Smart Nanomaterials Technology

Series Editors

Azamal Husen, Wolaita Sodo University, Wolaita, Ethiopia

Mohammad Jawaid, Department of Chemical and Petroleum Engineering, United Arab Emirates University, Al Ain, United Arab Emirates

Nanotechnology is a rapidly growing scientific field and has attracted a great interest over the last few years because of its abundant applications in different fields like biology, physics and chemistry. This science deals with the production of minute particles called nanomaterials having dimensions between 1 and 100 nm which may serve as building blocks for various physical and biological systems. On the other hand, there is the class of smart materials where the material that can stimuli by external factors and results a new kind of functional properties. The combination of these two classes forms a new class of smart nanomaterials, which produces unique functional material properties and a great opportunity to larger span of application. Smart nanomaterials have been employed by researchers to use it effectively in agricultural production, soil improvement, disease management, energy and environment, medical science, pharmaceuticals, engineering, food, animal husbandry and forestry sectors.

This book series in Smart Nanomaterials Technology (SNT) publishes cutting edge research monographs and edited volumes by top researchers and professionals across the globe. This book series aims to comprehensively cover topics in the fabrication, synthesis and application of these materials for applications in the following fields:

- Energy Systems—Renewable energy, energy storage (supercapacitors and electrochemical cells), hydrogen storage, photocatalytic water splitting for hydrogen production
- Biomedical—controlled release of drugs, treatment of various diseases, biosensors
- Agricultural—agricultural production, soil improvement, disease management, animal feed, egg, milk and meat production/processing
- Forestry—wood preservation, protection, disease management
- Environment—wastewater treatment, separation of hazardous contaminants from wastewater, indoor air filters

Review Process

The proposal for each volume is reviewed by the main editor and/or the advisory board. The chapters in each volume are individually reviewed single blind by expert reviewers (at least two reviews per chapter) and the main editor.

Ethics Statement for this series can be found in the Springer standard guidelines here - https://www.springer.com/us/authors-editors/journal-author/journal-author-helpdesk/before-you-start/before-you-start/1330#c14214

Jen-Tsung Chen
Editor

AI-Nano-Enabled Climate-Smart Agriculture and Sustainability

Springer

Editor
Jen-Tsung Chen
Department of Life Sciences
National University of Kaohsiung
Kaohsiung, Taiwan

ISSN 3004-8273 ISSN 3004-8281 (electronic)
Smart Nanomaterials Technology
ISBN 978-981-92-1693-2 ISBN 978-981-92-1694-9 (eBook)
https://doi.org/10.1007/978-981-92-1694-9

This Springer imprint is published by the registered company Springer Nature Singapore Pte Ltd.
The registered company address is: 152 Beach Road, #21-01/04 Gateway East, Singapore 189721, Singapore

Preface

This book is a frontier reference offering combinatorial technology by integrating two mainstream tools of artificial intelligence (AI) and nanotechnology in a wide range of agricultural applications, which can be highly expected to support smart, precision, and sustainable farming. These smart tools, so-called "AI-enabled nanotechnology", cover intelligent nanoparticles, working as nano-sensors, nano-fertilizers, nano-pesticides, and nano-carriers for targeted delivery of biostimulants such as plant growth-promoting rhizobacteria, and additionally, nucleic acids in advancing genetic engineering via CRISPR-Cas9 technology and epigenetic modulations via RNA technology. This book proposes promising crop-improving strategies constructed by AI-enabled nanotechnology with systems biology and omics tools, to obtain future crops with climate-ready capacities against temperature, water, and various potential environmental factors under a globally changing climate, and toward agricultural and environmental sustainability.

The readers include students, teachers, professors, researchers, and experts, in a wide range of agricultural sectors, in particular those who are interested in or learning materials science, nano-science, nanotechnology, agricultural biotechnology, bioinformatics involving AI technology, data science, plant physiology, and crop breeding.

Kaohsiung, Taiwan — Jen-Tsung Chen

Contents

Contributors

Hakima Achetoui Phytopathology Unit, Department of Plant Protection, Ecole Nationale d'Agriculture de Meknes, Meknes, Morocco

Debasmita Adak Centurion School of Smart Agriculture, Centurion University of Technology and Management, Vizianagaram, Andhra Pradesh, India

Arjun Kumar Agarwal Department of Genetics and Plant Breeding, Birsa Agricultural University, Kanke, Ranchi, Jharkhand, India

Purbita Aich Roy ICAR-Indian Agricultural Research Institute, New Delhi, India

Susmita Akhuli Microbiology Laboratory, Department of Botany (DST-FIST and UGC-DRS Funded), Institute of Science, Visva-Bharati (A Central University), Santiniketan, West Bengal, India

Sivaranjini Annamalai Department of Biotechnology, Sri Ramakrishna College of Arts & Science, Coimbatore, Tamil Nadu, India

Fatima Araish Department of Genetics and Plant Breeding, Birsa Agricultural University, Kanke, Ranchi, Jharkhand, India

Vijaya Anand Arumugam Department of Human Genetics and Molecular Biology, Bharathiar University, Coimbatore, Tamil Nadu, India

Nikita Arya Foundation of Healthcare Technologies Society, New Delhi, India

Alwin Robert Asirvatham Department of Endocrinology and Diabetes, Prince Sultan Military Medical City, Ministry of Defense, Riyadh, Kingdom of Saudi Arabia

Balamuralikrishnan Balasubramanian Department of Food Science and Biotechnology, Sejong University, Seoul, South Korea

Soham Barik Bidhan Chandra Krishi Viswavidyalaya, Mohanpur, Nadia, West Bengal, India

Madhumita Barooah Department of Agricultural Biotechnology, Assam Agricultural University, Jorhat, India;
DBT-NECAB, Assam Agricultural University, Jorhat, India

Farwa Basit Department of Biology, College of Science and Technology, Wenzhou-Kean University, Wenzhou, China

Nargiz Bayramova LANDAU School, Baku, Azerbaijan

Javaid Akhter Bhat Key Laboratory of Soybean Molecular Design Breeding, Northeast Institute of Geography and Agroecology, Chinese Academy of Sciences, Changchun, China;
Zhejiang Lab, Research Center for Life Science Computing, Hangzhou, China

Binty Department of Biotechnology, Jaypee Institute of Information Technology, Noida, Uttar Pradesh, India

Ishani Borthakur Department of Agricultural Biotechnology, Assam Agricultural University, Jorhat, India

Tahir Mohammad Chauhan CAE, F/ O Agricultural Sciences AMU, Aligarh, India

Bomba Dam Microbiology Laboratory, Department of Botany (DST-FIST and UGC-DRS Funded), Institute of Science, Visva-Bharati (A Central University), Santiniketan, West Bengal, India

Soumya Shephalika Dash Palli-Siksha Bhavana (Institute of Agriculture), Visva-Bharati, Sriniketan, West Bengal, India

Sandip Debnath Department of Genetics and Plant Breeding, Palli-Siksha Bhavana (Institute of Agriculture), Visva-Bharati (A Central University), Sriniketan, West Bengal, India

Sipika Devi Department of Plant Pathology, M S Swaminathan School of Agriculture, Centurion University of Technology and Management, Paralakhemundi, Gajapati, Odisha, India

Sukanya Dutta Faculty of Agriculture, JIS University, Kolkata, West Bengal, India

Pranjal Dwivedi Centre for Molecular Biology, Central University of Jammu, Bagla (Raya-Suchani), Samba, Jammu & Kashmir, India

Balaji Esakkimuthu Sudha Hospital, Hyderabad, Telangana, India

Rachid Ez-zouggari Phytopathology Unit, Department of Plant Protection, Ecole Nationale d'Agriculture de Meknes, Meknes, Morocco;
Laboratory of Biotechnology, Conservation and Valorization of Natural Resources (LBCVNR), Faculty of Sciences Dhar el Mehraz, Sidi Mohamed ben Abdallah University, Fez, Morocco

Xianzhong Feng Key Laboratory of Soybean Molecular Design Breeding, Northeast Institute of Geography and Agroecology, Chinese Academy of Sciences, Changchun, China

Thilagavathi Gangadharan Department of Information Technology, Coimbatore Institute of Technology, Coimbatore, Tamil Nadu, India

Bhoirob Gogoi DBT-NECAB, Assam Agricultural University, Jorhat, India

Rahul Gogoi Department of Agricultural Biotechnology, Assam Agricultural University, Jorhat, India

Subrata Goswami Banaras Hindu University, Uttar Pradesh, Varanasi, India

Shama Anzum Ismail Department of Biochemistry, PSG College of Arts and Science, Coimbatore, Tamil Nadu, India

Antony Prabhu Jeyabal Philomenathan Department of Human Genetics and Molecular Biology, Bharathiar University, Coimbatore, Tamil Nadu, India

Nie Jing Department of Biology, College of Science and Technology, Wenzhou-Kean University, Wenzhou, China

Payel Kar Bidhan Chandra Krishi Viswavidyalaya, Mohanpur, Nadia, West Bengal, India

Madhu Malathi Koduri Division of Agricultural Engineering, ICAR-IARI, New Delhi, India

Pradeep Kumar Rajasthan College of Agriculture, Maharana Pratap University of Agriculture and Technology, Udaipur, Rajasthan, India

Rajendra Kumar Division of Genetics, ICAR-Indian Agricultural Research Institute, New Delhi, India

Santanu Kundu ICAR-Indian Agricultural Research Institute, New Delhi, India

D. K. Kushwaha Division of Agricultural Engineering, ICAR-IARI, New Delhi, India

Rachid Lahlali Phytopathology Unit, Department of Plant Protection, Ecole Nationale d'Agriculture de Meknes, Meknes, Morocco

Vasantha Veerappa Lakshmaiah Department of Life Sciences, School of Sciences, Christ University, Bengaluru, India

Imane Lebyed Phytopathology Unit, Department of Plant Protection, Ecole Nationale d'Agriculture de Meknes, Meknes, Morocco

Trisha Majumder Department of Horticulture, Bangladesh Agricultural University, Maimansingh, Bangladesh

Sayanti Mandal Department of Chemistry and Biochemistry, Sharda School of Engineering and Sciences, Sharda University, Greater Noida, Uttar Pradesh, India

Tushar Kanti Mandal Mercedes-Benz Research and Development India Pvt. Ltd, Pune, India

Manoj Mani Department of Human Genetics and Molecular Biology, Bharathiar University, Coimbatore, Tamil Nadu, India

Samra Mirzayeva Science, Medicine and Education Department, Azerbaijan National Anti-Doping Agency, Baku, Azerbaijan

A. K. Mishra Division of Agricultural Engineering, ICAR-IARI, New Delhi, India

Bhagyashree Mishra School of Allied Health Science, The Apollo University, Chittoor, Andhra Pradesh, India

Ragini Mishra College of Community Science, Acharya Narendra Deva University of Agriculture and Technology, Kumarganj, Ayodhya, India

Sumit Kumar Mishra Department of Genetics and Plant Breeding, Birsa Agricultural University, Kanke, Ranchi, Jharkhand, India

Kinjal Mondal Rajasthan College of Agriculture, Maharana Pratap University of Agriculture and Technology, Udaipur, Rajasthan, India

Elshan Musazade Key Laboratory of Soybean Molecular Design Breeding, Northeast Institute of Geography and Agroecology, Chinese Academy of Sciences, Changchun, China

Praveen Nagella Department of Life Sciences, School of Sciences, Christ University, Bengaluru, India

Gouri Nanda Department of Biomedical Sciences, School of Health Sciences, The Apollo University, Chittoor, Andhra Pradesh, India

Aatika Nizam Department of Chemistry, School of Sciences, Christ University, Bengaluru, India

Nithya Palaniappan Department of Botany, Padmavani Arts and Science College for Women (Autonomous), Salem, Tamil Nadu, India

HimaBindu Pitaka Department of Biomedical Sciences, School of Health Sciences, The Apollo University, Chittoor, Andhra Pradesh, India

Wang Qian Department of Biology, College of Science and Technology, Wenzhou-Kean University, Wenzhou, China

Satyam Rai Department of Biotechnology, Centre for Molecular Biology, Central University of Jammu, Bagla (Raya-Suchani), Jammu & Kashmir, Samba, India

Ajay Rajan MS, Electrical and Computer Engineering, UMass Amherst, Amherst, MA, USA

Juhi Rani Department of Horticulture, Birsa Agricultural University, Kanke, Ranchi, Jharkhand, India

K. Baby Rani Department of Biomedical Sciences, School of Health Sciences, The Apollo University, Chittoor, Andhra Pradesh, India

Rajiv Ranjan Plant Biotechnology Lab, Department of Botany, Faculty of Science, Dayalbagh Educational Institute, Agra, India

Madhurya Ray Banaras Hindu University, Uttar Pradesh, Varanasi, India

Pravallika Sree Rayanoothala Department of Plant Pathology, M S Swaminathan School of Agriculture, Centurion University of Technology and Management, Paralakhemundi, Gajapati, Odisha, India

Prioty Bandhan Roop Department of Genetics and Plant Breeding, Palli-Siksha Bhavana (Institute of Agriculture), Visva-Bharati (A Central University), Sriniketan, West Bengal, India

Shweta Sachan College of Community Science, Acharya Narendra Deva University of Agriculture and Technology, Kumarganj, Ayodhya, India

Sanchayita Saha Microbiology Laboratory, Department of Botany (DST-FIST and UGC-DRS Funded), Institute of Science, Visva-Bharati (A Central University), Santiniketan, West Bengal, India

Palla Sahithi M S Swaminathan School of Agriculture, Centurion University of Technology and Management, Paralakhemundi, Odisha, India

Hridesh Harsha Sarma Department of Agronomy, Assam Agricultural University, Jorhat, India

Maksud Hasan Shah CAE, Faculty of Agricultural Science, Aligarh Muslim University, Aligarh, Uttar Pradesh, India;
Bidhan Chandra Krishi Viwavidyalaya, Mohanpur, India;
CAE, F/ O Agricultural Sciences AMU, Aligarh, India

Akash Shaikh Department of Life Sciences, School of Sciences, Christ University, Bengaluru, India

Aditya Pratap Singh Department of Genetics and Plant Breeding, Bidhan Chandra Krishi Viswavidyalaya, Mohanpur, Nadia, West Bengal, India

Longjam Boris Singh Department of Agriculture, Krishi Vigyan Kendra Thoubal, Manipur, Umiam, Meghalaya, India

Sadhna Singh College of Community Science, Acharya Narendra Deva University of Agriculture and Technology, Kumarganj, Ayodhya, India

Shubham Singh Rajmata Vijayaraje Scindia Krishi Vishwa Vidyalaya, Gwalior, Madhya Pradesh, India

Claire Copper Stone Department of Food Science, Nutrition and Dietetics, Faculty of Health Sciences, University of Malta, Msida, Malta

Desetty JMSNK Sri Veda Department of Plant Pathology, M S Swaminathan School of Agriculture, Centurion University of Technology and Management, Paralakhemundi, Gajapati, Odisha, India

Anisha Verma College of Community Science, Acharya Narendra Deva University of Agriculture and Technology, Kumarganj, Ayodhya, India

S. Vijay Kumar Uttar Banga Krishi Viwavidyalaya, Pundibari, Cooch Behar, India

Vikas Plant Biotechnology Lab, Department of Botany, Faculty of Science, Dayalbagh Educational Institute, Agra, India

K. K. Vishnupriya Department of Life Sciences, School of Sciences, Christ University, Bengaluru, India

Wang Wenhui Department of Biology, College of Science and Technology, Wenzhou-Kean University, Wenzhou, China

Neelam Yadav Centre of Food Technology, University of Allahabad, Allahabad, India

Hajar Zennouhi Phytopathology Unit, Department of Plant Protection, Ecole Nationale d'Agriculture de Meknes, Meknes, Morocco

Harnessing AI-Nano Approaches Towards Smart, Precision, and Sustainable Agriculture

Soham Barik, Shubham Singh, Soumya Shephalika Dash, Kinjal Mondal, Santanu Kundu, Maksud Hasan Shah, and S. Vijay Kumar

Abstract Agriculture remains a keystone of global economic stability and employment, yet it faces mounting challenges from population growth, climate change, urbanization, and the unsustainable use of agrochemicals. Excessive reliance on conventional fertilizers and pesticides has led to soil degradation, biodiversity loss, declining soil fertility, and reduced crop productivity worldwide. Precision agriculture (PA), also known as smart or site-specific crop management, has emerged as a sustainable solution by enabling accurate, data-driven management of agricultural inputs. The integration of artificial intelligence (AI) and nanotechnology significantly enhances the effectiveness of precision agriculture by improving accuracy, efficiency, and environmental performance. AI-driven systems leverage data from sensors, satellites, drones, and weather forecasts to optimize irrigation, fertilization, pest management, and yield prediction in real time. Nanotechnology complements these approaches through nano-enabled fertilizers, pesticides, biosensors, and targeted delivery systems that improve nutrient-use efficiency, enable early stress and disease detection, and minimize environmental contamination. Nano-biosensors play a critical role in real-time monitoring of soil health, crop physiology, contaminants, and microclimatic conditions, supporting informed decision-making in PA

S. Barik · M. H. Shah (✉)
Bidhan Chandra Krishi Viwavidyalaya, Mohanpur, India

S. Singh
Rajmata Vijayaraje Scindia Krishi Vishwa Vidyalaya, Gwalior, Madhya Pradesh, India

S. S. Dash
Palli-Siksha Bhavana (Institute of Agriculture), Visva-Bharati, Sriniketan, West Bengal, India

K. Mondal
Maharana Pratap University of Agriculture and Technology, Udaipur, India

S. Kundu (✉)
ICAR-Indian Agricultural Research Institute, New Delhi, India

M. H. Shah
CAE, Faculty of Agricultural Sciences, Aligarh Muslim University, Aligarh, Uttar Pradesh, India

S. Vijay Kumar
Uttar Banga Krishi Viwavidyalaya, Pundibari, Cooch Behar, India

J.-T. Chen (ed.), *AI-Nano-Enabled Climate-Smart Agriculture and Sustainability*,
Smart Nanomaterials Technology, https://doi.org/10.1007/978-981-92-1694-9_1

frameworks. Furthermore, the convergence of AI, nanotechnology, IoT, and cloud computing facilitates automated, scalable, and climate-resilient farming systems. Despite their transformative potential, challenges such as high initial costs, regulatory uncertainty, digital infrastructure gaps, and concerns over nanomaterial safety remain barriers to widespread adoption, particularly for smallholder farmers. This chapter highlights the synergies between AI, nanotechnology, and precision agriculture, discusses their applications, benefits, and limitations, and outlines future research directions needed to achieve sustainable, efficient, and resilient agricultural systems.

Keywords Climate change · Site-specific crop management · Biosensor · Economic stability

1 Introduction

The agricultural sector is regarded as one of the key drivers of employment in many nations throughout the world and as one of the pillars of the global economy. Growing urbanization, industrialization, and the world's population are the main obstacles to modern agriculture today, forcing the adoption of sustainable farming. However, the overuse of chemical fertilizers, the inevitable use of pesticides, herbicides, and other chemicals, as well as unsustainable farming practices, have resulted in soil degradation [48], the loss of beneficial microbes and biodiversity, a decline in the fertile status of productive lands, and ultimately a decrease in crop yield globally [28]. Precision agriculture, or smart farming, or site-specific crop management (SSCM), is a strategic computer-based technology that involves systematic monitoring, optimizing all types of inputs present in the field, and responsive actions to address the variability within and between fields in crop production with the help of different precision components towards sustainability [94, 92]. These precision farming techniques greatly aid in raising the overall crop production. They also lessen the amount of water, fertilizer, and pesticide wasted. Many technologies, such as GPS (Global Positioning System), GIS (Global Information System), and remote sensing, have been used to help identify temperature variations, soil properties, and climate changes in particular areas [112]. In agriculture, precision refers to the level of accuracy, and these days, integrating nanotechnology and artificial intelligence is becoming increasingly important to increase the level of accuracy in field crop production. Nanotechnology provides complementary innovations by introducing nanomaterials with a diameter of 100 nm or lower, which enhance plant health, facilitate early disease detection, and enable precise chemical delivery to targeted plant areas. This precision minimizes environmental pollution and maximizes resource usage efficiency by drastically lowering the use of pesticides [23]. Applications of nanotechnology also include gene editing tools to increase agricultural resilience, biosensors to detect food deterioration, and water disinfection. Additionally, nanoparticles improve knowledge of how plants control their interactions with environmental microbes and fend against illness [29]. Nanotechnology is regarded as an updated version of the conventional application method that provides more controlled and accurate pesticide distribution [58]. In agriculture, nanomaterials can be employed for remediation (pollutant remediation nanostructured systems), crop protection (nano-pesticides), plant nutrition (nano-fertilizers), and agricultural practice management (nanobiosensors) [31, 49].

Precision agriculture technologies driven by artificial intelligence maximize fertilization, irrigation, and pest control, resulting in lower input costs and less of an impact on the environment. Water conservation and increased crop yields are the outcomes of precise resource utilization made possible by technologies like variable rate technology (VRT) and artificial intelligence-driven irrigation systems [10]. One of the key applications of artificial intelligence is weather forecasting. It has also now become easy with the help of artificial intelligence for farmers to analyze the weather forecast and accordingly, they can plant the seeds, harvest the crops, and ensure proper irrigation [38].

Artificial intelligence, nanotechnology, and precision agriculture are three closely related fields that are transforming the agriculture industry. Supply chain management and agriculture could be completely transformed by artificial intelligence and similar technologies, especially when precision farming is used [8]. Each of these technologies has unique benefits for improving farming methods, and when combined, they have enormous potential to increase agricultural output, sustainability, and efficiency.

This chapter explores how artificial intelligence and nanotechnology can work together to support precision and sustainable agriculture. It looks at how these innovative technologies can solve important issues, transform farming methods, and promote an agrarian industry that is more productive and ecologically conscious. The chapter also discusses potential benefits, limitations, and socioeconomic and environmental consequences, as well as future research areas required for widespread adoption.

2 Synergies of Precision Agriculture and Nanotechnology for Sustainable Crop Growth

Precision agriculture is an advanced, knowledge-driven farming approach that combines geospatial tools, sensor-based monitoring technologies, and data-centric decision-support systems to enable accurate, site-specific crop management. The central aim of precision agriculture is to increase agricultural productivity while ensuring the efficient use of essential inputs such as water, fertilizers, and agrochemicals, thereby improving economic returns and minimizing environmental impacts. By integrating real-time sensor observations with predictive modeling and intelligent decision-support platforms, precision agriculture operationalizes the principle of applying the right input at the appropriate rate, location, and timing. This adaptive management strategy allows crops to respond optimally to spatial and temporal variability in field and environmental conditions, leading to improved yield stability, enhanced growth efficiency, and reduced resource losses [24].

Nanotechnology, a rapidly expanding interdisciplinary field, focuses on the manipulation of materials at the nanoscale, where unique physicochemical and biological properties become evident. In agricultural systems, nanotechnology offers innovative solutions that strengthen precision farming by improving sensing capabilities, enabling targeted delivery, and facilitating controlled release of inputs. For instance, nanosensors enable continuous, real-time monitoring of soil and plant health, allowing more precise and timely management interventions [111]. The convergence of precision agriculture and nanotechnology presents a promising pathway to address global food security challenges. By integrating nano-enabled farming systems with artificial intelligence and machine learning tools, researchers aim to mitigate the projected rise in global hunger, which is expected to affect nearly 840 million people by 2030 [116]. However, ensuring the safe, ethical, and responsible application of nanotechnologies in agriculture is essential. Further research is required to evaluate their long-term environmental and health implications, particularly concerning soil, water, and human safety.

The integration of nanotechnology with precision agriculture marks a transformative shift in modern agricultural production systems. This synergy enhances input efficiency, productivity, and environmental performance while supporting the transition toward climate-resilient and resource-efficient farming. Key dimensions of this integration are discussed below.

2.1 Nano-Enabled Nutrient Delivery Systems

Conventional fertilizers are characterized by poor nutrient use efficiency, with crops typically utilizing less than half of the applied nitrogen and approximately one-quarter of the phosphorus, while the remaining nutrients are lost through leaching, volatilization, or soil fixation [94, 92]. The rapid dissolution of soluble fertilizers in soil further exacerbates nutrient losses. Nano-fertilizers address these limitations by enabling precise, slow, and controlled nutrient release through nanoparticle encapsulation, thereby minimizing overapplication and nutrient wastage [19, 87, 88, 98]. Additionally, nanotechnology-enhanced precision agriculture employs nanosensors to monitor soil moisture, nutrient availability, pH, and plant physiological responses in real time, allowing high-resolution field mapping, early stress identification, and accurate irrigation and fertilizer scheduling [25].

2.2 Advanced Pest Management

Nanotechnology introduces novel approaches for pest and disease control through the use of nanoscale biosensors and nanoparticle-based formulations. These technologies enable early detection of pathogens and the targeted delivery of bioactive compounds, thereby reducing dependence on conventional chemical pesticides [53]. Nanopesticides improve the solubility, stability, and controlled release of active ingredients, which lowers environmental contamination and minimizes non-target impacts, ultimately supporting sustainable precision agriculture and improved food security [57, 86].

2.3 Environmental Sensing and Assessment

The extensive use of agrochemicals poses significant risks to environmental quality and agricultural productivity by contributing to soil degradation, water pollution, and biodiversity loss [90]. Nanomaterials and nanozymes, characterized by high surface area and enzyme-mimicking activity, offer enhanced reactivity and stability for environmental monitoring applications. Materials such as graphene, quantum dots, and metal nanoparticles exhibit strong potential for agricultural sensing, remediation, and diagnostic purposes [115].

2.4 Variable Rate Technology (VRT)

Variable rate technology (VRT) is a core component of precision agriculture, enabling the spatially optimized application of inputs such as fertilizers and herbicides based on crop condition and soil variability [83]. The incorporation of nanoparticles as delivery carriers further improves VRT by ensuring precise input placement at critical target zones. Nanocarrier-assisted VRT enhances crop performance by reducing degradation losses, minimizing repeated applications, and lowering environmental footprints while improving yield outcomes [114].

2.5 Data-Driven Interpretation and Analysis

Nanotechnology strengthens data analytics in precision agriculture by enhancing the accuracy, resolution, and timeliness of field-level data. Nano-enabled sensing platforms provide continuous, high-resolution measurements of soil moisture, nutrient levels, pH, and plant health, thereby supporting advanced analytical and predictive models [72]. Additionally, biosensors based on magnetic nanoparticles have demonstrated high sensitivity for detecting contaminants such as *Escherichia coli* in water systems [14].

3 Nano-Biosensors in Diagnostics under Precision Agriculture

The excessive application of fertilizers and agrochemicals has resulted in nutrient accumulation and toxic contamination of water bodies, leading to soil degradation, eutrophication, and increased economic and environmental costs. The adoption of precision agriculture systems supported by wireless networks and nanosensors enables site-specific crop management and improved sustainability outcomes [65]. Nanobiosensors facilitate real-time assessment of soil physicochemical parameters, including moisture, temperature, pH, nutrient status, and microclimatic conditions. They are also capable of detecting pollutants, plant pathogens, heavy metals, and pesticide residues [7]. A typical biosensor architecture includes a sensing element, signal processing unit, microprocessor, and wireless communication module for efficient data transmission [97].

3.1 Assessment of Crop Health

Smart nanobiosensors, particularly wearable plant sensors, are advancing crop diagnostics by enabling continuous monitoring of plant health and environmental conditions [82]. Flexible and low-cost sensor technologies, such as tape-based graphene sensors and stretchable strain sensors, are gaining widespread attention. Notable innovations include laser-induced graphene biosensors for pesticide detection and graphene-based sensors for monitoring volatile organic compounds released during plant stress responses [55, 118].

3.2 Monitoring Soil Nutrients

Nanobiosensors combine biological recognition elements with nanomaterials to deliver rapid, sensitive, and real-time measurement of soil pH and macro- and micronutrient concentrations. These capabilities support site-specific nutrient management, reduce fertilizer losses, and improve overall nutrient use efficiency within precision agriculture systems [33].

3.3 Detection of Metallic Contaminants

Soil contamination originating from domestic wastewater, industrial discharge, and agricultural activities negatively affects soil quality and beneficial microbial communities. Heavy metals such as lead, cadmium, mercury, and arsenic are traditionally detected using expensive analytical techniques, including HPLC, AAS, and ICP-AES. Nanosensor-based detection methods offer a cost-effective and user-friendly alternative. Carbon nanotube sensors can identify both organic and inorganic pollutants, while optical nanosensors employing gold and silver nanoparticles are effective for heavy metal detection [60]. Commercially available graphene-based nanosensors and metal–organic frameworks provide rapid response, high sensitivity, portability, and reduced costs, making them more practical than conventional analytical approaches [76].

3.4 Other Functions

Nanobiosensors generate real-time, quantifiable signals—typically electrical, optical, or mechanical—by detecting specific molecules within biological samples. These data are essential for optimizing agricultural productivity and ensuring food quality and safety [20]. Nanobiosensors have also demonstrated effectiveness in identifying contamination events in food processing facilities and in communicating safety information to consumers, regulatory bodies, and supply-chain stakeholders. Furthermore, biosensor-based platforms utilizing nanoparticles have been developed for allergen detection, with several technologies approaching commercial deployment [107] (Table 1).

4 Precision Agriculture and Cloud Computing

Current agricultural research emphasizes innovative technologies to enhance crop productivity while minimizing environmental impacts. Green nanotechnology and cloud computing support the development of sustainable farming systems [103]. Integrated with the Internet of Things (IoT), cloud computing enables automation, precision agriculture, and data-driven decision-making, improving input efficiency, pesticide management, and supply-chain performance [80]. Intelligent irrigation based on soil moisture monitoring is critical for crop quality and yield; however, high sensor costs, such as the ECHO-EC5 (~USD 169), limit adoption. Cost-effective alternatives like RFID-based GreenTag sensors address this challenge and support improved plant productivity [106].

Table 1 Nano biosensors in diagnostics under precision agriculture

Nanobiosensors	Role	Material used	References
Crop health biosensors	Assessment of crop health	Metal oxides	[9]
Soil moisture sensors	Monitoring the moisture levels present in the soil	Polymers	[36]
Metallic detection biosensors	Remediation of heavy metal compounds from wastewaters	Glutathione-functionalized	[104]
Plant disease-identifying biosensors	Detection of disease-causing pathogens in sugarcane	Gold nanomenzyme	[95]
Irrigation biosensors	Monitoring irrigation levels	Ceramics	[42]

5 Need for Nanotechnology in Precision Farming

Effective fertilizer management is critical for sustainable agriculture, and artificial intelligence (AI)-enhanced applications provide a viable solution to meet rising food demands driven by population growth. Overuse of chemical fertilizers has led to environmental problems such as soil degradation and water contamination. AI systems integrate data from soil sensors, satellite imagery, and weather forecasts to optimize fertilizer type, quantity, and application timing according to crop-specific requirements, thereby improving nutrient-use efficiency and reducing ecological impacts through precision agriculture. AI also plays a pivotal role in evaluating the long-term environmental effects of nanomaterials on soil, roots, and leaves, interpreting cumulative impacts on crop yield, soil fertility, and nutrient dynamics for sustainable deployment in precision farming [40]. Recent studies demonstrate AI's effectiveness in resource optimization; for example, an AI-based fertilizer recommendation model using real-time soil and crop data achieved a 15% reduction in fertilizer use without compromising yield, highlighting its potential to minimize nutrient runoff and soil degradation [59]. Furthermore, AI supports climate-resilient agriculture through precision farming, smart irrigation, and automation, particularly when integrated with IoT and renewable energy systems, and extends optimization to irrigation and pest management [62]. Talaviya et al. [99] report that artificial intelligence-enabled drones, robots, and automated sprayers improve irrigation efficiency and reduce pesticide and herbicide misuse through real-time sensing and image-based weed detection. Kumari et al. [47] further demonstrate how artificial intelligence, IoT and variable rate technologies enhance fertilizer application, soil health monitoring, and pest detection, although challenges such as high costs and data security remain barriers. Adewusi et al. [1] and Kumar et al. [45] reinforce artificial intelligence's role in precision agriculture through decision-support systems, remote sensing, and data analytics, enabling efficient water and nutrient management while reducing environmental impacts. Akintuyi [3] highlights adaptive artificial intelligence systems that leverage self-learning algorithms and IoT for real-time farm optimization, improving productivity and sustainability.

6 Reason behind AI and Nanotechnology Required for Precision Agriculture

AI-augmented fertilizer application plays a crucial role in advancing sustainable agriculture by improving nutrient-use efficiency and reducing environmental degradation. Growing population pressure has intensified fertilizer demand, which, although enhancing crop productivity, contributes to soil degradation, water pollution, and greenhouse gas emissions. Innovators and policymakers must address climate change and shrinking natural resources with adaptive and forward-looking strategies [81]. AI enables balanced fertilizer use by analyzing soil properties, crop growth stages, and climatic conditions, allowing farmers to improve yields while maintaining environmental responsibility.

Artificial intelligence supports soil health management by monitoring soil structure and nutrient retention and recommending sustainable practices such as crop rotation and organic inputs to maintain long-term fertility. Although initial investment costs may be high, AI reduces fertilizer wastage and increases profitability over time through precise nutrient delivery and automated data collection. However, limited accessibility for smallholder farmers remains a challenge, highlighting the need for training and collaboration among governments, technology providers, and agricultural institutions [3]. Traditional fertilizer practices often ignore crop-specific nutrient demand and soil absorption capacity, leading to overfertilization and ecosystem damage [85, 96]. AI-based precision agriculture integrates soil sensors, satellite imagery, and weather data to guide accurate fertilizer application in real time [87, 88]. Optimized fertilizer use also lowers emissions from fertilizer production [110] and minimizes nutrient runoff, thereby protecting water quality and reducing algal blooms [1].

6.1 AI and Nanotechnology as Pillars of Smart Farming Systems

Precision agriculture (PA) is transforming global farming by improving productivity, crop quality, profitability, sustainability, environmental protection, and rural livelihoods, making it central to modern agricultural strategies [83]. Nanotechnology, involving materials $\leq$ 100 nm, is increasingly integrated into PA due to its high surface area and unique physicochemical properties. Nanomaterials enhance nutrient management and plant protection, while nanoencapsulation enables controlled-release, non-persistent insecticides that reduce environmental contamination and improve agrochemical efficiency [75].

6.2 *Examples of Nanoparticles*

Precision farming in agriculture has made use of the synthesis of nanoparticles from several biological sources [26]. These nanoparticles include the following examples.:

- Silver nanoparticles: Silver nanoparticles have a greater antibacterial effect than commercial silver due to their vast surface area and atoms. The antibacterial qualities of silver nanoparticles have been used to treat a wide range of human ailments [71]. The antibacterial qualities of silver nanoparticles are therefore being used more often to treat plant ailments.
- Zinc oxide nanoparticles: Zinc deficiency is a common micronutrient problem that adversely affects agricultural output in alkaline soils that include calcium carbonate. Zinc fertilizers, including zinc sulfate and zinc oxides, are frequently used to address the zinc deficiency in soils.
- Titanium dioxide nanoparticles (TiO_2): According to Sang et al. [84], titanium is a strong and corrosion-resistant metal. According to Chen et al. [16], titanium boosts the production of additional carbohydrates, which accelerates photosynthesis and encourages plant growth.

7 Adoption Barriers of AI and Nanotechnological Innovations in Precision Farming

In order to improve crop management, resource efficiency, and environmental sustainability, precision agriculture integrates cutting-edge technologies like artificial intelligence (AI) and nanotechnology. Although the integration of nanotechnology with precision agriculture (PA) has great promise, there are still a number of technical, socioeconomic, ethical, and legal obstacles to overcome. To guarantee widespread, safe, and egalitarian adoption, particularly in the context of developing nations and smallholder farming systems, these issues must be resolved.

7.1 *Regulatory Uncertainty and Nanomaterial Safety*

Since many nations currently lack defined frameworks controlling the use of nanomaterials in food and farming systems, nanotechnology applications in agriculture suffer regulatory ambiguity. This ambiguity, together with a lack of long-term environmental and toxicity studies, causes regulators and farmers to be hesitant and slow in adoption [74].

7.2 High Costs and Economic Barriers

Even though commercial farms and controlled trials have shown that nanotechnology and PA tools are beneficial, many small and marginal farmers cannot afford the initial development, deployment, and maintenance costs of these technologies.

Low-income farming communities frequently cannot afford the high upfront costs of technologies like nanosensors, nano-formulated agrochemicals, and IoT systems [52].

7.3 Risk Assessment of Nanomaterials in Agro-ecosystems

The integration of artificial intelligence and nanotechnology in precision agriculture raises concerns regarding human health and environmental safety. Nano-enabled inputs may persist in ecosystems and harm beneficial organisms, while human exposure risks remain uncertain. Limited toxicological data and biased AI algorithms complicate risk assessment, highlighting the need for robust regulatory frameworks and environmental evaluations [79].

7.4 Digital Access and Infrastructure Challenges

Access to digital infrastructure, including sensor networks, data analytics platforms, and internet connectivity, is crucial to precision agriculture. But there is still a digital divide, particularly in rural and isolated areas with poor internet connection and technical capabilities [109]. The data-driven benefits of smart farming are not available to farmers without access to digital tools, which causes a gap in the adoption of technology between developed and poor nations (Table 2).

Table 2 Agriculture parameters are monitored and managed using AI

Parameter	Description	AI applications/techniques
Crop development and productivity	Tracking crop developmental phases, canopy expansion, and productivity levels	Machine vision, artificial neural networks, satellite-based sensing, UAVs
Soil condition assessment	Evaluation of soil physicochemical properties, nutrient availability, and fertility status	In-situ soil probes, hyperspectral sensors, data-driven learning models
Biotic stress identification	Early detection and diagnosis of insect pests and plant diseases	Pattern recognition, deep neural networks, vision-based analytics
Water resource optimization	Precision scheduling and regulation of irrigation to enhance water use efficiency	Sensor networks, climate data fusion, predictive analytics
Produce quality evaluation	Assessment of qualitative traits for grading, classification, and market value	Machine learning models, imaging systems, and spectral analysis
High-throughput crop Phenotyping	Measurement of morphological, physiological, and biochemical crop attributes	Multispectral/hyperspectral imaging, 3D reconstruction, AI algorithms
Agro-meteorological surveillance	Continuous monitoring and interpretation of weather variables affecting crop growth	Smart sensors, automated weather stations, big data analytics
Production forecasting	Predictive estimation of crop output for strategic planning and management	Satellite data analysis, regression models, and machine learning techniques
Automated harvest and post-harvest handling	Mechanization and optimization of harvesting and produce sorting operations	Intelligent vision systems, agricultural robotics, and AI-driven control systems

8 Application of Machine Learning and Artificial Intelligence in Agriculture

8.1 *AI Applications in Crop Management*

Artificial intelligence (AI) has emerged as a transformative tool in crop management by supporting precision-oriented and data-driven farming practices [1]. Using remote sensing platforms such as satellites and unmanned aerial vehicles, AI-enabled systems continuously assess crop conditions, identify physiological stress, and estimate yield potential. Machine learning algorithms process soil characteristics and climatic variables to improve decisions related to sowing time, irrigation scheduling, and fertilizer application. AI-supported crop suitability models further assist farmers in selecting crops that are best adapted to specific agro-climatic regions [12]. In parallel, AI-driven agricultural robotics improve harvesting efficiency, reduce dependence on manual labor, and contribute to environmentally sustainable crop production systems [93].

Recent developments highlight the role of AI in enhancing agricultural productivity and sustainability through the integration of machine learning, deep learning, computer vision, robotics, automation, and IoT-based sensor networks [1]. Learning-based approaches—including supervised, unsupervised, and reinforcement learning—are widely applied for yield estimation, crop planning, pest control, irrigation optimization, and adaptive greenhouse management by interpreting large and complex datasets. Computer vision techniques analyze imagery obtained from drones, satellites, and in-field sensors to assess crop and soil health, enable early detection of pests and diseases, and diagnose nutrient deficiencies for timely corrective measures [30, 78]. Robotics and automation technologies, such as robotic systems for seeding, weeding, and harvesting, enhance operational accuracy while reducing labor requirements. Furthermore, IoT-enabled precision agriculture and smart irrigation systems deliver real-time information on soil moisture, temperature, nutrient status, and weather conditions, enabling efficient input management and improved water-use efficiency. Collectively, these AI-based innovations promote data-informed decision-making, optimized resource utilization, and greater resilience of agricultural systems under changing environmental and economic conditions [1].

8.2 Pest and Disease Detection and Management

AI has become a key component of modern pest and disease management by enabling rapid and precise identification through image-based analysis and sensor-driven monitoring systems. Predictive modeling techniques help anticipate pest population dynamics and outbreak trends, thereby reducing excessive reliance on chemical pesticides and supporting environmentally friendly pest control strategies [66]. Interactive monitoring tools, including text-to-speech interfaces, further improve usability by providing real-time feedback to users. Deep learning methods, particularly Deep Convolutional Neural Networks and transfer learning models, have demonstrated high effectiveness in early detection of pests and diseases in crops such as banana and mung bean, with reported accuracy levels exceeding 93% [105].

8.3 Yield Prediction and Forecasting

AI has significantly improved crop yield prediction and forecasting, which are critical for economic planning, market analysis, and precision agriculture decision-making. Artificial neural networks and machine learning models integrate multiple data sources, including weather parameters and satellite-derived information, to generate accurate and reliable yield predictions.

8.4 AI in Drone Technology

The incorporation of AI into drone technology has transformed agricultural monitoring by providing detailed aerial imagery and actionable insights. Machine learning and computer vision algorithms enable drones to evaluate crop health, detect pest and disease incidence, monitor plant growth dynamics, and identify field zones requiring targeted management interventions [63].

8.5 Livestock Monitoring and Management

AI-based solutions in livestock management support continuous monitoring of animal health, behavior, and productivity. Wearable devices and RFID systems facilitate individual animal identification and tracking, while predictive analytics tools identify abnormal patterns and anticipate potential health disorders [4]. Multi-sensor platforms record physiological indicators such as body temperature, heart rate, and rumination activity, whereas camera-based systems analyze behavioral responses under varying environmental conditions [91]. Machine learning algorithms process these datasets to enable timely veterinary interventions, optimize feed formulations, and improve overall herd performance [101].

9 Sustainable and Precision Agriculture with the Internet of Everything (IoE)

The convergence of AI and the Internet of Things, collectively referred to as the Internet of Everything (IoE), is driving the adoption of sustainable and precision agriculture practices. These smart farming systems rely on real-time sensing, predictive analytics, and automated decision-making to enhance resource-use efficiency, lower production costs, and improve environmental sustainability. IoT sensors deployed across agricultural fields continuously collect data on soil moisture, temperature, and crop conditions, which AI systems analyze to optimize irrigation schedules, predict yields, and identify pest or disease outbreaks at early stages. This approach enables timely interventions, improves water-use efficiency, increases crop productivity, and reduces environmental impacts compared to conventional farming methods (Fig. 1) [6]. Through precision agriculture frameworks, the combined application of AI and IoT helps optimize the use of water, fertilizers, and pesticides, minimizing resource losses while supporting environmentally sustainable agricultural production systems [46] (Table 3).

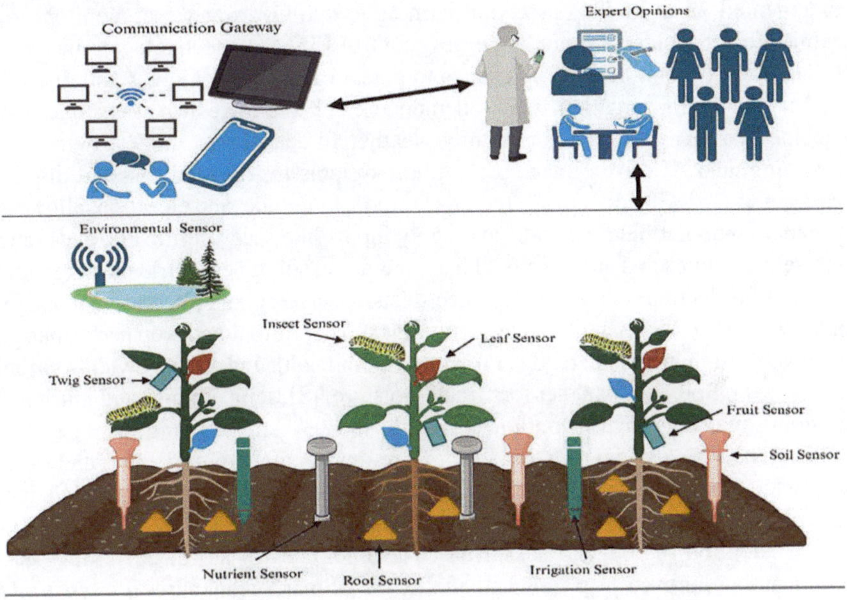

Fig. 1 An IoT-based network for smart farming

Table 3 Comparison of traditional farming and smart farming

Parameter	Traditional farming	IoT-based smart farming
Water consumption (liters per acre)	10,000	7000
Water savings (%)	0	30
Crop damage due to over/under irrigation (%)	15	5
Yield improvement due to smart irrigation (%)	0	12
Disease detection accuracy (%)	50	95
Reduction in pesticide use (%)	20	40
Crop loss due to disease (%)	0	5
Farmers adopting IoT-based disease detection (%)	10	60
Operational costs per acre (₹)	10,000	7000
Labor requirement (man-hours)	50	30
Profit per acre (₹)	20,000	26,000
Farmers experiencing cost benefits (%)	50	70

9.1 Implementation of AI and IoT in Precision Farming

In 2021, a project involving 200 farmers and an agricultural technology company implemented AI and IoT in precision farming to optimize water use, monitor crop health, predict yields, and automate pest control [35]. Key technologies included AI-based crop monitoring using drones to detect early signs of stress and disease, IoT-enabled smart irrigation with soil moisture sensors to reduce water wastage, machine learning models for real-time weather forecasting to guide sowing and harvesting, and AI-driven image recognition systems for targeted pest and disease management [5]. GPS accurately records latitude, longitude, and elevation, allowing farmers to pinpoint field locations and apply inputs like seeds, fertilizers, and water precisely. IS stores and analyzes field data, including soil types, nutrients, irrigation, and crop production, aiding in informed decision-making and alternative management strategies [108]. Satellite images monitor soil conditions and crop performance, using vegetation indices like NDVI to assess plant health and estimate yields before harvest [61]. Soil and plant sensors track moisture, pH, temperature, and nutrients, providing real-time data to optimize growth, manage stresses, and increase crop yields. Sensor technologies detect soil texture, nutrients, humidity, light, and other parameters, transmitting data to controllers for automated adjustments and efficient farm management [117].

The adoption of IoT and sensor-based technologies in advanced farming has significantly improved crop yield and quality [35]. Smart greenhouses use sensors to monitor humidity, temperature, light, and CO_2, enabling automated management and protection from environmental stresses. Hydroponic systems provide precise nutrient and water delivery to plants, monitored in real time via IoT, while vertical farming maximizes production with minimal land and water use. IoT-based phenotyping further aids crop breeding by linking plant traits with growth conditions, supporting efficient, high-quality, and sustainable agricultural practices [89]. IoT-based irrigation systems address water scarcity by continuously monitoring soil moisture and weather conditions to optimize water application [56]. Studies show that smart irrigation using IoT technology can save 30–40% of water compared to conventional methods [69]. The integration of AI algorithms with drip irrigation further enhances efficiency by delivering precise water quantities at the right time for crop needs [50].

IoT pest and disease detection systems use image recognition models to analyze high-resolution images captured by IoT-enabled drones and field cameras for early identification of crop threats [44]. These systems achieve up to 95% detection accuracy and reduce pesticide usage by nearly 40% [54]. They also help minimize crop damage while enhancing productivity and promoting environmentally friendly farming practices [2]. By leveraging AI-driven machinery, IoT sensors, and automated irrigation systems, farmers can reduce operational costs by approximately 25%. This increased efficiency not only lowers expenses but also boosts productivity, resulting in higher yields and improved farm management (Fig. 2).

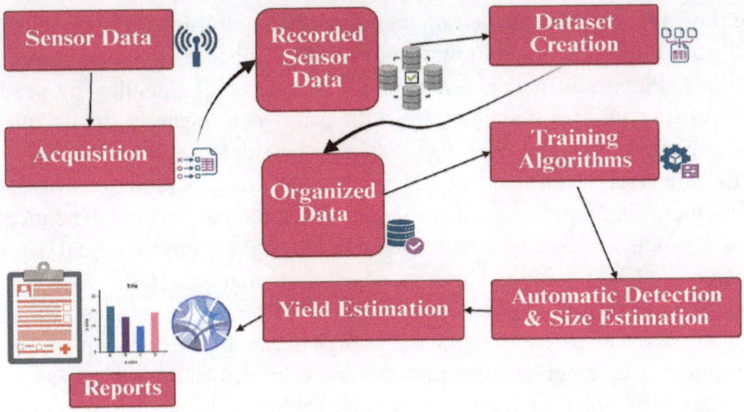

Fig. 2 Data collection and data extraction in precision farming

10 Decoding and Engineering the Phytobiome Communication for Smart Agriculture

The rapidly growing global population and climate change are intensifying pressure on agriculture, particularly through increasing water scarcity and inefficient agrochemical use [18]. Smart agriculture integrates IoT, robotics, and ML/AI to enhance resource efficiency and crop monitoring, yet most systems detect plant stress only after visible symptoms appear [113]. Emerging approaches based on plant electrophysiological signals and the Phytobiome concept enable early stress detection by considering plants as part of an interconnected biological ecosystem [77]. Integration of molecular communication and the Internet of Bio-Nano Things with AI facilitates precise, self-regulated interventions, promoting sustainable and high-efficiency agricultural management [11].

10.1 Overview of the Phytobiome Communication

Plants function as central communication hubs within the phytobiome, integrating internal and external signals to respond to biotic and abiotic stresses [77]. Signal transmission occurs through phytohormones such as jasmonic and salicylic acids, reactive oxygen species, calcium (Ca^{2+}) fluxes, electrical signals, and mobile RNAs, enabling rapid coordination among distant tissues [21, 67]. Localized stresses, such as herbivory, activate systemic defense responses in undamaged organs via jasmonate signaling [77]. Aboveground, plants emit volatile organic compounds that alert neighboring plants, while microorganisms communicate through quorum sensing to regulate collective behaviors. Insects and animals further mediate signaling through pheromones, shaping plant–herbivore and plant–pollinator interactions, highlighting the phytobiome as an interconnected, multilayered communication network [68].Interkingdom communication further enhances the complexity of the phytobiome. Plants can perceive and modulate microbial signaling by producing quorum-sensing mimics that interfere with pathogenic bacteria, while microbial volatile organic compounds (VOCs) activate plant defense responses [68]. Root exudates selectively recruit beneficial microorganisms, including nitrogen-fixing bacteria, improving nutrient acquisition and stress tolerance [119]. Chemical cues released by nematodes and insects also modify plant physiological signaling, reflecting coevolved, reciprocal interactions across kingdoms [68].

Modeling the phytobiome as a communication network provides a multiscale perspective. At the microscale, signaling through plasmodesmata, xylem, and phloem follows transmitter–receiver dynamics governed by diffusion and active transport [100]. Mesoscale interkingdom interactions resemble local area networks with plants as central hubs, while macroscale VOC-mediated links connect entire phytobiomes, analogous to wide area networks [43]. Integrating molecular communication, network theory, and machine learning can enable early stress detection, precision management, and sustainable, ecosystem-informed agriculture [34].

10.2 Engineering the Phytobiome Communication

(i) **Phytobiome monitoring**

Phytobiome monitoring leverages electrophysiological signals generated within plants in response to stress, such as drought, nutrient deficiency, or pest attacks [68]. These signals, which vary in amplitude, frequency, and temporal dynamics, are analyzed using ML/AI techniques to detect stress early [68]. Sensors capture these signals, which are then processed through bio-cyber interfaces for diagnostic purposes [68].

(ii) **Targeted delivery of agrochemicals and genes**

The Internet of Bio-Nano Things enables targeted delivery of agrochemicals and genes to specific plant regions [51]. Nanomachines carrying pesticides, fertilizers, or DNA payloads localize target cells using molecular communication and bind via engineered ligands [102]. This approach can operate in fully autonomous or semi-autonomous modes, allowing plants to manage nutrient deficiencies or pest infestations with minimal human intervention. Continuous monitoring ensures ML/AI algorithms improve stress diagnosis and delivery accuracy over time [32].

(iii) **Smart irrigation**

Smart irrigation systems integrate phytobiome monitoring with autonomous water management [68]. Plants' electrophysiological signals indicate water stress, prompting automated irrigation to prevent over- or underwatering [68]. Farmers can intervene remotely in semi-autonomous modes, while ML/AI algorithms continuously improve system efficiency by learning from real-time plant responses [22]. This approach enhances water use efficiency and increases crop yield while minimizing resource waste [37].

(iv) **Engineering phytobiome communication with genetically modified bacteria**

Genetically modified bacteria can enhance communication between plants and symbiotic microbes while suppressing pathogenic bacteria [13]. These engineered microbes improve nutrient uptake, disease resistance, and plant growth by strengthening molecular communication within the phytobiome [39]. Implementation of GM bacteria requires adherence to biosafety and regulatory guidelines to ensure responsible deployment in agricultural environments [15].

(v) **Inter-phytobiome communication**

Communication between separate phytobiomes occurs through wired and wireless channels, enabling the transfer of information such as pest infestations [11]. Plants receiving these signals can activate defense mechanisms, enhancing pest resilience [64]. Synthetic VOCs like methyl jasmonate can be applied via sprayer-based airborne molecular communication systems to efficiently transmit stress signals to neighboring plants [73].

(vi) **Optimization of communication parameters**

Inter-phytobiome communication systems can be optimized by adjusting parameters such as transmission time, modulation type, and signal power [27]. Wired channels can be enhanced using synthetic fungi to improve information flow [68]. Optimizing these communication pathways ensures faster, more accurate responses across agricultural fields, contributing to higher productivity and reduced chemical use [70].

(vii) **Integration of ML/AI in phytobiome applications**

ML/AI plays a critical role in interpreting electrophysiological signals, predicting plant stress, and managing automated interventions. The continuous feedback from monitoring and delivery systems enables models to learn and adapt, improving the accuracy of stress detection, water management, and targeted delivery of nutrients or pesticides [17]. This integration makes smart farming systems increasingly autonomous and efficient.

(viii) **Holistic agricultural impact**

By engineering phytobiome communication at multiple scales, viz., cellular, inter-kingdom, and inter-phytobiome, farmers can achieve precise, sustainable, and resilient agriculture [41]. Early stress detection, targeted interventions, optimized irrigation, and improved microbial interactions collectively enhance crop yield, reduce resource wastage, and minimize environmental impact, demonstrating the transformative potential of smart agriculture technologies [46].

11 Conclusion

Nanotechnology and artificial intelligence (AI) are revolutionizing agriculture by enhancing productivity, sustainability, and resource efficiency. Nanotechnology improves crop yields while minimizing chemical inputs and environmental impact, supporting food security and sustainable practices. AI integrates data from sensors, satellites, drones, and weather systems to optimize crop management, predict diseases, monitor nutrient status, and improve irrigation and supply chain efficiency. Together, these technologies enable precision farming, reduce resource wastage, mitigate climate and pest-related risks, and enhance profitability, offering transformative solutions for sustainable, high-quality, and resilient agricultural systems.

References

1. Adewusi AO, Asuzu OF, Olorunsogo T, Iwuanyanwu C, Adaga E, Daraojimba DO (2024) AI in precision agriculture: a review of technologies for sustainable farming practices. World J Adv Res Rev 21(1):2276–2285
2. Ahmed S, Marwat SNK, Brahim GB, Khan WU, Khan SN, Al-Fuqaha A, Kozieł S (2024) IoT based intelligent pest management system for precision agriculture. Sci Rep 14(1):31917
3. Akintuyi OB (2024) Adaptive AI in precision agriculture: a review: investigating the use of self-learning algorithms in optimizing farm operations based on real-time data. Res J Multidiscip Stud 7(02):016–030
4. AlZubi AA, Al-Zu'bi M (2023) Application of artificial intelligence in monitoring of animal health and welfare. Indian J Anim Res
5. Alazzai WK, Abood BSZ, Al-Jawahry HM, Obaid MK (2024) Precision farming: the power of AI and IoT technologies. E3S Web Conf 491:4006
6. Anshad AS (2025) Smart farming technologies: integrating AI and IoT for sustainable agriculture. Agric Innov Sustain J 1(1)
7. Antonacci A, Arduini F, Moscone D, Palleschi G, Scognamiglio V (2018) Nanostructured (bio)sensors for smart agriculture. TrAC Trends Anal Chem 98:95–103
8. Ashique S, Raikar A, Jamil S, Lakshminarayana L, Gajbhiye SA, De S, Kumar S (2025) Artificial intelligence integration with nanotechnology: a new frontier for sustainable and precision agriculture. Curr Nanosci 21(2):242–273
9. Ashraf S, Iqbal J, Abbasi BA, Haakim Z, Chalgham W, Mahmood T (2025) Nanobiosensors for real-time and high-resolution crop monitoring. In: Nanobiosensors for crop monitoring and precision agriculture. Springer Nature, Singapore, pp 281–305
10. Asritha M, Lokeshwar K, Gowtham G, Kotresh DJ, Kumar KN, Shahid MS (2024) The role of AI in modernizing Indian agriculture
11. Babar AZ, Akan ÖB (2024) Sustainable and precision agriculture with the internet of everything (IoE). arXiv (Cornell University)
12. Bera S, Dey T, Mukherjee A, Buyya R (2022) E-CropReco: a dew-edge-based multi-parametric Crop recommendation framework for internet of agricultural things. Research Square
13. Boo A, Toth T, Yu Q, Pfotenhauer AC, Fields BD, Lenaghan SC, Stewart CN, Voigt CA (2024) Synthetic microbe-to-plant communication channels. Nat Commun 15(1)
14. Bu T, Jia P, Liu J, Liu Y, Sun X, Zhang M, Tian Y, Zhang D, Wang J, Wang L (2019) Diversely positive-charged gold nanoparticles based biosensor: a label-free and sensitive tool for foodborne pathogen detection. Food Chem X 3:100052

15. Chemla Y, Sweeney CJ, Wozniak CA, Voigt CA (2025) Design and regulation of engineered bacteria for environmental release [Review of *design and regulation of engineered bacteria for environmental release*]. Nat Microbiol 10(2):281
16. Chen H, Seiber JN, Hotze M (2014) ACS select on nanotechnology in food and agriculture: a perspective on implications and applications. J Agric Food Chem 62(6):1209–1212
17. Chen E, Prakash S, Reddi VJ, Kim D, Rajpurkar P (2023) A framework for integrating artificial intelligence for clinical care with continuous therapeutic monitoring [Review of *a framework for integrating artificial intelligence for clinical care with continuous therapeutic monitoring*]. Nat Biomed Eng
18. Chukalla AD, Mekonnen MM, Gunathilake DMCC, Wolkeba FT, Gunasekara B, Vanham D (2025) Global spatially explicit crop water consumption shows an overall increase of 9% for 46 agricultural crops from 2010 to 2020. Nat Food 6(10):983
19. Cicek S, Nadaroglu H (2015) The use of nanotechnology in the agriculture. Adv Nano Res 3:207
20. Cipriano PE, Siueia Júnior M, da Silva DF, Mars G, da Silva RF, Rabêlo FHS (2025) Nanosensors in the agricultural system—the current understanding, role and prospects. In: Nanobiosensors for agricultural and other related sectors. Springer Nature, Singapore, pp 1–31
21. Coatsworth P, Gonzalez-Macia L, Collins A, Bozkurt TO, Güder F (2022) Continuous monitoring of chemical signals in plants under stress [review of *continuous monitoring of chemical signals in plants under stress*]. Nat Rev Chem 7(1):7
22. Coco MD, Leo M, Carcagnì P (2024) Machine learning for smart irrigation in agriculture: how far along are we? Information 15(6):306
23. De A, Bose R, Kumar A, Mozumdar S (2014) Targeted delivery of pesticides using biodegradable polymeric nanoparticles, vol 10. Springer, New Delhi, pp 978–981
24. Dhanaraju M, Chenniappan P, Ramalingam K, Pazhanivelan S, Kaliaperumal R (2022) Smart farming: internet of things (IoT)—based sustainable agriculture. Agriculture 12:1745
25. Dilshad M, Fatima H, Al-Amin M, Qureshi A, Ahmad I, Anum H et al (2025) Nanotechnology in precision agriculture applications of nanosensors in soil, crop and water management. Haya Saudi J Life Sci 10(8):298–310
26. Duhan JS, Gahlawat SK (2014) Biogenesis of nanoparticles: a review. Afr J Biotechnol 13(28)
27. Ergün D (2025) Decoding and engineering the phytobiome communication for smart agriculture. Agribio Innov 2(1):49
28. Fathi Qarachal J, Alizadeh M (2025) Harnessing precision agriculture and nanotechnology for sustainable farming: a review. Nanotechnol Environ Eng 10(2):37
29. Gad MA, Ming-ju L, Ahmed FK, Almoammar H (2020) Nanomaterials for gene delivery and editing in plants: challenges and future perspective. In: Multifunctional hybrid nanomaterials for sustainable agri-food and ecosystems. Elsevier, Berlin, pp 135–153
30. Ghazal S, Munir A, Qureshi WS (2024) Computer vision in smart agriculture and precision farming: techniques and applications. Artif Intell Agric 13:64
31. Ghormade V, Deshpande MV, Paknikar KM (2011) Perspectives for nanobiotechnology enabled protection and nutrition of plants. Biotechnol Adv 29(6):792–803
32. Giraldo JP, Wu H, Newkirk GM, Kruss S (2019) Nanobiotechnology approaches for engineering smart plant sensors [Review of *Nanobiotechnology approaches for engineering smart plant sensors*]. Nat Nanotechnol 14(6):541
33. Guerrero A, de Neve S, Mouazen AM (2021) Current sensor technologies for in situ and on-line measurement of soil nitrogen for variable rate fertilization: a review. Adv Agron 168:1–38
34. Gulec F, Awan H, Wallbridge N, Eckford AW (2025) Decoding and engineering the Phytobiome communication for smart agriculture
35. Hamrani A, Allouhi A, Bouarab FZ, Jayachandran K (2025) AI and robotics in agriculture: a systematic and quantitative review of research trends (2015–2025). Crops 5(5):75
36. Hashim A, Al-Khafaji Y, Hadi A (2019) Synthesis and characterization of flexible resistive humidity sensors based on PVA/PEO/CuO Nanocomposites. Trans Electr Electron Mater 20:530–536

37. Jaiswal N, Kumar TV, Shukla C (2025) Smart drip irrigation systems using IoT: a review of architectures, machine learning models, and emerging trends. Discov Agric 3(1)
38. Javaid M, Haleem A, Khan IH, Suman R (2023) Understanding the potential applications of artificial intelligence in agriculture sector. Adv Agrochem 2(1):15–30
39. Jing K, Wang B, Yoshikuni Y (2020) Microbiome engineering: synthetic biology of plant-associated microbiomes in sustainable agriculture [Review of *microbiome engineering: synthetic biology of plant-associated microbiomes in sustainable agriculture*]. Trends Biotechnol 39(3):244
40. Joshi H, Choudhary P, Mundra SL (2019) Future prospects of nanotechnology in agriculture. Int J Chem Stud 7(2):957–963
41. Karunathilake EMBM, Le AT, Heo S, Chung YS, Mansoor S (2023) The path to smart farming: innovations and opportunities in precision agriculture. Agriculture 13(8):1593
42. Kashyap B, Kumar R (2021) Sensing methodologies in agriculture for soil moisture and nutrient monitoring. IEEE Access 9:14095–14121
43. Kilic AB, Akan OB (2025) Information and communication theoretical foundations of the internet of plants, principles, challenges, and future directions
44. Kiobia DO, Mwitta C, Fue KG, Schmidt JM, Riley DG, Rains GC (2023) A review of successes and impeding challenges of IoT-based insect pest detection systems for estimating agroecosystem health and productivity of cotton [Review of *a review of successes and impeding challenges of IoT-based insect Pest detection Systems for Estimating Agroecosystem Health and Productivity of cotton*]. Sensors 23(8):4127
45. Kumar J, Chawla R, Katiyar D, Chouriya A, Nath D, Sahoo S et al (2023) Optimizing irrigation and nutrient management in agriculture through artificial intelligence implementation. Int J Environ Clim Change 13(10):4016–4022
46. Kumar, V., Sharma, K. V., Kedam, N., Patel, A. B., Kate, T. R., & Rathnayake, U. (2024). A comprehensive review on smart and sustainable agriculture using IoT technologies [review of *a comprehensive review on smart and sustainable agriculture using IoT technologies*]. Smart Agric Technol, 8, 100487. Elsevier BV
47. Kumari K, Mirzakhani Nafchi A, Mirzaee S, Abdalla A (2025) AI-driven future farming: achieving climate-smart and sustainable agriculture. AgriEngineering 7(3):89
48. Kundu S, Kumar SV, Pooniya V, Reddy BM, Basha NH, Das K, Barman A, Dutta S, Shah MH, Erapuram P (2025a) Soil conservation techniques and sustainable land management. In: Eco-resilience: climate change, land degradation and sustainable solutions. Springer, Cham, pp 381–404
49. Kundu S, Kumar SV, Reddy BM, Kasbe SS, Goud GMK, Pooniya V, Banoth SK, Shah MH, Ali S, Pramanik K (2025b) Effect of doses, particle shape, size, and composition of nanofertilizers on yield of crops. In: Nanofertilizers in agriculture: synthesis, mechanisms, and effect on plants. Springer, Cham, pp 291–310
50. Kunt YE (2025) Development of a smart autonomous irrigation system using Iot and AI
51. Kuşcu M, Unluturk BD (2021) Internet of bio-Nano things: a review of applications, enabling technologies and key challenges [review of *internet of bio-Nano things: a review of applications, enabling technologies and key challenges*]. ITU J Future Evolv Technol 2(3):1
52. Lal R (2020) Home gardening and urban agriculture for advancing food and nutritional security in response to the COVID-19 pandemic. Food Secur 12(4):871–876
53. Li X, Gao Y, Li H, Majoral JP, Shi X, Pich A (2023) Smart and bioinspired systems for overcoming biological barriers and enhancing disease theranostics. Prog Mater Sci 140:101170
54. Li WJ, Luo Y, Jiang P, Dong X, Tang K, Liang Z, Shi Y (2025) A sustainable crop protection through integrated technologies: UAV-based detection, real-time pesticide mixing, and adaptive spraying. Sci Rep 15(1):35748
55. Li Z, Zhou J, Dong T, Xu Y, Shang Y (2021) Application of electrochemical methods for the detection of abiotic stress biomarkers in plants. Biosens Bioelectron 182:113105
56. Liu X, Zhao Z, Rezaeipanah A (2025) Intelligent and automatic irrigation system based on internet of things using fuzzy control technology. Sci Rep 15(1):14577

57. Lu Y, Luo Q, Jia X, Tam JP, Yang H, Shen Y, Li X (2023) Multidisciplinary strategies to enhance therapeutic effects of flavonoids from Epimedii folium: integration of herbal medicine, enzyme engineering, and nanotechnology. J Pharm Anal 13(3):239–254
58. Ma Y (2019) Seed coating with beneficial microorganisms for precision agriculture. Biotechnol Adv 37(7):107423
59. Mahajan DA, Bhosale M, Nikam P (2024) Optimizing fertilizer application using AI for sustainable agricultural practices. Nanotechnol Percept 4517-4530:10.62441/nano-ntp.vi.5326
60. Mahajan MR, Nangare SN, Patil PO (2023) Nanosize design of carbon dots, graphene quantum dots, and metal–organic frameworks based sensors for detection of chlorpyrifos in food and water: a review. Microchem J 193:109056
61. Mamatkulov Z, Safarov E, Oymatov R, Abdurahmanov I, Rajapbaev M (2021) Application of GIS and RS in real time crop monitoring and yield forecasting: a case study of cotton fields in low and high productive farmlands. E3S Web Conf 227:3001
62. Mana AA, Allouhi A, Hamrani A, Rehman S, El Jamaoui I, Jayachandran K (2024) Sustainable AI-based production agriculture: exploring AI applications and implications in agricultural practices. Smart Agric Technol 7:100416
63. Manoj HM, Shanthi DL, Lakshmi BN, Archana K, Jyothi EVN, Archana K (2025) AI-driven drone technology and computer vision for early detection of crop disease in large agricultural areas. Sci Rep
64. Midzi J, Jeffery DW, Baumann U, Rogiers SY, Tyerman SD, Pagay V (2022) Stress-induced volatile emissions and signalling in inter-plant communication [review of *stress-induced volatile emissions and signalling in inter-plant communication*]. Plants 11(19):2566
65. Mondal R, Dam P, Chakraborty J, Paret ML, Katı A, Altuntas S et al (2022) Potential of nanobiosensor in sustainable agriculture: the state-of-art. Heliyon 8(12)
66. Mousavi SR, Hakemi A (2024) Artificial intelligence (AI) and insecticides in pest control. IntechOpen eBooks
67. Mudrilov M, Ladeynova M, Grinberg M, Balalaeva IV, Vodeneev V (2021) Electrical Signaling of plants under abiotic stressors: transmission of stimulus-specific information [Review of *electrical Signaling of plants under abiotic stressors: transmission of stimulus-specific information*]. Int J Mol Sci 22(19):10715
68. Najdenovska E, Dutoit F, Tran D, Rochat A, Vu B, Mazza M, Camps C, Plummer C, Wallbridge N, Raileanu LE (2021) Identifying general stress in commercial tomatoes based on machine learning applied to plant electrophysiology. Appl Sci 11(12):5640
69. Nsoh B, Katimbo A, Guo H, Heeren DM, Nakabuye HN, Qiao X, Ge Y, Rudnick DR, Wanyama J, Bwambale E, Kiraga S (2024) Internet of things-based automated solutions utilizing machine learning for smart and real-time irrigation management: a review [review of *internet of things-based automated solutions utilizing machine learning for smart and real-time irrigation management: a review*]. Sensors 24(23):7480
70. Otieno M (2023) An extensive survey of smart agriculture technologies: current security posture. World J Adv Res Rev 18(3):1207
71. Oves M, Khan MS, Zaidi A, Ahmed AS, Ahmed, F., Ahmad E et al (2013). Antibacterial and cytotoxic efficacy of extracellular silver nanoparticles biofabricated from chromium reducing novel OS4 strain of *Stenotrophomonas maltophilia*. PLoS One 8(3):e59140
72. Parameswari P, Belagalla N, Singh BV, Abhishek GJ, Rajesh GM, Katiyar D, Hazarika B, Paul S (2024) Nanotechnology-based sensors for real-time monitoring and assessment of soil health and quality: a review. Asian J Soil Sci Plant Nutrition 10(2):157–173
73. Pérez-Hedo M, Gallego-Giraldo C, Forner-Giner MÁ, Ortells-Fabra R, Urbaneja A (2024) Plant volatile-triggered defense in citrus against biotic stressors. Front Plant Sci 15
74. Priya G, Ramya B, Purandareswari D, Likhitha N, Hari A, Babu C (2025) Smart farming revolution: a comprehensive review of nanotechnology integration in precision agriculture. Int J Res Agron 8:192–197. https://doi.org/10.33545/2618060X.2025.v8.i6Sc.3163
75. Ragaei M, Sabry AH (2014) Nanotechnology for insect pest control. Int J Sci Environ Technol 3:528–545

76. Rathinavel S, Priyadharshini K, Panda D (2021) A review on carbon nanotube: an overview of synthesis, properties, functionalization, characterization, and the application. Mater Sci Eng B 268:115095
77. Reissig GN, de Oliveira TFC, de Oliveira RP, Posso DA, Parise AG, Nava DE, Souza GM (2021) Fruit herbivory alters plant electrome: evidence for fruit-shoot long-distance electrical Signaling in tomato plants. Front Sustain Food Syst 5
78. Rithesh L, Mohapatra S, Jose S, Debnath J, Nayak G, Rahman M et al (2025) Machine learning-enhanced plant disease detection and management. Mach Learn Plant Biol 131–154
79. Rizwan M, Ali S, Rehman MZU, Qayyum MF, Ok YS, Wang H (2022) Nanotechnology and plant health: current status and future prospects. Environ Res 212:113178
80. Sagar S, Birje MN (2025) Precision agriculture using internet of things and cloud computing: a review. SN Comput Sci 6(4):1–18
81. Saha S, Bhardwaj A (2025) A narrative review of artificial intelligence to optimize the use of fertilizers: a game changing opportunity. Crop Forage Turfgrass Manage 11(1):e70027
82. Saini P, Lone AF (2025) Nano sensors for real-time soil and crop health monitoring: a review. J Adv Biol Biotechnol 28(12):461–471
83. Samreen T, Rasool S, Kanwal S, Riaz S, Sidra-Tul-Muntaha, Nazir MZ (2022) Role of nanotechnology in precision agriculture. Environ Sci Proc 23(1):17
84. Sang L, Zhao Y, Burda C (2014) TiO_2 nanoparticles as functional building blocks. Chem Rev 114(19):9283–9318
85. Sarma HH, Das BC, Deka T, Rahman S, Medhi M, Kakoti M (2024) Data-driven agriculture: software innovations for enhanced soil health, crop nutrients, disease detection, weather forecasting, and fertilizer optimization in agriculture. J Adv Biol Biotechnol 27(8):878–896
86. Senapaty MK, Ray A, Padhy N (2023) IoT-enabled soil nutrient analysis and crop recommendation model for precision agriculture. Computers 12:61
87. Shah MH, Aktar SN, Barik S, Chowdhury A, Molla SR, Kundu S, Ahmad S, Mukherjee D, Mandal B, Hossain A (2025a) Effect of various Nanofertilizers on physiological activities of plants. In: Nanofertilizers in agriculture: synthesis, mechanisms, and effect on plants. Springer Nature, Cham, pp 311–354
88. Shah SS, van Dam J, Singh A, Kumar S, Kumar S, Bundela DS, Ritsema C (2025b) Impact of irrigation, fertilizer, and pesticide management practices on groundwater and soil health in the rice–wheat cropping system—a comparison of conventional, resource conservation technologies and conservation agriculture. Environ Sci Pollut Res 32(2):533–558
89. Sharma K, Shivandu SK (2024) Integrating artificial intelligence and internet of things (IoT) for enhanced crop monitoring and management in precision agriculture. Sens Int 5:100292
90. Shelar A, Salve S, Shende H, Mehta D, Chaskar M, Nile SH, Patil R (2024) Recent advances on highly sensitive plasmonic nanomaterial enabled sensors for the detection of agrotoxins: current progress and future perspective. Comput Electron Agric 227:109545
91. Shulkin JD, Vibhatasilpin A, Adhana V (2025) Pervasive sensing for livestock health and activity monitoring: current methods and techniques. arXiv (Cornell University)
92. Singh M, Goswami SP, Sachan P, Sahu DK, Beese S, Pandey SK (2024b) Nanotech for fertilizers and nutrients-improving nutrient use efficiency with nano-enabled fertilizers. J Experim Agric Int 46(5):220–247
93. Singh P, Sharma H, Kuntal P (2020) Agricultural robots for harvesting and planting. Türk Bilgisayar ve Matematik Eğitimi Dergisi 11(1):1112
94. Singh M, Upadhyay L, Badekhan A, Shil S, Gautam SK, Anand A (2024a) Leveraging remote sensing and nanotechnology to overcome barriers to agroforestry adoption by small holder farmers. J Sci Res Rep 30(5):113–140
95. Song Y, Wang Z, Wu Q, Su J, Liao J, Zhang X et al (2025) A dual-mode strategy for early detection of sugarcane pokkah boeng disease pathogen: a portable sensing device based on cross-N DNA framework and MoS2@ GDY. Biosens Bioelectron 267:116874
96. Sow S, Ranjan S, Seleiman M, Alkharabsheh H, Kumar M, Kumar N et al (2024) Artificial intelligence for maximizing agricultural input use efficiency: exploring nutrient, water and weed management strategies. Phyton 93(7):1569

97. Srinivasan B, Tung S (2015) Development and applications of portable biosensors. J Lab Autom 20(4):365–389
98. Stojanova MT, Djukic DA, Stojanova M (2025) Nano fertilizers: a new approach to horticultural. Int J Nanosci Nanotechnol 21(1):49–58
99. Talaviya T, Shah D, Patel N, Yagnik H, Shah M (2020) Implementation of artificial intelligence in agriculture for optimisation of irrigation and application of pesticides and herbicides. Artif Intell Agric 4:58–73
100. Tomkins M, HUGHES AP, Morris RJ (2021) An update on passive transport in and out of plant cells [review of *an update on passive transport in and out of plant cells*]. Plant Physiol 187(4):1973
101. Usigbe MJ, Uyeh DD, Park T, Ha Y, Mallipeddi R (2025) Many objective optimization and decision support for dairy cattle feed formulation. Sci Rep 15(1):13451
102. Vakilipoor F, Hirschmann N, Schladt J, Schwab S, Reineke A, Schober R, Castiglione K, Schaefer M (2025) MC for agriculture: a framework for nature-inspired sustainable Pest control. arXiv (Cornell University)
103. Vedantham AK (2024) Cloud and IoT technologies revolutionizing precision agriculture. Int J Sci Res Comput Sci Eng Inform Technol 10(6):860–867
104. Vidu R, Matei E, Predescu AM, Alhalaili B, Pantilimon C, Tarcea C, Predescu C (2020) Removal of heavy metals from wastewaters: a challenge from current treatment methods to nanotechnology applications. Toxics 8(4):101
105. Wang S, Xu D, Liang H, Bai Y, Li X, Zhou J, Su C, Wei W (2025) Advances in deep learning applications for plant disease and Pest detection: a review [Review of *advances in deep learning applications for plant disease and Pest detection: a review*]. Remote Sens 17(4):698
106. Wang J, Chang L, Aggarwal S, Abari O, Keshav S (2020) Soil moisture sensing with commodity RFID systems. In: Proceedings of the 18th international conference on mobile systems, applications, and services, pp 273–285
107. Warriner K, Reddy SM, Namvar A, Neethirajan S (2014) Developments in nanoparticles for use in biosensors to assess food safety and quality. Trends Food Sci Technol 40(2):183–199
108. Weraikat D, Šorić K, Žagar M, Sokač M (2024) Data analytics in agriculture: enhancing decision-making for crop yield optimization and sustainable practices. Sustainability 16(17):7331
109. Wolfert S, Ge L, Verdouw C, Bogaardt MJ (2017) Big data in smart farming–a review. Agric Syst 153:69–80
110. Xing Y, Wang X (2024) Precise application of water and fertilizer to crops: challenges and opportunities. Front Plant Sci 15:1444560
111. Yadav A, Yadav K, Ahmad R, Abd-Elsalam KA (2023) Emerging frontiers in nanotechnology for precision agriculture: advancements, hurdles and prospects. Agrochemicals 2(2):220–256
112. Yanes AR, Martinez P, Ahmad R (2020) Towards automated aquaponics: a review on monitoring, IoT, and smart systems. J Clean Prod 263:121571
113. Yousaf A, Kayvanfar V, Mazzoni A, Elomri A (2023) Artificial intelligence-based decision support systems in smart agriculture: Bibliometric analysis for operational insights and future directions. Front Sustain Food Syst 6
114. Zain M, Ma H, Rahman SU, Nuruzzaman M, Chaudhary S, Azeem I et al (2024) Nanotechnology in precision agriculture: advancing towards sustainable crop production. Plant Physiol Biochem 206:108244
115. Zeng S, Shakoor N, Rui Y (2025) Nanotechnology and agricultural sustainability: a review. Nano 15(23):1755
116. Zhang P, Guo Z, Ullah S, Melagraki G, Afantitis A, Lynch I (2021) Nanotechnology and artificial intelligence to enable sustainable and precision agriculture. Nat Plants 7(7):864–876
117. Zhang F, Li D, Li G, Xu S (2025) New horizons in smart plant sensors: key technologies, applications, and prospects [Review of *new horizons in smart plant sensors: key technologies, applications, and prospects*]. Front Plant Sci, 15:1490801
118. Zhao F, He J, Li X, Bai Y, Ying Y, Ping J (2020) Smart plant-wearable biosensor for in-situ pesticide analysis. Biosens Bioelectron 170:112636

119. Zhou X, Zhang J, Shi J, Rahman MK Liu H, Wei Z, Wu F, Dini-Andreote F (2024) Volatile-mediated interspecific plant interaction promotes root colonization by beneficial bacteria via induced shifts in root exudation. Microbiome 12(1)

Climate-Ready Agriculture: Integrating AI Models and Agricultural Biotechnology

Manoj Mani, Nithya Palaniappan, Balamuralikrishnan Balasubramanian, Thilagavathi Gangadharan, Alwin Robert Asirvatham, Antony Prabhu Jeyabal Philomenathan, Sivaranjini Annamalai, and Vijaya Anand Arumugam

Abstract Climate-smart agriculture requires the convergence of artificial intelligence and advanced biotechnologies to build resilient, productive, and sustainable food systems under climate change. This chapter synthesizes emerging innovations across AI-driven crop modeling, remote sensing, precision agriculture, and genome engineering to enhance stress tolerance and resource efficiency. AI-biotechnology integration accelerates crop improvement by enabling predictive breeding, multi-omics data analytics, and optimized trait selection, while digital agriculture platforms spanning IoT sensors, UAV imaging, digital twins, and smart irrigation provide real-time monitoring and adaptive field management. Genome editing tools such as CRISPR-Cas systems further strengthen climate resilience by enabling precise modification of stress-responsive pathways, complementing microbiome engineering and

M. Manoj · J. P. Antony Prabhu · A. Vijaya Anand (✉)
Department of Human Genetics and Molecular Biology, Bharathiar University, Coimbatore, Tamil Nadu, India
e-mail: avahgmb@buc.edu.in

P. Nithya
Department of Botany, Padmavani Arts and Science College for Women (Autonomous), Salem, Tamil Nadu, India

B. Balamuralikrishnan
Department of Food Science and Biotechnology, Sejong University, Seoul, South Korea

G. Thilagavathi
Department of Information Technology, Coimbatore Institute of Technology, Coimbatore, Tamil Nadu, India

A. Alwin Robert
Department of Endocrinology and Diabetes, Prince Sultan Military Medical City, Ministry of Defense, Riyadh, Kingdom of Saudi Arabia

A. Sivaranjini (✉)
Department of Biotechnology, Sri Ramakrishna College of Arts & Science, Coimbatore, Tamil Nadu, India
e-mail: sivaranjini@srcas.ac.in

J.-T. Chen (ed.), *AI-Nano-Enabled Climate-Smart Agriculture and Sustainability*, Smart Nanomaterials Technology, https://doi.org/10.1007/978-981-92-1694-9_2

soil health strategies. Ethical, economic, and regulatory frameworks are also examined to ensure equitable, safe, and scalable deployment of these technologies. Collectively, the chapter presents an integrated roadmap for developing climate-ready agriculture through converging digital, molecular, and ecological innovations.

Keywords Climate-smart agriculture · Artificial intelligence · Biotechnology · Genome editing · Precision agriculture

Abbreviations

APSIM	Agricultural production system simulator
CNN	Convolutional neural network
DL	Deep learning
DNN	Deep neural network
DT	Digital twins
IoT	Internet of things
ML	Machine learning
PGPR	Plant growth-promoting rhizobacteria
RF	Random forest
ROS	Reactive oxygen species
UAV	Unmanned aerial vehicle

1 Climate Change Stressors and Crop Adaptation

Abiotic stresses, including heat, drought, and salinity arising from climate change, seriously affect the physiology and productivity of crops, impairing key biochemical, physiological, and molecular processes. Such stress factors constrain biomass production and photosynthetic capacity, hence decreasing yield potential by altering growth pathways and constraining arable land use. Drought and salinity, which generally occur together in arid and semi-arid areas, cause severe reductions in grain yield and chlorophyll content in crops, especially for temperate cereal crops, for which osmotic stress and ion toxicity seriously impede metabolic homeostasis [56]. Responses to these factors include osmotic adjustment, activation of antioxidant defense pathways, and hormonal regulation, which, taken together, ultimately result in the development of stress tolerance in plants [2]. Artificial intelligence, precision agriculture, and modern biotechnological means are now increasingly considered an integral part of crop management frameworks toward improving resilience and resource-use efficiency [97]. Besides these approaches, genetic engineering and selection using genome resources have become highly essential for developing resistant genotypes because of the projected growth in population and resultant demand for food security worldwide. Elevated CO_2 may enhance photosynthetic rates and yields, yet this is mostly offset by global warming, and crop yields are expected to be adversely affected by ~ 10% for every 1 °C rise in temperature [22].

Due to increased climate variability, plants are more frequently facing combined and successive periods of stress, requiring multidimensional responses. For instance, under salt stress in combination with high irradiance, plants reprogram gene expression and reshape metabolic pathways through the action of transcription factors like MYB, WRKY, and bHLH to gain tolerance [82]. These molecular changes interact with cellular responses such as modulation of the levels of reactive oxygen species (ROS), which act not only as destructive agents but also as signaling molecules initiating defense pathways. Phytohormones again mediate cross-talk between abiotic and biotic stress signaling networks, thereby enabling plants to optimally allocate resources in dynamic environments [69]. The morpho-physiological traits comprise such basic tolerance attributes during episodes of drought, flooding, or heat: stomatal closure to prevent water loss; developing a better root system for enhanced water uptake; and structural changes in order to maintain cellular integrity. In coffee plants, for example, severe water deficit in association with heat stress provokes antioxidant enzyme activity and the build-up of light-protective pigments such as zeaxanthin and lutein, contributing to membrane stabilization and protecting photosystem II performance [76]. Transcriptomic studies further put in focus that the modulation of stress-responsive gene networks by plants is dynamic under successive and combined stress regimes.

Therefore, accurate predictions of crop responses to climate change require advanced modeling approaches that integrate physiological insights into environmental datasets. Combining machine learning (ML) methods with process-oriented crop models, such as Agricultural Production Systems Simulator (APSIM), significantly improves the predictive performance of crop models under extreme climate stresses. A hybrid APSIM-Random Forest (RF) framework improved wheat yield prediction in Australia by 33% compared to APSIM alone and emphasized the need to consider drought and heat stress during the critical growth stages of wheat [26]. Physiological determinants, such as stomatal conductance, enzyme stability, and photosynthetic efficiency, remain crucial for the mechanistic interpretation of the impact of abiotic stresses on yield [25]. Development of existing crop models, namely, ORYZA and CERES-Rice, that incorporate these physiological determinants enables improved simulations of rice responses to rising temperature, altered precipitation regimes, and increased CO_2. Physiology-dependent crop simulation models are also indispensable to extrapolate climate impact across agroclimatic zones and management frameworks, and thereby support the development of targeted adaptation strategies [9]. Such models have been widely applied to explore management interventions-particularly irrigation scheduling and planting dates-where genetic improvement strategies, although much needed for long-term resilience, comparatively received lesser emphasis [8].

2 AI Models for Predicting Crop Stress and Yield

ML and deep learning (DL) approaches have taken center stage in crop yield predictions under climate variability, as these techniques provide substantial enhancements compared to traditional statistical techniques. These models enable agricultural sustainability through optimization in resource allocation, early risk detection, and building resilience plans in the face of climate stress [81]. The performance of DL architectures such as convolutional neural networks (CNNs), long short-term memory networks, and deep neural networks (DNNs) stands out, handling nonlinear, large-scale, and multivariate spatiotemporal datasets, often superior to the performance of traditional ML algorithms, including support vector machines and RFs [34]. Integrating AI with geographic information systems and remote sensing technologies supports an increase in spatial and temporal resolution in yield prediction. Thus, models will become more robust against extreme climatic fluctuations [29]. Ensemble-based DL models employing stacking and boosting showed high predictive accuracy in environments characterized by high climate stress. Examples include Saudi Arabia, where models explain up to 96% of variability in crop yields at a minimum uncertainty [81]. Strong performances are also observable for hybrid frameworks that incorporate CNN, DNN, and recurrent neural network architectures combined with optimization algorithms. Benchmark datasets, such as the USCC dataset, revealed an improved capacity of machine and DL models to decode climate-crop performance interactions as foundational resources for developing predictive systems [29].

Parallel to yield prediction, the integration of AI with satellite, unmanned aerial vehicle (UAV), and multispectral imaging has transformed the detection of climate-induced stress in crops. Innovations in hyperspectral imaging, thermal sensing, and multimodal data fusion have hugely improved drought, heat, and nutrient stress detection by allowing for precise monitoring required for sustainable resource management [94]. Machine vision systems with multispectral and hyperspectral sensors are capable of non-invasive, real-time assessment of plant health by capturing diagnostic reflectance signatures associated with an early onset of stress, while AI-driven automation accelerates the analysis of extensive image datasets [32]. Low-cost computer vision platforms, in association with ML algorithms, have achieved more than 80% detection accuracy of drought stress in crops like corn and soybean, hence underlining the possibility of scalable tools for monitoring stress [71]. A multiscale monitoring framework that resolves key limitations of satellite-only approaches, including mixed pixel effects and temporal discontinuities, is enabled by fusing UAV-derived high-resolution imagery with satellite data. This, in turn, enhances the precision of early stress detection and field-level decision-making [78]. Vegetation Health Indices derived from AVHRR data have also played a prominent role in early drought warning and quantification of agricultural losses, thus demonstrating the global applicability of remote sensing toward climate impact assessment and food security monitoring [39].

With these developments, predictive modeling of climate-crop interactions increasingly utilizes hybrid AI frameworks that integrate environmental datasets with domain-based knowledge to support robust agricultural decision-making. Merging historical yield records, multisource satellite imagery, meteorological variables, and soil characteristics enables the development of high-precision models capable of capturing complex interactions across scales. Hybrid ensemble approaches that combine RF and gradient boosting with expert knowledge outperformed conventional techniques for yield prediction, underlining the added value brought by integrating empirical and heuristic components [75]. Embedding climate variables into ML frameworks such as multilinear regression, RFs, and XGBoost provided a more in-depth insight into crop responses under future climate scenarios; these predictions include yield forecasting for legume species subjected to different climatic pressures [51]. Besides this, hybrid stochastic-ML frameworks account explicitly for uncertainties inherent in agrometeorological processes, improving the robustness of the predictions made under environmental variability conditions [85]. The application of time series forecasting models using ensemble ML methods, such as RF and Bagging Regressor, further fortifies the accuracy and reliability of the predictions to dynamically support data-driven crop management strategies [101].

3 Digital Twins and Precision Agriculture

Agricultural digital twins (DT) are emerging as powerful tools for real-time simulation of crop growth and climate stress responses, thus promising a revolutionary paradigm for data-driven farming. A virtual twin of agricultural ecosystems integrates multimodal data streams from Internet of Things (IoT) sensors, remote sensing platforms, and precision farming machinery to provide dynamic simulations that accurately reflect actual conditions of crop growth [17]. Continuous data assimilation enables the DTs to offer high-fidelity predictions of crop developmental trajectories, hence enabling early detection of stress events and optimization of decision-making. Integration of DTs with IoT technologies further enhances their operational capabilities by way of continuous monitoring of crop health and provision of real-time analytics to support agronomic planning [15]. Recent advances involving the integration of the reasoning capability of large language models in DT platforms have ensured massive improvements in the prediction of vegetable crops through recognition of growth stages with high accuracy and simulation of long-run growth dynamics-skills necessary in optimizing inputs and mitigating risks associated with climate variability and biotic stressors [21]. DT frameworks will also be comprised of detailed representations of soil composition, microclimatic conditions, and field heterogeneity that will ultimately enable farmers themselves to make informed decisions on irrigation, fertilization, and pest management, leading to improvements in productivity with sustainability [64].

Complementing DT development, the use of remote sensing, AI, and precision farming in digital agriculture platforms enhances crop management and resource-use efficiency. For instance, Agri Sense systems utilize IoT sensors, AI-enabled analytics, and precision agriculture workflows to provide real-time insights into soil moisture dynamics, microweather patterns, and crop-specific irrigation requirements, thus enhancing yield potential with minimized environmental impacts [59]. Digital farming is increasingly embracing IoT networks, big data analytics, and remote sensing to allow for the precise site-specific management interventions that can reduce fertilizer and pesticide applications by up to 30%, while saving water by nearly 50% [74]. Centralized platforms extend this capability through the integration of soil moisture sensors and AI-based recommendation engines toward targeted crop advisory and improved resource-use efficiency [20]. Real-time detection of nutrient status, water stress, and infestation or infection of crops is enabled by integrating UAV-based imaging and hyperspectral sensing with ML, which in turn allows for responsive data-driven agronomic decisions [45].

These advances come together in DT architectures, which are increasingly seen for their potential to optimize irrigation scheduling, nutrient management, and climate-smart crop planning. By integrating IoT-generated field data, environmental sensor networks, and meteorological forecasts, DTs represent an integrated view of field conditions that enables the specification of precise irrigation and fertilization strategies to conserve resources and minimize environmental burdens [64]. Deployment of cloud-fog-edge computing infrastructures further increases the responsiveness of DTs by ensuring real-time processing and providing farmers with the capability to test "what-if" scenarios as a means of determining how to optimize field operations [35]. Beyond increasing productivity, DTs improve the sustainability of agriculture by reducing requirements for chemical inputs and enhancing traceability-critical attributes in light of increasing uncertainty in climate [11]. Case studies illustrate the opportunities for large reductions in water use, fertilizer inputs, and pesticide applications that accrue with DT-driven management, along with measurable improvements in the sustainability and efficiency of farm production [23].

4 Genome Editing an Biotech for Climate-Resilient Crops

The CRISPR/Cas-mediated genome editing approach has emerged as highly efficient in enhancing tolerance against drought, heat, and salinity, among the major constraints imposed by climate change on agricultural systems. This allows for precise and efficient modification of genes responsive to stresses, reducing dependency on conventional breeding strategies, which often prove slow and less accurate [33]. Some of the most striking examples of such an approach are represented by CRISPR/Cas9-mediated editing of OsRR22 and OsDST in rice, significantly enhancing salinity and drought tolerance, respectively [58]. Development of multiplex CRISPR/Cas platforms capable of targeting several genes at a time further accelerates the process of generating climate-resilient genotypes [103]. The new emerging genome editing tools, such as base editing and prime editing, enable a wider range of possible precise nucleotide modifications without inducing double-strand breaks, hence increasing efficiency and reducing off-target effects. These approaches become particularly critical in arid and semi-arid regions where high temperatures and scarcity of water may severely limit the productivity of crops [33]. Integration of CRISPR/Cas Technologies with Other Complementary Approaches. Epigenetic modulation, genome-wide association studies, and functional genomics provide a very robust framework for designing cultivars with multiple abiotic stressors [36, 58].

Genome editing, transgenic, and molecular breeding strategies complementing trait introgression across diverse crops make selection for climate-resilient characteristics possible. Improvement of adaptive traits for various abiotic stressors, such as heat tolerance, drought, and salinity, is targeted in several crops using breeding tools of different approaches: genomic selection, marker-assisted selection, and CRISPR-Cas systems [67]. Integrative genomics allows multi-omics, proteomics, and transcriptomics to enable the accelerated identification and functional characterization of genes involved in the stress signaling pathway that empowers even more precise and personalized decisions in breeding [54]. These innovative approaches support classical breeding to develop improved rice, wheat, and maize varieties with enhanced stress tolerance under climate variability [90]. Thus, tapping genetic diversity in underutilized crops, particularly pulses, will provide novel resources to help breed for temperature extremes and water deficit stress. Transgenics and genome editing could play an important role in the exploitation of such genetic variability [40]. Limitations related to the polygenic control of stress responses and regulatory constraints on breeders still remain; however, big data analytics, ML, and systems biology approaches continue to enhance actionable insights from large-scale genomic datasets, thereby facilitating climate-smart crop development even further [90].

Further reinforcement of crop resilience is imparted through the recent advances in the biotechnological modification of genes responsive to stresses, including enhancing the molecular pathways of stress perception, signaling, and tolerance. Due to trait improvements gained in drought tolerance, root architecture, and water-use efficiency by targeted manipulations of key regulatory genes using CRISPR/Cas9 and other genome-editing approaches, the ground condition has been remarkably altered [108]. Such interventions support biosynthesis and regulate various protective compounds, such as proline, glycine betaine, LEA proteins, and heat shock proteins, that maintain cellular homeostasis and reduce oxidative damage under extreme conditions. Transcription factors, including NAC, MYB, bZIP, and WRKY, act as master regulators in controlling stress responses, and their manipulation commonly aims at improving tolerance by perturbing physiological and biochemical pathways. These genome editing approaches are supplemented by various molecular breeding approaches, such as genomic selection and marker-assisted selection, which enable rapid screening and incorporation of the stress-adaptive allele into elite cultivars. When biotechnological innovations are combined with sustainable agricultural practices, beneficial soil microbes, and diversified cropping systems, the stress tolerance developed in the laboratory often translates into consistent field performance that is central to the pursuit of long-term resilience against climate-induced stressors [7].

5 Multi-omics for Climate Stress Trait Discovery

Collectively, genomics, transcriptomics, proteomics, and metabolomics represent an integrated framework that reinforces the dissection of the complex molecular architecture underlying plant responses to climate-induced abiotic stresses. Multi-omics methods achieve this by allowing for the characterization in one go of genetic variation, transcriptional regulation, protein dynamics, and changes in metabolite fluxes, which provide the systems-level understanding needed to identify resilient traits against abiotic stresses (Fig. 1). Genetic polymorphisms induced by abiotic stress, transcriptional reprogramming, and stress-responsive regulatory nodes are pointed out by in silico genomic and transcriptomic analyses. The latter identify tolerance mechanisms-related candidate genes and stress-responsive networks [6, 49]. By complementing these layers, proteomics and metabolomics capture functional outputs, such as enzymes and osmo-protectants induced by stress, antioxidants, and metabolic reconfigurations with a view to maintaining redox and cellular homeostasis under adverse conditions [53]. In this way, key pathways and biomarkers are discovered, which can be exploited in genetic engineering and precision breeding toward enhanced climate resilience [77].

Recent advances in high-throughput sequencing, spectral profiling, and computational biology have further empowered multi-omics network analysis to comprehensively decipher plant responses under drought, salinity, heat, and combined stresses. The integration of multi-layered omics datasets has enhanced resolution for the stress-responsive regulatory circuitry, hence allowing identification of transcriptional hubs, metabolic rewiring events, and cross-talk between signaling pathways [91]. Empirical studies in crops like potato have underlined the strength of this approach by pointing out a distinct molecular signature associated with co-stress imposed by heat and drought through suppression of photosynthetic machinery and accumulation of hormonal mediators like ABA and jasmonates [105]. These indeed provide a mechanistic lead for climate-smart development of crops through genome-assisted breeding and targeted engineering of stress-adaptive traits [77]. With this backdrop, biomarkers through omics are all set to occupy the center stage in the prediction of stress tolerance and speeding up trait discovery pipelines. The genes, proteins, and metabolites associated with stress, as identified through integrated omics analyses, enhance the precision of phenotyping and enable early-stage screening for tolerant genotypes [12, 49]. Multi-omics integration, especially the integration of transcriptomic and proteomic datasets with metabolomic signatures, enables reconstruction of stress-responsive metabolic networks and determines quantitative relationships between gene expression, protein abundance, and metabolic outputs [99]. Though the scale and complexity of multi-omics data pose a challenge for integration and interpretation, improved experimental design, normalization pipelines, and system biology analytical frameworks continuously drive rapid progress in the discovery of climate resilience traits [110].

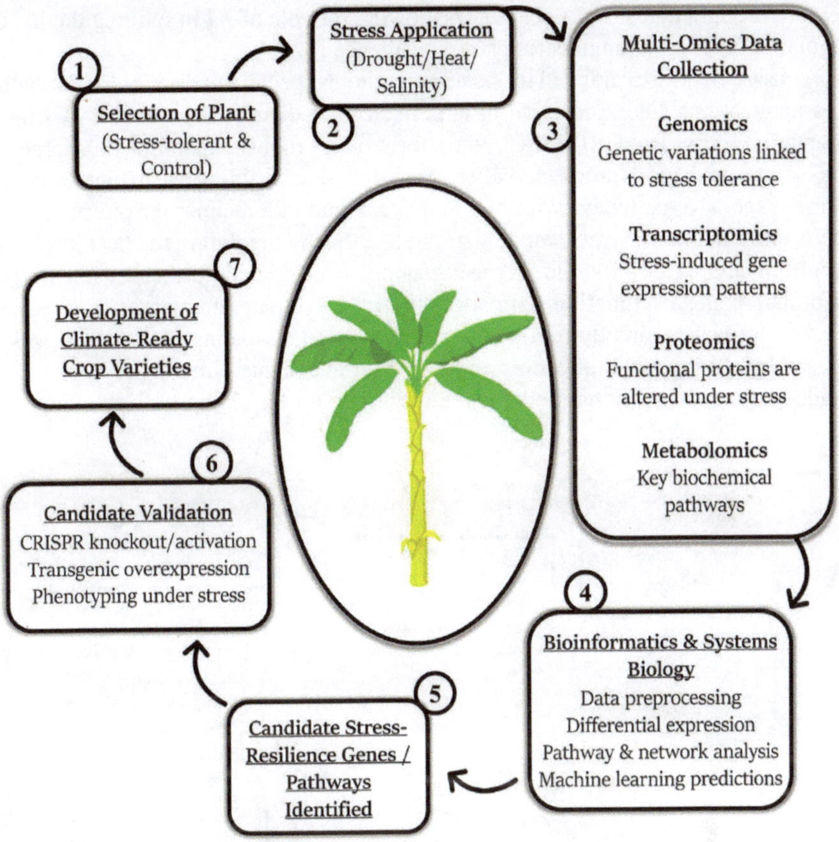

Fig. 1 Multi-omics integration for climate-stress trait discovery

6 AI-Guided Genome Editing Targets

The role of AI in target discovery and prioritization for the improvement of crops with climate resilience has become indispensable. By integrating multi-omics datasets, including genomics, transcriptomics, proteomics, metabolomics, and epigenomics, AI and ML systems identify genes, regulatory elements, and molecular pathways that underpin stress tolerance by deciphering complex regulatory networks [96]. In the concept of Breeding 5.0, AI-driven predictive analytics decodes germplasm architecture and measures genotype-environment interaction for accurate trait modeling and accelerates candidate selection suitable for genome editing interventions (Fig. 2) [28]. AI's ability to integrate genomic, phenotypic, and environmental datasets greatly improves the accuracy of target selection and enables data-driven breeding strategies toward the development of climate-resilient varieties [100]. Identification of regulatory small RNAs using in silico tools expanded the CRISPR target discovery pipeline; microRNAs like Os-miR408 and Os-miR535 have been validated for their roles in drought and salinity tolerance, showing the role of AI in refining the list of regulatory targets for engineering stress resilience [62].

AI-assisted prioritization of the gene targets holds special importance for CRISPR/Cas applications since precision in target selection dictates the success of stress-adaptation traits. The CRISPR/Cas9 platform offers robust capabilities for targeted knockouts, gene modulation, and allelic rewiring, thus enabling the manipulation of stress-responsive pathways with high efficiency and in a manner devoid of foreign DNA introduction [63]. ML models strengthen this by predicting the best targets to edit from large-scale genomic and post-genomic datasets and, further, by integrating information on gene function, network centrality, and environmental responsiveness. These approaches directly reinforce strategies aimed at abating global yield losses caused by abiotic and biotic stresses [60]. With increasing climate pressures, ML-guided target discovery has become indispensable for hastening the deployment of

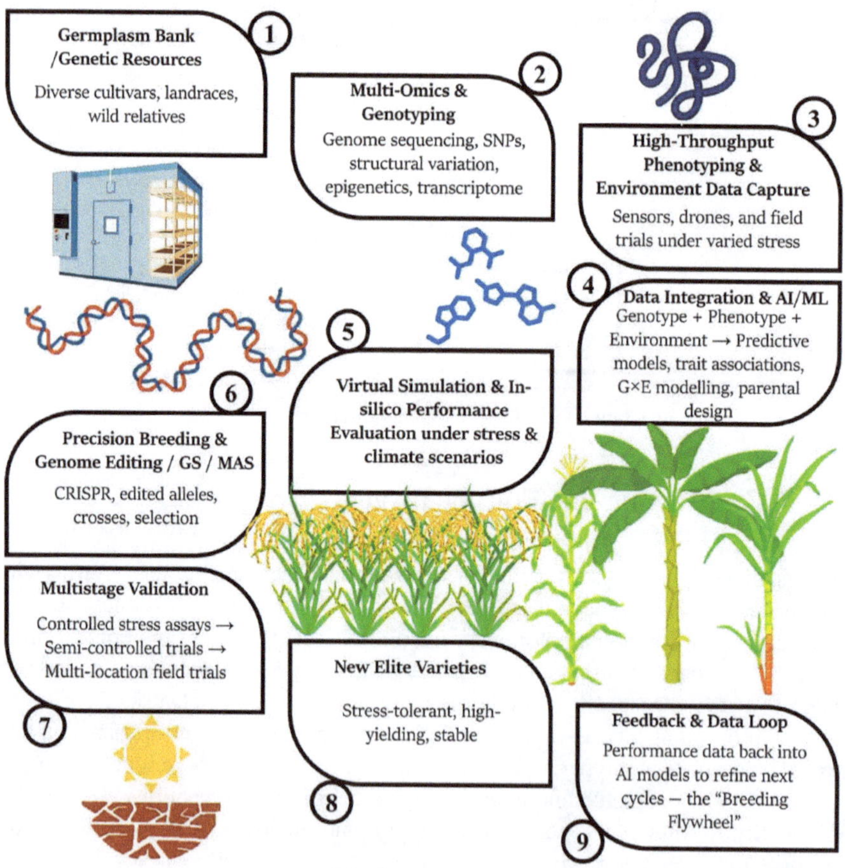

Fig. 2 Conceptual pipeline of breeding 5.0

stress-tolerant genotypes. However, regulations will have to evolve to accommodate the safe adoption of genome-edited crops [30].

AI-driven predictive modeling further strengthens target discovery by reconstructing and quantifying the dynamics of gene regulatory networks underlying abiotic stress responses. ML algorithms have shown excellent performance in the prediction of stress-responsive genes, for example, models developed in Chinese cabbage that classify genes linked to the cold, heat, drought, and salinity responses with excellent efficiency [102]. Network-based ML tools, such as mlDNA, enhance the detection of important transcription factors and regulatory nodes in species like Arabidopsis thaliana, outperforming traditional coexpression approaches in identifying stress-responding genes [42]. AI and DL frameworks also perform large-scale physiological and environmental data analysis for refining biomarker prediction and developing stress management protocols, including hormesis-based optimization strategies. ASRpro, a tool using support vector machines and other ML algorithms, allows the identification of genes responsive to multiple stresses and offers a computational alternative to time-consuming and work-intensive genetic screening [44].

7 Microbial Biotechnology and Soil Resilience

Plant growth-promoting rhizobacteria (PGPRs) represent an essential element in microbial biotechnology for the improvement of plant tolerance to drought and salinity, major abiotic stressors that limit productivity in arid and semi-arid ecosystems. Such microbes modulate plant responses through multi-faceted direct and indirect mechanisms that support plant growth under stress. Directly, PGPR enhances nutrient acquisition through biological nitrogen fixation, siderophore production, and the solubilization of key nutrients, thereby countering nutrient limitations that intensify in drought and salinity [73]. Indirectly, PGPR manipulate host physiological and biochemical pathways by enhancing antioxidant enzyme activities that partly mitigate ROS-induced oxidative damage. They further facilitate plant water relations through increased relative water content, reduced electrolyte leakage, and stabilization of chlorophyll concentration, collectively leading to improvement in photosynthetic performance [38]. Experimental evidence from barley systems reveals that strains such as Pseudomonas fluorescens and Pseudomonas putida significantly enhance seed germination, accumulation of biomass, and nutrient uptake under stress [106]. Similarly, the Rhizobium species has been reported to improve shoot growth and biomass and soil nutrient status under both drought and salinity adversities [5].

Building on the functional capacity of naturally occurring PGPR, microbiome engineering provides an advanced, programmable strategy for strengthening crop resilience via plant holobiont manipulation. By integrating synthetic biology, CRISPR interference, and computational modeling, it is possible to design engineered SynComs that can provide targeted, stress-mitigating functions in a manner that dynamically adapts to fluctuating environments [66]. The value contributed by these engineered microbiomes to crop performance is extended beyond enhancing tolerance to abiotic stresses, such as drought, heat, and salinity, through improving nutrient cycling efficiency and reducing dependency upon agrochemicals, thus aligning with broader sustainability goals for agricultural production [52, 95]. Indeed, manipulation of the root microbiota has been used in rice systems through inoculation and gene-editing approaches, demonstrating significant potential for generating cultivars that are resilient to climate change pressure [46], underlining translational prospects in staple crops.

Beyond those associated with plants, soil microbial communities are crucial to ecosystem-level resilience because they drive nutrient cycling, disease suppression, and growth promotion functions that have increasing importance under climate-induced stress regimes such as warming, drought, and salinity [84, 93]. Soil microbiome health affects carbon sequestration dynamics, soil fertility, and biocontrol potential, and thus it is critical for long-term agricultural stability. The diversity and complexity of plant-microbe-soil interactions require a systems-level framework to understand, predict, and modulate these interactions. Advanced tools like fabricated ecosystems, synthetic communities, and integrative genomic and biochemical profiling enable the mechanistic dissection of microbe-plant-soil interactions and promote the strategic design of microbial interventions to improve crop performance under changing climatic conditions [13].

8 IoT, Drones and Smart Farming Infrastructure

While a great step toward precision farming, the adoption of IoT sensor networks will enable continuous, real-time monitoring of microclimatic and soil conditions. IoT-enabled platforms are being deployed to acquire spatiotemporal high-resolution datasets unavailable with conventional sampling; these distribute sensor arrays that quantify the critical soil parameters of moisture, temperature, pH, electrical conductivity, and macronutrient concentrations [10]. In turn, the data streams are transmitted to cloud-based analytics, allowing adaptive irrigation scheduling and nutrient management decisions and the monitoring of crop health to optimize resource-use efficiency and support more sustainable agricultural intensification [57]. Further, long-range communication technologies such as LoRaWAN extend the coverage of these networks, therefore enabling monitoring over large production landscapes without appreciable energy demands [72]. Integration of weather sensors in addition to soil sensors further strengthens predictive capabilities by quantifying the interaction of atmospheric variables with soil processes, thus resulting in improved water management and conservation of soil [68]. Apart from this, self-powered sensor units based on solar harvesting increase the longevity of systems and reduce maintenance burdens, while analysis techniques such as EWMA improve the sensitivity to identify rapid changes in soil conditions relevant for the onset of stress [98].

The growth of UAVs and drone-based imaging systems further complements ground-based IoT networks in enabling the rapid, landscape-scale surveillance of crop stress dynamics. UAVs equipped with multispectral, thermal, and RGB sensors capture high-resolution imagery for detecting early indicators of abiotic stress, such as changes in canopy temperature, chlorophyll reflectance, and vegetation indices, which are crucial parameters for timely agronomic interventions [78]. Similarly, the incorporation of DL frameworks such as the enhanced Yolo-v8+ model considerably enhances the precision and recall of the detection of stress in crops like maize, thereby supporting real-time diagnostic capability [55]. Emerging aerial IoT platforms combine UAV mobility with hyperspectral imaging and capture physiological responses to drought, salinity, and thermal stresses at fine spatial and temporal scales [83]. Retrieval of vegetation indices-especially the normalized difference vegetation index, from drone imagery allows for the automated detection of hot spots of stress and enables autonomous assessments across fields with the help of advanced computer vision approaches [92]. In soybean systems, UAV-assisted assessments, when integrated with soil and microclimatic data, have proved to be effective in identifying the developmental windows sensitive to abiotic stress and underlying environmental drivers such as soil moisture deficit and elevated temperature as modulators of crop performance [43].

Complementing sensor networks and drone systems, smart irrigation and automation platforms are emerging as critical infrastructures for climate-adaptive agriculture. These systems integrate IoT, artificial intelligence, and data-driven decision frameworks to optimize water allocation with high temporal precision. The EcoStream SmartFlow System, for instance, uses temperature and humidity sensors to autonomously adjust irrigation based on the prevailing microclimatic conditions; this improves water-use efficiency and yield outcomes [80]. IoT-controlled irrigation architecture using microcontrollers like Raspberry Pis and workflow platforms like Node-RED can monitor in real time and automatically readjust irrigation schedules, thereby enabling sustainable water management, especially in water-scarce regions [1]. AI and ML further fine-tune these systems through the prediction of crop-specific water requirements under variable climate scenarios and by dynamically adjusting irrigation delivery [14]. Sensor-based irrigation systems drastically reduce the labor required while improving yield stability and water productivity, hence ultimately reinforcing agricultural resilience against increasing water scarcity driven by climate change [87].

9 Integrated AI-Biotechnology Pipelines

Integration of AI analytics with biotechnology for climate-smart agriculture calls for a holistic computational-biological platform that can enhance sustainability, productivity, and adaptive resilience. AI precision agriculture allows for high-resolution, real-time agroecosystem monitoring, while data analytics informs optimized irrigation, fertilization, and crop protection. Predictive analytics and modeling afford early detection of emerging stressors, and the modeling of climate-related scenarios improves preparedness and adaptive responses. The deployment of such analytical systems, often with other sensors and remote sensing platforms, offers continuous tracking of environmental fluctuations and contributes to climate-resilient agricultural management [47]. Complementing these digital technologies, biotechnology creates crop varieties that bear improved tolerance to drought and salinity conditions and heat stress, and underpins the creation of biofertilizers, biostimulants, and other biologically derived inputs that reduce chemical use and strengthen soil health. For Vietnam, this AI-biotechnology convergence was seen as one such path toward transitioning conventional agriculture into a technologically more advanced and environmentally sustainable production system, clearly emphasizing strategic investment in national research capacity and infrastructure [31].

This integration deepens further through advanced genetic and molecular technologies where AI accelerates predictive breeding, trait mapping, and stress-response modeling. With the use of IoT-driven data streams and high-throughput analytics, AI enhances efficiency in biotechnology pipelines, hence optimizing phenotyping, rapid genotype selection, and reducing resource-intensive trial-and-error approaches. Through precision breeding, genome editing, synthetic biology, and CRISPR-Cas9 technologies, biotechnology fortifies these pipelines with targeted modification of key traits participating in stress tolerance, biosynthesis of nutrients, and stability in yield [3]. Furthermore, AI and ML also assist in the interpretation of multi-omics data through the integration of genomics, transcriptomics, proteomics, and metabolomics with a view to deciphering complex biological pathways of climate resilience [96]. These digital-biological synergies reduce dependence on chemical pesticides and fertilizers, improve soil quality, and enhance the overall sustainability of production systems in tune with national and global climate-adaptation priorities [31].

A central example of such integrated pipelines is the development of data-driven, multi-scale platforms that integrate phenotyping, genomics, environmental data, and predictive modeling into unified breeding frameworks. The Genomes to Fields initiative epitomizes this approach through its compilation of a robust multi-year database encompassing genomic (G), environmental (E), and phenotypic (P) datasets integral to G $\times$ E modeling (Fig. 3). This G2F dataset integrates diverse maize genotypes; extensive, multi-trait phenotypic observations; and high-resolution environmental time-series data from field stations spanning the U.S. and Ontario, Canada. Notably, for analytical reliability, a stringent two-phase QC and CC pipeline is implemented to address environmental uncertainties; standardize data structures; impute missing values using ML algorithms; and remove rare genomic variants. Such multi-step processing generates the resulting, refined, multi-dimensional "CLIM4OMICS" database that follows the FAIR principles, enabling scalability for training and validation of climate-responsive models of crop growth. In this context, AI-biotechnology pipelines can play a critical role in rapid climate cultivar development and evidence-based agricultural decision-making by offering standardized integration with advanced predictive analytics.

Fig. 3 G2F: integrated genotype-environment-phenotype workflow for predictive agriculture and climate-resilience

10 Ethical, Economic and Policy Frameworks

The application of AI in conjunction with biotechnology in farming brings forth critical socioeconomic and ethical dimensions that determine their sustainable and equitable deployment. AI tools, comprising drones, sensor networks, and ML systems, could highly improve productivity by enhancing resource efficiency, accurate diagnostics of crop health, and real-time decision-making [4]. Their deployment, therefore, presents issues such as labor displacement, digital inequity, and problems revolving around data ownership and privacy [109]. Small-scale farmers have limited technological and financial abilities and, by virtue, risk intensifying the disparities between resource-poor producers and large commercial production entities [16]. Ethical frames emphasizing open data governance, equal access, and responsible innovation are critical in preventing technological exclusion and/or ensuring the fair distribution of benefits emanating from AI [24]. Moreover, this transition into smart farming systems embedded in the integration of IoT, robotics, and gene-editing technologies portrays broader ramifications for workforce restructuring and animal welfare, hence calls for an inclusive support program and adaptive policy mechanisms [86].

Regulatory architectures dealing with genome editing and AI applications in climate-smart agriculture are heterogeneous and continue to evolve in response to rapid technological progress. Genome editing, in particular through the CRISPR-Cas systems, has transformative potential to develop stress-resilient and high-yielding crops. However, the stringency of regulatory disposition is widely disparate between geopolitical regions. Some jurisdictions use product-based frameworks, whereby the characteristics of the final organism are assessed, while others use process-based regulations that place greater burdens on innovation. Chile presents a science-driven, flexible model for the regulation of genome-edited crops, which permits their development and commercialization with minimal additional regulatory hoops, placing this country at the frontier for agricultural biotechnology milestones toward sustainability and food security goals [79]. In Latin America and the Caribbean, the role of regulatory assessments is becoming more and more focused on how policy structures align with emerging technologies to help shape future governance, stimulate investment, and provide risk-proportionate oversight [107]. Recommendations emphasize the need for regulatory systems to learn from governance experience relating to GMOs but ensure that requirements imposed on gene-editing technologies are proportionate and no more onerous than necessary [18]. Alongside this, AI governance frameworks have to tackle key challenges, including high implementation costs and poor levels of digital literacy, especially in low-income countries. Targeted investments in digital infrastructure and enabling policy environments will be required to fully harness the potential of AI for climate-resilient agriculture [19].

Digital agriculture, therefore, rooted in the convergence of AI, IoT, and biotechnology, is going to have big implications for smallholder farmers and for national food security systems. Precision agriculture tools increase resource use efficiency and enable real-time crop monitoring, which is really useful in smallholder-dominated regions prone to climatic uncertainties and resource scarcity. Coupled with biotechnology innovations like precision breeding and genetic enhancement, these digital systems contribute to yield stability, environmental sustainability, and climate adaptation. Evidence from Odisha, India, during the COVID-19 pandemic illustrates the value of digital platforms like eKutir in maintaining supply chain connectivity and safeguarding farmer livelihoods by demonstrating the potential for the resilience of digital agriculture during systemic disruptions [27]. However, a myriad of persistent barriers, including weak policy ecosystems, the limited availability of contextually appropriate digital tools, and variable capacity for technology adoption, also continue to constrain the scale-up of digital innovations among smallholder farmers. Closing these gaps takes strategic policy interventions, capacity-building programs, and user-friendly digital app development that are commensurate with various agroecological and socioeconomic conditions [104]. Ethical and equitable adoption of digital and biotechnological innovations should, therefore, be ensured through inclusive governance frameworks that foster long-term viability and equitable benefits across all agricultural stakeholders [88].

11 Case Studies in Climate-Ready Agriculture

AI-assisted breeding programs are a transformative use case in accelerating the creation of climate-resilient varieties. ML and DL algorithms improve high-throughput phenotyping and genotyping pipelines for rapid identification of adaptive traits, whereas most breeding cycles are shortened. Coupling AI with multi-omics platforms enhances the resolution of genotype-environment interactions, thereby allowing breeders to decode intricate resilience mechanisms and design stress-tolerant varieties with greater precision [37]. Evidence from wheat improvement programs indicated that AI-driven predictive models resulted in superior yield, stress tolerance, and disease resistance forecasting compared to standard methods, thus accelerating selection gains and improving breeding pipelines [50]. Coupling AI with speed-breeding frameworks supports real-time monitoring of plant developmental responses under controlled environments, further improving the scope for generating climate-smart cultivars within compressed timeframes [70].

Parallel developments in CRISPR-Cas9-mediated genome editing provide another compelling example of climate-ready crop development. CRISPR allows the manipulation of genomic regions underlying drought and heat tolerance with unprecedented precision. This enables the fast generation of stress-tolerant varieties that thrive predictably across arid and semi-arid systems where conventional breeding has presented limited gains [33]. Demonstrations in maize and rice have realized yield gains of 20–30% under water-limited conditions, underscoring efficacy for climate-induced productivity losses. Similarly, genome-edited heat-tolerant versions realize stable yields under high temperature, lengthening growing seasons and strengthening system resilience [48]. Innovation around base editing and prime editing fosters more granular control over allelic changes, opening new windows of opportunity for tuning tolerance against multiple abiotic stresses, salinity, and cold [41]. Such new genome editing breakthroughs solidify the adaptive capacity of food systems while breaking linkages to costly chemical inputs and improving smallholder farmer incomes [48].

The use of AI in conjunction with biotechnology is further manifested in global climate-smart agriculture initiatives that put a premium on technological convergence for sustainability, productivity, and resilience. AI-enabled tools, from computer vision and predictive analytics to automated pest detection systems, support precision farming and optimal crop management that reduce input waste, especially in resource-constrained conditions [89]. In tandem with genetic engineering and synthetic biology, AI can accelerate the development of robust agri-biological solutions while considering ethical issues and equity concerns around access and benefit sharing [61]. Despite their proven value, AI-powered CSAT still faces some major barriers, including a lack of digital literacy, high costs of deployment, and infrastructural limitations-one of the most binding constraints in the Global South. Such constraints require focused policy support and investment for broader dissemination [19].

Real-world deployments of sensors and AI-integrated systems epitomize the role of enhanced climate readiness. Biosensor-enabled smart agriculture platforms apply real-time physiological and environmental measures to produce precise crop recommendations that minimize manual labor and improve efficiency [65]. Besides this, remote sensing integrated with AI enables the enhancement of predictive analytics, crop modeling, and continuous surveillance that may allow farmers to anticipate climate shocks and optimize resource use [47]. Together, these case studies show how AI, genome editing, and biotechnology collectively contribute toward the advancement of climate-ready agriculture through evidence-driven innovation and adaptive management from technological convergence.

12 Conclusion

The integration of AI, biotechnology, and sustainable agronomic practices represents the transformative pathway of climate-ready agriculture. AI-enabled analytics, including ML, DL, remote sensing, and predictive modeling, enabled high-resolution monitoring of crop performance, climate stress, and resource dynamics as a basis for rapid and informed decision-making at the field-to-landscape levels. Complementary advances in biotechnology-including CRISPR-mediated genome editing, precision breeding, and multi-omics-driven trait discovery-accelerate the generation of climate-resilient cultivars able to tolerate heat, drought, salinity, and emerging biotic stresses. Digital agriculture infrastructures comprise IoT sensor networks, UAV systems, smart irrigation, and DTs, which advance adaptive capacity by enabling continuous surveillance and optimization of water, nutrient, and soil management. At the same time, microbial biotechnology and soil health strategies strengthen ecosystem resilience through improved nutrient cycling, stress mitigation, and long-term productivity. At the same time, realizing these technologies' full potential requires supportive policy frameworks, ethical data governance, equitable access, and strong capacity-building efforts to avoid exacerbating the digital divide. Embedding scientific innovation within inclusive governance structures could dramatically enhance AI-biotechnology convergence for global food security and climate resilience.

Acknowledgement All the authors are thankful to their respective Universities and Institutions for their valuable supports.

Data Availability The authors confirm that the data supporting the findings of this study are available within the book/chapter.

References

1. Abdikadir NM, Hassan AA, Abdullahi HO, Rashid RA (2023) Smart irrigation system. Int J Electr Electron Eng 10(8):224–234
2. Afreen N, Jones I (2022) Physiological mechanisms of plant adaptation to abiotic stress. Plant Sci Arch 13:14
3. Ahlawat U, Naruka A, Changdeo WB, Rehsawla R, Sansanwal R, Mishra R, Singh VK (2024) A review of cutting-edge biotechnological solutions for next-generation farming. J Experim Agric Int 46(7):1187–1204
4. Akintuyi OB (2024) The role of artificial intelligence in US agriculture: a review: assessing advancements, challenges, and the potential impact on food production and sustainability. Open Access Res J Eng Technol 6(02):023–032
5. Al Mamun A, Islam SB, Mondal M, Al Hasan MA, Sultana S, Nazim MS (2025) Isolation and characterization of rhizobium strains for enhancing plant growth and stress tolerance under drought and salinity. https://doi.org/10.20944/preprints202504.2309.v1
6. Archana P, Ramamoorthy P, Swaminathan M, Muthurajan R, Alagarsamy S, Kathiresan PK (2024) Multi-omics approaches for abiotic stress tolerance in rice (*Oryza sativa* L.). Plant Sci Today 11:4843
7. Arif M, Shamshad J, Khalid F, Kuzyakov Y, Younas A, Li L (2025) Biotechnological and sustainable approaches to climate-resilient agriculture. J Agron Crop Sci 211(4):e70105
8. Asseng S, Cao W, Zhang W, Ludwig F (2009) Crop physiology, modelling and climate change: impact and adaptation strategies. Crop Physiol:511–543
9. Asseng S, Zhu Y, Wang E, Zhang W (2015) Crop modelling for climate change impact and adaptation. In: Crop physiology. Academic, pp 505–546
10. Awasthi P, Mishra S, Gupta N (2024) Integrated multi-parameter soil sensor technology for continuous real-time plant health analysis. In: 2024 International conference on augmented reality, intelligent systems, and industrial automation (ARIIA). IEEE, pp 1–6
11. Babanatsas T, Babanatis-Merce RM (2025) Study of the possibility of a conceptual framework for digital twin and cyber-physical systems to enhance sustainable agricultural management. https://doi.org/10.20944/preprints202507.1473.v1
12. Bagati S, Mahajan R, Nazir M, Dar AA, Zargar SM (2018) "Omics": a gateway towards abiotic stress tolerance. In: Abiotic stress-mediated sensing and signaling in plants: an omics perspective. Springer Singapore, Singapore, pp 1–45
13. Barnes EM, Tringe SG (2022) Exploring the roles of microbes in facilitating plant adaptation to climate change. Biochem J 479(3):327–335
14. Belghachi M (2025) Smart irrigation systems using AI to optimize water usage. In: Cases on AI-driven solutions to environmental challenges. IGI Global Scientific Publishing, pp 241–268
15. Benny B, Shirley CP, Jingle BJ, Vidhya K, Elakkiya E (2024) Leveraging IoT and digital twins to monitor crop growth and health in agriculture. In: 2024 10th International conference on advanced computing and communication systems (ICACCS), vol 1. IEEE, pp 2266–2271
16. Bhardwaj A, Gupta SP (2025) Empowering sustainable agriculture harnessing AI for enhanced yields, resource efficiency, and eco-friendly farming practices. In: AI and ecological change for sustainable development. IGI Global Scientific Publishing, pp 183–216
17. Bloch V, Palosuo T, Huitu H, Ronkainen A, Backman J, Pussi K, Pastell M (2023) Towards a digital twin for optimal field management. In: Precision agriculture'23, Wageningen Academic, pp 377–383
18. Brookes G, Smyth SJ (2024) Risk-appropriate regulations for gene-editing technologies. GM Crops Food 15(1):1–14
19. Chavula P, Kayusi F, Lungu G, Mambwe H, Uwimbabazi A (2024) AI application in climate-smart agricultural technologies: a synthesis study. LatIA 2:330–330
20. Chowdary CSA, Likith S, Maya BS, Lokesh SV, Manish Y, Asha T (2024) Centralized platform for sensor integration and AI models in precision agriculture. In: 2024 Second international conference on networks, multimedia and information technology (NMITCON). IEEE, pp 1–7

21. Chunjiang Z, Jingchen L, Huarui W, Yusen Y (2024) Vegetable crop growth modelling in digital twin platform based on large language model inference. Smart Agric 6(6):63
22. Dhir B (2018) Crop productivity in changing climate. In: Sustainable agriculture reviews 27. Springer International Publishing, Cham, pp 213–241
23. Escribà-Gelonch M, Liang S, van Schalkwyk P, Fisk I, Long NVD, Hessel V (2024) Digital twins in agriculture: orchestration and applications. J Agric Food Chem 72(19):10737–10752
24. Falola PB, Adeniyi AE, Madamidola OA, Awotunde JB, Olukiran OA, Akinola SO (2024) Artificial intelligence in agriculture: the potential for efficiency and sustainability, with ethical considerations. In: Exploring ethical dimensions of environmental sustainability and use of AI. IGI Global Scientific Publishing, pp 307–329
25. Feller U, Vaseva II (2014) Extreme climatic events: impacts of drought and high temperature on physiological processes in agronomically important plants. Front Environ Sci 2:39
26. Feng P, Wang B, Liu DL, Yu Q, Hu K (2022) Coupling machine learning with APSIM model improves the evaluation of climate extremes impact on wheat yield in Australia. In: Modelling processes and their interactions in cropping systems: challenges for the 21st century, pp 251–275
27. Ferguson R, Mishra S, Qureshi S, Injac L, Talukder B, Orbinski J (2024) Digital technologies and food security during crisis: COVID-19 experiences from smallholder farmers in Odisha, India. Front Sustain Food Syst 7:1150197
28. Fu J, Zheng S, Fan L, Zheng X, Qian Q (2025) Breeding 5.0: artificial intelligence (AI)-decoded germplasm for accelerated crop innovation. J Integr Plant Biol
29. Höhl A, Ofori-Ampofo S, Obadic I, Fernández-Torres MÁ, Salih Kuzu R, Zhu X (2023) USCC: A benchmark dataset for crop yield prediction under climate extremes. In: EGU General Assembly conference abstracts, p EGU-15540
30. Hossain A, Rahman MME, Ali S, Islam T, Syed MA, Syed T, Islam T (2022) CRISPR-Cas9-mediated genome editing technology for abiotic stress tolerance in crop plant. In: Plant perspectives to global climate changes, pp 331–354
31. Huyen PT (2025) Advancing sustainable agriculture in Vietnam through biotechnology and artificial intelligence. J Sci Univ Indus Trade 25(S2_ICA):26–46
32. Islam S, Reza MN, Samsuzzaman SA, Cho YJ, Noh DH, Chung SO, Hong SJ (2024) Machine vision and artificial intelligence for plant growth stress detection and monitoring: a review. Precis Agric 6(1):34
33. Jangir P, Khandelwal P, Soni P (2024) Genome editing for raising crops for arid lands: a perspective of increasing stress tolerance. In: Applications of genome engineering in plants, pp 369–393
34. Jayanthi M, Shanthi D (2023) Predicting crop yield with AI a comparative study of DL and ML approaches. In: International conference on soft computing and signal processing. Springer Nature, Singapore, pp 337–348
35. Kalyani Y, Vorster L, Whetton R, Collier R (2024) Application scenarios of digital twins for smart crop farming through cloud-fog-edge infrastructure. Future Internet 16(3):100
36. Kanth K, Mane RS, Prasad BD, Sahni S, Kumari P, Quaiyum Z, Chaudhary RK (2025) Editing the future: CRISPR/Cas9 for climate-resilient crops. IntechOpen
37. Khan MHU, Wang S, Wang J, Ahmar S, Saeed S, Khan SU, Feng X (2022) Applications of artificial intelligence in climate-resilient smart-crop breeding. Int J Mol Sci 23(19):11156
38. Kibret M, Devkota K, Bakrim WB, Ezzariai A, Terefe H, Karouach F, Kouisni L (2024) Plant growth promoting rhizobacteria mitigate drought and salinity stresses, and improve the physiological and agronomic performances in crops: a systematic review. CABI Rev 19(1)
39. Kogan F (2006) Early drought detection, monitoring, and assessment of crop losses from space: global approach. In: Disaster forewarning diagnostic methods and management, vol 6412. SPIE, pp 56–60
40. Kumar J, Choudhary AK, Gupta DS, Kumar S (2019) Towards exploitation of adaptive traits for climate-resilient smart pulses. Int J Mol Sci 20(12):2971
41. Kumar M, Prusty MR, Pandey MK, Singh PK, Bohra A, Guo B, Varshney RK (2023) Application of CRISPR/Cas9-mediated gene editing for abiotic stress management in crop plants. Front Plant Sci 14:1157678

42. Ma C, Xin M, Feldmann KA, Wang X (2014) Machine learning-based differential network analysis: a study of stress-responsive transcriptomes in Arabidopsis. Plant Cell 26(2):520–537
43. Machado ML, Simão MLR, Simão FR, Alexandrino RCS (2020) Stress conditions in soybean areas based on measurements of soil-plant-atmosphere system and UAV images. Pesquisa Agropecuária Tropical 50:e61785
44. Meher PK, Sahu TK, Gupta A, Kumar A, Rustgi S (2024) ASRpro: A machine-learning computational model for identifying proteins associated with multiple abiotic stress in plants. Plant Genome 17(1):e20259
45. Mishra S, Chavan BK, Dange MM, Sharma S, Kumari S, Shukla YK, Sumi K (2025) A review of transformative technologies in digital agriculture: integrating IoT, remote sensing, and AI for smart crop management. Int J Plant Soil Sci 37(6):449–459
46. Misu IJ, Kayess MO, Siddiqui MN, Gupta DR, Islam MN, Islam T (2025) Microbiome engineering for sustainable rice production: strategies for biofertilization, stress tolerance, and climate resilience. Microorganisms 13(2):233
47. Mmbando GS (2025) Harnessing artificial intelligence and remote sensing in climate-smart agriculture: the current strategies needed for enhancing global food security. Cogent Food Agric 11(1):2454354
48. Mohapatra R, Heena Kouser HM, Naan T, Chitra M, Ashwini R, Rout A, Lallawmkimi MC (2024) A review on biotechnological innovations in developing stress-tolerant crops for adverse environmental conditions. J Sci Res Rep 30(7):901–920
49. Morsy A, El-Sherif N (2024) Integrating omics approaches for abiotic stress tolerance in plants. In: Abiotic stress in crop plants-ecophysiological responses and molecular approaches. IntechOpen
50. Mushtaq MA, Ahmed HGMD, Zeng Y (2024) Applications of artificial intelligence in wheat breeding for sustainable food security. Sustainability 16(13):5688
51. Na MH, Na IS (2024) AI-powered predictive modelling of legume crop yields in a changing climate. Legum Res 47(8):1390–1395
52. Nadarajah K, Abdul Rahman NSN (2023) The microbial connection to sustainable agriculture. Plants 12(12):2307
53. Naik B, Kumar V, Rizwanuddin S, Chauhan M, Choudhary M, Gupta AK, Rustagi S (2023) Genomics, proteomics, and metabolomics approaches to improve abiotic stress tolerance in tomato plant. Int J Mol Sci 24(3):3025
54. Naqvi RZ, Siddiqui HA, Mahmood MA, Najeebullah S, Ehsan A, Azhar M, Asif M (2022) Smart breeding approaches in post-genomics era for developing climate-resilient food crops. Front Plant Sci 13:972164
55. Niu S, Nie Z, Li G, Zhu W (2024) Early drought detection in maize using UAV images and YOLOv8+. Drones 8(5):170
56. Ojeilua C, Onophurhi GA, Opeyemi BQ, Ayomide IP, Ojo TJ, Atumah PE, Olusesan OC (2025) Abiotic stress combinations in crops: a meta-analysis of drought and salinity responses across diverse climatic zones. Available at SSRN 5428054
57. Oyedokun JA, Okelola MO, Omidiora EO, Akanbi WB, Ajeigbe OA (2025) Development of an IoT-enabled in-situ soil monitoring system for enhancing precision in agricultural practices. Arch Curr Res Int 25(5):124–135
58. Pagnotta MA (2025) Molecular breeding for abiotic stress tolerance in crops: recent developments and future prospectives. Int J Mol Sci 26(18):9164
59. Pallavi K, Keerthana P, Devadharshini A (2024) Agrisense: MyRio and LabVIEW-powered precision farming for sustainable agriculture. In: 2nd International conference on computer vision and internet of things (ICCVIoT 2024), vol 2024. IET, pp 7–11
60. Pandita D (2023) CRISPR/Cas mediated genome editing for improving stress resilience in plants. In: New Frontiers in plant-environment interactions: innovative technologies and developments. Springer Nature, Cham, pp 143–167
61. Pasupuleti MK (2023) Agricultural revolution: AI and genomic breakthroughs in sustainable Agri-biotechnology. Int J Acad Indus Res Innov 3(7):55–66

62. Patil S, Joshi S, Jamla M, Zhou X, Taherzadeh MJ, Suprasanna P, Kumar V (2021) MicroRNA-mediated bioengineering for climate-resilience in crops. Bioengineered 12(2):10430–10456
63. Patra S, Chatterjee D, Basak S, Sen S, Mandal A (2024) CRISPR/Cas9 opens new horizon of crop improvement under stress condition. Biochim Biophys Acta Gen Subj 1868(10):130685
64. Peladarinos N, Piromalis D, Cheimaras V, Tserepas E, Munteanu RA, Papageorgas P (2023) Enhancing smart agriculture by implementing digital twins: a comprehensive review. Sensors 23(16):7128
65. Perumal B, Bargavram V, Atharsh MB, Kumar M (2023) Smart agriculture using bio-sensors and AI. In: 2023 2nd International conference on automation, computing and renewable systems (ICACRS). IEEE, pp 1071–1075
66. Portal-Gonzalez N, Wang W, He W, Santos-Bermudez R (2025) Engineering plant holobionts for climate-resilient agriculture. ISME J 19(1):wraf158
67. Prabhu KR, Kumar A, Yumkhaibam RS, Janeja HS, Krishna B, Talekar N (2023) A review on conventional and modern breeding approaches for developing climate resilient crop varieties. J Appl Nat Sci 15(3)
68. Qamar N, Ali MS, Siddique AM, Qureshi HU, Ali U, Khurram AA (2024) IoT-based soil and weather sensors for crop monitoring: a step towards smart agriculture. In: 2024 Global conference on wireless and optical technologies (GCWOT). IEEE, pp 1–7
69. Rafique S, Quadri SN, Abdin MZ (2024) Plant defense mechanism in combined stresses-cellular and molecular perspective. J Plant Stress Physiol:33–42
70. Rai KK (2022) Integrating speed breeding with artificial intelligence for developing climate-smart crops. Mol Biol Rep 49(12):11385–11402
71. Ramos-Giraldo P, Reberg-Horton C, Locke AM, Mirsky S, Lobaton E (2020) Drought stress detection using low-cost computer vision systems and machine learning techniques. IT Professional 22(3):27–29
72. Ramson SJ, León-Salas WD, Brecheisen Z, Foster EJ, Johnston CT, Schulze DG, Malaga MP (2021) A self-powered, real-time, LoRaWAN IoT-based soil health monitoring system. IEEE Internet Things J 8(11):9278–9293
73. Rani S, Sogarwal A, Mishra S, Kaur S (2025) Harnessing plant-bacterial interactions to enhance abiotic stress tolerance in plants: a review. Discover Plants 2(1):1–27
74. Raza A, Shahid MA, Safdar M, Tariq MAR, Zaman M, Hassan MU (2023) Exploring the impact of digital farming on agricultural engineering practices. Biol Life Sci Forum 27(01):10
75. Rebecca B, Padmini PS, Dhanakodi V, Gayathri M (2024) A novel hybrid method for predicting specific crop yield. Int J Sci Methods Comput Sci Eng 1(01):32–38
76. Rodrigues AP, Pais IP, Leitão AE, Dubberstein D, Lidon FC, Marques I, Ramalho JC (2024) Uncovering the wide protective responses in Coffea spp. leaves to single and superimposed exposure of warming and severe water deficit. Front Plant Sci 14:1320552
77. Roychowdhury R, Das SP, Gupta A, Parihar P, Chandrasekhar K, Sarker U, Sudhakar C (2023) Multi-omics pipeline and omics-integration approach to decipher plant's abiotic stress tolerance responses. Genes 14(6):1281
78. Sagan, V., Maimaitijiang, M., Sidike, P., Maimaitiyiming, M., Erkbol, H., Hartling, S., & Fritschi, F. U. A. V. (2019). UAV/satellite multiscale data fusion for crop monitoring and early stress detection. In: International archives of the photogrammetry, remote sensing and spatial information sciences—ISPRS archives, vol 42, no. 2/W13. International Society for Photogrammetry and Remote Sensing, pp 715–722
79. Sánchez MA (2024) The global advance of genome-edited plants to the market: the key role of Chile in its development. Plants 13(24):3597
80. Satasiya S, Patel AM, Jeyavathana RB, Ukrit MF (2024) EcoStream SmartFlow system. In: 2024 Second international conference on emerging trends in information technology and engineering (ICETITE). IEEE, pp 1–6
81. Sbai Z (2025) Deep learning models and their ensembles for robust agricultural yield prediction in Saudi Arabia. Sustainability 17(13):5807
82. Segarra-Medina C, Gómez-Cadenas A, Zandalinas SI (2025) Physiological, molecular, and metabolic adaptations of plants to combined salinity and high irradiance stress. Physiol Plant 177(2):e70164

83. Seralathan P, Edward S (2024) Revolutionizing agriculture: deep learning-based crop monitoring using UAV imagery—a review. In: 2024 International conference on expert clouds and applications (ICOECA). IEEE, pp 777–784
84. Shah AM, Khan IM, Shah TI, Bangroo SA, Kirmani NA, Nazir S, Biswas A (2022) Soil microbiome: a treasure trove for soil health sustainability under changing climate. Land 11(11):1887
85. Shahvar MP, Valenti D, Collura A, Micciche S, Farina V, Marsella G (2025) An integrated hybrid-stochastic framework for agro-meteorological prediction under environmental uncertainty. Stats 8(2):1–25
86. Sharma N, Rout S, Saritha B, Gautam SK (2024) A comprehensive review on future of smart farming and its role in shaping food production. J Exp Agric Int 46(5):486–493
87. Singh A, Pandey D, Iqbal MR, Nazir MR, Kumar S, Sonal D (2024) Smart irrigation system: optimizing water usage for sustainable agriculture. In: 2024 3rd International conference on automation, computing and renewable systems (ICACRS). IEEE, pp 421–424
88. Singh K, Kolar P, Abraham R (2024) Digital agricultural ecosystem: an introduction. In: Digital agricultural ecosystem: revolutionary advancements in agriculture, pp 1–15
89. Sivasamy S (2026) AI innovations in sustainable food production and agriculture. In: AI innovations for transforming food production. IGI Global Scientific Publishing, pp 1–24
90. Soni R, Ashaq M, Manjunath P, Pandey SK, Hosur S, Das R, Singh BV (2024) Harnessing genomic resources and molecular breeding techniques for advancing crop resilience and productivity. Plant Cell Biotechnol Mol Biol 25(11–12):124–136
91. Sonowal K, Sharma S, Kumar GA, Gupta V, Kumar G, Salvi P (2025) Exploring multi-omics tools and their advancement to study drought stress responses. Physiol Plant 177(5):e70520
92. Sriram GS, Aditya SVS, Teja GA, Uduthalapally V, Das D (2023) Drone vision based abiotic stress monitoring for smart agriculture. In: 2023 IEEE International symposium on smart electronic systems (iSES). IEEE, pp 218–223
93. Sultana T, Islam MR (2024) The potential of soil microbiomes in alleviating climate change-associated stresses on crop plants. In: Beneficial microbes for sustainable agriculture under stress conditions. Academic, pp 81–111
94. Swami P, Anusha M, David KM, Saivamsireddy G, Upadhyay L, Panigrahi CK (2025) Innovations in remote sensing techniques for monitoring agricultural crops and environmental stress conditions: a review. J Sci Res Rep 31(6):178–194
95. Tan C, Kalhoro MT, Faqir Y, Ma J, Osei MD, Khaliq G (2022) Climate-resilient microbial biotechnology: a perspective on sustainable agriculture. Sustainability 14(9):5574
96. Thingujam D, Gouli S, Cooray SP, Chandran KB, Givens SB, Gandhimeyyan RV, Mukhtar MS (2025) Climate-resilient crops: integrating AI, multi-omics, and advanced phenotyping to address global agricultural and societal challenges. Plants 14(17):2699
97. Ullah F, Ali S, Siraj M, Akhtar MS, Zaman W (2025) Plant microbiomes alleviate abiotic stress-associated damage in crops and enhance climate-resilient agriculture. Plants 14(12):1890
98. Vaish R, Al-Shaikh SM, Shah SK, Dhabliya D, Devi RV (2024) Automated smart crop protection utilising IoT: a comprehensive review. In: E3S web of conferences, vol 540. EDP Sciences, p 13012
99. Waseem M, Aslam MM (2024) Integrated approaches of "omics" for the improvement of stress tolerance in plants. In: Improving stress resilience in plants, pp 321–329
100. Wójcik-Gront E, Zieniuk B, Pawełkowicz M (2024) Harnessing AI-powered genomic research for sustainable crop improvement. Agriculture 14(12):2299
101. Yan Y, Wang Y, Li J, Zhang J, Mo X (2025) Crop yield time-series data prediction based on multiple hybrid machine learning models. arXiv preprint arXiv:2502.10405
102. You X, Shu Y, Ni X, Lv H, Luo J, Tao J, Feng S (2025) MLAS: machine learning-based approach for predicting abiotic stress-responsive genes in chinese cabbage. Horticulturae 11(1):44
103. Younas A, Riaz N, Rashid M, Tufail A, Hyder S, Noreen Z (2024) CRISPR/Cas system for achieving abiotic stress tolerance. In: OMICs-based techniques for global food security, pp 213–231

104. Yuan Y, Sun Y (2024) Practices, challenges, and future of digital transformation in smallholder agriculture: insights from a literature review. Agric Basel 14(12)
105. Zagorščak M, Abdelhakim L, Rodriguez-Granados NY, Široká J, Ghatak A, Bleker C, Knight MR (2024) Integration of multi-omics and deep phenotyping provides novel insights into multiple abiotic stress responses in potato. Change 26:27
106. Zaib S, Zubair A, Abbas S, Hussain J, Ahmad I, Shakeel SN (2023) Plant growth-promoting rhizobacteria (PGPR) reduce adverse effects of salinity and drought stresses by regulating nutritional profile of barley. Appl Environ Soil Sci 2023(1):7261784
107. Zarate S, Cimadori I, Roca MM, Jones MS, Barnhill-Dilling K (2023) Assessment of the regulatory and institutional framework for agricultural gene editing via CRISPR-based technologies in Latin America and the Caribbean. https://doi.org/10.18235/0004904
108. Zhang X, Ibrahim Z, Khaskheli MB, Raza H, Zhou F, Shamsi IH (2024) Integrative approaches to abiotic stress management in crops: combining bioinformatics educational tools and artificial intelligence applications. Sustainability 16(17):7651
109. Zhonghua LI (2025) The digital harvest: AI, rural transformation, and the new educational paradigm in agriculture. Artif Intell Educ Stud 1(2):79–94
110. Zhou R, Jiang F, Niu L, Song X, Yu L, Yang Y, Wu Z (2022) Increase crop resilience to heat stress using omic strategies. Front Plant Sci 13:891861

AI-Driven Nano-Informatics: Methodology and Agricultural Applications

Madhu Malathi Koduri, A. K. Mishra, D. K. Kushwaha, and Rajendra Kumar

Abstract Climate change poses unprecedented challenges to global agricultural systems, threatening food security, ecosystem stability and rural livelihoods. Climate-smart agriculture (CSA) has emerged as an integrative framework to address these challenges by simultaneously enhancing agricultural productivity, strengthening resilience to climate variability and reducing or mitigating greenhouse-gas emissions. In recent years, the convergence of artificial intelligence (AI) and nanotechnology has accelerated the transition towards next-generation CSA by enabling data-driven, precise and resource-efficient farming systems. This chapter examines the conceptual pathways and sustainability implications of AI–nano-enabled climate-smart agriculture. Complementarily, agricultural nanotechnology contributes material-level precision via nano-fertilizers, nano-pesticides, nano-sensors, seed nano-priming and soil and environmental remediation technologies, improving nutrient-use efficiency, crop protection and stress tolerance while minimizing ecological losses. Emphasis is placed on AI–nano integration mechanisms, where nano-sensor networks generate continuous field-level data that are processed through AI models to optimize input application, reduce risk and improve yield stability under climate stress. The chapter further evaluates the behaviour and fate of engineered nanomaterials in soil–plant–environment systems, addressing controlled-release dynamics, plant uptake pathways and sustainability considerations. Overall, the chapter concludes that AI–nano-enabled climate-smart agriculture represents a transformative and scalable pathway toward sustainable food systems. Its successful implementation will depend on interdisciplinary research, robust governance frameworks, life-cycle assessments, capacity building and inclusive policy support to ensure that technological advances translate into equitable, safe and climate-resilient agricultural development.

Keywords Climate-smart agriculture · Nanoparticles · Nanotechnology · Artificial intelligence · Sustainable agriculture

M. M. Koduri · A. K. Mishra · D. K. Kushwaha
Division of Agricultural Engineering, ICAR-IARI, New Delhi, India

R. Kumar (✉)
Division of Genetics, ICAR-IARI, New Delhi, India

J.-T. Chen (ed.), *AI-Nano-Enabled Climate-Smart Agriculture and Sustainability*, Smart Nanomaterials Technology, https://doi.org/10.1007/978-981-92-1694-9_3

1 Introduction

Climate-smart agriculture (CSA) has developed into a comprehensive framework aimed at ensuring food security and fostering sustainable growth, while simultaneously responding to the challenges posed by climate change [63]. CSA refers to a combination of farming strategies and innovative technologies that work together to steer agricultural output, strengthen the ability of farming systems towards environmentally friendly and climate-resilient development [26, 63]. Although CSA draws from traditional agricultural practices, scientific advances and sustainability concepts, it differs in several key aspects. To begin with, it directly targets climate change impacts within the agricultural and food sectors. Additionally, CSA intentionally evaluates the balance and interactions between increasing yields, adapting to climate variability and mitigating emissions [13]. Moreover, CSA incorporates a diverse set of methods customized to various ecological zones and socio-economic settings, such as climate-tolerant crop varieties, conservation tillage, agroforestry systems, precision agriculture tools, efficient water use strategies and better livestock practices [63]. When these methods are adopted, they generate three major benefits:

1. **Higher Productivity**: Increasing the quantity and quality of food without degrading natural resources, thereby improving nutritional outcomes and raising incomes—particularly for rural populations, where about 75% of the world's poor depend on farming for their livelihoods.
2. **Greater Resilience**: Minimizing exposure to climate-related hazards such as drought, pest outbreaks and disease, while enhancing the ability of agricultural communities to adapt to long-term pressures like shifting seasons and unpredictable weather conditions.
3. **Minimizing or Sequestering Emissions**: Cutting greenhouse gas outputs from agriculture, preventing forest loss caused by expanding cropland and promoting carbon storage in vegetation and soil.

Achieving these goals will rest on technologies that increase resource-use efficiency, enhance early detection of stress and reduce losses and leakages across the farm system [26]. To effectively manage climate fluctuations and advance sustainable agriculture, it is essential to pursue interdisciplinary research that integrates agronomy, environmental science and artificial intelligence (AI) [14]. Agriculture is increasingly adopting artificial intelligence (AI), which, together with other advanced computing technologies, has been applied in farming practices for many years [30]. Climate-smart agriculture (CSA) emphasizes the adoption of emerging technologies, including artificial intelligence (AI). This new agricultural approach positions AI as a key force in shaping future farming systems, given its strong potential to support strategies that address and mitigate the impacts of climate change [19, 30, 31, 39]. The incorporation of artificial intelligence (AI) into climate-smart agricultural technologies has prompted diverse viewpoints among scholars and scientists. Many experts emphasize that AI can transform the agricultural sector by improving resource efficiency, reducing vulnerabilities associated with climate variability and boosting overall yields. In contrast, some express concerns regarding barriers such as substantial investment requirements, limited digital skills among farming communities and inadequate technological infrastructure, which may hinder effective adoption [2, 14, 37].

As agriculture continues to advance, the incorporation of nanotechnology is opening new avenues for innovation. It provides modern, high-precision approaches to tackle worldwide challenges and has the potential to reshape farming practices for the future [40, 53]. Contemporary agricultural systems are increasingly integrating nanotechnology to advance the goals of precision farming. Various nanomaterials are being utilized to enhance farm operations, improve nutrient delivery and protect crops from pests, diseases and weeds [27, 28]. The presence and concentration of nanoparticles in the soil can significantly influence seed germination and subsequent plant development [27, 68].

2 Role of Nano-informatics in Climate-Smart Agriculture

Nano-informatics can be defined as the discipline that identifies information pertinent to nanoscale science and engineering and develops robust methods and tools for the systematic collection, validation, storage, sharing, analysis, modelling and practical utilization of this information [5]. Nano-informatics is crucial for analysing plant and ecosystem responses to newly developed nano-agriculture parameters and for optimizing these parameters to support safe, efficient and sustainable agricultural practices. Nano-informatics is the combination of nanotechnology and artificial intelligence, which has enhanced plant nutrient uptake and supported sustainable agricultural practices. The use of nanotechnology-based delivery systems for precise agrochemical application has reduced environmental contamination and minimized the negative impacts of agricultural activities on human health [5].

Agricultural nanotechnology refers to the use of nanoscale materials and innovative nanotech-based solutions to support different functions within the agriculture sector, including boosting crop productivity, enhancing soil properties and advancing sustainable farming systems. This emerging field encompasses a wide range of tools, such as nano-sensors used for real-time monitoring of plant physiological status and nano-fertilizers designed to improve the efficiency of nutrient supply [53, 65]. Nanotechnology also contributes significantly to crop protection through the use of nanoparticles and nanocomposites, typically sized between 1 and 100 nm [7, 46, 53]. These nano-enabled agrochemicals provide benefits like higher effectiveness, precise delivery to target sites and controlled release mechanisms, resulting in improved management of pests and diseases while reducing the environmental impact associated with conventional chemical inputs [1, 36, 53]. Furthermore, nanoscale carriers are increasingly important in biofortification strategies, enabling the controlled release of essential micronutrients. This enhances nutrient uptake by plants and ultimately strengthens the nutritional value of agricultural produce [53].

Artificial intelligence can significantly transform climate-smart agriculture by supporting accurate field monitoring, advanced prediction of crop and climate interactions and informed decision-making based on real-time data insights [6, 54, 56]. Artificial intelligence combines machine learning, computer vision and advanced analytics to support farmers in making better decisions, managing resources efficiently and reducing the uncertainties caused by climate change [9, 15, 56]. For instance, AI-based weather prediction systems enable farmers to foresee and respond to extreme weather conditions, while precision farming technologies help optimize the application of water, fertilizers and pesticides, thereby lowering input waste and reducing environmental harm. In addition, AI enables the creation of adaptive decision-support platforms tailored to the unique conditions of each farm. By bringing together information from satellite observations, IoT devices and long-term climate records, these systems generate practical recommendations. As a result, they not only increase farm productivity and profitability but also strengthen sustainability and resilience to climate-related challenges.

3 Nanotechnology in Climate-Smart Agriculture

Nanotechnology has emerged as an expanding multidisciplinary field with applications across medicine, agriculture, chemistry and several other sectors. In contemporary farming, the integration of nanotechnology is increasingly helping to advance and realize the goals of precision agriculture [27]. Nanoparticles applied in agriculture are commonly grouped into three major classes: organic, inorganic and carbon-based materials. Among recent advances, biologically synthesized nanomaterials—produced using natural resources such as plant extracts—have gained increasing attention as an environmentally responsible approach that minimizes waste and supports ecological sustainability [27, 57]. In the context of precision agriculture, a wide range of biological agents are utilized as green sources for nanoparticle synthesis, offering safer and more sustainable alternatives to conventional chemical methods [17, 27]. Table 1 shows the various nanoparticles with their functions and uses [27, 38]. Nanotechnology has broad classes in the CSA, which are listed in Fig. 1.

(a) Nano-fertilizers/Nano-nutrients for Sustainable Crop Nutrition

Nano-fertilizers are increasingly viewed as a promising substitute for traditional fertilizers. Their nanoscale dimensions and large surface area allow nutrients to be delivered more effectively and released in a controlled manner [23, 29]. By improving nutrient availability, these materials help enhance crop yield and quality, support better soil health and strengthen plants' ability to tolerate environmental stresses. They also lower input costs and promote safer, more precise nutrient management by ensuring nutrients are supplied slowly and consistently over time [29, 55]. As a result, smaller amounts of fertilizer are needed, which helps reduce ecological risks and overall environmental impact.

Table 1 Various nanoparticles with their functions and uses

Nanoparticle type	Functioning mechanism	Major uses
Silver nanoparticles	Release of ag$^+$ ions disrupts microbial cell membranes, enzymes, and DNA, leading to antimicrobial activity	– Antimicrobial coatings – Food packaging – Water purification – Plant disease management
Zinc oxide nanoparticles	Generate reactive oxygen species (ROS) and release Zn^{2+} ions; absorb UV radiation	– Nano-fertilizers – Antimicrobial agents – UV protection films – Crop protection
Titanium dioxide nanoparticles	Photocatalytic activity under UV/visible light produces ROS that degrade organic contaminants	– Self-cleaning surfaces – Pesticide degradation – Water treatment – Antimicrobial coatings
Iron oxide nanoparticles	Magnetic behaviour allows controlled movement and separation using external magnetic fields	– Targeted drug delivery – Soil remediation – Biosensing – MRI contrast agents
Silica nanoparticles	High surface area and porous structure enable adsorption and controlled release	– Carrier systems for fertilizers and pesticides – Food additives – Drug delivery
Polymeric nanoparticles	Biodegradable matrices allow slow and targeted release of active molecules	– Nano-fertilizers – Nano-pesticides – Drug delivery and tissue engineering
Copper nanoparticles	Release Cu^{2+} ions that interfere with microbial metabolism	– Fungicides – Plant growth promoters – Antimicrobial coatings
Quantum dots	Size-dependent fluorescence for sensitive optical detection	– Bioimaging – Plant stress detection – Biosensing – Labelling

(b) **Nano-pesticides/Nano-agrochemicals /Nano-formulations**

Pesticides are widely applied in agriculture to improve crop performance and productivity. Among the commonly used agrochemicals, urea-based fertilizers are frequently added to boost yields. However, urea releases urease, nitrite and nitrate compounds, which can eventually contribute to nutrient overload and eutrophication in water bodies. The primary objective of nano-pesticide development is to improve stability and effectiveness by preventing early degradation, enabling controlled and targeted release and lowering the quantity of active ingredients required for pest management. These systems are designed to reduce potential toxic effects on humans, beneficial organisms and the environment while enhancing overall pest control performance [59]. In addition, nano-pesticides can lower pest management costs, decrease the likelihood of resistance development, minimize environmental contamination and reduce the frequency of pesticide applications. Among the various nano-

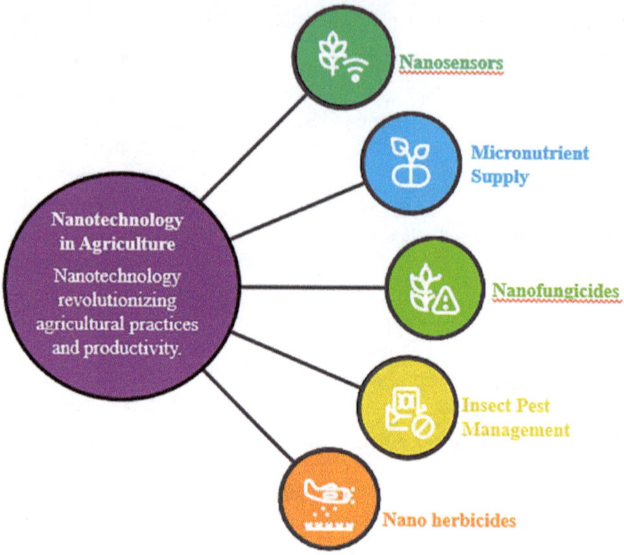

Fig. 1 Various classes of nanotechnology in CSA. *Source* Garg et al. [27], Kah et al. [38]

materials employed in agriculture, polymer-based nanoparticles are widely used as carriers for agrochemicals due to their high biocompatibility and reduced impact on non-target organisms. Metallic nanoparticles, such as silver and carbon-based nanomaterials, exhibit strong antimicrobial properties and have been reported to promote seed germination and plant growth. Furthermore, pesticide and herbicide nanoemulsions offer several advantages over conventional formulations, including broader applicability, improved penetration and enhanced adhesion to plant surfaces [24, 59]. Figure 2 shows the passage of the nano-pesticides and fertilizers through the plant stem and roots.

(c) **Nano-sensors/Nano-biosensor for Precision Agriculture**

Nano-biosensors (NBSs) exhibit markedly improved sensitivity and specificity when compared with conventional biosensors. They can detect pathogenic microorganisms either through direct interaction or via indirect recognition mechanisms, enabling continuous and real-time signal acquisition. In agricultural ecosystems, NBSs are developed for monitoring plant components, water quality and soil conditions. These sensors are commonly fabricated using nanostructures such as nanotubes, nanoparticles, nanowires or nanocrystals, typically with dimensions below 100 nm [59]. Owing to the unique physicochemical properties of nanomaterials, nano-biosensors represent an advanced analytical platform that outperforms existing biosensing technologies by effectively coupling biological recognition elements with physical and chemical transduction principles. These are helpful in real-time, sensitive monitoring of soil, water and crop health, which enables preci-

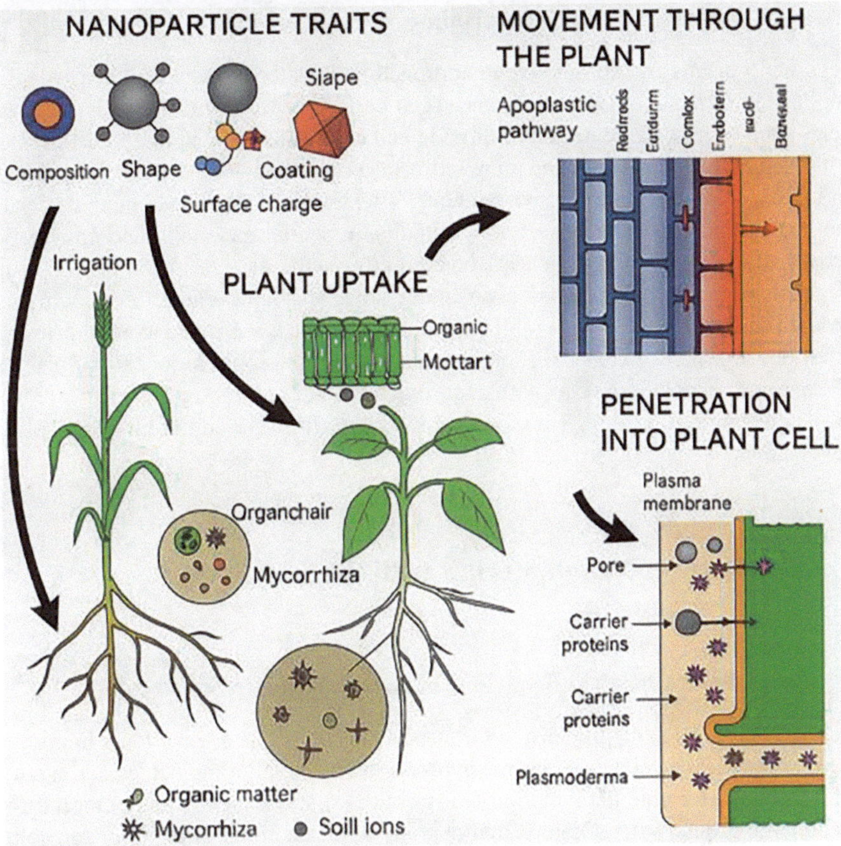

Fig. 2 Passage of the nano-pesticides and fertilizers through the plant stem and roots

sion agriculture—timely fertilization/irrigation/pesticide application; early detection of diseases or contamination supports data-driven decision-making [35, 59].

(d) **Soil and Environmental Remediation/Pollutant Management**

Metal-oxide nanoparticles such as nano-TiO_2, nano-Fe_3O_4 and doped nanocomposites have been investigated as adsorbents or photocatalysts to remove heavy metals, pesticide residues and organic pollutants from contaminated soils, drainage water or irrigation sources. Their high sorption capacity and, in some cases, light-activated degradation properties make them attractive for on-farm water treatment and rehabilitation of polluted fields [34]. Porous silica and clay-based nanostructures can likewise bind excess nutrients or contaminants, contributing to improved soil quality [49]. However, remediation applications must balance efficacy with the risk of introducing additional nanoparticles into the environment; therefore, current research focuses on immobilizing nanomaterials in composite matrices, designing magnetically recoverable particles and developing biodegradable or bio-synthesized nanomaterials to minimize long-term accumulation and ecotoxicity [49, 59].

(e) **Seed Treatment/Seed Nano-priming**

Seed nano-priming is an innovative approach in seed technology that involves the application of nanoparticles to enhance seed performance. In this method, nanoparticles such as iron oxide, titanium dioxide and carbon-based nanomaterials interact with seeds to promote faster and more uniform germination, improve seedling vigour and strengthen tolerance to environmental stresses [22, 49]. These nanomaterials facilitate enhanced water absorption, stimulate essential metabolic and enzymatic activities and support robust early-stage plant growth. As a result, improved crop establishment can be achieved even under adverse abiotic conditions, including drought and salinity. Research findings indicate that seeds subjected to nano-priming often develop more extensive root systems, higher chlorophyll concentrations [8, 49] and a greater capacity to withstand oxidative stress, highlighting the promise of nanotechnology in improving crop productivity from the initial phases of plant development.

4 Behaviour of Nanomaterials with the Soil, Plants and Environment

1. **Physical and Chemical Behaviour of Nanoparticles in Soil**

Nanoparticles (NPs) display distinct transport and persistence behaviours in porous media. Particle size, surface charge, hydrophobicity, coating molecules and aggregation state determine their mobility, adsorption to soil minerals and potential for leaching to groundwater. Organic matter content, pH and ionic strength of soil solution modulate NP fate. Encapsulation strategies (polymeric carriers, clay matrices) are used to tailor release kinetics and limit undesirable mobility [41].

2. **Plant Uptake and Translocation Pathways**

Plants can take up nanoparticles via root apoplast/apoplast bypass and, in some cases, via foliar stomata or cuticular pores. Smaller particles (typically < 20–50 nm) can pass cell walls or be endocytosed; surface functionalization (e.g. with peptides or polymers) alters affinity for root surfaces and phloem loading. Translocation may reach leaves, seeds or reproductive organs, depending on particle chemistry and plant species. Controlled carrier design seeks to maximize beneficial delivery (nutrients, RNAi) while minimizing accumulation in edible tissues [16, 64].

3. **Controlled-Release and Stimulus-Responsive Behaviours**

Polymeric nanocarriers and inorganic matrices (e.g. layered double hydroxides) can be engineered for pH, moisture or light-responsive release. Controlled-release limits nutrient pulses that cause leaching and enables synchronization of nutrient availability with crop demand, improving use efficiency and lowering emissions associated with fertilizer manufacture and loss. Stimulus-responsive nanocarriers can

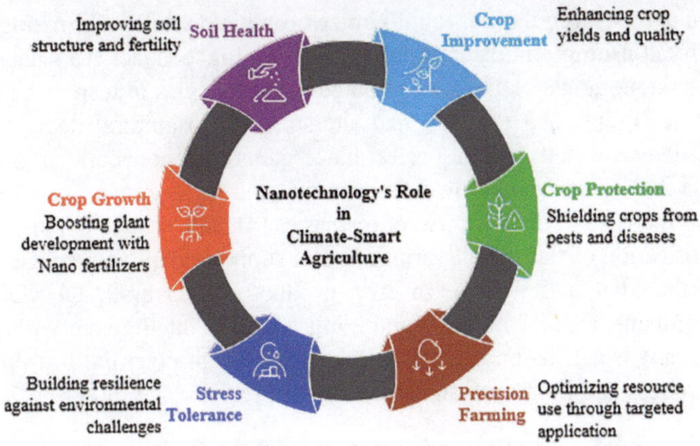

Fig. 3 Applications of nanotechnology in climate-smart agriculture

also be designed to release agrochemicals only in the presence of specific pathogen signals or under stress conditions [29]. Figure 3 demonstrates the applications of nanotechnology in climate-smart agriculture.

5 AI-Nano-integration in Climate-Smart Agriculture

The convergence of artificial intelligence and nanotechnology has emerged as a promising approach for enhancing sustainability and precision in agricultural systems. In this framework, real-time information collected by nano-sensors on soil parameters and crop health is analysed using advanced AI-driven models. Studies indicate that this synergy enables more reliable predictions, improves decision-making accuracy, minimizes inefficient use of inputs and substantially increases crop productivity. Despite these advantages, challenges remain in deploying sophisticated AI algorithms alongside nano-sensor networks under field conditions, highlighting the need for continued research to address technical integration and large-scale implementation constraints [4].

1. **Real-Time Crop Monitoring with Nano-sensor Using Artificial Intelligence Techniques**

Quantum-enhanced nano-sensor development integrated with reinforcement learning (QENDRL) is designed to support adaptive management of agricultural inputs and enhance crop performance. This approach integrates a deep Q-learning-based decision framework with quantum-enabled nano-sensors, enabling continuous, real-time assessment of key agronomic parameters such as soil moisture status, nutrient availability and surrounding environmental conditions. The reinforcement learning agent progressively learns optimal intervention policies for irrigation scheduling, fertilizer application and pest regulation, to maximize crop yield while minimizing unnecessary resource consumption. By continuously processing feedback from nano-sensors, the system autonomously adjusts its management strategies in response to evolving field conditions, ensuring efficient and site-specific agricultural decision-making. QENDRL demonstrated notable performance gains over conventional approaches, achieving a 25% improvement in data accuracy and measurement precision, along with a 30% increase in efficient use of resources [4]. The system's ability to enable early identification of field-level variations and support adaptive management practices contributed to a 25% rise in crop productivity. Overall, QENDRL represents a promising framework for combining artificial intelligence with quantum-enhanced nano-sensor technologies, offering significant potential to improve both the effectiveness and long-term sustainability of precision agriculture [12, 50].

2. **Robotics and Automation in Climate-Smart Agriculture**

Artificial intelligence integrated with robotic systems enables the automation of a wide range of agricultural operations. AI-controlled robots equipped with advanced sensors and vision systems can perform tasks such as sowing, weeding and harvesting with high accuracy and improved operational efficiency [3, 10, 11, 61]. Meanwhile, nanotechnology contributes by enabling the creation of stronger, lighter and more durable materials that enhance the performance and longevity of robotic components [4]. Figure 4 represents machine learning applications enhancing nano-enabled agricultural systems.

3. **AI with Nanotechnology in Autonomous Drones**

AI-driven autonomous vehicles and unmanned aerial systems are increasingly influencing modern agricultural practices. These systems are capable of independently

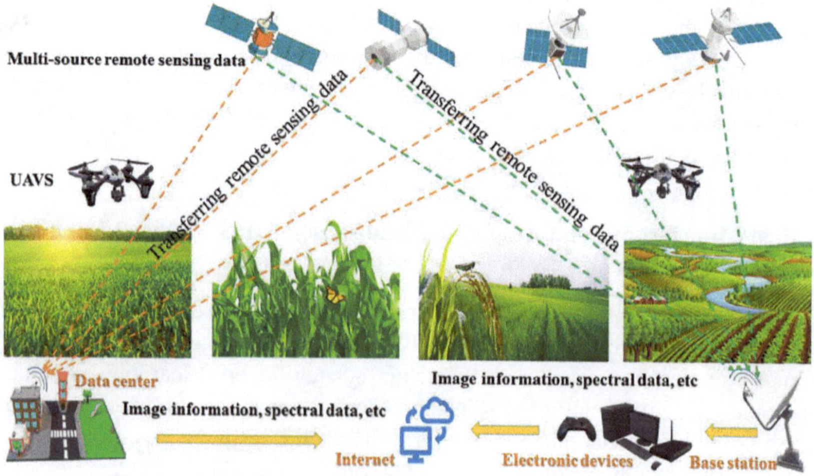

Fig. 4 Machine learning applications enhancing nano-enabled agricultural systems

moving through crop fields, acquiring real-time information and performing operations such as precision spraying, field mapping and crop health assessment [4]. The integration of nanotechnology further enhances their performance by enabling the development of light-weight, durable and energy-efficient structural materials, as well as compact nanoscale sensors for accurate and continuous data acquisition [25, 44, 47].

4. **Sustainable Agriculture with AI and Nanotechnology.**

Sustainable farming practices can be significantly strengthened through the integration of artificial intelligence and nanotechnology. AI-driven models support precise decision-making by optimizing irrigation schedules and minimizing excessive use of agrochemicals, thereby reducing environmental stress. At the same time, nanotechnology-based inputs such as nano-fertilizers and nano-pesticides enhance the efficiency of nutrient delivery while lowering the overall quantity of chemicals required [4, 33]. Nano-fertilizers are engineered delivery systems that improve plant nutrient uptake by encapsulating essential elements and releasing them in a controlled and gradual manner. This approach limits nutrient losses, ensures a steady supply to crops and supports long-term soil and ecosystem health. Commercial innovations, including nano-biomatters products and nano-pesticide solutions developed by crop enhancement companies, are designed to increase nutrient availability, reduce ecological impact and maximize crop productivity [4].

5. **Nano-pesticides and Nano-formulations for Pest Management.**

Nano-formulations increase solubility and stability of active ingredients, enable slow release, reduce photodegradation and can improve target specificity, reducing off-target impacts and the needed dose. Nanocarriers can integrate biological control agents or pheromones for mating disruption [64]. Several studies report improved efficacy at lower active ingredient doses and reduced non-target effects in controlled settings. Field adoption is hampered by regulatory questions, cost and the need for comparative long-term ecotoxicology studies [41].

6. **Nano-enabled Stress Tolerance and Plant Immunity Modulation**

Recent studies indicate that certain nanomaterials can prime plant antioxidant systems, modulate phytohormone signalling or deliver RNAi constructs for pest resistance. Nano-elicitors may therefore improve drought or heat resilience when applied in appropriate dosages. However, mechanisms are often species and formulation-dependent, requiring careful dose-response characterization [45, 64].

6 Challenges in Adoption of AI–Nano in Agriculture

Challenge category	Specific challenge	Explanation	References
Data and AI limitations	Data scarcity and poor quality	AI models require large, high-quality datasets, but agricultural data, especially from smallholder and marginal environments, are fragmented, sparse and often biased, reducing model accuracy and reliability	[48, 62]
	Lack of model transferability	AI algorithms trained in specific agro-climatic zones or for particular nano-formulations often fail when applied to new crops, soils, or climates due to system heterogeneity	[30, 42]
	Limited explainability	Many AI models operate as black boxes, making it difficult for farmers and policymakers to understand decision logic, which reduces trust and adoption	[52])
Nano-technology constraints	Inconsistent field performance	Nano-fertilizers and nano-pesticides show strong laboratory results, but their effectiveness varies widely under real field conditions due to soil chemistry, climate, and biological interactions	[32, 58]
	Scale-up and standardization issues	Producing nano-inputs at a commercial scale with consistent quality, stability, and cost-effectiveness remains technically challenging	[38, 67]
Environmental and health risks	Uncertain ecological impacts	The long-term fate of nanoparticles in soil–plant–water systems is not fully understood, raising concerns about accumulation, toxicity to soil organisms, and biodiversity loss	[38, 67]

(continued)

(continued)

Challenge category	Specific challenge	Explanation	References
	Food-chain and human exposure risks	Potential transfer of engineered nanoparticles into edible plant parts and subsequent dietary exposure remains insufficiently studied, affecting consumer acceptance and regulation	[18]
	Life-cycle environmental footprint	Potential transfer of engineered nanoparticles into edible plant parts and subsequent dietary exposure remains insufficiently studied, affecting consumer acceptance and regulation	(OECD 2022; Elsayed et al. 2025)
Infrastructure and access	Digital divide	AI-nano CSA depends on internet access, power supply, sensors, and computing infrastructure, which are often lacking in rural and developing regions most vulnerable to climate change	[21, 62]
	High cost of adoption	Nano-sensors, AI platforms, and precision equipment involve high initial investment, limiting adoption by small and resource-poor farmers	(Lowenberg-DeBoer et al. 2020)
Socio-economic challenges	Inequitable benefits	Early benefits of AI-nano-enabled systems tend to favor large-scale and capital-intensive farms, risking increased inequality in agricultural systems	[52, 66]
	Farmer skill and knowledge gaps	Successful use of AI tools and nano-inputs requires technical literacy and training, which many extension systems are not yet equipped to provide	[21, 48]
Climate and sustainability assessment	Difficulty in measuring climate benefits	Quantifying emissions reduction, carbon sequestration, or resilience gains attributable specifically to AI-nano interventions is methodologically complex	[21, 60]

(continued)

(continued)

Challenge category	Specific challenge	Explanation	References
	Rebound and trade-off effects	Efficiency gains from AI-nano technologies may unintentionally increase total input use or cultivated area, offsetting climate benefits	[60, 67]

7 Cost Economics Related to the AI–Nano with CSA

The cost economics of integrating artificial intelligence (AI) and nanotechnology within climate-smart agriculture (CSA) revolves around balancing higher initial investments with long-term efficiency gains, risk reduction and environmental benefits. Unlike conventional input-intensive agriculture, AI–nano-enabled CSA distributes costs and benefits over multiple production cycles, making economic evaluation inherently dynamic.

From a CSA perspective, AI-nano integration enhances economic resilience by reducing production variability and vulnerability to extreme weather events. Precision application of nano-inputs guided by AI forecasts lowers crop failure risk, which translates into reduced insurance costs and greater income predictability. When CSA co-benefits such as lower greenhouse-gas emissions, improved soil health and reduced environmental contamination are economically valued, overall system profitability improves substantially.

Economic benefits arise mainly through input optimization and productivity enhancement. AI-driven decision support systems improve the timing, dosage and spatial placement of nano-inputs, increasing nutrient-use efficiency and substantially reducing losses to soil, water and the atmosphere. As a result, total expenditure on fertilizers, pesticides, water and energy is often lower than in conventional systems, despite higher input sophistication. Yield stability and quality improvement further strengthen farm income, particularly under climate-induced stress conditions.

Although AI–nano-enabled CSA entails additional upfront and operational costs, these investments are frequently offset by higher yields, reduced input consumption, improved climate resilience and environmental co-benefits.

8 Conclusion

The chapter has comprehensively demonstrated that the integration of artificial intelligence and nanotechnology represents a transformative pathway for advancing climate-smart agriculture (CSA) in the context of global food security and sustainability challenges. AI-driven analytics, decision-support systems, robotics and autonomous platforms enable data-informed, site-specific management of crops and resources, thereby improving productivity while enhancing resilience to climate variability. At the same time, nanotechnology contributes precision at the material and molecular scale through nano-fertilizers, nano-pesticides, nano-sensors, seed nano-priming and soil remediation strategies, which collectively enhance input-use efficiency, reduce environmental losses and support ecosystem health. The convergence of these technologies, AI for intelligent decision-making and nanotechnology for precise intervention, creates a synergistic framework capable of delivering the core pillars of CSA: higher productivity, greater resilience and reduced greenhouse-gas emissions.

However, the chapter also highlights that large-scale adoption of AI–nano-enabled CSA is constrained by technical, environmental, socio-economic and governance-related challenges. Issues such as data scarcity, model transferability, high adoption costs, uneven access to digital infrastructure, uncertain long-term ecological impacts of nanoparticles and limited regulatory clarity must be addressed to ensure equitable and safe deployment. Bridging the digital divide, strengthening farmer-centric capacity building, improving explainable and transferable AI models and advancing life-cycle assessments of nano-inputs are critical priorities for future research and policy. Overall, AI–nano-enabled climate-smart agriculture holds substantial promise as a next-generation sustainable farming paradigm, but its success will depend on interdisciplinary collaboration, responsible innovation, supportive policies and inclusive implementation strategies that align technological progress with environmental and social equity.

References

1. Abd-Elsalam KA (2024) Special issue: agricultural nanotechnology Plants 13:489
2. Adesipo A, Fadeyi O, Kuca K, Krejcar O, Maresova P, Selamat A (2020) Smart and climate-smart agricultural trends as core aspects of smart village functions. Sensors 20(21):5977
3. Alreshidi E (2019) Smart sustainable agriculture (SSA) solution under pinned by internet of things (IoT) and artificial intelligence (AI)
4. Ashique S, Raikar A, Jamil S, Lakshminarayana L, Gajbhiye SA, De S, Kumar S (2025) Artificial intelligence integration with nanotechnology: a new frontier for sustainable and precision agriculture. Curr Nanosci 21:242–273
5. AzoNano (2022) Utilizing Nano-informatics in Agriculture. https://www.azonano.com/article
6. Balasundram SK, Shamshiri RR, Sridhara S, Rizan N (2023) The role of digital agriculture in mitigating climate change and ensuring food security: an overview. Sustainability 15(6):5325
7. Ball P (2002) Natural strategies for the molecular engineer. Nanotechnology 13:15

8. Basahi M (2021) Seed germination with titanium dioxide nanoparticles enhances water supply, reserve mobilization, oxidative stress, and antioxidant enzyme activities in pea. Saudi J Biol Sci 28:6500–6507
9. Behera D, Modak S (2022) Attitude and opinion of the farmers towards regulated market, a case of Balasore District. Indian Res J Extens Educ 22(2):169–172
10. Bini D, Pamela D, Shamia D, Prince S (2022) Intelligent agrobots for crop yield estimation using computer vision. Comput Assist Methods Eng Sci 29(1–2):161–175
11. Bogue R (2020) Fruit picking robots: has their time come. Industr Robot: Int J Robotics Res App, l47 (2):141–145
12. Cengiz A, Sama A (2025) Artificial intelligence techniques for quantum-enhanced nano-sensor development in precision agriculture and real-time crop monitoring. J Quantum Nano-Green Environ Syst (QNGES) 1(1):1–13
13. Chandra A, McNamara KE, Dargusch P (2018) Climate-smart agriculture: perspectives and framings. Clim Pol 18(4):526–541
14. Chavula P, Kayusi F, Lungu G, Mambwe H a UA (2025) AI application in climate-smart agricultural technologies: a synthesis study. LatIA 3:330
15. Das B (2014) ICTs adoption for accessing agricultural information: evidence from Indian agriculture. Agric Econ Res Rev 27(2):199–208
16. Deng P, Gao Y, Mu L, Hu X, Yu F, Jia Y, Wang Z, Xing B (2023) Development potential of nano-enabled agriculture projected using machine learning. Proc Natl Acad Sci 120(25):e2301885120
17. Duhan JS, Gahlawat SK (2013) Biogenesis of nanoparticles: a review. Afr J Biotechnol 13(28)
18. EFSA Scientific Committee, More S, Bampidis V, Benford, D, Bragard C, Halldorsson T, Hernández-Jerez A, Hougaard Bennekou S, Koutsoumanis K, Lambré C, Machera K, (2021) Guidance on risk assessment of nanomaterials to be applied in the food and feed chain: Human and animal health. EFSA J 19(8): e06768
19. Eli-Chukwu NC (2019) Applications of artificial intelligence in agriculture: a review. Eng Technol Appl Sci Res 9(4):4377–4383
20. Elsayed ME, Ayoub HA, Helal MI, Sang W, Shen Z, Abdelhafeez IA (2025) Nanotechnology-enabled soil management for sustainable agriculture: interactions, challenges, and prospects. Environmental Science: Nano. 12(4): 2128-2153
21. FAO, 2022. Climate-smart agriculture sourcebook
22. Faizan M, Sharma P, Sultan H, Alam P, Sehar S, Rajput VD, Hayat S (2024) Nano-priming: improving plant nutrition to support the establishment of sustainable agriculture under heavy metal stress. Plant Nano Biology 10:100096
23. Fatima F, H,shim A, Anees S (2021) Efficacy of nanoparticles as Nano fertilizer production: a review. Environ Sci Pollut Res 28:1292–1303
24. Feng J, Zhang Q, Liu Q, Zhu Z, McClements DJ, Jafari SM (2018) Application of nano-emulsions in formulation of pesticides, applications, and characterization
25. Fertu C, Dobrota LM, Balasan DL a, Stanciu S (2021) Monitoring the vegetation of agricultural crops using drones and remote sensing comparative presentation. Sci Pap Manag Econ Eng Agric Ru ral Dev 21:249–254
26. Food and Agricultural Organisation of the United States (2025). https://www.fao.org/climate-smart-agriculture/en/?utm_source.
27. Garg S, Rumjit NP, Roy S (2024) Smart agriculture and nanotechnology: technology, challenges, and new perspective. Adv Agrochem 3:115–125
28. Ghormade V, Deshpande MV, Paknikar KM (2011) Perspectives for nano-biotechnology enabled protection and nutrition of plants. Biotechnol Adv 29(6):792–803
29. González-Feijoo R, Martinez-Castillo C, Rodríguez-Seijo A, Pérez-Rodríguez P, Arenas-Lago D (2024) Nano-agrochemicals versus conventional fertilizers: a field case study with tailor-made Nano fertilizers for sustainable crop efficiency of *Brassica oleracea* L. convar. *capitata* var. *sabauda*. Agronomy 14:1885
30. Gryshova I, Balian A, Antonik I, Miniailo V, Nehodenko V, Nyzhnychenko Y (2024) Artificial intelligence in climate smart in agricultural: toward a sustainable farming future. Access Sci Business Innov Digital Econ 5(1):125–140

31. Gulzar M, Abbas G, Waqas M (2020) Climate smart agriculture: a survey and taxonomy. In: International conference on emerging trends in smart technologies, pp 1–6.
32. Gupta A, Rai P, Kumar V (2023) Nanotechnology applications in sustainable agriculture: challenges and opportunities. Sustain Chem Pharm 31:101012
33. Hajji-Hedfi L, Chhipa H (2021) Nano-based pesticides: challenges for pest and disease management. Euro-Medit J Environ Integr 6(3):69
34. Hamda AS, Areti HA, Gudeta RL, Abo LD, Jayakumar M (2025) Carbon-based nanomaterials for water treatment: a comprehensive review of recent advances and mechanisms. Chem Eng J Adv 23:100834
35. Hamers D, van Voorst Vader L, Borst JW, Goedhart J (2014) Development of FRET biosensors for mammalian and plant systems. Protoplasma 251(2):333
36. Jampílek J, Král'ová K (2017) 3—Nanopesticides: preparation, targeting, and controlled release. In: New pesticides and soil sensors, pp 81–127
37. Jariwala H, Santos R. M, Lauzon JD, Dutta A, Chiang WY (2022) Controlled release fertilizers (CRFs) for climate-smart agriculture practices: a comprehensive review. Environ Sci Pollut Res 29(36):53967–53995
38. Kah M, Kookana RS, Gogos A, Bucheli TD (2018) A critical evaluation of nanopesticides and nanofertilizers against their conventional analogues. Nat Nanotechnol 13:677–684
39. Klerkx L, Jakku E, Labarthe P (2019) A review of social science on digital agriculture, smart farming and agriculture 4.0: new contributions and a future research agenda. NJAS: Wageningen J Life Sci 90–91(1):1–16
40. Kumar A, Singh GA, Gourkhede P, Goutam P, Laxman T, Pandey S, Singh P (2023) Revolutionizing agriculture: a comprehensive review of nanotechnology applications. Int J Environ Climate Change 13:3586–3603
41. Kumari R, Suman K, Karmakar S, Mishra V, Lakra SG, Saurav GK, Mahto BK (2023) Regulation and safety measures for nanotechnology-based agri-products. Frontiers
42. Liakos KG, Busato P, Moshou D, Pearson S, Bochtis D (2018) Machine learning in agriculture: a review. Inform Process Agric 5(1):1–15
43. Lowenberg-DeBoer J, Huang, IY, Grigoriadis V, Blackmore S (2020) Economics of robots and automation in field crop production. Precision Agri 21(2): 278-299
44. Lukas V, Novák J, Neudert L, Svobodova I, Rodriguez Moreno F, Edrees M, Kren J (2016) The combination of UAV survey and landsat imagery for monitoring of crop vigor in precision agriculture. Int Arch Photogramm Remote Sens Spat Inf Sci 8:953–957
45. Masood HA, Zhang P, Noman M, Hafeez R, Ali S, Jabran K (2024) Recent advances in nano-enabled immunomodulation for enhancing plant resilience against phytopathogens. Front Plant Sci 15:1445786
46. Medina-Pérez G, Fernández-Luqueño F, Campos-Montiel RG, Sánchez-López KB, Afanador-Barajas LN, Prince L (2019) Nanotechnology in crop protection: status and future trends. In: Nano-Biopesticides Today and Future Perspectives, pp 17–45
47. Mitra M (2019) Robotic farmers in agriculture. Adv Robot Mech Eng 5:91–93
48. Mmbando GS, Baiyegunhi LJS a, Sekabira H (2025) Harnessing artificial intelligence and remote sensing for climate-smart agriculture: opportunities and challenges. Cogent Food Agric 11(1):2314567
49. Naziba AT, Kiruthika R (2025) Nanotechnology in modern agriculture: a mini review. Arch Food Nutrition Sci 9(1):020–023
50. Nizamani MM, Hughes AC, Zhang H, Wang Y (2024) Revolutionizing agriculture with nanotechnology: innovative approaches in fungal disease management and plant health monitoring. Sci Total Environ 928:172473
51. OECD (2022) OECD economic outlook, Volume 2022 Issue 2: confronting the crisis, OECD publishing, Paris. https://doi.org/10.1787/f6da2159-en
52. Omotayo AO, Adeniyi OR a, Aremu OJ (2025) Artificial intelligence in agriculture: ethics, impacts and governance challenges. Comput Electron Agric 219:108996
53. Pagano M, Lunetta E, Belli F, Mocarli G, Cocozza C, Cacciotti I (2025) Advancements in agricultural nanotechnology: an updated review. Plants 14:2939

54. Palanisamy A, Bharadwaj N (2017) Utilization of information disseminated through Mobile telephones by farmers in Tamil Nadu. J Extens Educ 29(3):5902–5909
55. Preetha PS, Balakrishnan N (2017) A review of Nano fertilizers and their use and functions in soil. Int J Curr Microbiol App Sci 6:3117–3133
56. Sahoo SK, Prangya S, Das KS (2025) AI-driven climate smart agriculture: enhancing resilience and productivity in a changing climate. In: Transforming agriculture: the role of effective extension education, pp 125–137
57. Sajid M, Płotka-Wasylka J (2020) Nanoparticles: synthesis, characteristics, and applications in analytical and other sciences. Microchem J 154:104623
58. Santos PA, Biraku X, Nielsen E, Ozketen AC, Ozketen AA, Hakki EE (2025) Agricultural nanotechnology for a safe and sustainable future: current status, challenges, and beyond. J Sci Food Agric 105(6):3159–3169
59. Shah MA, Shahnaz T, Zehab-ud-Din, Masoodi JH, Nazir S, Qurashi A, Ahmed GH (2024) Application of nanotechnology in the agricultural and food processing industries: a review. Sustain Mater Technol 39:e00809
60. Smith CJ, Al Khourdajie A, Yang P, Folini D (2023) Climate uncertainty impacts on optimal mitigation pathways and social cost of carbon. Environ Res Lett 18(9):094024
61. Su J, Yi D, Su B, Mi Z, Liu C, Hu X, Xu X, Guo L, Chen WH (2021) Aerial visual perception in smart farming: field study of wheat yellow rust monitoring. IEEE Trans Industr Inform 17(3):2242–2249
62. Wolfert S, Ge L, Verdouw C, Bogaardt MJ (2017) Big data in smart farming—a review. Agric Syst 153:69–80
63. World Bank Group (WB). Climate SMART agriculture. 2024. https://www.worldbank.org/en/topic/climate-smart-agriculture.
64. Wu H, Li Z (2022) Recent advances in nano-enabled agriculture for improving plant performance. Crop J 10:1–12
65. Yadav A, Yadav K, Abd-Elsalam KA (2023) Exploring the potential of nano fertilizers for a sustainable agriculture. Plant Nano Biol 5:100044
66. Yuan Y, Wu P, Zhang Y, Li X (2025) The values, challenges and strategies of artificial intelligence in empowering sustainable livelihoods for farmers. Front Sustain Food Syst 9:123456
67. Zeng Y, Wu J, Ma W, Wang C, Qi P, Zhang, D, and Wang P (2025) Toward triboelectric intelligence in AI-TENG systems. Cell Rep Phys Sci 6
68. Zheng L, Hong F, Lu S a LC (2005) Effect of nano-TiO_2 on strength of naturally aged seeds and growth of spinach. Biol Trace Elem Res 104:83–91

AI-Assisted Nano-Omics Approaches in Agriculture: Opportunities and Challenges

Vikas and Rajiv Ranjan

Abstract The convergence of artificial intelligence (AI), nanotechnology, and multi-omics has opened transformative pathways in agriculture, enabling precise, predictive, and sustainable food production systems. This chapter explores the emerging landscape of AI-assisted nano-omics approaches, detailing how they collectively enhance plant health monitoring, nutrient use efficiency, stress resilience, and microbiome dynamics. By integrating multi-layered biological data with nanodevices and AI-driven analytics, these technologies allow for real-time decision-making, early disease detection, and optimized input management across diverse agroecosystems. We begin by contextualizing the role of omics (genomics, transcriptomics, proteomics, metabolomics) in modern agriculture, highlighting their synergistic potential when paired with AI for big data processing and nanodevices for targeted delivery and diagnostics. The chapter then delves into technical advances in nanobiosensors, smart fertilizers, and precision phenotyping platforms, demonstrating how AI algorithms elevate the functionality of these tools through pattern recognition and predictive modeling. Despite these innovations, we underscore critical challenges, including data integration complexity, algorithm transparency, nano-toxicity concerns, and infrastructural disparities in low-income regions. To address these, the chapter provides policy and practice recommendations centered on ethical AI governance, biosafety regulation, interdisciplinary training, and inclusive technology access.

Keywords Artificial intelligence · Nano-omics · Sustainable agriculture · Plant–microbe interactions · Precision farming

Vikas · R. Ranjan (✉)
Plant Biotechnology Lab, Department of Botany, Faculty of Science, Dayalbagh Educational Institute, Agra, India
e-mail: rajivranjan@dei.ac.in

J.-T. Chen (ed.), *AI-Nano-Enabled Climate-Smart Agriculture and Sustainability*, Smart Nanomaterials Technology, https://doi.org/10.1007/978-981-92-1694-9_4

1 Introduction

Agriculture is increasingly challenged by climate change, land degradation, biodiversity loss, and the growing demand for food and bioresources, necessitating a fundamental shift toward more sustainable and resilient production systems. Conventional input-intensive practices are proving inadequate under rapidly changing environmental conditions. In response, recent advances in omics sciences have enabled deeper molecular understanding of plant, microbial, and soil processes, while nanotechnology and artificial intelligence have emerged as powerful enablers for translating these insights into field-level applications. The convergence of AI, nanotechnology, and multi-omics offers a promising framework for precision, climate-smart agriculture by integrating high-resolution biological data with intelligent analytics and targeted interventions.

1.1 The Agricultural Imperative in the Age of Climate Change

Global agriculture faces an unprecedented convergence of challenges, including climate volatility, land degradation, biodiversity loss, and rising demand for food, fiber, and bioenergy [10]. Conventional farming approaches, heavily reliant on synthetic inputs and intensive tillage, have exacerbated soil health deterioration, nutrient imbalances, and greenhouse gas emissions [32]. In this context, the pursuit of climate-smart and sustainable agriculture demands a paradigm shift toward precision, resilience, and ecological intensification [44].

1.2 Convergence of Omics, Nanotechnology, and AI

The past decade has witnessed rapid advances in omics sciences, genomics, transcriptomics, proteomics, metabolomics, and microbiomics that offer systems-level understanding of plant and microbial functions. These multi-dimensional datasets can decode the molecular architecture of traits such as stress tolerance, nutrient use efficiency, and plant–microbe interactions [40]. Various omics and their AI application in agriculture are listed in Table 1. However, the practical application of omics insights in real-world agroecosystems remains limited by issues of data complexity, poor integration, and suboptimal delivery systems.

Enter nanotechnology and artificial intelligence (AI), two enabling platforms that are revolutionizing agriculture. Nanotechnology offers tools for the precise detection,

Table 1 Comparative overview of omics techniques integrated with AI in agriculture

Omics type	Data output	AI applications	Agricultural use case	Challenges
Genomics	SNPs, gene sequences	Machine learning for trait prediction	Genotype selection, CRISPR design	Big data storage, variant annotation
Transcriptomics	mRNA expression profiles	Deep learning for expression pattern analysis	Stress response profiling	Temporal variability
Proteomics	Protein abundance and PTMs	Neural networks for protein function prediction	Disease resistance, enzyme optimization	Low reproducibility
Metabolomics	Metabolite fingerprinting	Random forest for biomarker detection	Nutrient cycling, stress diagnostics	Instrument calibration
Microbiomics	Community structure and function	AI for microbiome-based soil health scoring	Biofertilizer development	High taxonomic diversity

protection, and delivery of biomolecules at the nano-scale, improving the functionality and stability of omics-informed interventions [47]. Simultaneously, AI algorithms, particularly those in machine learning and deep learning, can extract actionable patterns from large, heterogeneous omics datasets and model complex biological–environmental interactions [42]. The synergy of nano-omics and AI thus holds the potential to unlock a new era of data-driven, precision agriculture.

2 Nano-Omics Technologies: An Overview

Nano-omics represents a transformative advancement in agricultural science by integrating nanotechnology with multi-omics platforms to enhance molecular-level understanding and field applicability. By leveraging engineered nanomaterials, nano-omics overcomes key limitations of conventional omics approaches, enabling ultra-sensitive detection, targeted delivery, and real-time monitoring of biological processes in crops and soils. The incorporation of artificial intelligence further strengthens this framework through predictive nanomaterial design and intelligent signal processing, ensuring accuracy in complex agroecosystems. Together, nano-omics tools in genomics, proteomics, metabolomics, and microbiomics provide a robust foundation for precision agriculture, sustainable crop improvement, and climate-resilient farming systems.

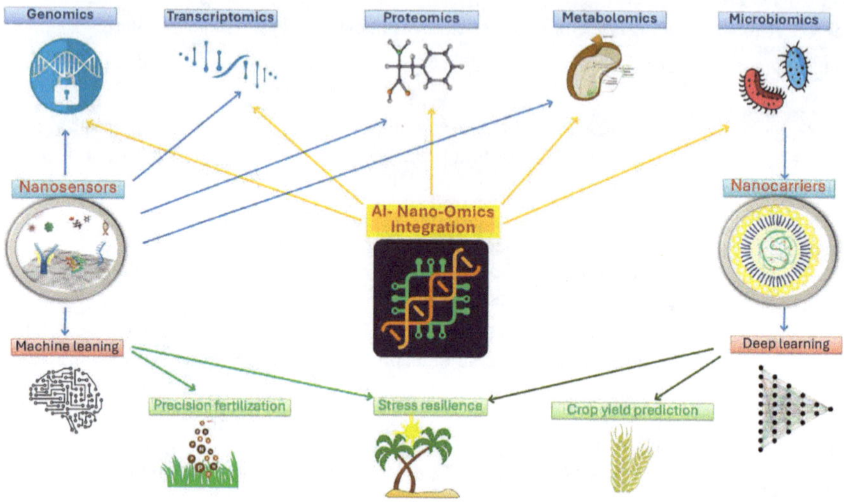

Fig. 1 Integrated framework showing how omics data, nanodevices, and AI models interact to enable real-time decisions in sustainable agriculture

2.1 Introduction to Nano-Omics in Agriculture

Nano-omics is an emerging interdisciplinary domain that fuses nanotechnology with various omics platforms, genomics, transcriptomics, proteomics, metabolomics, and microbiomics to improve the sensitivity, precision, and integration of molecular data across agricultural systems, as shown in Fig. 1. The use of nanomaterials in omics workflows enables ultra-sensitive detection, enhanced delivery of biomolecules, and the miniaturization of analytical tools, thereby addressing several limitations of conventional techniques [24, 31].

2.2 Nanomaterials in Agricultural Applications

Engineered nanomaterials such as carbon nanotubes, gold nanoparticles, graphene oxide, quantum dots, and metal–organic frameworks (MOFs) are leveraged for their high surface-area-to-volume ratios, chemical reactivity, and tunable functional groups [16, 22]. In genomics and transcriptomics, these nanostructures facilitate efficient DNA/RNA extraction, nucleic acid hybridization, and improved signal transduction in biosensing systems. For instance, nanoparticle-enhanced quantitative PCR (nano-qPCR) and portable lab-on-a-chip devices enable real-time monitoring of gene expression in crops under environmental stress [3, 31].

2.2.1 AI-Driven Nanomaterial Design

Artificial intelligence, particularly machine learning (ML), is increasingly being employed to rationally design engineered nanomaterials for agricultural nano-omics applications prior to their physical synthesis. ML models trained on large datasets comprising physicochemical descriptors such as particle size, surface charge, porosity, functional groups, and electronic properties can accurately predict key performance attributes, including stability, adsorption capacity, conductivity, and biological compatibility [47]. In the case of metal–organic frameworks (MOFs), supervised learning and regression-based approaches have been used to predict nutrient binding efficiency, ion selectivity, and structural resilience under variable soil pH and moisture conditions, allowing researchers to shortlist optimal MOF architectures before experimental validation [1]. Similarly, ML-guided optimization of carbon nanotubes and graphene-based nanomaterials has improved surface functionalization and electrical properties required for biosensing while minimizing phytotoxicity and environmental risks [11, 13]. This AI-driven predictive design significantly reduces experimental costs, accelerates innovation, and enhances the safety and effectiveness of nanomaterials intended for agricultural deployment.

2.2.2 AI-Based Signal Processing in Nano-Enabled Soil Sensing

In parallel with nanomaterial design, artificial intelligence plays a critical role in processing signals generated by nanosensors operating in complex and heterogeneous soil environments. Soil matrices are inherently noisy due to fluctuating moisture content, variable ionic strength, organic matter interference, and dynamic microbial activity, all of which can distort sensor outputs. Advanced AI algorithms, including convolutional neural networks, adaptive filtering models, and anomaly detection systems, are increasingly used to distinguish meaningful biological or chemical signals from background noise [38]. These models learn temporal and spatial signal patterns, enabling real-time correction of signal drift, suppression of environmental interference, and enhancement of signal-to-noise ratios. Consequently, high-fidelity data on nutrient availability, microbial metabolites, or plant stress biomarkers can be reliably extracted under field conditions [47]. Such AI-assisted signal processing is essential for ensuring the accuracy, robustness, and scalability of nano-omics platforms in precision agriculture.

2.3 Nano-Omics Tools in Proteomics and Metabolomics

In proteomics, nanotechnology enhances the isolation and quantification of proteins through magnetic nanoparticles and nanostructured affinity surfaces, while nano-electrospray ionization boosts the sensitivity of mass spectrometric detection [24]. In the metabolomics domain, electrochemical and optical nanosensors facilitate the detection of key phytohormones, stress metabolites, and signaling molecules, providing high-throughput, field-relevant metabolic profiling [27]. These innovations are crucial for understanding plant physiological responses and tailoring nutrient or stress management strategies.

2.4 Nano-Omics in Microbiome Studies

The integration of nanotechnology into microbiomics, particularly for rhizosphere and phyllosphere microbiome studies, has opened new avenues for non-invasive monitoring and intervention. Nanosensors embedded in soil or plant tissues can detect microbial metabolites and quorum-sensing signals with high spatial and temporal resolution [14]. Furthermore, nanoparticle-assisted metagenomic sequencing and single-cell analysis have enhanced the resolution and functional annotation of plant-associated microbial communities [35].

2.5 Functional Applications of Nano-Omics

Beyond diagnostics, nano-omics facilitates targeted functional interventions. Nanocarriers such as lipid nanoparticles, polymeric micelles, and mesoporous silica particles have been successfully used to deliver functional RNAs, CRISPR-Cas gene-editing complexes, and plant growth regulators [11]. These systems enhance the stability, targeting accuracy, and bioavailability of biomolecules while reducing off-target effects, enabling precision manipulation of genetic and metabolic pathways in crops.

2.6 Challenges and Future Perspectives

Despite the vast potential of nano-omics, key challenges remain. The uptake, transport, and biotransformation of nanoparticles in plant systems are influenced by multiple factors, including nanoparticle size, charge, coating, plant age, and environmental conditions [16, 35]. Moreover, potential toxicity, environmental accumulation, and regulatory uncertainties limit large-scale deployment. Advances in techniques such as single-particle inductively coupled plasma mass spectrometry (sp-ICP-MS), electron microscopy, and synchrotron-based imaging are helping to trace nanoparticle behavior in vivo [22], but a deeper systems-level understanding is necessary to optimize nano-enabled tools for sustainable, climate-resilient agriculture.

3 Role of Artificial Intelligence in Multi-Omics Data Processing and Integration

Artificial intelligence plays a pivotal role in transforming complex multi-omics datasets into meaningful biological and agronomic insights. By enabling advanced data integration, pattern recognition, and predictive modeling, AI overcomes the limitations of traditional analytical approaches. Its synergy with nano-omics supports real-time decision-making, network reconstruction, and precision interventions, forming the computational backbone of next-generation smart agriculture.

3.1 The Challenge of Multi-Omics Complexity

The generation of high-throughput omics datasets covering genomics, transcriptomics, proteomics, metabolomics, and microbiomics has exponentially increased the volume and dimensionality of agricultural data. However, integrating and interpreting these heterogeneous datasets to extract biologically meaningful insights remain a significant challenge. Each omics layer carries different data formats, noise levels, and dynamic ranges, making cross-platform alignment difficult [15, 26].

Traditional statistical methods fall short in capturing the nonlinear, high-dimensional relationships among genes, proteins, metabolites, and environmental variables. Moreover, many omics datasets are incomplete, sparse, or imbalanced conditions under which machine learning (ML) and deep learning (DL) models have demonstrated superior performance in biological data integration [21].

3.2 AI Algorithms in Agricultural Omics

Artificial intelligence encompasses a suite of algorithms capable of self-improvement and pattern discovery. In agricultural omics, several AI methods have proven particularly useful:

- **Supervised learning algorithms** (e.g., random forest, support vector machines) are widely used for classification tasks such as disease diagnosis, yield prediction, and stress phenotyping [7].
- **Unsupervised learning models** (e.g., k-means clustering, hierarchical clustering, t-SNE) are used for grouping genes or metabolites with similar expression patterns under stress or treatment conditions [26].
- **Deep learning architectures** (e.g., convolutional neural networks [CNNs], recurrent neural networks [RNNs], autoencoders) offer superior performance in modeling sequential and image-based omics data [5].

Table 2 Emerging AI models and algorithms in agricultural omics analysis

AI technique	Function in omics analysis	Example application	Advantages	Limitations
Machine learning (ML)	Pattern recognition, prediction	Genomic trait selection, soil microbiome classification	Fast, scalable	May require labeled data
Deep learning (DL)	Feature extraction from complex omics data	Transcriptomic response modeling under stress	High accuracy, nonlinear detection	"Black-box" nature
Random forest	Classification and regression	Proteomics-based disease marker identification	Handles large data, interpretable	Susceptible to overfitting
Support vector machine (SVM)	Binary classification tasks	Disease presence prediction using metabolomics	Effective in high-dimensional data	Requires tuning
Neural networks	Multi-omics integration	Crop yield modeling from genomic and environmental data	Strong nonlinear mapping	Computationally intensive
Natural language processing (NLP)	Literature mining for gene-function associations	Predictive annotations of uncharacterized genes	Leverages large corpora	Language/data bias

For example, CNNs have been employed to predict crop phenotypes directly from multi-omics profiles and environmental variables, enabling accurate genotype–phenotype mapping [23]. Meanwhile, DL-based integration of transcriptomics and metabolomics has uncovered regulatory modules in stress-adaptive crop responses [33]. Various emerging AI models and their role in agriculture are given in Table 2.

3.3 AI for Omics Data Fusion and Network Reconstruction

AI enables the **fusion of omics layers** into unified models that can capture complex regulatory, metabolic, and ecological networks. Approaches such as **multi-view learning, Bayesian networks**, and **graph neural networks (GNNs)** have been used to reconstruct gene regulatory networks and microbe–host interaction pathways [45].

In soil and rhizosphere microbiome studies, AI tools can identify keystone taxa, predict functional guilds, and simulate microbe–microbe or microbe–plant interactions under different nutrient or climate regimes [8]. These insights are critical for designing microbiome-informed interventions such as biofertilizers or biocontrol agents.

3.4 Predictive Modeling and Decision Support

The integration of AI with nano-omics platforms enables **real-time, predictive modeling** of plant physiological states, disease outbreaks, or nutrient fluxes. When nanosensors collect omics or environmental data in situ, AI models can process this information and generate actionable recommendations for precision farming ranging from irrigation schedules to microbial consortia deployment [19].

Decision support systems based on AI are also being integrated with mobile apps, drones, and Internet of Things (IoT) devices, making them accessible to farmers and agronomists in real time [14]. These systems benefit from continuous feedback and learning loops, improving their accuracy and relevance with ongoing use.

3.5 On-Device Processing and Edge AI in Nano-Enabled Agriculture

Edge artificial intelligence enables nano-enabled sensors to process and analyze data locally on devices such as drones, tractors, or handheld field units without reliance on continuous internet connectivity. By embedding lightweight machine learning models directly into sensor hardware, raw nanosensor signals can be filtered, interpreted, and converted into actionable insights in real time [47]. This on-device processing reduces data latency, enhances system reliability, and minimizes bandwidth requirements, which is particularly critical in remote rural and resource-limited agricultural regions [36]. Edge AI also improves data security and resilience by limiting cloud dependency while supporting rapid, site-specific decision-making in precision agriculture [37].

4 Applications of AI-Nano-Omics in Sustainable Crop Improvement and Soil Microbiome Engineering

The convergence of artificial intelligence and nano-omics has unlocked powerful applications for sustainable crop improvement and soil microbiome engineering. By integrating high-resolution molecular data with predictive analytics and nano-enabled sensing, these approaches enhance breeding efficiency, enable early stress diagnostics, and support precision nutrient management. Moreover, AI-nano-omics facilitates rational design of microbial consortia and climate-smart interventions, offering holistic solutions that improve productivity while maintaining soil health, ecosystem stability, and long-term agricultural sustainability.

4.1 Enhancing Crop Breeding and Trait Prediction

The integration of AI and nano-omics has greatly accelerated the discovery of key regulatory genes, quantitative trait loci (QTLs), and stress-responsive biomarkers in crops. Traditional breeding approaches are being replaced or enhanced by genomics-assisted breeding, where high-dimensional omics data (genotype, phenotype, transcriptome, epigenome) are processed using AI models to identify genotype–phenotype associations with high precision [13].

Machine learning (ML) and deep learning (DL) algorithms, such as random forests and deep neural networks, have successfully predicted complex traits like drought tolerance, nutrient use efficiency, and yield stability under variable environments [4]. These predictive tools help breeders prioritize parental lines and accelerate decision-making in breeding pipelines.

Nano-enabled omics platforms, such as nanopore sequencing and nanosensors for real-time metabolite detection, further enhance this predictive capability. For example, portable devices using nanopore technology allow field-based rapid genotyping, which, when coupled with AI models, can generate on-the-spot phenotype forecasts [46].

4.2 Precision Agriculture and Stress Diagnostics

AI-powered nano-omics tools are revolutionizing precision agriculture by allowing early detection and diagnosis of biotic and abiotic stressors in crops. Nanosensors functionalized with selective probes can detect specific plant hormones (e.g., ethylene, salicylic acid), pathogens, or reactive oxygen species, providing high-resolution spatial and temporal data [14]. These signals are then analyzed by AI algorithms to classify stress types, predict severity, and suggest tailored interventions. Many domains and associated AI-nano-omics applications are described in Table 3.

Deep learning-based image analysis systems also process hyperspectral and fluorescence imaging data to assess plant health, nutrient status, or pathogen load, enabling non-invasive diagnostics [28]. When combined with omics insights, these systems offer real-time and mechanistically grounded crop monitoring.

4.3 Soil Microbiome Engineering and Synthetic Community Design

Soil microbiomes are key to nutrient cycling, plant resilience, and ecosystem sustainability [43]. AI-nano-omics enables a system-level understanding of plant–microbe–soil interactions, helping design and deploy microbial consortia that promote sustainable crop productivity [47].

Table 3 AI-nano-omics applications across agricultural domains

Domain	Nano-enabled technology	AI functionality	Target output	Impact
Soil health	Nanosensors for nutrient detection	AI-driven nutrient modeling	N-P-K mapping	Precision fertilization
Plant health	Pathogen-detecting nano-biosensors	Disease prediction models	Early diagnosis	Reduced pesticide use
Crop breeding	Nano-assisted omics sampling	AI-assisted trait mining	High-throughput phenotyping	Faster breeding cycles
Stress tolerance	Smart nanoformulations	ML for resilience prediction	Drought/salinity markers	Climate adaptation
Microbiome engineering	Nanocarriers for probiotics	Network inference models	Functional microbial consortia	Soil regeneration

Multi-omics approaches, such as metagenomics, metatranscriptomics, and metabolomics, characterize microbial community composition and functional potential. AI tools then process this data to identify keystone taxa, infer functional guilds, and model microbe–plant associations [6]. For instance, ML models have been used to predict how microbial communities respond to fertilizers or climate conditions, improving soil amendment strategies.

Additionally, synthetic microbial consortia are being designed using AI-based simulations that assess microbial compatibility, stability, and plant-growth-promoting traits under varying environmental conditions. Such predictive designs enhance the field performance of biofertilizers and biocontrol agents while minimizing environmental risks.

4.4 Synthetic Microbial Communities (SynComs) and Modeling Interactions

It emerged as a powerful tool to enhance plant growth and resilience by introducing carefully designed consortia of beneficial microbes into the rhizosphere. Modeling the interactions within SynComs is critical, as microbial dynamics are highly complex. Artificial intelligence, particularly graph neural networks (GNNs), has been employed to capture these interactions. GNNs represent microbes as nodes and their interactions—such as competition, cooperation, and metabolic exchange—as edges in a network. By analyzing these graphs, AI can predict community stability, functional outcomes, and potential synergistic effects under varying environmental conditions [14, 43]. This approach enables the rational design of SynComs with optimized performance tailored to specific crops or soil types.

4.5 Nanocarrier Stability

Another key challenge in applying SynComs is ensuring microbial survival under harsh soil conditions, including fluctuations in pH, moisture, and temperature. Nanocarriers have been developed to address this issue by encapsulating synthetic microbes within protective matrices. These carriers buffer the microbes from environmental stressors, regulate their release, and enhance colonization efficiency. Studies have demonstrated that nanocarrier-stabilized SynComs maintain higher viability and functional activity in the field compared to unprotected microbes [20, 29]. Combining AI-guided modeling with nanocarrier technology thus represents a promising strategy for effective and sustainable SynCom deployment in agriculture.

4.6 Sustainable Nutrient Management and Carbon Sequestration

Nano-omics platforms equipped with nanosensors can monitor soil pH, nutrient fluxes (N, P, K), and microbial activity in real time. These data streams feed into AI models that optimize fertilization regimes, minimizing overuse and leaching [9]. For example, precision N-management tools based on real-time sensing and AI learning have reduced nitrogen application by up to 30% in field trials without compromising yield.

AI models also help predict carbon dynamics in soil microbiomes, supporting carbon sequestration initiatives through optimized tillage, cover cropping, and microbial management [14]. Such approaches contribute to climate-smart agriculture by coupling productivity with long-term ecosystem resilience.

5 Challenges, Limitations, and Ethical Considerations in AI-Nano-Omics Agriculture

The convergence of artificial intelligence, nanotechnology, and multi-omics science promises to transform agriculture into a more predictive, precise, and sustainable enterprise. However, this transformation is not without its complexities. As the applications of AI-nano-omics systems move from laboratory proof-of-concept to field-level implementation, a host of technical, ethical, ecological, and policy-related challenges emerge, as shown in Fig. 2.

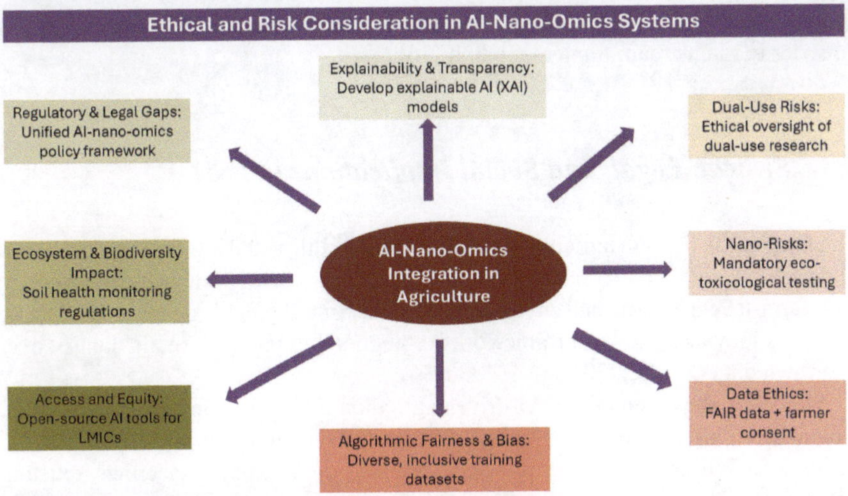

Fig. 2 Ethical risk wheel outlining AI-nano-omics challenges: nano-toxicity, fairness, biosafety, equity, transparency, and regulatory gaps

5.1 *Technical Barriers in Data Integration and Model Interpretability*

A central challenge in AI-nano-omics is the integration of heterogeneous omics datasets, genomics, transcriptomics, proteomics, metabolomics, and phenomics, each with distinct data formats, scales, and noise profiles. These datasets often suffer from sparsity and imbalance, complicating efforts to build robust predictive models [26]. Moreover, the lack of standardized protocols for data generation and annotation hinders interoperability across platforms and studies.

Equally important is the interpretability of AI models. Deep learning models, though powerful, often function as "black boxes," making it difficult for researchers to understand why certain predictions are made [34]. This lack of transparency can reduce trust and limit their adoption in regulatory and agronomic contexts.

5.2 *Limitations in Nano-Device Deployment*

The widespread application of nanotechnology in agriculture remains limited by challenges in scalability, cost, environmental persistence, and biosafety. Many nano-devices (e.g., biosensors, nano-fertilizers) are still in the prototype phase or are expensive for smallholder farmers to adopt at scale [39].

Furthermore, concerns over nanomaterial toxicity and accumulation in soil, plants, and food chains raise questions about long-term ecological impacts. Although several studies have shown promising outcomes in controlled environments, the lack of comprehensive field-scale assessments and environmental risk frameworks remains a barrier to mainstream implementation [16].

5.3 Ethical, Legal, and Social Implications (ELSI)

The integration of AI in agriculture raises complex ethical and legal questions. Issues of data ownership, privacy, and algorithmic bias are especially relevant when dealing with farm-level genomic and environmental data. Smallholder farmers in developing countries may lack the legal frameworks or negotiating power to retain control over their own data (Mark 2020).

Moreover, the potential for AI-driven agricultural technologies to displace labor, widen the digital divide, and create asymmetries in innovation access must be addressed. Without inclusive policies, AI-nano-omics may exacerbate existing socio-economic inequities in rural regions.

5.4 Regulatory Gaps and Standardization Issues

There is currently no unified regulatory framework governing the use of AI and nanotechnology in agriculture. Regulatory authorities in many countries have yet to define risk assessment protocols or standards for AI-generated predictions in crop development or for nano-enabled devices in soil and plant systems [47].

Moreover, the rapid pace of technological advancement far exceeds the rate at which regulations are updated, leading to ambiguities around accountability, liability, and safety. This lack of guidance creates hesitation among stakeholders and slows technology adoption.

6 Future Perspectives and Strategic Roadmap

The future of AI-assisted nano-omics lies in the convergence of edge computing, digital twins, and responsible innovation frameworks to enable real-time, field-level intelligence in agriculture. Integration of edge AI and advanced computing will allow decentralized processing of high-dimensional omics and sensor data, reducing latency and infrastructure dependence. Digital twins powered by nano-enabled biosensors and AI analytics will support personalized, site-specific crop and soil management. Equally important, policy innovation, ethical AI governance, and capacity building will be essential to ensure equitable access, biosafety, and sustainable deployment of AI-nano-omics for climate-resilient and biodiversity-rich agroecosystems.

6.1 Convergence of AI, Nano-Omics, and Edge Technologies

The future of AI-assisted nano-omics lies in converging innovations across synthetic biology, quantum computing, and edge AI. Integration of quantum machine learning could overcome current computational bottlenecks in processing ultra-high-dimensional omics data. Meanwhile, edge computing offers decentralized, real-time processing of sensor and omics data directly in the field, reducing latency and dependence on cloud infrastructure [36].

Additionally, bio-hybrid systems such as microfluidic biosensors embedded with living cells or engineered microbes could interface seamlessly with AI systems to provide real-time feedback on plant health, soil nutrient dynamics, or microbiome shifts.

6.2 Personalized Agriculture and Digital Twins

Emerging AI frameworks are enabling the development of "digital twins" for crops, fields, and microbial systems, dynamic virtual models that simulate biological and environmental responses under different scenarios [30]. These can predict crop responses to nutrient regimes, climate anomalies, or microbial inoculation strategies, enabling site-specific, tailored agricultural practices.

When powered by nano-enabled biosensors and real-time omics data, these systems will facilitate personalized agriculture: highly localized interventions based on genotype, phenotype, soil microbiome, and climate conditions.

6.2.1 Digital Twins for Predictive Agronomy

Digital twins are transforming modern agriculture by creating virtual replicas of crops, soils, and entire farm ecosystems. These digital representations integrate real-time data from sensors, weather forecasts, soil analyses, and crop models to simulate growth, nutrient uptake, and yield under varying conditions. One of the most powerful applications of digital twins is virtual stress testing, which allows researchers and farmers to predict long-term outcomes in a fraction of real time. For instance, a digital twin can simulate 10 years of climate change scenarios in just minutes, incorporating variations in temperature, precipitation, and extreme weather events. By introducing a specific nano-fertilizer into this virtual system, its effects on nutrient availability, crop growth, and resilience can be evaluated before field deployment [30].

This predictive capability enables the optimization of fertilization strategies, reduces resource wastage, and mitigates environmental impacts. Digital twins also facilitate scenario comparison, helping agronomists choose formulations and application schedules that maximize yield under future climate conditions. By combining high-resolution modeling with real-time farm data, digital twins provide a cost-effective, risk-free platform to test innovations such as nano-fertilizers, paving the way for precision agriculture that is adaptive, sustainable, and resilient [2].

6.2.2 Genotype–Environment Interaction (G × E)

Genotype-environment interaction plays crucial role in determining crop performance under varying environmental and management conditions. Understanding how different crop varieties respond to specific interventions is essential for precision agriculture. Artificial intelligence (AI) has emerged as a powerful tool to model these complex interactions, particularly when combined with nano-omics approaches—such as nano-enabled genomics, proteomics, and metabolomics. AI algorithms, including machine learning and deep learning models, can integrate multi-layered datasets comprising genotype information, environmental parameters (soil type, temperature, moisture), and management practices (fertilizer type, irrigation, nano-fertilizer application) to predict crop responses [37].

By simulating genotype × environment × management (G × E × M) interactions, AI can forecast how a specific nano-omics intervention will influence growth, nutrient uptake, stress tolerance, and yield for each crop variety. For instance, nano-fertilizer formulations can be virtually tested on multiple genotypes under varying climatic conditions to identify optimal combinations. This approach not only reduces experimental costs and time but also enhances crop resilience and productivity. Integrating AI with nano-omics enables precision agronomy, where interventions are tailored to the genetic and environmental context, maximizing efficiency while minimizing environmental impact [20, 29].

6.3 Policy Innovation and Responsible AI in Agriculture

The trajectory of AI-nano-omics must be guided by responsible innovation frameworks that prioritize transparency, fairness, and environmental sustainability. Future policies must address algorithmic governance, inclusive access to AI technologies, and ethical management of biological and agronomic data [12]. Some important gaps and recommendation policies are described in Table 4.

Capacity-building programs to train next-generation researchers in AI, omics, and agricultural ethics will be crucial. Public–private partnerships and international collaborations must prioritize open science, data sharing, and participatory innovation, particularly in the Global South [18].

Table 4 Governance and regulatory frameworks for AI-nano-omics in agriculture

Governance area	Current gaps	Recommended policy measures	Stakeholders involved
Nano-biosafety	Lack of environmental impact studies	Mandate nano-EIA and toxicity screening	Regulatory bodies, EPA-equivalents
AI ethics and bias	Algorithm opacity, lack of explainability	Promote transparent, explainable AI (XAI)	AI developers, agri-tech firms
Data sovereignty	Cross-border omics data transfers are unregulated	Enforce FAIR data principles + GDPR alignment	Farmers, bioinformatics platforms
Access and equity	Limited deployment in LMICs	Open-source tools, public AI platforms	Governments, NGOs
Intellectual property	Ambiguities around AI-generated omics innovations	Define AI-related IP laws in agriculture	Legal experts, WIPO, CGIAR
Education and training	Low awareness of integrated systems	Interdisciplinary curricula in AI-nano-omics	Universities, policy bodies

6.4 Toward Climate-Resilient and Biodiverse Agroecosystems

AI-nano-omics can serve as a cornerstone for climate-resilient agriculture, helping identify and deploy crop genotypes and microbial consortia adapted to fluctuating environmental stresses. Predictive analytics can optimize crop rotations, intercropping systems, and regenerative practices that enhance soil carbon sequestration, water-use efficiency, and agrobiodiversity [17].

By coupling precision intervention with systems-level ecological insight, AI-nano-omics can help shift global agriculture toward sustainability, resilience, and equity.

7 Summary of Key Policy and Practice Recommendations

1. **Promote Interdisciplinary Standardization and Open Data Sharing**
    - Establish interoperable data standards across omics, nano-devices, and AI platforms to ensure seamless integration and reproducibility.
    - Encourage the development of open-access databases and shared computational resources to democratize AI-nano-omics tools, especially for researchers in the Global South.
    - Foster interdisciplinary research consortia to bridge gaps between plant sciences, data science, nanotechnology, and sustainability policy.

2. **Implement Risk Assessment Frameworks for Nanomaterials and AI Tools**
    - Develop regulatory frameworks that assess long-term impacts of nanomaterials in agroecosystems, with protocols for biosafety, toxicity, and ecological persistence.
    - Institute ethical AI auditing mechanisms that assess transparency, explainability, and bias in machine learning algorithms used in agricultural decision-making.
3. **Strengthen Capacity-Building and Farmer-Centric Innovation**
    - Launch training programs for early-career scientists and extension workers on AI-omics-nano convergence, bioinformatics, and field-based deployment of smart sensors.
    - Involve local farmers and indigenous communities in participatory design of nano-enabled and AI-driven technologies, ensuring cultural appropriateness and usability.
    - Support digital literacy and infrastructure development (e.g., rural broadband) to reduce technological barriers in low-resource regions.
4. **Incentivize Climate-Smart Applications of AI-Nano-omics**
    - Provide public and private funding incentives for AI-nano-omics projects that enhance soil health, nutrient use efficiency, drought resilience, and biodiversity conservation.
    - Encourage precision agroecology approaches that use nano-biosensors and AI predictions to optimize sustainable practices (e.g., intercropping, microbial consortia, carbon farming).
5. **Embed Responsible Innovation in Agricultural Policy**
    - Institutionalize responsible innovation frameworks in national agriculture and science policies, focusing on inclusivity, equity, and long-term ecosystem stewardship.
    - Support multi-stakeholder platforms to co-develop guidelines for ethical use of agricultural AI and nanotechnologies, with input from farmers, scientists, policymakers, and civil society.

8 Conclusion

The integration of artificial intelligence (AI), nanotechnology, and multi-omics is transforming agricultural science into a predictive, precise, and proactive discipline. AI-assisted nano-omics offers unprecedented opportunities to enhance crop productivity, soil health, microbial efficiency, and climate resilience, key pillars of sustainable agriculture in the twenty-first century. By enabling high-resolution monitoring of plant and soil systems, predictive modeling of genotype-to-phenotype traits, and real-time decision-making through nano-enabled biosensors, this convergence promises to address major global challenges related to food security, biodiversity loss, and environmental degradation.

However, realizing this potential requires overcoming significant barriers. Technical limitations in data integration, lack of interpretability in AI models, and gaps in regulatory and ethical oversight demand urgent attention. Likewise, the deployment of nano-enabled technologies in real-world agroecosystems must be guided by robust risk assessments and sustainability metrics. If left unaddressed, these issues could undermine trust, safety, and equity in agricultural innovation.

Moving forward, the success of AI-nano-omics hinges on collaborative efforts across disciplines, sectors, and geographies. Policies must prioritize open science, inclusive access, and responsible governance. Equally important is the need for investments in training, digital infrastructure, and farmer-centric technology design. By anchoring innovation in ecological wisdom, equity, and long-term resilience, AI-nano-omics can catalyze the transition toward a climate-smart, biodiverse, and sustainable agri-food future.

References

1. Afantitis A, Tsoumanis A, Melagraki G (2020) Enalos suite of tools: enhancing cheminformatics and nanoinformatics through KNIME. Current Med Chem 27(38):6523–6535. https://doi.org/10.2174/0929867327666200727114410
2. Agelli M, Corona N, Maggio F, Moi PV (2024) Unmanned ground vehicles for continuous crop monitoring in agriculture: assessing the readiness of current ICT technology. Machines 12(11):750. https://doi.org/10.3390/machines12110750
3. Alharbi J, Jackson D, Usher K (2020) The potential for COVID-19 to contribute to compassion fatigue in critical care nurses. J Clinical Nursing 29(15–16):2762. https://doi.org/10.1111/jocn.15314
4. Ali T, Rehman SU, Ali S, Mahmood K, Obregon SA, Iglesias RC, Ashraf I (2024) Smart agriculture: utilizing machine learning and deep learning for drought stress identification in crops. Scient Reports 14(1):1–16. https://doi.org/10.1038/s41598-024-74127-8
5. Amiri Z, Heidari A, Navimipour NJ, Esmaeilpour M, Yazdani Y (2024) The deep learning applications in IoT-based bio-and medical informatics: a systematic literature review. Neural Comput Appl 36(11):5757–5797. https://doi.org/10.1007/s00521-023-09366-3
6. Arıkan M, Muth T (2023) Integrated multi-omics analyses of microbial communities: a review of the current state and future directions. Molecul Omics 19(8):607–623. https://doi.org/10.1039/D3MO00089C
7. Attri I, Awasthi LK, Sharma TP (2024) Machine learning in agriculture: a review of crop management applications. Multimedia Tools and Appl 83(5):12875–12915. https://doi.org/10.1007/s11042-023-16105-2
8. Bahram M, Hildebrand F, Forslund SK, Anderson JL, Soudzilovskaia NA, Bodegom PM, Bork P (2018) Structure and function of the global topsoil microbiome. Nature 560(7717):233–237. https://doi.org/10.1038/s41586-018-0386-6
9. Bhakta I, Phadikar S, Majumder K (2019) State-of-the-art technologies in precision agriculture: a systematic review. J Sci Food Agricul 99(11):4878–4888. https://doi.org/10.1002/jsfa.9693
10. Borah A, Sahu S, Srivastava RP, Singh M, Tyagi DB (2024) Exploring the economic challenges threatening global agriculture and food security. Ecol Environ Conserv 0971765X:30

11. Demirer GS, Zhang H, Matos JL, Goh NS, Cunningham FJ, Sung Y, Landry MP (2019) High aspect ratio nanomaterials enable delivery of functional genetic material without DNA integration in mature plants. Nature Nanotechnol 14(5):456–464. https://doi.org/10.1038/s41565-019-0382-5
12. Floridi L, Cowls J, Beltrametti M, Chatila R, Chazerand P, Dignum V, Vayena E (2018) AI4People—an ethical framework for a good AI society: opportunities, risks, principles, and recommendations. Mind Mach 28:689–707. https://doi.org/10.1007/s11023-018-9482-5
13. Gardner L, Kostarelos K, Mallick P, Dive C, Hadjidemetriou M (2022) Nano-omics: nanotechnology-based multidimensional harvesting of the blood-circulating cancerome. Nature Rev Clinical Oncol 19(8):551–561
14. Grunwald S (2022) Artificial intelligence and soil carbon modeling demystified: power, potentials, and perils. Carbon Footprints 1(1):N-A. https://doi.org/10.20517/cf.2022.03
15. Hasin Y, Seldin M, Lusis A (2017) Multi-omics approaches to disease. Genome Biol 18:1–15. https://doi.org/10.1186/s13059-017-1215-1
16. Kah M, Kookana RS, Gogos A, Bucheli TD (2018) A critical evaluation of nanopesticides and nanofertilizers against their conventional analogues. Nature Nanotechnol 13(8):677–684. https://doi.org/10.1038/s41565-018-0131-1
17. Khan MHU, Wang S, Wang J, Ahmar S, Saeed S, Khan SU, Feng X (2022) Applications of artificial intelligence in climate-resilient smart-crop breeding. Int J Molecul Sci 23(19):11156. https://doi.org/10.3390/ijms231911156
18. Kiosia A, Boylan S, Retford M, Marques LP, Bueno FTC, Kirima C, Wozencraft A (2024) Current data science capacity building initiatives for health researchers in LMICs: global and regional efforts. Front Public Health 12:1418382
19. Kumar P, Choudhury D (2025) Hybrid deep learning for predictive modelling of microbial biostimulants in precision agriculture. In: Artificial intelligence in microbial research, pp 351–383. https://doi.org/10.1007/978-981-96-3448-4_16
20. Kumar R, Choudhary DK, Debnath M (2020) Development of BSA conjugated on modified surface of quercetin-loaded lipid nanocarriers for breast cancer treatment. Mater Res Expr 7(1):015411
21. Libbrecht MW, Noble WS (2015) Machine learning applications in genetics and genomics. Nature Rev Genet 16(6):321–332. https://doi.org/10.1038/nrg3920
22. Liu R, Lal R (2015) Potentials of engineered nanoparticles as fertilizers for increasing agronomic productions. Sci Total Environ 514:131–139. https://doi.org/10.1016/j.scitotenv.2015.01.104
23. Mahmood U, Li X, Fan Y, Chang W, Niu Y, Li J, Lu K (2022) Multi-omics revolution to promote plant breeding efficiency. Front Plant Sci 13:1062952. https://doi.org/10.3389/fpls.2022.1062952
24. Majumdar S, Keller AA (2021) Omics to address the opportunities and challenges of nanotechnology in agriculture. Critical Rev Environ Sci Technol 51(22):2595–2636
25. Mark R (2019) Ethics of using AI and big data in agriculture: the case of a large agriculture multinational. The ORBIT J 2(2):1–27. https://doi.org/10.29297/orbit.v2i2.109
26. Misra BB, Langefeld C, Olivier M, Cox LA (2019) Integrated omics: tools, advances and future approaches. J Molecul Endocrinol 62(1):R21–R45
27. Mondal S, Priya S, Jha R, Bora J, Lata S, Hashem A, Malik S (2025) Nano-omics base plant genetic engineering: a key role in sustainable agriculture. In: Biogenic nanoparticles, Elsevier, pp 377–393. https://doi.org/10.1016/B978-0-443-33819-9.00019-6
28. Negi P, Anand S (2024) Plant disease detection, diagnosis, and management: recent advances and future perspectives. In: Artificial intelligence and smart agriculture: technology and applications, pp 413–436
29. Patel D, Patel B, Thakkar H (2021) Lipid based nanocarriers: promising drug delivery system for topical application. Europ J Lipid Sci Technol 123(5):2000264. https://doi.org/10.1002/ejlt.202000264
30. Peladarinos N, Piromalis D, Cheimaras V, Tserepas E, Munteanu RA, Papageorgas P (2023) Enhancing smart agriculture by implementing digital twins: a comprehensive review. Sensors 23(16):7128. https://doi.org/10.3390/s23167128

31. Qureshi MDA, Ramzan MF, Amjad F, Haider N (2024) Artificial intelligence in metabolomics for disease profiling: a machine learning approach to biomarker discovery. Indus J Biosci Res 2(02):87–96. https://doi.org/10.70749/ijbr.v2i02.146
32. Rai A, Tripathi S, Bahuguna A, Rai S, Rajput J, Gangwar A, Pandey S (2023) Conservation agriculture for soil health and carbon sequestration in the Indian Himalayan Region. In: Soil carbon dynamics in Indian Himalayan Region, Singapore, Springer Nature Singapore, pp 93–123
33. Roychowdhury R, Das SP, Gupta A, Parihar P, Chandrasekhar K, Sarker U, Sudhakar C (2023) Multi-omics pipeline and omics-integration approach to decipher plant's abiotic stress tolerance responses. Genes 14(6):1281. https://doi.org/10.3390/genes14061281
34. Samek W, Montavon G, Lapuschkin S, Anders CJ, Müller KR (2021) Explaining deep neural networks and beyond: a review of methods and applications. Proc IEEE 109(3):247–278. https://doi.org/10.1109/JPROC.2021.3060483
35. Sanzari I, Leone A, Ambrosone A (2019) Nanotechnology in plant science: to make a long story short. Front Bioeng Biotechnol 7:120. https://doi.org/10.3389/fbioe.2019.00120
36. Sarker IH, Kayes ASM, Badsha S, Alqahtani H, Watters P, Ng A (2020) Cybersecurity data science: an overview from machine learning perspective. J Big Data 7:1–29. https://doi.org/10.1186/s40537-020-00318-5
37. Sharma K, Shivandu SK (2024) Integrating artificial intelligence and internet of things (IoT) for enhanced crop monitoring and management in precision agriculture. Sensors Int 100292.https://doi.org/10.1016/j.sintl.2024.100292
38. Shaw DS, Honeychurch KC (2022) Nanosensor applications in plant science. Biosensors 12(9):675. https://doi.org/10.3390/bios12090675
39. Shawon ZBZ, Hoque ME, Chowdhury SR (2020) Nanosensors and nanobiosensors: agricultural and food technology aspects. In: Nanofabrication for smart nanosensor applications. Elsevier, pp 135–161. https://doi.org/10.1016/B978-0-12-820702-4.00006-4
40. Srikanth P, Sivakumar D, Sharma A, Kaushik N (2025) Recent developments in omics techniques for improving plant abiotic stress using microbes. Int J Environ Sci Technol 22(5):3787–3810. https://doi.org/10.1007/s13762-024-05957-2
41. Tariq A, Guo S, Farhat F, Shen X (2025) Engineering synthetic microbial communities: diversity and applications in soil for plant resilience. Agronomy 15(3):513. https://doi.org/10.3390/agronomy15030513
42. Tolani P, Gupta S, Yadav K, Aggarwal S, Yadav AK (2021) Big data, integrative omics and network biology. Adv Protein Chem Struct Biol 127:127–160. https://doi.org/10.1016/bs.apcsb.2021.03.006
43. Vikas, Ranjan R (2023) Biofertilizers and biofortification in future agriculture. In: Climate-resilient agriculture, Vol 1: crop responses and agro ecological perspectives, Cham, Springer International Publishing, pp 363–377. https://doi.org/10.1007/978-3-031-37424-1_18
44. Vishnoi S, Goel RK (2024) Climate smart agriculture for sustainable productivity and healthy landscapes. Environ Sci Policy 151:103600. https://doi.org/10.1016/j.envsci.2023.103600
45. Wu Y, Xie L (2024) AI-driven multi-omics integration for multi-scale predictive modeling of causal genotype-environment-phenotype relationships. arXiv preprint arXiv:2407.06405. https://doi.org/10.48550/arXiv.2407.06405
46. Yadav A, Yadav K, Ahmad R, Abd-Elsalam KA (2023) Emerging frontiers in nanotechnology for precision agriculture: advancements, hurdles and prospects. Agrochemicals 2(2):220–256
47. Zhang P, Guo Z, Ullah S, Melagraki G, Afantitis A, Lynch I (2021) Nanotechnology and artificial intelligence to enable sustainable and precision agriculture. Nature Plants 7(7):864–876. https://doi.org/10.1038/s41477-021-00946-6

AI-Nano Strategies for Developing Stress-Smart and Precision Farming

Purbita Aich Roy, Madhurya Ray, Soham Barik, Payel Kar, Subrata Goswami, Soumya Shephalika Dash, Tahir Mohammad Chauhan, Santanu Kundu, and Maksud Hasan Shah

Abstract Precision agriculture represents a paradigm shift from conventional, uniform field management to data-driven, site-specific decision-making aimed at enhancing productivity, profitability, and sustainability. Rooted in the concept of managing spatial and temporal variability, precision agriculture integrates advanced technologies such as remote sensing, GPS, GIS, automation, Internet of Things (IoT), and artificial intelligence (AI) to optimize resource use and reduce environmental impacts. Following the limitations of intensive chemical-based farming systems that

P. Aich Roy · S. Kundu
ICAR-Indian Agricultural Research Institute, New Delhi, India
e-mail: puaichroy@gmail.com

S. Kundu
e-mail: santanukundu120@gmail.com

M. Ray · S. Goswami
Banaras Hindu University, Uttar Pradesh, Varanasi, India
e-mail: raymadhurya999@bhu.ac.in

S. Goswami
e-mail: subrata.goswami@bhu.ac.in

S. Barik · P. Kar (✉)
Bidhan Chandra Krishi Viswavidyalaya, Mohanpur, Nadia, West Bengal, India
e-mail: kar.payel@bckv.edu.in

S. Barik
e-mail: bariksoham8@gmail.com

S. Barik
West Bengal, India

S. S. Dash
Palli Siksha Bhavana, Visva Bharati, Sriniketan, West Bengal, India
e-mail: soumyadashshephalika@gmail.com

T. M. Chauhan · M. H. Shah (✉)
CAE, F/ O Agricultural Sciences AMU, Aligarh, India
e-mail: maksudhasanshah@gmail.com

T. M. Chauhan
e-mail: drtmchauhan@gmail.com

J.-T. Chen (ed.), *AI-Nano-Enabled Climate-Smart Agriculture and Sustainability*,
Smart Nanomaterials Technology, https://doi.org/10.1007/978-981-92-1694-9_5

emerged after the Green Revolution—characterized by soil degradation, yield stagnation, and ecosystem disruption—precision agriculture has evolved as a sustainable alternative since the 1990s. Nanotechnology has further strengthened this approach through the development of nanosensors, which exploit unique physicochemical properties of nanomaterials (1–100 nm) such as high surface area, conductivity, and biocompatibility. Nano-enabled sensors, including optical, electrochemical, mechanical, thermal, electrical, electromagnetic, and biosensors, enable ultra-sensitive, real-time detection of soil nutrients, moisture, pH, temperature, pathogens, pesticides, heavy metals, and plant physiological signals. Materials such as carbon nanotubes, nanoparticles, nanowires, fullerenes, quantum dots, polymers, and metal oxides serve as key components in these sensing platforms. The operational mechanism of nanosensors involves selective recognition of target analytes, signal transduction, and amplification integrated with wireless communication and AI-based data analytics. Their deployment supports precision fertilization, irrigation scheduling, early stress and disease detection, environmental monitoring, and food safety assurance. Despite their immense potential, large-scale field adoption remains constrained by technical challenges, regulatory compliance, standardization, economic limitations, and environmental safety concerns. Nevertheless, the convergence of nanotechnology, precision agriculture, and digital farming systems positions nanosensors as transformative tools for achieving sustainable agriculture and global food security in the face of growing population pressure and climate variability.

Keywords Sustainability · Biocompatibility · Heavy metals · Food security

1 Introduction

Precision agriculture consists of two words: "precision" derived from "*praecisio*" (Latin) meaning "a cutting off", and agriculture derived from two Latin words "*ager*" meaning soil, and "*cultura*" meaning cultivation. Therefore, it streamlines the tedious maneuvering practices of conventional agriculture. The International Society for Precision Agriculture (2019) defined it as "*A management strategy that gathers, processes and analyzes temporal, spatial and individual data and combines it with other information to support management decisions according to estimated variability for improved resource use efficiency, productivity, quality, profitability and sustainability of agricultural production.*" In the era of globalization and artificial intelligence (AI), a new, smarter approach has replaced the traditional system to secure sustainability and stability. Integrating cutting-edge technologies like automation, vision technology, and Internet of Things (IoT) not only cuts out the unnecessary labor but also ensures an even distribution of technological advances. While conventional agriculture considers the field as a whole, precision agriculture identifies the intra-field variations and devises recommendations accordingly. This promises a better, more efficient, and effective system.

The field of agriculture has not been at a standstill, with the green revolutions in the 1960s, food safety had been ensured, but the environment was not taken into consideration. The intensive farming exhausted our lands, and production has reached a plateau. Chemical applications have increased exponentially, but the same has not reflected in the production in recent decades. Increasing population pressure has added more complications to the already fragile agroecosystems. Lands once fertile are turning unproductive, pollution is threatening mass extinctions, and ecosystem stability is disrupted. With the future of global farming collapsing, the need for a new system was realized. The introduction of RSs (Remote Sensing), GPSs (Global Positioning Systems), GISs (Geographic Information Systems), and yield monitors, "Precision agriculture" emerged in the 1990s as an avenue to address these issues. To meet the demands of the twenty-first century and satisfy the hunger for technology, farming systems have to undergo constant evolution. Manual systems paved the way for digital systems and finally smart systems [18]. So, precision agriculture is not just agriculture,it is an integration of novel technological branches, united in making the future of global farming a golden one. Nanotechnology is one such emerging branch; the utilization of particles of 1–100 nm (10^{-9} m or one billionth of a meter) is called nanotechnology. These nanomaterials (NMs) have unique properties like high conductivities, large surface area-to-volume ratio, and good biocompatibility [8].

Materials like metal oxides, chitosans, quantum dots, polymers, mesoporous silica, hydrogels, etc., are used for the preparation of nanoparticles. Their usage can vary from protectants, encapsulants, detectors, to biosensors and remote sensing devices [10]. Later, the infusion of NMs with sensing devices resulted in a boom in the variety of highly efficient and robust nanomaterial-enabled optical, electrochemical, and mass-sensitive detection platforms for sensing applications [3, 34, 35]. All these advances culminated in the birth of nanosensor technology.

2 Nano-Sensors: Unleashing the Potential of Nanotechnology

The word "sensor" is derived from the "*sentire*" (Greek) meaning "to perceive." A sensor is a device that detects a variable quantity, usually electronically, and converts the measurement into specific signals. These nanobiosensors are very cost-effective, simple, and rapid in real-time analysis of soil temperature, pH, moisture, pesticides, and detection of contaminants and microorganisms in soil [42].

> Nano-sensors are chemical or mechanical sensors that can be used to detect the presence of chemical species and nanoparticles, or monitor physical parameters such as temperature, on the nanoscale.

Nano-sensors must have at least one dimension within the nanometer range. Thus, physical, chemical, and biological changes can be perceived in a short period of time by nano sensors over traditional approaches. Also, their sensitivity differs according to environmental conditions, providing accurate site-specific responses. Additionally, owing to their sizes, nanoparticles can interact in vivo with biomolecules in a living system, providing us with real-time data where mechanical sensors cannot reach. Although a wide range of nanoparticles are used for manufacturing the nano-sensors, only a few of them are exploited extensively because of environmental concerns.

2.1 *Materials Used for Manufacturing Nanosensors*

2.1.1 Nanoparticles (NPs)

These are the building blocks of nanotechnology. Nanoparticles (NPs), typically characterized by dimensions between 1 and 100 nm, are regarded as advanced materials because of their exceptionally high surface area-to-volume ratio and distinctive physicochemical properties. These attributes make them highly promising for applications across diverse disciplines, especially in medicine and biomedicine [4]. These can be subdivided into carbon-based, metallic, semiconductors, and polymeric. Each one of them has its unique properties and uses.

2.1.2 Carbon Nanotubes

Discovered by Iijima in 1991, this is the most-used carbon-based nanomaterial. Extensive research indicates that CNTs may successfully pierce seed coatings and may interact with plant cells, affecting the amount of chlorophyll and improving the efficiency of light capture during photosynthesis [1]. Being sp2 hybridized, the free electron available contributes to excellent electronic, mechanical, and functional properties. Their easy customizability provides a means to interact selectively with different analytes, making them excellent Field-effect-transistors (FET).

2.1.3 Nanowires

These are wires in nanometric scale. Mostly Si is preferred because of its reproducibility by controlling the wire diameter and dopant concentration. These can also be made up of gold or silver NPs. These can help in identifying DNA changes or peptide nuclear detectors. These are also used for making gas sensors.

2.1.4 Fullerenes

Allomorphs of carbon are sp2 hybridized and highly symmetric. The building was named Fullerene in honor of an American architect because it was similar to his geodesic domes. Among the fullerene members, C60 is the most abundant archetype

[5]. These are often called "Bucky balls" after Buckminster Fuller. Fullerenes are very strong and bounce back to their initial shape after deformation, making them excellent electrochemical detectors [38].

Eight allotropes of carbon are shown below in Fig. 1.

Types of nanosensors:

The nanosensors can be classified into several types (Fig. 2) based on

A. **Interaction with the Analyte and Signal Transduction**:

A **Optical and Radiometric Sensors:** They use electromagnetic radiation for judging soil parameters like organic matter content, CEC (Cation Exchange Capacity), pH, mineralogy, etc. Nanoparticles, such as quantum dots, gold nanoshells, respond to specific analytes, like plant hormones, pesticides, or pathogens,

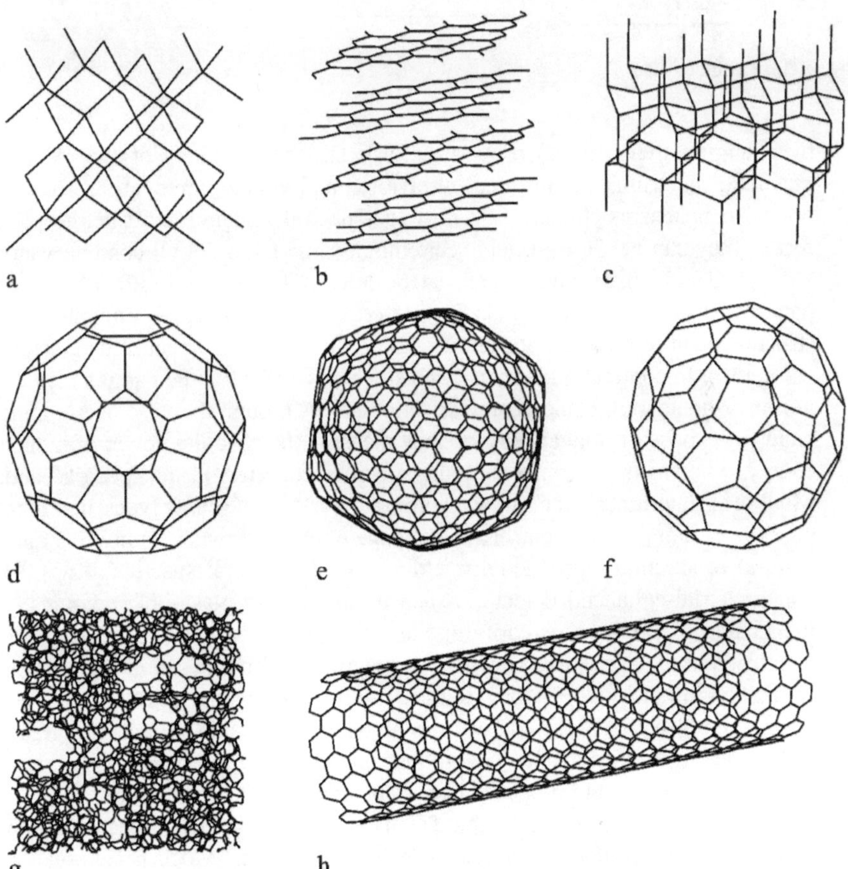

Fig. 1 Eight allotropes of carbon. **a** Diamond, **b** Graphite, **c** Lonsdaleite, **d** C_{60} buckminsterfullerene, **e** C_{540} fullerene, **f** C_{70} fullerene, **g** Amorphous carbon, **h** Armchair single-walled carbon nanotube. Adapted from Andel (2018), Wikimedia Commons, CC BY-SA 4.0

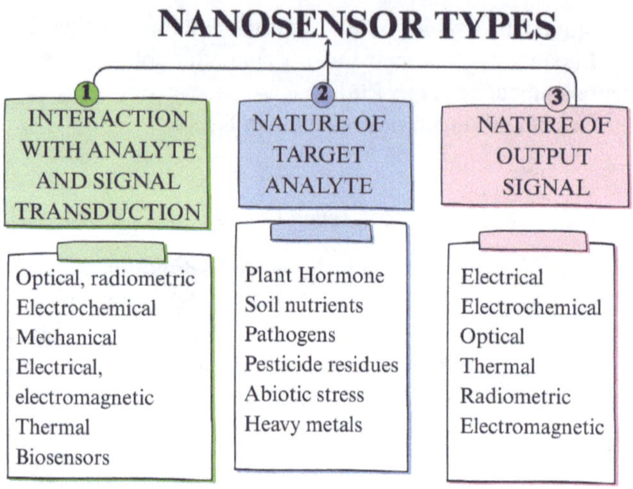

Fig. 2 Types of nanosensor

by changing their optical properties [44]. Different regions of the electromagnetic spectrum, like infrared, ultraviolet, and visible range, are utilized in optical nanosensors. Integrated with hyperspectral imaging and satellite platforms, they can perform in field-scale monitoring. Chlorophyll content, water stress, and early disease onset, etc., can be detected based on specific reflectance patterns in crop canopies [11]. Cutting-edge, highly sensitive nanophotonic structures can refine these techniques and reduce noise in spectral data acquisition [36]. In contrast, radiometric sensors exclusively use the gamma rays for non-invasive mapping and characterization [7]. Although they use overlapping bands like those of optical sensors, they measure the absolute power of radiant energy, whereas optical ones measure specific properties of the spectral band. Modern agricultural systems now incorporate a variety of sensor types, making it possible to monitor and control a wide range of characteristics, improving agricultural production's productivity, efficiency, and general sustainability [29]. Nanomaterial-enhanced detectors, such as those with plasmonic coatings or metamaterials, improve resolution and sensitivity under variable field conditions.

B **Electrochemical Sensors**: They detect specific ions in a system. Their action is mostly intracellular as they interact with cellular analytes. The electrochemical pulse is converted to an analytically favorable signal like Reactive Oxygen Species (ROS), phytohormone concentration, volatile organic compounds (VOC), etc. For instance, some can measure current, potential, or resistance based on a variable input [7]. They can also be used in heavy metal, pathogen, and pollutant detection, in plants, or in soil. Modern nanosensors work hand-in-hand with precision farming, are environmentally friendly, and can identify even minute changes in environmental conditions. Sensors used for site-specific fertilization reduce input costs by increasing application efficiency and minimizing contamination. Graphene-modified electrodes can detect nitrate concentrations efficiently due to their large surface area and fast electron transfer properties [45] and thus predict GHG emissions.

C **Mechanical sensors**: These can measure the mechanical properties of soil, like soil strength, compaction, plant growth, etc. They can determine wind load or stem deformation in plants. The different types of mechanical sensors are

- Strain Gauge sensors
- MEMs (Micro-Electro-Mechanical Systems)
- Capacitive Sensors
- Piezoresistive Sensors
- Resonant Sensors
- Ultrasonic Non-resonant Sensors

D **Electrical and Electromagnetic sensors**: These use electrical circuits and measure the electric charge or dielectric constant of a particular medium. The differences in conductivity/resistivity give an idea about the heterogeneity of the system. Electrical nanosensors work by detecting electrical resistance, current, voltage, or capacitance. For instance, CNT or nanowire-based Field-Effect Transistors (FETs) have been used to monitor phytohormones (like ethylene), which are critical markers of drought or pathogen stress. Similarly, graphene-enhanced electrodes can detect ions such as nitrate and potassium in soil with high selectivity and sensitivity, improving fertilizer use efficiency [45].

On the other hand, electromagnetic nanosensors detect changes in magnetic, dielectric, or radio-frequency (RF) fields in the environment as well as within the living organisms. Wireless field sensors use passive RF identification (RFID)-based nanotags and antennas that can transmit soil moisture or pH data without a power source [25]. Again, magnetoelastic nanosensors, composed of magnetically active thin films, can detect water stress in plants by responding to tissue hydration levels through shifts in resonant frequency without contact.

5. **Thermal sensors**: These are used to detect temperature changes in remote conditions efficiently. Being in the nanoscale, they are sensitive to subtle changes in temperature and can generate real-time micro-level temperature gradients in plants and soil. They are widely used for smart packaging in the food industry, plant stress detection via leaf temperature measurement, detecting temperature fluctuations in the root zone [2], thermal profiling in greenhouses [28], measuring photosynthetic efficiency through high-resolution thermal maps, etc. Now-a-days fluorescent nanothermometers employ rare earth-based probes to detect thermosensitive luminescent properties of objects [39].
6. **Biosensors**: A biosensor is a modest and compact investigative device that has the capability of producing definite systematic data either in a quantitative way or in a semi-quantitative form by employing a recognition component of biological origin, which is joined to a signal transformation unit [21, 26]. These are characteristics of the direct spatial combination of a matrix-bound biologically

active substance, the receptor, with an electronic device. The different sequences of processes in biosensors are

7. Specific recognition of the analyte.
8. The transduction of the physicochemical effect caused by the interaction with the receptor into an electric signal.
9. Signal processing and amplification.
10. **Nature of Target Analyte**:

The type of nanosensors varies according to the nature of the analyte that they are designed to detect (Table 1). They can be of the following types:

B. **Nature of Output Signal**:

Output is the signal that a system, device, or process transmits or sends out. Nanosensors can be classified based on the nature of the output they detect as follows (Table 2).

Mechanism of Action of Nanosensors in Precision Agriculture

The operational mechanism of nanosensors relies on their capacity to convert molecular interactions into measurable signals. These sensors typically consist of three essential components: a recognition element, a transducer, and a signal processing unit [19]. Together, they form an integrated system capable of detecting specific analytes, such as nutrients, pesticides, gases, or biomolecules, in agricultural matrices like soil, water, and plant tissues. The mechanism of action of nanosensors is shown in Fig. 3, and the three essential components are discussed below.

1. **Recognition Element**

The recognition element is the functional component that ensures selective interaction with a specific target molecule (Morales and Halpern 2018). It may consist of biological materials such as enzymes, antibodies, nucleic acids, or aptamers, or

Table 1 Target analytes, sensor types, and their applications in precision agriculture

Target analyte	Sensor type(s)	Purpose	References
Plant hormones	Electrical, Electrochemical	Drought/stress signaling	–
Soil nutrients	Electrochemical, Electrical	Precision fertilization	[45]
Pathogens	Optical, Colorimetric, Biosensors	Early disease detection	[28]
Pesticide residues	Electrochemical, Optical	Food and water safety	–
Abiotic stress (pH, T, etc.)	Optical, Thermal, Radiometric	Precision irrigation and crop monitoring	[36]
Heavy metals	Electrochemical	Soil and water toxicity monitoring	[28]

Table 2 Types of sensors, their output signals, and key agricultural applications

Sensor type	Nature of output signal	Typical agricultural application	References
Electrical	Resistance, voltage, or current changes	Monitoring nutrient ions, salinity, and plant hormone levels	[37]
Electrochemical	Current or potential from redox reactions	Detection of nitrates, pesticides, pH, and heavy metals	[45]
Optical	Fluorescence, absorbance, colour shifts	Detection of pathogens, plant stress, and toxins	[28]
Thermal	Temperature changes, thermal conductivity	Monitoring plant stress, leaf surface temperature	[14]
Radiometric	Infrared/microwave radiation intensity	Remote sensing of canopy temperature, evapotranspiration	[36]
Electromagnetic	Frequency, inductance, or RF signal	Wireless monitoring of moisture, hydration levels	[28]

of synthetic materials like molecularly imprinted polymers (MIPs). The recognition process is governed by molecular affinity, ensuring specificity even at very low analyte concentrations. Enzymatic nanosensors detect substrates through catalytic reactions that produce measurable by-products (e.g., hydrogen peroxide or protons), resulting in electrochemical or optical changes [23]. Antibody- and aptamer-based nanosensors rely on highly specific ligand-receptor interactions to recognize pathogens, hormones, or toxins [43]. MIP-based nanosensors replicate natural molecular recognition using engineered cavities that complement the shape and functional groups of the target analyte, ensuring durability under harsh field conditions [17].

2. **Transduction Process**

Once recognition occurs, the transducer converts the biochemical interaction into an electrical, optical, or mechanical signal. The main categories of transduction mechanisms include electrochemical transducers, which detect changes in current, potential, or impedance resulting from redox or ionic interactions [37]. Carbon nanotubes (CNTs), graphene, and conductive polymers enhance electron transfer and sensitivity, enabling the detection of nutrient ions or agrochemicals at trace levels. Also, there are optical transducers, which employ fluorescence, absorbance, or surface plasmon resonance (SPR) to translate binding events into light-based signals [16]. Quantum dots and gold nanoparticles provide high photostability and tunable optical responses, making them ideal for detecting stress-related metabolites or pollutants. Apart from these, there are mechanical and magnetic transducers [6].

3. **Signal Amplification and Processing**

The signals generated by nanosensors are often weak due to nanoscale interactions; thus, amplification is essential. Nanomaterials such as gold, silver, and graphene enhance signal intensity through plasmonic, catalytic, or conductive effects. The

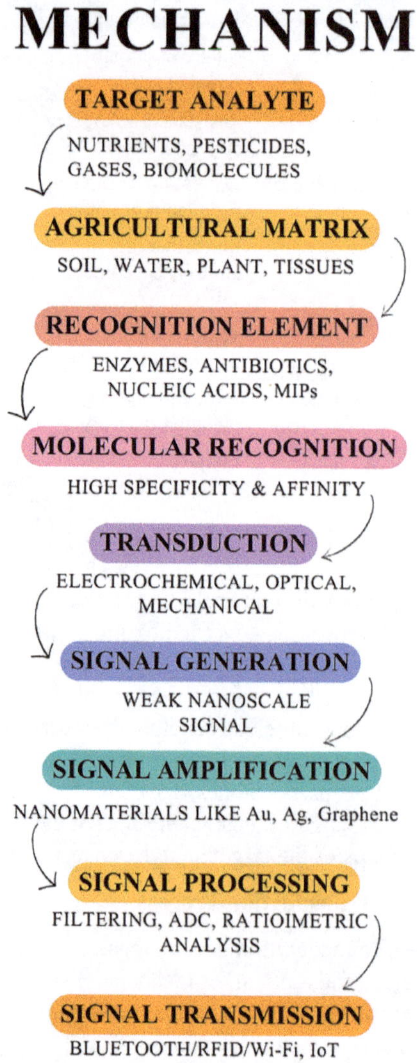

Fig. 3 Mechanism of action of nanosensors in precision agriculture

processed signal is then transmitted to data acquisition systems through wireless communication modules (e.g., Bluetooth, RFID, or Wi-Fi) [20]. Integration with Internet of Things (IoT) architectures enables real-time monitoring and cloud-based data analytics. Machine learning and artificial intelligence (AI) algorithms further enhance interpretation by recognizing patterns and predicting crop or soil behavior. This combination transforms nanosensors from passive detection devices into intelligent diagnostic systems [13].

The different mechanisms employed in the functioning of various types of nanosensors are discussed as follows:

1. **Förster Resonance Energy Transfer-Based Nanosensors**

Protein interactions, cell contents, and biophysical characteristics have all been extensively studied using optical nanosensors based on Léster resonance energy transfer. If the distance between fluorophores is within a nanometer-scale range, FRET relies on the non-radiative transmission of excited-state energy through dipole coupling. The donor's fluorescence emission is reduced, and the acceptor's fluorescence emission intensities are increased when energy is transferred from an excited donor to an acceptor molecule [30]. When examining the distance between two analytes or two places on a particular macromolecule, such as during protein conformational change, protein–protein interactions, and enzyme activity, FRET's narrow range and distance dependency make it a perfect instrument. FRET-based nanosensors can be introduced exogenously as externally synthesized chemicals or genetically encoded within the plant itself. Proteins, protein domains, nanoparticles, or chemical ligands that alter the separation between the donor and acceptor fluorescence domains of two fluorophores can be used by both genetically encoded and exogenously applied FRET-based nanosensors to report conformational changes. When a donor fluorophore is excited, resonance energy transfer transfers a portion of the energy to the acceptor fluorophore, causing the acceptor to glow. Ratiometric detection of tiny molecules is made possible by tracking the emission peaks of the donor and acceptor, which removes detection ambiguities through the self-calibration of two emission bands [31].

2. **Genetically Encoded FRET-Based Nanosensors**

A cyan fluorescent protein (CFP) and a yellow fluorescent protein (YFP) that work as a FRET pair are commonly utilized in genetically encoded FRET-based biosensors. The ratiometric readout is computed from (ExCFPEmYFP)/(ExCFPEmCFP) [27]. The plant-optimized pH-sensitive green fluorescent protein (pHGFP), which is based on the pH-sensitive pHluorin and redox oxidation-sensitive green fluorescent proteins (roGFPs), is an example of a single-fluorescent protein biosensor that can also be ratiometric when the fluorescent protein has two excitation wavelengths that react differently to an analyte. Protein interactions with nucleic acids, reactive oxygen species, and ion or metabolite levels have all been detected by genetically encoded FRET-based nanosensors [41].

3. **Electrochemical nanosensors**

Plant tissues are frequently homogenized in a known electrolyte solution in order to research plants using electrochemical nanosensors. Electrochemical nanosensors can also be used to investigate plants indirectly by concentrating on soil chemicals that are essential to plant growth and development. Plants can now be studied in situ without the requirement to homogenize plant tissues in electrolyte solution for additional examination, thanks to the invention of nanosensors made to live inside plants. One such sensor used photoactive nanomaterials (molybdenum-doped bismuth vanadate, or BiVO4) as a sensing unit in a photoelectrochemical platform to assess fruit antioxidant capability.

Advantages of Nanosensors in Precision Agriculture

(a) **Ultra-Sensitive and Real-Time Monitoring**: Nanosensors possess extremely high surface reactivity and signal-to-noise ratios, enabling detection of analytes at nanomolar or even picomolar levels. Their real-time response allows early diagnosis of plant stress, disease, or nutrient deficiencies, permitting timely corrective measures and minimizing yield losses. It also monitors the reactive oxygen bursts and helps in abiotic or biotic stress detection. Emerging nano-biohybrid systems integrate nanosensors directly into plant tissues, allowing non-invasive, real-time monitoring of physiological parameters like photosynthetic efficiency, transpiration, and hormonal flux. These plant-wearable sensors could revolutionize crop phenotyping and stress management [33].

(b) **Enhanced Resource Use Efficiency**: By providing accurate spatial and temporal data on soil moisture, nutrient availability, and crop status, nanosensors enable variable-rate input applications. This leads to optimized fertilizer and water use, reduced input wastage, and improved nutrient uptake efficiency—key pillars of sustainable precision agriculture. It is also used to monitor the plant water status, which is a key physiological parameter

(c) **Environmental Protection and Sustainability**: Precise control over agrochemical application reduces runoff, leaching, and contamination of surrounding ecosystems. Continuous monitoring of soil and water parameters contributes to environmental conservation and supports compliance with regulatory frameworks aimed at sustainable farming. By tracking nutrient dynamics and detecting contaminants during cultivation and post-harvest stages, nanosensors help ensure food safety and compliance with international standards. Their precision also enhances nutritional quality and post-harvest shelf life.

(d) **Integration with Digital and Smart Farming Technologies**: Nanosensors seamlessly interface with IoT, wireless sensor networks (WSN), and cloud computing platforms. Combined with AI and big data analytics, they provide predictive insights for adaptive management under dynamic climatic and environmental conditions. Due to their small size and low energy consumption, nanosensors can be deployed extensively across large fields or incorporated into portable devices, drones, and robotic systems, facilitating large-scale precision monitoring at minimal operational cost. Future nanosensors are expected to be multifunctional, capable of simultaneously measuring diverse parameters such as nutrient content, temperature, and pathogen presence. Incorporating energy-harvesting mechanisms, like piezoelectric or triboelectric nanogenerators, will produce self-powered, maintenance-free devices [12].

Upcoming Potentials in Nanosensor Usage:

Although there is a huge disparity between the technology of nanosensors that are theoretically possible, in contrast to the ones actually available in the field, there is a growing demand for nanotech devices, as they have proven to give consistent results. Their small size facilitates a huge capacity for data storage. A new aspect of the rapid modernization is the use of AI in nanotechnology. For instance, "intelligent sensors" coupled with IoTs have a feedback mechanism. In agriculture, the usage of nanomaterials has been growing annually. According to Nucleus, the Global Agricultural Nanotechnology Market is expected to be valued at 620 billion USD by 2028 (Fig. 4). And by 2032, the increment would be 11.7% from 2025. The next generation of nanosensors ranges from wireless, plant wearables, bio-compatibles, to nano biomarkers that can detect ultra-low concentrations of analytes through a physicochemical transducer. [40].

Also, super "novel nanomaterials" can increase the efficiency of biosensors in the near future. Converging nanotechnology, agriculture, plant macro and microclimate engineering can help us achieve all our Sustainable Development Goals 2030 with a high benefit–cost ratio.

Applications:

There have hardly been so many field-working sensors as laboratory-devised ones. This may be due to several scientific and non-scientific factors that hinder these nanosensors from proof-of-concept to fully commercialized products. The scientific parameters may include technical field difficulties overlooked in the laboratory and regulatory errors (validation and compliance). The non-scientific ones include IPR (Intellectual Property Rights) protection, standardization (legal), economic constraints, environmental (safety and security), and ethical issues. These factors are to be considered strategically for the better implementation of nanotechnological tools. But the demand for nanosensors has been ever-increasing to fulfill the demands of sustainability.

Case studies:

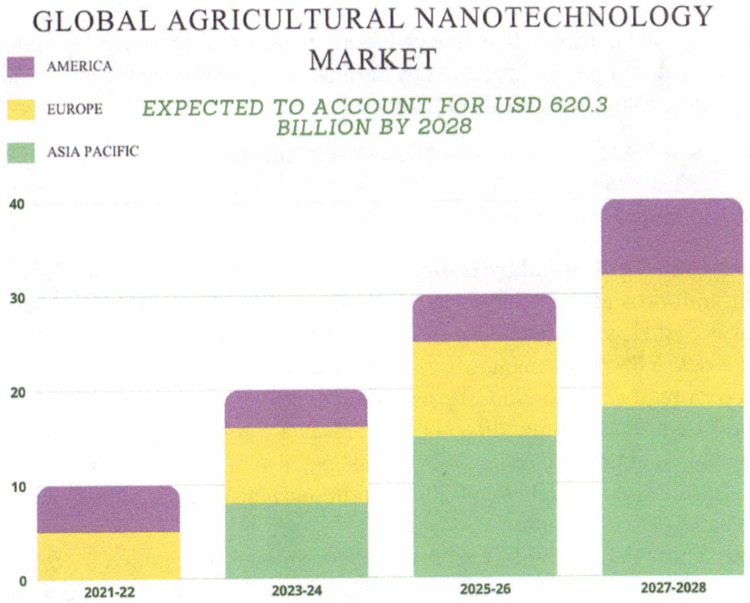

Fig. 4 Global agricultural nanotechnology market by 2028

1. **USDA Roadmap, "Nanoscale Science and Engineering for Agriculture and Food Systems"**: In December 2002, the United States Department of Agriculture (USDA) drafted the world's first "roadmap" for applying nanotechnology to agriculture and food. This was the first mapped approach to incorporate nanotechnology in the US agriculture and food system. Used for real-time monitoring of soil and plant health, food packaging for extended shelf life, and to enhance food safety
2. **Novel Approaches in Nanosensor Development**:

   - The US-based startup Razzberry developed portable chemical nanosensors to trace real-time chemical changes in water, soil, and the environment. These enhance on-site testing, offering rapid and sensitive detection of contaminants without reliance on laboratory infrastructure.
   - Italian startup Nasys invented a metal oxide-based nanosensor to detect air pollution and improve environmental surveillance systems.
   - nGageIT, a US-based startup, developed a "Digital Breath Biomarker" engine, a biological nanosensor for detecting biological contaminants in an agricultural setting.
   - Tracense, an Israeli security-tech startup implementing nanosensor technologies to detect biological and hazardous contaminants in agriculture and the food chain.

3. One of the Californian vineyards, Pickberry, in Sonoma County, has installed wi-fi systems with the help of the IT Company, Accenture. It enables the best grapes to be grown, which in turn produce finer wines that command a premium price. Potential applications across the food chain (in pesticides, vaccines, veterinary medicine, and nutritionally enhanced food), these nano- and micro-formulations are being developed and patented by agribusiness and food corporations such as Monsanto, Syngenta, and Kraft [22].

Challenges of Nano-sensor Application in Agriculture:

Despite their promise, several challenges limit the field-scale implementation of nanosensors (Fig. 5). A few are discussed below.

a. **Fabrication and Standardization**: Achieving reproducible sensor fabrication with uniform sensitivity and selectivity remains a technical bottleneck. Variability in nanoparticle synthesis methods can affect performance consistency. Moreover, the absence of global manufacturing standards hinders regulatory approval and interlaboratory comparison. Agricultural environments are dynamic, with fluctuating temperature, humidity, and soil pH. These conditions can degrade nanomaterial coatings or interfere with signal stability. Ensuring sensor robustness and longevity under field conditions requires advanced material engineering and protective encapsulation.
b. **Data Management and Interoperability**: Storage, integration, and interpretation of the massive amounts of data produced by nanosensor networks are

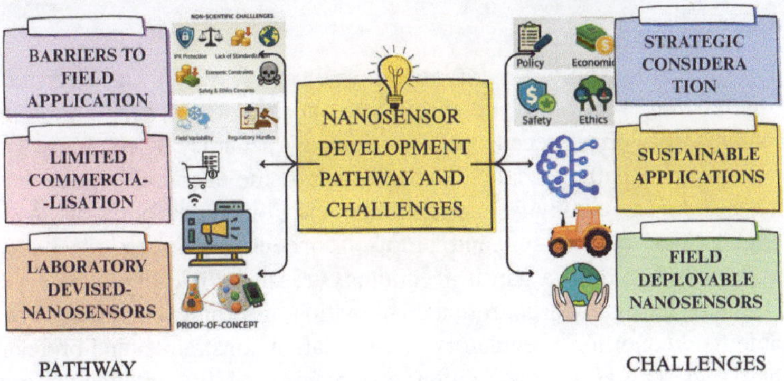

Fig. 5 Challenges of nanosensor application in agriculture

difficult. For digital integration to be effective, outstanding issues like cybersecurity, data standardization, and dependable connectivity in rural areas must be addressed.

c. **Economic Constraints**: The initial cost of nanosensor fabrication and deployment is relatively high. Additionally, many small-scale farmers lack access to the infrastructure and technical expertise required for sensor-based precision farming, posing socioeconomic barriers to adoption.
d. **Regulatory and Ethical Issues**: the absence of clear regulatory frameworks for the use of nanomaterials in agriculture raises safety and ethical concerns. Questions about data ownership, privacy, and equitable access to technology also require careful consideration.

Safety and Environmental Considerations

(a) **Nanotoxicity and Ecological Impact**: Concerns regarding possible toxicity and environmental persistence are raised when manmade nanoparticles are introduced into agricultural systems. Nanoparticles can bioaccumulate in aquatic or plant systems, modify nutrient cycles, and change soil microbial communities. Therefore, to quantify their ecological footprint, risk-assessment studies and life-cycle evaluations are crucial. Nanosensor residues may be harmful to human health if they get into edible plant parts or food chains. To prevent contamination and guarantee consumer safety, strict testing procedures, sensor retrieval systems, and adherence to safety standards (e.g., OECD, FAO, EFSA) are required.

(b) **Regulatory Framework and Responsible Innovation**: Developing harmonized international regulations is crucial for the safe deployment of nanosensors in agriculture. This includes defining acceptable exposure limits, labeling guidelines, and environmental monitoring protocols. Implementing "safe-by-design" strategies-where safety considerations are embedded from the early stages of sensor development—will promote public trust and sustainable innovation.

Conclusion

Nanosensor-based technologies are intricately linked to advancing precision agriculture by enabling real-time, site-specific monitoring of soil, plant, and environmental parameters with unprecedented sensitivity and accuracy. Their ability to detect nutrients, moisture, pathogens, and stress signals at the nanoscale optimizes input use, enhances crop productivity, and minimizes environmental footprints. Although substantial laboratory research and proof-of-concept demonstrations have been conducted, there has been a gap in technology-transfer. Field-deployable systems remain limited. This gap arises from a combination of technical challenges because of variable field conditions, regulatory and validation constraints, and broader non-scientific issues such as standardization, economic feasibility, intellectual property rights, and environmental and ethical concerns.

Addressing these challenges requires an integrated and strategic harmonization that will be essential for successful commercialization. Furthermore, nanosensors are pathways for sustainability goals—such as resource-use efficiency, climate resilience, and soil health conservation. With continued interdisciplinary research, policy frameworks, and stakeholder engagement, nanosensors will become key enablers of data-driven, sustainable, and resilient precision agriculture in the coming decades.

References

1. Abideen Z, Hanif M, Munir N, Nielsen BL (2022) Impact of nanomaterials on the regulation of gene expression and metabolomics of plants under salt stress. Plants 11(5):691
2. Agrawal J, Arafat MY (2024) Transforming farming: a review of AI-powered UAV technologies in precision agriculture. Drones (2504–446X) 8(11)
3. Akgönüllü S, Denizli A (2022) Nano-sensors and nano-devices for biological disaster monitoring (virus/disease epidemics/animal plagues detections). In: Nanotechnology-based smart remote sensing networks for disaster prevention, Elsevier, pp 43–57. https://doi.org/10.1016/B978-0-323-91166-5.00006-9
4. Atanda SA, Shaibu RO, Agunbiade FO (2025) Nanoparticles in agriculture: balancing food security and environmental sustainability. Discov Agric 3:26
5. Bakhsh S (2023) Fullerene, carbon nanotubes and graphene: a comprehensive review. Karbala Int J Modern Sci 9(3):1
6. Beegum S, Das S (2022) Nanosensors in agriculture. In: Agricultural nanobiotechnology, Woodhead Publishing, pp 465–478
7. Chugh B, Thakur S, Singh AK, Joany RM, Rajendran S, Nguyen TA (2022) Electrochemical sensors for agricultural applications. In: Nanosensors for smart agriculture, Elsevier, pp 147–164
8. Darwish MA, Abd-Elaziem W, Elsheikh A, Zayed AA (2024) Advancements in nanomaterials for nanosensors: a comprehensive review. Nanoscale Adv 6(2):455–478. https://doi.org/10.1039/D4NA00214H
9. Denizli A, Nguyen TA, Rajendran S, Yasin G, Nadda AK (Eds) (2021) In: Nanosensors for smart agriculture. Elsevier
10. Duhan JS, Kumar R, Kumar N, Kaur P, Nehra K, Duhan S (2017) Nanotechnology: the new perspective in precision agriculture. Biotechnol Reports 15:11–23. https://doi.org/10.1016/j.btre.2017.03.002. Electronics in Agriculture, 198, 107096

11. García-Vera YE, Polochè-Arango A, Mendivelso-Fajardo CA, Gutiérrez-Bernal FJ (2024) Hyperspectral image analysis and machine learning techniques for crop disease detection and identification: a review. Sustainability 16(14):6064
12. Gulia V, Dhull SS, Kamboj H, Kumar S, Deepali, Dhanda N, Rani J (2025) Pushing limits: integrating smart nanosensors and the internet of things in agriculture. Rendiconti Lincei. Scienze Fisiche e Naturali, 1–19
13. Gunasekaran K, Sreevardhan P, SamiKannu R (2025) Real-time soil fertility analysis, crop prediction, and insights using machine learning and deep learning algorithms. Front Soil Sci 5:1652058
14. Ikram M, Ameer S, Kulsoom F, Sher M, Ahmad A, Zahid A, Chang Y (2024) Flexible temperature and humidity sensors of plants for precision agriculture: current challenges and future roadmap. Comput Electron Agricul 226:109449
15. Javaid M, Haleem A, Singh RP, Rab S, Suman R (2021) Exploring the potential of nanosensors: a brief overview. Sensors Int 2:100130
16. Johnson MS, Sajeev S, Nair RS (2021) Role of nanosensors in agriculture. In: Proceedings of the 2021 international conference on computational intelligence and knowledge economy (ICCIKE), IEEE, pp 58–63
17. Kadhem AJ, Gentile GJ, Fidalgo de Cortalezzi MM (2021) Molecularly imprinted polymers in sensors for environmental and biomedical applications: a review. Molecules 26(20):6233
18. Karunathilake EMBM, Le AT, Heo S, Chung YS, Mansoor S (2023) The path to smart farming: innovations and opportunities in precision agriculture. Agriculture 13(8):1593
19. Lin J, Wu S (2025) Recent advances in agricultural sensors: towards precision and sustainable farming. Chemosensors 13(11):399
20. Long Y, Stahl Y, Weidtkamp-Peters S, Smet W, Du Y, Gadella TWJ Jr, Goedhart J, Scheres B, Blilou I (2018) Optimizing FRET–FLIM labeling conditions to detect nuclear protein interactions at native expression levels in living *Arabidopsis* roots. Front Plant Sci 9:639
21. Lu Y, Yang Q, Wu J (2020) Recent advances in biosensor-integrated enrichment methods for preconcentrating and detecting low-abundant analytes in agriculture and food samples. TrAC Trends in Anal Chem 132:115914
22. Manjunatha SB, Biradar DP, Aladakatti YR (2016) Nanotechnology and its applications in agriculture: a review. J Farm Sci 29(1):1–13
23. Metternich JT, Hill B, Wartmann JA, Ma C, Kruskop RM, Neutsch K, Herbertz S, Kruss S (2024) Signal amplification and near-infrared translation of enzymatic reactions by nanosensors. Angewandte Chemie Int Edition 63(9):e202316965
24. Morales MA, Halpern JM (2018) Guide to selecting a biorecognition element for biosensors. Bioconjug Chem 29(10):3231–3239
25. Ning Y, Wang Y, Zhang H, Li J (2021) Passive electromagnetic sensors for soil and plant monitoring: principles and perspectives. Biosens Bioelectron 180:113108. https://doi.org/10.1016/j.bios.2021.113108
26. Nosrati R, Golichenari B, Nezami A, Sahebkar A, Mohammadi A (2017) *Helicobacter pylori* point-of-care diagnosis: nano-scale biosensors and microfluidic systems. TrAC Trends in Anal Chem 97:428–444
27. Okumoto S, Jones A, Frommer WB (2012) Quantitative imaging with fluorescent biosensors. Annual Rev Plant Biol 63:663–706
28. Parisi SG, Di Nardo A, Santonico M, Pennazza G, D'Amico A (2020) Sensors for precision agriculture and environmental monitoring. Sensors 20(4):1145. https://doi.org/10.3390/s20041145
29. Paul K, Chatterjee SS, Pai P, Varshney A, Juikar S, Prasad V, Bhadra B, Dasgupta S (2022) Viable smart sensors and their application in data driven agriculture. Comput Electron Agricul 198:107096
30. Pham H, Hoseini Soflaee M, Karginov AV et al (2022) Förster resonance energy transfer biosensors for fluorescence and time-gated luminescence analysis of rac1 activity. Sci Rep 12:5291

31. Sagadevan S, Periasamy M (2014) Recent trends in nanobiosensors and their applications: a review. Rev Adv Mater Sci 36:62–69. https://doi.org/10.1002/9781118904923.ch2
32. Saleh Ahammad AJ, Lee JJ, Rahman MA (2009) Electrochemical sensors based on carbon nanotubes. Sensors 9(4):2289–2319
33. Shaw DS, Honeychurch KC (2022) Nanosensor applications in plant science. Biosensors 12(9):675
34. Singh A, Bedi S, Singh J (2017) Nanotechnology-based sensors. Sensors Actuators B: Chem 246:620–637
35. Singh M, Truong J, Reeves WB, Hahm JI (2017) Nanostructured sensors for biomedical and environmental applications. Sensors 17:428. https://doi.org/10.3390/s17020428
36. Singh RK, Berkvens R, Weyn M (2021) AgriFusion: an architecture for IoT and emerging technologies based on a precision agriculture survey. IEEE Access 9:136253–136283
37. Soussi A, Zero E, Sacile R, Trinchero D, Fossa M (2024) Smart sensors and smart data for precision agriculture: a review. Sensors 24(8):2647
38. Sudha PN, Sangeetha K, Vijayalakshmi K, Barhoum A (2018) Nanomaterials history, classification, unique properties, production and market. In: Emerging applications of nanoparticles and architecture nanostructures. Elsevier, pp 341–384. https://doi.org/10.1016/B978-0-323-51254-1.00012-9
39. Sun Y, Kong M, Ke J, Gu Y, Li F, Feng W (2025) Rare earth luminescent nanothermometers for biological thermal sensing. Coordinat Chem Rev 523:216222
40. Tshabalala ZP, Mokoena TP, Motaung DE (2022) Current commercial nanosensors and devices/products used in agriculture. In: Denizli A (ed) Nanosensors for smart agriculture. Elsevier
41. Uslu VV, Grossmann G (2016) The biosensor toolbox for plant developmental biology. Current Opinion in Plant Biol 29:138–147
42. Vishvakarma R, Siddiqui MA, Ansari N, Kumari P, Batool F, Parvez Z, Vimal A (2025) Nanosensors in agricultural applications: an overview. In: Handbook of nanotechnology in agriculture, pp 1–25
43. Wang W, He Y, He S, Deng L, Wang H, Cao Z, Yin Y (2025) A brief review of aptamer-based biosensors in recent years. Biosensors 15(2):120
44. Zhang C, Qiu M, Wang J, Liu Y (2023) Recent advances in nanoparticle-based optical sensors for detection of pesticide residues in soil. Biosensors 13(4):415
45. Zhou Y, Zhang H, Liu L, Li M, Zhang J (2019) Graphene-based electrochemical sensors for detecting nitrate in agricultural water. Sensors and Actuators B: Chem 296:126660

AI-Nano-Supporting Global Food and Nutrition Security: Technical Achievements and Prospects

Shweta Sachan, Sadhna Singh, Anisha Verma, Ragini Mishra, Ajay Rajan, Nikita Arya, Neelam Yadav, Claire Copper Stone, and Rajendra Kumar

Abstract Global food and nutrition security remains one of the greatest challenges of the twenty-first century, as over 828 million people face chronic hunger and more than two billion suffer from micronutrient deficiencies. With the global population projected to surpass nine billion by 2050, conventional agricultural methods alone are insufficient to ensure sustainable food systems. The convergence of Artificial Intelligence (AI) and Nanotechnology has emerged as a transformative force, offering innovative solutions to overcome resource limitations, climate change, and declining productivity. AI-driven predictive analytics, precision agriculture, and intelligent supply chain systems are revolutionizing resource management and reducing human effort, while nanotechnology enables breakthroughs in nutrient delivery, crop protection, smart nano-sensors, and food packaging that can detect spoilage and preserve nutritional value. These combined technologies have demonstrated measurable improvements in agricultural yield, food safety, and nutrient bioavailability across diverse environments. This chapter examines the synergistic role of AI and Nanotechnology as dual engines for global food and nutrition security. It highlights recent technical achievements, field applications, and policy developments from 2020–2024, while addressing ethical, regulatory, and sustainability

S. Sachan · S. Singh · A. Verma · R. Mishra
College of Community Science, Acharya Narendra Deva University of Agriculture and Technology, Kumarganj, Ayodhya, India

A. Rajan
MS, Electrical and Computer Engineering, UMass Amherst, Amherst, MA, USA

N. Arya
Foundation of Healthcare Technologies Society, New Delhi, India

N. Yadav
Centre of Food Technology, University of Allahabad, Allahabad, India

C. C. Stone
Department of Food Science, Nutrition and Dietetics, Faculty of Health Sciences, University of Malta, Msida, Malta

R. Kumar (✉)
Division of Genetics, ICAR-IARI, New Delhi, India
e-mail: rajendrak64@yahoo.co.in

J.-T. Chen (ed.), *AI-Nano-Enabled Climate-Smart Agriculture and Sustainability*, Smart Nanomaterials Technology, https://doi.org/10.1007/978-981-92-1694-9_6

considerations. The discussion concludes with prospects and strategic recommendations for equitable and responsible integration of AI–Nano systems into global food frameworks.

Keywords Food security · Nutrition security · Artificial intelligence · Nanotechnology · Precision agriculture · Sustainable food systems · Nutrient bioavailability · Climate resilience · Technological innovation

1 Introduction

The twenty-first century stands at a critical crossroads in ensuring global food and nutrition security. While the world produces sufficient food to feed its entire population, the distribution, accessibility, and nutritional adequacy of that food remain deeply unequal. According to *The State of Food Security and Nutrition in the World 2023* [8], approximately 828 million people continue to face chronic hunger, while over two billion individuals suffer from hidden hunger, also known as micronutrient deficiencies that impair growth, cognitive function, and overall well-being.

This crisis is compounded by multiple converging pressures: climate change, declining soil fertility, freshwater scarcity, and disruptions in global supply chains. Agricultural productivity has plateaued in many regions, with yield gaps persisting due to inefficient resource management and limited access to innovation. Simultaneously, malnutrition, including both undernutrition and overnutrition, poses a paradoxical challenge in the twenty-first century. "Around 45% of deaths among children under 5 years of age are linked to undernutrition." [31]. In the Lancet Commission report, Swinburn et al. [28] refer to the triple burden caused concomitantly by obesity, undernutrition, and climate change within the global population context, which are driving today's health and societal challenges, exacerbated by food and agriculture and land use, among other drivers.

To address these complex issues, technology must advance beyond traditional agricultural practices. Artificial Intelligence (AI) and Nanotechnology have emerged as revolutionary tools in this context. AI enables data-driven decision-making through precision monitoring, predictive analytics, and intelligent control systems. Nanotechnology, on the other hand, operates at the molecular scale to improve nutrient delivery, enhance soil health, protect crops, and reduce post-harvest losses.

Recent developments in India, China, and sub-Saharan Africa highlight that integrating AI and nano-based solutions has led to notable improvements in both productivity and nutritional outcomes. For instance, precision-guided nano-fertilizer deployment in paddy fields has increased the yield by 18–25%, while AI-driven irrigation scheduling has reduced water use by 30–40% [9]. Such outcomes signify that technology, when applied judiciously, can support the attainment of *Sustainable Development Goal (SDG)0.2 (Zero Hunger)*—ensuring access to sufficient, safe, and nutritious food for all, as well as supporting the implementation of objectives within other SDGs such as SDG 1 (No Poverty) and SDG 3 (Good Health and Well-Being) [30].

2 Global Food and Nutrition Security Challenges

2.1 Current Landscape

Food security exists when all people, at all times, have physical, social, and economic access to sufficient, safe, and nutritious food that meets their dietary needs and food preferences for an active and healthy life. Food security is determined by four pillars: availability, access, stability, and utilization. Global food systems are under unprecedented stress due to demographic expansion, climate variability, and unsustainable resource utilization. The population of the world is expected to reach 9.7 billion by 2050 [29]. The increased population and environmental stress pose a major challenge to access to nutritious and sustainable food. The 2025 EAT-Lancet Commission Report builds on the previous landmark report, reaffirming the need for healthy, sustainable, but also just food systems [24] and notes the increasing threats posed by more recent conflicts and climate change scenarios. The FAO [8] estimates that approximately 828 million individuals experience chronic undernourishment, while over two billion suffer from hidden hunger caused by micronutrient deficiencies, such as iron, vitamin A, and zinc. These deficiencies disproportionately affect women and children in low- and middle-income countries, resulting in impaired cognitive development, reduced productivity, and elevated morbidity rates [31].

Simultaneously, the dual burden of malnutrition, where undernutrition and obesity coexist, has emerged as a global paradox. In urbanizing societies, overconsumption of calorie-dense but nutrient-poor foods contributes to metabolic disorders, cardiovascular disease, and type-2 diabetes [22]. Further to this, the triple burden of the disease is now being recognized as the complex interactions between overnutrition, undernutrition, and climate change experienced in countries globally and which is giving rise cumulatively to barriers associated with health achievement, economic success, and social equity [28]. Meanwhile, rural communities often face inadequate access to nutrient-rich diets due to infrastructural gaps, market inefficiencies, and climate-induced crop failures. Realistic and practical solutions must therefore be presented to end the hunger burden and the related disease risk associated with all forms of malnutrition.

The environmental dimensions further exacerbate these challenges. Climate change poses a serious risk to crop production and overall food security. A serious reduction in production and provision of major crops will occur if the temperature rise is not curbed, together with issues associated with food distribution and access. Soil degradation affects more than 33% of the global land area, reducing the nutrient retention and water-holding capacity of soils [7]. Freshwater scarcity, particularly in arid and semi-arid regions, threatens irrigation-dependent agriculture, while unpredictable climatic events—such as droughts, floods, and heatwaves—disrupt planting and harvesting cycles. These compounding pressures necessitate transformative interventions beyond conventional agronomy.

2.2 *Need for Technological Intervention*

Traditional agricultural intensification through chemical fertilization, mechanization, and monocropping has reached its efficacy limits in many regions. Yield stagnation and environmental degradation underscore the urgent need for precision, efficiency, and sustainability in food production. Technological innovations, particularly at the intersection of artificial intelligence (AI) and nanotechnology, offer compelling solutions.

AI provides robust predictive capabilities, enabling farmers and policymakers to anticipate crop stress, optimize irrigation scheduling, and forecast yield outcomes with high accuracy [16]. Machine learning models, when integrated with satellite imagery, remote sensing data, and IT-enabled soil sensors, allow real-time monitoring and adaptive management across heterogeneous agro-ecological zones. However, the accuracy of these models is highly dependent on the availability, quality, and granularity of region-specific datasets,areas with sparse or inconsistent data often experience reduced model performance and less reliable predictions. As a result, strengthening local data infrastructure becomes essential for realizing the full potential of AI-driven agricultural solutions.

Nanotechnology (comment: provide a definition) complements these capabilities by enhancing nutrient delivery, crop protection, and post-harvest management at the molecular scale. Nano-fertilizers, nano-encapsulated pesticides, and nano-based food fortificants improve efficiency, reduce wastage, and elevate nutrient bioavailability [13]. For example, field trials in India demonstrated that nano-encapsulated zinc fertilizers increased grain zinc content by 35%, simultaneously reducing leaching and environmental runoff [9].

The synergistic integration of AI and nanotechnology thus represents a paradigm shift in sustainable agriculture. By combining data-driven decision-making with nano-enabled precision interventions, it is possible to optimize resource use, enhance nutritional quality, and mitigate environmental impact, creating a resilient food system capable of addressing both hunger and hidden hunger on a global scale.

3 Nanotechnology Applications in Agriculture and Nutrition

3.1 Enhancing Crop Productivity and Resource Efficiency

Nanotechnology offers transformative solutions to address the challenges of declining agricultural productivity and resource inefficiency. Recent studies have demonstrated that nanotechnology may benefit the agricultural industry by increasing the effectiveness of agricultural inputs and offering sustainable alternatives to traditional practices [11].

Nano-fertilizers and Nano-pesticides: The development of nano-fertilizers has shown promising results in enhancing nutrient uptake efficiency and reducing nutrient losses. For instance, a study by Kapoor et al. [12] reported that the application of nano-fertilizers led to a significant increase in nutrient use efficiency, resulting in improved crop yields and reduced environmental impact.

Nano-encapsulated Agrochemicals: Nano-encapsulation techniques have been employed to enhance the stability and controlled release of agrochemicals. Research by Zhang et al. [33] demonstrated that nano-encapsulated pesticides exhibited prolonged efficacy and reduced toxicity to non-target organisms, thereby contributing to sustainable pest management practices.

Soil Health Enhancement: Nanomaterials such as nano-zeolites and hydrogels have been utilized to improve soil structure and water retention capacity. A study by Zhang et al. [32] found that the application of nano-zeolites in soil improved nutrient retention and reduced leaching, leading to enhanced soil fertility and crop productivity.

3.2 Improving Nutrient Bioavailability and Food Fortification

Micronutrient deficiencies, or hidden hunger, affect a significant portion of the global population. Nanotechnology has emerged as a promising approach to enhance the bioavailability of essential micronutrients in food.

Nano-fortification Techniques: The use of nanoparticles for fortifying staple foods has shown potential in addressing micronutrient deficiencies. A review by Arshad et al. [3] highlighted that nanoparticles can improve the bioavailability of micronutrients such as vitamin B12, vitamin A, folic acid, and iron, thereby contributing to better nutritional outcomes. Researches have shown that vitamin spray-dispersed nanodroplets boost iron, curcumin, and folic acid absorption, and nanoparticles, especially nanocochleate structures, increase nutrition absorption [18].

Microencapsulation for Nutrient Stability: Microencapsulation techniques have been employed to protect sensitive nutrients from degradation during food processing and storage. Research by Ashraf et al. [4] demonstrated that microencapsulation improved the stability and bioavailability of nutrients in fortified foods, minimizing sensory changes and enhancing consumer acceptance. Nanoencapsulation helps in maintaining the stability of probiotics and delivery of bioactive compounds. This technology helps in enhancing the shelf life of food products through smart and intelligent and packaging, development of nano-sensors in detecting food pathogens and toxins, eventually helping in preventing food spoilage and safety [1, 20].

Targeted Delivery Systems: Nanotechnology enables the development of targeted delivery systems that release nutrients at specific sites within the gastrointestinal tract. A study by Altemimi et al. [2] explored the application of nanoparticles for the targeted delivery of bioactive compounds, enhancing their absorption and efficacy in the human body.

3.3 *Case Studies and Field Applications*

India: Field applications in India show that advanced Agri-technologies can improve productivity and resource efficiency. IFFCO's large-scale field trials report that nano-urea application leads to **up to an 8% increase in crop yields** across multiple regions Minello et al. [10]. Likewise, systematic reviews of precision and AI-enabled irrigation systems indicate **30–50% reductions in water use** compared with conventional irrigation practices [25]. These findings highlight the potential of nanotechnology and AI to enhance agricultural performance when supported by suitable local data and agro-ecological conditions.

China: In China, nano-encapsulated fertilizers have been utilized to enhance nutrient delivery efficiency. A study by Minello et al. [19] reported that the use of nano-fertilizers improved nutrient uptake and reduced environmental pollution, contributing to sustainable agricultural practices.

Sub-Saharan Africa: In Sub-Saharan Africa, nanotechnology has been employed to address soil fertility challenges. Research by Kapoor et al. [12] highlighted that the application of nano-fertilizers improved soil nutrient content and crop yields, offering a sustainable solution to soil degradation issues.

4 Integrated AI–Nano Strategies for Food and Nutrition Security

4.1 Precision Agriculture Through AI–Nanotechnology Convergence

The convergence of artificial intelligence (AI) and nanotechnology facilitates precision agriculture at an unprecedented level of granularity. AI systems, utilizing machine learning algorithms and predictive analytics, can process heterogeneous datasets derived from soil sensors, drones, satellite imagery, and IoT-enabled farm equipment. When coupled with nano-enabled inputs, such as nano-fertilizers and nano-pesticides, these systems enable site-specific management, reducing over-application of agrochemicals and optimizing nutrient delivery.

For instance, field trials in Punjab, India, demonstrated that AI-guided nano-fertilizer application increased wheat yield by approximately 22% while simultaneously reducing nitrogen leaching by 35% [26]. Similarly, in China, AI-controlled drip irrigation combined with nano-encapsulated micronutrients reduced water consumption by 38% and enhanced zinc and iron biofortification in staple crops [17]. These outcomes exemplify the capacity of integrated AI–Nano strategies to enhance productivity, nutritional quality, and environmental sustainability concurrently.

4.2 Real-Time Crop Monitoring and Stress Management

AI algorithms, particularly deep learning models, are increasingly applied to real-time crop monitoring for disease detection, pest management, and abiotic stress mitigation. High-resolution imaging combined with nano-sensors embedded in soil or plant tissues allows continuous monitoring of nutrient deficiencies, moisture levels, and pathogen presence. Nano-sensors, capable of detecting micronutrient concentrations or early-stage pathogen markers at the molecular level, feed data into AI models that predict stress events before they become yield-limiting. For example, deployment of AI–nano systems in Kenya's maize fields enabled early detection of fungal infestations, reducing crop losses by 20–25% [21]. These systems promote proactive interventions, enhancing resilience against climate variability and minimizing the need for broad-spectrum chemical applications.

4.3 *Optimization of Supply Chains and Post-Harvest Management*

Post-harvest losses remain a critical challenge for global food security, particularly in regions with inadequate storage infrastructure. AI-enabled supply chain management platforms, integrated with nanotechnology-based preservation methods, have demonstrated significant reductions in losses. Nano-coatings and nano-packaging materials enhance shelf life, prevent microbial contamination, and maintain nutrient integrity, while AI predicts optimal storage conditions and distribution pathways.

In Nigeria, nano-silver coatings in storage containers, coupled with AI-driven humidity and temperature control, reduced post-harvest maize losses by 30%, improving both food availability and quality [14]. These approaches address both the quantity and nutritional quality of food reaching consumers.

4.4 *Socio-Economic Implications*

The adoption of AI–nano strategies carries substantial socio-economic implications. Enhanced resource-use efficiency and yield can increase farmer incomes while reducing input costs. Moreover, biofortified crops produced through nano-enabled interventions help alleviate micronutrient deficiencies in vulnerable populations. Nonetheless, equitable access remains a concern, as smallholder farmers may face financial, technical, or infrastructural barriers. To maximize inclusivity, policymakers must implement supportive frameworks, including training programs, subsidies, and technology-sharing platforms.

4.5 *Ethical, Regulatory, and Sustainability Considerations*

Cloud-based AI platforms rely on continuous data sharing, where farmers' field images, soil profiles, crop health records, and yield statistics are uploaded to remote servers for analysis. While this enables real-time decision support, it also raises concerns about data ownership, consent, and potential misuse. Farmers may face risks such as unauthorized data mining, profiling by agri-companies, or exploitation through targeted pricing. Strong data-protection measures such as—secure encryption, clear user-control settings, anonymization protocols, and legally enforceable data-sharing agreements—are essential to ensure that farmers retain full rights over their information and are protected from commercial or institutional misuse (Table 1).

Table 1 Outcomes of integrated AI–nanotechnology strategies in agriculture and nutrition

Strategy/Region	Technology applied	Application area	Measurable outcome	References
India (Punjab)	AI-guided nano-fertilizer application	Precision agriculture, nutrient delivery	Increased 22% wheat yield; reduced 35% nitrogen leaching	[26]
China (Shandong Province)	AI irrigation + nano-encapsulated fertilizers	Water management, crop biofortification	Increased 22% rice yield; reduced 30% water use; improved Fe/Zn content	[17]
Kenya	AI + nano-sensors	Early pest and stress detection	Reduced 20–25% maize crop loss due to fungal infestations	[21]
Sub-Saharan Africa (Nigeria)	Nano-silver coatings + AI monitoring	Post-harvest storage, nutrient preservation	Reduced 30% maize post-harvest losses	[14]
India (Rajasthan)	Drone-assisted AI + nano-pesticides	Pest control, water efficiency	Reduced 40% pesticide use; reduced 35% water consumption	[27]
China (Xinjiang)	AI + nano-chemicals	Cotton farming	Reduced 20% pesticide use; reduced 25% water use	[23]
India (Vidarbha)	Nano-encapsulated micronutrients + AI	Nutrient biofortification	Increased 35% grain zinc content	[9]
China (Jiangsu Province)	Nano-fortification + AI monitoring	Micronutrient delivery	Increased 25% selenium; Increased 30% iron in grains	[19]
India (Uttar Pradesh)	AI + nano-fertilizers	Smallholder productivity and income	Increased 25% income; reduced input costs	[15]

5 Technical Achievements and Measurable Outcomes

The synergistic integration of AI and nanotechnology in agriculture has yielded demonstrable improvements in crop productivity, nutrient bioavailability, resource efficiency, and post-harvest management. Data from field trials, pilot projects, and large-scale applications between 2020 and 2025 underscore these measurable outcomes.

5.1 Crop Yield Enhancement

AI–nano strategies have consistently increased crop yields across multiple agro-ecological zones:

- **India (Punjab—Wheat)**: "AI-assisted decision-support system" nano-urea application improved wheat yield by **28%**, while simultaneously reducing nitrogen fertilizer use by **40%** [26].
- **China (Shandong Province—Rice)**: AI-controlled irrigation with nano-encapsulated fertilizers resulted in **22% higher yields** and a **30% reduction in water consumption** [17].
- **Sub-Saharan Africa (Kenya—Maize)**: Nano-fertilizers improved nutrient use efficiency by **30%**, increasing overall maize productivity [21].

These results indicate that precision nutrient management, guided by AI, enhances crop growth while optimizing input usage.

5.2 Nutrient Biofortification

Nano-enabled fortification has improved micronutrient content in staple crops:

- **India (Vidarbha—Paddy)**: Nano-encapsulated zinc fertilizers increased grain zinc content by **35%**, mitigating hidden hunger [9].
- **China (Jiangsu Province—Rice)**: Nano-fortification with selenium and iron enhanced grain selenium by **25%** and iron by **30%** [19].

Such interventions demonstrate the potential of AI–Nano strategies to simultaneously address both yield and nutritional quality.

5.3 Resource Efficiency and Environmental Sustainability

- **India (Rajasthan—Multi-crop farms)**: Drone-assisted AI applications with nano-pesticides reduced pesticide use by **40%** and water consumption by **35%**. Sharma and Meena [27].
- **China (Xinjiang—Cotton)**: AI combined with nano-chemicals reduced pesticide application by **20%** and water usage by **25%** [5].

These outcomes highlight the environmental benefits of integrated AI–Nano approaches, reducing chemical runoff and conserving water resources.

5.4 *Post-Harvest Loss Reduction*

- **Nigeria—Maize**: Nano-silver coatings in storage containers, guided by AI monitoring, reduced post-harvest losses by **30%**, improving food security [14].

Integration of AI-based monitoring with nanotechnology ensures that harvested crops maintain both quantity and quality until consumption.

These benefits illustrate the broader socio-economic potential of AI–Nano technologies in enhancing farmer livelihoods and food security (Table 2).

6 Challenges, Limitations, and Future Directions

6.1 *Technical and Operational Challenges*

Despite the demonstrable successes of integrated AI–Nano strategies, several technical and operational challenges remain. AI algorithms rely on high-quality, real-time data inputs, yet data scarcity and heterogeneity in rural regions constrain predictive accuracy. For instance, soil sensors and nano-sensors may be expensive or difficult to deploy at scale, particularly for smallholder farmers. Additionally, the calibration of AI models to account for region-specific soil types, climate variability, and crop varieties is a complex task that requires continuous validation and adaptive learning frameworks [16].

On the nanotechnology front, the synthesis, stability, and controlled release of nano-fertilizers or nano-encapsulated nutrients require precise engineering. Variations in nanoparticle size, surface charge, and solubility can significantly affect efficacy and environmental behavior [13]. Furthermore, integrating nano-enabled interventions with conventional agronomic practices demands sophisticated knowledge transfer and operational coordination, which is often limited in low-resource agricultural settings.

6.2 *Regulatory, Safety, and Ethical Considerations*

The rapid adoption of nanotechnology in agriculture raises important regulatory and ethical concerns. The potential toxicity of certain nanoparticles to non-target organisms and ecosystems remains under scrutiny. Studies indicate that long-term soil accumulation of metallic nanoparticles may alter microbial communities and affect soil fertility [32]. Consequently, regulatory frameworks must define permissible concentrations, environmental monitoring protocols, and risk assessment standards to ensure safe application.

Table 2 Quantitative outcomes of AI–Nano agricultural interventions

Region/Crop	Intervention	Yield increase	Water saving	Pesticide reduction	Nutrient Biofortification	References
India (Punjab—Wheat)	AI + Nano-urea	28%	–	–	–	[26]
China (Shandong—Rice)	AI irrigation + Nano fertilizers	22%	30%	–	Improved Fe/Zn	[17]
Kenya (Maize)	AI + Nano-fertilizers	–	–	–	–	[21]
India (Vidarbha—Paddy)	Nano-encapsulated Zn fertilizer	–	–	–	+ 35% Zn	[9]
China (Jiangsu—Rice)	Nano-fortification (Se, Fe)	–	–	–	+ 25% Se; + 30% Fe	[19]
India (Rajasthan—Multi-crop)	AI + Nano-pesticides	–	35%	40%	–	[27]
China (Xinjiang—Cotton)	AI + Nano-chemicals	–	25%	20%	–	[5]
Nigeria—Maize	Nano-silver coating + AI monitoring	–	–	–	–	[14]
India (Uttar Pradesh)	AI + Nano-fertilizers	–	–	–	–	[15]

Ethical considerations also include data privacy and ownership issues in AI-driven agricultural systems. Farmers' personal and production data, when collected via IoT and AI platforms, may be vulnerable to misuse or inequitable commercial exploitation [22]. Transparent policies, data-sharing agreements, and inclusive governance mechanisms are essential to maintain trust and promote equitable technology adoption.

6.3 Economic and Access Barriers

While AI–Nano solutions offer high efficiency and enhanced nutritional outcomes; their adoption may be constrained by cost and accessibility. High initial investment in sensors, drones, nano-formulations, and AI platforms may deter smallholder farmers in low- and middle-income countries. Moreover, disparities in digital literacy, infrastructure, and institutional support can exacerbate inequities in food systems [8]. To address these barriers, governments and development agencies should implement targeted subsidies, capacity-building initiatives, and cooperative frameworks to facilitate equitable access and technology transfer.

6.4 Environmental and Sustainability Implications

The sustainability of AI–Nano strategies hinges on careful consideration of ecological impacts. Nanomaterials, if not biodegradable or environmentally compatible, may pose risks to soil health, water quality, and biodiversity [4]. Additionally, the energy footprint of AI computation, data storage, and IoT deployment must be accounted for within broader sustainability goals. Research into eco-friendly nanomaterials, low-energy AI algorithms, and circular agricultural systems is critical to align with technological interventions and environmental stewardship and climate resilience objectives.

6.5 Future Research Directions Emerging Research Priorities Include:

- **Multi-scale Integration**: Linking farm-level AI–Nano interventions with regional and national food system modeling to optimize production, distribution, and nutritional outcomes.
- **Adaptive AI Models**: Development of AI systems capable of continuous learning from heterogeneous agro-ecological datasets, including satellite imagery, sensor networks, and socio-economic data.

- **Green Nanotechnology**: Advancement of biodegradable, non-toxic nano-fertilizers and nano-pesticides that maintain efficacy while minimizing ecological risk.
- **Nutrient Biofortification Platforms**: Innovative nano-delivery systems for essential micronutrients, vitamins, and bioactive compounds in staple crops and processed foods.
- **Policy and Governance Research**: Design of regulatory frameworks, ethical guidelines, and incentive mechanisms to facilitate safe, equitable, and sustainable adoption of AI–Nano strategies globally.

By addressing these technical, regulatory, and socio-economic challenges, the AI–Nano paradigm has the potential to transform global food systems, enhance nutrition security, and build resilience against climate-induced disruptions.

7 Policy Directions and Recommendations

7.1 Strengthening Regulatory Frameworks

To enable safe and equitable adoption of AI–Nano technologies, robust regulatory frameworks are imperative. Governments and international agencies should establish standardized protocols for:

- Risk assessment of nano-fertilizers, nano-pesticides, and nano-encapsulated nutrients.
- Environmental monitoring and long-term soil and water quality assessment.
- Data governance policies for AI-driven agricultural systems, ensuring privacy, ownership, and ethical usage.

For example, the European Union's REACH (Registration, Evaluation, Authorisation and Restriction of Chemicals) framework provides a model for regulating nanoparticle use while protecting environmental and human health [13]. Similar region-specific adaptations are needed in South Asia and Sub-Saharan Africa to balance innovation with safety.

7.2 Promoting Inclusive Technology Access

Equitable access is critical for the success of AI–Nano interventions. Policy measures should focus on:

- Subsidies and financial incentives for smallholder farmers to adopt precision farming tools and nano-enabled inputs.

- Public–private partnerships to disseminate cost-effective AI–Nano solutions and training programs.
- Strengthening digital infrastructure in rural regions, including high-speed connectivity, IoT deployment, and technical support centers.

7.3 *Encouraging Research, Development, and Innovation*

The Common Agricultural Policy (CAP) 2023–2027 provides the necessary vision linking farmers and consumers, as well as society in general, with main driving objectives which include promotion of the agricultural sector resilience and securing food for future generations. It addresses social, environmental, and economic goals and provides a strong framework to drive food systems changes. One main recommendation is that agricultural modernization solutions should be implemented through knowledge sharing and technological innovation [6].

Encouraging Research, Development, and Innovation, National and international research programs should prioritize:

- Development of eco-friendly nanomaterials and low-energy AI computation.
- Multi-disciplinary research integrating agronomy, materials science, data analytics, and nutrition science.
- Field-scale pilot projects that evaluate the socio-economic, environmental, and nutritional impacts of AI-Integrating **Food Security and Nutrition Policy** Nano interventions.

Government funding, innovation hubs, and collaborative research platforms can accelerate the translation of lab-scale breakthroughs into scalable, field-ready solutions [4].

7.4 *Integrating Food Security and Nutrition Policy*

AI–Nano technologies should be aligned with broader national food security and nutrition policies. Key recommendations include:

- Leveraging biofortified crops and nano-fortification programs to target micronutrient deficiencies in vulnerable populations.
- Incorporating AI-driven predictive analytics into supply chains and market regulation to minimize post-harvest losses.
- Monitoring and evaluation mechanisms to track the impact of technological interventions on food accessibility, dietary quality, and nutritional outcomes.

8 Conclusion

The integration of artificial intelligence and nanotechnology represents a transformative frontier in global food and nutrition security. Empirical evidence from India, China, and Sub-Saharan Africa demonstrates that AI–Nano strategies can simultaneously enhance crop yields, improve nutrient bioavailability, optimize resource use, and reduce post-harvest losses.

While technical, regulatory, and socio-economic challenges persist, the potential benefits are profound. With targeted policies, inclusive frameworks, and sustainable implementation strategies, AI–Nano technologies can contribute significantly to achieving Sustainable Development Goal 2—Zero Hunger— and other SDGs by ensuring equitable access to sufficient, safe, and nutrient-rich food.

Future research and policy initiatives must prioritize environmental safety, farmer inclusivity, and adaptive innovation to realize the full potential of this dual-engine approach. In doing so, AI–Nano convergence could offer a scientifically robust, socio-economically viable, and environmentally sustainable pathway to address the complex and interlinked challenges of 21st-century food systems for a better food and better future, the theme of FAO for the year 2025.

References

1. Ahmed H, Aya T (2025) Food 2050 concept: trends that shape the future of food. J Future Foods. https://doi.org/10.1016/j.jfutfo.2025.03.003
2. Altemimi A et al (2024) Targeted nutrient delivery using nanoparticles in staple crops: advances and prospects. J Agricul Nanotechnol 12(3):145–162. https://doi.org/10.1016/j.jagronano.2024.03.001
3. Arshad M, Khan R, Liu Y (2021) Nano-fortification techniques for improving micronutrient bioavailability in foods: a review. Food Chem 341:127897. https://doi.org/10.1016/j.foodchem.2020.127897
4. Ashraf S et al (2025) Microencapsulation of vitamins and minerals: Stability, bioavailability, and sensory attributes. Trends in Food Sci Technol 133:102–118. https://doi.org/10.1016/j.tifs.2024.11.008
5. Caiyu (2025) China's first 'super cotton field' project uses AI-driven tech, debunking lies of 'forced labor' in Xinjiang. Global Times
6. European Commission (2023) The common agricultural policy at a glance. Available at: https://agriculture.ec.europa.eu/common-agricultural-policy/cap-overview/cap-glance_en. (Accessed 06 Nov 2025)
7. FAO (2022) State of the World's land and water resources for food and agriculture. Food and Agriculture Organization of the United Nations, Rome
8. FAO (2023) The state of food security and nutrition in the world 2023: towards zero hunger. Food and Agriculture Organization of the United Nations, Rome
9. Gupta R, Sharma P, Singh A (2022) Nano-encapsulated zinc fertilizer improves grain zinc content in paddy: field trials in Vidarbha, India. Agritech Future 14(2):55–68
10. IFFCO (2023) Nano Urea Field Trial Summary Report. Indian Farmers Fertiliser Cooperative Limited
11. Kah M, Tufenkji N, White JC (2019) Nano-enabled strategies to enhance crop nutrition and protection. Nature Nanotechnol 14(6):532–540

12. Kapoor N, Chen X, Li J (2022) Nano-fertilizers and their role in sustainable agriculture: efficiency, environmental impact, and field applications. Sustain Agricul Rev 45:87–112. https://doi.org/10.1007/978-3-030-85089-0_4
13. Khan M, Zhang Y, Ahmad S (2022) Nanotechnology in agriculture: environmental, regulatory, and health perspectives. Environ Nanotechnol 17:100–115. https://doi.org/10.1016/j.envnano.2022.100115
14. Khan O, Adeyemi A, Nwankwo C (2023) AI-assisted post-harvest management with nano-silver coatings reduces maize losses in Nigeria. Int J Food Preserv 11(1):45–58
15. Kumar S, Srivastava A (2025) AI-enabled nano-fertilizer applications for enhancing small-holder productivity in Uttar Pradesh. Agronomy J 117(4):512–524
16. Li H, Zhang J, Wang Q (2023) AI-based predictive modelling for precision agriculture: applications in crop stress and yield forecasting. Comput Electron Agricul 208:107711. https://doi.org/10.1016/j.compag.2023.107711
17. Li X, Chen F, Min J (2024) AI-controlled irrigation combined with nano-encapsulated fertilizers enhances rice yield and micronutrient content in Shandong Province, China. Precis Agricul 25(4):987–1005. https://doi.org/10.1007/s11119-024-10145-7
18. McClements DJ (2020) Nano-enabled personalized nutrition: developing multicomponent bioactive colloidal delivery systems. Adv Colloid and Interface Sci 282:102211
19. Minello S, Zhao L, Tan Y (2025) Nano-fortification strategies for selenium and iron in rice: field application and nutritional outcomes. J Food Sci Nutrit 78(1):55–69. https://doi.org/10.1111/jfsn.2025.78.1.55
20. Mohammad ZH, Zaidi S, Ibrahim SA, Ahmad F (2022) Application of nanotechnology in different aspects of the food industry. Discover Food 2(1). https://doi.org/10.1007/s44187-022-00013-9
21. Mwangi P, Njoroge D, Otieno R (2024) Early detection of fungal infestations in maize using AI and nano-sensors in Kenya. African J Agricul Technol 15(2):102–117
22. Popkin B, Corvalan C, Grummer-Strawn L (2021) Dynamics of the global nutrition transition: the dual burden of undernutrition and obesity. Lancet 397(10278):1388–1394. https://doi.org/10.1016/S0140-6736(21)00212-4
23. Ray Nano Fertilizer (2024) Integrated AI and nano-chemical applications for sustainable cotton farming in Xinjiang. J Nanosci Agricul 6(2):45–59
24. Rockström J, Thilsted SH, Willett WC, Gordon LJ, Herrero M, Hicks CC, DeClerck F (2025) The EAT–lancet commission on healthy, sustainable, and just food systems. The Lancet 406(10512):1625–1700
25. Rodriguez D, Smith J, Patel R (2024) Precision irrigation technologies: a global systematic review. J Agricul Water Managem
26. Sharma R, Singh V, Mehta A (2023) AI-guided nano-urea application improves wheat yield and reduces nitrogen leaching in Punjab. Agronomy J 115(3):1254–1270. https://doi.org/10.1002/agj2.21520
27. Sharma R, Meena P (2025) Drone-assisted AI and nano-pesticide integration for sustainable pest management in semi-arid Rajasthan. Agronomy J 117(3):450–462
28. Swinburn BA, Kraak VI, Allender S, Atkins VJ, Baker PI, Bogard JR, Dietz WH (2019) The global syndemic of obesity, undernutrition, and climate change: the Lancet commission report. The Lancet 393(10173):791–846
29. UN (2019) World Population Prospects 2019: Highlights (ST/ESA/SER.A/423). Department of Economic and Social Affairs, Population Division
30. United Nations (2023) The sustainable development goals report 2023
31. WHO (2022). https://www.who.int/health-topics/malnutrition
32. Zhang X, Li Y, Huang T (2023) Nanomaterials for soil enhancement and water retention: mechanisms and field evaluation. J Soil Sci Environ Managem 14(1):33–48
33. Zhang Y, Sun W, Chen H (2024) Nano-encapsulated pesticides for controlled release and sustainable pest management. Pest Managem Sci 80(5):1120–1132. https://doi.org/10.1002/ps.6458

Improving Crop Quality and Yield by AI with Nanotechnological Approaches

Farwa Basit, Elshan Musazade, Wang Qian, Wang Wenhui, Nie Jing, Javaid Akhter Bhat, and Xianzhong Feng

Abstract The integration of artificial intelligence (AI) and nanotechnology in agriculture is revolutionizing traditional farming by enhancing precision, efficiency, and sustainability. AI technologies are increasingly used for soil monitoring, crop cultivation, pest control, farm optimization, and using real-time environmental data and advanced weather forecasting models to select the best seed types and determine optimal planting times. Simultaneously, nanotechnology offers innovative solutions, such as the targeted delivery of fertilizers and pesticides through nanoparticles (NPs), which enhance nutrient absorption and reduce resource waste. However, the integration of AI and nanotechnology enables smarter applications, including intelligent machinery for precise seed spacing, depth control, continuous crop monitoring, and early disease detection. This chapter explores how AI and nanotechnology together enhance crop productivity, reduce environmental impact, and promote sustainable practices, leading to significant progress in food and biofuel production.

Keywords Artificial intelligence (AI) · Crop monitoring · Soil health · Smart farming · Sustainable agriculture

---

Farwa Basit, Elshan Musazade: These authors contributed equally to this work.

F. Basit · W. Qian · W. Wenhui · N. Jing
Department of Biology, College of Science and Technology, Wenzhou-Kean University, Wenzhou, China

E. Musazade · J. A. Bhat (✉) · X. Feng (✉)
Key Laboratory of Soybean Molecular Design Breeding, Northeast Institute of Geography and Agroecology, Chinese Academy of Sciences, Changchun, China
e-mail: fengxianzhong@iga.ac.cn

J.-T. Chen (ed.), *AI-Nano-Enabled Climate-Smart Agriculture and Sustainability*,
Smart Nanomaterials Technology, https://doi.org/10.1007/978-981-92-1694-9_7

1 Introduction

The agricultural sector is often regarded as one of the foundations of the global economy and as the primary employer in many nations worldwide. However, the agriculture sector faces major obstacles in terms of agricultural sustainability, productivity, and quality, particularly in the context of global concerns like soil degradation, climate change, and growing food demand. Despite their traditional efficacy, conventional agricultural methods have drawbacks, particularly when it comes to resource exploitation, environmental contamination, and yield variability. These constraints necessitate more environmentally friendly methods that can boost crop quality and production [1].

Precision agriculture (PA), sometimes referred to as site-specific crop management (SSCM), satellite agricultural farming, or precision farming, is a strategic managerial framework that addresses crop production variability both within and between fields through responsive actions and systematic monitoring [36]. Precision agriculture (PA) enables the monitoring and management of crop production variability across fields using advanced technologies such as GPS, GIS, and remote sensing. Thus, it allows to collect data for real time on crop status, weather patterns, and soil health. PA aims to promote crop productivity while minimizing the waste of vital resources like water, pesticides, and fertilizers by facilitating more targeted and efficient management [54]. The overall goal of precision agriculture is to improve productivity while minimizing environmental impact and costs.

Alongside precision farming, nanotechnology has increasingly gained attention as a promising approach to enhance agricultural productivity. Through the manipulation of materials at the atomic scale, nanotechnology enables the development of innovative materials, electronics, and structures. Particles that have a diameter of 100 nm or less are included. Nanomaterials, such as nano-encapsulated fertilizers and pesticides, allow for more controlled and efficient delivery of agrochemicals, decreasing waste and environmental pollution [86]. Moreover, nanosensors embedded in soil and crops offer real-time monitoring of plant health, soil conditions, and pest presence at the microscale, enhancing the precision of farming practices.

Concurrently, the application of artificial intelligence (AI) has significantly transformed the agricultural industry by driving innovation in crop cultivation, real-time monitoring, processing, harvesting, and agricultural product marketing. The synergistic interaction between AI-technology and nanotechnology has the potential to revolutionize modern agriculture [36]. Crop health monitoring is one of the key AI applications. For example, Plantix, an app developed by PEAT that analyzes crop and soil images to detect nutrient deficiencies and plant diseases, provides farmers with tailored recommendations to optimize soil health, improve crop yield, and promote sustainable farming practices [31]. Thus, these AI-based sensors enable precise weed identification and targeted pesticide application, reducing chemical use and enhancing crop quality [54].

The agricultural industry is being revolutionized by the integration of AI and nanotechnology, which enhances efficiency, sustainability, and productivity. In agriculture, AI is essential for handling the massive amounts of data produced by precision farming and nanoscale sensors. With machine learning algorithms, AI can predict crop yields, detect diseases and pests early, and optimize the distribution of resources [86]. AI allows for more precise and effective farming methods when combined with nanotechnology. Embedded in crops and soil, nanosensors provide microscale real-time monitoring of insect infestation, soil conditions, and plant health. This integration enables the management of resources more precisely, which lowers the impact on the environment by lowering the demand for excessive water, fertilizer, and pesticides [80]. Furthermore, AI-nanotechnology integration offers several important benefits for the agricultural sector, including enhanced productivity, reduced environmental impact, and increased farmer profitability (Fig. 1). On the other hand, the ethical and legal implications of these technologies must be acknowledged. It's also necessary to handle issues like data security, privacy concerns, and closing the digital divide in rural areas. However, the integration of AI and nanotechnology holds significant promise for tackling the global challenges of food security and advancing the development of sustainable agricultural practices [58].

Despite substantial gains in agricultural productivity, these advances have often been accompanied by environmental degradation and unsustainable resource use. With limited opportunities for expanding arable land, sustainable intensification has become a central priority in modern agriculture. In this context, artificial intelligence has emerged as a powerful decision-support framework for precision agriculture, enabling data-driven optimization of inputs and management practices. This chapter critically examines the potential and limitations of precision farming and emphasizes the integration of AI with advanced nanotechnological approaches as a transformative strategy to enhance the precision, efficiency, and sustainability of crop quality and yield improvement.

2 Need for Artificial Intelligence in Modern Agriculture

The agriculture sector is undergoing numerous challenges caused by a rapidly growing population, urbanization, and changing consumer demands, which create predominant pressure on farmers to enhance crop productivity and quality by managing limited resources. Meanwhile, agriculture must operate under limited natural resources such as arable land, water, and nutrients, while reducing the environmental impacts. Also, farming is becoming more complicated due to climate change, soil degradation, unpredictable rainfall, pests, and diseases [73, 81]. However, traditional agricultural techniques are inefficient in coping with these challenges. Manual monitoring of crops, soil, and environmental conditions is time-consuming, labor-intensive, and highly susceptible to human error. Fertilizers and pesticides are often applied uniformly, as they require a lot of labor and basic management practices,

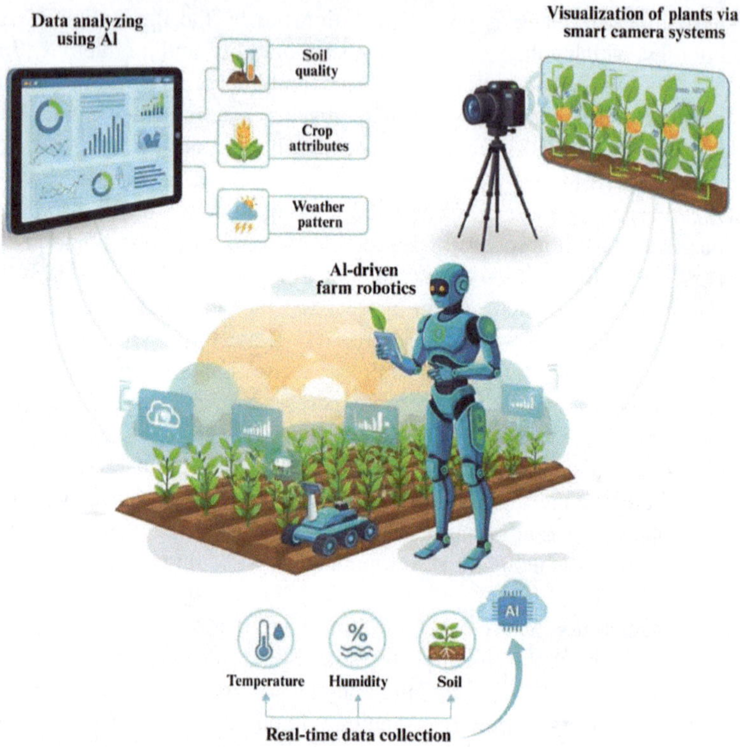

Fig. 1 Integration of artificial intelligence and nanotechnology in agricultural systems, illustrating applications across crop management, precision farming, and decision support

resulting in resource wastage, ineffective practices, reduced productivity, and environmental contamination. Similarly, utilizing traditional methods to forecast pest outbreaks, nutrient shortages, or crop diseases is outdated rather than proactive, which leads to lower yields and higher crop losses [73, 81].

Artificial intelligence (AI) offers a potent solution by enabling proactive, precise, and data-driven farm management. AI-technologies such as machine learning and computer vision can analyze large volumes of data from sources such as satellite imagery, drones, IoT sensors, and weather reports to monitor crop health, identify the early signs of stress or disease, and forecast yield with high accuracy (Table 1). Machine learning algorithms can predict crop yield, advise on optimal crop varieties, and support market planning, helping farmers to make informed decisions to maximize productivity and profit [72]. Simultaneously, AI may also optimize the time and quantity of inputs, such as water, fertilizer, and pesticides, to ensure that resources are used effectively and sustainably [10]. Furthermore, AI facilitates adaptive and predictive decision-making, assisting farmers in minimizing risks associated with market fluctuations, labor shortages, and climate change. AI reduces errors, improves productivity, and promotes sustainable farming practices by performing complex analysis and providing actionable insights [66].

In brief, environmental constraints, increasing production demands, and the shortcomings of traditional farming methods are the main reasons for the need for AI in modern agriculture. Therefore, AI-driven strategies are essential for achieving effective, resilient, and precision-based farming, forming a crucial foundation for the integration of cutting-edge technologies such as nanotechnology.

Table 1 Emerging applications of artificial intelligence (AI) in agriculture

Application domain	Applications of AI technologies	Key benefits	References
Precision irrigation	AI-driven sensors, machine learning, and real-time data analytics	Optimizes water usage, reduces waste, and enhances crop yields	[36]
Automated weed control	Machine learning, computer vision, robotics	Precision herbicide application, reduced chemical use, and labor cost savings	[37]
Soil health monitoring	AI algorithms, sensor networks, and data analytics	Optimized nutrient delivery, improved soil health, enhanced crop productivity	[68, 87]
Crop disease detection	Computer vision, drone imaging, and AI-powered sensors	Early identification of diseases, timely intervention, and reduced pesticide use	[34]
Climate-smart agriculture	AI models, satellite data, and machine learning	Enhanced resilience to climate change, optimized resource use, and sustainable farming practices	[28]
Local weather forecasting	AI-powered IoT devices, machine learning, and real-time data processing	Accurate microclimate predictions, informed irrigation, and planting decisions	[18]
Yield prediction	Machine learning, remote sensing, Internet of Things (IoT)	Accurate forecasting, informed decision-making, and resource optimization	[25]
Livestock health monitoring	Computer vision, AI sensors, behavioral analytics	Early disease detection, improved animal welfare, optimized breeding, and production	[5]
Regenerative agriculture monitoring	AI models, spectral probes, and automated soil analysis	Efficient carbon sequestration tracking, enhanced soil health management	[35]
Explainable crop recommendation	Explainable AI (XAI), machine learning, edge computing	Transparent decision-making, tailored crop suggestions, improved resource allocation	[4]

3 Nanotechnology in Agriculture: Materials and Properties

3.1 Classification and Characteristics of Nanoparticles Used in Agriculture

NPs, with their diverse classifications and characteristics, have demonstrated strong adaptability in the agricultural field. According to [39], the physical properties of NPs are influenced by morphological and topographical characteristics, such as size, shape, and localization, while composition and phase affect chemical properties. The commonly used nanomaterials in the agricultural field mainly include metal-based NPs, carbon-based NPs, and polymeric NPs. Each type of material has unique properties and is suitable for different agricultural application cases. With outstanding physical properties, nanomaterials can undergo chemical and biological interactions with plants, providing innovative solutions to challenges in agricultural production such as stress resistance improvement (alleviating drought stress, etc.), pest and disease control nutrient fortification (enhanced nitrogen fixation capacity, etc.), and crop yield increase such as promoting seed germination and improving photosynthetic capacity (Table 2) [39, 85]. Consequently, a comprehensive understanding of nanoparticle classification and characteristics provides a critical foundation for optimizing their application in plant science research and agricultural practice, thereby supporting sustainable crop productivity and quality improvement.

3.1.1 Polymeric Nanoparticles

Polymeric nanoparticles (PNPs) are nanosized particles composed of biodegradable polymers, such as polylactic acid-hydroxy acid (PLGA) or polycaprolactone (PCL) [52]. In agriculture, PNPs have emerged as promising tools for enhancing crop protection and nutrient delivery. For instance, pesticide-loaded PNPs can minimize environmental contamination by reducing leaching and volatilization while improving pest control efficiency [79]. The study also highlights the potential of PNPs in delivering plant growth promoters and biocontrol agents, thereby enhancing yield and stress resistance. Their biodegradability and low toxicity make them more environmentally friendly compared to conventional agrochemicals.

3.1.2 Metal-Based Nanoparticles

Common metal-based NPs are derived from substances such as aluminum (Al), lead (Pb), silver (Ag), and iron (Fe), and almost all metals can produce NPs [6]. Furthermore, the properties of metal-based NPs can be enhanced by adding corresponding oxides. For instance, common substances such as ferric oxide (Fe_2O_3) and silica (SiO_2) can be used to improve the reaction rate. In the field of agriculture, Zn NPs can promote the synthesis of photosynthetic pigments and improve the antioxidant

Table 2 Classification and characteristics of nanoparticles in plant sciences

NPs categories	NPs name	Key applications	Key properties	References
Metallic nanoparticles	Silver (Ag), Gold (Au), Zinc Oxide (ZnO), Iron Oxide (Fe_2O_3), Copper (Cu)	Antimicrobial activity, Nutrient delivery, Chlorophyll synthesis, Pathogen control, Enhanced micronutrient availability	High reactivity, Controlled release behavior, Nutrient uptake efficiency	[33, 67, 85]
Polymeric nanomaterials and composites	Chitosan-based NPs, Polymeric composites	Targeted and slow release of nutrients, pesticides, or growth regulators, and environmentally friendly pest management	Biodegradable, Biocompatible, Antimicrobial properties	[14, 85]
Carbon-based nanomaterials	Carbon Nanotubes (CNTs), Graphene Oxide, Fullerenes	Enhancing water and nutrient uptake, Modulating oxidative stress, Efficient agrochemical delivery, Biosensing applications	Structural stability, Electrical and mechanical properties, Creation of nanochannels in roots for better absorption	[14, 85]
Organic nanoparticles	Dendrimers, Liposomes, Ferritin	Nanopesticides, Nanosensors, Drug delivery systems	Non-toxic, Biodegradable, High encapsulation capacity	[67]
Metal oxide nanoparticles	Titanium Dioxide (TiO_2), Cerium Oxide (CeO_2), Nano Hydroxyapatite (NHAP)	Enhancing photosynthesis, improving seed germination and plant growth, Phytoremediation, and Stress tolerance	Photocatalytic activity, High surface area, Ability to alter plant physiological processes	[63, 76]
Metal–organic biohybrids	Copper High-Aspect Ratio Structures (CuHARS), Silver-Cystine Nanoparticles	Antimicrobial activity, Plant growth promotion, Disease resistance	Combination of metal ions and biological molecules, Scalable dimensions, High aspect ratio structures	[62]

response system; Cu NPs are often used to enhance resistance under non-biological stress; Fe NPs can promote plant growth and increase yield [64]. Metal NPs have been widely applied in agricultural production due to their unique and efficient effects.

3.1.3 Carbon-Based Nanoparticles

Carbon-based NPs, such as graphene, carbon nanotubes, and carbon black, are entirely composed of carbon. Due to the special structure and excellent performance of this material, it is widely used in agriculture [52]. Carbon-based NPs enhance the adaptability of plants and increase crop yields. For instance, carbon dots are used to enhance the salt tolerance of grapes. NPs with intrinsic fluorescence can be used for the detection of pesticides. For instance, graphene can be used to detect aromatic pesticides [88].

3.1.4 Comparative Overview of Nanoparticle Types

These three different types of NPs have distinct applications in agriculture. Understanding and leveraging this differentiated information can enhance their collaborative effectiveness in agriculture. Polymeric NPs, serving as carriers for delivering growth promoters and biological control agents, play a crucial role in enhancing agricultural productivity and protecting the environment due to their degradability and low spillage. Metal-based NPs are typically used to improve resistance under non-biological stress, and they are widely applied due to their high reaction rate and diverse types. However, these materials may pose a risk of low toxicity, which could affect their long-term use in agricultural production. Furthermore, carbon-based NPs can be used for pesticide tracking and have shown excellent performance in pesticide monitoring.

3.2 Physicochemical Properties Governing Nanoparticle Performance in Plant Systems

3.2.1 Surface Area and Chemical Reactivity

The surface area and chemical reactivity of NPs are critical factors influencing their interactions with plant systems. Due to their nanoscale dimensions, NPs exhibit a significantly higher surface area-to-volume ratio compared to bulk materials, enhancing their reactivity [55]. This increased surface area facilitates greater adsorption of nutrients, agrochemicals, and biomolecules, which can improve nutrient delivery and stress mitigation in plants. However, heightened reactivity may also lead to unintended consequences if NPs interact excessively with cellular components [48]. Surface modifications, such as coating with organic polymers, can modulate reactivity while maintaining beneficial effects. Thus, optimizing surface properties is essential for safe and effective NPs applications in agriculture.

3.2.2 Solubility, Size Distribution, and Dynamic Behavior

The solubility and size of NPs are critical determinants of their bioavailability and mobility in plant systems. Solubility influences how NPs release ions into the surrounding environment, which can affect nutrient delivery or toxicity levels [50]. Smaller NPs generally have higher solubility and greater mobility in soil and plant tissues, enabling easier penetration through root and leaf surfaces. This size-dependent behavior also influences translocation within the plant, with particles less than 50 nm often reaching systemic circulation [83]. However, overly high solubility might lead to rapid ion release, potentially causing toxicity. Conversely, poor solubility might limit the functional utility of NPs. Thus, tailoring the NPs' size and solubility profiles is essential for optimizing their agricultural applications, ensuring effective delivery while minimizing environmental risks.

4 Mechanisms Underlying Nanoparticle-Plant Interactions

The relationship between NPs and plants is complicated and includes changing the plant's metabolism, physiology, and overall welfare. They require plants to both absorb NPs from their leaves and roots and to transport them to different tissues. The uptake of NPs in plants may disrupt cellular structures and signal transduction, resulting in benefits related to growth, increased tolerance of stresses, and improved nutritional absorption by the plant [51]. Given that NPs can be helpful or harmful for plants depending on the type and dosage given, we need precise knowledge of these mechanisms to protect the environment and ensure they are used properly in agriculture [61].

4.1 Uptake, Transport, and Translocation of Nanoparticles

Through root absorption or foliar application, NPs can be taken up by plant roots or applied directly to the leaves and then moved to target tissues. There are two primary pathways for root uptake. One is the apoplastic pathway, where NPs travel in the spaces of the cell wall. The other is the simplistic pathway, which allows movement through the plasmodesmata after entering the cytoplasm. After passing through the epidermis and cortex via the apoplastic route, NPs encounter the Casparian strip. All molecules must move from the endodermal cells into the blood vessels to pass through this barrier [7]. After entering the root tissue, NPs reach the vascular system and are transported to leaves, flowers, and other above-ground parts via the xylem. For movement past the tissues, some NPs are subsequently carried by phloem and xylem, depending on their roles. The method of NP delivery and their own physical and chemical profiles influence their transport. Studies indicate that foliar-sprayed NPs can move through the phloem to reach roots and fruits, with their efficiency affected by size, charge, and the plant's health [7, 8]. In the xylem, NPs are carried upward with the water as it moves up, while the plant absorbs water through transpiration. Conversely, the primary function of foliar application is to facilitate the movement of NPs through the main xylem and phloem bundles. The transpiration-driven movement of moisture helps NPs travel through the plant's xylem. Additionally, NPs may enter leaves through either the cuticle or stomata. Smaller NPs (< 5 nm) can pass directly through the cuticle, whereas larger ones (> 10 nm) typically enter via stomata. Research has shown that NPs applied to leaves may traverse the cuticle or stomata and subsequently be moved by the vascular system to roots or fruits, bypassing the usual barriers found in the soil [82].

4.1.1 Foliar Application and Root Uptake Pathways

In soil-based systems, NPs primarily enter the plant through its roots, with two pathways for NPs to penetrate root tissues: apoplastic and symplastic. In the apoplastic pathway, substances move freely through the gaps in cell walls. Conversely, in the symplastic pathway, energy is required to transport water and solutes within cells via plasmodesmata. NPs may choose different pathways based on their size, charge, and surface properties, which influence both entry into root tissues and transport within cells [84]. After reaching the root cortex, NPs are transported to the xylem to move to the above-ground parts of the plant. This process has been shown to depend on transpiration-driven flow and root-to-shoot signaling mechanisms [74]. They can enhance nutrient absorption and strengthen roots, even when nutrients are scarce. Studies have demonstrated that appropriate concentrations of engineered NPs may stimulate root system development and improve macronutrient assimilation [29].

Foliar spraying is an effective method for delivering NPs, such as growth regulators and pesticides. When NPs are applied to leaves, they can penetrate the cuticle and stomata, thus avoiding soil-related obstacles. For instance, when nanoclusters of ZnO are sprayed on tomato leaves, they enhance both chlorophyll production and plant growth by facilitating nutrient uptake from the leaves. Studies have shown that foliar-applied ZnO NPs significantly increase nutrient absorption and biomass accumulation in tomato plants, thereby improving both photosynthetic efficiency and fruit quality without causing visible toxicity [3].

4.1.2 Translocation and Tissue-Level Distribution

Depending on NPs' functional properties and the site of application, they are transported within plants through different pathways. When applied to the roots, NPs move through the symplastic pathways to enter the xylem, then are transported upward to the shoots and leaves. In contrast, foliar application allows NPs to penetrate the leaf cuticle and stomata, entering the phloem for transport to non-photosynthetic tissues such as roots and fruits. The mobility and dispersion of NPs are influenced by factors such as their size, solubility, and the interactions they have with plant biopolymers.

Smaller particles tend to move more freely through cell walls and travel farther within vascular tissues, while larger ones are more likely to remain near the point of entry. If the NPs are highly soluble, they can break down into ions, making absorption and movement through plant fluids easier. By adjusting size and solubility, it's possible to control whether the NPs act locally or are delivered to specific parts of the plant.

4.2 Cellular and Molecular Effects of Nanoparticles

4.2.1 Impacts on Cell Structure and Integrity

NPs can have both positive and negative effects on plant cells. For example, copper NPs can induce structural and biochemical changes in the cell wall, and they help plants fight off microbes and handle physical stress better. These particles may accumulate in or near the cell wall and contribute to more protective barrier. Studies show that copper NPs help plants build stronger barriers against diseases [33, 84]. But if there are too many NPs that get inside the cells, they might break the cell wall and make it harder for things to move through. This can cause stress inside the cell and hurt important parts, such as the cell membrane and mitochondria. Too many NPs can lead to damage to the cells from harmful oxygen molecules, as ZnO NPs are known to generate reactive oxygen species (ROS) under certain conditions [20, 55, 77]. On the other hand, carbon-based NPs like carbon nanotubes (CNTs) help plants by improving water flow and nutrient uptake, especially when the plants are under stress like drought [46].

In the internal environment of the cell, NPs interact with key organelles, including chloroplasts, mitochondria, and nuclei, thereby influencing critical biological processes such as photosynthesis, respiration, and gene expression. For example, NPs help protect the photosynthetic system and regulate gene expression during drought stress [60]. Carbon-based NPs, including CNTs, facilitate the enhancement of cell membrane permeability and boost nutrient uptake through the formation of nanochannels in the cell membrane, thus advancing cellular activities under drought stress scenarios. In addition, NPs such as metal oxides and silicon-based types can reduce oxidative stress in plants. They do so by improving antioxidant responses and adjusting enzyme activities that combat ROS. As a result, ROS levels decrease, helping to protect plant cells from stress-induced damage [70]. This greatly enhances the stress resistance of plants under adverse environmental conditions.

4.2.2 Regulation of Nutrient Uptake and Plant Metabolism

NPs can really boost how effectively plants take up nutrients and how well their metabolism works. This means plants can grow better and become more resilient to stress. For example, recent studies show that NPs like zinc oxide (ZnO) and iron oxide (Fe_3O_4) are easy-to-absorb sources of important micronutrients. They deliver elements like zinc and iron more efficiently than traditional bulk fertilizers [69]. Because of their tiny size, these NPs can pass through plant cuticles and cell walls more easily, delivering nutrients directly to the parts of the cell where the work happens [59]. This helps boost enzyme functions and the production of chlorophyll, which is critical for photosynthesis and energy creation. As a result, plants grow more biomass and perform better, especially when faced with environmental stress like drought or salty soils [65].

In terms of plant metabolism, NPs help to regulate essential physiological and biochemical pathways, particularly those involved in energy metabolism and stress adaptation. For example, zinc and cerium oxide NPs have been shown to increase the activity of nitrate reductase, which enhances nitrogen assimilation, protein biosynthesis, and ultimately boosts plant biomass production [32]. In parallel, carbon-based NPs such as graphene oxide (GO) contribute to metabolic homeostasis by lowering the accumulation of ROS and mitigating oxidative damage during environmental stress [75]. These effects collectively support improved energy production, redox balance, and plant resilience under challenging conditions [11, 12].

5 Role of Nanotechnology in Precision Agriculture

The application of nanotechnology has the potential to transform precision agriculture by offering ground-breaking solutions to the various issues faced in the agricultural sector (Fig. 2). These are some key areas in which nanotechnology plays a significant role.

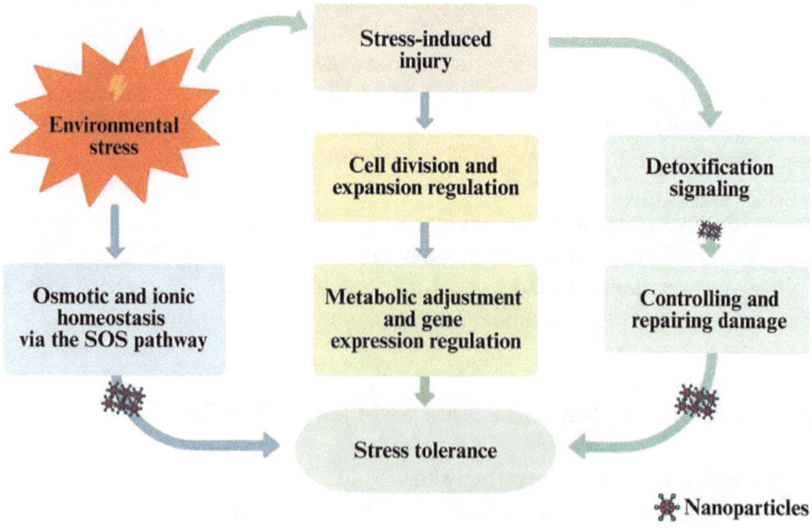

Fig. 2 Nanotechnology-based strategies for combating environmental stress in crop plants

5.1 Smart Nano-delivery Systems

Plant growth stimulants, insecticides, and mineral supplements are all precisely delivered using nanoparticles. These materials can be regulated and monitored for optimal efficacy by encapsulating them at the nanoscale. This minimizes the environmental impact of agrochemicals while lowering the total quantity needed [19].

5.2 Enhancement of Nutrient-Use Efficiency

Nanoparticles have the potential to improve plant nutrient availability and absorption. A consistent supply of nutrients can be ensured by slow-release mechanisms offered by nanostructured fertilizers. By increasing the surface area available for nutrient exchange, nanomaterials can improve plants' absorption of nutrients [71].

5.3 Crop Monitoring and Management

Nanosensors can be used to monitor pH balance, nutrient grade, soil humidity, and other vital variables in real-time. These sensors are able to provide accurate data about crop health, which makes it feasible to optimize pest control, fertilization, and irrigation strategies. This ultimately leads to increased crop productivity and resource efficiency. Various types of nanosensors used in living plants include sensors for growth, stress, and plant signaling. These nanosensors can identify a variety of plant signaling chemicals, including chemical cues, phytohormones, electrical signals, and gases. This technique has the potential to shed light on a number of aspects of plant physiology and biology [45]. A nanoscale sensor designed to detect and measure nitric oxide (NO) is a fluorescent ratiometric sensor that uses single-walled carbon nanotubes (SWCNTs). This sensor uses the unique characteristics of SWCNTs to detect NO, and it has been shown to be responsive in both the complex environment of plant leaves and controlled laboratory environments. This implies that the nanosensor can adjust to intricate and real-world circumstances [24]. The development and use of wearable nanosensors for a variety of purposes, such as tracking plant growth factors, has grown rapidly. These sensors have even been applied to the tracking of water transportation and distribution, sap flow, and nutrient distribution in the context of plant growth and development. A versatile electronic sensing device has been developed to facilitate continuous monitoring. For instance, the use of these nanosensors in watermelon production has shown interesting results. An alteration in the distribution of water between fruits and leaves during the day and night cycles has been noted [15]. In addition, researchers can quickly identify plant diseases with the use of portable, affordable, and precise nanosensors [41].

In short, nanosensors are essential for improving plant health monitoring, which significantly increases the effectiveness of plant management in agriculture. However, while incorporating these nanosensors into agricultural systems, we need to consider their stability. It also raises concerns about whether nanosensors currently offer the level of sensitivity and reliability needed for real-world agricultural applications. Even so, we are optimistic that ongoing research and technological advances will successfully overcome these challenges in the future.

5.4 *Precision Irrigation Strategies*

Moisture sensors at the nanoscale can improve irrigation procedures by providing precise data on soil moisture content. This enables farmers to accurately implement irrigation in the required location and timing, reducing water waste and improving the effectiveness of water usage [19].

5.5 *Soil Remediation and Sustainability*

Non-technological techniques may promote soil remediation by eliminating pollutants and enhancing soil quality. Heavy metals, pesticides, and other contaminants can be bound by nanoparticles, making it easier to remove them from the soil. Additionally, nanomaterials can improve the soil's ability to retain nutrients and water, making it healthier and more fertile [71].

5.6 *Disease and Pest Management*

Nanotechnology provides new methods for identifying and controlling pests and plant diseases. Proactive disease management is enabled by the early detection of pathogen presence using nanoscale biosensors [42, 49]. Additionally, nanoparticles can be engineered to produce bioactive substances that inhibit the growth of infections and pests, hence decreasing the need for traditional chemical treatments. Enhancing theranostic efficacy while reducing adverse effects is a promising application of nanomedicine [42]. The potential of bio-inspired systems to provide the precise and customized treatment required for each patient is astonishing. These strategies can address a variety of obstacles, frequently with intrinsically contradictory demands, when applied individually or in combination, promoting the development of successful disease theranostics. The majority of traditional delivery methods are frequently unable to accomplish this goal [47, 49]. Collectively, these advancements represent emerging applications of nanotechnology in modern agricultural systems (Fig. 3).

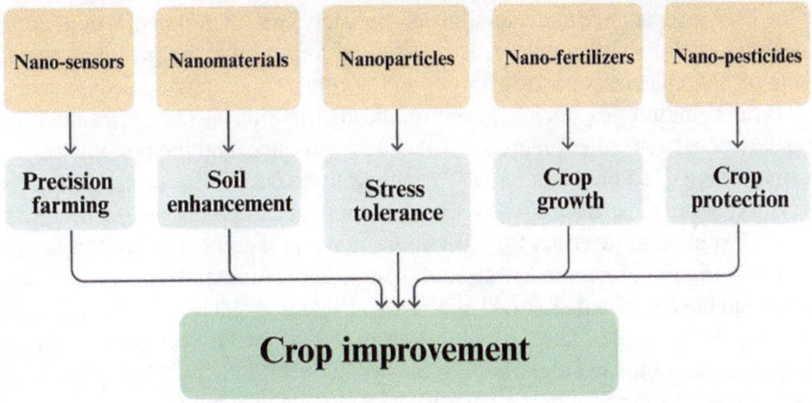

Fig. 3 Emerging applications of nanotechnology in modern agricultural systems

6 AI-Nanotechnology Integrated Approaches for Sustainable and Safe Crop Production

AI-Nanotechnology integration offers potential solutions to the escalating problems with global food security. Nearly 840 million people may suffer from hunger by 2030 according to UN predictions. In order to solve this problem, scientists are investigating how AI and farming driven by nanotechnology can transform agricultural sustainability and production [86]. Artificial intelligence (AI)-driven data analytics and machine learning techniques are increasingly applied to optimize agricultural operations. While nanotechnology permits precise material applications that improve soil health, fertilizer delivery, and insect control. The foremost challenge of nanotechnology application in agriculture is to ensure its safe and appropriate use.

According to Joshi [38], researchers emphasized the necessity of limiting the ongoing effects of NPs in agricultural settings, such as their impact on soil, roots, and leaves. AI plays a vital role in this scenario by allowing continuous monitoring of these impacts over time and predicting how repeated applications of nanoparticles impact soil health and ecosystem balance. Also, it assures that nanotechnology promotes sustainable farming practices rather than aggravating environmental degradation, this predictive accuracy is crucial to preventing any negative ecological impacts. Moreover, AI is also needed to examine the cumulative ecological impacts of nanomaterials in the agricultural ecosystem. AI interpreting the behavior of NPs in particular ecosystems requires a systems-level approach that makes use of the data now available on crop production, soil fertility, and nutrient-use efficiency (NUE). Through the utilization of AI and machine learning methodologies to examine the fundamental processes governing the conduct of NPs, precision agriculture can unleash their capabilities while guaranteeing ecological sustainability [38, 86].

However, an extensive understanding of the safety and environmental implications of nanomaterials is still required. It is essential to consider the potential negative effects of nanomaterials on both target and non-target organisms, their impact on ecological communities, their continuity and distribution in the environment, the breakdown or transformation products they may produce, and the possible residues they may leave in edible plant parts. To address these issues, biocompatibility and a safety-first approach are required when developing and applying nanoparticles. The safe-by-design nanomaterials for agriculture must be incorporated via research on the biodegradation of nanomaterials and their possible toxicity [13].

Through the combined use of AI and machine learning techniques, AI-technology makes it possible to evaluate cellular attachment and uptake, predict the formation of NPs corona, group and classify NPs, assess computational risk, and develop NPs safely by design. The goal of the AI-Nanotech combined application is to guarantee the sustainable and safe application of NPs in agriculture by utilizing these methods to improve our knowledge of their physicochemical characteristics and functionality [17]. For instance, microscopic photos of important species exposed to NPs have been evaluated using AI and machine learning algorithms to identify potential abnormalities or harmful consequences [40]. Another crucial breakthrough is the use of Bayesian networks for risk assessment of NPs exposure in the agriculture sector. These models assist in anticipating the outcomes of interactions with nanomaterials and set causal links within adverse outcome pathways (AOPs) [30]. Despite these developments, it is nevertheless challenging to integrate and curate the enormous datasets required for AI-Nanotech models. To increase the precision and reliability of AI predictions in nanomaterial applications, nanosafety data must be harmonized with other environmental, genetic, and geographic data types [2]. The use of conventional animal testing is gradually decreasing in the field of nanosafety as safe-by-design nanomaterials become more prevalent. In silico screening of nanomaterials to assess their safety profiles prior to their use in agriculture is now possible. AI-Nanotech models can direct the development of ecologically safe and efficient nanomaterials by integrating machine learning, predictive modeling, and experimental data. These AI-driven models are incorporated into frameworks such as Integrated Approaches to Testing and Assessment (IATA) to validate that the performance and safety are balanced while developing nanoparticles for agricultural systems [23].

For instance, AI-based simulations can determine optimal types and doses of NPs to improve crop quality and yield while lowering potential risks. Furthermore, as new data becomes available, dynamic models like Bayesian networks can be adjusted to provide real-time insights into the changing effects of nanomaterials. These advancements enable the use of AI-Nanotech as a feasible and cost-effective approach to advance precision agriculture, particularly as more data becomes available to improve nanomaterial design and refine predictions [9, 23].

7 Applications of AI-Nanotechnology Integration in Agriculture

In the agricultural sector, the integration of AI and nanotechnology is generating exciting new opportunities, particularly for making farming more intelligent, sustainable, and adaptable to global issues such as soil degradation, climate change, and food poverty. The targeted delivery of agrochemicals is one of the most promising applications. Traditionally, fertilizers and insecticides have been applied broadly across fields, leading to environmental pollution, resource waste, and the development of pest resistance. However, nanotechnology enables the packaging of these chemicals in nanoscale carriers, such as nano-encapsulated insecticides or fertilizers, which release their contents in a controlled manner. When integrated with AI, which can instantly analyze crop health, soil data, and weather patterns, these nanoformulations can work more efficiently. For instance, AI can evaluate real-time data from soil sensors, crop health monitors, and climate forecasting tools to optimize the timing, dosage, and placement of nano-enabled agrochemicals. This approach minimizes costs while promising that agrochemicals be used only when and where they are required, reducing adverse environmental effects [17, 56].

Precision farming is another important area where AI-Nanotech is playing its crucial role. Recent innovations in nanosensor technology have made it possible to develop microscopic sensors that can be attached to plants or located directly in the soil. These sensors can identify vital elements such as temperature, moisture content, nutrient shortages, and even plant disease symptoms. For example, sensors embedded in crops can identify stress from pests or diseases, while nanosensors placed in soil can provide real-time data on moisture and nutrient content. AI systems process the data when it is collected, giving farmers quick, useful insights. Farmers can use these insights to make data-driven decisions, such as early pest infestation alarms, ideal irrigation schedules, or accurate fertilizer recommendations. Precision farming systems can greatly boost crop yields, decrease waste, and enhance resource efficiency by combining nanosensors and AI-driven analytics. A prominent example of this integration is the implementation of AI-driven decision support systems, which may recommend real-time interventions based on collected data from sensor networks in fields [56].

In a nutshell, the combination of nanotechnology and AI is making agriculture a more advanced, data-driven sector. Real-time crop status monitoring, accurate agricultural input management, and predictive diagnostics for disease and pest outbreaks are all made possible by this confluence. When taken as a whole, these developments improve output, encourage environmentally friendly methods, and lessen the impact of contemporary farming.

7.1 Pest Identification

The use of AI and nanotechnology in pest identification offers a revolutionary approach to farming practices. AI-enhanced nanosensors enable major improvements in real-time monitoring systems. These nanosensors are used in fields to continually monitor plant health and environmental conditions. AI systems analyze the data and identify patterns and peculiarities leading to the presence of pests, allowing for prompt notifications and preventative measures. By using biochemical signatures, smart traps that incorporate nanosensors may identify and detect particular pest species. By analyzing the data from these traps, AI systems provide practical insights for focused pest control strategies, minimizing the environmental impact, and lowering the demand for broad-spectrum pesticides. AI-enhanced nanoencapsulated insecticides provide further advantages. Based on nanosensor data, AI algorithms can identify the best times and locations to apply pesticides, ensuring accurate application and minimizing environmental detrimental effects. Additionally, these pesticides can be designed to release their active ingredient in response to particular triggers, including enzymes associated with pests. AI forecasts environmental factors and pest behavior for efficient release [16, 34, 81].

7.2 Synergies of Nanotechnology and AI

AI enables intelligent decision-making beyond delivery. By combining AI-driven models with chitosan nanoparticles that contain certain biocontrol drugs, researchers may predict pathogen behavior based on environmental data. These nanoparticles minimize unnecessary applications by releasing their payload only when the model shows a high risk of infection [21]. According to Christiaens [16], AI also maximizes pest control by using intelligent nanocarriers that carry RNA molecules. AI algorithms direct dose estimations and carrier design for optimal impact and minimal environmental impact. Through extensive sensor networks, artificial intelligence also transforms precision management techniques [27]. AI algorithms provide farmers real-time information on crop health, soil moisture, and nutrient levels by optimizing sensor placement and data analysis. This information enables precision agricultural methods that are suited to certain field conditions. According to Parisi [56], even more futuristic uses integrate drone technology with nanotechnology. In this instance, AI-driven drones use nanoformulations of fertilizers or pesticides to target hotspots, minimizing total application quantities and increasing treatment effectiveness.

8 Applications of AI-Nanotechnology Integration in Agriculture

The integration of artificial intelligence (AI) and nanotechnology is opening new avenues for advancing intelligent, sustainable, and resilient agricultural systems capable of addressing challenges such as soil degradation, climate change, and food insecurity. One of the most impactful applications lies in the targeted delivery of agrochemicals. Conventional blanket application of fertilizers and pesticides often results in resource inefficiency, environmental contamination, and pest resistance. Nanotechnology enables the encapsulation of agrochemicals within nanoscale carriers that allow controlled and stimuli-responsive release. When coupled with AI, which rapidly analyzes crop health indicators, soil parameters, and weather dynamics, these nanoformulations can be applied with high spatial and temporal precision. AI systems integrate real-time data from soil sensors, crop monitoring platforms, and climate forecasting tools to optimize dosage, timing, and placement, ensuring that agrochemicals are applied only when and where they are needed, thereby reducing costs and environmental burden [56, 86].

Beyond delivery optimization, AI-nanotechnology integration plays a transformative role in precision farming through advanced sensing and decision-support systems. Recent developments in nanosensor technology have enabled the deployment of microscopic sensors directly in soil or attached to plant tissues to continuously monitor temperature, moisture, nutrient availability, and early stress or disease signals. These high-resolution data streams are processed by AI-driven analytics to generate actionable insights, such as early pest or disease warnings, optimized irrigation schedules, and precise fertilizer recommendations. The combination of nanosensors and AI-driven data interpretation substantially enhances resource-use efficiency, reduces waste, and improves crop productivity. A representative application is the use of AI-based decision-support systems that recommend real-time field interventions based on continuous sensor network inputs [56].

Pest monitoring and management further illustrate the synergistic potential of AI and nanotechnology. AI-enabled nanosensors facilitate continuous field surveillance by detecting biochemical and physiological signatures associated with pest infestation. Smart traps incorporating nanosensors can identify specific pest species, while AI algorithms analyze temporal and spatial patterns to guide targeted control strategies. This approach reduces reliance on broad-spectrum pesticides and mitigates non-target environmental effects. In parallel, nanoencapsulated insecticides can be designed to release active ingredients in response to pest-associated enzymes or environmental triggers, with AI models forecasting pest dynamics and environmental conditions to optimize release efficiency [44].

At a broader level, AI enhances the functional intelligence of nanotechnology-based systems beyond delivery and detection. Predictive AI models combined with nanocarriers, such as chitosan nanoparticles loaded with biocontrol agents, enable risk-based release strategies that activate only under high infection probability, thereby minimizing unnecessary applications [21]. AI-guided nanocarriers delivering RNA molecules further improve pest and pathogen control by optimizing dosage and carrier design for maximal efficacy and minimal ecological impact [16]. Through large-scale sensor networks, AI also refines precision management by optimizing sensor placement and data interpretation to deliver real-time insights into crop health, soil moisture, and nutrient dynamics, enabling field-specific management strategies [27]. More advanced implementations integrate AI-driven drones with nanoformulated fertilizers or pesticides to target localized hotspots, reducing overall chemical inputs while enhancing treatment effectiveness [56].

9 Challenges and Limitations of AI-Nanotechnology Integration

The application of nanotechnology and AI in agriculture holds promise for improvements in crop resilience, resource optimization, and precision farming. Nonetheless, a number of obstacles stand in the way of their broad acceptance and efficient application. Smallholder and resource-constrained farmers may find the high expenses of creating and implementing AI and nanotechnology technologies to be unaffordable. Furthermore, the usefulness and accessibility of these technologies are restricted in rural regions by a lack of infrastructure, such as dependable internet connectivity and technical support. The creation of scalable and reasonably priced solutions that are adapted to the requirements of various farming communities is necessary to guarantee equitable access [77]. When AI systems and nanotechnology are combined, hardware, software, and data must all be seamlessly integrated. While AI algorithms can optimize farming methods by analyzing data from nanosensors, the scalability of such integrated systems is still a difficulty, particularly in diverse and unstructured agricultural landscapes. Standardized AI-nanotechnology solutions are difficult to implement in many places due to factors including crop diversity, weather, and soil heterogeneity [43]. It is unclear how agricultural nanomaterials will affect the environment in the long run. Unintentional impacts on non-target organisms, bioaccumulation in the food chain, and possible toxicity to soil microbiota are among the worries. To reduce these risks, thorough risk evaluations and the creation of eco-friendly nanomaterials are crucial [22].

To successfully apply AI and nanotechnology in agriculture, a workforce with expertise in both fields is needed. There is currently a scarcity of interdisciplinary professionals, and farmers might not have the necessary skills to use these technologies efficiently. Initiatives to increase capacity and invest in education are essential for closing these knowledge gaps [78]. AI and nanotechnology in agriculture are developing at a faster rate than regulatory frameworks. Clear rules and regulations are required to guarantee the moral and secure application of these technologies. Ethical considerations also include the possible socioeconomic inequalities brought on by unequal access to cutting-edge agricultural technologies and data privacy, particularly concerning AI-driven data collection [53]. A cooperative strategy combining researchers, legislators, industry stakeholders, and farming communities is required to address these issues. To achieve sustainable and equitable growth, the agriculture industry may fully utilize AI and nanotechnology by supporting education and infrastructure development, encouraging interdisciplinary research, and creating inclusive regulations.

10 Conclusion and Future Directions

This book chapter elaborates on the thrilling prospects for further study in the domains of AI and nanotechnology to improve crop yields and promote sustainable agriculture. Advances in the field of nanotechnology have the potential to provide novel agricultural products, such as nanopesticides and fertilizers. These developments can maximize the availability of nutrients, encourage plant growth, and lessen the negative effects of conventional agrochemicals on the environment. However, it highlights the necessity of additional research in this area to improve the use of NPs in sustainable agriculture. The AI integration with nanotechnology offers a potent instrument for precision farming. To better understand the properties of NPs utilized in agriculture, the research community might explore nanoinformatics. It is essential to comprehend how these components affect plants and ecosystems in different environments and throughout different seasons. This coordinated strategy can reduce the hazards connected to traditional agrochemicals and promote sustainable and safe farming methods. However, the main goal of this particular area of research should be to examine the risks and difficulties that come with implementing nanotechnology in agriculture. Microbial populations and soil quality are especially vulnerable to the presence of engineered NPs. Their effects on these important elements of agricultural ecosystems need to be further investigated. Furthermore, studies on the phytotoxicity of various NPs should be conducted, taking into consideration their intricate relationships with plant growth.

In a nutshell, the prospects for further study have the ability to transform farming methods and support international initiatives to attain sustainable agriculture and food security by 2030. AI and nanotechnology combined in this multidisciplinary approach may lead to creative answers to the problems facing the agriculture industry. To further investigate these prospects and tackle the challenges of incorporating nanotechnological developments into agriculture while preserving the environment and public health, researchers, legislators, and stakeholders must work together.

Declarations **Authors' Contributions** F.B., and J.A.B. conceptualized the manuscript; F.B., and E.M. prepared the initial draft; F.B., E.M., W.Q., W.W., N.J., J.A.B., and X.F. reviewed and revised the manuscript. All authors have read and agreed to the published version of the manuscript.

Conflict of Interest Statement The authors declare that they have no conflict of interest.

Funding

Data Availability Statement All data generated or analyzed during this study are included in this published article.

Ethics Approval and Consent to Participate The use of plant parts in the present study complied with international, national, and/or institutional guidelines.

Consent for Publication Not applicable.

References

1. Abdo AI, Sun D, Shi Z, Abdel-Fattah MK, Zhang J, Kuzyakov Y (2025) Conventional agriculture increases global warming while decreasing system sustainability. Nat Clim Chang 15:110–117
2. Afantitis A (2020) Nanoinformatics: artificial intelligence and nanotechnology in the new decade. Comb Chem High Throughput Screening 23:4–5
3. Ahmed R, Uddin MK, Quddus MA, Samad MYA, Hossain MM, Haque ANA (2023) Impact of foliar application of zinc and zinc oxide nanoparticles on growth, yield, nutrient uptake and quality of tomato. Horticulturae 9:162
4. Akkem Y, Biswas SK, Varanasi A (2024) Streamlit-based enhancing crop recommendation systems with advanced explainable artificial intelligence for smart farming. Neural Comput Appl 36:20011–20025
5. Al-Ahmed SA, Ahamed T (2024) AI× IoT: increasing agricultural productivity of crops, orchards, and livestock management using smart agricultural space for achieving SDGs, IoT and AI in agriculture: smart automation systems for increasing agricultural productivity to achieve SDGs and Society 5.0. Springer, pp 481–490
6. Alam MJ, Abedin MJ, Rahman MZ (2024) Nanoparticles, nanocomposites, green/eco-composites, and hybrid composites and their applications in energy sectors
7. Ali S, Mehmood A, Khan N (2021) Uptake, translocation, and consequences of nanomaterials on plant growth and stress adaptation. J Nanomater 2021:6677616
8. Ali S, Mehmood A, Khan N (2021) Uptake, translocation, and consequences of nanomaterials on plant growth and stress adaptation. J Nanomater 2021:1–17
9. Ashique S, Raikar A, Jamil S, Lakshminarayana L, Gajbhiye SA, De S, Kumar S (2025) Artificial intelligence integration with nanotechnology: a new frontier for sustainable and precision agriculture. Curr Nanosci 21:242–273
10. Awasthi Y (2020) Press "A" for artificial intelligence in agriculture: a review. JOIV: Int J Inform Vis 4:112–116
11. Basit F, Bhat JA, Guan Y (2023a) Elucidation of synergistic interaction among metal oxide nanoparticles and PGPR on the plant growth and development, nanomaterials and nanocomposites exposures to plants: response, interaction, phytotoxicity and defense mechanisms. Springer, pp 311–325
12. Basit F, He X, Zhu X, Sheteiwy MS, Minkina T, Sushkova S, Josko I, Hu J, Hu W, Guan Y (2023) Uptake, accumulation, toxicity, and interaction of metallic-based nanoparticles with plants: current challenges and future perspectives. Environ Geochem Health 45:4165–4179
13. Bharadwaj A, Gupta M, Verma M, Gangwar P, Wahi N (2024) Advancements of nanobiotechnology in public health sector: benefits and challenges. Nanosci Nanotechnol Asia 14:E22106812316402
14. Carrillo-Lopez LM, Villanueva-Verduzco C, Villanueva-Sánchez E, Fajardo-Franco ML, Aguilar-Tlatelpa M, Ventura-Aguilar RI, Soto-Hernández RM (2024) Nanomaterials for plant disease diagnosis and treatment: a review. Plants 13:2634
15. Chai Y, Chen C, Luo X, Zhan S, Kim J, Luo J, Wang X, Hu Z, Ying Y, Liu X (2021) Cohabiting plant-wearable sensor in situ monitors water transport in plant. Adv Sci 8:2003642

16. Christiaens O, Whyard S, Vélez AM, Smagghe G (2020) Double-stranded RNA technology to control insect pests: current status and challenges. Front Plant Sci 11:451
17. Chugh G, Siddique K, Solaiman Z (2021) Nanobiotechnology for agriculture: Smart technology for combating nutrient deficiencies with nanotoxicity challenges. Sustainability. 13(4):1781
18. Das SK, Nayak P (2024) Integration of IoT-AI powered local weather forecasting: a game-changer for agriculture. arXiv preprint arXiv:2501.14754
19. Devarajan Y (2025) Investigation of emerging technologies in agriculture: an in-depth look at smart farming, nano-agriculture, AI, and Big Data. J Biosyst Eng 1–23
20. Dhiman V, Kondal N (2023) Bryophyllum pinnatum leaf extract mediated ZnO nanoparticles with prodigious potential for solar driven photocatalytic degradation of industrial contaminants. Environ Res 216:114751
21. Elmer WH, White JC (2016) The use of metallic oxide nanoparticles to enhance growth of tomatoes and eggplants in disease infested soil or soilless medium. Environ Sci Nano 3:1072–1079
22. El-Sayed M, Ayoub HA, Helal M, Sang W, Shen Z, Abdelhafeez IA (2025) Nanotechnology-enabled soil management for sustainable agriculture: interactions, challenges, and prospective. Environ Sci Nano
23. Garg D, Payasi DK (2020) Nanomaterials in agricultural research: an overview. Environ Nanotechnol 3:243–275
24. Giraldo JP, Landry MP, Kwak SY, Jain RM, Wong MH, Iverson NM, Ben-Naim M, Strano MS (2015) A ratiometric sensor using single chirality near-infrared fluorescent carbon nanotubes: application to in vivo monitoring. Small 11:3973–3984
25. Goel M, Pandey M (2024) Crop yield prediction using AI: a review. In: 2nd international conference on disruptive technologies (ICDT). IEEE, pp 1547–1553
26. Gogos A, Knauer K, Bucheli TD (2012) Nanomaterials in plant protection and fertilization: current state, foreseen applications, and research priorities. J Agric Food Chem 60:9781–9792
27. García L, Parra L, Jimenez JM, Lloret J, Lorenz P (2020) IoT-based smart irrigation systems: an overview on the recent trends on sensors and IoT systems for irrigation in precision agriculture. Sensors 20:1042
28. Gryshova I, Balian A, Antonik I, Miniailo V, Nehodenko V, Nyzhnychenko Y (2024) Artificial intelligence in climate smart in agricultural: toward a sustainable farming future. Access J 5:125–140
29. Guzmán-Báez GA, Trejo-Téllez LI, Ramírez-Olvera SM, Salinas-Ruíz J, Bello-Bello JJ, Alcántar-González G, Hidalgo-Contreras JV, Gómez-Merino FC (2021) Silver nanoparticles increase nitrogen, phosphorus, and potassium concentrations in leaves and stimulate root length and number of roots in tomato seedlings in a hormetic manner. Dose-Response 19:15593258211044576
30. Halappanavar S, Nymark P, Krug HF, Clift MJ, Rothen-Rutishauser B, Vogel U (2021) Non-animal strategies for toxicity assessment of nanoscale materials: Role of adverse outcome pathways in the selection of endpoints. Small 17:2007628
31. Hasteer N, Mallik A, Nigam D, Sindhwani R, Van Belle J-P (2024) Analysis of challenges to implement artificial intelligence technologies in agriculture sector. Int J Syst Assur Eng Manage 15:1841–1860
32. Hussain A, Rizwan M, Ali Q, Ali S (2019) Seed priming with silicon nanoparticles improved the biomass and yield while reduced the oxidative stress and cadmium concentration in wheat grains. Environ Sci Pollut Res 26:7579–7588
33. Jabran M, Ali MA, Muzammil S, Zahoor A, Ali F, Hussain S, Muhae-Ud-Din G, Ijaz M, Gao L (2024) Exploring the potential of nanomaterials (NMs) as diagnostic tools and disease resistance for crop pathogens. Chem Biol Technol Agric 11:75
34. Jafar A, Bibi N, Naqvi RA, Sadeghi-Niaraki A, Jeong D (2024) Revolutionizing agriculture with artificial intelligence: plant disease detection methods, applications, and their limitations. Front Plant Sci 15:1356260
35. Jagyasi B, Shyam M (2024) A review on AI and remote sensing based regenerative agriculture assessment. In: Harnessing data science for sustainable agriculture and natural resource management, pp 225–248

36. Javaid M, Haleem A, Khan IH, Suman R (2023) Understanding the potential applications of Artificial Intelligence in Agriculture Sector. Adv Agrochem 2:15–30
37. Jha K, Doshi A, Patel P, Shah M (2019) A comprehensive review on automation in agriculture using artificial intelligence. Artif Intell Agric 2:1–12
38. Joshi H, Choudhary P, Mundra S (2019) Future prospects of nanotechnology in agriculture. Int J Chem Stud 7:957–963
39. Joudeh N, Linke D (2022) Nanoparticle classification, physicochemical properties, characterization, and applications: a comprehensive review for biologists. J Nanobiotechnol 20:262
40. Karatzas P, Melagraki G, Ellis LJA, Lynch I, Varsou DD, Afantitis A, Tsoumanis A, Doganis P, Sarimveis H (2020) Development of deep learning models for predicting the effects of exposure to engineered nanomaterials on Daphnia magna. Small 16:2001080
41. Kashyap PL, Kumar S, Srivastava AK (2017) Nanodiagnostics for plant pathogens. Environ Chem Lett 15:7–13
42. Kaushik M, Kumar S, Singh M, Sharma H, Bhowmick M, Bhowmick P, Ashique S, Khatoon H, Pal R, Ansari MA (2024) Bio-inspired nanomaterials in cancer theranostics. In: Nanotheranostics for diagnosis and therapy, pp 95–123
43. Khattak WA, Ullah MW, Manan S, Islam SU, Khattak WA, Ul-Islam M (2024) Emerging applications and future trends of agri-nanotechnology. In: Revolutionizing agriculture: a comprehensive exploration of agri-nanotechnology, pp 429–458
44. Kumari A, Singh J, Yadav SK (2026) Harnessing nanobiotechnology for sustainable agriculture: from plant nanobionics to AI-driven precision farming. Biosci Nanotechnol 2:3
45. Kwak S-Y, Wong MH, Lew TTS, Bisker G, Lee MA, Kaplan A, Dong J, Liu AT, Koman VB, Sinclair R (2017) Nanosensor technology applied to living plant systems. Annu Rev Anal Chem 10:113–140
46. Lahiani MH, Khodakovskaya MV (2016). Concerns about nanoparticle hazard to human health and environment. In: Plant nanotechnology: principles and practices, pp 349–365
47. Li X, Hetjens L, Wolter N, Li H, Shi X, Pich A (2023) Charge-reversible and biodegradable chitosan-based microgels for lysozyme-triggered release of vancomycin. J Adv Res 43:87–96
48. Lu S, Duffin R, Poland C, Daly P, Murphy F, Drost E, MacNee W, Stone V, Donaldson K (2009) Efficacy of simple short-term in vitro assays for predicting the potential of metal oxide nanoparticles to cause pulmonary inflammation. Environ Health Perspect 117:241–247
49. Lu Y, Luo Q, Jia X, Tam JP, Yang H, Shen Y, Li X (2023) Multidisciplinary strategies to enhance therapeutic effects of flavonoids from Epimedii Folium: Integration of herbal medicine, enzyme engineering, and nanotechnology. J Pharmaceut Anal 13:239–254
50. Ma H, Williams PL, Diamond SA (2013) Ecotoxicity of manufactured ZnO nanoparticles—a review. Environ Pollut 172:76–85
51. Mahakham W, Sarmah AK, Maensiri S, Theerakulpisut P (2017) Nanopriming technology for enhancing germination and starch metabolism of aged rice seeds using phytosynthesized silver nanoparticles. Sci Rep 7:8263
52. Mandal M, Banerjee I, Mandal M (2022) Nanoparticle-mediated gene therapy as a novel strategy for the treatment of retinoblastoma. Colloids Surf, B 220:112899
53. Mathur A, Wadhwa S, Nagabooshanam S, Roy S (2021) Journey of agricultural sensors—from conventional to ultra-modern. In: Biosensors in agriculture: recent trends and future perspectives, pp 217–240
54. Mridha K, Hasan SMA (2021) Artificial intelligence (AI) for agricultural sector, 2021 international conference on control, automation, power and signal processing (CAPS). IEEE, pp 1–6
55. Nel A, Xia T, Madler L, Li N (2006) Toxic potential of materials at the nanolevel. Science 311:622–627
56. Parisi C, Vigani M, Rodríguez-Cerezo E (2015) Agricultural nanotechnologies: what are the current possibilities? Nano Today 10:124–127
57. Rai M, Khan S, Belgamwar A (2023) Nanotechnology principles in drug targeting and diagnosis. Elsevier

58. Raj EFI, Appadurai M, Athiappan K (2022) Precision farming in modern agriculture, Smart agriculture automation using advanced technologies: data analytics and machine learning, cloud architecture, automation and IoT. Springer, pp 61–87
59. Rajput VD, Minkina T, Upadhyay SK, Kumari A, Ranjan A, Mandzhieva S, Sushkova S, Singh RK, Verma KK (2022) Nanotechnology in the restoration of polluted soil. Nanomaterials 12:769
60. Rasheed A, Li H, Tahir MM, Mahmood A, Nawaz M, Shah AN, Aslam MT, Negm S, Moustafa M, Hassan MU (2022) The role of nanoparticles in plant biochemical, physiological, and molecular responses under drought stress: A review. Front Plant Sci 13:976179
61. Rico CM, Majumdar S, Duarte-Gardea M, Peralta-Videa JR, Gardea-Torresdey JL (2011) Interaction of nanoparticles with edible plants and their possible implications in the food chain. J Agric Food Chem 59:3485–3498
62. Ruiz-Hitzky E, Darder M, Aranda P (2008) An introduction to bio-nanohybrid materials. Bio-inorganic hybrid nanomaterials, pp 1–40
63. Saleem H, Zaidi SJ (2020) Recent developments in the application of nanomaterials in agroecosystems. Nanomaterials 10:2411
64. Santás-Miguel V, Arias-Estévez M, Rodríguez-Seijo A, Arenas-Lago D (2023) Use of metal nanoparticles in agriculture. A review on the effects on plant germination. Environ Pollut 334:122222
65. Seleiman MF, Ahmad A, Battaglia ML, Bilal HM, Alhammad BA, Khan N (2023) Zinc oxide nanoparticles: a unique saline stress mitigator with the potential to increase future crop production. S Afr J Bot 159:208–218
66. Shadrin D, Menshchikov A, Ermilov D, Somov A (2019) Designing future precision agriculture: detection of seeds germination using artificial intelligence on a low-power embedded system. IEEE Sens J 19:11573–11582
67. Shao C, Zhao H, Wang P (2022) Recent development in functional nanomaterials for sustainable and smart agricultural chemical technologies. Nano Convergence 9:11
68. Sharma S, Verma K, Hardaha P (2023) Implementation of artificial intelligence in agriculture. J Comput Cogn Eng 2:155–162
69. Sheta MH, Abd El-Wahed AH, Elshaer MA, Bayomy HM, Ozaybi NA, Abd-Elraheem MA, El-Sheshtawy A-NA, El-Serafy RS, Moustafa MM (2024) Green synthesis of zinc and iron nanoparticles using Psidium guajava leaf extract stimulates cowpea growth, yield, and tolerance to saline water irrigation. Horticulturae 10:915
70. Singh AP, Chetry S, Singh H, Ghosh T, Atta K, Rout RK, Das U, Acharya D (2025) Realistic application of biogenic nanoparticles: future perspectives and myth. In: Biogenic nanoparticles. Elsevier, pp 25–53
71. Singh S (2025) Optimizing nutrient uptake: nano-fertilizers and soil amendments, food and Industry 5.0: transforming the food system for a sustainable future. Springer, pp 229–248
72. Skvortsov E (2020) Prospects of applying artificial intelligence technologies in the regional agriculture. Ekonomika Regiona Econ Regions 563
73. Subeesh A, Mehta C (2021) Automation and digitization of agriculture using artificial intelligence and internet of things. Artif Intell Agric 5:278–291
74. Sun X-D, Ma J-Y, Feng L-J, Duan J-L, Xie X-M, Zhang X-H, Kong X, Ding Z, Yuan X-Z (2023) Magnetite nanoparticle coating chemistry regulates root uptake pathways and iron chlorosis in plants. Proc Natl Acad Sci 120:e2304306120
75. Taha NA, Hussein AA, El-Hak HNG, El-Shenawy NS (2025) A mini-review of nanoparticle therapeutics targeting oxidative stress and inflammation in diabetes. J Basic Appl Zool 86:29
76. Trivedi R, Upadhyay TK, Mujahid MH, Khan F, Pandey P, Sharangi AB, Muzammil K, Nasir N, Hassan A, Alabdallah NM (2022) Recent advancements in plant-derived nanomaterials research for biomedical applications. Processes 10:338
77. Upadhyay SK, Kumar S, Rout C, Vashistha G, Aggarwal D (2022) Risks and concerns of use of nanoparticles in agriculture. In: The role of nanoparticles in plant nutrition under soil pollution: nanoscience in nutrient use efficiency. Springer, pp 371–394

78. Usman M, Farooq M, Wakeel A, Nawaz A, Cheema SA, ur Rehman H, Ashraf I, Sanaullah M (2020) Nanotechnology in agriculture: Current status, challenges and future opportunities. Sci Total Environ 721:137778
79. Vemula M, Reddy AVB (2023) Polymeric nanoparticles as effective delivery systems in agriculture sustainability. Nanotechnol Environ Eng 8:805–814
80. Vijayakumar M, Surendhar G, Natrayan L, Patil PP, Ram PB, Paramasivam P (2022) Evolution and recent scenario of nanotechnology in agriculture and food industries. J Nanomater 2022:1280411
81. Vijayakumar V, Balakrishnan N (2021) RETRACTED ARTICLE: Artificial intelligence-based agriculture automated monitoring systems using WSN. J Ambient Intell Humaniz Comput 12:8009–8016
82. Wang P, Lombi E, Zhao F-J, Kopittke PM (2016) Nanotechnology: a new opportunity in plant sciences. Trends Plant Sci 21:699–712
83. Wang S, Liu Z, Wang W, You H (2017) Fate and transformation of nanoparticles (NPs) in municipal wastewater treatment systems and effects of NPs on the biological treatment of wastewater: a review. RSC Adv 7:37065–37075
84. Wang X, Xie H, Wang P, Yin H (2023) Nanoparticles in plants: uptake, transport and physiological activity in leaf and root. Materials 16:3097
85. Zaman W, Ayaz A, Park S (2025) Nanomaterials in agriculture: a pathway to enhanced plant growth and abiotic stress resistance. Plants 14:716
86. Zhang P, Guo Z, Ullah S, Melagraki G, Afantitis A, Lynch I (2021) Nanotechnology and artificial intelligence to enable sustainable and precision agriculture. Nature Plants 7:864–876
87. Zhang X, Yang P, Lu B (2024) Artificial intelligence in soil management: The new frontier of smart agriculture. Adv Resour Res 4:231–251
88. Zhu L, Chen L, Gu J, Ma H, Wu H (2022) Carbon-based nanomaterials for sustainable agriculture: their application as light converters, nanosensors, and delivery tools. Plants 11:511

Pathogen-Resilient Crops: Highlighting the Impact of AI-Driven Nanotechnology

Rachid Ez-zouggari, Hajar Zennouhi, Hakima Achetoui, Imane Lebyed, and Rachid Lahlali

Abstract Climate change is accelerating the emergence and spread of plant pathogens, threatening global food security and ecosystem stability. The development of pathogen-resilient crops is therefore essential to meet the growing demand for sustainable agricultural systems. This chapter explores how the convergence of artificial intelligence (AI) and nanotechnology can revolutionize plant protection by enabling early detection, intelligent intervention, and adaptive disease management. Nanotechnology offers a powerful toolbox for the delivery of nanoscale materials with antifungal, antibacterial, and antiviral properties. Engineered nanoparticles, ranging from metallic to polymeric and carbon-based forms, can directly inhibit pathogens or enhance plant innate immunity. Meanwhile, AI technologies such as machine learning, computer vision, and predictive modeling are transforming how we diagnose plant diseases, analyze big data, and optimize agricultural decisions. When integrated, AI-driven nanotechnology enables the development of smart nanosensors capable of detecting pathogens in real time, intelligent delivery systems that release nano-biopesticides in response to biotic signals, and adaptive systems that learn and improve disease control strategies over time. This chapter reviews the current state of AI-nanotech innovations applied to phytopathology, presents successful case studies, and analyzes their potential in the context of climate-smart agriculture. The chapter concludes by addressing key challenges, including biosafety, environmental impact, regulatory frameworks, and ethical considerations. It advocates for interdisciplinary collaboration to develop scalable, low-risk, and accessible solutions that strengthen crop resilience while supporting ecological sustainability.

Keywords Smart nanosensors · Pathogen detection · Precision agriculture · Sustainable crop protection · Climate-smart agriculture

---

R. Ez-zouggari · H. Zennouhi · H. Achetoui · I. Lebyed · R. Lahlali (✉)
Phytopathology Unit, Department of Plant Protection, Ecole Nationale d'Agriculture de Meknes, Meknes, Morocco
e-mail: rlahlali@enameknes.ac.ma

R. Ez-zouggari
Laboratory of Biotechnology, Conservation and Valorization of Natural Resources (LBCVNR), Faculty of Sciences Dhar el Mehraz, Sidi Mohamed ben Abdallah University, Fez, Morocco

J.-T. Chen (ed.), *AI-Nano-Enabled Climate-Smart Agriculture and Sustainability*, Smart Nanomaterials Technology, https://doi.org/10.1007/978-981-92-1694-9_8

1 Introduction

Modern agriculture faces unprecedented challenges, exacerbated by climate change, population growth, and rapidly evolving plant pathogens. Whether fungal, bacterial, or viral, these pathogens severely compromise agricultural productivity, alter crop quality, and threaten global food security [55, 132]. According to the FAO, 20–40% of global agricultural production is lost each year due to diseases, pests, and weeds, causing economic losses estimated at $220 billion for plant diseases and $70 billion for invasive insects. This burden, exacerbated by climate change, increased international trade, and ecosystem degradation, seriously jeopardizes global food security. While plants are the basis of human food and livelihoods for millions of people, particularly in rural areas, their health remains largely neglected. However, once established, pathogens are difficult to eradicate and costly to control, making prevention and strengthening crop resilience essential in the face of growing biotic threats [50].

In this context, the convergence of nanotechnology and AI is opening up new opportunities to modernize agriculture and overcome the limitations of conventional approaches. Initially introduced in developed countries, the integration of these technologies was aimed at improving productivity while addressing environmental challenges. Given these promising results, many developing countries are now exploring these innovations, particularly to strengthen crop protection and the sustainability of agricultural systems [34]. Plant resilience to disease traditionally relies on strategies such as varietal selection, plant protection products, cultural practices, and biocontrol. However, these approaches have limitations: increased pathogen resistance, ecotoxic effects, high costs, and limited accessibility for small farmers. Furthermore, the complexity of plant epidemics, often influenced by variable environmental factors, now calls for predictive, intelligent, and adaptive solutions capable of responding in real time to biotic threats [5, 78]. It is in this context that nanotechnology, defined as the engineering of materials at the nanoscale (1–100 nm), is a promising lever for strengthening plants' resistance to disease. Nanomaterials (metallic, polymeric, or carbon-based) have unique properties, such as high specific surface area, chemical reactivity, and improved bioavailability that enable them to induce plant defenses (SAR, ISR, PR proteins) or act directly on pathogens [20, 111]. They can also deliver active agents (fungicides, interfering RNAs, bio stimulants) in a targeted and controlled manner, thereby increasing their effectiveness at low doses [71]. More recently, the integration of smart nanosensors enables early detection of infections and triggers an automated response, while the use of nanoparticles as non-viral vectors improves the targeted delivery of gene editing systems (e.g., CRISPR/Cas) to create more resistant varieties without transgenesis [76]. These innovations pave the way for precision agriculture that is sustainable and better equipped to deal with emerging pathogens [33, 118]. At the same time, AI offers new prospects for improving crop resilience to disease. Thanks to its ability to analyze vast sets of genomic, transcriptomic, and environmental data, AI can identify multivariate correlations between plant molecular profiles and their response to pathogens, which are often invisible to conventional human analysis [18, 116]. Machine learning models, such as deep neural networks (DNNs), random forests, and support vector machines (SVMs), can predict plant–pathogen interactions and infection risks by taking into account multiple variables such as meteorological data, soil characteristics, and phenotypic data [104]. This predictive capability facilitates marker-assisted selection and rapid identification of resistant genotypes in plant breeding programs [57]. Furthermore, the integration of multi-omics approaches through intelligent algorithms facilitates understanding of the complex interactions between the plant, the pathogen, and the environment, which provides a valuable basis for the development of pathogen-resilient crops. Thus, AI represents a strategic lever for addressing the challenges posed by new epidemics, climate change, and pressure on chemical inputs [139]. Thus, the combined use of AI and nanotechnology is revolutionizing plant pathology by offering predictive, real-time, and sustainable tools [160]. AI enables early and accurate disease diagnosis through deep learning models and phenotypic image recognition, while nanotechnology enhances plant defenses and enables smart delivery systems. Together, these approaches enable precision agriculture systems in which nanosensors provide continuous crop health data, interpreted by AI for targeted and timely interventions fostering resilient, high-yield agriculture under increasing global pressures [140, 144].

This chapter explores how the application of nanotechnology in combination with artificial intelligence could strengthen crop resistance to plant diseases. This text discusses the use of major nanomaterials in agriculture, such as metal, polymer, and carbon nanoparticles. It reveals their properties, including their ability to destroy pathogenic structures, generate reactive oxygen species, and trigger plant defense mechanisms. It also discusses their role in the targeted and controlled release of active agents. The combination of nanotechnology and artificial intelligence can lead to intelligent crop protection systems. Nanotechnologies use nanosensors, reactive nano-delivery platforms, and adaptive feedback systems. These innovative concepts have proven their effectiveness in real-world practice through practical case studies, such as the detection of fusarium wilt in bananas or the monitoring of tomato late blight. Finally, the chapter addresses the environmental, technical, and ethical issues that arise when using these technologies and emphasizes how important it is for scientists, engineers, and agricultural stakeholders to work together across disciplines to ensure sustainable and inclusive implementation. AI-based nanotechnology is a revolutionary strategy for producing pathogen-resistant crops and building a stronger, more flexible, and more sustainable future for agriculture.

2 Nanotechnology for Plant Disease Management

2.1 *Overview of Nanomaterials Used in Agriculture*

Nanotechnology is based on the manipulation of materials at the nanometric scale, generally ranging from 0.1 to 100 nm, and offers a wide array of potential applications in various fields of everyday life. It highlights the use of submicron particles, molecules, or even isolated atoms within biological, chemical, and physical systems [45]. Nanoparticles (NPs) are characterized by their large specific surface area and nanometric size, making them increasingly important for many applications [17]. In agriculture, nanotechnology finds applications in diverse domains, notably through the development of synthetic nano fertilizers and nano pesticides that enhance the biological efficiency of active compounds, as well as the design of alternative nano formulations, as well as nano vectors and nanosensors, which offer an innovative, environmentally friendly, and highly effective solution for combating plant pathogens [31]. Their simple synthesis and ease of structural modification enhance their potential as sustainable bioremediation tools (Abd-Elsalam et al. 2020). The adoption of nanotechnology-based plant protection strategies has grown rapidly, leading to significant improvements in agricultural productivity. Advances in research have led to the development of new formulations incorporating a variety of metallic nanoparticles, while polymeric nanoparticle-based carriers and carbon-based nanomaterials have emerged as key components of these innovations. Together, these nanostructures play an essential role in modern crop protection strategies. Their diverse applications are illustrated in Fig. 1, which shows the main functions of NPs in plant protection [125].

Metallic NPs, especially those composed of silver (AgNPs), copper (CuNPs), copper oxide (CuO), zinc (ZnNPs), or titanium dioxide (TiO_2) and metal oxides, including zinc oxide (ZnO), are attracting increasing interest in agriculture due to their outstanding antimicrobial and antifungal properties [123]. These NPs can interact with photons. When enough energy is received, an electron is stimulated to a higher energy state, inducing the formation of highly reactive hydroxyl radicals. These radicals then participate in the lysis of pathogens. This mechanism shows that NPs are an effective solution for combating resistance to conventional antibiotics [124]. Fan et al. demonstrated that foliar application of ZnO nanoparticles at a concentration of 0.25 mg/mL effectively controlled *Pseudomonas syringae* by suppressing its biofilm formation pathway [48]. For example, several polysaccharide-based formulations are listed in Table 1 and have been described in the literature for their ability to ensure efficient loading and controlled release of active ingredients [60]. Among these polysaccharides, poly-lactic-co-glycolic acid (PLGA) and chitosan (CH) stand out, with the latter appearing as a particularly promising candidate, notably due to the non-toxicity of its degradation products [74]. ChNPs are distinguished by their significantly superior inhibition of fungal growth compared

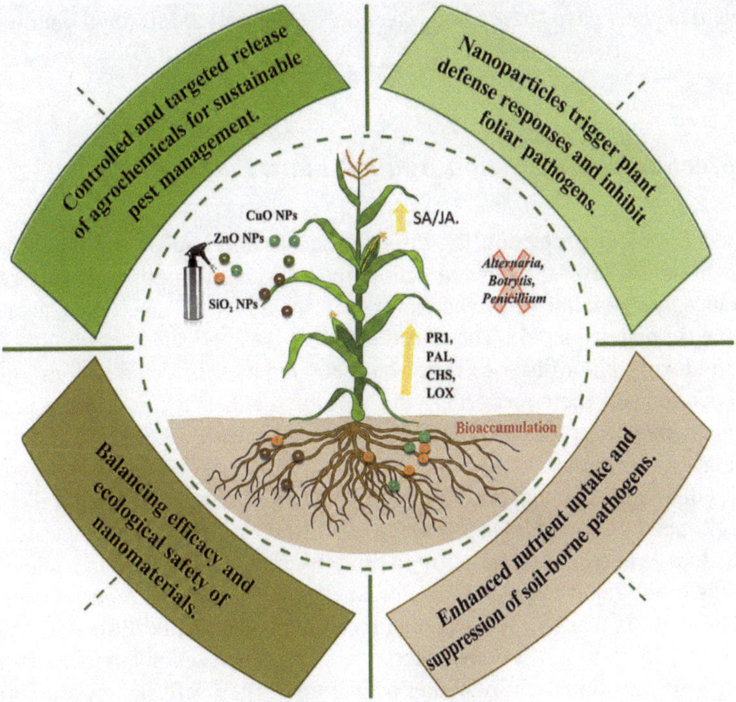

Fig. 1 Nanoparticle applications in agriculture

with bulk chitosan [87]. They represent a promising alternative to chemical pesticides. In addition to their remarkable antifungal activity, they also possess antibacterial, antiviral, nematocidal, and insecticidal properties [103]. Similarly, PLGA-polymeric NPs are attracting growing interest in the agricultural sector for the controlled release of biologically active molecules. Their appeal lies in their exemplary biocompatibility and natural biodegradability: They break down into lactic acid and glycolic acid, which are metabolized without toxic effect [26]. In tomato plants, García-Sánchez et al. reported that the application of polymer-based nanospheres at 70 ppm led to a marked reduction in the characteristic symptoms caused by *Candidatus Liberibacter solanacearum*, along with a significant decrease in bacterial DNA levels after 25 days of treatment [56]. Carbon-based nanomaterials are another class of nanomaterials investigated for the management of plant diseases, including graphene oxide (GO) and carbon dots. GO has a large specific surface area and oxidizing functional groups, which give it high reactivity and the ability to act in a targeted and prolonged manner against pests and diseases [54]. Yi Hao reported that soil application of fullerene (C_{60}) and reduced graphene oxide (rGO) at a concentration of 500 mg·kg^{-1} in rice cultivation resulted in the elimination of the Proteobacteria phylum [62]. In addition, inoculation of C60 fullerenes into *N.*

benthamiana for 5 days suppressed infection by cucurbit chlorophyll yellows virus [6].

2.2 Mechanisms of Action Against Plant Pathogens

NPs used for plant protection act through various mechanisms. Their protective effect is based on several modes of action: the physical disruption of pathogen cell membranes, the generation of reactive oxygen species (ROS), and the activation of plant defense pathways [75]. The effectiveness of these NPs in preventing, controlling, or reducing plant diseases relies on a combination of these actions and may vary depending on their properties, the type of plant, and the specific pathogen with which they interact [22, 77]. NPs can mechanically damage the cell walls of pathogenic fungi due to their small size and large specific surface area, which enhances their adhesion and penetration. This action disrupts the structural integrity of the cell membranes [65]. Once attached to the cell surface, they penetrate microbial membranes and bind to the sulfur in proteins or the phosphorus in nucleic acids. This leads to the rapid degradation of organelles and enzymes, reduces membrane permeability, and causes the leakage of intracellular components, ultimately resulting in cell death [90]. Some NPs do not exert direct antifungal action but rather stimulate the plant's natural defenses by acting as elicitors. They activate several signaling pathways, including those involving ethylene, jasmonic acid, and salicylic acid. This activation triggers a cascade of defense responses, including the synthesis of antimicrobial metabolites, the strengthening of cell walls, and increased expression of defense-related genes, which enhances both induced systemic resistance (ISR) and systemic acquired resistance (SAR). In parallel, these NPs promote the accumulation of pathogenesis-related (PR) proteins, phytoalexins, and key defense enzymes such as peroxidase, catalase, superoxide dismutase (SOD), phenylalanine ammonia-lyase (PAL), and β-1,3-glucanase, which collectively strengthen the plant's innate immunity [38]. Furthermore, NPs induce the production of ROS in pathogen cells, including superoxide anions, hydroxyl radicals, and hydrogen peroxide. These highly reactive molecules cause severe oxidative stress, damaging essential cellular components such as lipids, proteins, and DNA. The resulting processes, lipid peroxidation, protein oxidation, and degradation of genetic material, ultimately lead to cell death [88] (Table 2).

2.3 Nanocarriers for Controlled Release

Nanotransporters are an innovative technology for the targeted, controlled delivery of plant protection agents. They ensure a gradual, tailored release of active ingredients, optimizing their efficacy while limiting losses. Encapsulating fungicides, insecticides, or biocontrol agents in nanostructured matrices protects these compounds from

Table 1 Examples of nanoparticles and their protective effects on plants

Exemple	Application	Pathogen	Results	References
AgNP	Antifungal	*Ustilaginoidea virens/rice*	Better inhibition of the pathogen at 100 ppm (67.11%)	[23]
Fe_3O_4	Antiviral	*Tobacco mosaic virus (TMV)/ tabac*	Induction of antiviral resistance by strengthening plant defense mechanisms, as well as control and prevention of viral infections	[28]
ChNPs	Nematicide	*Meloidogyne incognita/ tomato*	Suppression of egg hatching; reduction in the number of adult females, egg masses, galls, and second-stage juveniles in roots/soil	[96]
PLGA	Antifungal	*P. infestans/S. lycopersicum*	Inhibition of fungal spore release in vitro; suppression of foliar infection in vivo	[54]
SiO_2	Herbicide	*Amaranthus retroflexus/ amaranth, dandelion*	At high doses (2000–4000 mg L^{-1}), they inhibit germination, growth, pigments, proteins, and amino acids of weeds	[126]
CS-Ag NC	Antiviral	*Alfalfa Mosaic Virus (AMV)/ bell pepper (Capsicum annuum)*	Inhibition of the virus up to 91%; increased levels of proline, capsaicin, and phenol in treated plants; protective effect following inoculation	[46]
AgNP	Antifungal	*Bipolaris sorokiniana/* wheat	Inhibits conidial germination and foliar infection	[92]
MWCNT	Antifungal	*A. solani/tomato (Solanum lycopersicum)*	Foliar applications reduce disease incidence and severity by stimulating the antioxidant system	[58]
AgNP	Antibacterial	*Pectobacterium carotovorum/ bell pepper (Capsicum annuum)*	A minimum inhibitory concentration of 0.0625 mg/mL was sufficient to limit the disease to 15%	[15]
AgNP	Insecticides	*Tribolium castaneum/ cereals*	Mortality rates of 55.2, 52.8, and 47.4% at 150 $\mu g/cm^2$	[142]
Cu–ChNPs	Antifungal	*A. Solanum, F. oxysporum/ tomato (Solanum lycopersicum)*	Demonstrated strong antifungal activity, with inhibition rates of up to 73.5% for mycelial, as well as control efficacy against late blight (87.7%) and fusarium wilt (61.1%) under semi-controlled conditions	[121]

(continued)

Table 1 (continued)

Exemple	Application	Pathogen	Results	References
NiO	Insecticides	*Callosobruchus maculatus*/ *Vigna unguiculata*	Significant reduction in fecundity and prolongation of developmental duration in a dose-dependent manner (5–40 ppm)	[112]
GO, CNT	Antimicrobial	*Pectobacterium carotovorum*,	Induce oxidative stress, inhibit fungal and bacterial growth, and stimulate antioxidant defenses	[16]

Table 2 Modes of action of certain NPs against plant pathogens

Mode of action	NPs	Descriptions
Physical disruption of pathogen cell walls	AgNPs	Affects the membrane permeability of *F. oxysporum* spores, leading to their destruction [101]
	CS–Mg	Damages the cell wall and internal organelles of *A. oryzae* and *R. solani*, causing loss of cellular integrity and death of the pathogens [4]
	Ag-TNTs	They cause lysis of Botrytis cinerea conidia by disrupting their membrane integrity [120]
	GO–CuO	They induce the formation of pores and cavities in the membranes of fungal cells *F. oxysporum*, leading to leakage of cellular components and cell death [44]
Generation of reactive oxygen species	CuO NPs	They induce oxidative stress causing lipid peroxidation and organelle destruction in *Aspergillus niger* (A. [131])
	TiO_2 NPs	Generate ROS under light exposure, inducing oxidative stress that damages cellular components, leading to growth inhibition or pathogen death [83]
	AgNPs	They catalyze the production of O_2^-, OH, H_2O_2 causing oxidation of proteins and membrane lipids of *F. oxysporum* and *A. flavus* [38]
Induction of plant defense pathways	CuO	They induce oxidative stress in *A. niger*, leading to lipid peroxidation and cell death [24]
	ChNP	They stimulated the expression of catalase and polyphenol oxidase in banana plants infected with banana streak virus [2]
	CuZn	They induced a strong production of SOD and PAL in pea (*Pisum sativum*), enhancing resistance against *F. solani* [113]
	CS-SiO_2	Activates the natural defenses of tomato plants against *Phytophthora infestans*, reducing late blight symptoms [98]

degradation, improves their bioavailability, and reduces their dispersal to non-target organisms [9, 86]. Designed to react to specific signals (pH variations, temperature changes or biochemical signals emitted by pathogens), these systems release active ingredients only when infection conditions are met. This precision of action makes it possible to reduce the doses applied, limit off-target dispersion, and reduce environmental impact, while maintaining or even enhancing the effectiveness of treatments [11]. Several studies highlight the potential of this technology: Encapsulating hexaconazole in chitosan NPs enabled prolonged release and enhanced antifungal efficacy against *F. oxysporum* [65]. Alginate NPs loaded with carbendazim improved antifungal activity against *A. solani* [122]. Mesoporous silica NPs incorporating prochloraz provided long-lasting protection against *M. oryzae* [11], halloysite nanotubes served as effective carriers for the controlled release of pesticides, thereby reducing their dispersion in the environment [91]. Moreover, ch-AgNP nanocomposites exhibited significant inhibition of *Penicillium expansum* [1]. Regarding stimuli-responsive controlled release, NPs engineered to respond to acidic pH successfully targeted validamycin delivery against *R. solani* [43]. Additionally, pH-sensitive PLGA-based nano formulations enabled controlled release of fungicides in the event of infection. [38]. Concurrently, polymeric nano capsules delivering RNA interference (RNAi) molecules have inhibited the expression of pathogenic genes [88], and Fe_3O_4/ZnO/AgBr-photoactivated nanocomposites have released antifungal ions effective against *F. graminearum* [7]. In *mesoporous organosilica* NPs (MON) encapsulated the fungicide cymoxanil (CYM) was encapsulated with salicylic acid (SA). This encapsulation increased the photostability of CYM and boosted its antifungal efficacy, achieving 86.22% protection against cucumber downy mildew [154]. Kashyap et al. also show that chitosan NPs encapsulating mancozeb have shown superior efficacy against *Phytophthora infestans* on tomato [73].

3 Artificial Intelligence in Plant Pathogen Detection and Management

3.1 Machine Learning and Computer Vision in Disease Detection

3.1.1 Convolutional Neural Networks (CNNs)-Based Image-Based Diagnosis

Conventional human visual inspection-based plant disease diagnosis is subjective as well as time-consuming and error-prone in the case of large area cropping. Deep learning, in particular, convolutional neural networks (CNNs), has revolutionized this procedure, providing automated and accurate detection of plant pathogens from leaf and stem images [51]. CNNs can learn hierarchical features from raw images, such as color variation, texture, and lesion patterns. These are more efficient than handcrafted methods. Research has revealed classification accuracies of more than 95% for several crops and diseases, which include wheat rust, grapevine powdery mildew, and potato late blight [95]. More recently, [82] tested CNN models in the field, demonstrating application robustness to real-world variations, such as noise, shading, occluded leaves, and environmental variation.

These can, and are now, largely being incorporated into mobile apps, which can be used by farmers who capture images of leaves and provide immediate diagnosis and management advice [51]. Efficient CNN architectures like Res Net, Inception, and Efficient Net have the potential to enable disease detection to be democratized even inside resource-poor farming systems with limited resources, as they can be run efficiently on mobile phones, deep learning, and computer vision in plant disease detection: a comprehensive review of techniques, models, and trends in precision agriculture[162].

3.1.2 Real-Time Drone and Satellite Surveillance

Multispectral cameras are widely employed in agricultural analytics, as they capture images with high-spatial resolution and measure near-infrared reflectance, thereby improving biomass estimation [36]. Following this, drones equipped with thermal and hyperspectral sensors are the most powerful techniques for detecting and identifying early-stage pathogen contagions in crops, facilitating large-scale disease detection and monitoring [66]. These technologies are progressively combined with computer vision models capable of analyzing anomalies in plant physiology, such as chlorophyll fluorescence, canopy temperature, or variations in Normalized Difference Vegetation Index (NDVI), which can indicate the early signal of pathogen attack [143]. For instance, [159] showed that deep learning applied to drone images detected grapevine downy mildew with over 92% accuracy several days before visible symptoms appeared. Similarly, satellite-based imagery, while offering lower resolution, provides regional or national surveillance of epidemics, supporting governmental monitoring programs and early warning systems. These approaches extend AI-based diagnostics from individual farms to landscape-level plant health surveillance [130].

3.2 Predictive Modeling for Epidemic Outbreaks

3.2.1 Epidemiological Models Enhanced by AI Interactions

As it concerns the hybrid models which integrate AI and machine learning, the application of algorithmic methods such as Bayesian networks, random forests, and recurrent neural networks is instrumental to the detection of patterns in complex datasets which are critical to the impact of an epidemic (Journal & [70]). Moreover, these methods can be further improved through real-time data and making predictions based on previous data trends [114].

- Data on Human Mobility: Acquire and integrate human mobility datasets which include travel, commuting, and migration patterns as they are critical to modeling disease transmission and progression at subnational and global levels [157].

- Social and Economic Statistics: Collect economic and demographic data including population density, health care, and public health facilities, as well as hospital admissions associated with early-symptom reports. These data can impact the likelihood of infection as well as disease transmission [109]. These data assists health officials in identifying priority areas for targeted interventions (Rama Krishna Debbadi & Obed [115]).
- Accumulation of Epidemiology Datasets: To create an essential reference for understanding chronological patterns of disease evolution, it is necessary past datasets on epidemics events, including patients' files, geographical data, social statistics, and outbreak chronologies [157].
- Acquisition of Ecological Datasets: It is vital to analyzing the potential effect of multiple factors on pathogen propagation, such as climate conditions, pollution levels, hydric sources, and land exploitation [10].

Recurrent neural networks (RNNs) and long short-term memory (LSTM) models are effective in the time forecasting of epidemics. LSTM models are capable of capturing temporal dependencies in sequential data and thus are appropriate for the prediction of diseases over time [29]. For example, K means clustering aids public health officials in the identification and intervention in the appropriate locations by grouping areas with similar epidemiological features [115]. There is a growing interest in forecasting epidemics using hybrid models which are AI mechanistic with epidemiological models. These models utilize the interpretability of the compartmental models and the accuracy of the ML techniques; for example, some hybrid AI-epidemiological models utilize social mobility indicators alongside case incidence reports to forecast the transmission of COVID-19 [157]. In AI models, interpretability and case-by-case analytics are helpful to epidemiologists and policymakers while accurate hinges on regular updates of high-caliber data [10]. Trust in AI-based outbreak prediction systems will require the mitigation of biases in the training data and foster explainability [53]. Another study we can cite by [94] reports a machine learning model for detecting a pathogen outbreak in wheat crops projecting a performance of 80% for model prediction and well outperforming the traditional models which averaged 30%. These systems can predict epidemic outbreaks and simulate scenario management, including the strategic application of fungicides.

3.2.2 Disease Risk Forecasting Using Weather and Crop Data

The nature of numerous plants disease has the nature processes, and their dependence on multiples variables can disrupt our comprehension. Therefore, the predictive capacity of numerous statistics and mathematics techniques is naturally limited [13]. This encouraged researchers to initially study solutions based on analytical and simulation models using weather and crop data, which play an important role in predicting crop diseases and their economic impacts [21]. To understand and evaluate innovative models and interaction among plants, recent research aimed to focus increasingly on ML and AI methods which can be defined with defined by mathematical formulation [84, 152]. Commonly, to generate more precise estimation and to detect innovative dynamics, which are practical to make sense for the complicated mechanisms, the researchers prefer to use per hour meteorological data in place of day-to-day meteorological data.

For example, the researchers created a combination of the support vector machine (SVM) and logistic regression (LR) algorithms to forecast fungal mildew infection in tomato crops and reduce the errors in data through the adaptive sampling-based noise reduction (ANR) technique. Nevertheless, the study did not apply any variable selection algorithm to detect the most important parameters [152]. Additionally applied the artificial neural network analysis (ANN) algorithm to discriminate and categorize fungal diseases in oil palm trees at the preliminary stage using first [152] and second derivative spectroradiometer datasets [158], additional research confirmed the development of strawberry leaf anthracnose through examining multispectral images and identifying a group of spectral ranges that function to separate between a normal and an asymptomatic leaf, achieving to classification results of 88% rate via a progressive classification technique [158], in this setting, a significant obstacle is to identify and measure infections across a plant crop before the propagation throughout a greenhouse to allow control in a brief timeframe. Usually, this preventive intervention is performed out via laboratory analysis [39]. The purpose of the scientific study was to establish if it was possible to substitute current biological and laboratory diagnostic systems with an electronic nose.

3.3 Decision Support Systems

3.3.1 AI Platforms for Farm-Level Disease Control Recommendations

AI platforms can be used to propose targeted strategies such as localized fungicide spraying, adoption of tolerant cultivars, or adjustments in irrigation and plant density [79]. A notable example is represented by decision support systems (DSSs), which provide a platform that synthesizes detection and forecasting data, by converting complex artificial intelligence algorithms into practical field-scale guidelines for farmers [153]. Liakos et al. highlighted that DSS based on AI minimized pesticide application by as much as 30% within Mediterranean cereal systems without reducing yield. Smartphone-based DSS tools also offer smallholder farmers with real-time alerts and customized guidelines, reducing the gap between advanced digital tools and practical agricultural management [81].

3.3.2 Integration with Smart Farming and Robotics

Crop protection by ecofriendly methods is an important advancement in AI, with smart agriculture systems. Pesticides use and costs can be decreased by using robotics

Table 3 Applications of AI in plant pathogen detection and management

Domain	AI techniques	Example application	Agricultural impact	References
Detection	CNNs, computer vision	Automatic leaf-based disease diagnosis	Faster, objective disease detection	[105]
Surveillance	Drones, satellites + deep learning	Early downy mildew detection in vineyards	Prevention of epidemic spread	[64]
Predictive modeling	Random forest, Bayesian networks, LSTM	Forecasting wheat rust and potato blight	Improved epidemic preparedness	[68]
DSS	Mobile apps, AI platforms	Farm-level disease control recommendations	Reduced pesticide use	[156]
Robotics	AI-guided sprayers	Selective fungicide application	Cost savings, environmental protection	[19]

sprayers and computer vision guided to identify diseased plants and precisely apply treatments, reducing pesticides usage and costs [27]. Moreover, monitoring temperature, humidity, and soil composition continuously feed data into AI models by Internet of Things (IoT) sensors, ensuring real-time adaptation of disease management strategies. These systems coupled with robotics and DSS platforms, form the backbone of smart farming, enabling adaptive and environmentally friendly agriculture (Table 3) [72].

4 AI-Driven Nanotechnology: Smart Systems for Crop Protection

The integration of nanotechnology with AI is reshaping modern agriculture by enabling smarter, data-driven, and environmentally responsible crop protection. Through real-time sensing, intelligent diagnostics, and targeted nano-delivery systems, AI-driven nanotechnology allows crops to be protected in a precise, adaptive, and sustainable manner. These interconnected technologies form a feedback ecosystem capable of detecting pathogens early, delivering agrochemicals only when and where they are needed and continuously optimizing agricultural decisions [141].

4.1 Smart Nanosensors and AI-Powered Diagnostics

The advent of smart nanosensors, which incorporate AI, is revolutionizing the field of plant disease detection and diagnosis. Unlike conventional approaches, which are often hampered by their slow speed and lack of sensitivity, these systems enable real-time, on-site monitoring with high accuracy [102]. These biosensors, which rely on a variety of nanomaterials such as gold (AuNPs), silver (AgNPs), carbon nanotubes (CNTs), graphene, and metal oxides (SnO_2, TiO_2), transform biorecognition interactions (such as antibody-antigen binding, detection of pathogenic proteins, metabolites, or volatile organic compounds) into measurable electrochemical, optical, or mechanical signals [117]. The integration of AI, particularly through the use of machine learning and deep learning algorithms, offers the possibility of thoroughly analyzing complex signals from sensors, detecting subtle patterns, and reducing experimental noise, leading to improved reliability of the diagnostic process. This combination of nanosensors and AI offers new opportunities for the early detection of fungal, bacterial, and viral pathogens, reducing the need for chemical use and improving crop resistance (Table 4).

Combining nanomaterial-based biosensors with AI, such as machine learning and deep learning, allows for autonomous interpretation of complicated datasets, making it easier to make quick diagnoses and predictions with little human involvement [135]. These solutions work with edge computing and cloud platforms to allow ongoing monitoring, real-time analysis, and scalability across agricultural fields [161]. In general, smart nanosensors powered by AI can rapidly and accurately detect plant infections on-site, offering a promising approach to improve disease management and enhance agricultural sustainability.

4.2 Intelligent Nano-Delivery Platforms

Intelligent nano-delivery systems complement nanosensor networks by ensuring that plant protection agents are released in a controlled, site-specific, and responsive manner. Among these, chitosan nanoparticles (ChNPs) are particularly promising carriers due to their cationic nature, biodegradability, and high solubility under slightly acidic conditions [35]. These properties improve adhesion to leaf surfaces, prolong contact time with plant tissues, and facilitate the uptake of active ingredients through cell membranes [146]. Various synthesis methods such as ionic gelation, emulsion cross linking, and coacervation allow precise tuning of particle size, stability, and release profiles to suit agricultural applications [69]. Both hydrophilic and hydrophobic compounds can be encapsulated, offering versatility for pesticide, fungicide, or nutrient delivery, and the release behavior of these systems depends on pH, particle size, cross-linking density, and enzymatic degradation, often exhibiting a biphasic pattern: An initial burst followed by sustained release [128]. When coupled with AI, these nanocarriers can become "smart delivery systems." For example, data

Table 4 Examples of smart nanosensors applied to plant pathogen detection

Application	Type of nanosensors/ nanomaterial	Target detected	Key performances	References
Bacterial detection (maize)	AuNP-based electrochemical immunosensor	*Pantoea stewartii*	Wide detection range; LOD 20 × lower than ELISA	[30]
Detection of PthA protein (citrus)	AuNP-modified immunosensor	PthA protein (citrus canker)	Range 0.03–100 nM; LOD 0.01 nM; excellent selectivity and stability	[61]
Viral disease detection (citrus, plum)	FET and EGOFET biosensors	SDE1 protein (HLB citrus), Plum Pox Virus (PPV)	Rapid and selective in planta virus detection	[138]
VOC detection (strawberry)	SnO_2 and TiO_2 nanoparticles on carbon electrodes	*Phytophthora cactorum* (p-ethylguaiacol)	LOD of 35 and 62 nM; high selectivity	[49]
Bio-e-nose with SWCNTs and ssDNA	Bioelectronic sensor based on carbon nanotubes + AI	VOCs (e.g., 4-ethylphenol, marker of *P. cactorum*)	Detection from 0.25%–100% saturated vapor; accurate AI modeling (NN + GPR)	[148]
Fungal detection via VOC fingerprints	Carbon nanostructure-modified e-noses	*Rhizopus, Aspergillus*	Differentiation of fungal species by VOC profiles	[59]
Plant virus detection	Graphene oxide-based immunosensor	Groundnut bud necrosis virus (GBNV)	Detection range 0.5–150 ng/mL; LOD ≈ 5.7 ng/mL	[32]
Bacterial pathogen detection	Gold nanoparticles (AuNPs)—colorimetric sensor	*Xanthomonas campestris*	Detection from ~ 100 cells, 100% specificity, 10^2–10^6 CFU	[108]
Viral pathogen detection	Gold nanoparticles (AuNPs) —colorimetric assay	Tomato yellow leaf curl virus (TYLCV)	Sensitivity 5 ng μL^{-1}, detection range 0.75–200 ng μL^{-1}, eliminates need for PCR amplification	[119]

(continued)

Table 4 (continued)

Application	Type of nanosensors/ nanomaterial	Target detected	Key performances	References
Fungal pathogen detection	AuNPs-streptavidin particles—STNPCR-LFBA biosensor	*Alternaria panax Whetz* (ginseng)	Sensitivity 0.01 pg μL^{-1}, 100 × more sensitive than conventional PCR-LFBA, range 0.01 pg μL^{-1}–1 ng μL^{-1}	[149]
Oomycete detection	Cys-functionalized AuNPs/ AuNRs—colorimetric sensor	*P. infestans* (tomato late blight)	$\geq$ 95% detection (lab and field), LOD 0.4 ppm (VOC), smartphone-based platform	[80]

collected by nanosensors can instruct the AI model to trigger or modulate the release of agrochemicals only when a pathogen is detected or environmental stress exceeds a certain threshold. This real-time responsiveness reduces excessive chemical use, limits environmental contamination, and enhances treatment efficiency [97]. Ultimately, AI-integrated ChNP systems exemplify precision nanomedicine for plants delivering the right compound, in the right place, at the right time.

4.3 Data-Driven Feedback Loops

At the heart of AI-driven nanotechnology lies the concept of data-driven feedback loops continuous cycles of sensing, analysis, and actuation. By merging the Internet of Things (IoT), the Internet of Nano-Things (IoNT), and adaptive machine learning systems, modern agriculture can achieve unprecedented precision in real-time decision-making [3].

Nanosensors and nanoactuators monitor pathogen–plant interactions at the molecular level, providing data on soil nutrients, moisture, and pH that guide automatic adjustments in fertilization and irrigation [85]. Smart imaging tools mounted on drones or satellites enhance this capability by enabling large-scale surveillance and early detection of biotic or abiotic stresses [150].

IoT-enabled devices such as pest traps, environmental cameras, and soil probes feed continuous streams of information into AI-driven predictive models [41]. These systems can generate early warnings of disease outbreaks, automate irrigation schedules, and optimize water and nutrient management for climate-resilient farming [47]. In this closed-loop framework, every sensor reading refines the algorithm's predictive accuracy, allowing the system to "learn" and self-correct over time. By linking nanoscale monitoring with real-time analytics, farms evolve into autonomous, adaptive ecosystems that make precise interventions while conserving resources and reducing environmental impact [134].

5 Challenges, Risks, and Ethical Considerations

While the prospects remain positive, the use of nanosensors and AI in agriculture poses some challenges and environmental, technical, economic, legal and ethical risks (Fig. 2). For responsible and sustainable adoption, the consideration of all these dimensions is necessary.

5.1 Environmental and Health Risks

A major challenge lies in understanding the environmental fate and behavior of engineered nanomaterials once introduced into agricultural systems. Recent studies show that factors such as soil pH, texture, and organic matter content can significantly influence nanoparticle mobility, aggregation, and bioavailability [67]. For instance, the dissolution and transformation of ZnO and TiO_2 nanoparticles vary according to soil properties and incubation time, which can affect metal ion release and potential phytotoxicity [151]. Moreover, long-term accumulation of metallic nanoparticles in soil can disrupt microbial diversity, modify nutrient cycling, and affect beneficial interactions such as mycorrhization or nitrogen fixation. Beyond soil interactions, the toxicological effects of nano-enabled agrochemicals on non-target organisms

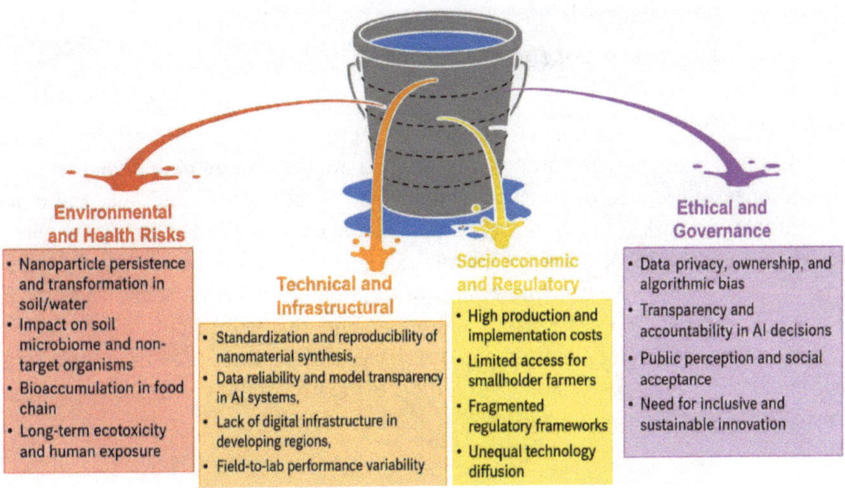

Fig. 2 Major challenges and risks in AI–nanotechnology integration for sustainable agriculture

remain a pressing concern [40]. A systematic review by [107] demonstrated that even at sublethal concentrations, nano pesticides can induce oxidative stress, DNA damage, and cytotoxic responses in aquatic and terrestrial species. Human exposure through food chains or contaminated water also poses potential health risks, including neurotoxicity and endocrine disruption. Despite these findings, data on long-term bioaccumulation and chronic exposure remain scarce, underscoring the need for comprehensive life-cycle assessments and standardized ecotoxicological testing protocols [99].

5.2 Technical and Infrastructural Challenges

From a technological standpoint, reproducibility and scalability remain significant bottlenecks. The synthesis of nanomaterials with consistent physicochemical properties size, morphology, surface charge, and functionalization is difficult to achieve at industrial scale. Small variations can drastically alter their antimicrobial activity or toxicity profiles. Furthermore, the transition from controlled laboratory conditions to open-field environments often results in unpredictable outcomes due to fluctuating temperature, humidity, soil chemistry, and microbial interactions [63]. The integration of AI further complicates implementation. Reliable AI-driven nanosensing and decision systems depend on robust data collection infrastructures, stable connectivity, and advanced computational capacity [102]. However, many agricultural regions, particularly in developing countries, lack such digital infrastructures, creating a technological gap that may slow or limit the adoption of precision nanotechnologies [37].

5.3 Socioeconomic and Regulatory Barriers

These technologies also possess an arguably critical socioeconomic dimension. For example, smallholder farmers are limited to the extent to which they can adopt pioneering technologies, such as nanosensors, advanced materials, and data-gathering frameworks, due to their high production costs [137]. Disparities between industrial and traditional farming systems can generate inequities, as Allan's recent perspective explains. Also, many countries continue to function without cohesive regulatory frameworks to assess and authorize nano-enabled agricultural goods [8]. Moreover, the absence of clearly defined international expectations around labeling, risk evaluation, and post-market surveillance greatly impairs efforts to promote transparency and protect consumers.

5.4 Data Governance and Algorithmic Bias

AI systems used for pathogen prediction, nanomaterial optimization, or precision applications rely heavily on big data analytics. Yet, concerns about data ownership, privacy, and misuse are increasingly prominent [89]. Many current datasets used to train AI models come from large-scale, data-rich farms in temperate regions, resulting in algorithmic bias when applied to small or resource-limited farming contexts. This may lead to inaccurate predictions, unsuitable treatment recommendations, or unintended ecological impacts. Ensuring transparency in model design, traceability in data collection, and equitable access to digital tools is thus vital to avoid reinforcing existing inequalities [136]. Moreover, responsibility and liability in AI-driven decision-making remain unclear. When an automated system recommends a nanoparticle-based treatment that later causes crop damage or environmental contamination, accountability between developers, suppliers, and end-users becomes a legal and ethical dilemma. Establishing clear governance frameworks and participatory data policies will be crucial to address these issues [145].

5.5 Ethical and Sustainability Considerations

Ethical considerations extend beyond technical safety and regulation. The responsible development of AI-guided nanotechnologies must align with principles of transparency, inclusivity, and sustainability. The concept of "technology for good" emphasizes that innovations should benefit not only productivity but also ecological balance, biodiversity preservation, and social equity [106]. This requires integrating diverse stakeholder perspectives of researchers, policymakers, farmers, and local communities throughout the research-to-application pipeline. Equally, public perception and trust play decisive roles in the adoption of such advanced technologies. Insufficient communication about potential risks or exaggerated claims of benefits can erode social acceptance. Participatory risk communication, education, and community engagement are therefore essential to bridge the gap between scientific innovation and societal values [42].

5.6 Toward Responsible Innovation

Addressing the challenges associated with AI-driven nanotechnology requires a coordinated, interdisciplinary approach. Regulatory agencies, research institutions, and industry stakeholders must collaborate to develop international standards for nanomaterial testing, AI transparency, and ethical data management. Long-term monitoring programs should assess environmental persistence, bioaccumulation, and socio-economic impacts under real agricultural conditions [102]. Ultimately, the integration of nanotechnology and AI in agriculture must be guided by a precautionary and participatory framework one that balances innovation with environmental stewardship and human well-being. Only by embedding ethical reflection and sustainability at the core of technological development can AI-driven nanotechnology truly fulfill its promise of creating pathogen-resilient crops for a sustainable future [52].

6 Future Prospects and Opportunities

The future of integrating nanotechnology and AI into agriculture opens a wealth of opportunities to strengthen crop resilience and ensure sustainable food security. The convergence of these two fields promises unprecedented interdisciplinary innovations, particularly through the integration of nanomaterials into image recognition systems, enabling the design of ultra-sensitive sensors capable of providing rapid and accurate diagnoses of plant health [155]. Combined with predictive models and autonomous robotics, these advances pave the way for intelligent platforms for detection and targeted intervention. Concrete experiments are already confirming this potential, such as the use of robots equipped with AI and nanosensors to control powdery mildew in vineyards through the targeted application of nano formulated fungicides, reducing both the incidence of the disease and the excessive consumption of chemicals [100]. Similarly, autonomous robots using computer vision are now capable of selective weeding, helping to reduce the use of herbicides and promote more sustainable agricultural practices [147].

In the short and medium term, nanorobots and biohybrid devices capable of diagnosing and delivering protective agents at the cellular level offer unprecedented therapeutic precision. These tools, combined with machine learning algorithms for optimizing administration routes and doses, could significantly reduce the use of chemical inputs and limit exposure to non-target organisms [93]. These technologies extend beyond plant protection to include other strategic applications, such as artificial pollination. For example, one study demonstrated the effectiveness of microrobots carrying pollen NPs in pollinating apple trees. Other applications include precision agriculture and selective harvesting, improving the sustainability and efficiency of agricultural systems [110]. Optimizing the deployment of smart sensors is a crucial lever for maximizing the efficiency of data collection while reducing costs. AI, via reinforcement learning, genetic algorithms, and particle swarms, is already improving coverage and energy efficiency compared to traditional methods [129]. In this context, AI embedded in low-cost sensors and mobile platforms, coupled with robust nanosensors, could enable real-time local monitoring and provide recommendations tailored to local constraints (cultivated varieties, available resources, climate) [133].

The combination of nanotechnology with genetic engineering, particularly through CRISPR Cas technology, opens up the possibility of developing crops that are more resistant to biotic and abiotic stresses, while optimizing resource use through nano fertilizers, nano pesticides, and nanosensors [14]. The delivery of CRISPR/Cas systems via nanotechnological vectors polymeric, lipid, or carbon-based NPs already offers "DNA-free" solutions for directly introducing RNA–protein complexes (RNPs), promoting targeted and transient modifications to plant resistance. The integration of AI tools to optimize the design of these vectors (size, charge, coating) and predict off-target effects improves the safety and efficacy of these approaches, which are now considered a priority in the literature [127].

The sustainable development of nanotechnology-based AI solutions will depend on enhanced interdisciplinary collaboration between nanomaterials engineers, plant pathologists, AI specialists, sociologists, and policymakers. Recent studies highlight the need for public–private partnerships and standardized pilot tests, including assessments of the fate of NPs in soils and their impacts on non-target microbiomes, before any large-scale deployment [12]. Innovative business models, such as sensor rental, open access platforms, and local training, will be essential to avoid a worsening of technological inequalities and democratize access to these tools [133].

In short, the synergy between nanotechnology and AI points toward a new generation of agriculture based on intelligent, sustainable, and scalable solutions. Realizing these opportunities on a large scale will require close coordination between AI engineers, nanotechnologists, and plant pathologists, as well as a strong commitment from public and private stakeholders, in order to build an agricultural future that is resilient to climate and disease challenges [155].

7 Conclusion

The combination of artificial intelligence and nanotechnology is taking modern agriculture in a new direction, offering a smarter, more precise, and sustainable approach to crop protection. While nanoscale delivery systems ensure the economically viable and environmentally friendly use of agrochemicals, smart nanosensors linked to AI algorithms work together to detect infections at an early stage, forecast outbreaks, and direct focused treatments. With the help of these technologies, plant protection is no longer a reactive process but rather a proactive, self-adaptive system that can react quickly to biological and environmental changes. AI-driven nanotechnology improves crop resilience and lessens reliance on traditional pesticides by encouraging early diagnosis, regulated distribution, and continual learning. However, ethical data use, responsible execution, and fair access to innovation are necessary for its success. In conclusion, this contribution provides a route to intelligent, resilient, and sustainable agriculture that can guarantee food security amidst worldwide.

Acknowledgements This study was financially supported by the Phytopathology Unit of the Ecole Nationale d'Agriculture de Meknès, Morocco.

References

1. Abd-Elsalam KA, Alghuthaymi MA, Shami A, Rubina MS, Abramchuk SS, Shtykova EV, Yu. Vasil'kov A (2020) Copper-chitosan nanocomposite hydrogels against Aflatoxigenic Aspergillus flavus from dairy cattle feed. J Fungi 6(3):112. https://doi.org/10.3390/jof6030112
2. Abdelbaset TE, Alahmari AS, Elkammar HF, Alothaim T, Ahmed NG, Abd El Moneim D, Negm S, Alshaharni MO, El-kott AF, Abou-Zeid MA (2024) Stimulating banana tree resistance to banana streak virus (BSV) disease by chitosan nanoparticles. Eur J Plant Pathol 169(4):825–840. https://doi.org/10.1007/s10658-024-02872-7
3. Aggarwal M, Khullar V, Goyal N, Gautam R, Alblehai F, Elghatwary M, Singh A (2023) Federated transfer learning for rice-leaf disease classification across multiclient cross-silo datasets. Agronomy 13(10):2483. https://doi.org/10.3390/agronomy13102483
4. Ahmed T, Noman M, Luo J, Muhammad S, Shahid M, Ali MA, Zhang M, Li B (2021) Bioengineered chitosan-magnesium nanocomposite: a novel agricultural antimicrobial agent against Acidovorax oryzae and Rhizoctonia solani for sustainable rice production. Int J Biol Macromol 168:834–845. https://doi.org/10.1016/j.ijbiomac.2020.11.148
5. Akhtar H, Usman M, Binyamin R, Hameed A, Arshad SF, Aslam HMU, Khan IA, Abbas M, Zaki HEM, Ondrasek G, Shahid MS (2024) Traditional strategies and cutting-edge technologies used for plant disease management: a comprehensive overview. Agronomy 14(9):2175. https://doi.org/10.3390/agronomy14092175
6. Al-Zaban MI, Alhag SK, Dablool AS, Ahmed AE, Alghamdi S, Ali B, Al-Saeed FA, Saleem MH, Poczai P (2022) Manufactured nano-objects confer viral protection against cucurbit chlorotic yellows virus (CCYV) infecting Nicotiana benthamiana. Microorganisms 10(9):1837. https://doi.org/10.3390/microorganisms10091837
7. Alghuthaymi MA, Kalia A, Bhardwaj K, Bhardwaj P, Abd-Elsalam KA, Valis M, Kuca K (2021) Nanohybrid antifungals for control of plant diseases: current status and future perspectives. J Fungi 7(1):48. https://doi.org/10.3390/jof7010048
8. Allan J, Belz S, Hoeveler A, Hugas M, Okuda H, Patri A, Rauscher H, Silva P, Slikker W, Sokull-Kluettgen B, Tong W, Anklam E (2021) Regulatory landscape of nanotechnology and nanoplastics from a global perspective. Regul Toxicol Pharmacol 122:104885. https://doi.org/10.1016/j.yrtph.2021.104885
9. Ameen F (2022) Optimization of the synthesis of fungus-mediated bi-metallic Ag-Cu nanoparticles. Appl Sci 12(3):1384. https://doi.org/10.3390/app12031384
10. Ameh B (2024) Sustainable supply chains as strategic instruments for environmental protection, public health, and economic resilience. Int J Res Publ Rev 5(12):1192–1208. https://doi.org/10.55248/gengpi.5.1224.3428
11. An C, Sun C, Li N, Huang B, Jiang J, Shen Y, Wang C, Zhao X, Cui B, Wang C, Li X, Zhan S, Gao F, Zeng Z, Cui H, Wang Y (2022) Nanomaterials and nanotechnology for the delivery of agrochemicals: strategies towards sustainable agriculture. J Nanobiotechnol 20(1):1–19. https://doi.org/10.1186/s12951-021-01214-7
12. Ansari MM, Shin M, Kim M, Ghosh M, Kim S-H, Son Y-O (2024) Nano-enabled strategies in sustainable agriculture for enhanced crop productivity: a comprehensive review. J Environ Manage 372:123420. https://doi.org/10.1016/j.jenvman.2024.123420
13. Arun R, Bhuvaneswari R (2019) Buying behavior of meat consumption relates to food safety from north and south part of the Coimbatore city. Int J Recent Technol Eng 7(5):429–433
14. Ashique S, Raikar A, Jamil S, Lakshminarayana L, Gajbhiye SA, De S, Kumar S (2024) Artificial intelligence integration with nanotechnology: a new frontier for sustainable and precision agriculture. Curr Nanosci 21(2):242–273. https://doi.org/10.2174/0115734137275111231206072049
15. Ayisigi M, Cokislerel A, Kucukcobanoglu Y, Yalcin T, Aktas LY (2020) Green synthesized silver nanoparticles for an effective control on soft rodisease pathogen pectobacterium carotovorum and growth stimulation in pepper. Bulgarian J Agr Sci 26(3):574–584

16. Azizi-Lalabadi M, Hashemi H, Feng J, Jafari SM (2020) Carbon nanomaterials against pathogens; the antimicrobial activity of carbon nanotubes, graphene/graphene oxide, fullerenes, and their nanocomposites. Adv Coll Interface Sci 284:102250. https://doi.org/10.1016/j.cis.2020.102250
17. Baker S, Volova T, Prudnikova SV, Satish S, Prasad MN (2017) Nanoagroparticles emerging trends and future prospect in modern agriculture system. Environ Toxicol Pharmacol 53:10–17. https://doi.org/10.1016/j.etap.2017.04.012
18. Banerjee S, Mondal AC (2023) An intelligent approach to reducing plant disease and enhancing productivity using machine learning. Int J Recent Innov Trends Comput Commun 11(3):250–262. https://doi.org/10.17762/ijritcc.v11i3.6344
19. Barbosa Júnior MR, dos Santos RG, de Sales LA, de Oliveira LP (2024). Advancements in agricultural ground robots for specialty crops: an overview of innovations, challenges, and prospects. Plants 13(23). https://doi.org/10.3390/plants13233372
20. Bayda S, Adeel M, Tuccinardi T, Cordani M, Rizzolio F (2019) The history of nanoscience and nanotechnology: from chemical-physical applications to nanomedicine. Molecules 25(1):112. https://doi.org/10.3390/molecules25010112
21. Bhagawati R, Bhagawati K, Singh AKK, Nongthombam R, Sarmah R, Bhagawati G (2015) Artificial neural network assisted weather based plant disease forecasting system. Int J Rec Innov Trends Comput Commun 3(6):4168–4173
22. Bhanja SK, Samanta SK, Mondal B, Jana S, Ray J, Pandey A, Tripathy T (2020) Green synthesis of Ag@Au bimetallic composite nanoparticles using a polysaccharide extracted from Ramaria botrytis mushroom and performance in catalytic reduction of 4-nitrophenol and antioxidant, antibacterial activity. Environ Nanotechnol Monitor Manage 14:100341. https://doi.org/10.1016/j.enmm.2020.100341
23. Bhargava P, Kumar A, Kumar S, Azad CS (2018) Impact of fungicides and nanoparticles on Ustilaginoidea virens causing false smut disease of rice. J Pharmacogn Phytochem 7(1):1541–1544
24. Bhavyasree PG, Xavier TS (2020) Green synthesis of copper oxide/carbon nanocomposites using the leaf extract of Adhatoda vasica Nees, their characterization and antimicrobial activity. Heliyon 6(2):e03323. https://doi.org/10.1016/j.heliyon.2020.e03323
25. Blessing A (2024). Advancing national security and economic prosperity through resilient and technology-driven supply chains. World J Adv Res Rev 24(3):483–500. https://doi.org/10.30574/wjarr.2024.24.3.3723
26. Botanica EDE, De Angelis G (2022) Fisiche e naturali. PhD thesis Tutor
27. Bottacini M, Scollo A, Edwards SA, Contiero B, Veloci M, Pace V, Gottardo F (2018) Skin lesion monitoring at slaughter on heavy pigs (170 kg): welfare indicators and ham defects. PLoS ONE 13(11):1–16. https://doi.org/10.1371/journal.pone.0207115
28. Cai L, Cai L, Jia H, Liu C, Wang D, Sun X (2020) Foliar exposure of Fe_3O_4 nanoparticles on Nicotiana benthamiana: evidence for nanoparticles uptake, plant growth promoter and defense response elicitor against plant virus. J Hazard Mater 393:122415. https://doi.org/10.1016/j.jhazmat.2020.122415
29. Calandra D, Favareto M (2020) Artificial intelligence to fight COVID-19 outbreak impact: an overview. Eur J Soc Impact Circular Econ 1(3):84–104. https://doi.org/10.13135/2704-9906/5067
30. Cardoso RM, Pereira TS, Facure MHM, dos Santos DM, Mercante LA, Mattoso LHC, Correa DS (2022) Current progress in plant pathogen detection enabled by nanomaterials-based (bio)sensors. Sens Actuat Rep 4:100068. https://doi.org/10.1016/j.snr.2021.100068
31. Carrillo-Lopez LM, Villanueva-Verduzco C, Villanueva-Sánchez E, Fajardo-Franco ML, Aguilar-Tlatelpa M, Ventura-Aguilar RI, Soto-Hernández RM (2024) Nanomaterials for plant disease diagnosis and treatment: a review. Plants 13(18):2634. https://doi.org/10.3390/plants13182634
32. Chaudhary M, Verma S, Kumar A, Basavaraj YB, Tiwari P, Singh S, Chauhan SK, Kumar P, Singh SP (2021) Graphene oxide based electrochemical immunosensor for rapid detection of groundnut bud necrosis orthotospovirus in agricultural crops. Talanta 235:122717. https://doi.org/10.1016/j.talanta.2021.122717

33. Chausali N, Saxena J, Prasad R (2023) Nanotechnology as a sustainable approach for combating the environmental effects of climate change. J Agric Food Res 12:100541. https://doi.org/10.1016/j.jafr.2023.100541
34. Chavula P, Kayusi F (2025) Systematic review on the application of nanotechnology and artificial intelligence in agricultural economics. LatIA 3:322. https://doi.org/10.62486/latia2025322
35. Chenxi Z, Hemmat A, Thi N, Afrand M (2025) Nanoparticle-enhanced drug delivery systems: an up-to-date review. J Mol Liq 424:126999. https://doi.org/10.1016/j.molliq.2025.126999
36. Chew R, Rineer J, Beach R, O'neil M, Ujeneza N, Lapidus D, Miano T, Hegarty-Craver M, Polly J, Temple DS (2020) Deep neural networks and transfer learning for food crop identification in UAV images. Drones 4(1):1–14. https://doi.org/10.3390/drones4010007
37. Choruma DJ, Dirwai TL, Mutenje MJ, Mustafa M, Chimonyo VGP, Jacobs-Mata I, Mabhaudhi T (2024) Digitalisation in agriculture: a scoping review of technologies in practice, challenges, and opportunities for smallholder farmers in sub-Saharan Africa. J Agric Food Res 18:101286. https://doi.org/10.1016/j.jafr.2024.101286
38. Cruz-Luna AR, Cruz-Martínez H, Vásquez-López A, Medina DI (2021) Metal nanoparticles as novel antifungal agents for sustainable agriculture: current advances and future directions. J Fungi 7(12). https://doi.org/10.3390/jof7121033
39. Cui S, Ling P, Zhu H, Keener HM (2018) Plant pest detection using an artificial nose system: a review. Sensors (Switzerland) 18(2):1–18. https://doi.org/10.3390/s18020378
40. da Silva Júnior AH, Mulinari J, de Oliveira PV, de Oliveira CRS, Reichert Júnior FW (2022) Impacts of metallic nanoparticles application on the agricultural soils microbiota. J Hazard Mater Adv 7:100103. https://doi.org/10.1016/j.hazadv.2022.100103
41. Domingues T, Brandão T, Ferreira JC (2022) Machine learning for detection and prediction of crop diseases and pests: a comprehensive survey. Agriculture 12(9):1350. https://doi.org/10.3390/agriculture12091350
42. Dong L, Herweijer M, Mcdermott R, Van Den Berg C, Zhang X (2024) The role of trust and risk communication in determining residents' flood risk perception and preparedness intention: evidence from Wuhan, China. https://doi.org/10.22541/au.173438190.09628114/v1
43. Dutta P, Kumari A, Mahanta M, Biswas KK, Dudkiewicz A, Thakuria D, Abdelrhim AS, Singh SB, Muthukrishnan G, Sabarinathan KG, Mandal MK, Mazumdar N (2022) Advances in nanotechnology as a potential alternative for plant viral disease management. Front Microbiol 13. https://doi.org/10.3389/fmicb.2022.935193
44. El-Abeid SE, Ahmed Y, Daròs J-A, Mohamed MA (2020) Reduced graphene oxide nanosheet-decorated copper oxide nanoparticles: a potent antifungal nanocomposite against fusarium root rot and wilt diseases of tomato and pepper plants. Nanomaterials 10(5):1001. https://doi.org/10.3390/nano10051001
45. El-Baky NA, Amara AAAF (2021) Recent approaches towards control of fungal diseases in plants: an updated review. J Fungi 7(11):900. https://doi.org/10.3390/jof7110900
46. El-Ganainy SM, Soliman AM, Ismail AM, Sattar MN, Farroh KY, Shafie RM (2023) Antiviral activity of chitosan nanoparticles and chitosan silver nanocomposites against Alfalfa Mosaic Virus. Polymers 15(13):2961. https://doi.org/10.3390/polym15132961
47. Eze VHU, Eze EC, Alaneme GU, Bubu PE, Nnadi EOE, Okon MB (2025) Integrating IoT sensors and machine learning for sustainable precision agroecology: enhancing crop resilience and resource efficiency through data-driven strategies, challenges, and future prospects. Discov Agric 3(1):83. https://doi.org/10.1007/s44279-025-00247-y
48. Fan G, Xiao Q, Li Q, Xia Y, Feng H, Ma X, Cai L, Sun X (2023) Antimicrobial mechanisms of ZnO nanoparticles to phytopathogen Pseudomonas syringae: Damage of cell envelope, suppression of metabolism, biofilm and motility, and stimulation of stomatal immunity on host plant. Pestic Biochem Physiol 194:105455. https://doi.org/10.1016/j.pestbp.2023.105455
49. Fang Y, Umasankar Y, Ramasamy RP (2014) Electrochemical detection of p-ethylguaiacol, a fungi infected fruit volatile using metal oxide nanoparticles. Analyst 139(15):3804–3810. https://doi.org/10.1039/C4AN00384E
50. FAO (2020) International year of plant health

51. Ferentinos KP (2018) Deep learning models for plant disease detection and diagnosis. Comput Electron Agric 145(January):311–318. https://doi.org/10.1016/j.compag.2018.01.009
52. Feroz F, Mehraj M, Kumar S, Wani AW, Saini N, Chaudhuri M (2025). Nanotechnology and artificial intelligence to enable sustainable and precision agriculture. In: Nanomaterials for enhanced plant-based food production. Elsevier, pp 227–241. https://doi.org/10.1016/B978-0-443-23688-4.00006-3
53. Francisca CU, Nwankwo EI, Igwama GT, Olaboye JA (2023) Real-Time data integration in diagnostic devices for predictive modeling of infectious disease outbreaks. Comput Sci IT Res J 4(3):525–545. https://doi.org/10.51594/csitrj.v4i3.1502
54. Fukamachi K, Konishi Y, Nomura T (2019) Disease control of Phytophthora infestans using cyazofamid encapsulated in poly lactic-co-glycolic acid (PLGA) nanoparticles. Colloids Surf, A 577:315–322. https://doi.org/10.1016/j.colsurfa.2019.05.077
55. Gai Y, Wang H (2024) Plant disease: a growing threat to global food security. Agronomy 14(8):1615. https://doi.org/10.3390/agronomy14081615
56. García-Sánchez AN, Yáñez-Macias R, Hernández-Flores JL, Álvarez-Morales A, Valenzuela-Soto JH, Guerrero-Sanchez C, Guerrero-Santos R (2021) Exogenous application of polycationic nanobactericide on tomato plants reduces the candidatus liberibacter solanacearum infection. Plants 10(10):2096. https://doi.org/10.3390/plants10102096
57. Garrett KA, Bebber DP, Etherton BA, Gold KM, Plex Sulá AI, Selvaraj MG (2022) Climate change effects on pathogen emergence: artificial intelligence to translate big data for mitigation. Annu Rev Phytopathol 60(1):357–378. https://doi.org/10.1146/annurev-phyto-021021-042636
58. González-García Y, Cadenas-Pliego G, Alpuche-Solís ÁG, Cabrera RI, Juárez-Maldonado A (2021) Carbon nanotubes decrease the negative impact of Alternaria solani in Tomato crop. Nanomaterials 11(5):1080. https://doi.org/10.3390/nano11051080
59. Greenshields M, Cunha B, Coville N, Pimentel I, Zawadneak M, Dobrovolski S, Souza M, Hümmelgen I (2016) Fungi active microbial metabolism detection of Rhizopus sp. and Aspergillus sp. section Nigri on strawberry using a set of chemical sensors based on carbon nanostructures. Chemosensors 4(3):19. https://doi.org/10.3390/chemosensors4030019
60. Gupta A, Dhiman A, Sood A, Bharadwaj R, Silverman N, Agrawal G (2023) Dextran/eudragit S-100 based redox sensitive nanoparticles for colorectal cancer therapy. Nanoscale 15(7):3273–3283. https://doi.org/10.1039/D3NR00248A
61. Haji-Hashemi H, Norouzi P, Safarnejad MR, Larijani B, Habibi MM, Raeisi H, Ganjali MR (2018) Sensitive electrochemical immunosensor for citrus bacterial canker disease detection using fast Fourier transformation square-wave voltammetry method. J Electroanal Chem 820:111–117. https://doi.org/10.1016/j.jelechem.2018.04.062
62. Hao Y, Ma C, Zhang Z, Song Y, Cao W, Guo J, Zhou G, Rui Y, Liu L, Xing B (2018) Carbon nanomaterials alter plant physiology and soil bacterial community composition in a rice-soil-bacterial ecosystem. Environ Pollut 232:123–136. https://doi.org/10.1016/j.envpol.2017.09.024
63. Herdiana Y, Wathoni N, Shamsuddin S, Muchtaridi M (2022) Scale-up polymeric-based nanoparticles drug delivery systems: development and challenges. OpenNano 7:100048. https://doi.org/10.1016/j.onano.2022.100048
64. Hernández I, Silva R, Melo-Pinto P, Gutiérrez S, Tardaguila J (2025) Early detection of downy mildew in vineyards using deep neural networks for semantic segmentation. Biosyst Eng 252:15–31. https://doi.org/10.1016/j.biosystemseng.2025.02.007
65. Ingle AP (ed) (2021) Nanotechnology in plant growth promotion and protection. Wiley. https://doi.org/10.1002/9781119745884
66. Iqbal Z, Khan MA, Sharif M, Shah JH, ur Rehman MH, Javed K (2018) An automated detection and classification of citrus plant diseases using image processing techniques: a review. Comput Electron Agric 153:12–32. https://doi.org/10.1016/j.compag.2018.07.032
67. Irumva O, Twagirayezu G, Xia A, Uwimpaye F, Nizeyimana JC, Nizeyimana I, Uwimana A, Birame CS (2025) Environmental fate, transport, impacts, and future perspectives of engineered nanoparticles in surface waters. Environ Res 285:122267. https://doi.org/10.1016/j.envres.2025.122267

68. Jensen A, Brown P, Groves K, Morshed A (2025) Next generation crop protection: A systematic review of trends in modelling approaches for disease prediction. Comput Electron Agric 234(February):110245. https://doi.org/10.1016/j.compag.2025.110245
69. Jha R, Mayanovic RA (2023) A review of the preparation, characterization, and applications of chitosan nanoparticles in nanomedicine. Nanomaterials 13(8):1302. https://doi.org/10.3390/nano13081302
70. Journal I, Ahamed G (2025) Advancements in machine learning and deep learning for plant disease detection: a comprehensive survey of techniques. Datasets Fut Direct. https://doi.org/10.17577/IJERTV14IS020069
71. Kah M, Kookana RS, Gogos A, Bucheli TD (2018) A critical evaluation of nanopesticides and nanofertilizers against their conventional analogues. Nat Nanotechnol 13(8):677–684. https://doi.org/10.1038/s41565-018-0131-1
72. Kamilaris A, Prenafeta-Boldú FX (2018) Deep learning in agriculture: a survey. Comput Electron Agric 147:70–90. https://doi.org/10.1016/j.compag.2018.02.016
73. Kashyap P, Shirkot P, Das R, Pandey H, Singh D (2023) Biosynthesis and characterization of copper nanoparticles from Stenotrophomonas maltophilia and its effect on plant pathogens and pesticide degradation. J Agric Food Res 13:100654. https://doi.org/10.1016/j.jafr.2023.100654
74. Khan A, Alamry KA (2021) Recent advances of emerging green chitosan-based biomaterials with potential biomedical applications: a review. Carbohyd Res 506:108368. https://doi.org/10.1016/j.carres.2021.108368
75. Khan AU, ur Rahman A, Yuan Q, Ahmad A, Khan ZUH, Mahnashi MH, Alyami BA, Alqahtani YS, Ullah S, Wirman AP (2020) Facile and eco-benign fabrication of Ag/Fe_2O_3 nanocomposite using Algaia Monozyga leaves extract and its' efficient biocidal and photocatalytic applications. Photodiagn Photodyn Therapy 32:101970. https://doi.org/10.1016/j.pdpdt.2020.101970
76. Kim M, Hwang Y, Lim S, Jang H-K, Kim H-O (2024) Advances in nanoparticles as non-viral vectors for efficient delivery of CRISPR/Cas9. Pharmaceutics 16(9):1197. https://doi.org/10.3390/pharmaceutics16091197
77. Krishnani KK, Boddu VM, Chadha NK, Chakraborty P, Kumar J, Krishna G, Pathak H (2022) Metallic and non-metallic nanoparticles from plant, animal, and fisheries wastes: potential and valorization for application in agriculture. Environ Sci Pollut Res 29(54):81130–81165. https://doi.org/10.1007/s11356-022-23301-4
78. Lahlali R, Ezrari S, Radouane N, Kenfaoui J, Esmaeel Q, El Hamss H, Belabess Z, Barka EA (2022) Biological control of plant pathogens: a global perspective. Microorganisms 10(3):596. https://doi.org/10.3390/microorganisms10030596
79. Lamichhane JR, You MP, Barbetti MJ (2020) Benefits of fungicide seed treatment rate and volume of fungicides used for seed treatment. Plant 3:1–14
80. Li Z, Paul R, Ba Tis T, Saville AC, Hansel JC, Yu T, Ristaino JB, Wei Q (2019) Non-invasive plant disease diagnostics enabled by smartphone-based fingerprinting of leaf volatiles. Nature Plants 5(8):856–866. https://doi.org/10.1038/s41477-019-0476-y
81. Liakos KG, Busato P, Moshou D, Pearson S, Bochtis D (2018) Machine learning in agriculture: a review. Sensors (Switzerland) 18(8):1–29. https://doi.org/10.3390/s18082674
82. Liu J, Sun H, Katto J (2023) Learned image compression with mixed transformer-CNN architectures. In: Proceedings of the IEEE computer society conference on computer vision and pattern recognition, pp 14388–14397. https://doi.org/10.1109/CVPR52729.2023.01383
83. Malandrakis AA, Kavroulakis N, Chrysikopoulos CV (2022) Metal nanoparticles against fungicide resistance: alternatives or partners? Pest Manag Sci 78(10):3953–3956. https://doi.org/10.1002/ps.7014
84. Malicdem AR, Fernandez PL (2015) Rice blast disease forecasting for northern philippines college of information technology. WSEAS Trans Inf Sci Appl 12(September):120–129
85. Malik JA, Saleem M, Almutairi S, Qadri S, Raza MA, Alsanoosy T, Chaudhary UMD (2024) Implementing fog computing in precision agriculture for real-time soil health monitoring and data management, pp 371–400. https://doi.org/10.1007/978-3-031-67450-1_14

86. Malik MA, Albeladi SS, Al-Maaqar SM, Alshehri AA, Al-Thabaiti SA, Khan I, Kamli MR (2023) Biosynthesis of novel Ag-Cu bimetallic nanoparticles from leaf extract of salvia officinalis and their antibacterial activity. Life 13(3):653. https://doi.org/10.3390/life13030653
87. Maluin FN, Hussein MZ (2020) Chitosan-based agronanochemicals as a sustainable alternative in crop protection. Molecules 25(7):1611. https://doi.org/10.3390/molecules25071611
88. Manisha R, Karanwal R, Singh A, Jagtap PD, Panigrahi CK, Das N, Sharavanan PT, Panotra N (2025) Nanotechnology in plant disease management–functional roles of nanoparticles in pathogen detection and control. Plant Cell Biotechnol Molecular Biol 26(7–8):175–189. https://doi.org/10.56557/pcbmb/2025/v26i7-89426
89. Miglietta L, Rawson TM, Galiwango R, Tasker A, Ming DK, Akogo D, Ferreyra C, Aboagye EO, Gordon NC, Garcia-Vidal C, Rodriguez-Manzano J, Holmes AH (2025) Artificial intelligence and infectious disease diagnostics: state of the art and future perspectives. Lancet Infect Dis. https://doi.org/10.1016/S1473-3099(25)00354-8
90. Mikhailova EO (2020) Silver nanoparticles: mechanism of action and probable bio-application. J Funct Biomater 11(4):84. https://doi.org/10.3390/jfb11040084
91. Minchitha KU, Seshagiri S, Kumar H, Shivashankara NG, Surynan S (2024) Agricultural application of nanotechnology in plant growth and protection: a review. Lett Appl NanoBioSci 13(2):82. https://doi.org/10.33263/LIANBS132.082
92. Mishra S, Singh BR, Singh A, Keswani C, Naqvi AH, Singh HB (2014) Biofabricated silver nanoparticles act as a strong fungicide against Bipolaris sorokiniana causing spot blotch disease in wheat. PLoS ONE 9(5):e9788, 1. https://doi.org/10.1371/journal.pone.0097881
93. Mittal D, Kaur G, Singh P, Yadav K, Ali SA (2020) Nanoparticle-based sustainable agriculture and food science: recent advances and future outlook. Frontiers in Nanotechnology 2. https://doi.org/10.3389/fnano.2020.579954
94. Modgil S, Singh RK, Hannibal C (2022) Artificial intelligence for supply chain resilience: learning from Covid-19. Int J Logist Manag 33(4):1246–1268. https://doi.org/10.1108/IJLM-02-2021-0094
95. Mohanty SP, Czakon J, Kaczmarek KA, Pyskir A, Tarasiewicz P, Kunwar S, Rohrbach J, Luo D, Prasad M, Fleer S, Göpfert JP, Tandon A, Mollard G, Rayaprolu N, Salathe M, Schilling M (2020) Deep learning for understanding satellite imagery: an experimental survey. Front Artif Intell 3(November):1–21. https://doi.org/10.3389/frai.2020.534696
96. Mouniga R, Anita B, Lakshmanan A, Shanthi A, Karthikeyan G (2023) Nematicidal properties of chitosan nanoformulation. J Nematol 55(1). https://doi.org/10.2478/jofnem-2023-0033
97. Naik GG, Jagtap VA (2024) Two heads are better than one: unravelling the potential impact of artificial intelligence in nanotechnology. Nano TransMed 3:100041. https://doi.org/10.1016/j.ntm.2024.100041
98. Nguyen TN, Huynh TN, Hoang D, Nguyen DH, Nguyen QH, Tran TH (2019) Functional nanostructured oligochitosan–silica/carboxymethyl cellulose hybrid materials: synthesis and investigation of their antifungal abilities. Polymers 11(4):628. https://doi.org/10.3390/polym11040628
99. Nthunya LN, Mosai AK, López-Maldonado EA, Bopape M, Dhibar S, Nuapia Y, Ajiboye TO, Buledi JA, Solangi AR, Sherazi STH, Ndungu PN, Mahlangu OT, Mamba BB (2025) Unseen threats in aquatic and terrestrial ecosystems: nanoparticle persistence, transport and toxicity in natural environments. Chemosphere 382:144470. https://doi.org/10.1016/j.chemosphere.2025.144470
100. Oberti R, Marchi M, Tirelli P, Calcante A, Iriti M, Tona E, Hočevar M, Baur J, Pfaff J, Schütz C, Ulbrich H (2016) Selective spraying of grapevines for disease control using a modular agricultural robot. Biosys Eng 146:203–215. https://doi.org/10.1016/j.biosystemseng.2015.12.004
101. Ogunyemi SO, Luo J, Abdallah Y, Yu S, Wang X, Alkhalifah DHM, Hozzein WN, Wang F, Bi J, Yan C, Li B (2024) Copper oxide nanoparticles: an effective suppression tool against bacterial leaf blight of rice and its impacts on plants. Pest Manag Sci 80(3):1279–1288. https://doi.org/10.1002/ps.7857

102. Olawade DB, Ige AO, Olaremu AG, Ijiwade JO, Adeola AO (2024) The synergy of artificial intelligence and nanotechnology towards advancing innovation and sustainability—a mini-review. Nano Trends 8:100052. https://doi.org/10.1016/j.nwnano.2024.100052
103. Oluoch G, Nyongesa M, Mamati EG, Matiru V (2022) Thymol and eugenol nanoparticles elicit expression of Ralstonia solanacearum virulence and potato defense genes and are potential bactericides against potato bacterial wilt. Arch Phytopathol Plant Protect 55(9):1136–1157. https://doi.org/10.1080/03235408.2022.2081766
104. Otero M, Velasquez LF, Basile B, Onrubia JR, Pujol AJ, Pijuan J (2022). Data driven predictive models based on artificial intelligence to anticipate the presence of Plasmopara viticola and Uncinula necator in Southern European winegrowing regions. https://doi.org/10.3233/FAIA220333
105. Pai P, Amutha S, Patil S, Shobha T, Basthikodi M, Shafeeq BMA, Gurpur AP (2025) Deep learning-based automatic diagnosis of rice leaf diseases using ensemble CNN models. Sci Rep 15(1):1–13. https://doi.org/10.1038/s41598-025-13079-z
106. Papagiannidis E, Mikalef P, Conboy K (2025) Responsible artificial intelligence governance: a review and research framework. J Strateg Inf Syst 34(2):101885. https://doi.org/10.1016/j.jsis.2024.101885
107. Paz-Trejo C, Flores-Márquez AR, Gómez-Arroyo S (2023) Nanotechnology in agriculture: a review of genotoxic studies of nanopesticides in animal cells. Environ Sci Pollut Res 30(25):66473–66485. https://doi.org/10.1007/s11356-023-26848-y
108. Peng H, Chen IA (2019) Rapid colorimetric detection of bacterial species through the capture of gold nanoparticles by chimeric phages. ACS Nano. https://doi.org/10.1021/acsnano.8b06395
109. Pham QV, Nguyen DC, Huynh-The T, Hwang WJ, Pathirana PN (2020) Artificial Intelligence (AI) and big data for coronavirus (COVID-19) pandemic: a survey on the state-of-the-arts. IEEE Access 8:130820–130839. https://doi.org/10.1109/ACCESS.2020.3009328
110. Pinheiro I, Aguiar A, Figueiredo A, Pinho T, Valente A, Santos F (2023) Nano aerial vehicles for tree pollination. Appl Sci 13(7):4265. https://doi.org/10.3390/app13074265
111. Pramanik P, Krishnan P, Maity A, Mridha N, Mukherjee A, Rai V (2020) Application of nanotechnology in agriculture, pp 317–348. https://doi.org/10.1007/978-3-030-26668-4_9
112. Rahman MA, Parvin A, Khan MSH, War AR, Lingaraju K, Prasad R, Das S, Hussain B, Bhattacharyya A (2021) Efficacy of the green synthesized nickel-oxide nanoparticles against pulse beetle, Callosobruchus maculatus (F.) in black gram (Vigna mungo L.). Int J Pest Manage 67(4):306–314. https://doi.org/10.1080/09670874.2020.1773572
113. Rai M, Ingle AP, Paralikar P, Anasane N, Gade R, Ingle P (2018) Effective management of soft rot of ginger caused by Pythium spp. and Fusarium spp.: emerging role of nanotechnology. Appl Microbiol Biotechnol 102(16):6827–6839. https://doi.org/10.1007/s00253-018-9145-8
114. Raissi M, Perdikaris P, Karniadakis GE (2019) Physics-informed neural networks: a deep learning framework for solving forward and inverse problems involving nonlinear partial differential equations. J Comput Phys 378:686–707. https://doi.org/10.1016/j.jcp.2018.10.045
115. Rama Krishna D, Obed B (2025) Enhancing cognitive automation capabilities with reinforcement learning techniques in robotic process automation using UiPath and automation anywhere. Int J Sci Res Arch 14(2):733–752. https://doi.org/10.30574/ijsra.2025.14.2.0450
116. Ramcharan A, Baranowski K, McCloskey P, Ahmed B, Legg J, Hughes DP (2017) Deep learning for image-based cassava disease detection. Front Plant Sci 8. https://doi.org/10.3389/fpls.2017.01852
117. Rawat S, Phogat P, Shreya, Chand B (2025) Advances in nanomaterial-based biosensors: innovations, challenges, and emerging applications. Mater Today Commun 48:113334. https://doi.org/10.1016/j.mtcomm.2025.113334
118. Ray MK, Mishra AK, Mohanta YK, Mahanta S, Chakrabartty I, Kungwani NA, Avula SK, Panda J, Pudake RN (2023) Nanotechnology as a promising tool against phytopathogens: a futuristic approach to agriculture. Agriculture 13(9):1856. https://doi.org/10.3390/agriculture13091856

119. Razmi A, Golestanipour A, Nikkhah M, Bagheri A, Shamsbakhsh M, Malekzadeh-Shafaroudi S (2019) Localized surface plasmon resonance biosensing of tomato yellow leaf curl virus. J Virol Methods 267:1–7. https://doi.org/10.1016/j.jviromet.2019.02.004
120. Rodríguez-González V, Domínguez-Espíndola RB, Casas-Flores S, Patrón-Soberano OA, Camposeco-Solis R, Lee S-W (2016) Antifungal nanocomposites inspired by titanate nanotubes for complete inactivation of Botrytis cinerea isolated from tomato infection. ACS Appl Mater Interfaces 8(46):31625–31637. https://doi.org/10.1021/acsami.6b10060
121. Saharan V, Sharma G, Yadav M, Choudhary MK, Sharma SS, Pal A, Raliya R, Biswas P (2015) Synthesis and in vitro antifungal efficacy of Cu–chitosan nanoparticles against pathogenic fungi of tomato. Int J Biol Macromol 75:346–353. https://doi.org/10.1016/j.ijbiomac.2015.01.027
122. Salim SA, Taha AA, Khozemy EE, EL-Moslamy SH, Kamoun EA (2022) Electrospun zinc-based metal organic framework loaded-PVA/chitosan/hyaluronic acid interfaces in antimicrobial composite nanofibers scaffold for bone regeneration applications. J Drug Deliv Sci Technol 76:103823. https://doi.org/10.1016/j.jddst.2022.103823
123. Sánchez-López E, Gomes D, Esteruelas G, Bonilla L, Lopez-Machado AL, Galindo R, Cano A, Espina M, Ettcheto M, Camins A, Silva AM, Durazzo A, Santini A, Garcia ML, Souto EB (2020) Metal-based nanoparticles as antimicrobial agents: an overview. Nanomaterials 10(2):292. https://doi.org/10.3390/nano10020292
124. Sehgal N, Kumari A et al (2021) Latest developments and applications of nanotechnology in agriculture sector: a review. Agric Rev. https://doi.org/10.18805/ag.R-2175
125. Shang Y, Hasan MK, Ahammed GJ, Li M, Yin H, Zhou J (2019) Applications of nanotechnology in plant growth and crop protection: a review. Molecules 24(14):2558. https://doi.org/10.3390/molecules24142558
126. Sharifi-Rad J, Sharifi-Rad M, Teixeira Da Silva JA (2016) Exposed to SiO2 nanoparticles. Int J Agric Sci Tech 18
127. Sharma P, Lew TTS (2022) Principles of nanoparticle design for genome editing in plants. Front Genome Edit 4. https://doi.org/10.3389/fgeed.2022.846624
128. Shin J, Cole BD, Seyedmohammad M, Lim SI, Jang Y (2024) Protein nanocarriers capable of encapsulating both hydrophobic and hydrophilic drugs, pp 143–150. https://doi.org/10.1007/978-1-0716-3469-1_10
129. Shukla R, Dubey G, Malik P et al (2021) Detecting crop health using machine learning techniques in smart agriculture system. J Sci Ind Res 80(8). https://doi.org/10.56042/jsir.v80i08.44034
130. Silva JCF, Teixeira RM, Silva FF, Brommonschenkel SH, Fontes EPB (2019) Machine learning approaches and their current application in plant molecular biology: a systematic review. Plant Sci 284(April):37–47. https://doi.org/10.1016/j.plantsci.2019.03.020
131. Singh A, Rawat S, Rajput VD, Minkina T, Mandzhieva S, Eloyan A, Singh RK, Singh O, El-Ramady H, Ghazaryan K (2024) Nanotechnology products in agriculture and environmental protection: advances and challenges. Egypt J Soil Sci 64(4). https://doi.org/10.21608/ejss.2024.300047.1802
132. Singh BK, Delgado-Baquerizo M, Egidi E, Guirado E, Leach JE, Liu H, Trivedi P (2023) Climate change impacts on plant pathogens, food security and paths forward. Nat Rev Microbiol 21(10):640–656. https://doi.org/10.1038/s41579-023-00900-7
133. Soussi A, Zero E, Sacile R, Trinchero D, Fossa M (2024) Smart sensors and smart data for precision agriculture: a review. Sensors 24(8):2647. https://doi.org/10.3390/s24082647
134. Spanaki K, Karafili E, Sivarajah U, Despoudi S, Irani Z (2022) Artificial intelligence and food security: swarm intelligence of AgriTech drones for smart AgriFood operations. Prod Plan Control 33(16):1498–1516. https://doi.org/10.1080/09537287.2021.1882688
135. Taha BA, Ahmed NM, Talreja RK, Haider AJ, Al Mashhadany Y, Al-Jubouri Q, Huddin AB, Mokhtar MHH, Rustagi S, Kaushik A (2024) Synergizing nanomaterials and artificial intelligence in advanced optical biosensors for precision antimicrobial resistance diagnosis. ACS Synth Biol 13(6):1600–1620

136. Taheri Hosseinkhani N (2025) Artificial intelligence applications and economic impacts in sustainable agriculture. Available at SSRN 5474048
137. Touch V, Tan DKY, Cook BR, Liu DL, Cross R, Tran TA, Utomo A, Yous S, Grunbuhel C, Cowie A (2024) Smallholder farmers' challenges and opportunities: Implications for agricultural production, environment and food security. J Environ Manage 370:122536. https://doi.org/10.1016/j.jenvman.2024.122536
138. Tran T-T, Clark K, Ma W, Mulchandani A (2020) Detection of a secreted protein biomarker for citrus Huanglongbing using a single-walled carbon nanotubes-based chemiresistive biosensor. Biosens Bioelectron 147:111766. https://doi.org/10.1016/j.bios.2019.111766
139. Tzachor A, Devare M, King B, Avin S, Ó Éigeartaigh S (2022) Responsible artificial intelligence in agriculture requires systemic understanding of risks and externalities. Nat Mach Intell 4(2):104–109. https://doi.org/10.1038/s42256-022-00440-4
140. Upadhyay K, Katel S, Mandal H, Yadav S, Kharel A, Dahal R (2021) A review on Nano technology application in agriculture. Fundam Appl Agric 1. https://doi.org/10.5455/faa.127296
141. Usman M, Farooq M, Wakeel A, Nawaz A, Cheema SA, ur Rehman H, Ashraf I, Sanaullah M (2020) Nanotechnology in agriculture: current status, challenges and future opportunities. Sci Total Environ 721:137778. https://doi.org/10.1016/j.scitotenv.2020.137778
142. Vadlapudi V, Amanchy R (2017) Phytofabrication of silver nanoparticles using Myriostachya wightiana as a novel bioresource, and evaluation of their biological activities. Braz Arch Biol Technol 60. https://doi.org/10.1590/1678-4324-2017160329
143. van Dijk ADJ, Kootstra G, Kruijer W, de Ridder D (2021) Machine learning in plant science and plant breeding. IScience 24(1):101890. https://doi.org/10.1016/j.isci.2020.101890
144. Vardhan PNH, Badavath A, Srivalli P (2025) Artificial intelligence and its applications in agriculture: a review. Environ Conserv J 26(1):274–280. https://doi.org/10.36953/ECJ.28802904
145. Varghese VM, Kavitha O, Poulose G, Joseph N (2025) Future trends and challenges in agricultural nanotechnology, pp 547–584. https://doi.org/10.1007/978-3-031-90506-3_22
146. Venkidasamy B, Shelar A, Dhanapal AR, Nile AS, Patil R, Zhang Y, Kuksal K, Nile SH (2025) Emerging biopolymer nanocarriers for controlled and protective delivery of food bioactive compounds- current status and future perspective. Food Hydrocolloids 160:110769. https://doi.org/10.1016/j.foodhyd.2024.110769
147. Visentin F, Cremasco S, Sozzi M, Signorini L, Signorini M, Marinello F, Muradore R (2023) A mixed-autonomous robotic platform for intra-row and inter-row weed removal for precision agriculture. Comput Electron Agric 214:108270. https://doi.org/10.1016/j.compag.2023.108270
148. Wang H, Wang Y, Hou X, Xiong B (2020) Bioelectronic nose based on single-stranded DNA and single-walled carbon nanotube to identify a major plant volatile organic compound (p-Ethylphenol) released by phytophthora cactorum infected strawberries. Nanomaterials 10(3):479. https://doi.org/10.3390/nano10030479
149. Wei S, Sun Y, Xi G, Zhang H, Xiao M, Yin R (2018) Development of a single-tube nested PCR-lateral flow biosensor assay for rapid and accurate detection of Alternaria panax Whetz. PLoS ONE 13(11):e0206462. https://doi.org/10.1371/journal.pone.0206462
150. Wilhelm RC, van Es HM, Buckley DH (2022) Predicting measures of soil health using the microbiome and supervised machine learning. Soil Biol Biochem 164:108472. https://doi.org/10.1016/j.soilbio.2021.108472
151. Wu P, Cui P, Du H, Alves ME, Zhou D, Wang Y (2021) Long-term dissolution and transformation of ZnO in soils: the roles of soil pH and ZnO particle size. J Hazard Mater 415:125604. https://doi.org/10.1016/j.jhazmat.2021.125604
152. Xiao Q, Li W, Kai Y, Chen P, Zhang J, Wang B (2019) Occurrence prediction of pests and diseases in cotton on the basis of weather factors by long short term memory network. In: BMC bioinformatics. Springer International Publishing, vol 20. https://doi.org/10.1186/s12859-019-3262-y

153. Xing Y, Wang X (2024) Precision agriculture and water conservation strategies for sustainable crop production in arid regions. Plants 13(22). https://doi.org/10.3390/plants13223184
154. Xue Z, Liu F, Wang B, Shi X, Liu P, Liang Y, Liu X (2025) Fabrication of a dual-stimuli-responsive mesoporous organosilica nanoplatform for co-delivery of fungicide and plant immune inducer toward synergistic disease management. J Nanobiotechnol 23(1). https://doi.org/10.1186/s12951-025-03605-6
155. Yadav A, Yadav K (2025) Harnessing nanotechnology and artificial intelligence to revolutionize sustainable smart agriculture, pp 123–147. https://doi.org/10.1007/978-981-96-7756-6_7
156. Yang H, Jin Y, Jiang L, Lu J, Wen G (2025) AI roles in 4R crop pest management—a review. Agronomy 15(7):1–15. https://doi.org/10.3390/agronomy15071629
157. Yang H, Zhang S, Liu R, Krall A, Wang Y, Ventura M, Deflitch C (2022) Epidemic Informatics and control: a review from system informatics to epidemic response and risk management in public health. In: Springer proceedings in business and economics. https://doi.org/10.1007/978-3-030-75166-1_1
158. Yeh YH, Chung WC, Liao JY, Chung CL, Kuo YF, Lin TT (2016) Strawberry foliar anthracnose assessment by hyperspectral imaging. Comput Electron Agric 122:1–9. https://doi.org/10.1016/j.compag.2016.01.012
159. Zang H, Wang C, Zhao Q, Zhang J, Wang J, Zheng G, Li G (2025) Constructing segmentation method for wheat powdery mildew using deep learning. Front Plant Sci 16(May):1–15. https://doi.org/10.3389/fpls.2025.1524283
160. Zhang P, Guo Z, Ullah S, Melagraki G, Afantitis A, Lynch I (2021) Nanotechnology and artificial intelligence to enable sustainable and precision agriculture. Nat Plants 7(7):864–876. https://doi.org/10.1038/s41477-021-00946-6
161. Zhang X, Cao Z, Dong W (2020) Overview of edge computing in the agricultural internet of things: key technologies, applications, challenges. IEEE Access 8:141748–141761
162. Zulfiqar A, Izquierdo E, Chandramouli K (2025) Harnessing deep learning for plant species classification: a comprehensive review. Comput Electron Agric 237(PC):110663. https://doi.org/10.1016/j.compag.2025.110663

Ways to Develop Environmentally Stress-Resilient Crops by AI-Powered Nanobiotechnology

Pranjal Dwivedi, Binty, Satyam Rai, Tushar Kanti Mandal, and Sayanti Mandal

Abstract The more frequent occurrences of environmental stresses such as drought, salinity, heat, and heavy metal toxicity have become increasingly severe challenges in crop yield and food security worldwide. Conventional farming systems can no longer meet these demands sustainably. This chapter discusses the convergence of AI and nanobiotechnology, an innovative strategy for synergistically promoting environmentally stress-tolerant crops. The unique physicochemical properties of ENMs, like metal-based, carbonaceous, and polymeric nanoparticles, have been utilized for modulation of antioxidant defense systems, osmolyte accumulation, hormonal regulation, and photosynthetic efficiency in plants. In combination with AI-based predictive modeling, ML algorithms will be able to optimize nanoparticle synthesis, delivery, and their environmental safety on one side, and nutrient use efficiency on the other end, thereby decreasing toxicity accordingly. Precision agriculture through nanosensor networks, remote sensing, and data-driven irrigation and pest control systems. In addition, the combination of nanocarrier-based gene editing with an AI-supported genomic selection expedites stress-resistant breeding. Although several challenges like nanotoxicity, data standardization, and economic scalability remain to be addressed, this interdisciplinary paradigm is in harmony with the United

P. Dwivedi
Centre for Molecular Biology, Central University of Jammu, Bagla (Raya-Suchani), Samba, Jammu & Kashmir, India

Binty
Department of Biotechnology, Jaypee Institute of Information Technology, Noida, Uttar Pradesh, India

S. Rai
Department of Biotechnology, Centre for Molecular Biology, Central University of Jammu, Bagla (Raya-Suchani), Jammu & Kashmir, Samba, India

T. K. Mandal
Mercedes-Benz Research and Development India Pvt. Ltd, Pune, India

S. Mandal (✉)
Department of Chemistry and Biochemistry, Sharda School of Engineering and Sciences, Sharda University, Greater Noida, Uttar Pradesh, India
e-mail: mandalsayanti@gmail.com

J.-T. Chen (ed.), *AI-Nano-Enabled Climate-Smart Agriculture and Sustainability*, Smart Nanomaterials Technology, https://doi.org/10.1007/978-981-92-1694-9_9

Nations Sustainable Development Goals (SDGs), which will provide a way forward to resilient agricultural ecosystems and global food security.

Keywords Artificial intelligence (AI) · Climate-smart agriculture · SDG · Nanobiotechnology · Precision agriculture · CRISPR

1 Principles of Nanobiotechnology in Crop Stress Resilience

Environmental changes like climate change, soil degradation, pollution, and population growth have exerted immense pressure on the agricultural system. Continuous exposure to abiotic stresses (drought, salinity, heavy metal, heat) and biotic stresses decreases global yield by more than 50% in a few decades [1–5]. Rapid, precise, and sustainable solutions cannot be attained by traditional approaches like selective breeding and agrochemical applications. Nanobiotechnology is a combination of nanotechnology with biological systems, which can be used to enhance plant resilience through nano-enabled stress management. Engineered nanomaterials (ENMs) with special physicochemical properties, like size (< 100 nm), catalytic activity, high surface-to-volume ratio, and tunable surface charge, allow them to interact indirectly with plant cells, signaling pathways, and biomolecules [6]. Nanomaterials can help plants tolerate abiotic stress by modulating antioxidant defense systems, hormonal balance, osmolytes accumulation, and photosynthetic efficiency [7]. For combating biotic stress and ensuring sustainable agriculture, nano-enabled nutrient delivery systems, gene-editing platforms, and nano-pesticides can be utilized [8].

1.1 Nanomaterials and Plant Stress Mechanisms

Carbon-based nanomaterials (e.g., graphene, carbon nanotubes), metal and metal-oxide nanoparticles (e.g., ZnO, TiO_2, Fe_2O_4), and lipid/polymeric/organic nanoparticles are the ENMs, and they are the spotlight of improving plant resilience to environmental stresses [9–11]. They exhibit properties like a high surface area to volume ratio, along with tunable size, surface charge, coating properties, and shape. These properties allow them to bore into the plant tissue efficiently and link with important biochemical pathways in a controlled manner [12].

Once they are inside the cell, ENMs show different stress mitigation mechanisms, like the enhancement of antioxidant defenses by ZnO-NPs. ZnO-NPs enhance the activity of antioxidant enzymes, which helps reduce the oxidative damage under drought and salinity conditions [13]. Another stress mitigation mechanism is osmolyte accumulation and hormonal modulation by TiO_2 nanoparticles. TiO_2

nanoparticles modulate phytohormone (e.g., enhancing GA, IAA; reducing ABA) levels by increasing the production of osmoprotectants such as soluble sugars, proline, and phenolics; overall enhancing membrane stability, water retention, and drought tolerance [14]. Photosynthetic enhancement is also an important stress mitigation mechanism in which CuO, ZnO, and other metal nanoparticles increase PS-II efficiency and stabilize photosystem degradation under stress by increasing photosynthetic performance by preserving chlorophyll and carotenoids [15].

Researchers have observed practical examples in the case of salinity stress that ENMs can efficiently minimize the adverse effects of abiotic stresses on crops. For example, in salinity stress, if ZnO-NPs are applied through seed priming and foliar spray, they can enhance different physiological and biochemical factors responsible for plant resilience. Those plants that are treated with these ENMs show improved photosynthetic pigments, higher relative water content, and increased stomatal conductance, which together can improve photosynthetic efficiency during saline conditions (Fig. 1). Furthermore, the performance of antioxidant enzyme activities was increased, which is helpful in the mitigation of reactive oxygen species (ROS), hence lowering the stress-induced membrane damage markers such as hydrogen peroxide (H_2O_2) and malondialdehyde (MDA) [16]. Another ionic imbalance, which is caused by excessive sodium accumulation, can also be corrected by ZnO-NP treatment, which shows that it has an important role in the maintenance of physiological stability and cellular homeostasis under salt stress [17].

Studies on semi-arid wheat showed that even during drought stress conditions, ZnO-NPs have an important role in promoting stress tolerance [3–5]. There was a great improvement in grain zinc content, total biomass, yield attributes, and grain protein levels as compared to untreated control when ZnO-NPs are applied at a concentration of 150 ppm under water-inadequate conditions. These studies suggest that their role as safe, yield-oriented, and nutrient-enriching nanofertilizers [3–5].

Together, these studies suggest that these ENMs work through different ways, but they are linked together—like modulation of osmolytes and stress hormones, activation of antioxidant defense systems, stabilization of photosynthetic machinery, and ionic balance regulation—which help plants tolerate different environmental stresses [18]. Recent research also shows the importance of fine-tuning nanoparticle physicochemical properties (e.g., charge, coating, size, and solubility) to enhance the benefits and, in turn, reduce the risk of environmental accumulation or phytotoxicity [19]. So, we can say that nanomaterials hold a promise in improving resilient, high-yield crop systems under challenging climate scenarios.

1.2 Nanobiotechnology for Nutrient and Pest Management

Nanobiotechnology has an important role in improving nutrient delivery and pest management in agriculture, specifically under abiotic stress like drought, nutrient deficiencies, and salinity. Traditional strategies like fertilizer and pesticides are not that effective, and can cause soil contamination, ecological harm, and nutrient

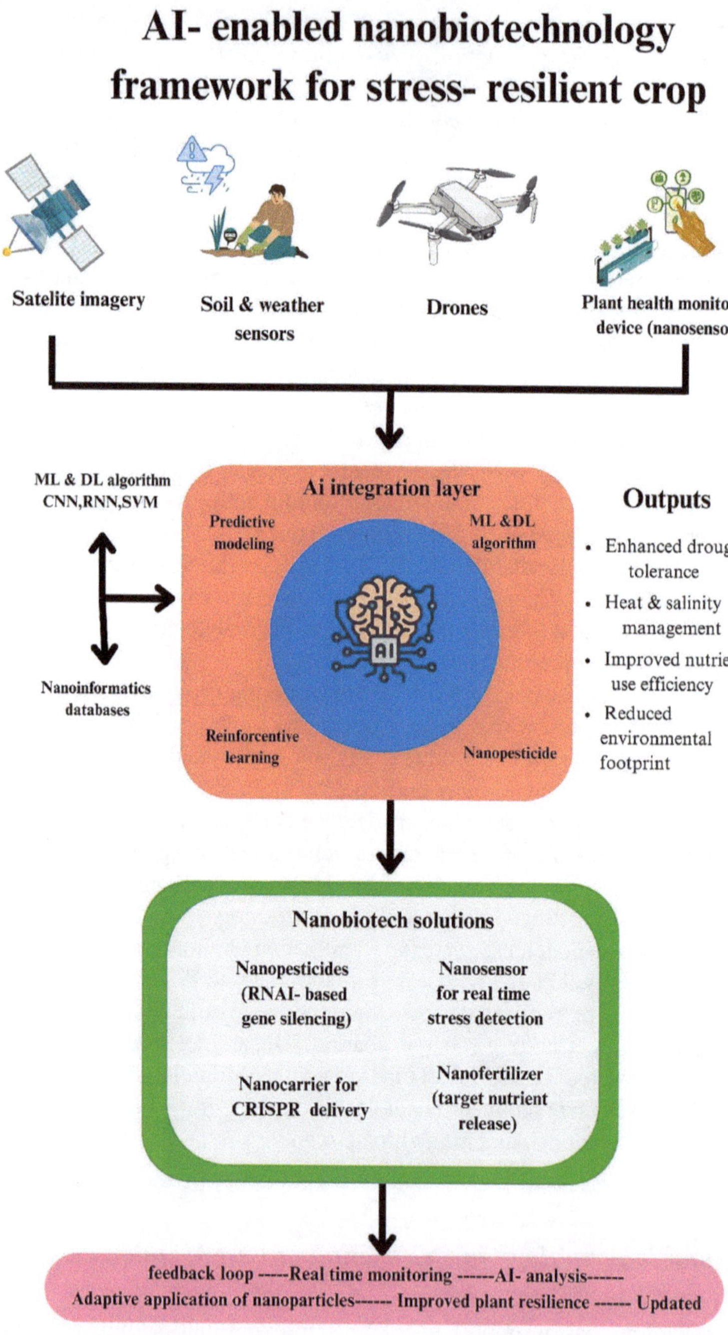

Fig. 1 Biotechnology framework for stress-resistant crops that is enabled by artificial intelligence

leaching. ENMs are a more resilient alternative because they allow target-specific, slow-release, and site-specific delivery of important nutrients and agrochemicals, therefore reducing environmental loss and enhancing plant uptake efficiency. Nano-packaging of nutrients like phosphorus, nitrogen, and potassium fertilizers can increase their solubility, assure controlled release, and decrease volatilization, which can assure a stable nutrient supply during stress conditions [20]. Similarly, micronutrients like zinc, copper, and iron can also be delivered by the use of nanoparticles, which helps in overcoming deficiencies that sometimes worsen the stress responses in crops. ZnO-NPs have improved antioxidant activity, chlorophyll synthesis, and overall production during salinity and drought stress in maize and rice crops [21].

Nanobiotechnology not only has a role in nutrient delivery, but in fact nanobiotechnology is also explored in pest and pathogen management because it offers a precise and very eco-friendly solution to traditional pesticides. The stability and bioavailability of biopesticides can be improved by nanobiotechnology-based formulation, allowing lower dosage and increased protection against pests [22]. Examples, silver and silica nanoparticles have strong antifungal and antimicrobial properties against plant pathogens like *Alternaria* and *Fusarium*. Another example can be chitosan nanoparticles, which act as an activator of systemic resistance pathways and plant defense responses [23]. RNA interference (RNAi)-based pest control is another strategy that involves nanoparticle-assisted delivery of (double-stranded RNA) dsRNA [24]. This allows silencing of genes in pathogens and insect pests, promising very specific, eco-friendly, and safe pest suppression. dsRNA can be protected from degradation by nanosheets and polymeric nanoparticles, and they can also facilitate its uptake by target organisms, leading to more effective pest mortality and gene knockdown [25].

Furthermore, nanobiotechnology is also intersecting with modern biology to promote plant genome editing for increased stress resilience. Several nanoparticles, like mesoporous silica nanoparticles, carbon nanotubes, and lipid-based nanocarriers, are coming out as an excellent vector for delivering DNA templates and CRISPR-Cas9 ribonucleoproteins into plant cells, skipping conservative transformation ways which are mostly species-specific and technically challenging [26, 27]. This nanoparticle-based gene editing also reduces dependency on biolistic-based systems and Agrobacterium, offering a common, universal, and less effective method of crop improvement. Traits like drought tolerance, pest resilience, and salinity resistance can be precisely engineered by CRISPR-Cas9, hence supporting the development of next-generation climate-smart crops [28].

Together, these advanced nanoparticles-mediated nutrient delivery, genome editing, and pest management play an important role in making resilient agriculture systems [9–11]. By combining all the factors like RNAi-based suppression, nutrient release, and precise genome editing, nanobiotechnology not only increases crop performance under stress conditions but also helps in making more eco-friendly agriculture practices that can decrease the load on chemicals and environmental footprints [29].

2 Artificial Intelligence for Nanobiotechnology and Agriculture

The rise of artificial intelligence (AI) in agricultural science has turned out to be a transforming step into the traditional farming practices that are providing remarkable opportunities for nanobiotechnology-driven solutions. AI includes deep learning (DL), machine learning (ML), and data-driven predictive modeling, helpful in correct decision making, high-throughput analysis, and real-time monitoring (Figs. 1 and 2), which is important for the production of high crop yield and stress-resistant plants [30]. When combined with nanobiotechnology, AI can refine nanoparticle delivery, design, and application, making a more optimal, eco-friendly, sustainable, and safe strategy for crop management. This section tells you about the multidisciplinary role of AI in modern agriculture, especially in agriculture, material design, nanosensor-based monitoring, and nanoinformatics.

2.1 AI in Precision Agriculture

Resource utilization, prediction of environmental stress, monitoring plant health, significantly decreasing waste, improving crop yield, and many more are the areas

Fig. 2 Applications of nanobiotechnology in agriculture

in which precision agriculture can utilize AI. Large datasets from different sources like satellite imagery, IoT sensors, drones, and weather stations can be analyzed by ML algorithms, providing actionable insights [31]. Early identification of drought, pathogen attacks, and nutrient deficiencies before visible symptoms can be done by image-based stress detection techniques using convolutional neural networks (CNNs). Early detection is important because delayed information can lead to irreversible crop damage and yield loss [32].

Soil moisture variability and weather patterns can be predicted by an AI model, allowing flexible irrigation scheduling. Farmers can anticipate heat waves, drought, or flooding events through recurrent neural networks (RNNs) and long short-term memory (LSTM) networks excelling in time-series prediction [33]. Drone-assisted imaging, coupled with AI-driven analysis, can generate high-resolution crop maps that reveal heterogeneity in plant health, canopy coverage, and pest infestations. Site-specific interventions, like variable rate nanofertilizer application and target NP-based pesticide delivery, can be guided by AI models [34].

Furthermore, AI increases predictive modeling for stress response, yield estimation, and resource use efficiency. Environmental variables, NP treatment outcomes, and plant phenotypic traits can be used by Random Forest algorithms and gradient boosting models to generate precise recommendations for crop management. This predictive capability is important for executing nanobiotechnology in agriculture at scale, as it makes sure that nanomaterials are effective, safe, and applied only when required [35].

2.2 AI in Nanoinformatics and Material Design

Nanoinformatics means the application of informatics to nanomaterials, and this field heavily depends on AI to predict, analyze, and use nanoparticle properties. AI models can work on large-scale datasets of ENMs together, like shape, size, surface chemistry, functionalization, and charge, to predict plant uptake, toxicity, and bioavailability. Prediction of stress mitigation efficacy and NP-plant interaction can be done by ML algorithms like support vector machine (SVM), artificial neural network (ANNs), and k-nearest neighbors (k-NN) [36].

Researchers have developed an ML-based model that can link nanoparticle characteristics with responses produced by plants. This model can analyze parameters like NP composition, surface modification, concentration, and predict antioxidant activity, photosynthetic efficiency, and osmolyte accumulation under stress conditions [37]. This model allows the development of nanoparticles that can maximize stress tolerance and minimize environmental risks and potential phytotoxicity.

High-throughput virtual screening and optimization of nanoparticle formulation can be facilitated by AI models. Generative models and reinforcement learning can also be used to develop new nanomaterials designed especially for crop types and the stress conditions they usually suffer. For example, ML algorithms can be used to make changes in surface functionalization that can be helped in improving the delivery of nucleic acids for CRISPR-based genome editing and uptake in roots or leaf tissues [38]. These advanced approaches help reduce experimental trial-and-error, and speed up the translation of nanobiotechnology innovations from the lab to the real-life fields [39].

AI is also important for evaluating nanoparticle safety. Integration of toxicogenomics datasets can help predictive models in estimating potential off-target effects on non-target plants, aquatic organisms, and soil microbiomes. This confirms regulatory compliance and environmental sustainability, tackling one of the main concerns linked with broad NP application in agriculture [40].

2.3 *Machine Learning for Nanosensors and Monitoring*

Nanosensors play an important role in real-time monitoring of soil chemistry, plant health, and environmental conditions. AI has an important role in analyzing complex signals generated by these sensors; it can convert raw data into operational knowledge. ROS, like H_2O_2, is a key indicator of oxidative stress in plants, which can be detected by carbon nanotube (CNT)-based nanosensors [41, 42]. Machine learning algorithms, ensemble methods, and support vector regression (SVR) interpret spatial and temporal variation in sensor signals to forecast stress onset, progression, and severity [43].

Similarly, the "bioristor" is an important in vivo nanosensor inserted in plant xylem, used in the measurement of ionic fluctuations such as Na^+ under salinity stress. AI models can manage these high-frequency datasets to construct predictive maps of ion accumulation and water balance across the plant. Precise actions can be made, like foliar application of nanoparticle-based osmoprotectants or targeted irrigation through these AI models [43].

Furthermore, if a nanosensor network is combined with AI models, it can perform multiple functions, like combining data from environmental sensors, phenotypic imaging, and NP tracking. Undetectable and complex stress signatures can be revealed by the fusion of these heterogeneous datasets by deep learning models like autoencoders and recurrent neural networks [44]. For example, AI can forecast ROS-mediated leaf senescence and drought-induced stomatal closure before any visible symptoms, allowing timely nanobiotechnological intervention [45]. AI can also improve adaptive decision-making. Reinforcement learning frameworks can continuously improve NP application strategies based on sensor feedback, flexibly adjusting concentration, delivery routes, and timing to increase maximum efficacy while decreasing waste and toxicity. So, all of this represents the future of AI-driven nanobiotechnology in precision agriculture, allowing sustainable crop production under rapidly changing environmental conditions.

3 AI-Enhanced Nano-agri Applications and Platforms

3.1 AI-Driven Nanomaterial Optimization

AI has now become one of the most important instruments in science and is the driving force behind advancements in a broad range of fields and will have a significant impact on how mankind will develop in the future. An emerging field with unlimited potential in this world of quick changes and innovation is the integration of AI and nanotechnology. These two fields, which have revolutionized science by their own means, have been integrated to form a synergy that has the potential to solve numerous global problems [46].

To ensure global food security, it is required to tackle the problems of climate change, growth in population, increased demands for land to produce fuels and unfavorable conditions of soil. Hence, to achieve sustainable solutions, the integration of multidisciplinary knowledge is required. There is encouraging potential for sustainable food production by integrating disciplines like precision agriculture, nanotechnology, and artificial intelligence (AI) [47].

Due to their complementary advantages, AI and nanotechnology are naturally convergent. AI and nanotechnology have transformed numerous procedures in a variety of fields. The role of AI in agriculture mainly emphasizes relieving the effects of natural calamities, preparing for climate change, managing water, and also in precision farming. Nanotechnology had an impact on seeds, fertilizers, pesticides, herbicides, irrigation systems, solar energy, and forecasting. By the use of this integration of AI and nanotechnology, it is now possible for farmers to maximize the distribution of their resources, which helps in making crucial decisions to rapidly respond in real-time for better production of crops [48]. Using ML models along with this integration of AI and nanotechnology can help in making agriculture more sustainable by enhancing important factors like precision targeting, effective absorption, effective distribution, and optimized nutrient assimilation.

3.1.1 Machine Learning-Assisted Engineering of Nanoparticles

Numerous tactics, including surface modification and functionalization, have been tried to improve the distribution and effectiveness of nanoparticles and verify their uses. In many of the solutions discussed, experimentation and conventional trial-and-error methods are prevalent, expensive, and erratic. But lately, ML-based techniques have become more popular for analyzing and assessing drug delivery systems. Data-driven nanoparticles have lately gained attention in this field because of artificial intelligence (AI)-based techniques, including machine learning (ML) and deep learning [49]. Data-driven nanoparticles have established a reputation for being grounded in both experimental data and mathematical modeling and evaluations, in contrast to other approaches that rely primarily on theory. AI techniques apply to a broad range of applications and do not have a size limit, in contrast to other computing approaches. AI has been used, for instance, to manage the physicochemical features of formulations (e.g., size, shape, drug loading efficiency, etc.) and to design tiny therapeutic molecules.

3.2 Precision Delivery Technologies

Nanoparticles that can provide us with precise delivery of bioactive compounds, nutritional components, and genetic materials can help us make agriculture less prone to environmental changes and ultimately more sustainable. It might be feasible to increase the efficiency and efficacy, and crop yield by improving the nanoparticle's ability to precisely carry and deliver the components at site-specific locations [50].

The combination of precision farming and AI, specifically drones and autonomous vehicles that can carry nano-formulations, GPS, remote sensing for crop monitoring, data analytics, variable rate technology, precision irrigation systems, automation, and crop health monitoring tools, has the potential to transform agriculture into more sustainable forms. Precision agriculture, with the help of drones and other technological advancements, helps to address the challenge of increasing crop productivity while reducing environmental impact. UAV-mounted sensors collect multispectral and geographic data, analyzed by AI drone systems for targeted pesticide application, using ML algorithms. By the use of proper data analysis and prediction models, real-time predictions for the correct pesticide, dosage, and the schedule of treatment can be generated by the system [51]. IoT-based systems are smart agriculture systems that aim at integrating real-time monitoring, automation, and decision-making that is data-driven, which ultimately optimizes crop production. The data is sent to a central processing unit by a network of sensors that can analyze the spread of diseases, temperature, humidity, etc. [52]. Using these analytics and advanced ML models, accurate decisions can be taken for proper and sustainable crop production.

3.3 Integrated Genomic and Phenotypic Selection

To meet the demands of a growing population and ensure continued food production, crop development rates must rise. One promising breeding technique that has been effectively used in plant breeding is genomic selection (GS). By quickly selecting superior genotypes, GS promises to speed up breeding cycles. Training population design, model construction, prediction, and selection are its four main phases.

When GS is utilized together with other platforms of crop enhancement, such as AI, breeding is hastened in order to increase genetic gain. Using data science technologies like machines and deep learning can help boost the accuracy of the models used to make predictions. The use of ML is gradually replacing the conventional models in plant-omics data analysis because it is able to handle big and complicated data. The advent of high-throughput sequencing technology has brought a plethora of new information and biologists are able to investigate complex associations, decode plant stress responses, and untangle mysteries of genomic responses [53. ML, more so DL, has been effective in addressing such challenges and producing precise analyses of the plant properties that are influenced by interaction between genotypes and environment [54]. In the field of plant systems biology, unsupervised and semi-supervised ML have (although without large labeled training sets) been used to enable the analysis of big data [55]. The use of ML is also involved in enhancing the plant agronomic characteristics by using big omics data [56]. It is hoped that the advancements in the field of artificial intelligence will allow the integration of various types of data into prediction models. The reduced breeding period and the increase in the capability of GS to study larger populations have seen it widely applied in breeding plants.

DL has shown incredible qualities in crop management, including different crops such as rice, wheat, tomato, and potato. Li et al. [57] were the first to present a system of video detection using DL that would combat plant diseases and pests in crops. Their main aim was to detect the presence of plant diseases and pests using video analysis as fast as possible, based on the powerful models, Faster R-CNN and YOLO v3, as the analytical systems in the video in real time. The technique consisted of converting videos into single frames, with the Faster R-CNN framework for detection to perform that work. They also proposed video-based evaluation measures that would be used to measure the quality of detection and their custom backbone system was proven to be the best compared to the previous systems in detecting videos of rice that were not trained. Concentrating on the wheat stripe rust, which is a common wheat disease, Mi et al. [58] proposed a new DL network named C-DenseNet. In this network, the convolutional Block Attention Module (CBAM), which is a densely connected convolutional network (DenseNet), outperforms classical DenseNet and ResNet networks on wheat stripe rust severity grading with a test error of 97.99%. Finally, through these AI algorithms, the diagnostic screening of individual stress-resistance genes can be achieved on a large genomic and phenomics dataset, exposing its full potential in managing targeted alterations on the plant genome to improve agricultural productivity and sustainability [59]. Owing to their unparalleled ability to induce precise nucleotide changes, CRISPR-Cas systems have emerged as a powerful tool to improve crop yields and stress resistance. For example, disrupting Gn1a, DEP1, and GS3 genes in rice led to larger grain size and higher yield [60], targeted mutations in the MLO allele produced bread wheat resistant to powdery mildew [61], and editing the promoter region of host sucrose transporter genes generated rice lines with broad-spectrum resistance against bacterial blight [62].

3.4 Monitoring and Data Assimilation

An important aspect of crop health monitoring is the precise and on-time identification of plant stresses, which would ensure the prevention of food loss and promise food security by inhibiting the spread of diseases and providing efficient management. A typical diagnosis of crop disease is done by using direct or indirect methods, like the study of symptoms and serological testing. At the asymptomatic stage, these methods are effective but less dependable. They also take a lot of time, need expensive equipment and skilled operators, and are not field-applicable [63].

Smart nanosensors are biological signal-sensing and transducer devices that link living things to digital data in the Internet of Things (IoT) era. The use of nanosensors could allow for automatic water and nutrient allocation in addition to the real-time detection of plant chemical responses for plant health surveillance. For instance, it was recently possible to monitor plant health and identify biotic or abiotic stressors using a nanosensor based on single-walled carbon nanotubes functionalized with a hemin-decorated DNA aptamer [64]. By integrating these near-infrared (nIR) fluorescent nanosensors with *Arabidopsis thaliana* plant leaves, it was possible to visualize H_2O_2, a crucial signaling chemical linked to the development of plant stress in distant models. Plant stress communication can be better understood with the use of these sensors. By attaching a camera to multispectral, hyperspectral, or thermal sensors, unmanned aerial vehicles (drones) or crop dusters, which are already utilized for irrigation, spraying, or sowing, can be utilized to monitor or detect the reflective qualities of crops. Crop reflective qualities reveal common farming problems such as nutrient deficiencies, disease prevalence, and weed density. Larger trends at the landscape level can be observed via satellites [41, 42]. Compared to airborne drones, which can study the earth's characteristics and regional weather patterns for forecasting and vegetation index detection, this monitoring is often conducted on a broader spatial scale and with lower resolutions. Open sources and services, like the Copernicus Land Monitoring Service, can provide the satellite data.

Cross-scale precise assessment of crucial metrics like crop yield, pest, and disease thresholds is made possible by combining remote sensing data with ground-based biomass forecasts produced by crop growth models based on AI and ML models that have monitored the crops for the collection of data via ground methods, drones, or satellites. This increases the economic and environmental benefits of agricultural production by enabling agricultural decision-makers to respond quickly to a variety of hazards and put precision management techniques into place [65, 66]. Additionally, exploring the real-world implementations of these assimilation techniques in precision agriculture has significant practical implications for improving agricultural production systems and strengthening food security defenses, in addition to helping to modernize and transform agricultural paradigms. This is mainly attributed to its special advantages that comprise timeliness, synchronization, continuous coverage, and broad areas of observation [67]. With the recent development of technology, several quantitative inversion products, which are directly correlated to crop physiological and biochemical traits, have emerged. These products are a source of rich and useful data used to monitor agriculture, analyze yields, and give early warning.

4 Nanobiotechnological Strategies for Major Stressors

The management of the majority of Agricultural stressors, both Biotic (pests, pathogens) and Abiotic (drought, salinity, temperature, heavy metals), is being revolutionized by Nanobiotechnological strategies that use nanomaterials, conjugated biomolecules, and advanced delivery mechanisms for crop resilience and productivity [9–11]. These strategies aim at bringing advanced sustainability and resilience to modern agriculture by using engineered nanoparticles and bio-nanocomplexes that have the potential to provide multitarget solutions for biotic and abiotic stressors.

4.1 *Drought Stress*

Drought Stress is one of the most harmful abiotic factors that stop plant growth and development, lowering its yield and causing financial losses. The positive effects of carbon-based nanomaterials (NMs), including metal-based nanoparticles (NPs) (Ag, Au, Cu, Fe_2O_3, TiO_2, and ZnO), graphene, fullerene, multiwalled carbon nanotubes (MWNTs), and single-walled carbon nanotubes (SWNTs), on plants in extreme conditions such as drought have been proven by different studies [68, 69]. By increasing yield and enhancing plant growth indices, NPs help plants in countering drought. It enhances the uptake and utilization of nutrients and water. By changing the cell membranes and controlling stomatal closure, it aids in water retention. As the pigment content and rate of photosynthesis increased in NP-treated plants, the photosynthetic metrics apparently improved as well. It has been observed that NP exposure has increased the activation of both enzymatic and nonenzymatic antioxidants [70]. These antioxidants are important components of the stress-reduction mechanism.

Additionally, it has been documented that NPs control the production of phytohormones such as cytokinin, auxin, gibberellin, and abscisic acid, which aid plants in fending off drought stress, as well as the expression of genes that respond to drought. For example, according to a study's findings, TiO_2 NPs enhanced plant growth during drought and operated effectively. However, when 40 ppm calcium phosphate and 40 ppm TiO_2 NPs were applied to plants together, the best outcomes were achieved [71]. In comparison to the control, they boosted the wheat's fresh weight and dry weight, as well as its root and shoot lengths. Additionally improved were the physiological measures such as the osmolyte (sugar and proline) content, membrane stability index, relative water content, and chlorophyll content [72, 73].

While avoiding over- and under-irrigation that can result from droughts, it is preferable to irrigate to satisfy plant water demands at the appropriate time. For a duration of typically less than a week, this often entails scheduling and controlling irrigation activities on an hourly or daily basis. The goal of precision irrigation is to precisely identify and measure the water requirements of plants. The irrigation volume and timing are based on measurements of soil, plant, and environmental variables from which the plant water need is inferred using AI and ML models [74]. By utilizing developments in sensor, control, and modeling technology, precision irrigation has been demonstrated to increase agricultural output, decrease energy consumption, and improve water usage efficiency.

4.2 Heat Stress

Recent studies have found that the nanotechnology strategies, especially the application of nanoparticles nano-silica (SiO_2) and titanium dioxide (TiO_2), can enhance the ability of leaves and canopies to reflect infrared radiation, which in turn can directly help in lowering their surface temperatures during intense sunlight and heat stress [75]. Experimental validation of these studies has proved that the leaves that had been coated with nanoparticles have shown lower temperatures than the leaves that were uncoated, proving their cooling effect. Studies on transparent surfaces using silica nanopillars show that these features aid in maintaining lower surface temperatures in greenhouse and field settings by reflecting more solar infrared wavelengths [9–11]. Silica-bridged inorganic–organic hybrid membranes have been designed to increase the dissipation of thermal energy and protect the crop canopies against the immense heat stress, providing a solution that is both efficient and scalable against the heat stress mitigation of crops.

Applying different phytohormones to plants is one of the most promising methods of protecting them. Salicylic acid (SA) regulates systemic acquired resistance (SAR) signaling pathways. To shield plants from biotic and abiotic stresses, the expression of genes coding for antioxidant enzymes can be boosted. Systems for the delivery and gradual release of SA in plants must be developed in order to sustain ideal SA concentrations throughout time [76]. For this, a variety of carrier-based nanoparticles (NPs) made from both organic and inorganic substances are employed. In thermogenic and non-thermogenic plants, SA also naturally stimulates the production of heat, encourages the assimilation of CO_2, controls homeostasis, inhibits the activity of catalase and peroxidase, and stomatal conductance, which can help counter heat stress. Treatments with exogenous SA can induce an acclimation effect, enabling plants to adapt to various types of abiotic stress [77]. Improvement in rice yield under high-temperature conditions has been reported when plants were treated with exogenous SA [78].

Climate models using AI, when combined with real-time local sensors, allow very accurate short and long-term prediction of heat waves. These predictions have the ability to transform farm management by allowing growers to take appropriate measures, such as applying nanocoatings to foliage or adjusting shade and irrigation schedules to counter the predicted heat waves and immense sunlight [48].

4.3 *Salinity Stress*

Crop productivity might be cut by 50% due to salinity stress. Soil salinity development is frequently linked to sodium, magnesium, calcium, chlorides, and sulfates. The salinity of the soil affects seed germination. This is because salt causes seeds to absorb more water and have lower gibberellic acid levels while simultaneously having higher amounts of abscisic acid. Salinity stress affects roughly 20% of farmed land worldwide, and its prevalence is rising year by year [79]. Therefore, to boost crop yield, the salt content of the soil or water must be decreased. Nanotechnology might be a significant factor in this issue.

Badawy et al. [80] evaluated the efficacy of selenium nanoparticles on reducing salinity hazard in two varieties of rice (salt-sensitive variety Giza 177 and salt-tolerant variety Giza 178) cultivated in saline soil. The results found that using selenium nanoparticles improves salinity stress tolerance in Giza 177 by increasing its ion selectivity and increasing the crop yield. Kiumarzi et al. [81] estimated the ability of selenium nanoparticles (foliar application) on reducing the harmful effects of salinity stress in pineapple mint, *Mentha suaveolens*. Different selenium nanoparticle concentrations (10 and 20 mg/L) were used in different levels of salinity (30, 60, and 90 mM NaCl). The obtained results showed that selenium nanoparticles, especially 20 mg/L, improve the essential oil content in pineapple mint under salinity stress. The soil salinity management is advancing rapidly by integrating AI and ML models. AI systems combine local sensor networks and remote sensing imagery to generate high-resolution soil salinity maps. These maps can precisely predict salinity hotspots in the field, which allows for the localized applications of the nanoparticles to mitigate saline stresses [82].

4.4 *Pathogens and Pests (Biotic Stress)*

The agricultural sector is facing a severe crisis due to the fast-changing climate, decreased soil fertility, deficiency of macro and micro nutrients, presence of heavy metals in soil, and improper and excessive use of fertilizers and pesticides. Farmers are integrating nano-based materials with advanced agricultural methods to increase the productivity and yield of the crops. For sustainable crop growth, nanoparticles like silver, copper, chitosan, and silica can be used, which have been found to have antimicrobial and insecticidal properties, making them appropriate agents for biological pest control and nano-pesticides [83]. With half the resources used, the productivity has doubled due to the use of nano-based materials. Controlled-release nanofertilizers increase crop productivity, yield, and growth.

Several reports suggest the combination of metal nanoparticles like Fe, Zn, Cu, nano-N, nano-P, nano-K, etc., with commercial fertilizers as foliar spray drastically improved plant growth, economic yield, and NPK nutrient use efficiency (NUE), among other parameters [84, 85]. Additionally, slow-release nanofertilizers are used

Table 1 AI and sensor-integrated nano-delivery systems in smart agriculture

Study	Sensor types	ML trigger for nano-delivery
Smart agriculture systems using IoT and nanoparticles	Soil moisture, temp	ML-driven control of nano-fertilizer, irrigation
Efficient use of nano fertilizers using sensor-based technology	Soil fertility, moisture	Sensors trigger nano-fertilizer release
Nanobiosensors and nanoformulations in agriculture	Nanobiosensors(multi)	Real-time stress cue triggers, nanoformulation
Nanosensors for plant science insights	Nano/micro sensors	ML/AI triggers nano-nutrient application
Proximal soil moisture sensing for real-time water delivery	Soil moisture	Real-time water (or nano-input) delivery

to compensate for inorganic fertilizer waste. Plants might then absorb many of their nutritional needs without losing any vital elements. Field studies on nutrient usage efficiency of engineered nanofertilizers in rice showed much higher usage efficiency than traditional fertilizer application, resulting in higher crop yield (7.9 ton/ha) by only providing half the necessary nitrogen-nutrient value compared to traditional application (7.3 ton/ha) [86].

With the help of ML and DL-based image recognition models, AI surveillance on drones and camera-captured images is becoming more efficient at diagnosing early pest and disease outbreaks in crops. By the use of this technology, early and precise information is made possible, which makes taking rapid decisions for timely intervention easier. To predict pathogen evolution, ML algorithms are increasingly being used, which helps in guiding the design of nanoencapsulated multi-action biocontrol agents for effective crop protection [87]. Recent studies explore how ML models understand genomic, proteomic, and evolutionary data of pathogens to predict mutations, resistance traits, and virulence shifts, enabling and enhancing targeted formulation of nanoparticles loaded with multiple biocontrol agents or antimicrobials (Table 1).

5 Crop-Type and Regional Applications

The application of AI-enhanced nanobiotechnology or any climate-smart strategies for stress resilience is not a "one size fits all" paradigm. Effectiveness of the strategies must be tailored to particular crop species, environments, and regional climate conditions so that specific types of physiological responses must be stimulated in plants [88]. By associating nano-enabled interventions and AI-derived monitoring instruments, specific crop types and geography can be targeted, resulting in achieving both sustainability and productivity on a large scale.

5.1 Major Food Crops

Nanotechnology research is primarily based on the globally produced staple crops for over half of the entire population of the world. This includes the cereal (such as wheat, rice, and maize), legumes, as well as oilseeds (such as soybeans, lentils) [1]. As reported in cereal crops, in wheat and rice, it has been proven that using ZnO and TiO_2 nanoparticles, tolerance to drought and salinity has been enhanced. This happens due to the stimulation of photosynthetic pigments and by maintaining relative water content (RWC) as well as by improvement of antioxidant defense [3]. In addition to this, the usage of nano-silica and chitosan nanoparticles applies with stress adaptation properties through stabilizing membranes and stomatal conductance regulation [89]. Apart from this, the efficacy of nanoparticles for the improvement of stress-resilient properties of cereals like wheat, rice, and maize has also been documented in several studies. In a recent study conducted, it has also been reported that in wheat, zinc oxide nanoparticles (ZnO-NPs) showed higher root and shoot biomass even when kept in saline environments by improving antioxidant enzyme activities and also by maintaining ionic homeostasis [90]. On the other hand, titanium dioxide nanoparticles are found to assist in improving drought resilience in wheat, showing that when higher is the relative water content (RWC), chlorophyll concentration gets elevated along with improving photosynthetic efficacy [72, 73]. For a different perspective, here's another example of ZnO nanoparticles. ZnO nanoparticles have proven to be better for the improvement of resilience during severe drought conditions as well as the increased rate of melatonin synthesis and antioxidant activity [91]. Various other nanoparticles, such as nano-silica, are also proven to have a reliable effect on the growth and yield of cereals like rice and maize even under adverse conditions [92]. With AI to design prediction maps, in cereals, by using techniques such as satellite imaging, including NDVI for the distribution of nanofertilizers for the areas detected for deficiency [93]. Besides this, by narrowing down the use of AI-guided UAVs (unmanned aerial vehicles), crop phenology could be supplemented, thereby improving recovery at critical growth stages in adverse/stress conditions [94].

5.2 Horticultural and Vegetable Crops

Usage of nanoparticles in the treatment of horticulture crops is very substantial as it is of high value, which, when improved, can boost both productivity and quality even in saline conditions or any pathogen attack [95]. With the intervention of AI, we can actually ensure crop-specific protection and prevent over-applications. As an example, computer vision systems are now trained to monitor stress indicators in tomato or grapevine leaves earlier, this enables us to predict specific nano-treatments [96]. Fruits and vegetables like tomatoes, strawberries, and brassicas have shown enhanced stress resilience and improved post-harvest quality using nanoparticles.

For example, in tomato plants, ZnO-NPs have been reported to increase antioxidant enzyme activity and biomass, and Ag NPs have improved seedling vigor and better germination rates in tomatoes under saline environments [97]. Even nanomaterials like silicon and chitosan NPs are widely studied for their ability to strengthen cell walls, enhance shelf-life, and induce systemic resistance against pathogens in a variety of fruit crops like citrus and brassica. Intervention with AI-guided scheduling ensures crop-specific protection. As an example, computer vision systems trained to monitor early stress indicators in tomato or grapevine leaves help to enable predictive nano-treatments [98]. Medicinal and regional crops incorporated with nanoparticles, which act as elicitors of secondary metabolites, are important for their medicinal and aromatic properties. For example, alkaloid and flavonoid production is enhanced by nanocarrier-delivered methyl jasmonate in plant tissue cultures [8, 99].

An AI model is being trained on climate datasets that support region-specific interventions. For example, in Mediterranean regions, AI systems are built to enhance stress-tolerant tomatoes, and in coastal Asia, they are built to support salt-tolerant rice with the intervention of nanoparticles. By the integration of nanomaterial with adapted AI models, one can support both ecological and socio-economic needs that, in the long term, support food security and small farmers [100].

6 Challenges and Future Directions

Combining AI and nanotechnology shifted a paradigm in agriculture from laboratory to large-scale implementation, which is fraught with challenges. Keeping these issues in mind is to ensure the technology is reliable, effective, equitable, and essential, while also providing long-term solutions for future technologies.

6.1 Toxicity and Environmental Safety

Properties of nanoparticles, like small size, high reactivity, and high surface-to-volume ratio when engineered, create a highly positive impact on agricultural agents, but sometimes raise concerns like environmental and health issues at the same time. Because of their size, which is usually smaller than 20 nm, they can effectively penetrate the tissue part of the plant, which results in bioaccumulation in edible portions. This further enters the food chain and potentially impacts the soil microbiota, various types of pollinators, and other useful organisms, as indicated by toxicological studies [101]. Underlining the requirement for nano-ecotoxicology, this duality further enhances the bioavailability, persistence, and modification of nanomaterials under different field conditions. This domain uses predictive capabilities to correlate various nanoparticle physicochemical features, such as size, charge, coatings, and reactivity, to biological outcomes. Also, by associating artificial intelligence, we can create "safe-by-design" nanomaterials. Regulatory frameworks are evolving globally on the governance front. Particularly for genetic-engineering applications such as NP-mediated gene delivery, the EU, US, and China are currently developing guidelines for nano-enabled agro-products that ensure harmonized international regulations and consistent safety evaluation, which will be critical for responsible adoption [102].

6.2 *Data and Modeling Limitations*

Broadening the AI application in nano-agriculture requires fundamental constraints like the availability of large, high-quality, and well-annotated datasets. However, the current field of nano-agriculture is very sparse, with scattered and non-standardized data that are often limited to laboratory or small-scale trials. This data also limits the development of machine learning (ML) models, which can be used to predict nano-agriculture [103]. The problems become more sophisticated when results from different studies are compared due to the variation in experimental conditions, NP characterization, and reporting standards.

6.3 *Scalability and Cost*

AI-nanotech solutions having the potential to create a real-world impact are required to have economic viability and scalability. The nanoformulations have an expensive synthesis, functionalization, and quality control as compared to conventional agrochemicals. The current economic and engineering challenge that persists is the consistent production of tons of stable nanomaterial for agricultural use. To combat cost barriers, identification of the minimal effective dosage and optimizing the application schedule to maximize efficiency and improve cost-effectiveness are such challenges that can be easily achieved by AI optimization. However, across diverse agro-ecological zones, there are extensive field trials and demonstration projects in developing countries, ranging from large-scale commercial farms to smallholder settings in developing countries—these are essential. These field trials and projects are very important to validate lab results in real-world conditions, analyze the economic benefits for farmers, and develop possible business models that check whether these advanced tools do not worsen the global digital and technological divide [104].

6.4 *Interdisciplinarity*

The development of effective and impactful AI-driven nanobiotech solutions needs authentic, well-founded, and deep collaboration of the niches that were earlier used in isolation. Thus, to develop such platforms, plant physiologists, data scientists, nano-material scientists, agricultural engineers, economists, and social scientists should team up to effectively address necessary concerns. The aim behind this is to establish a "digital agriculture stack"—a coordinated and integrated platform that combines real-time data from nanosensors and IoT devices, phenomics and genomics datasets, AI-based predictive tools, and nanoinformatics design systems [105]. In making such compatible and interconnected platforms, a few challenges exist, it is important to convert theoretical innovations into practical solutions for real-world agricultural needs.

6.5 Future Opportunities

Apart from some present challenges, these cutting-edge technologies introduce exciting avenues to further improve this integration:

- **Synthetic Biology**: Genetically modified microbes to produce and combine directly within the rhizosphere, targeted delivery systems could be improvised, decreasing both manufacturing and application costs while enhancing precision [106].
- **Blockchain Technology**: The unification of blockchain with IoT sensors could set up transparent and tamper-proof records across the nanoagriculture supply chain. This would monitor nano-formulation usage from production to application, assuring credibility and encouraging greater consumer confidence [107].
- **Digital Twin Technology**: Developing "digital twin farms"—exact virtual replicas of real agricultural systems—would equip researchers and farmers to model and refine AI-driven nano-intervention strategies in a risk-free environment before applying them in actual field conditions [108].

7 Conclusion

Synergy between artificial intelligence (AI) and nanobiotechnology in agriculture is a game-changer that would transition agriculture from reactive to proactive, data-driven, sustainable systems. This convergence sets the scientific base for producing robust crops, which are able to grow under multiple environmental stresses without disrupting the ecology and food security across the world. This nanotechnology and AI combination brings forth not only precision and predictability in agriculture but also promotes advancement in crop improvement, nutrient management, and stress resistance.

The dependent application of nanotechnology to reduce abiotic stress factors (e.g., drought, salinity, and heat) by engineered nanomaterials (ENMs), e.g., ZnO, TiO_2, and carbon-based nanoparticles, has good potential. With a combination of AI-athrophagic models, it is possible with astonishing precision to design, refine, and implement these nanoparticles. The prediction of the nanoparticle plant interaction could be performed using machine learning (ML)- and deep learning (DL)-based algorithms, which could reveal environmental risk assessment and propose a safe dose that would help to reduce the toxicity and increase the efficacy.

Moreover, AI-powered nanosensors maintain both the well-being and the nutrient circulation and reaction of a crop to strain and ensure that the designated farmers can make informed choices at any given moment that enhance productivity without depleting resources. Precise gene editing with CRISPR-Cas through nanoparticle-mediated delivery is another potential of AI-based nanobiotechnology that can be used to encourage the creation of stress-tolerant crop strains. AI-based genomic selection models have the potential to function with massive omics measurements and identify gene variants related to drought-tolerance, pest-resistance, and yield characteristics.

The combination of these technologies will provide a strong infrastructure for the fast cycle and sustainable enhancement of genetics. However, to realize its global talent, there are a few steps that should be successfully overcome. Strict regulatory processes and the collaboration of the international community should discuss issues related to the toxicity of nanoparticles, their environmental half-life, harmonization of data, and the overall economic viability. To continue, more databases on nanoinformatics should be open access, nanomaterials should be designed to be safer, and eco-toxicological risk assessments should be integrated into AI algorithms. Moreover, equitable access to AI-nano technologies, especially among small-scale farmers in developing countries, will be necessary to avoid the formation of another digital divide. The future of agriculture is believed to be witnessed in the form of Digital-Nano-Bio (DNB) agriculture, in which the entire scene at a field level, like nanotechnology-based sensors, farm data, and machine learning analytics, works closely with biotechnological breakthroughs.

References

1. Lesk C, Rowhani P, Ramankutty N (2016) Influence of extreme weather disasters on global crop production. Nature 529(7587):84–87. https://doi.org/10.1038/nature16467
2. Mahanta S, Kundu R, Sen P, Muhammed Mubarak KP, Shaw S, Hossain A, Dhar A, Sarkar S, Santra SC, Moulick D (2025) Heavy metal accumulation in vegetable crops: a threat to sustainable agroecosystem. Abiotic Stress Undergr Veg: Response Mitig, 265–292. https://doi.org/10.1016/B978-0-443-23961-8.00018-8
3. Raza MAS, Muhammad F, Farooq M, Aslam MU, Akhter N, Toleikienė M, Binobead MA, Ali MA, Rizwan M, Iqbal R (2025) ZnO-nanoparticles and stage-based drought tolerance in wheat (*Triticum aestivum* L.): effect on morpho-physiology, nutrients uptake, grain yield and quality. Sci Rep 15(1):5309. https://doi.org/10.1038/s41598-025-89718-2
4. Raza A, Khare T, Zhang X, Rahman MM, Hussain M, Gill SS, Chen H, Zhou M, Hu Z, Varshney RK (2025) Novel Strategies for Designing Climate-Smart Crops to Ensure Sustainable Agriculture and Future Food Security. J Sustain Agric Environ 4(2):e70048. https://doi.org/10.1002/sae2.70048
5. Raza MA, Muhammad F, Farooq M, Aslam MU, Akhter N, Toleikienė M, Binobead MA, Ali MA, Rizwan M, Iqbal R (2025) ZnO-nanoparticles and stage-based drought tolerance in wheat (*Triticum aestivum* L.): Effect on morpho-physiology, nutrients uptake, grain yield and quality. Sci Rep 15(1):1–14. https://doi.org/10.1038/s41598-025-89718-2

6. Kant R, Shukla DR, Mandal S (2025) Nanocarriers for delivering plant growth promoting microorganisms. In Chen J (ed) Nanocarriers in plant science and agriculture. IGI Global Scientific Publishing, pp 235–272. https://doi.org/10.4018/979-8-3693-9869-2.ch009
7. Khose S, Bhausaheb D, Shekhar S (2023) The role of precision farming in sustainable agriculture: advancements and impacts article. ResearchGate. https://www.researchgate.net/publication/373166415_The_Role_of_Precision_Farming_in_Sustainable_Agriculture_Advancements_and_Impacts_Article
8. Singh Y, Kumar U, Panigrahi S, Balyan P, Mehla S, Sihag P, Sagwal V, Singh KP, White JC, Dhankher OP (2023) Nanoparticles as novel elicitors in plant tissue culture applications: current status and future outlook. Plant Physiol Biochem 203:108004. https://doi.org/10.1016/j.plaphy.2023.108004
9. Raza A, Zaman QU, Hu Z (2024) Leveraging a new thermosensor for heat-smart future agriculture. Plant commun 5(9):101007. https://doi.org/10.1016/j.xplc.2024.101007
10. Raza A, Bhardwaj S, Rahman MA, García-Caparrós P, Copeland RG, Charagh S, Rivero RM, Gopalakrishnan S, Corpas FJ, Siddique KH, Hu Z (2024) Fighting to thrive via plant growth regulators: green chemical strategies for drought stress tolerance. Physiol Plant 176(6):e14605. https://doi.org/10.1111/ppl.14605
11. Raza A, Gangurde SS, Singh Sandhu K, Lv Y (2024) Omics-assisted crop improvement under abiotic stress conditions. Plant Stress 14:100626. https://doi.org/10.1016/j.stress.2024.100626
12. Ali Z, Yaqoob S, Yu J, D'Amore A (2024) Critical review on the characterization, preparation, and enhanced mechanical, thermal, and electrical properties of carbon nanotubes and their hybrid filler polymer composites for various applications. Comp Part C: Open Access 13:100434. https://doi.org/10.1016/j.jcomc.2024.100434
13. Mendoza RP, Brown JM (2019) Engineered nanomaterials and oxidative stress: current understanding and future challenges. Curr Opin Toxicol 13:74–80. https://doi.org/10.1016/j.cotox.2018.09.001
14. Rasheed A, Li H, Tahir MM, Mahmood A, Nawaz M, Shah AN, Aslam MT, Negm S, Moustafa M, Hassan MU, Wu Z (2022) The role of nanoparticles in plant biochemical, physiological, and molecular responses under drought stress: a review. Front Plant Sci 13:976179. https://doi.org/10.3389/fpls.2022.976179
15. Carvalho FEL, Ware MA, Lima Neto MC, Aranjuelo I (2022) Editorial: photosynthetic efficiency under multiple stress conditions: prospects for increasing crop yields. Front Plant Sci 13:893730. https://doi.org/10.3389/fpls.2022.893730
16. Hasanuzzaman M, Fujita M (2022) Plant responses and tolerance to salt stress: physiological and molecular interventions. Int J Mol Sci 23(9):4810. https://doi.org/10.3390/ijms23094810
17. Assaha DVM, Ueda A, Saneoka H, Al-Yahyai R, Yaish MW (2017) The role of Na+ and K+ transporters in salt stress adaptation in glycophytes. Front Physiol 8:509. https://doi.org/10.3389/fphys.2017.00509
18. Chauhan J, Prathibha M, Singh P, Choyal P, Mishra UN, Saha D, Kumar R, Anuragi H, Pandey S, Bose B, Mehta B, Dey P, Dwivedi K, Gupta N, Singhal RK (2023) Plant photosynthesis under abiotic stresses: damages, adaptive, and signaling mechanisms. Plant Stress 10:100296. https://doi.org/10.1016/j.stress.2023.100296
19. Verma SK, Das AK, Patel MK, Shah A, Kumar V, Gantait S (2018) Engineered nanomaterials for plant growth and development: a perspective analysis. Sci Total Environ 630:1413–1435. https://doi.org/10.1016/j.scitotenv.2018.02.313
20. Ijaz M, Khan F, Ahmed T, Noman M, Zulfiqar F, Rizwan M, Chen J, Siddique KHM, Li B (2023) Nanobiotechnology to advance stress resilience in plants: current opportunities and challenges. Materials Today Bio 22:100759. https://doi.org/10.1016/j.mtbio.2023.100759
21. Pal K, Khurana SM, Debnath N, Das S, Pal K, Debnath N, Das S (2022) Micronutrient nanoparticles: synthesis, properties and application in agriculture. ResearchGate. https://www.researchgate.net/publication/365700555_Micronutrient_Nanoparticles_Synthesis_Properties_and_Application_in_Agriculture_15_Sustainable_agriculture_Green_synthesis_Precision_farming_Nanofertilizers_Slow_release

22. Arshad R, Gulshad L, Haq IU, Farooq MA, Al-Farga A, Siddique R, Manzoor MF, Karrar E (2021) Nanotechnology: a novel tool to enhance the bioavailability of micronutrients. Food Sci Nutr 9(6):3354–3361. https://doi.org/10.1002/fsn3.2311
23. Mussin J, Giusiano G (2022) Biogenic silver nanoparticles as antifungal agents. Front Chem 10:1023542. https://doi.org/10.3389/fchem.2022.1023542
24. Mukherjee S, Sikdar B, Roy A, Hossain A, Moulick D (2025) Small interfering RNAs: potent silencing tools for plant protection. Funct RNAs Plants, 111–126. https://doi.org/10.1016/B978-0-443-33341-5.00022-7
25. Wu Q, Gao G, Li Y, Zhang C, Yan H, Xu N, Tan Y (2025) Nanoparticle-loaded dsRNA delivery system for pesticides: Status and perspective. Adv Agrochem 4(3):235–248. https://doi.org/10.1016/j.aac.2025.08.003
26. Akhter Bhat T, Rather RA, Bashir S, Mishra N, John R (2025) Nanoparticles in plant systems: omics-based perspectives on stress adaptation and toxicological implications. Plant Nano Biol 13:100181. https://doi.org/10.1016/j.plana.2025.100181
27. Jash S, Das S, Hurrah IM, Kumari A, Mandal S (2025) CRISPR-edited plants for resistance to viral diseases: methods and applications. CRISPR Technol Combat Plant Pests Pathog, 129–160. https://doi.org/10.1016/B978-0-443-43920-9.00007-5
28. Vats S, Kumawat S, Brar J, Kaur S, Yadav K, Magar SG, Jadhav PV, Salvi P, Sonah H, Sharma S, Deshmukh R (2022) Opportunity and challenges for nanotechnology application for genome editing in plants. Plant Nano Biol 1:100001. https://doi.org/10.1016/j.plana.2022.100001
29. Cunningham FJ, Goh NS, Demirer GS, Matos JL, Landry MP (2018) Nanoparticle-mediated delivery towards advancing plant genetic engineering. Trends Biotechnol 36(9):882–897. https://doi.org/10.1016/j.tibtech.2018.03.009
30. Mana A, Allouhi A, Hamrani A, Rehman S, El Jamaoui I, Jayachandran K (2024) Sustainable AI-based production agriculture: exploring AI applications and implications in agricultural practices. Smart Agric Technol 7:100416. https://doi.org/10.1016/j.atech.2024.100416
31. Hoque A, Padhiary M (2024) Automation and AI in precision agriculture: Innovations for enhanced crop management and sustainability. Asian J Res Comput Sci 17:95–109. https://doi.org/10.9734/ajrcos/2024/v17i10512
32. Rosa A, Barão L, Chambel L, Cruz C, Santana M (2023) Early identification of plant drought stress responses: changes in leaf reflectance and plant growth promoting rhizobacteria selection—the case study of tomato plants. Agronomy 13(1):183. https://doi.org/10.3390/agronomy13010183
33. Maity R, Srivastava A, Sarkar S, Khan MI (2024) Revolutionizing the future of hydrological science: impact of machine learning and deep learning amidst emerging explainable AI and transfer learning. Appl Comput Geosci 24:100206. https://doi.org/10.1016/j.acags.2024.100206
34. Mallik S, Chakraborty A, Podder K, Talukdar S, Rahman A, Mishra U (2025) Enhancing soil moisture prediction with explainable AI: integrating IoT and multi-sensor remote sensing data through soft computing. Appl Soft Comput 180:113406. https://doi.org/10.1016/j.asoc.2025.113406
35. Jabed MA, Murad MAA (2024) Crop yield prediction in agriculture: a comprehensive review of machine learning and deep learning approaches, with insights for future research and sustainability. Heliyon 10(24):e40836. https://doi.org/10.1016/j.heliyon.2024.e40836
36. Maojo V, Fritts M, de la Iglesia D, Cachau RE, Garcia-Remesal M, Mitchell JA, Kulikowski C (2012) Nanoinformatics: a new area of research in nanomedicine. Int J Nanomed 7:3867–3890. https://doi.org/10.2147/IJN.S24582
37. Yang L, Wang H, Leng D, Fang S, Yang Y, Du Y (2024) Machine learning applications in nanomaterials: recent advances and future perspectives. Chem Eng J 500:156687. https://doi.org/10.1016/j.cej.2024.156687
38. Basu S, Murmu S, Murmu K, Choudhury A, Upadhyay MK, Hossain A, Moulick D (2025) Agronomic approaches to deal with the abiotic stress tolerance in root and tuber crops. Abiotic Stress Undergr Veg: Response Mitig, 215–225. https://doi.org/10.1016/B978-0-443-23961-8.00015-2

39. Gimeno A, Ojeda-Montes MJ, Tomás-Hernández S, Cereto-Massagué A, Beltrán-Debón R, Mulero M, Pujadas G, Garcia-Vallvé S (2019) The light and dark sides of virtual screening: what is there to know? Int J Mol Sci 20(6):1375. https://doi.org/10.3390/ijms20061375
40. Ahmadi M, Ayyoubzadeh SM, Ghorbani-Bidkorpeh F (2024) Toxicity prediction of nanoparticles using machine learning approaches. Toxicology 501:153697. https://doi.org/10.1016/j.tox.2023.153697
41. Sharma M, Mahajan P, Alsubaie AS, Khanna V, Chahal S, Thakur A, Yadav A, Arya A, Singh A, Singh G (2025) Next-generation nanomaterials-based biosensors: real-time biosensing devices for detecting emerging environmental pollutants. Mater Today Sustain 29:101068. https://doi.org/10.1016/j.mtsust.2024.101068
42. Sharma H, Sidhu H, Bhowmik A (2025) Remote sensing using unmanned aerial vehicles for water stress detection: a review focusing on specialty crops. Drones 9(4):241. https://doi.org/10.3390/drones9040241
43. Adam T, Gopinath SCB (2022) Nanosensors: recent perspectives on attainments and future promise of downstream applications. Process Biochem 117:153–173. https://doi.org/10.1016/j.procbio.2022.03.024
44. Yin S (2025) Artificial intelligence-assisted nanosensors for clinical diagnostics: current advances and future prospects. Biosensors 15(10):656. https://doi.org/10.3390/bios15100656
45. Olawade DB, Ige AO, Olaremu AG, Ijiwade JO, Adeola AO (2024) The synergy of artificial intelligence and nanotechnology towards advancing innovation and sustainability: a mini-review. Nano Trends 8:100052. https://doi.org/10.1016/j.nwnano.2024.100052
46. Moulick D, Hossain A, Barek V et al (2024) Heavy metal stress in the agro-environment: consequences, adaptations and remediation. Int J Environ Sci Technol 21:9297–9340. https://doi.org/10.1007/s13762-024-05657-x
47. Pandey M, Srivastava AK, Penna S (2024) Molecular insights for improving plant resilience to salinity. In: Al-Khayri JM, Ingle KP, Jain SM, Penna S (eds) Plant molecular breeding in genomics era. Advances in plant breeding strategies, vol 4. Springer, Cham. https://doi.org/10.1007/978-3-031-68598-9_3
48. Singh SB, Sonone N, Tavare N, Suprasanna P (2025) Transgenic breeding for improving abiotic stress tolerance in rice. Rice Cultiv Abiotic Stress, 395–406. https://doi.org/10.1016/B978-0-443-21793-7.00025-8
49. Laib I, Gheraissa N, Benaissa A, Benkhira L, Azzi M, Benaissa Y, Abdelaziz AG, Tian F, Walsh M, Bechelany M, Barhoum A (2025) Tailoring innovative silver nanoparticles for modern medicine: the importance of size and shape control and functional modifications. Materials Today Bio 33:102071. https://doi.org/10.1016/j.mtbio.2025.102071
50. Unde SS, Kurkute VK, Chavan SS et al (2025) The expanding role of multirotor UAVs in precision agriculture with applications AI integration and future prospects. Discov Mech Eng 4:38. https://doi.org/10.1007/s44245-025-00132-4
51. Ahmed N, Shakoor N (2025) Advancing agriculture through IoT, Big Data, and AI: a review of smart technologies enabling sustainability. Smart Agric Technol 10:100848. https://doi.org/10.1016/j.atech.2025.100848
52. Sharma K, Shivandu SK (2023) Integrating artificial intelligence and internet of things (IoT) for enhanced crop monitoring and management in precision agriculture. Sens Int 5:100292. https://doi.org/10.1016/j.sintl.2024.100292
53. Singh S, Praveen A, Dudha N, Bhadrecha, P (2024) Integrating physiological and multi-omics methods to elucidate heat stress tolerance for sustainable rice production. Physiology and molecular biology of plants: Int J Funct Plant Biol 30(7), 1185–1208. https://doi.org/10.1007/s12298-024-01480-3
54. Arsenovic M, Karanovic M, Sladojevic S, Anderla A, Stefanovic D (2019) Solving current limitations of deep learning based approaches for plant disease detection. Symmetry 11:939. https://doi.org/10.3390/sym11070939
55. Yan J, Wang X (2022) Unsupervised and semi-supervised learning: the next frontier in machine learning for plant systems biology. Plant J: Cell Mol Biol 111(6):1527–1538. https://doi.org/10.1111/tpj.15905

56. Isewon I, Apata O, Oluwamuyiwa F et al. (2022) Machine learning algorithms: their applications in plant omics and agronomic traits' improvement. F1000Research 11:1256.https://doi.org/10.12688/f1000research.125425.1
57. Li D, Wang R, Xie C, Liu L, Zhang J, Li R, Wang F, Zhou M, Liu W (2020) A recognition method for rice plant diseases and pests video detection based on deep convolutional neural network. Sensors 20:578. https://doi.org/10.3390/s20030578
58. Mi Z, Zhang X, Su J, Han D, Su B (2020) Wheat Stripe rust grading by deep learning with attention mechanism and images from mobile devices. Frontiers Plant Sci 11:558126. https://doi.org/10.3389/fpls.2020.558126
59. Singh J, Varshney V, Saddhe AA, Manuka R, Penna S, Jha S (2024) Molecular techniques for the improvement of abiotic stress tolerance in plants. In: Al-Khayri J, Alnaddaf LM, Jain SM, Penna S (eds) Innovative methods in horticultural crop improvement. advances in plant breeding strategies, vol 1. Springer, Cham. https://doi.org/10.1007/978-3-031-61081-3_3
60. Li L, Snyder, JC, Pelaschier, IM, Huang J, Niranjan U-N, Duncan P, Rupp M, Müller K-R, Burke K, (2016). Understanding machine-learned density functionals. Int J Quantum Chem 116(11):819–833. https://doi.org/10.1002/qua.25040
61. . Wang J, Chen L, Wang Y, Zhang J, Liang Y, Xu D (2013) A computational systems biology study for understanding salt tolerance mechanism in rice. PloS one 8(6):e64929. https://doi.org/10.1371/journal.pone.0064929
62. Oliva R, Ji C, Atienza-Grande G, Huguet-Tapia JC, Perez-Quintero A, Li T, Eom JS, Li C, Nguyen H., Liu, B., Auguy F, Sciallano C, Luu, VT, Dossa GS, Cunnac S, Schmidt SM, Slamet-Loedin IH, Vera Cruz, C, Szurek B, Frommer WB, … Yang B (2019) Broad-spectrum resistance to bacterial blight in rice using genome editing. Nat Biotechnol37(11):1344–1350. https://doi.org/10.1038/s41587-019-0267-z
63. Rahman MA, Karthikeyan M, Johnson I, Raja K, Sekar C, Mary XA, Basha JS (2025) Ensuring food security through rapid and in-field detection of diseases in food crops using real time and portable sensors. Anal Biochem 705:115925. https://doi.org/10.1016/j.ab.2025.115925
64. Bharti A, Jain U, Chauhan N (2024) From lab to field: nano-biosensors for real-time plant nutrient tracking. Plant Nano Biol 9:100079. https://doi.org/10.1016/j.plana.2024.100079
65. Wang J, Wang Y, Qi Z (2024) Remote sensing data assimilation in crop growth modeling from an agricultural perspective: new insights on challenges and prospects. Front Plant Sci 15:1256792. https://doi.org/10.3390/agronomy14091920
66. Wang J, Wang Y, Qi Z (2024) Remote sensing data assimilation in crop growth modeling from an agricultural perspective: new insights on challenges and prospects. Agronomy 14(9):1920. https://doi.org/10.3390/agronomy14091920
67. Abiri R, Rizan N, Balasundram SK, Shahbazi AB, Abdul-Hamid H (2023) Application of digital technologies for ensuring agricultural productivity. Heliyon 9(12):e22601. https://doi.org/10.1016/j.heliyon.2023.e22601
68. Chandrashekar HK, Singh G, Kaniyassery A, Thorat SA, Nayak R, Murali TS, Muthusamy A (2023) Nanoparticle-mediated amelioration of drought stress in plants: a systematic review. Front Plant Sci 14:10491566. https://doi.org/10.1007/s13205-023-03751-4
69. Chandrashekar HK, Singh G, Kaniyassery A, Thorat SA, Nayak R, Murali TS, Muthusamy A (2023) Nanoparticle-mediated amelioration of drought stress in plants: a systematic review. 3 Biotech 13(10):336. https://doi.org/10.1007/s13205-023-03751-4
70. Pandya P, Kumar S, Sakure AA, Rafaliya R, Patil GB (2023) Zinc oxide nanopriming elevates wheat drought tolerance by inducing stress-responsive genes and physio-biochemical changes. Curr Plant Biol 35–36:100292. https://doi.org/10.1016/j.cpb.2023.100292
71. Mustafa H, Ilyas N, Akhtar N, Raja NI, Zainab T, Shah T, Ahmad A, Ahmad P (2021) Biosynthesis and characterization of titanium dioxide nanoparticles and its effects along with calcium phosphate on physicochemical attributes of wheat under drought stress. Ecotoxicol Environ Saf 223:112519. https://doi.org/10.1016/j.ecoenv.2021.112519
72. Rehman M, Salam A, Ulhassan Z, Ali B, Haider Z, Ahmad I, Yasin MU, Javaid MH, Yang C, Fayyaz M, Gan Y (2025) Titanium dioxide nanoparticles in crop stress management: Mechanisms, applications, and abiotic stress mitigation. Plant Nano Biol 5:100207. https://doi.org/10.1016/j.plana.2025.100207

73. Rehman SU, Yang J, Zhang J, Zhang L, Hao X, Song R, Chen S, Wang G, Hua L (2025) Salt stress in wheat: a physiological and genetic perspective. Plant Stress 16:100832. https://doi.org/10.1016/j.stress.2025.100832
74. Umutoni L, Samadi V (2024) Application of machine learning approaches in supporting irrigation decision making: a review. Agric Water Manag 294:108710. https://doi.org/10.1016/j.agwat.2024.108710
75. Dilnawaz F, Kalaji MH, Misra AN (2023) Nanotechnology in improving photosynthesis under adverse climatic conditions: cell to canopy action. Plant Nano Biol 4:100035. https://doi.org/10.1016/j.plana.2023.100035
76. Mishra S, Roychowdhury R, Ray S, Hada A, Kumar A, Sarker U, Aftab T, Das R (2024) Salicylic acid (SA)-mediated plant immunity against biotic stresses: an insight on molecular components and signaling mechanism. Plant Stress 11:100427. https://doi.org/10.1016/j.stress.2024.100427
77. Torun IE, Kılınc YB, Kilinc E (2022) Endogenous and exogenous serotonin, but not sumatriptan, ameliorate seizures and neuroinflammation in the pentylenetetrazole-induced seizure model in rats. Arquivos de neuro-psiquiatria, 80(1):48–55. https://doi.org/10.1590/0004-282X-ANP-2021-0101
78. Yang W, Hao Q, Liang J, Tan Q, Luan X, Lin S, Zhu H, Bu S, Liu Z, Liu G, Wang S and Zhang G (2022) Fine mapping of two major quantitative trait loci for rice chalkiness with high temperature-enhanced additive effects. Front. Plant Sci. 13:957863. https://doi.org/10.3389/fpls.2022.957863
79. Hasan MM, Rahman MA, Corpas FJ, Rahman MM, Jahan MS, Liu X, Ghimire S, Alabdallah NM, Wassem M, Alharbi BM, Raza A, Fang X (2024) Salt stress tolerance in rice (*Oryza sativa* L.): a proteomic overview of recent advances and future prospects. Plant Stress 11:100307. https://doi.org/10.1016/j.stress.2023.100307
80. Badawy SA., Zayed BA, Bassiouni SMA, Mahdi AHA, Majrashi A, Ali EF, Seleiman MF (2021) Influence of Nano Silicon and Nano Selenium on Root Characters, Growth, Ion Selectivity, Yield, and Yield Components of Rice (Oryza sativa L.) under Salinity Conditions. Plants (Basel, Switzerland), 10(8):1657. https://doi.org/10.3390/plants10081657
81. Kiumarzi F, Morshedloo MR, Zahedi SM, Mumivand H, Behtash F, Hano C, Chen JT, Lorenzo JM (2022) Selenium nanoparticles (se-nps) alleviates salinity damages and improves phytochemical characteristics of pineapple mint (mentha suaveolens Ehrh.). Plants (Basel, Switzerland), 11(10):1384. https://doi.org/10.3390/plants11101384
82. Parra-López C, Ben Abdallah S, Garcia-Garcia G, Hassoun A, Trollman H, Jagtap S, Gupta S, Aït-Kaddour A, Makmuang S, Carmona-Torres C (2025) Digital technologies for water use and management in agriculture: recent applications and future outlook. Agric Water Manag 309:109347. https://doi.org/10.1016/j.agwat.2025.109347
83. Ahmed N, Zhang B, Chachar Z, Li J, Xiao G, Wang Q, Hayat F, Deng L, Narejo M, Bozdar B, Tu P (2023) Micronutrients and their effects on horticultural crop quality, productivity and sustainability. Sci Hortic 323:112512. https://doi.org/10.1016/j.scienta.2023.112512
84. Rahman S, Hoque M, Uddin M (2021) Problems faced by the crop farmers related to extension services provided by department of agricultural extension in Bangladesh. Int J Agri Ext 9(3):373–388. https://doi.org/10.33687/ijae.009.03.3606
85. Abd El-Azeim MM, Sherif, MA., Hussien MS, Tantawy, IAA, Bashandy SO (2020) Impacts of nano- and non-nanofertilizers on potato quality and productivity. Acta Ecologica Sinica, 40(5):388–397. https://doi.org/10.1016/j.chnaes.2019.12.007
86. Carmona FJ, Guagliardi A, Masciocchi N (2022) Nanosized calcium phosphates as novel macronutrient nano-fertilizers. Nanomaterials (Basel, Switzerland), 12(15):2709. https://doi.org/10.3390/nano12152709
87. Majdalawieh M, Martins C, Radi M, Alaraj M, Khan S (2025) Precision agriculture in the age of AI: a systematic review of machine learning methods for crop disease detection. Smart Agric Technol 12:101491. https://doi.org/10.1016/j.atech.2025.101491
88. Edo GI, Mafe AN, Ali AB, Akpoghelie PO, Yousif E, Isoje EF, Igbuku UA, Zainulabdeen K, Owheruo JO, Essaghah AEA, Umar H, Ahmed DS, Alamiery AA (2025) Eco-friendly

nanoparticle phytosynthesis via plant extracts: mechanistic insights, recent advances, and multifaceted uses. Nano TransMed 4:100080. https://doi.org/10.1016/j.ntm.2025.100080
89. Saberi Riseh R, Vatankhah M, Hassanisaadi M, Kennedy JF (2023) Chitosan/silica: a hybrid formulation to mitigate phytopathogens. Int J Biol Macromol 239:124192. https://doi.org/10.1016/j.ijbiomac.2023.124192
90. Kale SS, Chauhan R, Nigam B, Gosavi S, Chaudhary IJ (2024) Effectiveness of nanoparticles in improving soil fertility and eco-friendly crop resistance: a comprehensive review. Biocatal Agric Biotechnol 56:103066. https://doi.org/10.1016/j.bcab.2024.103066
91. Dey S, Mohanty DL, Divya N, Bakshi V, Mohanty A, Rath D, Das S, Mondal A, Roy S, Sabui R (2025) A critical review on zinc oxide nanoparticles: synthesis, properties and biomedical applications. Intell Pharm 3(1):53–70. https://doi.org/10.1016/j.ipha.2024.08.004
92. McLean B, Yarovsky I (2024) Structure, properties, and applications of silica nanoparticles: recent theoretical modeling advances, challenges, and future directions. Small 20(51):e2405299. https://doi.org/10.1002/smll.202405299
93. Alami Machichi M, El Mansouri L, Imani Y, Bourja O, Hadria R, Lahlou O, Benmansour S, Zennayi Y, Bourzeix F (2022) CerealNet: a hybrid deep learning architecture for cereal crop mapping using sentinel-2 time-series. Informatics, 9(4):96. https://doi.org/10.3390/informatics9040096
94. Hossin MA, Yin S, Dan R, Islam MS, Khatun MS, Hossin MA (2025) Integrating artificial intelligence in unmanned vehicles: navigating uncertainties, risks, and the path forward for the fourth industrial revolution. Hum Soc Sci Commun 12:312. https://doi.org/10.1057/s41599-025-04554-z
95. Abou El-Nasr MK, Hassan KM, Abd-Elhalim BT, Kucher DE, Rebouh NY, Ansabayeva A, Abdelkader M, Ali MA, Nasser MA (2025) The emerging roles of nanoparticles in managing the environmental stressors in horticulture crops—a review. Plants 14(14):2192. https://doi.org/10.3390/plants14142192
96. Sidhu AK, Sharma M, Bhickchand Agrawal S, Kapoor R, Kumar A, Kumar V (2024) Nanomaterial strategies for enhancing plant resilience in the face of temperature stress. CABI Agric Biosci 5:60. https://doi.org/10.1186/s43170-024-00255-w
97. Rahaman MM, Hossain R, Herrera-Bravo J, Islam MT, Atolani O, Adeyemi OS, Owolodun OA, Kambizi L, Daştan SD, Calina D, Sharifi-Rad J (2023) Natural antioxidants from some fruits, seeds, foods, natural products, and associated health benefits: an update. Food Sci Nutr 11(4):1657–1670. https://doi.org/10.1002/fsn3.3217
98. Rezić I (2022) Nanoparticles for biomedical application and their synthesis. Polymers 14(22):4961. https://doi.org/10.3390/polym14224961
99. Singh A, Mazahar S, Chapadgaonkar SS, Giri P, Shourie A (2023) Phyto-microbiome to mitigate abiotic stress in crop plants. Front Microbiol 14:1210890. https://doi.org/10.3389/fmicb.2023.1210890
100. Camps-Valls G, Fernández-Torres MÁ, Cohrs KH, Giffard-Roisin S, Kadow C, Boucher MA, Chantry M, Ioannou I, Kladny KF, Lguensat R, Lessig C, Linares MC, Marras M, Morel T, O'Gorman PA, Ouala S, Pérus M, Prando G, Rojas AG et al (2025) Artificial intelligence for modeling and understanding extreme weather and climate events. Nat Commun 16:1919. https://doi.org/10.1038/s41467-025-56573-8
101. Eker F, Duman H, Akdaşçi E, Bolat E, Sarıtaş S, Karav S, Witkowska AM (2024) A comprehensive review of nanoparticles: from classification to application and toxicity. Molecules 29(15):3482. https://doi.org/10.3390/molecules29153482
102. Dedman CJ (2022) Nano-ecotoxicology in a changing ocean. SN Appl Sci 4:264. https://doi.org/10.1007/s42452-022-05147-0
103. Zhang P, Guo Z, Ullah S, Melagraki G, Afantitis A, Lynch I (2021) Nanotechnology and artificial intelligence to enable sustainable and precision agriculture. Nat Plants 7(7):864–876. https://doi.org/10.1038/s41477-021-00946-6
104. Patil M, Perumal C, Choudhari P, Patil S, Chavan S, Singh HV (2025) Differential impacts of regenerative agriculture practices on soil organic carbon: a meta-analysis of studies from India. Sci Rep 15:33470. https://doi.org/10.1038/s41598-025-12149-6

105. Papastratis I, Konstantinidis D, Daras P, Kordas G, Rallis I, Vovoli E, Vretos N (2024) AI nutrition recommendation using a deep generative model and ChatGPT. Sci Rep 14:14620. https://doi.org/10.1038/s41598-024-65438-x
106. Yan Q, Ming Y, Liu J, Yin H, He Q, Li J, Huang M, He Z (2025) Microbial biotechnology: from synthetic biology to synthetic ecology. Adv Biotechnol 3(1):1. https://doi.org/10.1007/s44307-024-00054-4
107. Sizan NS, Dey D, Layek MA, Uddin MA, Huh E-N (2025) Evaluating blockchain platforms for IoT applications in Industry 5.0: a comprehensive review. Blockchain: Res Appl 6(3):100276. https://doi.org/10.1016/j.bcra.2025.100276
108. Peladarinos N, Piromalis D, Cheimaras V, Tserepas E, Munteanu RA, Papageorgas P (2023) Enhancing smart agriculture by implementing digital twins: a comprehensive review. Sensors 23(16):7128. https://doi.org/10.3390/s23167128

Temperature-Smart Agriculture Constructed by AI Models with Nanotechnology

Manoj Mani, Balaji Esakkimuthu, Shama Anzum Ismail, Thilagavathi Gangadharan, Alwin Robert Asirvatham, Antony Prabhu Jeyabal Philomenathan, Sivaranjini Annamalai, and Vijaya Anand Arumugam

Abstract Temperature-smart agriculture (TSA) stands for the application of artificial intelligence (AI) and nano in such ways that promote crop resilience through optimization of temperature controls for sustainability in the face of global challenges. Predictive analytics integrating AI, IoT sensors, and digital twins enable real-time monitoring and adaptive microclimate control through improved water, nutrient, and energy use efficiency. In the same way, nanotechnology brings in various multifunctional approaches, including nano-fertilizers, nanocomposites, and nanophotonic coatings, to further enhance thermotolerance and reduce oxidative stress and balance phytohormones. The synergy of AI and nano enhances the development of temperature-responsive materials and autonomous systems, hence guaranteeing precision environmental management. Ideas about quantum AI, federated learning, and green nanotechnology have been proposed to ensure sustainable innovation with reduced ethical and environmental impacts. In general, integration of AI and nano

M. Mani · A. P. Jeyabal Philomenathan · V. A. Arumugam (✉)
Department of Human Genetics and Molecular Biology, Bharathiar University, Coimbatore, Tamil Nadu, India
e-mail: avahgmb@buc.edu.in

B. Esakkimuthu
Sudha Hospital, Hyderabad, Telangana, India

S. A. Ismail
Department of Biochemistry, PSG College of Arts and Science, Coimbatore, Tamil Nadu, India

T. Gangadharan
Department of Information Technology, Coimbatore Institute of Technology, Coimbatore, Tamil Nadu, India

A. R. Asirvatham
Department of Endocrinology and Diabetes, Prince Sultan Military Medical City, Ministry of Defense, Riyadh, Kingdom of Saudi Arabia

S. Annamalai (✉)
Department of Biotechnology, Sri Ramakrishna College of Arts & Science, Coimbatore, Tamil Nadu, India
e-mail: sivaranjini@srcas.ac.in

J.-T. Chen (ed.), *AI-Nano-Enabled Climate-Smart Agriculture and Sustainability*, Smart Nanomaterials Technology, https://doi.org/10.1007/978-981-92-1694-9_10

offers a transformational framework toward climate-adaptive, resource-efficient, data-driven agricultural systems resilient against thermal fluctuations.

Keywords Temperature-smart agriculture · Artificial intelligence · Nanotechnology · Thermal regulation

Abbreviations

3D-NoC	3D Network-on-Chip
ACO	Ant Colony Optimization
AI	Artificial Intelligence
ANFIS	Adaptive Neuro-Fuzzy Inference Systems
ANNs	Artificial Neural Networks
BTSAM	Balanced Thermal-State-Aware Mapping
CEA	Controlled Environment Agriculture
CSA	Climate-Smart Agriculture
DL	Deep Learning
ENMs	Engineered Nanomaterials
ETS	Ethical, Transparent, and Sustainable
GESs	Genetically Encoded Sensors
GNPs	Gold Nanoparticles
HFL	Hierarchical Federated Learning
HT-NRs	Heater-Thermometer Nanorobots
IoT	Internet of Things
IR	Infrared Radiation
MCG-RE	Multisensory Crop Guidance REALTIME Engine
MMSE	Multi-Model Super Ensemble
NMs	Nanomaterials
PAR	Photosynthetically Active Radiation
PCMs	Phase Change Materials
PNIPAM	Poly(N-isopropylacrylamide)
PPO	Proximal Policy Optimization
PSO	Particle Swarm Optimization
PTC	Positive Temperature Coefficient
PTM	Photothermal Microscopy
QAI	Quantum AI
RL	Reinforcement Learning
ROS	Reactive Oxygen Species
SERS	Surface-Enhanced Raman Scattering
SVR	Support Vector Regression
T-NN	Trust Neural Networks
TPQ	Thermal Pure Quantum
TSA	Temperature-Smart Agriculture

UV	Ultraviolet
VOCs	Volatile Organic Compounds
VQE	Variational Quantum Eigensolver
XAI	Explainable AI

1 Introduction to Temperature-Smart Agriculture

Temperature-smart agriculture (TSA) is an evolving paradigm that extends the basic fundamentals of climate-smart agriculture (CSA), emphasizing mitigation and adaptation aspects in view of rising temperature stresses associated with global climate change. While CSA has a broad mandate for enhancing agricultural resilience, reducing environmental impact, and ensuring food security through integrated strategies, TSA zeroes in on thermal adaptation mechanisms for safeguarding crop productivity under heat stress. This refined approach places greater emphasis on physiological, molecular, and ecological plant responses to a high-temperature environment with the use of modern technological interventions aimed at optimizing productivity and sustainability [128].

Increased global temperature has affected all facets of the physiology of plants, from vital processes like photosynthesis and respiration to reproductive growth, thereby affecting yields and food security. High temperatures hinder photosynthetic efficiency, chlorophyll and carotenoids synthesis, and photorespiration, leading to a decline in biomass production and grain quality. Detrimental effects are very detrimental during the sensitive phases of plant growth: germination, flowering, and grain filling, during which time heat stress diminishes pollen viability, seed germination rate, and duration of grain filling. For instance, in the case of wheat, sharp declines in yield are observed when temperatures exceeded 30–32 °C due to thermal damage to photosynthetic pigments directly linked to energy assimilation and yield stability [23]. Moreover, high temperatures elevate vapor pressure deficits, thereby increasing transpiration stress and decreasing water-use efficiency. Although an increase in atmospheric CO_2 may facilitate photosynthetic rates, physiological benefits from these increases are usually negated by thermal and hydric stresses as C_3 and C_4 crops show variable responses under these conditions [46]. These nonlinear and threshold-dependent effects make it very important to continue refining the models of crop simulation and to further validate their performance under realistic field climate conditions.

Recent progress in the field of artificial intelligence (AI) and nanotechnology has been able to shed new light on mitigation strategies against temperature stresses in crops and the induction of thermotolerance. Integration of AI with nanotechnology will not only facilitate monitoring and instantaneous management of crop environments for efficient use of inputs, but also modulate the response of the plants towards abiotic stress [14, 160]. Thus, nanotechnology increases the abiotic stress tolerance of crops through engineering metal and metalloid nanoparticles such as silver, silicon, titanium, and selenium, modulating gene expression, stimulating antioxidative defences, and inducing the accumulation of osmoprotectants. Such nanoparticles can penetrate cellular barriers and trigger a protective biochemical cascade that stiffens a plant's defence against high temperatures. Integration of AI allows for intelligent deployment of nanoscale agrochemicals and fertilizers that improve nutrient use efficiency, reduce environmental losses, and ensure sustainability in soil health.

AI-driven temperature regulation systems are increasingly integrated into the smart agriculture frameworks of various countries for better climate control and resource management. With the addition of Internet of Things (IoT)-enabled sensors, AI in smart greenhouses will be able to provide real-time regulation of microclimatic variables, hence enhancing water and energy efficiency at optimum growth conditions [62]. AI will continue to support climate-smart agricultural technologies (CSAT) through the provision of predictive analytics for irrigation scheduling, early stress detection, and climate advisory services that become particularly important in regions prone to temperature extremes. At the same time, its wider dissemination is confronted by specific barriers to be overcome, such as implementation costs, generally low levels of digital infrastructure, and low levels of digital literacy in many developing economies [32]. Further challenges related to data involve ensuring data quality, algorithmic transparency, and interoperability among systems, which creates the need for targeted policy support, capacity building, and ethical governance frameworks that ensure fairness in technological diffusion [34, 137].

Alongside AI, nanotechnology also continues to revolutionize sustainable agriculture in light of temperature stress. Nano-fertilisers and nano-pesticides were developed, which improve nutrient delivery and pest control efficiently, hence reducing the amount of chemical inputs and environmental contamination [27]. Nano sensors enable the continuous monitoring of soil moisture, nutrient levels, and plant physiological responses to provide actionable data for site-specific decisions. Carbon-based nanostructures enhance seed vigor, root development, and overall thermal tolerance, adding to climate-resilient crop systems [44]. In addition, the use of nanomaterials (NMs) for soil remediation supports the long-term fertility and detoxification of pollutants therein. Notwithstanding these benefits, environmentally safe deployment of nanotechnology in agriculture requires rigorous environmental risk assessment and the establishment of robust regulatory frameworks to avoid potential ecotoxicological impacts [139].

2 Artificial Intelligence Models for Temperature Optimization

AI also represents an approach in which agricultural temperature optimization and prediction come with robust frameworks for precision farming and climate resilience. Deep learning (DL) models have been found capable of predicting microclimate temperature variation in diverse agricultural contexts with a high degree of accuracy. Several models, such as Bi-LSTM, ANN, GBM, and RF, were tested for the microclimate of greenhouses, with the GBM and ANN obtaining R^2 values of 0.9998 and 0.999999, respectively. This underscores their precision in estimation both for temperature and associated environmental parameters [35]. Hybrid architectures, for example, ConvLSTM*CNN-BP, which integrates a convolutional and a back-propagation neural network, are effective for multi-horizon, multifactor forecasts without using IoT data, thus, present cost-effective options for developing regions [83]. Similarly, the use of EEMD-Conv3D, leveraging Ensemble Empirical Mode Decomposition to extract complex spatiotemporal features, yielded high R^2, with reduced RMSE and MAE, enhancing the accuracy of soil temperature prediction [157]. The MSLKSTNet model, driven by a multi-scale spatiotemporal attention mechanism, surmounts this with an improved MSE of 42% over ConvLSTM. It globalizes and captures localized temperature variations, therefore leading to a massive improvement over most existing works [51]. In all these ways, DL enhances sustainable and climate-smart agriculture through enhanced predictiveness of the models, optimization of input use, and enhancement of farm resilience [69].

Transformer-based architectures have taken temperature forecasting in smart agriculture to the next level by harnessing the power of attention mechanisms that model complex spatial–temporal dependencies across multimodal data. The AgriTransformer model is a prime example, incorporating tabular agricultural data with vegetation indices, leading to improved yield forecasts through dynamic feature fusion [68]. MTTF uses a multimodal transformer structure that incorporates reanalysis temperature data from ERA5, allowing it to successfully and efficiently forecast temperature [30]. Similarly, TempCast used self-attention mechanisms and decoupled multimodal fusion strategies that avoid autoregressive accumulation of errors, especially improving prediction reliability over highland and mountainous areas [162]. Domain adaptation methodologies using transformer-based deep neural networks maintain regional variations, which ensure consistency and accuracy of temperature forecasts across heterogeneous agroecological environments [149]. Such transformer models show enormous potential for adaptive climate analytics and will support precision temperature management across different farming systems.

Reinforcement learning (RL) is another important AI paradigm for dynamic temperature regulation and control in agricultural systems. RL algorithms apply adaptive feedback mechanisms to optimize environmental parameters like temperature and humidity for better crop yields and resource efficiency. Gorantla and Veluchamy developed a Proximal Policy Optimization (PPO)-based RL agent that efficiently managed a virtual greenhouse by improving crop health through adaptive environmental control [56]. Similarly, Li et al. proposed a multi-agent transfer deep RL system that reduced energy consumption and adjustment time while maintaining precise thermal and humidity regulation [89]. Integrating IoT with RL further enhances these benefits by facilitating autonomous greenhouse management with minimal human intervention, leading to significant energy savings [116]. Zhang et al. extended this work by embedding deep RL with computer vision and crop growth models, presenting optimized setpoints that improve yield outcomes with lower operational costs. Overall, RL-driven systems enable a more scalable and adaptive framework for real-time temperature optimization to foster more resilient and sustainable agricultural ecosystems [161].

Explainable AI (XAI) helps to add the interpretability and transparency needed in temperature-based agricultural decision-making, complementing advanced models. Frameworks like LIME and SHAP are useful in explaining the internal logics of predictive models, determining the key environmental drivers, especially temperature, which most affect the performance of crops [85, 133]. For instance, rice yield prediction across India, the SHAP applied on LSTM networks identified that temperature explained 42% of the yield variability and thus presented actionable insights for climate-resilient agricultural planning [129]. The use of XAI methodologies beyond yield modelling in precision agriculture enhances the interpretability of AI systems for irrigation scheduling, nutrient management, and stress forecasting. By making model outputs transparent and interpretable, XAI enables informed decision-making by farmers and policymakers to ensure the ethical and trustworthy adoption of AI in temperature-sensitive agricultural systems [96, 108].

Federated AI introduces a decentralized approach to thermal analytics. It enables collaborative model training across distributed agricultural networks in a privacy-preserving manner. In this concept, farms share model parameters rather than transferring raw data, allowing robust and private knowledge aggregation. It enhances the accuracy of temperature optimization models under data-scarce conditions and ensures regional data protection standards [40]. Extensions of this power are found in Hierarchical Federated Learning (HFL), which can enable context-aware model personalization and inter-farm adaptability [57]. Other frameworks, such as the FL-DDQN method, have demonstrated scalability and efficiency in variable climatic and soil environments while integrating deep RL to make better decisions on temperature regulation [22]. Finally, the integration of energy-efficient AI techniques, such as model compression and adaptive federated learning, may help reduce computational demands for the sustainable deployment of this technology [119]. Some of the key challenges ahead include addressing statistical heterogeneity and data imbalance, which is definitely a challenging task, and only a few attempts have been made in multi-farm federated systems so far [57].

Finally, quantum AI (QAI) is the new frontier in temperature optimization, integrating quantum computing paradigms with AI for better predictive efficiency and decision accuracy. These models leverage quantum-enhanced learning to process complex multi-omics, IoT, and environmental datasets and have achieved up to 83% accuracy in predicting crop responses to temperature fluctuations [76, 167]. By integrating genomic, phenotypic, and climatic data, QAI models can enhance crop stress tolerance by 12–15%, thus helping in developing thermotolerant cultivars [76]. The amalgamation of QAI with Remote Sensing (RS) and IoT permits a real-time assessment of risks, adaptive irrigation scheduling, and automated environmental control [132]. Despite some challenges identified with regard to computational intensity, handling, and accessibility, merging QAI with conventional AI frameworks could transform precision temperature management and achieve global agricultural sustainability [17].

3 Nanotechnology-Enabled Systems for Temperature Regulation

Nanotechnology is furnishing new and multi-functional approaches toward temperature control in agriculture through the improved resistance of crops to thermal stresses, while further improving soil–plant-atmosphere interaction. Because of their high ratio of surface area to volume and novel physicochemical properties, NMs can interact at cellular and tissue levels, modulating physiological and biochemical processes involved in thermotolerance. Translocation of NMs through cellular barriers triggers stress-responsive pathways that regulate growth, maintenance of water balance, and photosynthetic efficiency at elevated temperatures. It stabilizes the cellular pro-/antioxidant equilibrium, hence reducing oxidative damage mediated by reactive oxygen species (ROS) under thermal stress conditions [131, 165]. Further, NMs trigger the expression of genes related to the synthesis of heat shock proteins and aquaporins, thus maintaining membrane integrity and osmotic homeostasis during heat exposure. These nanoparticles also influence phytohormone levels, hence enabling the sustenance of photosynthetic activity and metabolic balance under fluctuating temperature regimes [141]. Targeted delivery and controlled-release mechanisms of NMs improve nutrient absorption and water retention important for plant survival under heat stress conditions [165]. However, the biological efficacy and ecotoxicological safety of NMs were found to depend on concentration, composition, and size, hence, controlled application and continuous monitoring will be required for sustainable deployment in agriculture [131].

Recent innovations in thermo-responsive nanocarriers have introduced intelligent temperature-sensitive delivery systems for fertilizers and pesticides. These nanocarriers regulate the release of agrochemicals in response to ambient temperature fluctuations, matching nutrient availability to plant demand while minimizing environmental losses. For example, naturally derived coatings, such as castor oil and polycaprolactone composites, have been designed for temperature-responsive urea fertilizers that show thermal switches at about 32.5 °C, extending nutrient release up to 70 days and hence enhancing nutrient use efficiency and environmental sustainability [164]. Likewise, poly(N-isopropylacrylamide) (PNIPAM)-based nanocarriers modulate pesticide release rates with variations in temperature, hence optimizing the efficacy of pest control under minimum chemical residues [136]. Multi-compartment liposome nanovesicles further enhance co-delivery of pesticides or fertilizers with temperature-dependent release profiles, hence improving efficiency under high thermal conditions [43]. Although these systems hold immense potential, full-scale deployment based on economic viability, uncertainty over regulatory issues, and those factors involving environmental risk all point toward the need for further research into safety and scalability by interdisciplinary approaches.

Other promising avenues in mitigating temperature stress include the improvement of soil thermal buffering capacity and water retention with nanostructured hydrogels. This hydrophilic three-dimensional polymer network absorbs a large volume of water that it slowly releases, maintaining both soil moisture and temperature constant, especially very important in arid and semi-arid regions [114]. Continuous hydration has the added benefit of minimizing irrigation frequency and further protects roots against heat-induced desiccation. Water retained this way in soil will thus act as thermal mass by absorbing excess heat during the day and releasing it at night, therefore moderating fluctuation in soil temperature [106]. Besides, hydrogels improve the structure of the soil by enhancing porosity and reducing bulk density, hence promoting root penetration and nutrient uptake [87]. The application of hydrogel enhances the cation exchange capacity, thus retaining nutrients and promoting long-term fertility. However, overall application requires strict environmental assessment of nanostructured hydrogels in order to warrant their biodegradability and their nontoxicity to avoid pollution in the soil or other ecological mal-effects [9].

Nanophotonic coatings offer the promise for optical modulation of temperature through tunable light absorption and reflection on the surface of plants or greenhouses. This may include selectivity in photosynthetically active radiation (PAR) transmission while reflecting harmful ultraviolet (UV) and infrared (IR) radiation. Such coatings may serve as low-, high-, or band-pass filters to further optimize the light spectra for plant growth upon proper tuning of optical parameters such as layer thickness and nanoparticle size [152]. SafeWax is a superhydrophobic coating bioinspired from plants, which acts like a plant cuticle to provide significant protection against UV radiation, offers anti-fungus properties, increases captured water, and reduces pesticide use [117]. Likewise, reflective films based on kaolin maintain photosystem II integrity in grapevine leaves under high light and temperature, thus increasing photochemical efficiency and preventing photoinhibition [42]. The state-of-the-art nano-photonic structures include photonic crystals and plasmonic nanostructures with the capabilities for enhancing the spectral quality and energy efficiency inside greenhouses by manipulating light at the nanoscale [120]. Material cost, long-term stability, and translation into large-scale agriculture remain the major challenges.

The improvement of nutrient uptake, modulation of gene expression, and physiological stability by bioinspired nanocomposites further mitigate thermal stresses in plants. Composites based on ZnO, silica, and carbon nanotubes upregulate stress-responsive genes that encode for heat shock proteins and aquaporins, leading to enhanced cellular thermotolerance [151]. There is also enhancement in the activity of antioxidant enzymes, hence amelioration of oxidative damages triggered by heat stress. Nano-encapsulated agrochemicals allow controlled nutrient release and minimize environmental contamination, hence increasing their efficiency [138]. Besides this, nanocarrier-mediated delivery of CRISPR-Cas9 enables the precision genome editing for the development of climate-resilient crops with better temperature tolerance [67]. Despite these advantages, ecological safety, regulatory oversight, and ethical use of large-scale nanocomposite applications are still critical issues to be evaluated owing to species-specific responses and environmental variability.

Green nanotechnology is the new face of environmental sustainability for thermal management in agriculture. The biogenic nanoparticles derived from both plants and microorganisms are an environmentally friendly alternative to fertilizers, pesticides, and biosensors by reducing the ecological footprint of conventional agricultural inputs [84]. Such biogenic nanoparticles help improve nutrient uptake, enhance growth, and alter the phytohormones of plants, thus improving productivity and thermal resilience of plants [145]. Smart Nano sensors made of green NMs monitor in real-time a variety of soil and plant physiological parameters for early detection of thermal and water stress [111]. Moreover, green NMs help in remediation of soil and water by adsorption of pollutants and catalytic degradation of harmful substances capable of restoring soil fertility and environmental balance [24]. Green nanotechnology, while offering green alternatives, has also raised several issues related to bioaccumulation, phytotoxicity of nanoparticles, and partial regulatory frameworks [84].

4 AI-Nano Synergy for Precision Temperature Management

AI-powered nanofabrication is a breakthrough in developing temperature-regulating materials through precision in nanoscale structural control, hence optimizing thermal management and energy efficiency. Machine learning and predictive modelling by AI refine the synthesis and functionality of NMs in adaptive responses to environmental fluctuation, hence contributing to sustainable agricultural performance. Most importantly, phase change materials (PCMs) have been optimized by AI, with nano-enhancements being guided by AI algorithms to achieve superior heat storage, adaptive thermal regulation, and controlled energy release. Consequently, integrating AI with nanoengineered PCMs achieves autonomous temperature stabilization in solar-based agricultural systems. Such a development in regard to smart self-adaptive materials is represented by metafabrics that integrate radiative cooling and phase-change cycling to maintain stable crop micro-environments against variable temperatures [102, 153].

AI-assisted fabrication also enhances the thermal conductivity and phase stability of flexible temperature-regulating materials, such as positive temperature coefficient (PTC) composites that can self-regulate heat flow in sensitive agricultural processes. In the same light, hierarchical lignocellulosic thermostats illustrate a prospect of AI-driven nanomanufacturing for engineering robust thermally adaptive materials for use within agricultural infrastructure [156]. Beyond solid-state materials, AI enables the design for bidirectional thermo-responsive hydrogels to maintain crops and livestock within their comfort zones through programmed cycles of swelling-deswelling. In any case, large-scale application of AI-guided nanofabrication will be difficult owing to a set of challenges related to economic viability, environmental sustainability, and regulatory considerations, which need to be overcome to ensure equity in application at the agricultural level [113].

This is accomplished by encouraging precision thermoregulation through real-time monitoring and intelligent environmental control. These Nano-IoT sensors constantly provide high-resolution information related to temperature, humidity, and soil moisture. The information is converted into actionable insights through AI-driven analytics, supporting proactive management and predictive intervention against several modules of agriculture (Fig. 1) [14]. This will help optimize irrigation and fertilizer schedules and reduce resource use up to 40%, with a yield increase of approximately 30% [2]. Real-time feedback makes it possible to adapt to thermal fluctuations without much delay, thus reducing crop stress, enabling early disease detection, and yield forecasting, which is considered important in precision agriculture [86]. Though the prospect of deploying the AI-Nano-IOT systems appears bright, several challenges have to be surmounted, including high operation cost, data security, and environmental implications, hence, more innovation and regulations are required [14].

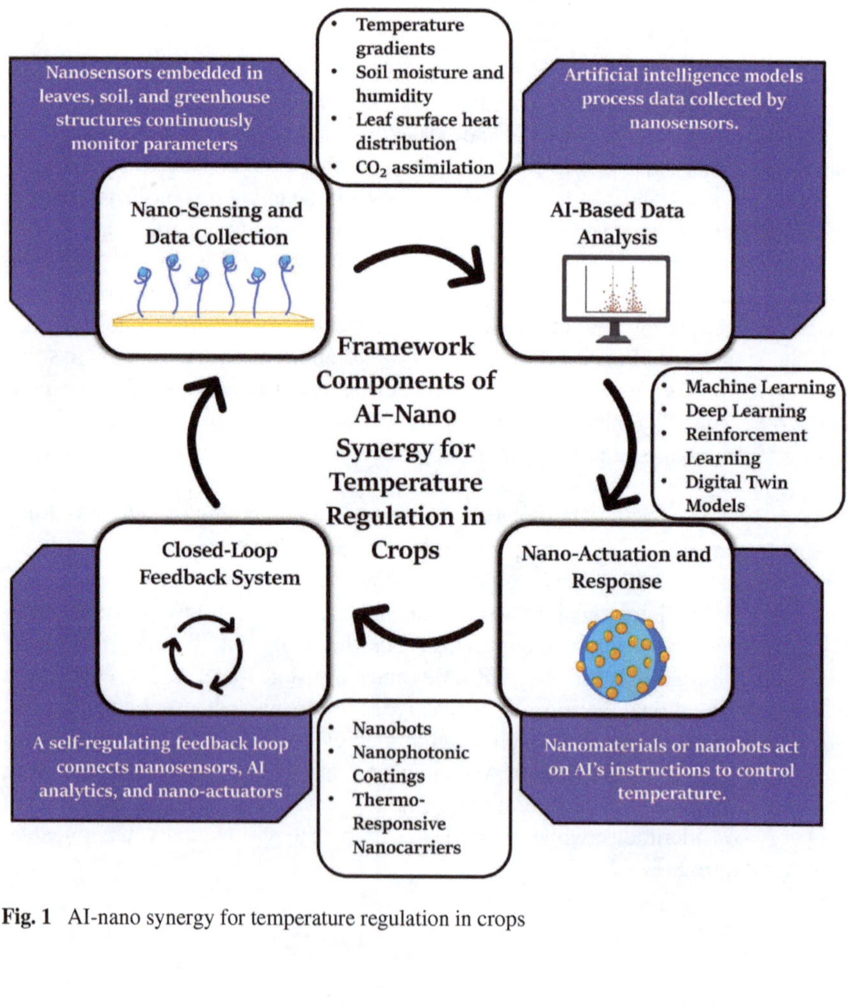

Fig. 1 AI-nano synergy for temperature regulation in crops

However, this convergence of AI, IoT, and nanotechnology has catalyzed the development of smart irrigation capable of mitigating heat-induced water stress. Supported by AI models like deep neural networks that analyze the dynamics of soil moisture-temperature to decide on the time and quantity of irrigation, IoT-based monitoring allows real-time adjustment while minimizing loss of water. Predictive AI allows inclusion of meteorological data that minimizes thermal stress and averts water inefficiency in crops [7]. Complementing these technologies, NMs modulate phytohormone activities, improve the expression of stress-responsive genes, and provide slow-release water or nutrient delivery, thus collectively improving crop resilience under extreme conditions [165]. While AI-enabled nano-irrigation systems improve water use efficiency and yield, costs and technical challenges remain too high for resource-poor farmers. Targeted policy support and capacity-building programs will be essential for wider dissemination [103].

The precision of thermal monitoring in multi-sensor networks of agricultural systems is further enhanced by AI-driven data fusion approaches. Weighted averaging and Kalman filtering reduce sensor noise and improve the predictive accuracy of thermal parameters [148], while fuzzy logic and Trust Neural Networks (T-NN) enhance reliability in uncertain field conditions [166]. DL models assess fused multispectral datasets of thermal, visible, and near-infrared data with high precision regarding plant stress, disease, and water deficits [73]. Overall, these multi-modal data fusion systems enhance the reliability of the AI-driven thermal regulation framework, however, redundancy and real-time processing constraints remain challenges for large-scale deployment [104].

Beyond monitoring, closed-loop AI-nano systems can autonomously regulate crop temperature by integrating sensing, analytics, and actuation. For instance, the Greenhouse Intelligence System utilizes NASA Earth observation data together with AI analytics to achieve up to 95% accuracy in temperature forecasting, thus enabling dynamic environmental adjustment [4]. The AI-GECS framework leverages Multi-Model Super Ensemble (MMSE) forecasting for proactive regulation of cooling systems in greenhouses, while the Multisensory Crop Guidance REALTIME Engine (MCG-RE) system integrates nano sensors with cloud analytics for real-time irrigation and nutrient optimization. Deep RL architectures further fine-tune set point control by allowing energy economy and responsiveness of the system [74]. Scalability into open-field systems is one of the dominant research frontiers for these, despite their noted effectiveness within controlled greenhouse environments [88].

Nano-enhanced environmental virtual simulation or agricultural digital twin is a recent but powerful tool for precision temperature control. The integrated IoT sensing, nanomaterial data, and predictive algorithms in these AI-integrated models simulate and optimize the microclimatic conditions. Digital twins can thus use model predictive control in hydroponic systems to optimize trade-offs between temperature and energy use [29]. In vertical farming, nano-enhanced material simulations showed enhanced thermal efficiency and yield [16]. Such virtual models are improved with refinement by machine learning methods like RL, genetic algorithms, and neural networks for better decision making and adaptive control of such systems [54].

5 Advanced Nano-sensing and Monitoring Technologies

Recent developments on graphene-based nano sensors have greatly improved the detection of ultrafast thermal gradients in plants. Being made of a highly thermally conductive, electrically sensitive, and mechanically flexible material, graphene nano sensors are used for high-resolution, fast-temperature monitoring at the cellular and tissue levels. Thermocouples in the form of graphene-coated textiles have excellent electrothermal performance: 1.5 s response time and 2.5 mV voltage output for a temperature gradient of 32 °C that offers reliable real-time thermal monitoring in plant systems [47]. These advantages are further extended by Photothermal microscopy (PTM), offering submicron spatial and 10 μs temporal resolution, able to detect temperature variations of only 6 mK, which allows nanoscale mapping of heat diffusion within plant tissues [33]. Other studies documented that plants themselves can provide measurable voltages generated by root-leaf temperature differentials because of strong thermoelectric coupling, further pointing at the potential of graphene-based nano sensors in plant thermophysiology and associated adaptation to stress conditions. Despite these advantages, integration into agricultural environments is more complicated since, due to variability in the system, field calibration is complicated and will require further optimization for the realization of robust deployment.

Meanwhile, parallel development of nano-biosensing technologies is in progress with a view to enhancing the accuracy of soil and plant temperature monitoring. Carbon nanotube- and graphene-based sensors, owing to their ultra-high sensitivity and fast response rate, can detect subtle microclimatic fluctuations across the plant-soil interface [39]. At nanoscale dimensions, dense deployment achieves high-resolution thermal mapping over extended surfaces of agricultural fields. Further integration into wireless communication networks can enable continuous real-time environmental monitoring/data transmission on cloud platforms for advanced analytics [112]. These sensor networks will enable dynamic decision-making—for example, changes in irrigation timing with records of concurrent soil temperature and moisture. Plasmonic nanostructure-based electrochemical nanowire sensors have also been developed, retaining measurement accuracy under fluctuating UV radiation and pH levels in the soil for long-term reliability. However, scaling up deployment, production costs, and ecotoxicological effects remain looming issues; hence, emphasis is given to applying biodegradable and biocompatible designs for sensors toward sustainable agricultural integration [41].

Incorporation of AI into the nano sensor platforms has increased the resolution, flexibility, and interpretability of temperature monitoring in agriculture. AI algorithms like artificial neural networks (ANNs) and DL models further advance nano sensor calibration by managing nonlinearities, filtering out the noise in the signal, and granting stability against changes in the field conditions [91]. Bivariate Gaussian mixture models improve sensor calibration within noisy agricultural contexts, while deep-wavelet neural networks allow thermocouple signal enhancements through the suppression of background interference and maximizing of the signal-to-noise ratio

[72]. However, together with the high-resolution real-time thermal and nutrient monitoring, such AI nano sensing systems have enabled predictive management of crop stress and resource allocation. Overfitting, heterogeneous data, and environmental interference are concerns that bound model robustness and generalizability [41].

Added to this, nano sensor-based agricultural monitoring is further being reinforced by emerging architectures in cloud-edge interoperability. Reducing latency, preserving bandwidth, and support for immediate responses to thermal and environmental changes are possible through the hybrid model of combining real-time edge computing with cloud-level analytics. Local edge processing guarantees fast adjustments in irrigation or nutrient delivery, while centralized cloud analysis makes large-scale forecasting possible with enhanced pattern recognition [55, 130]. Improved disease detection, increased precision in irrigation, and yield prediction under continuous adaptive learning are thus enabled in this kind of distributed architecture. Thus, this enables edge-level processing so as to minimize the demand on data transmission and hence ensures system reliability while reducing operational costs in geographical areas where network infrastructure remains minimal [143]. Low-power event-driven detection with a scope for sustainable precision agriculture ensures integration of IoT-edge systems with TinyML [75]. Besides the enhancement of responsiveness and security through local information processing native to these architectures, challenges that still call for additional optimization and involve not limited to infrastructure resilience and adapting the system to regions are persisting [55].

Neuromorphic architecture introduced a new frontier in processing data from nano sensors in AI Nano farms, especially for temperature regulation. Integration of bioinspired computing systems with memristor-based hardware and 3D Network-on-Chip (3D-NoC) has demonstrated high computational efficiency and low thermal buildup. It has been observed that Balanced Thermal-State-Aware Mapping (BTSAM) decreases the peak temperature by as much as 12.4 K compared to traditional approaches in neuromorphic 3D-NoC systems, which might extend operational lifespan fourfold at a practically negligible power overhead. Similarly, the cluster-based, thermal-aware mapping methodology using genetic algorithms achieves up to a reduction of 39 °C in hotspot temperatures and can support thermal uniformity and computational stability [93]. Memristor-SNN systems provide even further efficiency, reducing power consumption by up to 60% and tripling the speed of processing over CMOS architectures [97]. Thermal neuristors operating under conduction mechanisms ruled by temperature offer scalable and low-power options for real-time AI inference in agriculture. Other techniques like dynamic spike encoding and synaptic pruning enhance adaptability by 45% to realize faster processing in a variable agricultural microclimate. Reconfigurable neuristor arrays allow cascaded information flow of thermal information toward scalable adaptive nanosensing. Although considerable progress has been achieved, scalability and hybrid integration remain the main challenges [121].

6 Nano-robotics and Autonomous Temperature Control

Nanobots, therefore, represent a revolutionary concept in intraleaf temperature regulation for the realization of precise thermal modulation toward optimization of photosynthesis and mitigation of abiotic stress. These nanosystems are engineered to enable dynamic responses to environmental stimuli through the delivery of stress modulators or by promoting adaptation in intracellular metabolic activity, hence making plants resilient and improving crop yield. Temperature-sensing systems in controlled environments attained a 28–43% yield in tomato cultivation and evidenced the potential of nanoscale thermal regulation as an input for precision agriculture [142]. Temperature-responsive nanobots that release the bioactive molecule spermidine enhance tolerance to heat-light stress, providing longer protection—up to 15 days—less physiological damage caused by climate change, protecting them from global warming [163]. Gold nanoparticles (GNPs) have been showcased as efficient photothermal transducers of remote induction of biological processes such as heat-shock gene expression that confers increased thermal tolerance. Their uptake and translocation into *Arabidopsis* really herald very promising applications in both temperature modulation and molecular regulation [82]. This, however, brings into play the grave aspects of environmental toxicity, high costs of fabrication, and ecologically risky long-term implications that restrain immediate large-scale deployment and demand further assessment for the sustainable implementation of such technology [151].

By incorporating robotics, thermography, and the principles of bioheat transfer, AI-driven nano-drones extend the functionalities of nanorobotics to macro-scale environmental monitoring and precision in temperature management. These autonomous drones apply infrared thermography, allowing non-invasive mapping of the heat distribution and zones of stress within crop canopies, which reduce dependencies on fixed sensor arrays. These, when combined with environment sensors, enable real-time monitoring of air temperature and humidity for interventions targeted at adaptive microclimate control [65]. Beyond that, AI-enabled drones maintain localized thermal comfort by means of detecting human state and applying adaptive airflow or water-spraying systems, thus allowing energy-efficient alternatives over conventional cooling devices [79]. This will further enable nanoscale thermal modulation with the integration of nanoparticle-based heating systems, and thermally resilient materials such as polyimide aerogels help enhance durability under extreme environmental conditions. However, due to challenges related to payload capacity, energy efficiency, and ethical issues, including privacy and data governance, there is a need to develop lightweight architecture and robust regulatory frameworks [26, 59].

The swarm intelligence algorithms have put forward a new paradigm of decentralized, adaptive, and self-organized thermal optimality in the coordination of nanorobotics. Algorithms in modelling collective biological behaviour, such as Particle Swarm Optimization (PSO) and Ant Colony Optimization (ACO), allow ways to reach nanodevice-to-nanodevice heat distribution and self-regulation efficiently. For instance, PSO models dynamics of collective motion that can be adapted for positioning nanoscale robots to enhance coordinated thermal control [15], while ACO makes it possible to find optimal paths of heat dissipation by biomimetic foraging behaviour [140]. A specific example concerns Heater-Thermometer Nanorobots (HT-NRs with swarming action for targeting certain regions, they may make use of nanorobot systems to reach real-time thermal feedback while modulating hyperthermia by implementing reactive, stateless rules for autonomous assembly, self-repair, and adaptive function in time-varying thermal gradients [92]. While swarm coordination generally reduces entropy and increases system reliability, there are challenges that persist, including those about communication fidelity, scalability, and ease of programming with respect to appropriate nanoscale communication frameworks [126].

Self-organization of nanorobotic systems further enhances the mitigation of temperature stress in plants through the use of engineered NMs, which activate plant molecular defence pathways. These materials upregulate the stress-responsive genes encoding heat shock proteins and aquaporins, modulating ROS, enhancing water transport, and improving structural stability under fluctuating temperatures [151, 165]. NMs are also used in targeted delivery into plants of nutrients and stress-alleviating compounds that reduce agrochemical consumption with a view towards environmental sustainability [5]. However, incomplete information on nanoparticle toxicity, their persistence, and ecosystem impact creates a need to formulate comprehensive regulatory frameworks in order for deployment of nanoparticles to be environmentally safe in agriculture [52].

AI-Nano feedback architectures represent the next step in the evolution of autonomous temperature regulation that integrates data from nano sensors with real-time predictive analytics and modelling of the digital twin for the self-regulating farm environment. Other systems, such as AI-GECS, use hybrid DL frameworks that enable the prediction of microclimatic fluctuations and hence possibly pre-emptive control of cooling systems, with better optimization in energy use and efficiency in greenhouses [88]. The PSO-BP-PID control framework enhances sensor calibration accuracy and further stabilizes interactions between humidity and temperature, reducing deviation from target conditions [154]. In other platforms, such as Digi-HortiRobot, the digital twin integrates monitoring and prediction of environmental variables with autonomous task execution through a closed-loop system continuously optimizing the temperature [48]. Dynamic adjustment of control strategies concerning growth stages enabled by deep RL further refines this process and enables efficient low-energy climate regulation [3]. Newer controllers, such as LI-RBF-PID, introduce an element of self-tuning intelligence that further enhances the accuracy and convergence characteristics of these systems where there is considerable variation in environmental parameters [158]. System integration, infrastructural cost, and operational complexity are still some of the challenges that need to be met in the diffusion of this technology for wider agricultural applications. Cross-disciplinary innovations at the level of design itself and, therefore, implementation of AI-nano systems are still in need.

7 Digital Twins, Simulation Models, and Computational Frameworks

Digital twin technology in agriculture can be described as an advanced integration of real-time sensing, data-driven modelling, and control systems to simulate temperature dynamics with the view to optimize microclimate regulation for productivity with sustainability. Coupling IoT-based environmental sensing with predictive modelling, digital twins execute state-of-the-art environmental controls of temperature, humidity, and CO_2 concentration in greenhouses and vertical farming systems (Fig. 2). IoT sensor networks continue collecting real-time data on microclimates, maintaining temperature deviation below 0.1 °C and humidity error within 2%, thus maintaining stable environmental controls and uniform crop growth [29]. Data-driven algorithms like Support Vector Regression (SVR) predict internal temperature fluctuations in greenhouse environments with high accuracy for the timely regulation of cooling systems such as fans and evaporative wet walls [66]. Neural networks and genetic programming enhance intelligent control by virtual testing, evaluation, and optimization of adaptive responses of digital twins against environmental perturbations. The population-based model predictive control system maintains temperature uniformity under dynamic conditions, hence improving energy efficiency and crop health [16].

Controlled Environment Agriculture (CEA) digital twin frameworks combine physical and data-driven modelling approaches to balance yield optimization and resource efficiency. Their architecture allows for fine-tuned control of environmental variables, with aims towards sustainability, crop quality, and food security [31]. A set of complementary simulation platforms including PlantSimEngine, GEOSPACE, SOILSIM, and SiSPAT further develop the capabilities in modelling plant-soil-atmosphere interactions at a range of scales. PlantSimEngine is an easily accessible, high-level platform for the simulation of processes that involve transpiration, energy flux, and soil moisture without advanced programming skills [150]. Other similar platforms include the modularly designed object-oriented GEOSPACE, which can model the soil–plant-atmosphere continuum while allowing easy inclusion of new processes, hence enabling modelling at multiple scales [37], SOILSIM, focused on energy and moisture fluxes for large-scale evapotranspiration estimation using inputs from remote sensing [49], and SiSPAT, which offers validated surface energy balance and hydrothermal modelling from high-resolution field data, enabling the establishment of sustainable management of soil and water under various climatic conditions [8]. The choice of simulation frameworks should be based on specific research objectives, system complexity level, and prevailing environmental conditions, modularity, and integration of data with respect to adaptive agricultural modelling.

AI-driven simulations improve the accuracy and flexibility of the digital twin, with hybrid modelling approaches to analyze crop-environment interactions under temperature variability. Integration of IoT, meteorological, and environmental data develops virtual models of agricultural systems that could even simulate crop response to thermal oscillations. Such integration backs predictive irrigation, active

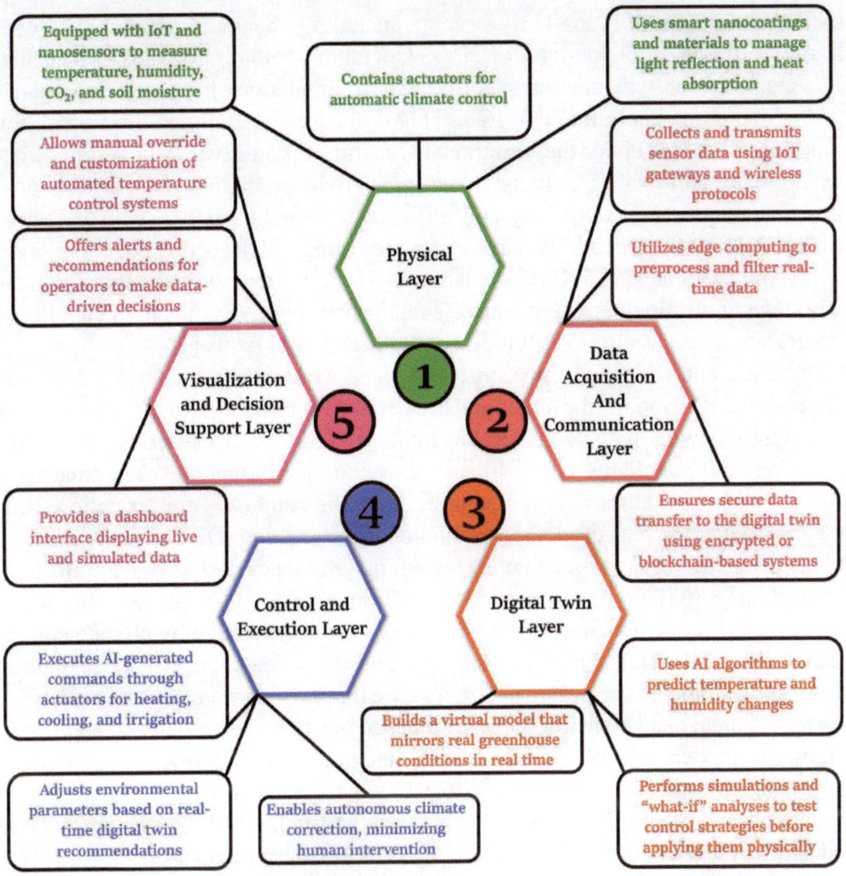

Fig. 2 Digital twin-driven smart greenhouse architecture

nutrient management, and pest control methods—all with real-time environmental data serving as input [115]. Hybrid models will integrate process-based crop growth models into DL frameworks in order to represent nonlinear impacts at extreme climatic events on crop yield across variable growth stages. Dynamic refinement of model parameters allows optimized resilience with minimum environmental impacts possible [134]. AI-powered models can predict temperature-dependent root zone moisture in enhancing irrigation precision and water-use efficiency in arid areas toward climate-resilient agricultural systems [21]. At the same time, this advanced simulation platform also includes a few challenges in relation to data interoperability, protection of data privacy, and requirements for large quantities of high-quality data essential for reliable performance [115].

Quantum computing is a new frontier in nano-enabled temperature modelling. It represents an avenue for dealing with the severe quantum mechanical complexity underlying thermal behavior at nanoscale levels. Quantum algorithms allow for the efficient simulation of many-body systems, including phonon interactions, influential on the thermal conductivity of NMs. Quantum computing avoids exponential state sampling for scalable and precise thermal simulations by generating canonical thermal pure quantum (TPQ) states [118]. Accurate modelling of multi-phonon scattering processes under the Variational Quantum Eigensolver (VQE) framework hence updates nanoscale heat transfer and conductivity prediction [146]. These capabilities make quantum computing a promising complement to digital twin and materials optimization applications. Challenges regarding qubit decoherence and cryogenic stability remain [125] Hybrid quantum–classical integration, though still in the early stage of development, may enable even further improvements in the fidelity of temperature prediction models within agricultural digital twins [107].

Big Data analytics, IoT infrastructures, and AI frameworks further facilitate convergence to support the emergence of temperature-intelligent ecosystems in which digital twins are central orchestrators of real-time monitoring, predictive modelling, and adaptive control. In such a system, environmental data emanating from IoT devices that monitor temperature, humidity, and CO_2 concentration feeds into AI models for real-time analysis and decision-making. Digital twins model a variety of operating scenarios that ensure energy efficiency and resource utilization in greenhouses [63]. For the prediction of environmental load, DL and RL algorithms are adopted, against which an optimum strategy for energy-efficient climate control is devised [10]. Big Data analytics can provide anomaly detection, predictive maintenance, and automated correction of temperature through continuous pattern analysis. The deployment and extension of the system become possible through standardized, parametric models and plug-and-play IoT integration on scalable frameworks. It seems that it is actually challenging to take on fragmented data, interoperability limitations, and privacy concerns [144]. Integrating Big Data, IoT, and AI provides a real pathway toward intelligent, scalable, and sustainable computational frameworks for precision environmental management [10].

8 Applications in Temperature-Smart Agriculture

AI-nano systems reflect a very new concept within temperature smart agriculture; these integrate AI, IoT, and networks of nano sensors to allow for far more precise control of environmental parameters. These hybrid systems are able to monitor and alter the amount of temperature, humidity, and CO_2 in real-time, thus helping to maximize plant growth, raise yields, and reduce energy consumption together with operational costs. Controllers based on AI using DL algorithms, together with microcontrollers and distributed sensor networks, conduct automatic control over fan speed, frequency of irrigation, and systems of ventilation in order to keep optimum microclimate conditions for the development of the crop [105]. Besides that, AI optimizes the consumption of energy by integrating renewable sources, like photovoltaics, with dynamic pricing algorithms, which make cost-effectiveness and sustainability go together in tune in controlled environment agriculture [95, 158]. Despite the advantages mentioned above, some practical challenges in the form of high implementation cost, limited technical expertise, and regional inequalities in technological infrastructure remain pivotal bottlenecks toward large-scale deployment of AI-nano systems in agriculture [105].

Apart from thermoregulation, AI-driven irrigation and cooling systems greatly enhance crop resilience to heat stress through their data-driven irrigation management. These systems, with the help of real-time data on soil moisture, atmospheric temperature, and weather forecasting, decide on irrigation timing and volume, hence optimizing water use efficiency. Machine learning models further improve the accuracy by changing the irrigation parameters dynamically to keep up a soil-temperature balance and reduce evaporative losses. An intelligent valve control system that is tailored for each micro- and piped irrigation network adapts to soil and crop conditions [7]. On large scales, AI-driven irrigation cooling has been noted to increase maize yield by up to 81% in the US Midwest for different climate change scenarios, hence its impressive agronomic and economic potentials [94]. In the meantime, there is potential for precision irrigation to improve environmental sustainability by improving soil health, reducing runoff of nutrients, and increasing carbon sequestration while reducing operating costs and resource wastage [103]. However, this variability of the performance across climatic zones, as influenced by differences in data accuracy, model robustness, and sensor calibration, calls for optimization and adaptation at local levels [71].

Nano-enabled amendments provide additional mechanisms for temperature stabilization at the soil interface, especially in arid and semi-arid ecosystems. Incorporation of nanoparticles, nanoclays, and biochar improves the structural integrity, water retention, and nutrient availability of the soil, reduces diurnal temperature fluctuation, and, as such, reduces thermal stress. Biodegradable radiative cooling mulches with high solar reflectance and thermal emissivity showed a 60% reduction in soil moisture loss and 30% improvement in plant growth by efficiently reflecting the incident solar radiation and emitting heat [90]. In like manner, biochar prepared from agricultural residues reduces soil bulk density and thermal conductivity, while in coal-mined arid regions, application of 2% biochar reduced soil temperature by 0.93 °C at a depth of 15 cm, hence, showing its potential to buffer extreme surface heating [155]. Nano-structured soil conditioners, therefore, bring about sustainable land management and improved capacity to withstand climate variability. However, unresolved issues relating to persistence, leaching, and long-term ecological impacts of nanoparticles require thorough environmental risk assessment for safe agricultural applications [45].

AI also significantly speeds up the development of temperature-resilient crops through multi-omics data integration, including genomic, transcriptomic, proteomic, and phenomic data, to uncover the genetic determinants of heat tolerance. Machine learning models allow for high-throughput analysis of genotype–phenotype relationships, thus enabling biomarker and regulatory network discovery with respect to thermal resilience [77]. AI-operated drones and hyperspectral imaging platforms further enable the rapid phenotyping of crops under thermal stress conditions that underpin the selection of genotypes with superior adaptation capacities [147]. This multi-omics-AI synergy brings greater efficiencies in breeding and supports predictive crop management across variable climate scenarios. However, inconsistent data quality, limited computational infrastructure, and ethical concerns with respect to data privacy and equity are severe constraints to wider adaptation [58].

The ingenious solutions that nanotechnology proposes, in relation to the control of thermal regulation and welfare management within the livestock system, involve nanocoatings, which improve the process of heat dissipation from the structure. It should be able to maintain favourable temperature and humidity conditions so as to reduce the incidence of heat stress. Moreover, it allows the embedded nano sensors to monitor all the microclimatic and physiological parameters continuously and to send real-time data controlling the environment automatically [36]. Nutritional nanotechnology improves the bioavailability of nutrients while integrating feed ingredients promoting thermotolerance [18]. The integration of IoT-nano enables precision livestock farming because several projects have used nano sensors and thermographic imaging with the intention of monitoring animal responses with the aim of optimizing comfort and productivity—such projects include AWARTECH [13]. Despite the encouraging results, the use of NMs in animal systems needs to be assessed for safety with great care, and their use regulated so as to avoid accumulation of these substances in living organisms and their exposure to the environment [135].

These AI-nano hybrid frameworks drive innovative, region-specific creations globally in the field of TSA, using data intelligence, material science, and automation. For example, AI-based systems such as CropX and NetBeat in Israel integrate soil and climatic data to deduce optimum irrigation schedules, hence improving water-use efficiency [12]. Nano sensors and predictive AI systems in Indian agricultural systems enable the early detection of pests and diseases, reduce yield losses, and make farming resilient [61]. An integrated model of AI-nano in the Netherlands deploys biochar, nano sensors, and smart analytics for nutrient, water, and carbon loops to advance sustainable and circular agricultural ecosystems [81]. In turn, IoT-enabled US companies like Moocall and Ida facilitate real-time monitoring towards enhanced welfare and productivity. These examples suggest that at the core of the global relevance is the integration of AI-nano into agriculture. Still, there are some challenges: high technology costs, skill gaps, and infrastructure deficits remain the critical barriers to adoption. It is in this regard that such obstacles call for multi-stakeholder coordinated innovation, capacity building, and policy support to scale up sustainable TSA [12].

9 Data Governance, Sustainability, and Ethical Dimensions

Blockchain technology ensures more traceability and transparency in integrated AI-nanotechnology agricultural systems through a decentralized, immutable ledger system where the accuracy of product origin, quality, and safety information is ensured. Blockchain ensures a tamper-free recordkeeping system that verifies data integrity; thus, it guarantees traceability along the whole agri-food supply chain, providing very fast contamination responses and verifiable provenance data that engender consumer trust in accountability within the market [1]. Blockchain works together with AI and IoT; this enhances seamless interconnectivity and transparency in data flow, while AI enhances analytics along the supply chain through real-time interpretation of blockchain-verified data sets, and nano sensors enable precision monitoring of product quality and safety parameters by connecting molecular-scale data into a larger traceability network [123]. Aside from its role in transparency, blockchain enhances sustainability in reduction of transaction costs, fraud cases, and enabling ethical sourcing based on verifiable records of production and distribution practices [28]. Besides benefits, large-scale adoption has infrastructural deficits, interoperability concerns, and lack of stakeholder alignment; hence, policies and targeted technical interventions are necessary in realizing full benefits in AI-nanotech agriculture [1].

While blockchain adds a layer of robustness to the data integrity, the greater proliferation of engineered NMs (ENMs) brings on board some very difficult environmental and moral problems that raise a need for strict governance. ENMs in nano-enabled fertilizers, pesticides, and conditioners may cause ecotoxicological effects in terrestrial ecosystems, causing soil contamination and bioaccumulation in trophic chains [70]. The environmental fate of these materials, governed by particle size, surface charge, and the composition of the medium, controls their mobility, transformation, and potential toxicity. It has been evidenced that the buildup of ENMs can alter the physicochemical properties of soil, a process that, astonishingly, significantly reduces soil fertility and changes microbial communities, including more distressing secondary reaction products from deteriorating quality of the soil [80]. Long-term bioaccumulation and biomagnification threatens food safety and ecosystem health due to a lack of information on combined interactions with other co-contaminants [53]. Thus, their fate, transformation, and persistence, modelled through lifecycle and exposure modelling, form the backbone of secure and sustainable use in agriculture.

AI technologies complement the environmental risk assessment to further improve lifecycle evaluation and carbon footprint modelling of nano-enabled agricultural practices. Adaptive computational models, such as Adaptive Neuro-Fuzzy Inference Systems (ANFIS) and ANNs, can be applied to enhance the accuracy of sustainability analyses for prediction of the environmental indicators of emissions, energy consumption, and resource use [101]. On the basis of existing research outputs, the hybrid model ANFIS-FCM showed higher accuracy in the prediction of environmental metrics and hence provided enhanced reliability in the lifecycle analysis. AI improves functionality through the enhancement of controlled delivery and release mechanisms of bioactive nutrients in nano-fertilizers and nano-pesticides, thus limiting the amount of wastes and ecological impacts. Combined with nanoinformatics, AI accelerates the design of safer, functionally optimized nanoscale agrochemicals and may help contribute to mitigating carbon footprints from agriculture [160]. Novel AI-based platforms, such as ParallelFarm, apply multisource data streams in intelligent decision-making for real-time monitoring toward predictive sustainability management that may enable transition toward carbon–neutral and resource-efficient farming [25]. Scalability challenges persist, along with issues on data standardization and regulatory integration, and there is a need for harmonized frameworks which could ensure appropriate and ethical deployment.

The integration of AI and nanotechnology into agriculture brings immense ethical considerations that should be managed through strong governance frameworks for accountability, transparency, and the division of benefits. The Ethical, Transparent, and Sustainable (ETS) framework puts much emphasis on the responsible acceptance of AI in agriculture by aligning innovations with principles of equity, data safety, and environmental care [11]. Green nanotechnology furthers this perspective by focusing on an eco-friendly synthesis and application of degradable, nontoxic NMs that reduce ecological and health hazards in the manufacture of traditional NMs [84]. Safety guidelines aligned with defined rules on toxicity testing, monitoring, and lifecycle analysis are important guides toward responsible nanotechnology deployment. At the same time, striking a balance between innovation and ethical and social protections by accelerating technological evolution tends increasingly to mark the core of all the challenges determined by adaptive governance mechanisms and interdisciplinary collaboration for data privacy, algorithm bias, and ecological resilience [6].

Complementary policy frameworks are instrumental in developing temperature-smart and sustainable agriculture by establishing legal, financial, and organizational mechanisms that favour climate adaptation and technological innovation. Good governance enhances resilience through the integration of smart technologies such as IoT, AI, and precision agriculture into the country's mitigation and sustainable production strategies [98]. Long-term outcome monitoring is assured while credible environmental safety and performance indicators are put in place, political and societal commitment is also enhanced towards sustainable transformation. Removal of other socio-economic and infrastructural obstacles—technology access would further push both the adoption and equitable benefit of CSA across diverse farming systems [100]. Moreover, improving institutional coordination and multi-stakeholder engagement and capacity-building measures support the coherent development and implementation of the policies of CSA. A holistic approach to governance, bringing to the forefront issues on sustainability, inclusivity, and technological adaptability, would thus become crucial in minimizing the impacts of climate change by assuring ethical innovation and assured resilience in global food systems [20].

10 Future Directions and Visionary Concepts

AI-driven bio-nano interfaces will, therefore, form a new frontier both in plant communication and temperature sensing because of how nanoscale sensor technologies integrate into the physiological processes of plants. The systems allow plants to become active biosensors that produce continuous signals related to environmental and physiological events, potentiating adaptive responses against thermal oscillations, enhancing resilience, and productivity. New-generation genetically encoded sensors (GESs) coupled with wearable sensors allow, in high resolution and real-time, for the monitoring of responses from the plants to microclimatic variation for better strategic decisions in precision agriculture. New classes of sensors, such as surface-enhanced raman scattering (SERS) nano sensors showing extraordinary sensitivity upon contact with volatile organic compounds (VOCs) released under conditions of temperature stress, enable the early detection of abiotic stresses [60]. Designs such as VERDANT TERRA, based on ultrasonic sensors and machine learning, decode the acoustic emissions from the plants in order independently to detect the rise in temperature increase-related stress and to autonomously mitigate this factor, while the system Plant-Keeper utilizes IoT-based ion transmission monitoring to assess the plant physiological state and the changes in ion fluxes resulting from temperature variations by contactless monitoring of plant physiology [110].

However, despite these, AI-driven bio-nano interfaces face persistent challenges in sensor integration, interpretation of large-scale data, and ethical issues related to the genetic alteration of plants. Overcoming such challenges requires close collaboration among interdisciplinary areas of specialization with the express aim of further improving AI modelling, nanomaterial synthesis, and sensor design in scalable ways [159]. QAI and nano-robotics have tremendous potential in enhancing efficiency, sustainability, and resiliency in adaptive climate agriculture. This is because QAI enhances the computational power needed to enhance precision in yield forecasting, pest detection, and soil analysis; meanwhile, nano-robotics allows for the control of nutrients at the molecular level in nutrient delivery, as well as the targeted management of pests (Fig. 3). This indicates that irrigation logistics and supply chains will be more effectively optimized for the overarching objectives of sustainable and CSA [132]. Nano-fertilizers and nano-pesticides, along with smart nano sensors, ensure nutrition with high precision and allow for early detection of stressors for climate resilience and soil health [38, 78]. On the other hand, stringent regulatory mechanisms are needed vis-à-vis access to quantum-enabled agricultural systems, data governance, and ethical issues to ensure fair and responsible adoption [17].

Another paradigm shift in precision agriculture is toward the emergence of self-learning agroecosystems that integrate AI-driven feedback mechanisms, predictive modelling, and real-time data analytics for autonomous regulation of crop temperature and environmental parameters. Convergence of AI with IoT, edge computing, and optimization algorithms further enhances adaptive greenhouse management to minimize energy consumption while improving resilience against climate variability. For this reason, data-driven predictive control frameworks [64], and deep RL algorithms [3] ensure higher sample efficiency, adaptiveness, and safety under extreme environmental conditions. Systems such as the AI-powered Greenhouse Environmental Control System, Lee et al. make use of hybrid DL architectures to predict microclimatic changes and, thus, allow proactive cooling system control that maintains optimal crop temperature conditions [88]. Physics-informed neural networks model complex nonlinear interactions between environmental variables, enhancing prediction accuracy at lower computational error rates. Other architectures, such as AGRARIAN, integrate AI, IoT, and satellite networks, enabling large-scale data integration in real-time for decision-making, yield forecasting, and efficient resource use, among others [19].

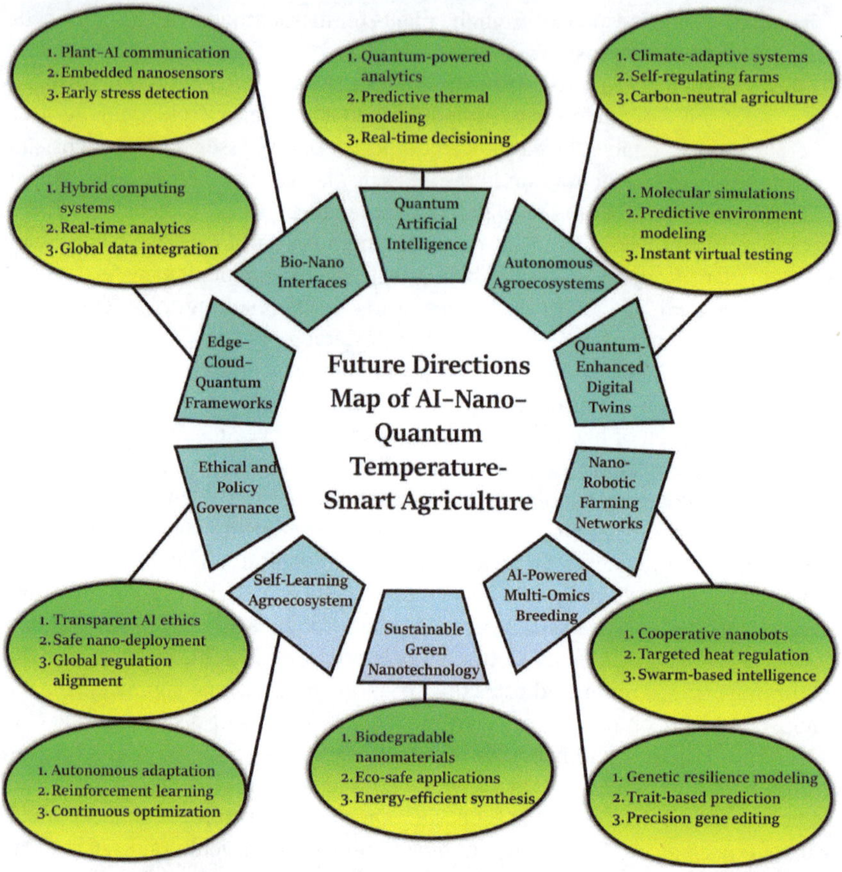

Fig. 3 Future directions map of AI–nano–quantum temperature-smart agriculture

With these developments, the AI-assisted thermo-biohybrid development of crops focuses on the design of plant systems that have adaptive thermal responses programmed at the genetic level. Using machine learning, AI is now capable of integrating genomics, transcriptomics, and metabolomics in decoding the architecture of germplasm and its plasticity owing to environmental factors for the identification of thermally resilient genotypes [50]. Integration of data in multiple modes and XAI-derived actionable insights hence allows optimization of parental selection toward accelerating temperature-resilient breeding [124]. Field phenotyping in combination with cultivar-specific temperature response models further sharpens the phenological predictions and deepens the knowledge of genotype-environment interaction under heat stress [127]. However, decoding of complex gene-environment dynamics remains a central challenge that further calls for interdisciplinary research in AI, molecular biology, and environmental modelling.

By 2050, fully autonomous AI-IoT-robotic agroecosystems are foreseen to transform food production in pursuit of sustainability, efficiency, and resilience. These will be integrated systems that apply real-time analytics to precision resource management, predictive crop monitoring, and autonomous operations, contributing to reduced human labour and resource waste [99]. AI-driven frameworks may help buffer the negative impacts of climate change through dynamic adaptation to extreme weather and pest outbreaks and hence assure global food security [122]. However, this challenge will also remain to be addressed on socioeconomic fronts, ethics, and possible labour displacement, for which comprehensive policy support and multi-sectoral collaboration need to be initiated to ensure that the adoption is equitable [109].

11 Conclusion

TSA embodies the radical shift toward climate-resilient and sustainable food systems through the integration of AI and nanotechnology. AI-driven models could provide predictive temperature management, precision irrigation, and resources optimization, whereas nanotechnology itself enhances thermotolerance with engineered materials, nano sensors, and temperature-responsive agrochemicals. Their convergence could allow the development of intelligent, adaptive frameworks that will enable dynamic environmental fluctuation assessment in real-time with reduced energy use and increased yield stability. QAI, federated learning, and bio-nano interfaces might take it even further into autonomous self-learning agro-ecosystems. This, however, would demand consideration of some critical challenges in the form of regulatory oversight, data governance, cost barriers, and ecological safety. Ethical frameworks orientated toward transparency, sustainability, and equitability of access are, therefore, crucial for responsible innovation. Merging AI-nano systems with green nanotechnology, models of the circular economy, and global digital infrastructures are more promising developments. A new path of multidisciplinary collaboration and governance will thus place TSA in a position to redefine the role of agriculture in climate adaptation, food security, and the long-term resilience of the planet.

Acknowledgements All the authors are thankful to their respective Universities and Institutions for their valuable support.

Data Availability The authors confirm that the data supporting the findings of this study are available within the book/chapter.

References

1. Adewusi AO, Chikezie NR, Eyo-Udo NL (2023) Blockchain technology in agriculture: enhancing supply chain transparency and traceability. Finance Acc Res J 5(12):479–501

2. Ahmed RM (2024) Integration of wireless sensor networks, Internet of Things, artificial intelligence, and deep learning in smart agriculture: a comprehensive survey: integration of wireless sensor networks, Internet of Things. J Innov Intell Comput Emerg Technol (JIICET) 1(01):8–19
3. Ajagekar A, Mattson NS, You F (2023) Energy-efficient ai-based control of semi-closed greenhouses leveraging robust optimization in deep reinforcement learning. Adv Appl Energy 9:100119
4. Ajina A, Nithin VS, Devika P, Chinmaya BS, Prithviraj BM (2025) An integrated system for real-time greenhouse environmental forecasting and adaptive crop management. In: 2025 3rd International conference on sustainable computing and data communication systems (ICSCDS). IEEE, pp 1923–1930
5. Akter R, Rahman MH, Chowdhury MAR, Manirujjaman M, Elshenawy SE (2022) Advances of nanotechnology in plant development and crop protection. In: Applications of computational intelligence in multi-disciplinary research. Academic Press, pp 143–157
6. Alam MW, Junaid PM, Gulzar Y, Abebe B, Awad M, Quazi SA (2024) Advancing agriculture with functional NM:"pathways to sustainable and smart farming technologies." Discover Nano 19(1):197
7. Alex B, Jignasa G, Madhubabu K, Gopi A (2024).AI-driven smart irrigation: enhancing agricultural water efficiency through intelligent valve regulation in piped and micro irrigation networks. In: 2024 First international conference on pioneering developments in computer science and digital technologies (IC2SDT). IEEE, pp 76–81
8. Alharbi A, Ghonimy M (2025) Modelling soil-atmosphere interactions to support sustainable soil management and agricultural resilience in temperate europe using the SiSPAT model. Sustainability 17(18):8114
9. Ali K, Asad Z, Agbna GH, Saud A, Khan A, Zaidi SJ (2024) Progress and innovations in hydrogels for sustainable agriculture. Agronomy 14(12):2815
10. Amangeldy B, Tasmurzayev N, Imankulov T, Baigarayeva Z, Izmailov N, Riza T, Zhumagulov B (2025) AI-powered building ecosystems: a narrative mapping review on the integration of digital twins and LLMs for proactive comfort, IEQ, and energy management. Sensors 25(17):5265
11. Anisul Islam Opy AIO (2025) Bridging renewable energy and smart agriculture: a review of ai-driven innovations for sustainable growth. Bridg Renew Energy Smart Agric: Rev AI-Driven Innov Sustain Growth 2(11):47–57
12. Anwar I (2025) Shaping the future of agriculture with intelligent systems. J Comput Sci Appl Eng (JOSAPEN) 3(2):41–45
13. Arora CH, Autade SD, Avhad AM, Deokar RD (2024) Smart cattle care: an IOT based monitoring and management system. Int J Innov Sci Res Technol 9(4):1928–1931
14. Ashique S, Raikar A, Jamil S, Lakshminarayana L, Gajbhiye SA, De S, Kumar S (2025) Artificial intelligence integration with nanotechnology: a new frontier for sustainable and precision agriculture. Curr Nanosci 21(2):242–273
15. Asrar A, Servatkhah M, Yasrebi M (2020) Providing a bird swarm algorithm based on classical conditioning learning behavior and comparing this algorithm with sinDE, JOA, NPSO and D-PSO-C based on using in nanoscience. J Optoelectron Nanostructures Summer 5(3)
16. Ates C, Bicat D, Yankov R, Arweiler J, Koch R, Bauer HJ (2023) Model predictive evolutionary temperature control via neural-network-based digital twins. Algorithms 16(8):387
17. Bansod PJ, Usharani R, Oliver AS, Prasad SS, Sharma DM, Myilsamy S (2024) Quantum computing-powered agricultural transformation: optimizing performance in farming. In: Technological advancements in data processing for next generation intelligent systems. IGI Global Scientific Publishing, pp 169–195
18. Barwant MM, Rasool A, Mannam VK, Hussain A, Abdullah M, Nisa Memon Z, Khan UA (2024) Nanotechnology applications in livestock farming and animal health. In: Harnessing nanoomics and nanozymes for sustainable agriculture. IGI Global Scientific Publishing, pp 413–442

19. Batistatos MC, De Cola T, Kourtis MA, Apostolopoulou V, Xilouris GK, Sagias NC (2025) AGRARIAN: a hybrid AI-driven architecture for smart agriculture. Agriculture 15(8):904
20. Behnassi M, Boussaid M, Gopichandran R (2014) Achieving food security in a changing climate: the potential of climate-smart agriculture. Environmental cost and face of agriculture in the Gulf cooperation council countries: fostering agriculture in the context of climate change. Springer International Publishing, Cham, pp 27–42
21. Belghachi M (2025) Smart irrigation systems using AI to optimize water usage. In: Cases on AI-driven solutions to environmental challenges. IGI Global Scientific Publishing, pp 241–268
22. Benameur R, Dahane A, Boudiaf A, Laouadi S, Belharizi N (2025) FL-DDQN: a federated double deep Q-learning framework for enhancing irrigation management across family multi-farms environments. J Ambient Intell Smart Environ 17(4):414–439
23. Bernacchi CJ, Ruiz-Vera UM, Siebers MH, DeLucia NJ, Ort DR (2023) Short-and long-term warming events on photosynthetic physiology, growth, and yields of field grown crops. Biochemical Journal 480(13):999–1014
24. Bhandari G, Dhasmana A, Chaudhary P, Gupta S, Gangola S, Gupta A, Slama P (2023) A perspective review on green nanotechnology in agro-ecosystems: opportunities for sustainable agricultural practices and environmental remediation. Agriculture 13(3):668
25. Bi G, Zhang Y, Xue Q, Li Y, Yuan Z, Liu K, Yang P (2024) ParallelFarm: An AI-enabled sustainable farming management system for carbon neutrality. In: 2024 IEEE 22nd international conference on industrial informatics (INDIN). IEEE, pp 1–6
26. Bischof JC, Diller KR (2018) From nanowarming to thermoregulation: new multiscale applications of bioheat transfer. Annu Rev Biomed Eng 20(1):301–327
27. Bist DR, Chapagaee P, Kunwar A, Pant BD, Khatri L, Mandal A (2025) Nanotechnology in agriculture: a review of innovations in crop protection and food security. Adv Agric 2025(1):8892001
28. Bosona T, Gebresenbet G (2023) The role of blockchain technology in promoting traceability systems in agri-food production and supply chains. Sensors 23(11):5342
29. Bukhtoyarov VV, Nekrasov IS, Timofeenko IA, Gorodov AA, Kartushinskii SA, Trofimov YV, Lishik SI (2025) Hybrid digital twin for phytotron microclimate control: integrating physics-based modelling and iot sensor networks. AgriEngineering 7(9):285
30. Cao Y, Zhai J, Zhang W, Zhou X, Zhang F (2024) MTTF: a multimodal transformer for temperature forecasting. Int J Comput Appl 46(2):122–135
31. Chaux JD, Sanchez-Londono D, Barbieri G (2021) A digital twin architecture to optimize productivity within controlled environment agriculture. Appl Sci 11(19):8875
32. Chavula P, Kayusi F, Lungu G, Mambwe H, Uwimbabazi A (2024) AI application in climate-smart agricultural technologies: a synthesis study. LatIA 2:330–330
33. Chen Z, Shan X, Guan Y, Wang S, Zhu JJ, Tao N (2015) Imaging local heating and thermal diffusion of nanomaterials with plasmonic thermal microscopy. ACS Nano 9(12):11574–11581
34. Chitengi S, Sakapaji JJ (2023) Harnessing AI for climate-resilient agriculture: opportunities and challenges. Eur J Theor Appl Sci 1:1144–1158
35. Cletus F, John AE (2024) Comparative analysis of machine learning models for greenhouse microclimate prediction. Brill: Res Artif Intell 4(1):162–175
36. da Cruz VF, Baptista F, Rico J, Coelho D (2022) AWARTECH project: from laboratory experiments to farm conditions. In: Practical precision livestock farming. Wageningen Academic, pp 137–153
37. D'Amato C, Tubini N, Rigon R (2025) A component-based modular treatment of the soil-plant-atmosphere continuum: the GEOSPACE framework (v. 1.2. 9). Geosci Model Dev 18(20):7321–7355
38. Danai-Varsou D, Zhang P, Afantitis A, Guo Z, Lynch I, Melagraki G (2023) Nanoinformatics and artificial intelligence for nano-enabled sustainable agriculture. In: Nano-enabled sustainable and precision agriculture. Academic Press, pp 503–531

39. Das H, Singh SK, Priyadarshini E, Maurya CL, Tiwari AK, Gulaiya S, Singh BV (2024) Nanotechnology facilitated real time soil monitoring for optimized crop production. Asian J Soil Sci Plant Nutr 10(2):582–607
40. Devaraj H, Sohail S, Ooi M, Li B, Hudson N, Baughman M, Rana O (2024) RuralAI in tomato farming: Integrated sensor system, distributed computing, and hierarchical federated learning for crop health monitoring. IEEE Sens Lett 8(5):1–4
41. Dilshad M, Fatima H, Al-Amin M, Qureshi A, Ahmad I, Anum H, Akbar A (2025) Nanotechnology in precision agriculture applications of nano sensors in soil, crop and water management. Haya: Saudi J Life Sci 10(8):298–310
42. Dinis LT, Ferreira H, Pinto G, Bernardo S, Correia CM, Moutinho-Pereira J (2016) Kaolin-based, foliar reflective film protects photosystem II structure and function in grapevine leaves exposed to heat and high solar radiation. Photosynthetica 54(1):47–55
43. Du Q, Ding XQ, Gao F, Cui B, Wang TY, Chen FY, Zeng ZH (2024) Thermo-responsive liposome nano-vesicles for co-delivery of emamectin benzoate and nitenpyram with synergistic pest control. Chem Eng J 479:147548
44. El-Ramady H, Brevik EC, Abowaly M, Ali R, Saad Moghanm F, Gharib MS, Prokisch J (2024) Soil degradation under a changing climate: management from traditional to nano-approaches. Egypt J Soil Sci 64(1):287–298
45. El-Ramady H, Prokisch J, El-Mahrouk ME, Bayoumi YA, Shalaby TA, Brevik EC, Solberg SØ (2024) Nano-food farming approaches to mitigate heat stress under ongoing climate change: a review. Agriculture 14(5):656
46. Fang SB, Shen B, Tan KY, Gao XN (2010) Effect of elevated CO_2 concentration and increased temperature on physiology and production of crops. Chin J Eco-Agric 18(5):1116–1124
47. Farooq U, Ali B, Bidsorkhi HC, D'Aloia AG, Sarto MS (2024) Graphene-based flexible thermal sensor and thermopile approach for enhancing thermometry sensitivity in next-generation wearables. IEEE Sens Lett 8(10):1–4
48. Fernández R, Navas E, Rodríguez-Nieto D, Rodríguez-González AA, Emmi L (2025) DigiHortiRobot: an AI-driven digital twin architecture for hydroponic greenhouse horticulture with dual-arm robotic automation. Future Internet 17(8):347
49. Field RT (1990) Documentation for program SOILSIM: a computer program for the simulation of heat and moisture flow in soils and between soils, canopy and atmosphere (No. NAS 1.26: 186953)
50. Fu J, Zheng S, Fan L, Zheng X, Qian Q (2025) Breeding 5.0: artificial intelligence (AI)-decoded germplasm for accelerated crop innovation. J Integr Plant Biol, 1–13.
51. Gao F, Fei J, Ye Y, Liu C (2024) MSLKSTNet: multi-scale large kernel spatiotemporal prediction neural network for air temperature prediction. Atmosphere 15(9):1114
52. Gautam R, Manisankar G, Bhuyan S, Kurdekar AK (2023) Exploring the future of agriculture through nanotechnology: a review. Int J Environ Clim Chang 13(10):215–228
53. Gilbertson LM, Pourzahedi L, Laughton S, Gao X, Zimmerman JB, Theis TL, Lowry GV (2020) Guiding the design space for nanotechnology to advance sustainable crop production. Nat Nanotechnol 15(9):801–810
54. Goldenits G, Mallinger K, Raubitzek S, Neubauer T (2024) Current applications and potential future directions of reinforcement learning-based digital twins in agriculture. Smart Agric Technol 8:100512
55. Gong R, Zhang H, Li G, He J (2025) Edge computing-enabled smart agriculture: technical architectures, practical evolution, and bottleneck breakthroughs. Sensors 25(17):5302
56. Gorantla S, Veluchamy S (2023) Automated crop growth monitoring and optimizing the yield with reinforcement learning. In: 2023 3rd international conference on mobile networks and wireless communications (ICMNWC). IEEE, pp 1–6
57. Grataloup A, Jonas S, Meyer A (2025) Wind turbine condition monitoring based on intra-and inter-farm federated learning. IEEE Access
58. Gul D, Banday RUZ (2024) Transforming crop management through advanced AI and machine learning: insights into innovative strategies for sustainable agriculture. AI Comput Sci Robot Technol. https://doi.org/10.5772/acrt.20240030

59. Häusermann D, Bodry S, Wiesemüller F, Miriyev A, Siegrist S, Fu F, Kovač M (2023) FireDrone: multi-environment thermally agnostic aerial robot. Adv Intell Syst 5(9):2300101
60. He T, Wang J, Chae E, Lee C (2025) In situ plant sensors: toward real-time, high-resolution monitoring. ACS Sens 10(9):6275–6291
61. Hooda R (2025) Harnessing artificial intelligence in agriculture: in-depth case studies and strategic insights. Int J Adv Res Comput Sci 16(4)
62. Hoque A, Mazumder AS, Roy S, Saikia P, Kumar K (2025) Transformative approaches to agricultural sustainability: automation, smart greenhouses, and AI. Int J Res Appl Sci Eng Technol 13(1):1011–1023
63. Howard DA, Ma Z, Veje C, Clausen A, Aaslyng JM, Jørgensen BN (2021) Greenhouse Industry 4.0-digital twin technology for commercial greenhouses. Energy Informat 4(Suppl 2)
64. Hu G, You F (2023) Energy management for controlled environment agriculture based on physics informed neural networks and adaptive linearization based data-driven robust model predictive control with AI. In: Computer aided chemical engineering, vol 52. Elsevier, pp 2205–2210
65. Hu X, Assaad RH (2024) Automated heat stress monitoring and water-spraying robotic system for improving work conditions using drone (UAV) infrared thermography. In: Construction research congress 2024, pp 748–756
66. Hull K, Booysen MJ, Mabitsela M, Phiri E (2023) Using a digital twin for greenhouse tunnel temperature management and prediction. In: 2023 IEEE AFRICON. IEEE, pp 1–6
67. Imran, Ortas I (2024) Nano biotechnology: exploring the potential of nano materials in enhancing crop health and combating abiotic stress. Commun Soil Sci Plant Anal 55(16):2509–2528
68. Jácome Galarza L, Realpe M, Viñán-Ludeña MS, Calderón MF, Jaramillo S (2025) AgriTransformer: a transformer-based model with attention mechanisms for enhanced multimodal crop yield prediction. Electronics 14(12):2466
69. Jahana K (2025) Integrating AI and machine learning into agricultural meteorology for sustainable and climate-smart farming. J Exp Agric Int 47(9):300–311
70. Jamal NN, Duncan E, Owens G (2021) Application of metal oxide nanomaterials in agriculture: benefit or bane? Nanomaterial biointeractions at the cellular, organismal and system levels. Springer, Cham, pp 231–248
71. Jog S, Bhole J, Gaikwad A, Modi A, Jadhav C (2023) The green Revolution 2.0: weather forecasting and precision irrigation reshaping crop yield enhancement. Int J Sci Res Sci Eng Technol 10(6):83–89
72. Julian J, Wahyuni F, Dewantara AB, Winarta A, Putra N (2024) Enhancing the accuracy of low-cost thermocouple devices through deep-wavelet neural network calibration. Int J Electr Comput Eng 14(3):2088–8708
73. Kamarudin MH, Ismail ZH, Saidi NB (2021) Deep learning sensor fusion in plant water stress assessment: a comprehensive review. Appl Sci 11(4):1403
74. Kamboj A, Malik A, Agarwal A, Ghorpade J, Sumathy G (2024) Multisensor crop guidance realtime engine with cloud analytics and recommendation system. In: 2024 International conference on advances in computing, communication and applied informatics (ACCAI). IEEE, pp 1–7
75. Karras A, Karras C, Giannaros A, Giotopoulos KC, Tsolis D, Oikonomou K, Sioutas S (2023) TinyML-based event detection: an edge-cloud approach for smart agriculture over LoRa WSNs. In: 2023 8th south-east europe design automation, computer engineering, computer networks and social media conference (SEEDA-CECNSM). IEEE, pp 1–10
76. Kaushik P (2025) The quantum-AI genomic frontier platform: revolutionizing vegetable crop resilience in a changing climate. https://doi.org/10.20944/preprints202503.1073.v1
77. Khan MHU, Wang S, Wang J, Ahmar S, Saeed S, Khan SU, Feng X (2022) Applications of artificial intelligence in climate-resilient smart-crop breeding. Int J Mol Sci 23(19):11156
78. Khobragade P, Dhankar PK, Titarmare A, Dhone M, Thakur S, Saraf P (2024) Quantum-enhanced AI robotics for sustainable agriculture: pioneering autonomous systems in precision

farming. In: 2024 International conference on artificial intelligence and quantum computation-based sensor application (ICAIQSA). IEEE, pp 1–7
79. Kim H, Kim G, Jain A, Lee D (2024) Mobile robot based personalized thermal comfort scheme. In: 2024 21st International conference on ubiquitous robots (UR). IEEE, pp 84–91
80. Kokilavani S, Janani B, Balasurya S, Khan SS (2023) Accumulation of engineered nanomaterials in soil, water, and air. In: Agricultural and environmental nanotechnology: novel technologies and their ecological impact. Springer, Singapore, pp 551–582
81. Kong M, Liu X, Chen D, Xu Y, Zhang J, Chen X, Xu HL (2025) Biochar synergy with smart agriculture and environment: from soil amendment to precision regulation systems. https://doi.org/10.5772/intechopen.1011847
82. Koo Y, Lukianova-Hleb EY, Pan J, Thompson SM, Lapotko DO, Braam J (2016) In planta response of Arabidopsis to photothermal impact mediated by gold nanoparticles. Small 12(5):623–630
83. Kow PY, Lee MH, Sun W, Yao MH, Chang FJ (2022) Integrate deep learning and physically-based models for multi-step-ahead microclimate forecasting. Expert Syst Appl 210:118481
84. Kumar S (2025) fundamental principles of green nanotechnology in agriculture. In: Eco-friendly nanotechnology: harnessing small-scale technologies for a cleaner and healthier planet. Deep Science Publishing, pp 01–12
85. Kumar S, Kumar M (2024) Enhancing agricultural decision-making through an explainable AI-based crop recommendation system. In: 2024 International conference on signal processing and advance research in computing (SPARC), vol 1. IEEE, pp 1–6
86. Kumari K, Mirzakhani Nafchi A, Mirzaee S, Abdalla A (2025) AI-driven future farming: achieving climate-smart and sustainable agriculture. AgriEngineering 7(3):89
87. Laxmi S, Chanu PH, Rani P, Rai S, Prasad SK, Singh RK (2019) Effect of hydrogel on soil moisture stress. J. Pharmacogn Phytochem 8(6):316–320
88. Lee MH, Yao MH, Kow PY, Kuo BJ, Chang FJ (2024) An artificial intelligence-powered environmental control system for resilient and efficient greenhouse farming. Sustainability 16(24):10958
89. Li Y, Sun Y, Liu Z, Zhang X, Guo D, ZhangY (2023) Indoor temperature and humidity control system based on transfer deep reinforcement learning. In: 2023 China automation congress (CAC). IEEE, pp 5720–5725
90. Lin C, Hur J, Zhang M, Zhang Y, Zhang L, Huang Z, Li W (2023) Alleviating heat stress and water scarcity in cultivation with biodegradable radiative cooling mulch. https://doi.org/10.21203/rs.3.rs-2922832/v1
91. Liu C, Zhao C, Wang Y, Wang H (2023) Machine-learning-based calibration of temperature sensors. Sensors 23(17):7347
92. Liu J, Li L, Cao C, Feng Z, Liu Y, Ma H, Mou F (2023) Swarming multifunctional heater-thermometer nanorobots for precise feedback hyperthermia delivery. ACS Nano 17(17):16731–16742
93. Maatar M, Wang Z, Dang KN, Abdallah AB (2024) BTSAM: balanced thermal-state-aware mapping algorithms and architecture for 3D-NoC-Based neuromorphic systems. IEEE Access 12:126679–126692
94. Mahan JR, Burke JJ (2016) U.S. Patent No. 9,247,689. U.S. Patent and Trademark Office, Washington, DC
95. Maraveas C (2022) Incorporating artificial intelligence technology in smart greenhouses: current State of the Art. Appl Sci 13(1):14
96. Martin RJ, Mittal R, Malik V, Jeribi F, Siddiqui ST, Hossain MA, Swapna SL (2024) XAI-powered smart agriculture framework for enhancing food productivity and sustainability. IEEE Access
97. Mehonic A, Sebastian A, Rajendran B, Simeone O, Vasilaki E, Kenyon AJ (2020) Memristors-From in-memory computing, deep learning acceleration, and spiking neural networks to the future of neuromorphic and bio-inspired computing. Adv Intell Syst 2(11):2000085
98. Milovanova M, Khambulatova Z, Ramazan K (2024) Climate-smart agriculture legal strategies for enhancing environmental sustainability in farming production. In: Bio web of conferences, vol 140. EDP Sciences, p 02009

99. Mim MI, Sultana F, Hasan MR (2025) AI-powered autonomous farming: the future of sustainable agriculture. Eur J Theor Appl Sci 3(1):11–31
100. Mishra H, Mishra D (2024) Sustainable smart agriculture to ensure zero hunger. In: Sustainable development goals. CRC Press, pp 16–37
101. Mohammadi Kashka F, Tahmasebi Sarvestani Z, Pirdashti H, Motevali A, Nadi M, Valipour M (2023) Sustainable systems engineering using life cycle assessment: application of artificial intelligence for predicting agro-environmental footprint. Sustainability 15(7):6326
102. Mohammadpour J, Lee A, Timchenko V, Taylor R (2022) Nano-enhanced phase change materials for thermal energy storage: a bibliometric analysis. Energies 15(9):3426
103. Mole M (2025) Revolutionizing agricultural water management through AI-driven irrigation systems: a comprehensive framework for sustainable farming practices. Int J Comput Eng 7(21):1–11
104. Navaneethan S, Sampath JL, Kiran SS (2023) Development of a multi-sensor fusion framework for early detection and monitoring of corn plant diseases. In: 2023 2nd international conference on automation, computing and renewable systems (ICACRS). IEEE, pp 856–861
105. Noureddine S, Mohammed B, Abdelmadjid MK (2024) Smart technology-based microclimate control system for greenhouse. Stud Eng Exact Sci 5(2):e10122–e10122
106. Omar H, Alsharaeh E (2024) Improving water retention in sandy soils with high-performance superabsorbents hydrogel polymer. ACS Omega 9(22):23531–23541
107. Otgonbaatar S, Nurmi O, Johansson M, Mäkelä J, Kocman T, Gawron P, Dumitru CO (2023) Quantum computing for climate change detection, climate modelling, and climate digital twin. https://elib.dlr.de/198760/
108. Pai DG, Balachandra M, Kamath R (2025) Explainable AI in agriculture: review of applications, methodologies, and future directions. Eng Res Express 7(3):032202
109. Pandey A (2025) The transformative potential of artificial intelligence in building sustainable agricultural systems: innovations, challenges, and future directions. In: AI and ecological change for sustainable development. IGI Global Scientific Publishing, pp 161–182
110. Pandey U, Vashisth A, Mishra A, Kumari M (2025) AI and IoT-driven soil health restoration: a machine learning approach for sustainable agriculture. https://doi.org/10.21203/rs.3.rs-6490610/v1
111. Paneru P, Singh SP, Deepak D, Kumar M (2025) Development and application of eco-friendly nanomaterials in plant science for sustainable agriculture. In: Eco-friendly nanotechnology: harnessing small-scale technologies for a cleaner and healthier planet, pp 13–28
112. Parameswari P, Belagalla N, Singh BV, Abhishek GJ, Rajesh GM, Katiyar D, Paul S (2024) Nanotechnology-based sensors for real-time monitoring and assessment of soil health and quality: a review. Asian J Soil Sci Plant Nutr 10(2):157–173
113. Park G, Park H, Seo J, Yang JC, Kim M, Lee BJ, Park S (2023) Bidirectional thermo-regulating hydrogel composite for autonomic thermal homeostasis. Nat Commun 14(1):3049
114. Patra SK, Poddar R, Brestic M, Acharjee PU, Bhattacharya P, Sengupta S, Hossain A (2022) Prospects of hydrogels in agriculture for enhancing crop and water productivity under water deficit condition. Int J Polym Sci 2022(1):4914836
115. Peladarinos N, Piromalis D, Cheimaras V, Tserepas E, Munteanu RA, Papageorgas P (2023) Enhancing smart agriculture by implementing digital twins: a comprehensive review. Sensors 23(16):7128
116. Platero-Horcajadas M, Pardo-Pina S, Cámara-Zapata JM, Brenes-Carranza JA, Ferrández-Pastor FJ (2024) Enhancing greenhouse efficiency: integrating IoT and reinforcement learning for optimized climate control. Sensors 24(24):8109
117. Polishchuk I, Prudnikov E, Abu-Hamad H, Ben-Arie N, Sklar J, Kellermeier M, Pokroy B (2025) SafeWax: a bio-inspired multifunctional coating for sustainable crop protection. Small e05360
118. Powers C, Bassman Oftelie L, Camps D, de Jong WA (2023) Exploring finite temperature properties of materials with quantum computers. Sci Rep 13(1):1986
119. Prabakeran S, Sethukarasi T, Indumathi V (2025) Federated learning frameworks for energy-efficient ai in distributed data centres. In: Energy efficient algorithms and green data centers for sustainable computing. IGI Global Scientific Publishing, pp 399–426

120. Prasad A (2023) Nanophotonic for efficient light harvesting in greenhouse agriculture. Knowl Res Multidisc J 1(11):55–61
121. Qiu E, Zhang YH, Ventra MD, Schuller IK (2024) Reconfigurable cascaded thermal neuristors for neuromorphic computing. Adv Mater 36(6):2306818
122. Rampalli NS, Sri YG, Bhuvaneshwari KS (2024) Autonomous agriculture and food production: agritech revolution. In: The convergence of self-sustaining systems with AI and IoT. IGI Global Scientific Publishing, pp 40–63
123. Rana RL, Tricase C, De Cesare L (2021) Blockchain technology for a sustainable agri-food supply chain. British Food J 123(11):3471–3485
124. Raza A, Bashir S, Khare T, Karikari B, Copeland RG, Jamla M, Varshney RK (2024) Temperature-smart plants: a new horizon with omics-driven plant breeding. Physiol Plant 176(1):e14188
125. Reich A, Carroll D, Carlson T, Reich M, Disalvo R (2023) The potential and challenges of quantum computing for engineering applications. In: Pressure vessels and piping conference, vol 87455. American Society of Mechanical Engineers, p V002T02A020
126. Requicha A (2012) Swarms of self-organized nanorobots. In: Nanorobotics: current approaches and techniques. Springer, New York, pp 41–49
127. Roth L, Binder M, Kirchgessner N, Tschurr F, Yates S, Hund A, Walter A (2024) From neglecting to including cultivar-specific per se temperature responses: extending the concept of thermal time in field crops. Plant Phenom 6:0185
128. Safdar M, Shahid MA, Yang C, Rasul F, Tahir M, Raza A, Sabir RM (2024) Climate smart agriculture and resilience. In: Emerging technologies and marketing strategies for sustainable agriculture. IGI Global Scientific Publishing, pp 28–52
129. Saket PK (2025) Explainable AI for assessing climate change impacts on rice yield in India: a Hybrid LSTM-SHAP approach. https://doi.org/10.20944/preprints202506.1108.v1
130. Santhosh S, Gurrapu O, Dimri N, Vladimirovna ON, Karthik B (2025) Integrating IoT and edge computing for smart agriculture with real-time data analytics. In: 2025 3rd International conference on sustainable computing and data communication systems (ICSCDS). IEEE, pp 1–6
131. Sarraf M, Vishwakarma K, Kumar V, Arif N, Das S, Johnson R, Hasanuzzaman M (2022) Metal/metalloid-based nanomaterials for plant abiotic stress tolerance: an overview of the mechanisms. Plants 11(3):316
132. Senthil GA, Monica KM, Prabha R, Prinslin L, Elavarasi R (2025) Quantum AI: a cognitive machine learning technique based on nurturing food security sustainability predictive analysis for life science-bioengineering in healthcare. In: Bio web of conferences, vol 172. EDP Sciences, p 02002
133. Shams MY, Gamel SA, Talaat FM (2024) Enhancing crop recommendation systems with explainable artificial intelligence: a study on agricultural decision-making. Neural Comput Appl 36(11):5695–5714
134. Shan B, Qian H, Guan X, De Michele C (2025) Enhancing crop yield simulations under extreme climate events using a hybrid model. In: EGU General assembly conference abstracts, pp EGU25–12074
135. Sharma M, Tirupathi Y, Singh RK, Das A, Singh TU, Tarafdar A (2025) Emerging applications of nanotechnology in livestock: updates and perspectives. Ann Animal Sci. https://doi.org/10.2478/aoas-2025-0081
136. Shen Y, An C, Jiang J, Huang B, Li N, Sun C, Wang Y (2022) Temperature-dependent nanogel for pesticide smart delivery with improved foliar dispersion and bioactivity for efficient control of multiple pests. ACS Nano 16(12):20622–20632
137. Shilpa K (2024) Frontiers of artificial intelligence in agricultural sector: trends and transformations. J Sci Res Rep 30(10):970–980
138. Sidhu AK, Sharma M, Bhickchand Agrawal S, Pradip Bhavsar P, Samota MK (2024) Nanomaterial strategies for enhancing plant resilience in the face of temperature stress. CABI Agric Biosci 5(1):60

139. Singh SP, Paneru P, Deepak D (2025) Nanotechnological approaches for enhancing climate resilience and sustainable adaptation in agriculture. In: Eco-friendly nanotechnology: harnessing small-scale technologies for a cleaner and healthier planet, 189–198
140. Sissodia R, Rauthan MS, Barthwal V, Dwivedi V (2025) Evolutionary algorithms for optimization and swarm intelligence-based optimization. In: Optimization tools and techniques for enhanced computational efficiency. IGI Global Scientific Publishing, pp 17–42
141. Sodhi GK, Wijesekara T, Kumawat KC, Adhikari P, Joshi K, Singh S, Mitra D (2025) Nanomaterials-plants-microbes interaction: plant growth promotion and stress mitigation. Front Microbiol 15:1516794
142. Son KH, Sim HS, Lee JK, Lee J (2023) Precise sensing of leaf temperatures for smart farm applications. Horticulturae 9(4):518
143. Sreedevi V, Kulkarni V, Rinku DR, Jyothi V, Kanakaprabha S, Rao PV (2025) Edge computing for agricultural monitoring: a bandwidth-efficient iot approach for real-time sensor data analysis. In: 2025 8th International conference on computing methodologies and communication (ICCMC). IEEE, pp 361–366
144. Stogia M, Naserentin V, Dimara A, Eleftheriou O, Tzitzios I, Papaioannou C, Krinidis S (2024) A scalable and user-friendly framework integrating IoT and digital twins for home energy management systems. Appl Sci 14(24):11834
145. Sundararajan N, Habeebsheriff HS, Dhanabalan K, Cong VH, Wong LS, Rajamani R, Dhar BK (2024) Mitigating global challenges: harnessing green synthesized nanomaterials for sustainable crop production systems. Global Chall 8(1):2300187
146. Tan X (2024) Quantum computing for phonon scattering effects on thermal conductivity. arXiv preprint arXiv:2407.15808
147. Thingujam D, Gouli S, Cooray SP, Chandran KB, Givens SB, Gandhimeyyan RV, Mukhtar MS (2025) Climate-resilient crops: Integrating AI, multi-omics, and advanced phenotyping to address global agricultural and societal challenges. Plants 14(17):2699
148. Torres AB, da Rocha AR, da Silva TLC, de Souza JN, Gondim RS (2020) Multilevel data fusion for the internet of things in smart agriculture. Comput Electron Agric 171:105309
149. Tran V, Septier F, Murakami D, Matsui T (2024) Spatial-temporal temperature forecasting using deep-neural-network-based domain adaptation. Atmosphere 15(1):90
150. Vezy R (2023) PlantSimEngine: a simulation engine for the soil-plant-atmosphere system. J Open Source Softw 8(86):5371
151. Vinay JU, Kurjogi M, Uppar A, Vanti G (2023) Engineered nanomaterials in crop plants temperature and or heat stress management. In: Engineered nanomaterials for sustainable agricultural production, soil improvement and stress management. Academic Press, pp. 227–239
152. Weber M, Hoobler R, Cope D, Gibbons P (2009) U.S. Patent Application No. 11/827,911
153. Wei W, Tian Y, Shao Z, Bao B, Chen G, Chen D, Wang H (2025) Self-adaptive metafabric enabling thermal rectification and radiative-phase change energy cycling for all-weather thermal regulation. Adv Funct Mater, e15535
154. Wu W, Yao B, Huang J, Sun S, Zhang F, He Z, Gao R (2023) Optimal temperature and humidity control for autonomous control system based on PSO-BP neural networks. IET Control Theory Appl 17(15):2097–2109
155. Xiong J, Yu R, Islam E, Zhu F, Zha J, Sohail MI (2020) Effect of biochar on soil temperature under high soil surface temperature in coal mined arid and semiarid regions. Sustainability 12(19):8238
156. Yu A, Xi Y, Xu W, Wang A, Liu L, Zhang L (2023) Fabrication and performance evaluation of the flexible positive temperature coefficient material for self-regulating thermal control. J Appl Polym Sci 140(34):e54311
157. Yu F, Hao H, Li Q (2021) An Ensemble 3D convolutional neural network for spatiotemporal soil temperature forecasting. Sustainability 13(16):9174
158. Yu J, Sun C, Zhao J, Ma L, Zheng W, Xie Q, Wei X (2025) Prediction and control of greenhouse temperature: methods, applications, and future directions. Comput Electron Agric 237:110603

159. Zhang J, Wei Z, Sun J, Guo M, Huang X, Chen X, Peng Y (2022) Plant keeper: towards wireless sensing to ion transmission of plants. In: Proceedings of the 1st ACM workshop on no power and low power internet-of-things, pp 14–20
160. Zhang P, Guo Z, Ullah S, Melagraki G, Afantitis A, Lynch I (2021) Nanotechnology and artificial intelligence to enable sustainable and precision agriculture. Nat Plants 7(7):864–876
161. Zhang T, Li J, Li J, Wang L, Li F, Liu J (2019) Model embedded DRL for intelligent greenhouse control. arXiv preprint arXiv:1912.00020
162. Zhang W, Cao Y, Zhai JH, Mu Z, Zhang S (2022) TempCast: a multi-modal transformer for short-term temperature forecasting. In: 2022 IEEE 8th International conference on cloud computing and intelligent systems (CCIS). IEEE, pp 469–473
163. Zhang Y, Fu L, Martinez MR, Sun H, Nava V, Yan J, Lowry GV (2023) Temperature-responsive bottlebrush polymers deliver a stress-regulating agent in vivo for prolonged plant heat stress mitigation. ACS Sustain Chem Eng 11(8):3346–3358
164. Zhao J, Li X, Liu F, Zhao W, Chen S (2025) Smart fertilizer with temperature-responsive behavior coated using a naturally derived material. J Appl Polym Sci 142(15):e56721
165. Zhao W, Wu Z, Amde M, Zhu G, Wei Y, Zhou P, Lynch I (2023) Nanoenabled enhancement of plant tolerance to heat and drought stress on molecular response. J Agric Food Chem 71(51):20405–20418
166. Zhao X, Wu J, Wang M, Li G, Yu H, Feng W (2020) Multi-sensor data fusion algorithm based on adaptive trust estimation and neural network. In: 2020 IEEE/CIC international conference on communications in China (ICCC). IEEE, pp 582–587
167. Zidan F, Febriyanti DE (2024) Optimizing agricultural yields with artificial intelligence-based climate adaptation strategies. IAIC Trans Sustain Dig Innov (ITSDI) 5(2):136–147

Water Resilient Farming Contributed by AI Models with Nanotechnology

Bhagyashree Mishra, Gouri Nanda, HimaBindu Pitaka, K. Baby Rani, and Rajendra Kumar

Abstract Water stresses are a critical criterion for agricultural landscapes, which shows a greater impact not only on climate change but also on water scarcity, extreme temperatures, soil degradation, crop productivity, and protectivity, and its impact converts agricultural land into barren land, constraints threatening food security and farming practices, which need immediate attention to develop strategies on an immediate basis to enhance resilience. This comprehensive study explores the capacity of smart nano-systems with Artificial Intelligence as a transformative solution to enhance water stress, maintain soil health, and provide sustainable alternatives to conventional agricultural systems. This study demonstrates that the smart Nano-system integrated with Artificial Intelligence shows an innovative and sustainable solution to address the agricultural challenges. The nano-encapsulated PGPR system, particularly using ZnO nanoparticles as carriers, significantly improves crop resilience to water stress, protects the microbiome from environmental stress, while ensuring targeted delivery to plant root zones where needed. Nano-bio hybrid systems enhance water uptake efficiency, improving soil structure with nutrient availability, and strengthening plant defense mechanisms against abiotic stress using a hydrogel system. This integrated approach of smart nano-systems with AI represents a transformative shift towards sustainable agriculture with a practical pathway to maintaining food security, conserving water, and adapting to climate change.

Keywords PGPR · Nanotechnology · Nano-hydrogel · Quantum sensor · ZnO nanoparticles

---

B. Mishra
School of Allied Health Science, The Apollo University, Chittoor, Andhra Pradesh, India

G. Nanda · H. Pitaka · K. B. Rani
Department of Biomedical Sciences, School of Health Sciences, The Apollo University, Chittoor, Andhra Pradesh, India

R. Kumar (✉)
Division of Genetics, ICAR-Indian Agricultural Research Institute, New Delhi, India
e-mail: rajendrak64@yahoo.co.in

J.-T. Chen (ed.), *AI-Nano-Enabled Climate-Smart Agriculture and Sustainability*,
Smart Nanomaterials Technology, https://doi.org/10.1007/978-981-92-1694-9_11

1 Introduction

Water stresses are important criteria for agricultural landscapes that have a greater impact not only on climate change but also on water scarcity, extreme temperatures, soil degradation, crop productivity, and protectivity, and their impact converts agricultural land into barren land constraints, threatening food security and farming practices, which need immediate attention to strategies on an immediate basis to enhance resilience. This comprehensive study explored the capacity of smart nano-systems with Artificial Intelligence as a transformative solution to enhance water stress, maintain soil health, and provide sustainable alternatives to conventional agricultural systems. The potential of smart nano-systems integrates more efficient and advanced technologies, resource-conserving, and long-term food production, showcasing them as sustainable alternatives to traditional farming systems.

Problems such as poor shelf life, environmental instability, and consistent field performance lead to Efficacy with Nano-encapsulated PGPR transformative solutions with biodegradable smart nano-system carriers that safeguard microbial viability with controlled integration under soil-specific conditions. To optimize the situation, AI smart modeling such as machine learning algorithms including random forest and deep neural networks, is being used to analyze multi-spectral data, soil parameters, and plant physiological responses, which detect the nano-PGPR responses, enabling data driven decision making for precision agriculture [1].

Nano-bio hybrid gives a dual role by protecting microbial efficacy and precise, responsive delivery near the plant rhizosphere. Nanoparticles, such as ZnO, are being used as carriers to protect PGPR activity and improve stress management. Innovative ideas that show a good result on PGPR plant productivity, protectivity, mineral solubilization, and siderophores formation with increasing crop yield, which has a significant impact on crop production; however, it has some drawbacks, reducing its performance potential for targeted delivery with limited diffusion. To overcome these difficulties, Nano-emulsion smart technology can improve the delivery system, enhance the stability of PGPR with controlled release, and improve bioavailability. Climate change is the most common challenge in modern agricultural activities, changing environmental conditions that crops have evolved to tolerate and cause unexpected stress to agricultural systems worldwide [2]. With increasing global temperature, erratic precipitation patterns, and extreme weather conditions that create abiotic stresses, crop productivity and food security. Over the past century, global water withdrawal rates have exceeded population growth rates by more than 1.7 times, with approximately 70% of the freshwater consumption dedicated to agricultural purposes. To maintain sustainable and environmentally friendly enhancement of crop resilience, PGPR inhibit the rhizosphere and narrow the zone of soil surrounding plant roots to obtain the benefits of microbes and host plants [2].

Therefore, the role of PGPR in abiotic stress has a beneficial role in colonizing the rhizosphere and promoting growth-producing phytohormones, solubilizing specific nutrients by triggering stress-specific responsive genes. PGPR increases root surface area to enhance water uptake by producing exopolysaccharides that improve soil aggregation and water retention, inducing systemic resistance (ISR) in plants, and modulating ethylene with specific Nano-carriers that increase survival rate and persistence in the soil environment, mitigating soil abiotic stress. Nano-materials such as ZnO, silica, chitosan, carbon nanotubes, and polymeric matrices are being used for the encapsulation of PGPR with different functional groups and show controlled release and target-specific delivery; the inability to maintain PGPR availability under extreme pH, salinity, and temperature, and its porosity enhances the interaction with plant root exudates, minimizing water loss and runoff [3].

Coating PGPR with ZnO nanoparticles using the sol–gel method or electrospinning methods results in unique properties such as biocompatibility, antimicrobial properties, protection from environmental stresses such as UV radiation, extreme temperature, pH fluctuation, and enzymatic degradation, while maintaining their viability and biological activity. However, the extensive use of nanomaterials in consumer products has raised concerns regarding the interaction of nanoparticles with living organisms (including microorganisms). Zinc oxide nanoparticles (ZNo-NPs) are well known for their bioactivity due to the generation of reactive oxygen species (ROS). Therefore, their release into the environment is expected to raise major concerns regarding their ecotoxicity [4].

Due to the overproduction of Reactive Oxygen species (ROS), plants face oxidative stress. To overcome this issue, PGPR are being used to maintain cellular homeostasis, enhance the activity of antioxidant enzymes, and promote the synthesis of non-enzymatic antioxidant properties, thereby maintaining stress tolerance as a global challenge nowadays. Nano-material formulations not only benefit plant growth but also maintain real-time soil health and yield crop production by detecting early environmental factors. Ion homeostasis is important for stress tolerance, where PGPR helps the plants maintain proper sodium and potassium ratios and often exhibits specialized morphological adaptation with succulent leaves and waxy outer layers, which enhances the natural defense mechanism in plants. The integration of synthetic biology Quantum nano-sensors technology detects plant stress signals through soil moisture and enhances water uptake by modulating hydrogel matrices based on algorithmic prediction at the docking site [5]. Quantum nano-sensor molecules enabled the PGPR Delivery system to engineer microbial strains with enhanced traits and improved drought-responsive gene expression to optimize the root colonization pattern, persistence, and activity of beneficial bacteria in the rhizosphere with its attachment to the docking site. Microbiome-based solutions for healthy soil enhance plant resilience to abiotic and biotic stress and increase the potential of the soil microbiome by promoting resilience to climate change, facilitating adaptive strategies to address the issue of land degradation and food security [6].

Recent advancements in hydrogel encapsulation nanotechnology have introduced stimuli-responsive systems that adjust the target delivery system characteristics based on environmental conditions by incorporating pH-responsive polymers, temperature-sensitive materials, and moisture-activated components to adjust the rate of bacterial release to match the plant's release water stress levels. It releases specific plant stress hormones through bacterial signaling molecules, creating a biological synthetic system that optimizes plant health. Smart Nano-delivery system enables detection of real-time monitoring of soil and plant health parameters, including soil moisture content, temperature, pH levels, nutrient availability, and microbial activity, by minimizing sensors using electrochemical, optical, and mechanical responses and wireless communication and data processing algorithms to create an everyday monitoring network that detects small changes in the soil and plant health condition and its physiological status. Plant stress nano-sensors can monitor plant transpiration rate, stomatal conductance, chlorophyll fluorescence, and electrical impulses at the molecular level, enabling the early detection of symptoms and stress-induced damage to plants [7].

This study demonstrates that the smart nano-system integrated with Artificial Intelligence provides innovative and sustainable solutions to address agricultural challenges and that the nano-encapsulated PGPR system, particularly using ZnO nanoparticles as carriers, significantly improves crop resilience to water stress and protects the microbiome from environmental stress while ensuring targeted delivery to plant root zones where needed. Nano-bio hybrid systems enhance water uptake efficiency, improving soil structure with nutrient availability, and strengthening plant defense mechanisms against abiotic stress using a hydrogel system. This integrated approach of smart nano-systems with AI represents a transformative shift towards sustainable agriculture with a practical pathway to maintain food security, conserving water, and adapting to climate change [8].

2 Nanotechnology in Agriculture: Transformative Potential for Sustainable Crop Production

Nanotechnology is emerging as a vital tool in modern agriculture, offering new ways to improve resource efficiency, protect crops, and adapt to environmental stresses. By manipulating materials at the nanoscale (typically 1–100 nm), scientists can engineer particles and systems that behave differently from their larger counterparts. These unique properties open doors to innovations such as slow-release nano-fertilizers, targeted nano-pesticides, sensor-based monitoring, and seed coatings that boost germination and stress tolerance [9].

One of the key applications lies in nutrient delivery. Traditional fertilizers often face losses by leaching or volatilization, while nano-fertilizer formulations can improve uptake and retention. For example, nanoparticles provide increased surface area and improved interactions with plant roots, enabling nutrients to be delivered more precisely and in smaller amounts. This not only reduces input cost and environmental runoff but also supports plant growth under marginal water conditions [10].

Another major domain is pest and disease management. Nano-pesticides or nano-encapsulated active ingredients permit controlled release, higher stability under field conditions, and targeted action against pathogens or insects. This precision reduces collateral damage to beneficial organisms and lessens the overall chemical burden on soils and ecosystems. In addition, nano-coatings applied to seeds can protect emerging roots against pathogens and harsh conditions, giving plants an early advantage [11].

Sensor and monitoring technologies are also rapidly advancing. Nano-sensors embedded in soil or plant tissues can detect moisture levels, nutrient status, pH shifts, or even early stress biomarkers before visible symptoms appear. Harnessed within an Internet-of-Things (IoT) framework, these tiny devices can feed real-time data into decision-support systems, enabling farmers or automated controllers to respond quickly by adjusting irrigation, applying nutrients, or triggering protective treatments [12].

When nanotechnology is combined with plant stress-management strategies, its benefits become more evident. For instance, nanoparticles have been explored to enhance root hydraulic conductivity, modify soil wettability, or act as carriers for bioactive molecules that boost antioxidant defense in plants. These traits are particularly valuable under drought or heat stress, where plants must optimize water uptake, minimize transpiration losses, and maintain cellular health.

However, the promise of nano-agriculture comes with important considerations. Because nanoparticles are highly reactive and mobile, their interactions with soil microbes, plant tissues, and ecosystems must be carefully assessed. Potential toxicity, accumulation in food chains, and long-term soil health implications remain under investigation. Moreover, scaling nanomaterials from lab to field cost-effectively, ensuring uniform application and meeting regulatory standards, poses non-trivial challenges [10].

Looking toward the future, the integration of nanotechnology with precision agriculture tools, remote sensing, and AI represents a smart farming paradigm. In such systems, nano-sensors detect stress signatures, nanoparticles deliver corrective agents to specific zones, and algorithms optimize timing and quantity of interventions—all aimed at maintaining yield while minimizing inputs and environmental impact. Full adoption will likely depend on demonstration trials, cost reduction of nanomaterials, robust safety frameworks, and farmer training [13]. In summary, nanotechnology holds significant potential to transform agriculture. By enabling targeted delivery of nutrients and pesticides, enhancing plant resilience to stress, and facilitating real-time monitoring, nano-enabled systems offer a path toward higher efficiency and sustainability. As global agriculture grapples with climate change, water scarcity, and soil degradation, such innovations may become essential components of future crop production systems that are both productive and environmentally responsible.

3 Nano-sensors and Biosensors for Real-Time Monitoring in Agriculture

In modern agriculture, the ability to monitor soil and plant health in real-time is rapidly becoming a game-changer—especially under the growing pressures of climate variability, water scarcity, and changing nutrient dynamics. Traditional sampling methods (soil cores, lab analyses) are slow, labor-intensive, and unable to capture fast changes across a field. That is where nano-sensors and biosensors step in, offering high-resolution, continuous data streams that can be used to inform precision decisions on irrigation, fertilization, microbe delivery, and stress mitigation [14].

At their core, nano-sensors are tiny devices built with nanomaterials (for example, graphene, carbon nanotubes, metal-oxide nanoparticles) that respond to specific physical, chemical, or biological changes—such as moisture drop, ion shifts, pH change, or microbial activity. Biosensors build on that by incorporating a biological recognition element—an enzyme, aptamer, or antibody—that binds a target (say, a plant stress hormone or pathogen marker), converting that binding event into an electrical or optical signal. The advantage is twofold: high sensitivity (they can detect very low concentrations) and fast response (sub-minute or real-time).

In practical terms, imagine a smart network buried in a field: nano-sensors at several depths pick up dropping soil moisture, increasing salinity, or elevated root-zone temperature. A biosensor on a leaf surface detects early signs of pathogen attack or drought stress hormones. These sensors transmit data via low-power wireless links (LoRa, NB-IoT) to a cloud platform or edge computer, which then triggers alerts, irrigation adjustments, or targeted delivery of beneficial microbes. Real-time monitoring transforms reactive farming into proactive management [15].

The sensor architectures vary, and electrochemical sensors use nanowire electrodes that change current when ions or metabolites bind; optical/fluorescent sensors use quantum dots or nanoscale fluorophores whose emission changes upon analyte binding; and surface-plasmon-resonance (SPR) sensors detect refractive-index changes at metal surfaces coated with recognition molecules. These diverse platforms allow detection of soil nutrients (K^+, NO_3^-), heavy metals, moisture/potential changes, plant stress markers, and even microbial quorum-sensing molecules.

Why is this especially important for drought or water-limited agriculture? Because plants do not wait for a full wilting to signal distress. Early shifts in root exudate chemistry, ion balance, or leaf potential may occur hours before visible symptoms. Nano-sensors can catch these "silent alarms" and feed them to machine learning models (in a broader smart-farm system), which then decide: "Trigger microbe release now," or "Delay irrigation for zone B." In essence, these sensors provide the nervous system of a smart farm [16].

Still, several challenges remain. Deploying nanodevices in soil means dealing with dust, varying pH, biofouling (microbial build-up on sensors), power supply (for remote sensors), and cost. Moreover, although lab-scale sensors show excellent sensitivity, field ruggedness, and long-term stability, calibration across seasons remains under-researched. There are also questions of nanomaterial safety in soils, sensor disposal, and data management infrastructure in rural settings.

Looking ahead, the convergence of nano-sensors with bio-functionalized surfaces (for microbes or plant-hormone detection), wireless IoT networks, and AI/machine-learning analytics offers tremendous promise. Farms of the future could have self-learning systems: sensors detect early plant stress → AI determines optimal action → controlled release of beneficial microbes or nutrients occurs → sensors monitor outcome and refine next step. This closed-loop, adaptive system has the potential to make agriculture more resilient, efficient, and precise [17].

Nano-sensors and biosensors are enabling a leap from periodic manual measurements to continuous, fine-grained monitoring of agro ecosystems. By capturing subtle changes in soil, plant, and microbial health, these technologies offer farmers actionable insights at the right time and place. Combined with decision-support systems, they promise smarter, more sustainable agriculture in an age of increasing environmental stress.

4 System Architecture: Integration of Nano, Bio, and AI Layers

Designing a farming system that is truly resilient to water stress means weaving together multiple technologies so they operate as one coordinated unit. At the heart of such a system are three intertwined layers: the nano/micro delivery layer (materials and microbes), the sensing/communication layer (data capture and transmission), and the intelligence/actuation layer (AI decision-making and execution). When these layers are aligned, the farm becomes a living network rather than isolated parts.

First, the nano/bio layer comprises the carrier materials (for example, nano-hydrogels, nano-clays, or polymer-encapsulated microbes) and the beneficial microbes themselves (such as drought-tolerant PGPR). These materials are placed in the soil or root zone in a spatially mapped layout. Their job is to protect microbial cells, control release timing, and deliver nutrients or stress-relief agents exactly when and where plants need them.

Above that lies the sensing and communication layer. Here, a network of sensors monitors real-time soil parameters (moisture, temperature, ion concentrations), plant physiological signals (leaf water potential, chlorophyll fluorescence), and microbial activity. These sensors are connected via low-power wireless systems (LoRa, NB-IoT, or ZigBee) to edge computing nodes and cloud servers. This layer acts like the nervous system of the farming field—capturing signals, transmitting data, and enabling real-time awareness.

The third layer is the AI/actuation layer. Data from the sensing grid arrives and is processed in models that learn temporal patterns and spatial heterogeneity. Machine learning or reinforcement-learning engines predict when plants face water stress, identify which zone needs intervention, and determine the optimal delivery volume and timing of nano-bio carriers. Once a decision is made, actuators trigger the release mechanism (via swelling of a hydrogel, pH change, or thermal activation) to deliver microbes or nutrients. The effect is then sensed again—forming a closed-loop: sense → decide → act → sense.

Integration happens when the spatial mapping of nano-bio carriers is fed into the AI system, sensors confirm microbial colonization and plant response, and the system adapts over time. For example, if Zone 5 shows early signs of drying roots, the AI schedules a micro-release of PGPR in that zone, sensors verify improved root hydration, and future models adjust by lowering release amounts or extending intervals. Over seasons, the architecture becomes more refined, requiring fewer interventions.

Such a system architecture requires careful design of each interface: durable nanomaterials so microbes survive, rugged sensors for field environments, robust communications for connectivity, and AI models that generalize across conditions. It also demands user interfaces so farmers or agronomists can monitor the system, receive alerts, and override decisions if needed. In essence, this three-tiered architecture forms a cyber-physical-biological system that adapts dynamically to water stress, rather than simply reacting.

In summary, by integrating nano-bio delivery mechanisms, continuous sensing and wireless networking, and advanced AI decision-making, the system architecture transforms the field into an adaptive, intelligent environment—capable of sensing incipient stress, intervening precisely, and learning for next time. System Architecture: Integration of Nano, Bio, and AI Layers.

5 Nanoencapsulation and Smart Hydrogels

Nanoencapsulation and smart hydrogels represent advancements in materials science that are transforming how bioactive agents are delivered and their efficacy in agriculture. Nanoencapsulation refers to the process of enclosing helpful substances like plant growth-promoting rhizobacteria (PGPR), nutrients, or pesticides within nanoscale carriers, typically measuring from 10 to 1000 nm. These nanocarriers can be made from polymers like chitosan, alginate, polyethylene glycol (PEG), or inorganic materials such as zinc oxide and silica nanoparticles. Smart hydrogels consist of three-dimensional, hydrophilic polymer frameworks that can absorb significant amounts of water or biological fluid; they can be specially designed with properties that allow to respond to stimuli for controlled release of their contents based on environmental triggers. In nanoencapsulation, sometimes the active agent is physically in the nanogel's mesh or chemically bound via degradable linkers.

The controlled release mechanisms of nanoencapsulation and smart hydrogels rely on systems that respond to stimuli, which can detect changes in environmental conditions such as pH, temperature, moisture, or the presence of specific enzymes. These stimuli can alter the physical and chemical properties (swelling, shrinking, network degradation), and this change causes the release of the encapsulated agent. For example, pH-sensitive release, which tends to change the local pH, where the polymers with ionizable groups will swell or shrink, and thermo-responsive release, which contains polymers that collapse or swell at a critical temperature. Additionally, Enzyme/redox-responsive release, which is broken in the presence of specific enzymes, and composite/remote-stimulus release embedding magnetic or light-responsive nanoparticles, which allows external triggers like magnetic field. The mechanisms typically involve (i) diffusion of the active agent through the gel mesh, (ii) swelling/deswelling of the gel, altering the mesh size and permeability, and (iii) degradation or erosion of the gel matrix itself.

Several notable examples highlight the practical use and effectiveness of these systems. Nanoparticles made from chitosan have been used to encapsulate PGPR, providing protection to microbes from harmful environments such as UV, exposure, drying out, and predation microfauna. Microbes encapsulated in zinc oxide nanoparticles have shown an enhanced drought stress resilience in crops by improving the nutrient uptake and modulating stress hormones. Smart hydrogels, which have been loaded with micronutrients and beneficial microbes, create a moist microenvironment with a steady release pattern, which reduces nutrient leakage and improves plant resilience to abiotic stress. These nano-systems function with plant root exudates and microbial communities, adapting release kinetics based on real-time soil conditions.

The efficiency of this system is maintained by several design features. It has a high loading capacity, where the nanoscale size and functional group allow more drugs to be loaded and retained until triggered. By designing the networks that remain intact until the specific stimulus is encountered, the payload isn't released prematurely into any healthy tissue, thus it improves targeting and reduces the side effects. It has controlled kinetics, i.e., by tuning the polymer composition, cross-link density, and stimulus sensitivity, the release rate can be tailored rather than a burst release. The system is made with safe materials, hence biocompatible and biodegradable. The stimuli (such as pH or redox) are specific to pathological sites (tumor, inflammation) so the release happens wherever needed. This just localizes the effect and increases efficacy per dose.

Therefore, by combining nano-scale encapsulation with smart, stimuli-responsive hydrogels, one obtains delivery systems that are adaptable, targeted, and efficient—improving therapeutic outcomes and reducing side-effects.

6 AI and ML Integration in Agriculture

Artificial Intelligence (AI) refers to the computer systems that perform tasks that normally require human intelligence, such as reasoning, problem-solving, and pattern recognition. Machine Learning (ML), a subset of AI, refers to the algorithms that learn from data, improving their performance on specific tasks without explicit programming. Artificial Intelligence (AI) and Machine learning (ML) are transformative technologies increasingly incorporated into modern agriculture, offering extraordinary abilities for observation, prediction, and decision making. Both combined enable data driven precision agriculture, especially in managing challenges like water stress and drought.

The input data that drives the AI and ML models in agriculture is diverse, including environmental and biological parameters. In order to make useful predictions, ML systems draw on a variety of data sources, including soil moisture measurements and soil temperatures, which tell us how wet/dry and warm the soil is. Air temperature, humidity, rainfall/ precipitation, evapotranspiration (weather data), all these are collected because the crops and microbes respond to the climatic conditions. The drone/ UAV (unmanned aerial vehicle) images, satellite imagery (crop canopy, vegetation indices, NDVI) give us a complete view of plant health, stress, coverage. Furthermore, the historical field data including the past yields, crop type, planting dates, irrigation events, microbial treatments and geospatial data such as topography, soil type, texture, all these data streams are increasingly used in smart- farming/ precision- agriculture systems driven by AI.

Once all this data is collected and cleaned, ML algorithms learn to predict outcomes. Some commonly used models in agriculture involve Random Forest, which is an algorithm that builds a decision from samples of data, then averages their predictions. For example, RF has been used for drought index/soil moisture prediction. Neural networks/Deep neural networks are layered models that tend to learn from complex relationships in data (e.g. hidden layers). They are used when data inputs(images) are large, and relationships are highly non-linear. Deep learning has been used for drought prediction and crop yield. Further models include Support Vector Machines (SVM), Gradient Boosting Machine, etc. All these models get trained on labeled data (for example, when drought occurred vs when it did not), learn to map input features (moisture, temperature, drone images) to an outcome. Once trained, the model can predict for new input conditions.

These systems can collect the sensor data (soil moisture, temperature) continuously and the ML models detect signs of drought onset, for example, falling soil moisture despite irrigation, rising soil temperature, and changing vegetation index. With such a prediction, the system may take decisions like releasing drought-tolerant microbes, applying microbial inoculants, adjusting irrigation, or even nutrient management. As the time goes it learns from outcomes. This feedback loop allows the model to refine its predictions. Thus, AI/ML supports determining the timing and location for interventions to proactively manage drought stress, instead of responding to damage after it occurs.

A best example is an AI based platform that combines soil sensors, multispectral imaging and weather data to model plant water stress in real time the system makes use of the neural network trained on past drought responses and current conditions to trigger timely release of microbial inoculants. Field trials have indicated increased crop yields and reduced water consumption due to this adaptive strategy. Such technologies represent a big step towards autonomous farming, where the AI not only diagnoses stress but also prescribes responsive solutions, enhancing resource efficiency and sustainability.

7 Mechanism of Resilience

Smart agricultural systems, which include AI and nanotechnology, confer resilience by supporting many important plant responses to water stress. These systems promote stronger and deeper root growth, which enables plants to access moisture from lower soil layers. As there is increased soil contact, there would be more nutrient uptake. By combining materials like hydrogels, the soil's capacity to retain moisture is drastically enhanced, leading to less water loss and prolonged plant hydration during drought conditions. This retained moisture creates a more stable environment for both roots and microbes, ultimately helping crops remain productive even with decreased irrigation [18].

Nanoparticle-based systems also deliver essential nutrients directly to plant roots in a controlled and targeted manner. When these nutrient particles are encapsulated or attached to nanocarriers, their gradual release matches with the crop's need, i.e., it improves the nutrient uptake and minimizes the waste production. As a result, plants experience more growth, healthier leaves, and consistent yields even in water-scarce conditions [19].

Moreover, exposure to specific nanoparticles activates the antioxidant defense systems in plants. This activation helps in neutralizing harmful compounds known as reactive oxygen species, which can build up during environmental stress and potentially harm plant tissues. By reducing the oxidative damage, these systems help maintain stable physiological functions such as photosynthesis during drought conditions [20].

Early stress detection through hormone signaling, sensor responses give us timely adjustment, like stomatal closure, root-architecture modification which helps in minimizing damage from stress onset. For example, hormones such as abscisic acid (ABA) help detect water deficit and trigger root changes or stomatal responses. All this builds resilience so plants stay functional or just as they are under challenging environments [21].

Together, these interconnected functions create a strong framework that not only helps plants cope with immediate water shortages but also helps them survive when environmental changes happen. This resilience strategy minimizes reliance on external water sources and chemical treatments, supporting more sustainable and resource-efficient agricultural practices. As the challenges related to climate variability and water scarcity increase, such emerging technology ensures promising solutions to protect food security while conserving natural resources. The ongoing development and application will be crucial as it helps in shaping the future of resilient and sustainable agriculture worldwide [22].

8 Role of Plant Growth-Promoting Rhizobacteria (PGPR) in Drought Conditions

Drought stress is basically one of the most important environmental factors which will be the reason for limiting crop productivity across the world. Because of the limited water availability, plants will be experiencing physiological and biochemical disruptions such as reduced photosynthetic activity, impairment of nutrient uptake, oxidative stress, and the accumulation of the toxic reactive oxygen species (ROS). In recent years, plant growth-promoting rhizobacteria have emerged as a promising, eco-friendly strategy to enhance the plant tolerance to the drought through multiple direct and indirect mechanisms [23].

PGPR has many benefits such as soil microorganisms, and these will colonize the plant roots and will promote growth by producing various metabolites and signaling molecules. The studies demonstrated that inoculation with specific PGPR strains improved the drought resilience in wheat by enhancing the biochemical and physiological parameters such as chlorophyll content, relative water content, and antioxidant enzyme activity. These beneficial effects arise from mechanisms such as phytohormone modulation, osmolyte accumulation, and improved root system architecture, which collectively support water and nutrient acquisition under stress [24].

One of the most well-studied traits of PGPR in drought conditions is the production of 1-aminocyclopropane-1-carboxylate (ACC) deaminase. This enzyme lowers ethylene levels in the plants, which is a stress hormone that will inhibit the root growth under drought conditions. The studies reported that maize inoculated with ACC deaminase-producing rhizobacteria showed enhanced root elongation and improved drought tolerance compared to the untreated controls. The bacteria will help these plants to maintain their growth by preventing ethylene-induced senescence and sustaining photosynthetic efficiency even under water scarcity [24].

PGPR will also contribute to drought tolerance by producing exopolysaccharides (EPS) that will improve the soil aggregation and water retention around the root zone. This microenvironmental improvement will enhance moisture availability to the plant roots during the dry periods. PGPR will synthesize Osmo protectants such as proline, glycine betaine, and trehalose, which will stabilize the cellular membranes and maintain the osmotic balance. According to Frontiers in Plant Science (2024), maize plants co-inoculated with multiple PGPR strains exhibited stronger drought and heat tolerance than those treated with a single strain, indicating that synergistic interactions among the microbial consortia may offer superior protection.

Moreover, PGPR promote the accumulation of antioxidant enzymes like catalase, peroxidase, and superoxide dismutase, reducing oxidative damage caused by ROS accumulation under water—limited conditions. They also influence the hormonal signaling pathways, especially through the synthesis of indole-3-3acetic acid (IAA), cytokinin's, and gibberellins, which regulate root growth and the stomatal behavior—Key adaptations for water conservation. The review by Plant Growth Regulation (2024) says that such microbial inoculants enhance not only physiological drought tolerance but also contribute to the soil microbial diversity and long-term soil health.

Recent studies are also exploring the integration of PGPR- based bioformulations with nanocarriers and biosensors to monitor the plant stress signals and enable controlled microbial release. These nano-enabled delivery systems can be further enhanced bacterial survival and activity in harsh soil environments, so the PGPR play a crucial role in helping plants withstand the drought by modulating stress—related hormones, improving water-use efficiency, and protecting against oxidative stress. Their integration into modern agricultural systems—especially when combined with nanotechnology and AI-driven monitoring—holds great promise for developing climate-resilient, sustainable crop production strategies [25].

9 AI and ML Integration in Agriculture

Artificial Intelligence (AI) refers to the computer systems that perform tasks that normally require human intelligence, such as reasoning, problem-solving, and pattern recognition. Machine Learning (ML), a subset of AI, refers to the algorithms that learn from data, improving their performance on specific tasks without explicit programming. Artificial Intelligence (AI) and Machine learning (ML) are transformative technologies increasingly incorporated into modern agriculture, offering extraordinary abilities for observation, prediction, and decision making. Both combined enable data-driven precision agriculture, especially in managing challenges like water stress and drought [26].

The input data that drives the AI and ML models in agriculture is diverse, including environmental and biological parameters. To make useful predictions, ML systems draw on a variety of data sources, including soil moisture measurements and soil temperatures, which tell us how wet/dry and warm the soil is. Air temperature, humidity, rainfall/ precipitation, and evapotranspiration (weather data) are collected because the crops and microbes respond to the climatic conditions. The drone/ UAV (unmanned aerial vehicle) images, satellite imagery (crop canopy, vegetation indices, NDVI), give us a complete view of plant health, stress, coverage. Furthermore, the historical field data, including the past yields, crop type, planting dates, irrigation events, microbial treatments, and geospatial data such as topography, soil type, texture, all these data streams are increasingly used in smart- farming/ precision-agriculture systems driven by AI [27].

Once all this data is collected and cleaned, ML algorithms learn to predict outcomes. Some commonly used models in agriculture involve Random Forest, which is an algorithm that builds decisions from samples of data and then averages their predictions. For example, RF has been used for drought index/soil moisture prediction. Neural networks / Deep neural networks are layered models which actually tend to learn from complex relationships in data (e.g. hidden layers). They are used when data inputs (images) are large and relationships are highly non-linear. Deep learning has been used for drought prediction and crop yield. Further models include Support Vector Machines (SVM), Gradient Boosting Machine, etc. All these models get trained on labeled data (for example, when drought occurred vs when it did not), learn to map input features (moisture, temperature, drone images) to an outcome. Once trained, the model can predict for new input conditions [28].

These systems can collect the sensor data (soil moisture, temperature) continuously, and the ML models detect signs of drought onset, for example, falling soil moisture despite irrigation, rising soil temperature, and changing vegetation index. With such a prediction, the system may take decisions like releasing drought-tolerant microbes, applying microbial inoculants, adjusting irrigation, or even nutrient management. As time goes it learns from outcomes. This feedback loop allows the model to refine its predictions. Thus, AI/ML supports in determining the timing and location for interventions to proactively manage drought stress, instead of responding to damage after it occurs [29].

A best example is an AI based platform that combines soil sensors, multispectral imaging, and weather data to model plant water stress in real time the system makes use of the neural network trained on past drought responses and current conditions to trigger timely release of microbial inoculants. Field trials have indicated increased crop yields and reduced water consumption due to this adaptive strategy Such technologies represent a big step towards autonomous farming, where the AI not only diagnoses stress but also prescribes responsive solutions, enhancing resource efficiency and sustainability [30].

10 Conclusion

Water scarcity continues to be one of the most serious challenges affecting global agriculture, especially under changing climate conditions. Traditional methods like improved irrigation, mulching, and drought-tolerant crop varieties have provided some relief, but they often fall short when water availability becomes critically low. The integration of nanotechnology, beneficial microbes, and artificial intelligence (AI) offers a more adaptive and intelligent solution to this persistent problem.

Through nano-encapsulation, plant growth-promoting rhizobacteria (PGPR) and bioactive compounds can be protected from environmental stress and released precisely when and where plants need them. Nano-sensors and biosensors make it possible to monitor soil and plant health in real-time, giving farmers detailed insights into changing field conditions. AI-based decision systems then analyze these inputs, predict drought stress, and guide timely interventions—creating a smart feedback loop that improves efficiency and sustainability.

While challenges remain—such as nanotoxicity, cost, and long-term field validation—the potential benefits are immense. When developed responsibly, these integrated systems can not only strengthen plants' drought tolerance but also preserve soil health and reduce water waste. Ultimately, the convergence of nano, bio, and AI technologies represents a promising step toward climate-resilient, resource-efficient agriculture, ensuring food security and environmental balance for future generations.

References

1. Akter R, Rahman MdH, Chowdhury MAR, Manirujjaman M, Elshenawy SE (2022) Advances of nanotechnology in plant development and crop protection. In: Applications of computational intelligence in multi-disciplinary research. Elsevier, pp 143–157
2. Ahmad HM, Fiaz S, Hafeez S, Zahra S, Shah AN, Gul B et al (2022) Plant growth-promoting rhizobacteria eliminate the effect of drought stress in plants: a review. Front Plant Sci 11:13
3. Vurukonda SSKP, Vardharajula S, Shrivastava M, SkZ A (2016) Enhancement of drought stress tolerance in crops by plant growth promoting rhizobacteria. Microbiol Res 184:13–24
4. Vurukonda SSKP, Vardharajula S, Shrivastava M, Skz A (2016) Enhancement of drought stress tolerance in crops by plant growth promoting rhizobacteria. Microbiol Res 184:13–24
5. Raza A, Charagh S, Salehi H, Abbas S, Saeed F, Poinern GEJ et al (2023) Nano-enabled stress-smart agriculture: can nanotechnology deliver drought and salinity-smart crops? J Sustain Agric Environ 2(3):189–214
6. Kumar A, Ansari MZ, Banjara RA, Romano R, De Luca L, Santini A et al (2025) Nanotechnology for sustainable biotic stress management in plants. Plant Stress 18:101064
7. Akter R, Rahman MdH, Chowdhury MAR, Manirujjaman M, Elshenawy SE (2022) Advances of nanotechnology in plant development and crop protection. In: Applications of computational intelligence in multi-disciplinary research. Elsevier, pp 143–57
8. Ashique S, Raikar A, Jamil S, Lakshminarayana L, Gajbhiye SA, De S et al (2024) Artificial intelligence integration with nanotechnology: a new frontier for sustainable and precision agriculture. Curr Nanosci 21(2):242–273
9. Rasheed A, Li H, Tahir MM, Mahmood A, Nawaz M, Shah AN et al (2022) The role of nanoparticles in plant biochemical, physiological, and molecular responses under drought stress: a review. Front Plant Sci 24:13
10. Chaudhary S, Prasanna D, Trivedi P, Nandy R, Prasad SS, Sunitha S et al (2025) Nanotechnology applications in agriculture for crop protection and nutrient delivery: a review. Int J Plant Soil Sci 37(9):106–119
11. Khundi Q, Jiang Y, Sun Y, Rui Y (2025) Nanofertilizers for sustainable african agriculture: a global review of agronomic efficiency and environmental sustainability. Nanomaterials 15(5):390
12. Prasad R, Bhattacharyya A, Nguyen QD (2017) Nanotechnology in sustainable agriculture: recent developments, challenges, and perspectives. Front Microbiol 20:8
13. Kumar C, Singh AK, Tomar A, Amami R, Ghazouani H (2026) Artificial intelligence in sustainable and smart agrochemicals. In: Artificial intelligence in chemical engineering. Elsevier, pp 443–470
14. Miguel-Rojas C, Pérez-de-Luque A (2023) Nanobiosensors and nanoformulations in agriculture: new advances and challenges for sustainable agriculture. Emerg Top Life Sci 7(2):229–238
15. Shaw DS, Honeychurch KC (2022) Nanosensor applications in plant science. Biosensors (Basel) 12(9):675
16. Sarabandi M, Zargar M, Ghorbani A, Chen M (2025) Smart and sustainable nano-biosensing technologies for advancing stress detection and management in agriculture and beyond. Ind Crops Prod 226:120713
17. Kwak SY, Wong MH, Lew TTS, Bisker G, Lee MA, Kaplan A et al (2017) Nanosensor technology applied to living plant systems. Annu Rev Anal Chem 10(1):113–140
18. Zhang P, Guo Z, Ullah S, Melagraki G, Afantitis A (2021) Lynch I nanotechnology and artificial intelligence to enable sustainable and precision agriculture. Nat Plants 7(7):864–876
19. Kumar A, Ansari MdZ, Banjara RA, Romano R, De Luca L, Santini A et al (2025) Nanotechnology for sustainable biotic stress management in plants. Plant Stress 18:101064
20. Al-Khayri JM, Rashmi R, Surya Ulhas R, Sudheer WN, Banadka A, Nagella P et al (2023) The role of nanoparticles in response of plants to abiotic stress at physiological, biochemical, and molecular levels. Plants. 12(2):292

21. Logeshwaran J, Senthil A, Anitha K, Moorthy PS, Raghu R, Karthikeyan R et al (2025) Nanomaterial based drought mitigation in crops is through antioxidant defense system. Plant Sci Today 9
22. Rasheed A, Li H, Tahir MM, Mahmood A, Nawaz M, Shah AN et al (2022) The role of nanoparticles in plant biochemical, physiological, and molecular responses under drought stress: a review. Front Plant Sci 24:13
23. Mikiciuk G, Miller T, Kisiel A, Cembrowska-Lech D, Mikiciuk M, Łobodzińska A et al (2024) Harnessing beneficial microbes for drought tolerance: a review of ecological and agricultural innovations. Agriculture 14(12):2228
24. Chieb M, Gachomo EW (2023) The role of plant growth promoting rhizobacteria in plant drought stress responses. BMC Plant Biol 23(1):407
25. Danish S, Zafar-ul-Hye M, Mohsin F, Hussain M (2020) ACC-deaminase producing plant growth promoting rhizobacteria and biochar mitigate adverse effects of drought stress on maize growth. PLoS One 15(4):e0230615
26. Liakos K, Busato P, Moshou D, Pearson S, Bochtis D (2018) Machine learning in agriculture: a review. Sensors 18(8):2674
27. Gupta G, Kumar Pal S (2025) Applications of AI in precision agriculture. Discov Agric 3(1):61
28. Kumar C, Singh AK, Tomar A, Amami R, Ghazouani H (2026) Artificial intelligence in sustainable and smart agrochemicals. In: Artificial intelligence in chemical engineering. Elsevier, pp 443–470
29. Zeng Q, Yu S, Fan Z, Huang Y, Song B, Zhou T (2022) Nanocone-array-based platinum-iridium oxide neural microelectrodes: structure, electrochemistry, durability and biocompatibility study. Nanomaterials 12(19):3445
30. Ashique S, Raikar A, Jamil S, Lakshminarayana L, Gajbhiye SA, De S et al (2024) Artificial intelligence integration with nanotechnology: a new frontier for sustainable and precision agriculture. Curr Nanosci 21(2):242–273

Salinity-Stress Resilience of Crops Mediated by AI Models with Nanotechnology

Akash Shaikh and Praveen Nagella

Abstract Soil salinization poses a major challenge to global agriculture, affecting crop productivity, soil fertility, and food security. This chapter explores the integration of artificial intelligence (AI) and nanotechnology as a transformative approach to enhance crop resilience against salinity stress. It begins by outlining the physiological, molecular, and anatomical effects of salt stress on plants, emphasizing the dual impact of osmotic and ionic toxicity on growth, metabolism, and gene regulation. The limitations of conventional remediation methods underscore the need for precision technologies. Nanotechnology offers promising solutions through nanofertilizers, soil remediation materials, and nanosensors that enable targeted delivery, improved nutrient use efficiency, and real-time monitoring. However, these technologies alone face challenges related to scalability, environmental safety, and data complexity. AI complements nanotechnology by providing predictive modeling, data analytics, and intelligent decision-making to optimize nano-interventions and accelerate breeding for salt-tolerant cultivars. The synergistic integration of AI-driven nanosensor networks and machine learning-based analytics enables the development of adaptive, autonomous systems capable of real-time stress detection and response. The chapter concludes with a discussion on implementation challenges and outlines strategies for sustainable deployment through open-source innovation, policy reform, and participatory research. Together, AI and nanotechnology represent a next-generation framework for achieving climate-resilient, resource-efficient, and sustainable agriculture in the face of expanding soil salinity.

Keywords Salinity stress · Artificial intelligence · Nanotechnology · Precision agriculture · Nanosensors

---

A. Shaikh · P. Nagella (✉)
Department of Life Sciences, School of Sciences, Christ University, Bengaluru, India
e-mail: praveen.n@christuniversity.in

J.-T. Chen (ed.), *AI-Nano-Enabled Climate-Smart Agriculture and Sustainability*, Smart Nanomaterials Technology, https://doi.org/10.1007/978-981-92-1694-9_12

1 Introduction

Soil salinization represents one of the most formidable threats to global agricultural sustainability and food security. Over 833 million hectares of land worldwide are currently affected by salinity, encompassing approximately 8.7% of the planet's surface and representing a critical constraint on agricultural productivity [22, 42]. The magnitude of this challenge extends beyond mere statistics, as salinity affects 20% of total cultivated land and 33% of irrigated agricultural areas globally, with devastating consequences for crop yields and farmer livelihoods [71]. The economic implications of soil salinization are staggering, with global losses estimated at $27.3 billion annually [15]. In regions heavily dependent on agriculture, such as South Asia, where 52 million hectares are salt-affected, the impact on food security and rural economies is particularly severe. The problem is expected to intensify significantly due to climate change, rising sea levels, and poor irrigation practices, with projections suggesting that more than half of cultivable land worldwide will be salinized by 2050 [42].

Traditional approaches to salinity management have proven inadequate for addressing this increasing crisis. Conventional breeding programs are limited by the complex, multigenic nature of salt tolerance traits, whereas agronomic interventions offer only temporary relief [30]. The limitations of current methodologies have created an urgent need for innovative, technology-driven solutions that can provide both immediate intervention capabilities and long-term sustainability.

The emergence of Artificial Intelligence (AI) and nanotechnology as disruptive agricultural technologies offers unprecedented opportunities to revolutionize salinity stress management. AI enables real-time monitoring, predictive analytics, and data-driven decision-making through advanced sensor networks and machine learning algorithms [64]. Simultaneously, nanotechnology and the subsequent usage of nanomaterials provide enhanced delivery systems, targeted interventions, and precision-controlled applications that optimize resource utilization while minimizing environmental impact [32, 48]. Moreover, the convergence of these technologies will undoubtedly create a synergistic framework for addressing salinity stress that transcends the limitations of individual approaches. Thus, an integrated AI-nanotechnology system aligns with the principles of climate-smart agriculture and precision farming methodologies, offering scalable solutions for diverse agricultural systems and varying environmental conditions. In this chapter, we examine the scientific foundations, technological applications, and transformative potential of this emerging framework.

2 Physiological and Molecular Basis of Salinity Stress

2.1 *Physiological Impacts of Salt Stress on Plants*

It is essential to understand the basis of how salinity stress affects plants before attempting to address the issue using AI models or other intervention methods. While it is common knowledge that soil salinity stress affects plant growth and crop production, the nuances in how excess salt impacts plant physiology and molecular metabolism enable us to address the root causes of these negative effects. Salt stress imposes severe physiological constraints on plant growth and development through two primary mechanisms: osmotic stress and ionic toxicity [92].

Osmotic stress occurs immediately upon salt exposure, creating a water potential gradient that reduces the plant's ability to absorb water from the soil. This osmotic imbalance results in reduced water uptake, stomatal closure, and physiological drought conditions that persist even in environments with abundant water [12, 92].

Ionic toxicity, particularly from sodium (Na^+) and chloride (Cl^-) ions, represents a more extreme and long-term threat to plant health. Excessive Na^+ influx disrupts cellular ion homeostasis by interfering with essential nutrients, particularly potassium (K^+), calcium (Ca^{2+}), and magnesium (Mg^{2+}). This ionic disruption compromises enzyme activation, membrane integrity, and metabolic processes, ultimately leading to cellular dysfunction and plant death [33, 92]. The chloride component of salt stress has emerged as an equally critical factor in plant toxicity. Research demonstrates that chloride accumulation in chloroplasts can reach levels that inhibit the activities of photosynthetic enzymes and disrupt electron transport chains. High chloride concentrations cause chlorophyll degradation, resulting in characteristic chlorotic symptoms and a reduction in photosynthetic capacity. The accumulation of Cl^- in sensitive tissues triggers oxidative stress through the generation of reactive oxygen species (ROS), further exacerbating cellular damage [26, 84]. However, other studies also highlight the potential benefits of chloride salts in plants, specifically in improving drought tolerance [25], and biotic challenges like bacterial pathogen infection [87]. Therefore, there is a need to delicately balance the positives and negatives of salt ions in the soil. A conglomeration of AI models and nanotechnology can provide the minute mechanisms needed to monitor ion levels in real-time, allowing agriculture to thrive in places that had earlier been rendered non-arable due to rising salt levels.

2.2 *Morphological and Anatomical Adaptations*

Salt stress induces a cascade of morphological and anatomical changes in plants, mostly negative, that precede and ultimately necessitate adaptation. High concentrations of soluble salts in the rhizosphere alter both osmotic and ionic equilibria,

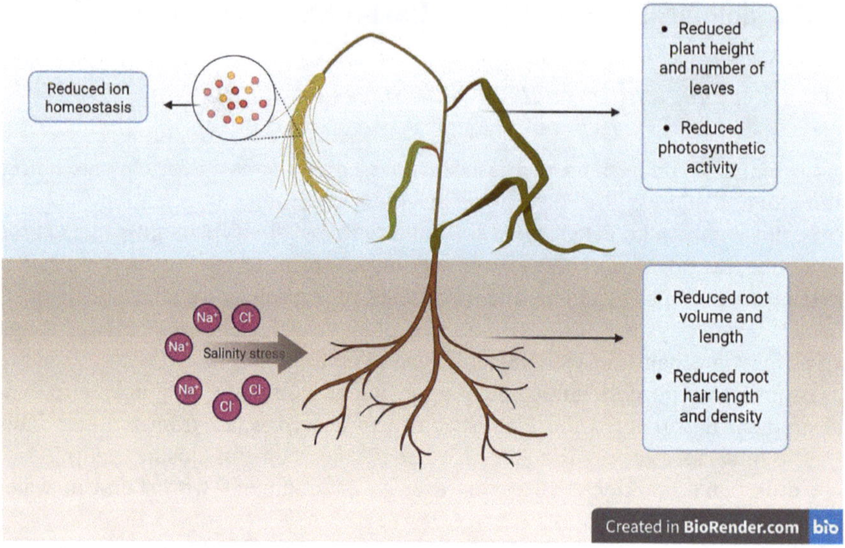

Fig. 1 Morphological response to salt stress in plants. Created in https://BioRender.com

through mechanisms discussed above, leading to significant modifications in plant growth, structure, and overall physiology (Fig. 1). These morphological and anatomical alterations are not merely passive consequences of abiotic stress but represent active physiological readjustments that enable plants to withstand and survive in saline environments.

High salinity in the soil matrix triggers an osmotic imbalance that substantially reduces the water potential gradient between the soil and root cells, thereby diminishing the plant's capacity to absorb water. This reduction in water availability induces a pseudo-drought condition, resulting in reduced cell turgor and inhibition of cell expansion [14]. This is often followed by excess cell lignification, which results in stunted primary growth of roots, leading to a cascade effect that affects all other parts of the plant, including reduced leaf area, shoot length, and photosynthetic activity [17, 54].

These significant changes in the plant's anatomy due to external stress also trigger several adaptive mechanisms, primarily beginning at the roots, to try and minimize the uptake of salt ions at the source. Salinity leads to modifications in root diameter and root surface area by reducing the length and density of root hairs, with the added consequence of lowered ion uptake from the soil [10, 63]. However, aside from excluding Na^+ and Cl^-, lower ionic uptake also restricts the accumulation of essential ions, such as Ca^{2+}, Fe^{2+}, Mg^{2+}, and Zn^{2+}, which are crucial for root growth [20]. Slight adaptations are also observed in shoot and leaf structures, primarily aimed at ionic compartmentalization and water conservation [40, 47].

The balance of injury and adjustment varies across species and genotypes, reflecting genetic diversity in salt-tolerance mechanisms. Understanding these different responses is crucial for integrating AI-based predictive modeling and nanotechnological interventions that can simulate, diagnose, and ultimately mitigate the morphological consequences of salt stress in real time.

2.3 Molecular Responses and Gene Regulation

The molecular basis of salt tolerance involves complex genetic networks that regulate ion transport, osmotic adjustment, and cellular protection mechanisms. The interplay between distinct signaling modules, notably the Salt Overly Sensitive (SOS) pathway, the High-Affinity Potassium Transporter (HKT) system, and vacuolar Na^+/H^+ exchangers (NHX), constitutes the primary molecular framework facilitating salt resilience across diverse plant species [5, 29, 34, 35]. The Salt Overly Sensitive (SOS) pathway, initially characterized in *Arabidopsis thaliana*, is one of the most extensively studied regulatory systems in plant ion homeostasis. This pathway comprises the SOS1, SOS2, and SOS3 components, which collectively mediate sodium efflux and calcium-dependent signaling under high salinity conditions [16]. Under saline stress, elevated cytosolic Ca^{2+} binds to SOS3 (or SCaBP8), activating SOS2, which in turn phosphorylates SOS1 to enhance Na^+ efflux activity. Recent insights have revealed additional molecular components that fine-tune the activation and repression dynamics of the SOS module, depending on the intensity and duration of stress. The cumulative effect of these signaling events is the restoration of cytosolic ion balance and sustained osmotic stability [5, 16, 35].

Complementing the SOS system, High-Affinity K^+ Transporters (HKT) play a critical role in mediating the selective movement of sodium and potassium ions between root and shoot tissues. HKT isoforms function within xylem parenchyma cells to retrieve Na^+ ions from xylem sap, preventing excessive translocation to photosynthetically active tissues. This selective reabsorption safeguards chloroplast function and photosynthesis under saline conditions by maintaining a favorable Na^+/K^+ ratio in leaves [6, 31].

The third integral system in salinity tolerance involves vacuolar Na^+/H^+ exchangers, encoded by NHX genes, which compartmentalize sodium ions into vacuoles as a primary detoxification strategy [34]. Research in *A. thaliana* has established that the AtNHX1 gene mediates the sequestration of excess cytosolic Na^+ into vacuoles, thereby preserving cytoplasmic enzyme activity and osmotic balance [37]. Moreover, the upregulation of NHX transcripts in response to hyperosmotic and ionic stress underscores their responsiveness within broader stress-related transcriptional networks that intersect with ABA and ROS signaling cascades [2].

Collectively, the integration of SOS, HKT, and NHX modules represents an adaptive molecular strategy that coordinates sodium extrusion, xylem reabsorption, and vacuolar sequestration. This multi-level regulation ensures efficient maintenance of ionic equilibrium while minimizing cytotoxic effects at both cellular and organ levels [86]. Modern genomic and transcriptomic analyses are increasingly revealing that these pathways are not discrete but interdependent systems regulated by calcium signatures, phosphorylation events, and hormonal crosstalk [68]. Such understanding forms the foundation for AI-driven modeling efforts aimed at predicting gene regulation patterns under saline environments and identifying molecular targets for engineering salt-tolerant crops.

3 The Need for AI Models in Conjunction with Nanotechnology

Traditional approaches to ameliorating soil salinity, including gypsum application, leaching with high-quality irrigation water, and organic amendments such as compost or biochar, have long been employed to displace sodium ions, improve soil structure, and enhance crop productivity [13, 56]. However, these methods often require large input volumes, extended equilibration periods, and intensive labor, limiting their practicality and sustainability, particularly in resource-constrained rural areas, where the need for preventative techniques is the most [75]. Gypsum dissolution rates are hampered by low surface area, necessitating high application rates to achieve modest reductions in exchangeable sodium percentage, while conventional biochar amendments, though beneficial for improving cation exchange capacity, lack the specificity and reactivity to address dynamic salinity fluctuations [13]. Moreover, traditional breeding for salt tolerance remains slow and arduous, constrained by the polygenic nature of stress tolerance traits, making it harder for scientists to zero in on a particular set of genes that can act universally against all forms of ionic stress [38, 59]. These limitations create a critical gap in the ability to deliver precise, rapid, and cost-effective interventions tailored to evolving field conditions. The convergence of artificial intelligence and nanotechnology presents a compelling solution to address this shortcoming. AI-driven analytics can provide real-time salinity diagnostics and predictive insights, while nanomaterials enable the targeted delivery of amendments and stimulants at the molecular scale, forming an integrated framework that can overcome the shortcomings of conventional methods [90]. The sections below will outline some of the advancements made in both nanotechnological and artificial intelligence-mediated areas with respect to agricultural monitoring, and how combining these two can be useful for implementing salinity stress resilience in crops.

4 Role of Nanotechnology in Alleviating Salinity Stress

4.1 Nano-Fertilizers for Targeted Nutrient Delivery

Under saline conditions, plants experience specific deficiencies in essential micronutrients, particularly zinc, iron, silicon, and copper, due to competitive inhibition by sodium and chloride ions at uptake sites and reduced bioavailability in alkaline saline soils. Nano-fertilizers address these constraints through their nanoscale dimensions (typically 1–100 nm), which confer superior solubility, reactivity, and cellular penetration compared to conventional fertilizers [72, 83].

Silicon nanoparticles (SiNPs) have emerged as effective ameliorants for salinity stress. Silicon, though not classified as an essential nutrient, functions as a quasi-essential element that modulates gene expression, enhances structural rigidity, and mitigates oxidative stress. Under saline conditions, SiNPs improve osmotic adjustment, enhance antioxidant enzyme activities (including superoxide dismutase, catalase, ascorbate peroxidase, and guaiacol peroxidase), and increase the synthesis of osmoprotectants such as proline, phenolics, and flavonoids [7, 21]. Studies on *Ocimum basilicum* have demonstrated that SiNP application increases chlorophyll a and b content, as well as fresh and dry biomass, while maintaining photosynthetic efficiency under NaCl stress through enhanced osmotic regulation [36]. The superior performance of SiNPs relative to bulk silicon salts is attributed to their enhanced bioavailability and prolonged residence time in plant tissues, which sustain protective effects throughout the stress period. In the same manner, Zinc nanoparticles (ZnNPs) have also emerged as a potent elicitor in improving salt tolerance in plants, owing to its ability to function as a cofactor for over 300 enzymes essential for protein synthesis and membrane stability. Application of ZnNPs to salt-stressed *Solanum melongena* has resulted in increased growth vigor and chlorophyll activity, attributed to the increased concentrations of total soluble sugars, proteins, and free amino acids [9]. Iron nanoparticles (FeNPs), including iron and its oxides, have also been shown to alleviate chlorosis in crops grown in saline soils by enhancing chlorophyll synthesis and scavenging reactive oxygen species through catalase-like activities, as well as by reducing Na^+ concentrations [8, 49]. Hence, there is sufficient precedent to support the use of nano-fertilizers in the form of nanoparticles to alleviate saline stress, an approach which can be enhanced by integrating it with AI.

4.2 Soil Remediation Using Nanomaterials

Aside from attempting to enhance the salt-resistance ability of the plant, nanomaterials can be used to directly reclaim saline soil, much like biochar. Biochar-nanoparticle composites can synergistically combine biochar's high surface area and cation exchange capacity with nanoparticle functionalities [18, 74]. These composites adsorb Na^+ and Cl^- ions, reducing bioavailability, while embedded nanoparticles provide catalytic degradation of pollutants. The application of *Broussonetia papyrifera*-derived biochar combined with selenium NPs increased soil organic carbon, available nutrients, and microbial biomass, while reducing electrical conductivity and the exchangeable sodium percentage, and also lead pollution [53]. Nano-gypsum is another promising candidate in this regard, exhibiting dramatically improved dissolution kinetics with surface areas far exceeding those of conventional gypsum [65]. In general, nanocomposites and nanomaterials have exhibited nutrient mobilization in the rhizosphere under saline conditions, as well as reduction of ionic (Na^+) toxicity in the soil, making it an effective agent against soil salinization [67].

4.3 Nanocarriers for Gene and Hormone Delivery

Perhaps the most advanced and direct method through which nanotechnology can be utilized in combating rising salinity stress in the soil is by employing nanomaterials as targeted delivery systems for precise genetic and hormonal modifications [27]. Research has already demonstrated the use of nanomaterial-based targeted delivery systems for amelioration of stress and growth enhancement in plants using salicylic acid, a known biostimulant effective against saline stress [58], Trung [77]. Additionally, CRISPR/Cas9 nanodelivery systems can enable transient gene editing without stable transformation [70], allowing for a novel way to impart salinity stress tolerance in plants.

4.4 Nanosensors for Real-Time Monitoring

The deployment of nanosensors represents a pivotal advancement in real-time monitoring of soil salinity, enabling farmers to implement precision interventions that address spatial and temporal variability in salt distribution [57, 88]. Electrochemical nanosensors fabricated from carbon nanotubes, graphene oxide, and metal oxide nanoparticles achieve detection sensitivities at the micromolar level for critical ions including Na^+ and Cl^-, far exceeding the capabilities of conventional soil testing methods [55]. These sensors can leverage voltametric techniques that resolve ionic concentrations across biologically susceptible ranges within seconds, providing near-instantaneous feedback on soil salinity status [39]. The integration of such nanosensor arrays with wireless sensor networks (WSNs) and Internet of Things (IoT) infrastructure can facilitate continuous, spatially distributed monitoring across agricultural landscapes [46]. This also opens up the option of introducing an on-board artificial intelligence component to such nanosensors, which can help make real-time decisions to mitigate rising salt salinity, a possibility that will be discussed in-depth later in this chapter.

4.5 Limitations and Environmental Concerns

Despite the transformative potential of nanotechnology-enabled salinity management, several critical limitations constrain standalone nano-approaches. The accumulation of engineered nanoparticles in soil systems raises concerns regarding long-term ecological impacts, potential phytotoxicity at elevated concentrations, and bioaccumulation through trophic chains [51, 79]. Nanosensor networks, while providing high-resolution spatial data, generate massive data streams that overwhelm manual interpretation capabilities and require sophisticated analytics to extract actionable insights [46]. Furthermore, optimal nanoparticle application rates, timing, and formulations remain empirically determined through costly trial-and-error approaches lacking predictive frameworks, leading to high initial costs.

The integration of artificial intelligence addresses these fundamental gaps by providing self-regulatory monitoring and intelligent decision-making capabilities that optimize nano-intervention strategies [90]. AI-driven models will enable the rational design of nanomaterials with enhanced efficacy and reduced toxicity, while adaptive feedback loops can continuously refine intervention protocols based on observed outcomes. This highlights the importance of artificial intelligence in combating salinity stress in soil, a topic that will be further discussed below.

5 Applications of Artificial Intelligence in Salinity Stress Management

Artificial Intelligence or AI has had an unprecedented, meteoric rise in the day-to-day lives of human beings in the past few years, and many may not have realized that the technological implementation of AI starts right from the way humans grow their food for sustenance. The implementation of AI in agriculture has been ongoing for almost a decade, with machine learning-based algorithms integrated with monitors and sensors, enabling minute control of cultivation conditions that is impossible with simple human interaction [1]. Everything, from optimizing irrigation schedules in relation to ambient weather conditions to calculating the correct amount of pesticides and herbicides to be added to the soil based on soil microbiota, can be fine-tuned by AI [76]. In the context of soil salinity, AI's capabilities span early detection of salt stress symptoms, precise prediction of crop performance under saline conditions, and optimizing irrigation and nutrient management for resilience. Rather than requiring farmers to possess advanced technical expertise, modern AI systems present recommendations through accessible interfaces like mobile apps, web dashboards, and automated alerts that are capable of guiding management practices in real-time [3].

5.1 Early Detection of Salinity Stress

One of the most practical applications of AI in agriculture is identifying salt stress symptoms before they become visible to the human eye. Crops subjected to salinity undergo physiological changes, modifications in leaf color, water stress responses, and growth patterns that can be detected through machine learning analysis of field images captured by hyperspectral imaging [80]. AI algorithms with supervised learning can become particularly effective at analyzing images, helping in detecting crop diseases and stress symptoms when applied to agricultural imagery [80]. These systems learn to recognize the subtle visual signatures of salt-induced stress, such as chlorosis (yellowing of leaves), wilting, and growth stunting, that appear days or even weeks before a farmer would notice them through conventional field inspection.

The advantage of early detection extends beyond simply identifying problems; it can also map out soil salinity across vast regions, allowing for proper planning and management of agricultural land even before the crops are sown. Studies in China have already utilized Multi-Spectral Imaging (MSI) data to identify the varying degrees of salinity between higher and lower altitude areas, even going so far as to correctly attribute the reason for higher salinity in lower regions to the runoff of salts from the higher regions due to gravity [82, 91]. The most appealing aspect of such studies is that the results were obtained solely by analyzing imaging data from satellites, thereby eliminating the need for human inspection of different soil zones and saving the cost of on-field studies [82]. Some researchers have also been successful in utilizing Machine Learning-based algorithms to not only map but also predict regions of soil salinity, bypassing traditional lab-based detection techniques altogether [85].

5.2 *Precision Irrigation and Water Management*

Water management represents another critical AI application for addressing salinity in arid and semi-arid regions. AI-driven irrigation systems integrate data from soil moisture sensors, weather forecasts, and plant water requirements to generate precise irrigation schedules that minimize both water waste and salt accumulation in the root zone [76]. Rather than farmers applying fixed irrigation volumes based on guesswork or convention, AI algorithms can analyze historical data, current soil conditions, and weather patterns to predict exactly how much water each field section needs.

Practical field demonstrations effectively illustrate the approach's effectiveness. In a Michigan study involving corn, blueberry, and tomato production, an IoT-based system equipped with soil moisture sensors and AI-driven recommendations achieved a 30% reduction in irrigation water use while maintaining or exceeding crop yields compared to farmer-managed conventional practices [19]. More importantly, the system prevented overirrigation, which is a primary cause of salt accumulation in soils, by providing real-time feedback on soil water availability at multiple depths. Remote monitoring capabilities enabled farmers to make irrigation decisions from their homes or offices rather than requiring frequent field visits, reducing labor requirements while improving response times to changing conditions [28].

The integration of weather forecasting into AI irrigation systems also enables proactive management that exploits natural rainfall for salt leaching. Systems can predict precipitation events days in advance and adjust irrigation schedules accordingly, reducing reliance on expensive artificial irrigation while allowing salts to be flushed from the root zone through natural leaching [19, 28]. This capability is particularly valuable in regions where rainfall is seasonal or episodic, allowing farmers to anticipate and take advantage of favorable conditions.

5.3 Genomic Selection and Cultivar Recommendation for Salinity Tolerance

The integration of artificial intelligence with genomic science has revolutionized crop improvement programs by dramatically accelerating the identification and selection of salt-tolerant cultivars. Traditionally, breeding for salinity tolerance, a complex trait governed by multiple genes, required 15–20 years of field evaluation across diverse environments. Machine learning algorithms now enable researchers to identify the specific genetic markers associated with salt tolerance within years rather than decades, fundamentally transforming the breeding pipeline [23, 52].

Machine learning algorithms, particularly, Support Vector Machines (SVMs) and Artificial Neural Networks (ANNs), have proven exceptionally effective in identifying genes that control salt resistance. Researchers have utilized SVM-based models to predict salt resistance genes in rice [81], and a similar ANN-based analysis has been used to identify effective genetic indices for predicting salt-tolerant wheat varieties among Iranian germplasm [61]. These algorithms analyze Genome-Wide Association Study (GWAS) data, which are bits of information from thousands of single-nucleotide polymorphisms (SNPs) distributed across crop genomes, to pinpoint which specific genetic variants associate most strongly with salt tolerance traits [52]. The advantage of machine learning and AI over traditional statistical approaches lies in its ability to detect complex, non-linear relationships between genetic markers and phenotypic traits. A single crop typically contains millions of genetic variants; traditional methods struggle with this dimensionality. Machine learning algorithms, by contrast, efficiently sift through massive datasets to identify the signal [23].

5.4 Multi-omics Integration for Deeper Understanding

Contemporary breeding programs increasingly employ AI-assisted analysis of multi-omics data; information spanning genomics (genes), transcriptomics (gene expression), proteomics (proteins), and metabolomics (small chemical compounds). The end goal is to gain a comprehensive understanding of salt tolerance mechanisms. Newer models like GenPhenML are already being utilized to develop stress-resistant cultivars of barley, with reported accuracies of predicted models being as high as 97% [4]. This translates to much shorter times for breeding programs, effectively allowing researchers to move from lab-controlled environments to production-ready crops in record time. This is because machine learning algorithms trained on multi-omics datasets learn to recognize the molecular signature of salt-tolerant plants,a specific pattern of gene expression, protein abundance, and metabolite accumulation that distinguishes tolerant from susceptible genotypes [52]. This integrated approach enables breeding programs to select not just for salt tolerance per se, but for the underlying physiological mechanisms conferring this tolerance, creating more robust and durable resistant cultivars.

6 AI-Nanotechnology Synergy: A Convergent Approach

The true transformative potential of the above two discussed technologies emerges when artificial intelligence and nanotechnology operate in concert rather than isolation. This convergence creates feedback loops where AI optimizes nano-interventions while nanosensors provide the data that AI algorithms require for intelligent decision-making [11]. In practical terms, this means farmers gain access to self-regulating systems that detect salinity problems, prescribe nano-enabled solutions, and continuously refine their approaches based on observed outcomes.

6.1 Intelligent Monitoring and Adaptive Response

The first and foremost area where AI and nanotechnology can be combined to combat against salinity stress is by continuous monitoring and responding to stresses in real time. Nanosensors embedded throughout a field can continuously measure soil salinity, moisture, nutrient levels, and plant stress indicators; data that traditional sensors cannot easily capture at such spatial resolution or sensitivity [46]. Such sensors can transmit real-time information through wireless networks to cloud-based platforms where AI algorithms can process the data streams. As discussed earlier, machine learning models, trained on years of field observations, recognize patterns that indicate emerging stress conditions weeks before visible symptoms appear [80]. Rather than requiring farmers to interpret complex data, AI systems can automatically trigger appropriate responses, such as adjusting irrigation schedules to prevent salt accumulation, deploying nano-fertilizers precisely where needed, or alerting farmers to take protective measures [76]. This autonomous capability transforms salinity management from reactive problem-solving into proactive prevention. The practical advantage lies in eliminating the gap between detection and response. Conventional field monitoring might identify a salinity problem after observable crop damage has already begun. AI-enabled nanosensor networks can trigger interventions days before symptoms manifest, fundamentally changing the trajectory of crop health. A multi-layered smart agriculture framework integrating nanosensors, fog computing for local data processing, and cloud platforms for centralized analytics enables real-time and medium-term decision-making simultaneously, accelerating the transition from data collection to actionable management [73].

6.2 Optimization of Nanomaterial Application

Integration of AI with nanomaterial application will be useful since it will no longer be a matter of simply deploying nanomaterials; AI models can optimize when, where, and how much to apply. Machine learning algorithms analyze field heterogeneity, including parameters like variations in soil type, salinity distribution, and crop requirements, to generate site-specific nano-amendment recommendations. Rather than applying uniform nano-fertilizer across an entire field, AI can identify zones of highest need and allocate resources accordingly, improving cost-effectiveness while minimizing unnecessary nanoparticle accumulation [43, 60]. This precision prevents both under-treatment (where salt stress continues) and over-treatment (where excess nanoparticles might accumulate). The system will continuously learn: if a particular nano-intervention fails to achieve the expected results, AI algorithms can adjust the dosage, timing, or formulation for future applications, creating a continuously improving management system [44].

6.3 Current Landscape and Emerging Research

The integration of artificial intelligence with nanotechnology has transitioned from theoretical promise to practical field deployment, with documented case studies demonstrating tangible improvements in agriculture. The synergy between these technologies creates intelligent systems capable of real-time sensing, autonomous decision-making, and targeted nano-enabled interventions that respond to field conditions continuously [46]. Practical trials have already been done to achieve breakthroughs in AI systems integrated into nanobiosensors, albeit for use in detection against biotic stress agents, namely phytonematodes [45]. Such advancements can be easily translated into AI-nanobiosensors specifically geared towards detecting soil salinity and making appropriate decisions to combat stress in real-time. Moreover, contemporary research increasingly explores how machine learning enhances the design and deployment of nanomaterials themselves. Extensive research has documented that nanotechnology offers potential solutions including improvement in nutrient use efficiency, efficacy of pest management, and plant protection when integrated with advanced monitoring and decision systems [89]. AI-driven approaches enable rational design of nanomaterials through predictive modeling: machine learning algorithms analyze vast databases of nanoparticle properties and biological responses to predict optimal characteristics that maximize agricultural efficacy while minimizing environmental persistence [90].

Nano-biosensing technologies integrated with AI represent an emerging capability combining molecular-level detection with intelligent interpretation [45]. AI-driven nano-biosensors can analyze real-time data, enhance detection precision, and provide actionable insights for farmers, enabling proactive and targeted interventions before stress becomes apparent. The convergence of AI, biosensors, and advanced sensing technologies opens new possibilities for autonomous stress management systems wherein plants autonomously signal their physiological needs while synthetic nano-components deliver responsive interventions [66].

7 Challenges and Limitations

Although the future of integrating AI models with nanotechnology to combat salinity stress in plants seems exciting and promising, it is not without its challenges. The real-world deployment faces substantial barriers that cannot be ignored if we hope to see these technologies reach the farmers who need them most. Understanding these challenges is not pessimistic; rather, it is a prerequisite for developing realistic solutions.

7.1 Practical and Economic Barriers

The most immediate challenge confronting widespread adoption is cost. Comprehensive AI-IoT-nanosensor systems can typically require capital investments of approximately $10,000 per hectare, a prohibitive sum for smaller farmers who manage the majority of global agricultural land and are most vulnerable to salinity stress [24, 50]. A global survey of 5,500 farmers revealed that 52% of North American farmers and 48% of European farmers cite high costs as their primary barrier to adopting precision agriculture technology, with an additional 40% concerned about unclear returns on investment [24]. This economic burden intertwines with infrastructure deficits that exacerbate inequality. Reliable internet connectivity, which is essential for cloud-based AI analytics, remains absent across much of rural agricultural territory, particularly in developing countries where salinity problems are most severe. Additionally, unreliable electrical grids threaten sensor operation, while the absence of technical support services leaves farmers without practical guidance on installation, maintenance, and troubleshooting [78].

7.2 Data Quality and Technical Complexity

Contemporary AI systems face a fundamental constraint: they require large, high-quality datasets for effective training. Yet agricultural data collection faces three interconnected challenges termed data noise, data fog, and data islands. Data noise refers to errors and inconsistencies from imperfect collection processes; data fog emerges from difficulty integrating diverse data sources with different formats and standards; data islands occur when systems cannot communicate or share information [62]. A recent comprehensive analysis demonstrated that data noise significantly impacts precision agriculture predictions, leading to biased decisions and wasted resources, while data fog complicates decision-making processes and data islands hinder data sharing essential for collaborative management [62].

7.3 Regulatory and Safety Uncertainties

Regulatory frameworks for agricultural nanotechnology remain fragmented globally, creating uncertainty that inhibits investment and deployment. In the United States, various agencies regulate different aspects. The Environmental Protection Agency (EPA) governs pesticides under the Toxic Substances Control Act (TSCA) and Federal Insecticide, Fungicide, and Rodenticide Act (FIFRA), while the Food and Drug Administration (FDA) regulates food safety, agencies whose regulations cannot be avoided when introducing nanofertilizers and any other nanotechnology to edible crop cultivation [41]. Similarly, the European Union employs the Registration, Evaluation, Authorisation and Restriction of Chemicals (REACH) regulations for monitoring nanotechnology usage, while China utilizes multiple agency oversight structures. This lack of international harmonization increases compliance costs and slows innovation [41].

7.4 Environmental Concerns

The long-term ecological fate of engineered nanoparticles in soil systems remains poorly characterized, raising legitimate concerns about potential groundwater contamination or food chain accumulation, thus opening the door for regulatory roadblocks as mentioned above. Recent years have seen a growing body of research dedicated to understand the environmental implications of nanotechnology-based systems in agriculture, particularly in relation to nanoparticles and nano-fertilizers. Studies have shown that nanomaterials may interact differently with soil, water, and living organisms compared to their conventional counterparts owing to their small particle size and enhanced reactivity [32]. Concerns have been raised about the long-term accumulation and persistence of engineered nanoparticles in soil ecosystems. Experimental evidence suggests that excessive or repeated application can lead to changes in soil microbial diversity, potential bioaccumulation in food chains, and disruption of soil enzymatic activities [32, 69]. While acute toxicity is generally rare at recommended application rates, chronic and subtle effects, such as oxidative stress in non-target soil fauna or alteration of beneficial microbial populations, warrant caution. As regulatory evaluations lag behind rapid field deployment, ongoing research emphasizes the necessity for lifecycle and fate studies to inform safe and sustainable nanotechnology adoption in agriculture.

8 Future Prospects

The convergence of artificial intelligence and nanotechnology represents one of the most promising developments in contemporary agriculture, offering integrated solutions to the escalating challenge of soil salinization. Throughout this chapter, we have examined how these technologies operate synergistically; AI systems provide intelligent decision-making, while nanomaterials enable precision interventions at the molecular scale, functions that can address salinity stress through mechanisms that conventional agriculture cannot achieve. From early stress detection weeks before visible symptoms, to genomic selection accelerating the development of salt-tolerant cultivars, to real-time irrigation optimization preventing salt accumulation, this integrated framework demonstrates measurable improvements across multiple dimensions of agricultural sustainability.

The path forward demands coordinated action across multiple stakeholder groups. Researchers must continue to develop modular, cost-effective technologies designed for smallholder contexts, rather than exclusively for industrial-scale operations. Technology developers should prioritize open-source platforms and knowledge-sharing mechanisms that reduce barriers to adoption. Policymakers must invest in rural infrastructure, broadband connectivity, electrical grids, and regional innovation centers that create enabling environments for technology deployment. Educational institutions should modernize curricula to prepare agricultural professionals with interdisciplinary competencies spanning plant science, data science, and nanotechnology. Regulatory agencies should develop adaptive frameworks balancing innovation incentives with rigorous safety stewardship through international harmonization efforts. Most importantly, farmers themselves must be positioned as co-designers rather than passive recipients of technology. Participatory approaches, wherein practitioners contribute insights about local problems and help shape solutions, enhance relevance and increase the likelihood of adoption. The most sophisticated AI algorithm or innovative nanomaterial matters little if it fails to address real challenges faced by those working on the land.

As climate change intensifies agricultural stress and global population demands substantial increase in food production, the integration of AI and nanotechnology provides viable pathways towards agricultural resilience and sustainability. The technologies discussed throughout this chapter are tools requiring wisdom, careful stewardship, and equitable access to unlock their transformative potential. By addressing current barriers through coordinated innovation and inclusive development, the international agricultural community can ensure that these powerful technologies contribute meaningfully to alleviate the harmful effects of soil salinity on crop production, thus ensuring global food security, environmental protection, and improved livelihoods for farming communities worldwide, particularly those confronting the escalating threat of soil salinization in the twenty-first century.

9 Conclusion

The convergence of AI and nanotechnology represents a transformative paradigm for addressing salinity stress through molecular-level precision guided by field-scale intelligence. This integrated approach enables autonomous agricultural systems to detect, diagnose, and respond to stress conditions in real-time. Evidence demonstrates measurable improvements, as early stress detection extends intervention windows by weeks, genomic selection reduces breeding timelines from decades to years, precision irrigation achieves water savings while maintaining yields, and nano-enabled remediation sustains productivity in previously marginal environments. However, widespread implementation faces substantial barriers. Data quality challenges, infrastructure deficits, high adoption costs, and regulatory uncertainties hinder diffusion, particularly in rural environments where the impacts could be greatest. Overcoming these roadblocks requires coordinated efforts spanning technology development, capacity building, infrastructure investment, and governance innovation. Open-source platforms, modular architectures, and collaborative networks offer democratization pathways. Adaptive frameworks and international harmonization can maintain safety while enabling innovation.

The urgency to adopt such conglomerate systems stems from the accelerating effects of climate change and population growth, which intensify pressures on agricultural systems already stressed by degradation and scarcity. The window for developing transformative technologies narrows as salinity expands and affected lands exhaust productive capacity. The combination of AI and nanotechnology presents a viable pathway toward climate-resilient, resource-efficient agriculture. Realizing this potential requires sustained commitment from research communities, policymakers, innovators, and farming populations working toward shared objectives of agricultural sustainability, environmental stewardship, and global food security.

References

1. Abdullahi HO, Mahmud M, Hassan AA, Ali AF (2023) A bibliometric analysis of the evolution of IoT applications in smart agriculture. Ingénierie Des Systèmes D Information 28(6):1495–1504. https://doi.org/10.18280/isi.280606
2. Adabnejad H, Kavousi HR, Hamidi H, Tavassolian I (2015) Assessment of the vacuolar Na+/H+ antiporter (NHX1) transcriptional changes in *Leptochloa fusca* L. in response to salt and cadmium stresses. Mol Biol Res Commun 4(3):133–142. https://pmc.ncbi.nlm.nih.gov/articles/PMC5019205/

3. Aijaz N, Lan H, Raza T, Yaqub M, Iqbal R, Pathan MS (2025) Artificial intelligence in agriculture: advancing crop productivity and sustainability. J Agric Food Res 20(101762):101762. https://doi.org/10.1016/j.jafr.2025.101762
4. Akbari M, Sabouri H, Sajadi SJ, Yarahmadi S, Ahangar L (2024) Classification and prediction of drought and salinity stress tolerance in barley using GenPhenML. Sci Rep 14(1):17420. https://doi.org/10.1038/s41598-024-68392-w
5. Ali A, Petrov V, Yun D-J, Gechev T (2023) Revisiting plant salt tolerance: novel components of the SOS pathway. Trends Plant Sci 28(9):1060–1069. https://doi.org/10.1016/j.tplants.2023.04.003
6. Ali A, Raddatz N, Aman R, Kim S, Park HC, Jan M, Baek D, Khan IU, Oh D-H, Lee SY, Bressan RA, Lee KW, Maggio A, Pardo JM, Bohnert HJ, Yun D-J (2016) A single amino-acid substitution in the sodium transporter HKT1 associated with plant salt tolerance. Plant Physiol 171(3):2112–2126. https://doi.org/10.1104/pp.16.00569
7. Almutairi ZM (2016) Effect of nano-silicon application on the expression of salt tolerance genes in germinating tomato (*Solanum lycopersicum* L.) seedlings under salt stress. Plant Omics J 9(1):106–114. https://www.pomics.com/almutairi_9_1_2016_106_114.pdf
8. Alsamadany H, Anayatullah S, Zia-Ur-Rehman M, Usman M, Ameen T, Alharby HF, Alharbi BM, Abdulmajeed AM, Yong JWH, Rizwan M (2024) Residual efficiency of iron-nanoparticles and different iron sources on growth, and antioxidants in maize plants under salts stress: life cycle study. Heliyon 10(7):e28973. https://doi.org/10.1016/j.heliyon.2024.e28973
9. Anwar T, Qureshi H, Fatimah H, Siddiqi EH, Anwaar S, Moussa IM, Adil MF (2023) Elucidating effect of ZnO-nanoparticles and melatonin on physiological adjustments and growth of *Solanum melongena* under salinity stress. Sci Hortic 322(112455):112455. https://doi.org/10.1016/j.scienta.2023.112455
10. Arif MR, Islam MT, Robin AHK (2019) Salinity stress alters root morphology and root hair traits in *Brassica napus*. Plants 8(7):192. https://doi.org/10.3390/plants8070192
11. Ashique S, Raikar A, Jamil S, Lakshminarayana L, Gajbhiye SA, De S, Kumar S (2025) Artificial intelligence integration with nanotechnology: a new frontier for sustainable and precision agriculture. Curr Nanosci 21(2):242–273. https://doi.org/10.2174/0115734137275111231206072049
12. Balasubramaniam T, Shen G, Esmaeili N, Zhang H (2023) Plants' response mechanisms to salinity stress. Plants 12(12):2253. https://doi.org/10.3390/plants12122253
13. Bello SK, Alayafi AH, Al-Solaimani SG, Abo-Elyousr KAM (2021) Mitigating soil salinity stress with gypsum and bio-organic amendments: a review. Agronomy (Basel, Switzerland) 11(9):1735. https://doi.org/10.3390/agronomy11091735
14. Byrt CS, Munns R, Burton RA, Gilliham M, Wege S (2018) Root cell wall solutions for crop plants in saline soils. Plant Sci: Int J Exp Plant Biol 269:47–55. https://doi.org/10.1016/j.plantsci.2017.12.012
15. Chele KH, Tinte MM, Piater LA, Dubery IA, Tugizimana F (2021) Soil salinity, a serious environmental issue and plant responses: a metabolomics perspective. Metabolites 11(11):724. https://doi.org/10.3390/metabo11110724
16. Chen C, He G, Li J, Perez-Hormaeche J, Becker T, Luo M, Wallrad L, Gao J, Li J, Pardo JM, Kudla J, Guo Y (2023) A salt stress-activated GSO1-SOS2-SOS1 module protects the *Arabidopsis* root stem cell niche by enhancing sodium ion extrusion. EMBO J 42(13):e113004. https://doi.org/10.15252/embj.2022113004
17. Chourasia KN, Lal MK, Tiwari RK, Dev D, Kardile HB, Patil VU, Kumar A, Vanishree G, Kumar D, Bhardwaj V, Meena JK, Mangal V, Shelake RM, Kim J-Y, Pramanik D (2021) Salinity stress in potato: understanding physiological, biochemical and molecular responses. Life (Basel, Switzerland) 11(6):545. https://doi.org/10.3390/life11060545
18. Das R, Panda SN (2022) Preparation and applications of biochar based nanocomposite: a review. J Anal Appl Pyrol 167(105691):105691. https://doi.org/10.1016/j.jaap.2022.105691
19. Dong Y, Werling B, Cao Z, Li G (2024) Implementation of an in-field IoT system for precision irrigation management. Front Water 6. https://doi.org/10.3389/frwa.2024.1353597

20. EL Sabagh A, Islam MS, Skalicky M, Ali Raza M, Singh K, Anwar Hossain M, Hossain A, Mahboob W, Iqbal MA, Ratnasekera D, Singhal RK, Ahmed S, Kumari A, Wasaya A, Sytar O, Brestic M, Çig F, Erman M, Habib Ur Rahman M, Ullah N, Arshad A (2021) Salinity stress in wheat (*Triticum aestivum* L.) in the changing climate: adaptation and management strategies. Front Agron 3. https://doi.org/10.3389/fagro.2021.661932
21. Elsheery NI, Helaly MN, El-Hoseiny HM, Alam-Eldein SM (2020) Zinc oxide and silicone nanoparticles to improve the resistance mechanism and annual productivity of salt-stressed mango trees. Agronomy (Basel, Switzerland) 10(4):558. https://doi.org/10.3390/agronomy10040558
22. FAO (2021) World soil day: fao highlights the threat of soil salinization to global food security. FAO.org; Food and Agriculture Organization of the United Nations. https://www.fao.org/newsroom/detail/world-soil-day-fao-highlights-threat-of-soil-salinization-to-food-security-031221/en
23. Farooq MA, Gao S, Hassan MA, Huang Z, Rasheed A, Hearne S, Prasanna B, Li X, Li H (2024) Artificial intelligence in plant breeding. Trends Genet: TIG 40(10):891–908. https://doi.org/10.1016/j.tig.2024.07.001
24. Fiocco D, Ganesan V, de la Serrana Lozano MG, Sharifi H (2023) Agtech: breaking down the farmer adoption dilemma. McKinsey & Company. https://www.mckinsey.com/industries/agriculture/our-insights/agtech-breaking-down-the-farmer-adoption-dilemma
25. Franco-Navarro JD, Díaz-Rueda P, Rivero-Núñez CM, Brumós J, Rubio-Casal AE, de Cires A, Colmenero-Flores JM, Rosales MA (2021) Chloride nutrition improves drought resistance by enhancing water deficit avoidance and tolerance mechanisms. J Exp Bot 72(14):5246–5261. https://doi.org/10.1093/jxb/erab143
26. Geilfus C-M (2018) Chloride: from nutrient to toxicant. Plant Cell Physiol 59(5):877–886. https://doi.org/10.1093/pcp/pcy071
27. Gholizadeh F, Gohari G, Pál M, Szalai G, Khan I, Janda T (2025) Enhancing wheat resilience to salt stress through an integrative nanotechnology approach with chitosan proline and chitosan glycine. Sci Rep 15(1):11126. https://doi.org/10.1038/s41598-025-91496-w
28. Gupta G, Kumar Pal S (2025) Applications of AI in precision agriculture. Discov Agric 3(1). https://doi.org/10.1007/s44279-025-00220-9
29. Hamamoto S, Horie T, Hauser F, Deinlein U, Schroeder JI, Uozumi N (2015) HKT transporters mediate salt stress resistance in plants: from structure and function to the field. Curr Opin Biotechnol 32:113–120. https://doi.org/10.1016/j.copbio.2014.11.025
30. Hoang T, Tran T, Nguyen T, Williams B, Wurm P, Bellairs S, Mundree S (2016) Improvement of salinity stress tolerance in rice: challenges and opportunities. Agronomy (Basel, Switzerland) 6(4):54. https://doi.org/10.3390/agronomy6040054
31. Horie T, Hauser F, Schroeder JI (2009) HKT transporter-mediated salinity resistance mechanisms in Arabidopsis and monocot crop plants. Trends Plant Sci 14(12):660–668. https://doi.org/10.1016/j.tplants.2009.08.009
32. Iavicoli I, Leso V, Beezhold DH, Shvedova AA (2017) Nanotechnology in agriculture: opportunities, toxicological implications, and occupational risks. Toxicol Appl Pharmacol 329:96–111. https://doi.org/10.1016/j.taap.2017.05.025
33. Isayenkov SV, Maathuis FJM (2019) Plant salinity stress: many unanswered questions remain. Front Plant Sci 10:80. https://doi.org/10.3389/fpls.2019.00080
34. Jiang X, Leidi EO, Pardo JM (2010) How do vacuolar NHX exchangers function in plant salt tolerance? Plant Signal Behav 5(7):792–795. https://doi.org/10.4161/psb.5.7.11767
35. Ji H, Pardo JM, Batelli G, Van Oosten MJ, Bressan RA, Li X (2013) The salt overly sensitive (SOS) pathway: established and emerging roles. Mol Plant 6(2):275–286. https://doi.org/10.1093/mp/sst017
36. Kalteh M, Alipour ZT, Ashraf S, Aliabadi MM, Nosratabadi AF (2014) Effect of silica Nanoparticles on Basil (*Ocimum basilicum*) under salinity stress. J Chem Health Risks 4(2):49–55. https://www.jchr.org/index.php/JCHR/article/view/408/410
37. Khan OU, Ahmad B, Mamoona Z, Irfan Z, Ismail A, Saleem S, Shaheen M, Khan M, Zafar M, Bashir MW (2024) The *Arabidopsis* nhx1 gene: overexpression and its impact on salt tolerance

in transgenic plants. Biol Clin Sci Res J 2024(1):1148. https://doi.org/10.54112/bcsrj.v2024i1.1148
38. Kotula L, Zahra N, Farooq M, Shabala S, Siddique KHM (2024) Making wheat salt tolerant: what is missing? Crop J. https://doi.org/10.1016/j.cj.2024.01.005
39. Kumar AA, Kumar SKN, Fernandez RE (2021) Real time sensing of soil Potassium levels using zinc oxide-multiwall carbon nanotube-based sensors. IEEE Trans Nanobiosci 20(1):50–56. https://doi.org/10.1109/TNB.2020.3027863
40. Kumari P, Gupta A, Chandra H, Singh P, Yadav S (2021) Effects of salt stress on the morphology, anatomy, and gene expression of crop plants. In: Physiology of salt stress in plants. Wiley, pp 87–105. https://doi.org/10.1002/9781119700517.ch6
41. Kumari R, Suman K, Karmakar S, Mishra V, Lakra SG, Saurav GK, Mahto BK (2023) Regulation and safety measures for nanotechnology-based agri-products. Front Genome Ed 5:1200987. https://doi.org/10.3389/fgeed.2023.1200987
42. Kumar P, Sharma PK (2020) Soil salinity and food security in India. Front Sustain Food Syst 4. https://doi.org/10.3389/fsufs.2020.533781
43. Mahajan A, Bhosale M, Nikam P (2024) Optimizing fertilizer application using AI for sustainable agricultural practices. Nanotechnol Percept 4517–4530. https://doi.org/10.62441/nano-ntp.vi.5326
44. Majumdar P (2025) Revolutionising the fertiliser industry through strength-focused AI integration. Am J Inf Sci Technol 9(3):242–255. https://doi.org/10.11648/j.ajist.20250903.17
45. Mani J, Nagachandrabose S, Somasundaram P, Deenan S (2025) Artificial intelligence integrated nano biosensor technology: a breakthrough in early detection and sustainable management of phytonematodes. Physiol Mol Plant Pathol 139(102756):102756. https://doi.org/10.1016/j.pmpp.2025.102756
46. Mansoor S, Iqbal S, Popescu SM, Kim SL, Chung YS, Baek J-H (2025) Integration of smart sensors and IOT in precision agriculture: trends, challenges and future prospectives. Front Plant Sci 16:1587869. https://doi.org/10.3389/fpls.2025.1587869
47. Mishra P, Mishra J, Arora NK (2021) Plant growth promoting bacteria for combating salinity stress in plants—recent developments and prospects: a review. Microbiol Res 252(126861):126861. https://doi.org/10.1016/j.micres.2021.126861
48. Mridha D, Lamsal B, Antonangelo JA (2025) Nanotechnology in agriculture: Innovations for sustainability and greenhouse gas mitigation—a review. Sci Total Environ 995(180065):180065. https://doi.org/10.1016/j.scitotenv.2025.180065
49. Mukhtiar A, Zia MA, Alawadi HFN, Naqve M, Seleiman MF, Mahmood A, Majeed MI, Hafeez MB, Khan BA, Khan N (2024) Role of iron oxide nanoparticles in maize (*Zea mays* L.) to enhance salinity stress tolerance. Notulae Botanicae Horti Agrobotanici Cluj-Napoca 52(3):13695. https://doi.org/10.15835/nbha52313695
50. Munz J, Maurmann I, Schuele H, Doluschitz R (2024) Digital transformation at what cost? A case study from Germany estimating the adoption potential of precision farming technologies under different scenarios. Smart Agric Technol 9(100585):100585. https://doi.org/10.1016/j.atech.2024.100585
51. Murali M, Gowtham HG, Singh SB, Shilpa N, Aiyaz M, Alomary MN, Alshamrani M, Salawi A, Almoshari Y, Ansari MA, Amruthesh KN (2022) Fate, bioaccumulation and toxicity of engineered nanomaterials in plants: current challenges and future prospects. Sci Total Environ 811(152249):152249. https://doi.org/10.1016/j.scitotenv.2021.152249
52. Murmu S, Sinha D, Chaurasia H, Sharma S, Das R, Jha GK, Archak S (2024) A review of artificial intelligence-assisted omics techniques in plant defense: current trends and future directions. Front Plant Sci 15:1292054. https://doi.org/10.3389/fpls.2024.1292054
53. Murtaza G, Hassan NE, Usman M, Zaman QU, Rizwan M, Deng G, Ahmed Z, Majeed A, Iqbal J, Elshikh MS, AlKubaisi N, Ahmad S, Gurbanova L, Iqbal R, Malaga-Tobola U, Gancarz M (2025) Combine effects of *Broussonetia papyrifera*-derived biochar and selenium nanoparticles for lead-polluted saline soils remediation during barley cultivation. Sci Rep 15(1):26837. https://doi.org/10.1038/s41598-025-12279-x

54. Neves GYS, Marchiosi R, Ferrarese MLL, Siqueira-Soares RC, Ferrarese-Filho O (2010) Root growth inhibition and lignification induced by salt stress in soybean. J Agron Crop Sci 196(6):467–473. https://doi.org/10.1111/j.1439-037x.2010.00432.x
55. Ogunsola SS, Oladipo ME, Oladoye PO, Kadhom M (2024) Carbon nanotubes for sustainable environmental remediation: a critical and comprehensive review. Nano-Struct Nano-Objects 37(101099):101099. https://doi.org/10.1016/j.nanoso.2024.101099
56. Ondrasek G, Rathod S, Manohara KK, Gireesh C, Anantha MS, Sakhare AS, Parmar B, Yadav BK, Bandumula N, Raihan F, Zielińska-Chmielewska A, Meriño-Gergichevich C, Reyes-Díaz M, Khan A, Panfilova O, Seguel Fuentealba A, Romero SM, Nabil B, Wan CC, Shepherd J, Horvatinec J (2022) Salt stress in plants and mitigation approaches. Plants 11(6):717. https://doi.org/10.3390/plants11060717
57. Parameswari P, Belagalla N, Singh BV, Abhishek GJ, Rajesh GM, Katiyar D, Hazarika B, Paul S (2024) Nanotechnology-based sensors for real-time monitoring and assessment of soil health and quality: a review. Asian J Soil Sci Plant Nutr 10(2):157–173. https://doi.org/10.9734/ajsspn/2024/v10i2272
58. Polyakov V, Bauer T, Butova V, Minkina T, Rajput VD (2023) Nanoparticles-based delivery systems for salicylic acid as plant growth stimulator and stress alleviation. Plants 12(8):1637. https://doi.org/10.3390/plants12081637
59. Qin H, Li Y, Huang R (2020) Advances and challenges in the breeding of salt-tolerant rice. Int J Mol Sci 21(21):8385. https://doi.org/10.3390/ijms21218385
60. Rajak P, Ganguly A, Adhikary S, Bhattacharya S (2023) Internet of Things and smart sensors in agriculture: scopes and challenges. J Agric Food Res 14(100776):100776. https://doi.org/10.1016/j.jafr.2023.100776
61. Ravari SZ, Dehghani H, Naghavi H (2016) Assessment of salinity indices to identify Iranian wheat varieties using an artificial neural network. Ann Appl Biol 168(2):185–194. https://doi.org/10.1111/aab.12254
62. Ren Y, Qu Y, Gao R (2025) Data quality challenges of AIGC application in smart agriculture. Front Artif Intell 8(1640805):1640805. https://doi.org/10.3389/frai.2025.1640805
63. Robin AHK, Matthew C, Uddin MJ, Bayazid KN (2016) Salinity-induced reduction in root surface area and changes in major root and shoot traits at the phytomer level in wheat. J Exp Bot 67(12):3719–3729. https://doi.org/10.1093/jxb/erw064
64. Ryan M, Isakhanyan G, Tekinerdogan B (2023) An interdisciplinary approach to artificial intelligence in agriculture. NJAS: Impact Agric Life Sci 95(1). https://doi.org/10.1080/27685241.2023.2168568
65. Santhosh Kumar MP, Thiyageshwari S (2017) Performance of nano-gypsum on reclamation of sodic soil. https://doi.org/10.1002/ldr.v14:3/issuetoc
66. Sarabandi M, Zargar M, Ghorbani A, Chen M (2025) Smart and sustainable nano-biosensing technologies for advancing stress detection and management in agriculture and beyond. Ind Crops Prod 226(120713):120713. https://doi.org/10.1016/j.indcrop.2025.120713
67. Sári D, Ferroudj A, Abdalla N, El-Ramady H, Dobránszki J, Prokisch J (2023) Nano-management approaches for salt tolerance in plants under field and in vitro conditions. Agronomy (Basel, Switzerland) 13(11):2695. https://doi.org/10.3390/agronomy13112695
68. Shah WH, Rasool A, Saleem S, Mushtaq NU, Tahir I, Hakeem KR, Rehman RU (2021) Understanding the integrated pathways and mechanisms of transporters, protein kinases, and transcription factors in plants under salt stress. Int J Genomics 2021:1–16. https://doi.org/10.1155/2021/5578727
69. Shang Y, Hasan MK, Ahammed GJ, Li M, Yin H, Zhou J (2019) Applications of nanotechnology in plant growth and crop protection: A review. Molecules (Basel, Switzerland) 24(14):2558. https://doi.org/10.3390/molecules24142558
70. Sharma P, Lew TTS (2022) Principles of nanoparticle design for genome editing in plants. Frontiers in Genome Editing 4:846624. https://doi.org/10.3389/fgeed.2022.846624
71. Shoukat A, Maryam U, Pitann B, Zafar MM, Nawaz A, Hassan W, Seleiman MF, Saqib ZA, Mühling KH (2025) Efficacy of nano and conventional zinc and silicon fertilizers for nutrient use efficiency and yield benefits in maize under saline field conditions. Plants 14(5). https://doi.org/10.3390/plants14050673

72. Singh RP, Handa R, Manchanda G (2021) Nanoparticles in sustainable agriculture: an emerging opportunity. J Control Release: Off J Control Release Soc 329:1234–1248. https://doi.org/10.1016/j.jconrel.2020.10.051
73. Sneha, Singh PD, Tripathi V, Maurya S (2025) Nano-enhanced IoT sensors and hybrid scheduling algorithms for smart agriculture: a multi-layered framework towards sustainability. J Nano-Electron Phys 17(4):04014-1–04014-04016. https://doi.org/10.21272/jnep.17(4).04014
74. Sultan H, Li Y, Ahmed W, Yixue M, Shah A, Faizan M, Ahmad A, Abbas HMM, Nie L, Khan MN (2024) Biochar and nano biochar: enhancing salt resilience in plants and soil while mitigating greenhouse gas emissions: a comprehensive review. J Environ Manage 355(120448):120448. https://doi.org/10.1016/j.jenvman.2024.120448
75. Sultan MT, Mahmud U, Khan MZ (2023) Addressing soil salinity for sustainable agriculture and food security: innovations and challenges in coastal regions of Bangladesh. Futur Foods: Dedic J Sustain Food Sci 8(100260):100260. https://doi.org/10.1016/j.fufo.2023.100260
76. Talaviya T, Shah D, Patel N, Yagnik H, Shah M (2020) Implementation of artificial intelligence in agriculture for optimisation of irrigation and application of pesticides and herbicides. Artif Intell Agric 4:58–73. https://doi.org/10.1016/j.aiia.2020.04.002
77. Trung Luu D, Zwiazek JJ, Amodeo G, Kwon E-H, Adhikari A, Imran M, Lee D-S, Lee C-Y, Kang S-M, Lee I-J (2023) Exogenous SA applications alleviate salinity stress via physiological and biochemical changes in St John's Wort plants. Plants 12(2):310. https://doi.org/10.3390/PLANTS12020310
78. United Nations Development Programme (2021) Precision Agriculture for Smallholder Farmers. UNDP Global Centre for Technology, Innovation and Sustainable Development. https://www.undp.org/sites/g/files/zskgke326/files/2021-10/UNDP-Precision-Agriculture-for-Smallholder-Farmers.pdf
79. Wahab A, Muhammad M, Ullah S, Abdi G, Shah GM, Zaman W, Ayaz A (2024) Agriculture and environmental management through nanotechnology: eco-friendly nanomaterial synthesis for soil-plant systems, food safety, and sustainability. Sci Total Environ 926:171862. https://doi.org/10.1016/j.scitotenv.2024.171862
80. Walsh JJ, Mangina E, Negrão S (2024) Advancements in imaging sensors and AI for plant stress detection: a systematic literature review. Plant Phenomics (Washington, D.C.) 6(0153):0153. https://doi.org/10.34133/plantphenomics.0153
81. Wang J, Chen L, Wang Y, Zhang J, Liang Y, Xu D (2013) A computational systems biology study for understanding salt tolerance mechanism in rice. PLoS ONE 8(6):e64929. https://doi.org/10.1371/journal.pone.0064929
82. Wang J, Peng J, Li H, Yin C, Liu W, Wang T, Zhang H (2021) Soil salinity mapping using machine learning algorithms with the Sentinel-2 MSI in arid areas China. Remote Sens 13(2):305. https://doi.org/10.3390/rs13020305
83. Wang L, Ning C, Pan T, Cai K (2022) Role of silica nanoparticles in abiotic and biotic stress tolerance in plants: a review. Int J Mol Sci 2022(4):1947. https://doi.org/10.3390/ijms23041947
84. Wang L, Xu J-Y, Jia W, Chen Z, Xu Z-C (2020) Chloride salinity in a chloride-sensitive plant: focusing on photosynthesis, hormone synthesis and transduction in tobacco. Plant Physiol Biochem 153:119–130. https://doi.org/10.1016/j.plaphy.2020.05.021
85. Xiao C, Ji Q, Chen J, Zhang F, Li Y, Fan J, Hou X, Yan F, Wang H (2023) Prediction of soil salinity parameters using machine learning models in an arid region of northwest China. Comput Electron Agric 204(107512):107512. https://doi.org/10.1016/j.compag.2022.107512
86. Yang Y, Guo Y (2017) Elucidating the molecular mechanisms mediating plant salt-stress responses. New Phytol 217(2):523–539. https://doi.org/10.1111/nph.14920
87. Yang Y-B, Yang C, Zheng J-R, Xu L-Z, Yao N (2024) Chloride salt enhances plant resistance to biotic stresses. Front Plant Sci 15:1385164. https://doi.org/10.3389/fpls.2024.1385164
88. Zain M, Ma H, Nuruzzaman M, Chaudhary S, Nadeem M, Shakoor N, Azeem I, Duan A, Sun C, Ahamad T (2023) Nanotechnology based precision agriculture for alleviating biotic and abiotic stress in plants. Plant Stress 10(100239):100239. https://doi.org/10.1016/j.stress.2023.100239

89. Zain M, Ma H, Ur Rahman S, Nuruzzaman M, Chaudhary S, Azeem I, Mehmood F, Duan A, Sun C (2024) Nanotechnology in precision agriculture: Advancing towards sustainable crop production. Plant Physiol Biochem 206(108244):108244. https://doi.org/10.1016/j.plaphy.2023.108244
90. Zhang P, Guo Z, Ullah S, Melagraki G, Afantitis A, Lynch I (2021) Nanotechnology and artificial intelligence to enable sustainable and precision agriculture. Nature Plants 7(7):864–876. https://doi.org/10.1038/s41477-021-00946-6
91. Zhao J, Nurmemet I, Muhetaer N, Xiao S, Abulaiti A (2023) Monitoring soil salinity using machine learning and the polarimetric scattering features of PALSAR-2 data. Sustainability 15(9):7452. https://doi.org/10.3390/su15097452
92. Zhou H, Shi H, Yang Y, Feng X, Chen X, Xiao F, Lin H, Guo Y (2024) Insights into plant salt stress signaling and tolerance. Yi Chuan Xue Bao (J Genet Genomics) 51(1):16–34. https://doi.org/10.1016/j.jgg.2023.08.007

Sustainable Farming Contributed by Smart-Delivering Nanofertilizers

Rahul Gogoi, Bhoirob Gogoi, Hridesh Harsha Sarma, Ishani Borthakur, and Madhumita Barooah

Abstract Sustainable agriculture aims to harmonize productivity with environmental stewardship, fostering long-term food security while minimizing ecological harm. In recent years, nanotechnology has emerged as a transformative force in agriculture, particularly through the development of nanofertilizers. Smart delivery nanofertilizers represent an innovative class of nutrient inputs designed to enhance nutrient usage efficiency (NUE), minimize losses via leaching and volatilization and mitigate the environmental consequences of conventional fertilization techniques. These nanoformulations are engineered for enhanced and targeted release, reacting to certain environmental stimuli such as pH, moisture and root exudates. Consequently, the nutrients will be accessible precisely when and where the crops necessitate them. This chapter explores the design, synthesis and mechanisms of smart nanofertilizers, highlighting their application in enhancing crop yield, soil quality and resource efficiency in smart agriculture. Emphasis is placed on environmental sustainability achieved by reduced application rates, less nutrient runoff and lowered greenhouse gas emissions. Furthermore, regulatory issues, cost efficiency and enduring ecological impacts are meticulously examined. The integration of smart nanofertilizers with precision farming operations offers a promising avenue for a resilient and sustainable agricultural future. This chapter concludes by identifying knowledge gaps and proposing future research areas essential for the secure advancement and widespread implementation of nanotechnology in agriculture.

Keywords Smart nanofertilizers · Nutrient use efficiency · Controlled release · Precision farming · Environmental sustainability · Soil health · Crop productivity · Agricultural nanotechnology

---

R. Gogoi · I. Borthakur · M. Barooah (✉)
Department of Agricultural Biotechnology, Assam Agricultural University, Jorhat, India
e-mail: madhumita.barooah@aau.ac.in

B. Gogoi · M. Barooah
DBT-NECAB, Assam Agricultural University, Jorhat, India

H. H. Sarma
Department of Agronomy, Assam Agricultural University, Jorhat, India

J.-T. Chen (ed.), *AI-Nano-Enabled Climate-Smart Agriculture and Sustainability*,
Smart Nanomaterials Technology, https://doi.org/10.1007/978-981-92-1694-9_13

1 Introduction

The need to provide food resources to a highly growing population, accompanied by offsetting the massive environmental impact of the conventional agronomy, is a world issue that requires groundbreaking paradigm shifts in the way agriculture is practiced. This should be a transition that would balance agricultural output with the strict environmental stewardship to enhance long-term resilience in the intricate agroecosystems [47]. In turn, sustainable agriculture has been conceptualized as a wise combination of operations that optimize the output and minimize the ecological impact, thus being able to sustain the viability of food-production systems [98]. The Green Revolution, which is based on the intensive use of agrochemicals, is already at a point of crisis; modern agriculture demands high-performing inputs, which reduce the wastage of resources, preserve limited resources and keep within the limits of the planet. These needs are in line with the United Nations Sustainable Development Goals, which refer to the reduction of hunger and the mitigation of climate change, and it is important to state that the essential change in the agronomic paradigm is crucial [38].

High yields have been based on traditional mineral fertilizers, including urea and diammonium phosphate, for a number of decades. However, their widespread use creates strong inefficiencies and environmental destruction. One of its main weaknesses is the poor nutrient use efficiency (NUE) and in this case, the fertility of nitrogen, whereby less than 40% of the applied nitrogen is utilized by the target crop, with estimates showing that as much as 80% goes to waste [1]. Surplus nitrogen is washed away into the environment through runoff, leaching and volatilization; and added to pervasive water pollution (eutrophication due to nitrate and phosphate runoff), soil erosion (salinization, loss of microbial diversity) and increased atmospheric pollution via greenhouse gases like nitrous oxide (N_2O). The associated economic and environmental impacts are enormous and increase production expenses and worsen climate change [91].

Nanotechnology has become a powerful force in the field of agricultural science in response to the apparent restrictions the conventional inputs place upon it. The ability to work at the nanoscale, a range of 1–100 nm, allows one to produce materials with distinct physicochemical properties, such as a significantly greater surface-to-volume ratio and reactivity [92]. In the agricultural sector, such ability is utilized to come up with nanofertilizers that connect with plant systems at the molecular scale, contributing to unprecedented improvements in nutrient absorption and usefulness. Such nano-enabled substance engineering also constitutes an innovative way of developing a more efficient, resource-saving agrochemical [97].

The future of nanotechnology in agriculture consists of smart-delivered nanofertilizers. They are not only nanosized nutrients but also sophisticated nanoformulations which are designed to serve as controlled- and sustained-release delivery systems. The major task of such advanced constructs is to align nutrient release systems with the exact agronomic demands of the crop in order to maximize the NUE and reduce environmental losses [75]. These formulations with such properties are often stimulus-responsive, with their cargo being released when certain environmental factors are detected (e.g. pH changes, change in water content, change in temperature or even the presence of root exudates) [127].

This chapter will provide an in-depth discussion of these new contributions that outline the general design principles, synthesis procedures and launch systems that belong to smart nanofertilizers. It will also explain how they have been utilized to increase crop production rates, raise the health of their soil and record high resource utilization. Special attention will be given to their direct contribution to environmental sustainability by lowering the rate of application and runoff, which also correlates with the spirit of precision farming. Finally, the chapter will make a case for the potentially powerful and indispensable possibility of a sustainable, high-performance and truly sustainable agricultural future made possible through the incorporation of smart nanofertilizers with advanced digital technologies.

2 Nanofertilizers: An Overview

The radically increasing global population, which is estimated to hit almost 10 billion in the middle of the twenty-first century, poses a significant challenge to the agricultural systems on a global scale, whereby there is a need to significantly and sustainably increase the productivity and, at the same time, reduce the ecological footprint of food production. Traditional methods of fertilization that have been the main pillars of the Green Revolution have shown that there are serious inherent inefficiencies and overall negative environmental effects, and this is mainly due to the large rates of application and resultant losses of nutrients. It is on this critical background that the nanotechnology research has provided a groundbreaking solution in the nanofertilizer [87]. These represent a paradigm shift in nutrient delivery, providing a specific, highly efficient and proven sustainable alternative to a conventional source of nutrients. This part is devoted to the detailed, extensive description of nanofertilizers, which include their definition in the form of a fundamental one, their classification according to the type of nutrient and structure, the main synthesis procedures used, the main physicochemical characteristics that determine their functionality, and lastly, the intrinsic and major benefits of the product relative to the traditional products. Introduction of nanofertilizers into contemporary farming is not only an incremental enrichment, but it can be termed as the re-engineering of the nutrient supply chain with respect to the key tenets of sustainable agriculture [5].

The variety of nanofertilizers can be differentiated in a number of different ways, and this offers a detailed structure by which one can consider the diversity of the nanofertilizers and the way in which they can be used. The most used classification systems are either according to the nature of the nutrient being transported or the structural purpose the nanoscale component will have in the end.

2.1 *Classification According to Type of Nutrient*

This paradigm rationally resembles the traditional fertilizers, but in contrast to them, the role of the nanoscale delivery system or the size of the particles in improving the functionality of the element is considered.

2.1.1 Nano-macronutrient Fertilizers

These solutions are used to provide all the necessary factors needed in the greatest amounts to promote the growth of the plants, leading to nitrogen (N), phosphorus (P) and potassium (K). A typical business case is bulk urea granules that are thoroughly covered with the ultra-thin layers of nanopolymers, which is a membrane that can be controlled to ensure that nitrogen is released to the soil solution. On the same note, Phosphorus, which is also highly fixable in the soil, can be provided in the form of nano-hydroxyapatite $(Ca_{10})(PO_4)_6(OH)_2$ or nano-phosphates. It is the nanoscale coating or matrix that essentially gives the coating its properties of being smart or slow release to ensure that there is a reduction of major losses through leaching and volatilisation, which are the major culprits in the reduction of the nitrogen use efficiency of bulk fertilizers [54].

2.1.2 Nano-micronutrient Fertilizers

These contain the essential factors that, although they are only in minute quantities, are important to the important metabolic processes of plants, such as zinc (Zn), iron (Fe), copper (Cu) and manganese (Mn). The deficiencies in micronutrients are common worldwide, either due to the ease with which the elements get bound or bound in the soil or because they are not readily available to be absorbed by the roots. The solubility and intrinsic mobility of these elements radically increase in the form of nanoformulations, e.g. zinc oxide nanoparticles (ZnO NPs) or nano-ferrous sulphate ($FeSO_4$). This guarantees sufficient quantities of the nutrient in the rhizosphere, which is plant-available even with greatly reduced application rates compared to conventional chelates or sulphates and, therefore, is used to tackle the common micronutrient deficiencies in staple crops as well as improve the nutritional quality [24].

2.1.3 Nano-biofertilizers Bio-based

These are androids of products in which nanomaterials are used strategically to entrap, physically shield or add value to the delivery and colonization of useful microorganisms. These microbes, including nutrient fixation and mobilization, are important in natural nutrient fixation and mobilization, e.g. nitrogen-fixing Rhizobium or phosphate-solubilising Azotobacter. The carrier materials in nanoscale (e.g. nanopolymers or nanoclays) are meant to form a microenvironment that would protect these delicate microbes against the adverse environmental stressors (e.g. UV radiation, desiccation, thermal stress) and toxicants. This protection enhances the viability and shelf life of the bio-inoculant dramatically until the bio-inoculant is transported to and established effectively within the rhizosphere, thus encouraging highly effective biological nutrient cycling of soil [67].

2.2 *Strategic Classification by Structure and Function*

A different and more descriptive division depends on the design and purpose of the nanocomponent in architecture.

2.2.1 Nanomaterials that Are Encapsulated (Core Shell)

In this design, the nutrient load is fully enclosed in a protective and functional shell of an inert substance, which is usually a biodegradable polymer, a metal oxide or a clay derivative. The fundamental nutrient is thus removed from the environment, and its release is determined by only the characteristics of the shell, like its rate of degradation or its ability to absorb water. These core shells are the ideal examples of smart delivery systems [90].

2.2.2 Surface-Coated Nanomaterials

In this case, the nutrient is adsorbed or physically coated onto the outer part of an already existing nanoparticle, and the high surface area for the so-called loading is used quickly. This may also consist of a layer of nanoparticle coating a bulk fertilizer particle (e.g. a nano-silica coating of bulk urea). The role of the coating is usually to alter the surface energy, enhance dispersibility in water or be the responsive component in a stimuli-mediated release [44].

2.2.3 Matrix Nanomaterials (Embedded)

In this structure, the nutrient is impregnated into the material or dispersed uniformly across a porous and usually three-dimensional, nanoscale porous structure. Specialized nanoclays, zeolites or polymeric hydrogel materials are usually used. The nutrient release process is usually a slow progressive process that is controlled by slow diffusion of the nutrient through the matrix pores or progressive degradation of the matrix structure through time due to chemical/biological processes [37].

2.2.4 Self-assembled Nanomaterials According to the Intrinsic Nature of the Nutrient

In its most basic sense, even the necessary nutrient element is produced as a nanoparticle. The examples are zinc oxide nanoparticles (ZnO NPs) or copper nanoparticles (Cu NPs). Here, the nutrient itself acts as the active delivery agent, as it is the small particle size that positively affects the nutrient solubility and bioavailability [101, 102].

2.3 *Methods of Synthesis*

The exact mode of synthesis to be used constitutes a deciding factor of the properties of the resulting nanofertilizer, which directly affects the homogeneity of its size, shape, inherent stability, purity, working efficiency and most importantly, the ultimate product cost. Synthesis methods are loosely and rationally divided into physical (top-down), chemical (bottom-up) and biological (green) methods [70].

2.3.1 Top-Down (Physical) Approach

The top-down methodology is a physical process where the macroscopic or mechanical breakdown of bulk materials into particles in the desirable nanoscale is carried out. It is considered to be the least accurate among the existing methods.

High-Energy Ball Milling (Mechanical Milling)

This method is perhaps the most scalable, least costly and industrially applicable large-scale approach. It is the exposure of bulk fertilizer material or pure nutrient compounds to high-energy impact, grinding and intense shear forces in a specialized mill, and in effect, causing mechanical fracture and attrition of particles until they achieve the nanoscale. An advantage of the method is its simplicity and its ability to provide high throughput, but its major disadvantage is that mechanical milling may introduce contaminants that have been washed off the milling media; besides, it is often characterized by a relatively broad particle size distribution, and the shape of the particles is often irregular and not of the ideal shape [73].

Thermal Decomposition/Evaporation

In this process, a mass material is vaporized to high temperatures, and then it is quickly cooled down; in this process, the vapour condenses to finely grained nanoparticles. The method has the capability of generating high-purity particles, but it is intensive in energy and not easily scaled at a cost-effective scale to commodity agricultural products [118].

2.3.2 Bottom-Up (Chemical Methods) Approach

The bottom-up method is defined as a constructive assembly of atoms or molecules into pre-defined structures of nanoscale. This approach has much better control over the purity, size and shape of the resulting nanoparticles, which results in high-quality and more viable products [4].

Chemical Precipitation

The key chemical mechanism is the accurate reaction of two or more soluble molecular precursors in a liquid-phase under strictly controlled conditions (e.g. pH, temperature, concentration). The product of the reaction is an insoluble product, which then comes out of solution as highly pure nanoparticles. The technique is most suitable for the synthesis of a variety of inorganic nanofertilizers, such as numerous metal-oxide, as well as metal-phosphate nanomaterials.

Sol-Gel Process

One of the most versatile methods is whereby a colloidal suspension of precursors (usually metal or metal salts) is first formed. This sol is then left to undergo a process of hydrolysis and condensation reaction to form a continuous, solid, network-like structure referred to as a gel. The gel is then dried and subjected to treatment. Sol-gel is especially useful in the synthesis of high-purity, structurally homogeneous metal-oxide nanoparticles and more complex, porous silica-based carriers that are best suited to loading them with nutrients [59].

Microemulsion/Emulsion Polymerization

This method is mostly used in the production of complex polymeric nanocapsules, which are the main aspects of smart delivery systems. The polymerization process is strictly limited to nanoscale droplets (the microemulsion), which are stabilized by surfactants. This is a very well-controlled environment whereby the final polymer shell thickness can be fine-tuned and creates a narrow distribution in particle size, which is essential in achieving uniform release kinetics [41].

2.3.3 Green Synthesis (Biological Methods)

Green synthesis is also becoming prominent and the most environmentally oriented method of application in agriculture. It is essentially non-toxic, much more cost-efficient and does not involve any use of dangerous chemicals or extensive energy requirements that the two above techniques do.

Phytosynthesis Is Plant-Mediated Synthesis

It utilizes naturally occurring plant extracts (leaf, root or seed extracts) that are easy to obtain and inexpensive, which represent a natural arsenal of biomolecules (polyphenols, alkaloids and terpenoids) in plant tissues. Both of these natural compounds act as reducing agents, which transform metal ions into elemental nanoparticles and as stabilizing/capping agents, which inhibit agglomeration. Indicatively, even a basic leaf extract can be used to prepare silver or zinc oxide nanoparticles of their respective metal salts at ambient conditions [57].

Microbial Synthesis

The Microbial Synthesis or Myco- or Bacteriosynthesis is a type of synthesis in chemistry and biochemistry.

In this method, different biological agents are applied, such as certain strains of bacteria, fungi or yeasts, and used to make nanoparticles. These organisms spontaneously release certain enzymes and other metabolites into the medium in which they are cultured that effectively reduce metal ions into stable nanoparticles, usually at room temperature and standard atmospheric pressure. It is considered the most sustainable, non-polluting and scalable route to the mass production of safe, biodegradable nanofertilizers, and hence the industrial requirement of both efficacy and environmental responsibility [53].

2.3.4 Physicochemical Properties that Affect the Performance of the Fertilizers

The final performance of any nanofertilizer after being introduced into the complex soil-plant system is unavoidably tied to the accurate physicochemical characteristics of the same. These properties are carefully designed and manipulated in the course of synthesis [78].

Surface Area and Particle Size

As has been mentioned above, it is the most crucial defining characteristic. An exponential increase in the specific surface area to volume ratio goes with the reduction in particle size. It is the surface area thus maximized which is the ultimate cause of all the performance advantages: it causes very large increases in solubility, very large increases in the chemical reactivity with the soil solution and physical contact over the minute structures of plant roots and soil constituents. In its turn, it determines how fast and effectively the nutrient dissolution and the following absorption by the plant occur [40].

Surface Charge and Zeta Potential

The electric charge at the particle liquid interface, which is quantified by the Zeta potential, is essential in two aspects. To begin with, the Zeta potential is a high magnitude (either negative or positive) that is necessary to create a powerful repulsion between particles caused by the electrostatic repulsion, which is crucial in preventing agglomeration (clumping) of particles; this is important in maintaining the long-term stability of the fertilizer suspension, which is crucial to shelf life. Secondly, the nanofertilizer surface charge affects the critical interaction of the overall negatively charged soil particles (e.g. clay minerals and organic matter) with the surface charge, thus affecting the mobility and uptake pathway of the nanofertilizer [112].

Crystal Structure and Purity

The crystalline structure of atoms (the internal structure) and the general chemical purity of the nanofertilizer have a direct influence on the kinetics of their nutrient release and their possible ecotoxicity. In most cases, an amorphous (non-crystalline) structure dissolves and sheds its nutrient content more rapidly compared to a highly crystalline structure. In agricultural inputs, high chemical purity is uncompromisingly crucial since it excludes the possibility of introducing poisonous contaminants in the food chain, or unwanted byproducts, of toxic heavy metals [19].

Shape and Morphology

Besides size, the geometric structure of a particle (e.g. spherical, rod, cubic or sheet) will play a role in a variety of biological and physical processes, including the uptake and transportation of materials to plants. Research suggests that they may utilize different mechanisms of uptake into plant cells when using anisotropic (non-spherical) nanoparticles as compared to traditional spherical nanoparticles. The use of layered structures, such as specialized nanoclays, provides an abundant number of unique intercalation sites that can facilitate the physical loading and delivery of large amounts of nutrient ions in a controlled manner [130].

Stability in Suspension (Colloidal Stability)

Nanoparticle suspensions created for use on liquids (drip irrigation or foliar spraying) rely on the long-term colloidal stability of the nanoparticle suspension to remain stable. This means that the nanoparticle suspension must be able to resist sedimentation (settling) and aggregation (clumping) and therefore should be managed appropriately. Non-ionic surfactants and protective polymeric surface coatings can be used strategically to achieve a uniform dose delivery over an entire field during application [105].

2.3.5 Comparative Advantages Over Conventional Fertilizers

Smart nanofertilizers (NFs) are specifically designed to offer enhanced performance for crop production and sustainability compared to traditional fertilizer products. Several of these benefits include increased efficacy when delivering nutrients to plants, creating a controlled-release effect, having less negative impact on the environment and improving the overall quality of crops grown using them. Details of these benefits are outlined in Table 1.

3 Mechanisms of Smart Delivery in Nanofertilizers

The paradigm change in the form of smart delivery nanofertilizers lies in their enhanced, designed capacity to control the release of nutrients, making the wasteful character of burst release characteristic of standard fertilizers be substituted with a highly mechanical and focused delivery system. In this section, the key mechanisms underlying this intelligent functionality have been outlined in detail and include the principles of controlled release, the types of stimuli that are used to induce nutrient availability, the various types of nanocarriers that are used and the key biological interactions that occur at the root-soil interface. These smart delivery systems form the technological core to support high nutrient use efficiency (NUE) at low environmental cost in modern-day precision agriculture [46].

3.1 Controlled- and Sustained-Release Mechanisms

The main principle behind smart delivery is the controlled and constant release, developed to match the nutrient supply rate to the dynamic needs of the developing

Table 1 Table comparing smart nanofertilizers and conventional fertilizers across key agricultural and environmental outcomes

Feature/Outcome	Smart nanofertilizers (NFs)	Conventional fertilizers	References
Nutrient use efficiency (NUE)	High (up to 20–30% more efficient; >40% in some cases)	Low (often <35%)	Nongbet et al. [70]; Seleiman et al. [94]
Nutrient release	Slow, controlled and targeted (40–50 days)	Rapid, less controlled (4–10 days)	Kekeli et al. [46] Upadhyay et al. [116]
Crop yield and quality	Increased yield (up to 93% improvement), better quality, enhanced stress tolerance	Moderate yield, less quality improvement	Azam et al. [12]; Sadati Valojai et al. [88]
Environmental impact	Reduced leaching, volatilization and soil/water pollution	High nutrient losses, environmental pollution	Babu et al. [13]; Sadati Valojai et al. [88]
Application dosage	Lower required amounts can reduce conventional fertilizer use by 25–50%	Higher dosages are needed for a similar effect	Madzokere et al. [55]; Taha et al. [110]
Cost and efficiency	Potentially lower long-term costs, higher energy efficiency	Higher input costs, lower energy efficiency	Upadhyay et al. [116]; Kumar et al. [49–51]
Plant stress tolerance	Enhanced tolerance to biotic/abiotic stresses	Limited stress mitigation	Seleiman et al. [94]; Kekeli et al., [46]
Soil health	Supports soil microbial activity and reduces degradation	Can degrade soil health over time	Zulfiqar et al. [132]; Kumar et al. [49–51]
Risks/limitations	Potential nanoparticle toxicity, need for further safety research	Well-known risks, but less concern for nanotoxicity	El-Saadony et al.[29]; Verma et al. [117]

crop, a principle commonly referred to as feeding the plant on demand. Unlike in traditional fertilizers, where a large proportion of nutrients is immediately released in the soil, leading to serious losses through leaching, volatilisation or microbial degradation, nanofertilizer uses physical barriers to control the release of nutrients to the soil [101, 102]. The controlled-release process is not immediate, but controlled by one or a combination of three main physical processes.

3.1.1 Diffusion-Controlled Release

Under this mode, the nutrient is trapped by a porous or semi-permeable polymeric or inorganic shell (the nanocarrier) wrap. The rate of release is given by the speed at which the soil water permeates the shell and dissolves the inner nutrient core, and then the diffusion of the dissolved nutrient out of the shell through the pores of the shell into the soil around it. The parameters that are engineered to control the rate of diffusion are shell thickness, porosity and tortuosity at the nanoscale. A denser, less porous shell has a slower and more extended-release profile, best preferred when the release duration of the crop is long. This will help in maintaining a constant and low-level supply of nutrients within a few weeks or months, which will make sure that the plant does not undergo any stages of deficiency or excessive nutrient over-supply [123].

3.1.2 Degradation-Controlled Release

In this case, the dissolution of nutrients is directly connected with the dissolution of the nanocarrier matrix itself. The carrier itself is produced out of a substance that will disintegrate over time within the environment of the soil. Soil microbes, hydrolysis (reaction with water) or photodegradation (reaction with light) can mediate degradation. As an illustration, coating urea with biodegradable polymer like poly-lactic acid (PLA) or chitosan would make the nutrient dissolve when the polymer chain is broken down to form harmless, non-toxic subunits. The mechanism would be crucial in the full eradication of the residual synthetic nanocarrier wastes in the field and, hence, overcome a significant environmental concern that is related to nanotechnology [80].

3.1.3 Osmotic Pressure/Swelling-Controlled Release

Commonly used in nano-hydrogel formulations, the mechanism in question consists of incorporating the nutrient into the network of the polymer that is highly sensitive to water. In the case of increased soil moisture (e.g. after irrigation or rainfall), the water enters the hydrogel matrix, resulting in significant swelling. The swelling phenomenon increases the pore volume and pushes the nutrients that are enclosed in the pore into the soil solution. This process is terminated when the soil becomes dry, and the hydrogel shrinks, forming a very sensitive system that is directly proportional to nutrient availability and water availability, which is a key component of water stress management in arid and semi-arid agriculture [6].

3.2 Stimuli-Responsive Release Mechanisms

The idea behind the smartness of delivery goes beyond sustained delivery by offering mechanisms that react to special environmental or biological stimuli, ensuring a high level of specificity in nutrient delivery. These stimulus-sensitive mechanisms represent an on-demand mechanism, significantly increasing the efficiency and effectiveness of fertilization by adjusting the supply to the short-term requirement.

3.2.1 pH-Responsive Release

One of the most common stimuli that is used is soil solution pH. The rhizosphere, a zone directly surrounding the root, is acidic as a rule, and the pH can often decrease by one unit because of active H^+ ion and organic acid secretion by the plant root as an attempt to mobilize nutrients. The polymers employed in the development of nanocarriers may have pH-sensitive functional groups (e.g. amine or carboxylic group) which swell, dissolve or change conformation at lower pH values. This makes a large portion of the fertilizer non-reactive in the bulk soil, though selectively released only in the region of physiological activity around the roots where nutrients are required [27].

3.2.2 Moisture/Water-Responsive Release

The moisture-responsive systems are closely associated with the osmotic mechanism in that they release the nutrients only when the moisture content in the soil reaches a critical level. The nanoscale super-absorbent polymers (SAPs) are used to store the nutrients until adequate water makes the contents swell and burst out. The process is critical in water management where the nutrients are not lost during dry seasons when the plant is not actively uptaking water, and only when the water, which is the medium of metabolism and uptake, is present releases the nutrients [58].

3.2.3 Temperature-Responsive Release (Thermo-Responsive)

Though soil temperature does not show much dynamic variation compared to pH or moisture, it is closely related to the metabolism rate of the plant and root development, which makes it an effective trigger. Nanocarriers may be prepared using materials of a particular Lower Critical Solution Temperature (LCST) or Upper Critical Solution Temperature (UCST). As an example, an LCST polymer could be designed to fold and release the nutrient over the temperature range that is favourable to plant development, or expand under it. The mechanism can be used especially to control nutrient release in certain stages of development or in conditions with extreme variation of temperature throughout the year [62].

3.2.4 Enzymatic Activity and Root Exudates-Responsive Release

This is the most advanced form of intelligent delivery, which is the most accurate in targeting biologicals. The rhizosphere environment is actively remodelled through the release of different enzymes (e.g. phytase, cellulase) and low-molecular-weight organic compounds (root exudates) by the active secretion of plant roots. Nanofertilizers may be functionalised with biodegradable linkers or shells that are only cleaved by the enzymes released by plants. As an example, a nutrient could be attached to a carrier through a bond that may only be broken by an enzyme made by the plant when it is lacking in phosphorus. The use of biological intelligence means the nutritional release is not only regulated but also guaranteed. The plant will receive the nutrient as a direct transactional response to its recorded lack of phosphorus [119].

3.3 *Nanocarriers: The Building Blocks of Smart Systems*

The nanocarriers are the main embodiment of the functionality of the smart delivery systems that envelop, protect and control the release of the nutrient payload. The nature of the controlled-release mechanism and sensitivity to environmental stimuli depend on the choice of carrier material.

3.3.1 Nanocapsules and Nanospheres (Polymeric Carriers)

These carriers are one of the most versatile. Nanocapsules have a core-shell structure, with a liquid or solid core (i.e. nutrient) surrounded by a polymeric membrane (i.e. PLA or poly-caprolactone). Nanospheres are systems of the matrix, where a nutrient is evenly distributed in a solid polymer. The use of polymers is very popular due to the possibility of easily chemically functionalizing with pH-sensitive groups, enzyme-cleavable bonds or specific cross-linking agents to provide a variety of stimulus-responsive actions. Further, agricultural polymers, including chitosan (chitin-based polymers) and alginate (seaweed-based polymers), are biodegradable and biocompatible, which makes sure that the carrier does not cause the soil to be contaminated in the long term [131].

3.3.2 Nanoclays, Zeolites and Mesoporous Silica

These inorganic substances are highly surface area based, layered, or porous based in their internal organization, which makes them superior nutrient loading hosts.

Nanoclays

Nanoclays (e.g. montmorillonite) have great cation exchange capacity (CEC) and extensive internal surface area. Nutrient ions (including K^+ or NH_4^+) are located between the clay layers and are released gradually by ion exchange with the soil solution. The diffusion-controlled and release are mainly maintained [99].

Mesoporous Silica Nanoparticles (MSNs)

Mesoporous Silica Nanoparticles (MSNs) possess pores that are highly uniform, tunable and can hold a large quantity of fertilizer. The pores may be topped by pH-sensitive or enzyme-cleavable gates, which will convert the MSN into a proper smart delivery vehicle [68].

Zeolites

The Microporous, aluminosilicate minerals, and zeolites, are molecular sieves and powerful ion exchangers and provide a high-capacity, long-lasting and economical platform to deliver nutrients [121].

3.3.3 Bio-nanocomposites and Carbon-Based Carriers

Bio-nanocomposites

Bio-nanocomposites represent the best qualities of organic and inorganic materials, often incorporating a biodegradable polymer with a high-strength inorganic scaffold (e.g. cellulose nanocrystals or nanoclays). Such integration increases the structural stability without loss of biodegradability [66].

Carbon-Based Carriers

Carbon-based carriers (e.g. carbon nanotubes (CNTs) and graphene oxide (GO)) have an exceptionally large surface area to load and physical adsorption. Specifically, CNTs may be functionalized to include nutrients and have been demonstrated to possibly invade plant cell walls, making it possible to introduce nutrients directly; although their safety in the environment is currently under intense research [31].

3.4 Interaction of Nanofertilizers with Plant Root Systems and Rhizosphere

The last crucial process in the intelligent delivery system is the process of effective communication between the nanofertilizer and the plant. The effect of nanofertilizers is a fundamental alteration of the nutrient uptake dynamics in two key avenues: altered interaction with soils and altered biological pathways of uptake [16].

3.4.1 Rhizosphere-Targeted Delivery

As explained in the pH-responsive mechanism, intelligent nanofertilizers are structured around the concentration of their release activity in the rhizosphere, which is the slim layer of soil that is directly impacted by the root secretions and related microbiota. They increase the probability that the nutrient will be captured at this zone, where the nutrient demand is greatest, and minimize the chances that the nutrient was previously chemically absorbed by other soil constituents or was consumed by other non-target organisms, a significant challenge to conventional fertilizers. They are also small enough to allow increased uniformity of distribution in this important zone [61].

3.4.2 Enhanced Biological Uptake Pathways

The nanoscale dimensions allow the potential uptake routes to be inaccessible to the conventional chelates of micronutrients and bulk ions.

Endocytosis

It is theorized that plant root cells can internalize whole nanoparticles using a process called endocytosis, in which the cell membrane surrounds the particle. This takes the nutrient directly into the cell, where it avoids the complex and at times limiting transport of the free ions [104].

Apoplastic and Symplastic Transport

Nanoparticles or the released ions, once internalized, may travel through the cell walls and intercellular spaces (apoplastic pathway) or through the cytoplasm via plasmodesmata (symplastic pathway). Their tiny size enables their easy movement through these internal plant transport systems, resulting in excellence of translocation to roots, shoots and leaves relative to bulk-sized particles [95].

Stomatal Uptake (Foliar Application)

In the case of nanofertilizers used on the leaves (foliar feeding), the small size will presumably allow them to penetrate the stomata and cuticular layer more efficiently than the larger droplets or the traditional foliar sprays [26].

4 Contribution to Sustainable Farming

The integration of smart-delivering nanofertilizers into contemporary agricultural systems represents a shift in paradigm towards a more sustainable way of food production (Fig. 1). This has had a wide range of positive effects, which have, in turn, aided in farming in various ways. These have been discussed in detail, herewith.

4.1 Enhancing Nutrient Use Efficiency

Nutrient use efficiency (NUE) is one of the most important parameters with regard to the evaluation of agricultural sustainability. NUE is represented by the ratio of nutrients harvested in crop biomass to the total nutrients applied. Conventional fertilizers typically exhibit low NUE values, with nitrogen efficiency often ranging between 30 and 35%, phosphorus at 18 and 20%, and potassium at 35 and 40% [30]. This inefficiency translates to substantial economic losses and environmental degradation, making NUE a paramount objective in sustainable agriculture.

Smart-delivering nanofertilizers can transform the dynamics of nutrient delivery through their unique physicochemical properties and controlled-release mechanisms. The nanoscale dimensions of these materials typically range from 1 to 100 nm [46]. And they provide a high surface-area-to-volume ratio, which facilitates enhanced interaction with plant root systems and soil components. This increased surface area enables more efficient nutrient dissolution and availability to plants compared to conventional bulk fertilizers [61].

The controlled-release system, which is in nanofertilizers, ensures that nutrients are delivered to plants in a synchronized manner with their physiological demands throughout different growth stages. During the vegetative phase, when

Fig. 1 The integration of smart-delivering nanofertilizers into contemporary agricultural systems

nitrogen demand is the highest, smart nanofertilizers can be engineered to release higher nitrogen concentrations. Similarly, during reproductive stages when phosphorus and potassium requirements increase, the release profile can be tailored as needed. This temporal matching of nutrient supply with crop demand minimizes luxury consumption and maximizes utilization efficiency [36].

Encapsulation technologies employed in the design of nanofertilizer create protective barriers around nutrient cores. This helps in shielding them from premature degradation, chemical transformation or immobilization in the soil matrix [35]. For example, nitrogen encapsulated within polymer nanoparticles remains protected from microbial nitrification and denitrification processes that typically result in substantial nitrogen losses. Research has demonstrated that nano-encapsulated urea can increase nitrogen use efficiency by 40–50% compared to conventional urea applications, representing a significant advancement in sustainable nitrogen management [85].

The site-specific delivery capability of nanofertilizers amplifies the NUE by ensuring that nutrients reach the precise locations where the uptake by roots is maximum. Functionalized nanoparticles can be designed with certain surface modifications that can enhance their mobility towards root zones. This is done with the help of electrochemical gradients or recognition of root exudates. This targeted approach minimizes nutrient dispersion into non-productive soil regions, concentrating fertilizer resources where they can be most effectively utilized by crops. Moreover, nanofertilizers facilitate enhanced penetration through plant cuticles and cell walls when applied as foliar treatments. This enables direct nutrient delivery to photosynthetically active tissues [70].

4.2 Reducing Fertilizer Losses

One of the most significant contributions of nanofertilizers to sustainable agriculture is their capacity to dramatically reduce nutrient losses. This is done through various environmental pathways. Pathways involving conventional fertilizer systems suffer from three primary loss mechanisms, viz. leaching, volatilization and surface runoff. Each type of loss can contribute to economic losses and also an overall reduction in productivity [46].

Leaching is the downward movement of water-soluble nutrients beyond the root zone, aided by gravitational pull. This movement makes them unavailable for plant uptake vis-à-vis contaminating groundwater resources. One such element, nitrogen in the form of nitrate, is particularly susceptible to leaching. This is due to its high solubility as well as its negative charge. This prevents their adsorption to soil particles. In conventional systems, leaching can account for 15–30% of applied nitrogen [2]. This is particularly prevalent in sandy soils that have low cation exchange capacity or in regions where there is high rainfall. Smart nanofertilizers combat this challenge through various mechanisms. These so-called control-release mechanisms sustain the nutrients in the root zone for extended periods. The gradual release pattern ensures that nutrient concentrations in soil solution remain moderate, preventing the accumulation of excess soluble nutrients that would otherwise leach during irrigation or rainfall events. Field studies have documented that leaching loss is reduced by 30–70% when nanofertilizers substitute conventional fertilizers. The magnitude of reduction, however, depends on the soil type, climate and crop system [122].

Volatilization losses primarily affect nitrogen fertilizers. The most common ones are urea- and ammonium-based compounds. They can convert to gaseous ammonia under alkaline soil conditions or high temperatures. These losses have been found to range from 10 to 20% of applied nitrogen in conventional systems [83]. It represents both economic waste and atmospheric pollution. The encapsulation of nitrogen within nanocarriers creates a physical barrier that substantially reduces volatilization rates. Polymer-coated nano-urea and zeolite-encapsulated ammonium formulations have documented to achieve volatilization reductions exceeding 50% compared to conventional applications. Additionally, the controlled-release nature of nanofertilizers allows for gradual nitrogen availability. This helps to prevent the sudden spikes in soil ammonia concentration, which are known to be the main drivers of volatilization losses [49].

Surface runoff and erosion transport nutrients from the agricultural fields to the water bodies. This is a contributing factor towards eutrophication, which in turn leads to the degradation of aquatic ecosystems. Phosphorus is particularly susceptible to surface losses due to its strong adsorption to soil particles that can easily be mobilized during events like runoff and or erosion. Conventional phosphorus fertilizers often exceed their requirements, which leads to their accumulation in surface soils [107]. Due to this, they become vulnerable to runoff. Nano-phosphate fertilizers improve the retention of phosphorus in the soil by enhancing their adsorption characteristics and their controlled release. Their nanoscale size allows them to better integrate with the soil matrices, and their controlled release prevents accumulation at the soil surface. Research has shown that nano-phosphate formulations can reduce surface phosphorus losses by 40–60% compared to the application of conventional phosphate fertilizers like superphosphate or diammonium phosphate [79].

The cumulative effect of reduced nutrient losses through leaching, volatilization and runoff translates into environmental benefits. Decrease in nitrogen leaching keeps the groundwater quality in check and thereby reduces the risk of nitrate contamination in drinking water supplies. Decrease in volatilization losses means decrease in the atmospheric ammonia concentrations, which reduces its contributions to acid rain and particulate matter formation. Minimization of runoff prevents eutrophication of surface waters and protects aquatic ecosystems and biodiversity. These strategies of environmental protection align directly with sustainable farming principles and demonstrate how nanofertilizer technology can make agricultural productivity go hand in hand with maintaining ecological balance [72].

4.3 Maintaining Soil Fertility and Structure

Soil health represents the foundation of sustainable agricultural systems, which encompass physical, chemical and biological properties. It is not only essential for crop production but also for the maintenance of ecosystem functions. Conventional fertilizer application practices, when applied intensively over extended periods, have the capacity to degrade the quality of soil quality. This is done in many ways that include acidification, salinization, organic matter depletion and disruption of microbial communities. Smart-delivering nanofertilizers have significant advantages in maintaining and enhancing the soil fertility and structure [111].

There is a significant reduction in the application rates of fertilizers due to the use of nanofertilizers. Their high efficiency weakens the accumulation of salts that contribute to soil salinization. Conventional fertilizers often contain large concentrations of chloride, sulphate and other salt-forming ions that accumulate in soil profiles. This is particularly prevalent under irrigation systems where leaching is inadequate. By delivering nutrients more efficiently, nanofertilizers allow farmers to reduce total fertilizer inputs by 20–40%, all the while maintaining or improving crop yields. This consequently reduces salt accumulation in agricultural soils [100].

Soil acidification results from excessive use of nitrogenous fertilizers, especially ammonium-based products, which is a threat to soil fertility in the long term. During nitrification, ammonium is converted to nitrate, and this process releases hydrogen ions. This, in turn, gradually lowers the soil pH to levels that lower the nutrient availability and also impair the activity of beneficial microbes. The moderate ammonium release rates of Nano-encapsulated nitrogen fertilizers enable their gradual nitrification and thereby reduce the level of acidification. Some nanofertilizer formulations also include pH-buffering components in their carrier matrices, providing additional protection against acidification [128].

The physical structure of soil, such as aggregate stability and porosity, is a vital component for water infiltration, root penetration and aeration. Indiscriminate application of conventional fertilizers disrupts soil aggregation through mechanisms like sodium-induced clay dispersion and accelerated organic matter decomposition. Nanofertilizers help preserve soil structure through a reduction in the volume of fertilizer that needs to be applied and also enables a balanced nutrient delivery in support of natural soil-forming processes. Biodegradable nanofertilizers, which are made from organic polymers, chitosan or lignin, also contribute to organic matter as they are degraded. This enhances the aggregate stability and helps water retention [11].

Soil microbial communities are essential when it comes to nutrient cycling, organic matter decomposition and the suppression of diseases. High rates of conventional fertilizers can shift microbial composition towards fast-growing copiotrophic organisms. But this is done at the expense of diverse oligotrophic populations. The gradual, balanced nutrient supply from nanofertilizers aids in the promotion of microbial diversity by avoiding the ‘feast-famine’ cycles that are an outcome of conventional fertilization. Studies have indicated that nanofertilizer management maintains higher microbial biomass and greater functional diversity. This is a reflection of improved biological soil health [7].

4.4 *Minimizing Environmental Pollution and Greenhouse Gas Emissions*

The environmental footprint of conventional fertilizer use extends beyond direct nutrient losses. It also has broader impacts on air quality, climate change and the overall health of the ecosystem. Nanofertilizers have potential for mitigating these environmental concerns. In turn, they contribute towards sustainable agricultural systems and reduce the detrimental impacts on ecology [126].

One of the major contributors to global warming is greenhouse gas emissions from agricultural systems. Nitrous oxide (N_2O), in particular, is said to pose a critical environmental threat. The global warming potential is approximately 300 times greater than that of carbon dioxide, and it significantly contributes to ozone depletion in the stratosphere [69]. Agricultural soils are the primary anthropogenic source of N_2O. Their emissions are mainly driven by microbial nitrification and denitrification processes. These processes are intensified when nitrogen is highly available. Conventional nitrogen fertilizers create conditions conducive to N_2O production through localized zones of high nitrogen concentration, which are combined with anaerobic microsites [103].

Nanofertilizers lower emissions of N_2O in several ways. Controlled release eliminates the hotspots of nitrogen, which are a cause of denitrification, keeping relatively even values of nitrogen in soils. Studies proved that using nitrogen fertilizers encapsulated in nanoparticles reduces emissions of N_2O by 30–50% in comparison with using conventional urea fertilizer [49–51]. Together with enhanced efficiency, which decreases total nitrogen use, overall reductions of more than 60% of total emissions of N_2O can be obtained, which has a great potential for mitigating climate change [60].

The industrial processing and transportation of conventional fertilizers, especially those with a high amount of nitrogen, produced using the energy-demanding Haber-Bosch process, are major contributors to carbon emissions in agriculture. Indirectly, nanofertilizers help in mitigating carbon emissions, as they make it feasible to use lower amounts of fertilizer without adversely affecting their productivity. Studies on life cycle analyses of nanofertilizers indicate that even with their increased production costs, their environmentally adverse effects are reduced [86].

Another important aspect of the environmental advantages of nanofertilizers is the protection of water quality. Nitrogen and phosphorus pollution of surface and groundwater resources poses a severe environmental problem worldwide. Nutrient-based eutrophication of water bodies contributes to toxic algae blooms, 'dead' zones, and a decrease in aquatic biodiversity. Nitrate pollution of freshwater sources poses a severe hazard to human health. The substantial decrease in nutrient pollution of groundwater and surface water, which results from the application of nanofertilizers, contributes significantly to providing superior water quality [14].

4.5 Supporting Precision and Climate-Smart Agriculture

The development of sustainable agriculture has become reliant on precision farming techniques, which optimize resources using site-specific management decisions informed by data. Smart delivery nanofertilizers are a technology that would ideally be used in conjunction with a precision agriculture management system, which would provide unprecedented control over fertilizer delivery in terms of space and time [124].

Precision agriculture involves using sensors, remote sensing, geographic information systems and variable rate application technology to provide insight into site-level variability of soils, crop conditions and yields. This would help farmers target the application of inputs where needed, without applying where there are sufficient nutrients. Nanofertilizers would, in this regard, find application in this technology as a means of targeted nutrient delivery that has been determined by precise crop monitoring [96].

The integration of nanofertilizers with digital platforms in agriculture opens opportunities for the development of intelligent management systems for fertilizers that are intelligent. These systems would help slow down the application of nanofertilizers if there are forecasts of heavy rain in order to cut down on leaching. Another area would involve speeding up nutrient release when crops are growing in periods when nutrients are most needed. These smart technology approaches in agriculture are areas that are part of the sustainable intensification of agriculture [34]. The role of smart delivery nanofertilizers in sustainable agriculture, therefore, has a wide range of dimensions that cover important issues related to efficiency in resource use, environmental conservation, soil management and strategies for mitigating the effects of climate change. Increasing pressures from a growing population, declining resources and a changing environment pose serious threats to agricultural systems, which make it important for farmers to take advantage of new technology that provides alternatives for developing a sustainable food chain [61].

5 Applications and Case Studies

5.1 Nanofertilizers in Cereal Crops (Rice, Wheat, Maize)

Sahoo et al. [89] experimented to evaluate integrated nano-formulated DAP applications on rice growth, yield and nutrient use efficiency compared with conventional fertilization [89]. The experimental design was RBD with three replications in total.

Results revealed that treatment T_{10} consisting of 50% of the soil test recommended dose (STD) of N and P, seedling root dipping (SRD) with nano-DAP at 5 mL L^{-1} along with two foliar sprays (FS) of nano-DAP at 4 mL L^{-1} at 25 and 45 DAT demonstrated highest growth parameters which included plant height, panicle length, number of tillers $hill^{-1}$, total grains $panicle^{-1}$ and root length. Although the treatment of 100% STD (T_2) showed maximum mean grain yields (4.12 and 4.05 Mg ha^{-1}) and nutrient uptake, these values were statistically comparable to those of T_{10}. T_{10} also recorded the highest benefit-cost ratio (2.26) 3.5% higher than T_2. Moreover, recovery efficiency varied from 30 to 94.2% (N) and 11.2 to 90.4% (P) and agronomic use efficiency for N and P ranged from 7.5 to 31.5% and 15 to 63%, respectively, post-harvest soil pH and available N and P were significantly higher in T_{10}, while soil MBC, MBN, MBP, urease and phosphatase activities were greatest in T_2, followed by T_{10}.

Kashyap et al. [45] conducted a study to evaluate innovative nitrogen (N) and zinc (Zn) management strategies using a factorial RBD which included four treatments of N control, 100% recommended nitrogen dose (RDN) through urea, 50% RDN through urea + two nano-urea sprays at the six- and ten-leaf stages, and 33.33% RDN through urea + two nano-urea sprays at the same stages and a total of four Zn treatments which are control, soil application of $ZnSO_4$, foliar application of $ZnSO_4$ and foliar application of nano-Zn. A total of three replications were taken. Findings indicated that applying 50% RDN through urea combined with two foliar sprays of nano-urea produced optimum yields, which were comparable to the full 100% RDN applied solely through urea fertilizer. For Zn management, application of $ZnSO_4$ as both soil and foliar applications demonstrated similar green and dry fodder yields, improving fodder quality was recorded in foliar application. Moreover, integration of optimized nitrogen and zinc strategies increases protein content and reduces fibre fractions in maize [45].

Mohsen et al. [63] conducted a field experiment to evaluate the effect of four wheat varieties, *Baraka*, *Tammuz-2*, *Latifia* and *Ibaa*-95, respectively, on different concentrations of Fe nanofertilizer (0, 1, 2 and 3 g L^{-1}). The research was based on factorial arrangement within a randomized complete block design (RCBD) with a total of three replications. Results showed that the *Baraka* variety surpassed others in both grain yield and biological yield, reaching 4.892 and 12.943 tons ha^{-1}, respectively, which can be attributed to its efficient and higher growth traits and yield components. All measured characteristics were influenced significantly by Fe nanofertilizer (3 g L^{-1}), producing the highest mean grain yield (4.966-ton ha^{-1}). In terms of interactions, Baraka treated with 3 g L^{-1} Fe nanofertilizer showed the maximum grain yield (5.800-ton ha^{-1}), demonstrating its enhanced yield parameters under the same treatment [63].

5.2 Horticultural Applications (Fruits, Vegetables, Ornamentals)

Davarpanah et al. [23] experimented on the yield and fruit quality of pomegranate (*Punica granatum* cv. Ardestani) via foliar nanofertilizers of zinc (Zn) and boron (B) using a factorial experiment in a completely randomized block design. A total of nine treatments and four replications were taken into consideration. 0, 60, 120 mg Zn L^{-1} of Nano-Zn chelates and 0, 3.25, 6.5 mg B L^{-1} of nano-B chelate were supplied only once before full bloom at 5.3 L $tree^{-1}$. By August, both nutrients enhanced leaf Zn and B concentrations, indicating better tree nutritional status. Moreover, enhanced fruit yield, highly through an increase in fruit number in treatments of lower application rates (about 34 mg B $tree^{-1}$ and 636 mg Zn $tree^{-1}$), with B showing a higher influence than Zn. The highest levels of Zn and B effectively improved several quality attributes, producing 4.4–7.6% increases in total soluble solids, 9.5–29.1% reductions in titratable acidity, 20.6–46.1% increases in maturity index and 0.28–0.62 unit rises in juice pH, while physical fruit traits remained unaffected. Only slight changes were observed in total sugars and total phenolics, and neither antioxidant activity nor total anthocyanin content was meaningfully altered by the treatments [23].

El-Desouky et al. [28] studied the impact of nano-based iron fertilizer application as a precision strategy to increase greenhouse tomato (*Solanum lycopersicum* cv. Bigdena F1) production in an environmentally sustainable and economical manner. The design of the experiment was factorially randomized complete block design with a total of four replications to compare a total of three Fe sources—conventional Fe ($FeCl_3{\cdot}6H_2O$; Conv-Fe), chelated Fe (6% Fe; Che-Fe) and nano-Fe_2O_3 (Nano-Fe; alpha, 99%, 30–50 nm) that was applied at 0, 50 and 100 mg Fe kg^{-1} soil. Across Fe application rates, Nano-Fe consistently increased plant growth parameters, including height, leaf number and area, and both shoot and root fresh and dry biomass, along with increased SPAD values. The 100 mg kg^{-1} Nano-Fe treatment produced the strongest physiological responses, significantly increasing photosynthetic rate (Pn), stomatal conductance (gs), transpiration (E) and overall yield and yield components relative to other Fe sources and rates. Although simple and interaction effects of Fe on physiological parameters were not statistically significant, Nano-Fe maintained superior Pn with lower E and higher gs. At 100 mg kg^{-1}, Nano-Fe also resulted in the greatest improvements in fruit diameter, fruit number per plant, total fruit weight per plant, average fruit weight and corresponding yield metrics per hectare, followed by Nano-Fe at 50 mg kg^{-1}. It was concluded that Nano-Fe outperformed other Conv-Fe and Che-Fe during both seasons by increasing photosynthate accumulation and translocation, resulting in higher tomato yield by approximately 11% and offering sustainable advantages for grower profitability [28].

Srisha and Prasad [106] experimented to study the response of gladiolus (*Gladiolus grandiflorus* L.) under different doses of NPK and nanofertilizer application. Randomized block design (RBD) was the experimental design with a total of eight different treatments and three replications. Findings revealed prominent variation among treatments, with the cultivar named American Beauty performing particularly well. Maximum plant height of 116.22 cm, leaf number (14.0), early spike emergence (76.8 days), first flower opening (118.44 days), 50% floret opening (106.88 days), 50% sprouting (19.73 days), spike length (72.33 cm), spike weight (93.81 g) and floret diameters (6.27 and 6.01 cm) were demonstrated by treatment T_5, nano-iron at 0.25 g L^{-1}. The same treatment also resulted in the highest number of corms per plant (2.33) and per plot (20.31), longest vase life (11.5 days), most spikes per plant (1.83) and greatest number of florets per spike (12.77). Economically, T_5 generated the highest net income of ₹1,563,017 ha^{-1}, gross return of ₹2,187,165 ha^{-1} and benefit-cost ratio of 1:2.5, confirming nano-iron at 0.25 g L^{-1} as the most effective treatment for enhancing gladiolus growth, productivity and commercial profitability [106].

5.3 *Integration with Organic and Biofertilizers*

Al-Waeli and Al-Janabi [10] evaluated the effects of biofertilizers, organic fertilizers and nanofluids on a French black grape cultivar in a private orchard in the Al-Hamzah AlGharbi region of Babylon province using a three-factor, three-replicate randomized complete block design (RCBD). The study applied Bacillus megaterium and Mycorrhiza glomus at 0- and 10-ml L^{-1} to the soil; liquid organic fertilizer at 0-, 15- and 30-ml L^{-1}; and foliar-applied organic nanofluid at 0.1- and 3-ml L^{-1} throughout 2021. Data were analysed using GenStat with the least significant difference (L.S.D.) test at 5% probability. The experiment included 54 vines arranged randomly across three sectors. However, results revealed that soil biofertilization at 10 ml L^{-1} promptly enhanced most quantitative yield traits, increasing total yield to 36.2–38.6 kg $vine^{-1}$. Application of organic nanofertilizer (foliar spray) at 3 ml L^{-1} further enhanced the average number of clusters per vine to 51.6–53.9 and showed the highest total yield of 37.0 kg $vine^{-1}$. Additionally, soil-applied organic fertilizer at 30 ml L^{-1} positively affected most yield parameters, producing total yields of 36.6–37.5 kg $vine^{-1}$, indicating that integrated bio- and nano-fertilization strategies can properly improve grape productivity [10].

Dahham et al. [21] experimented on cucumber hybrids to evaluate their growth characteristics and productivity under protected cultivation using organic fertilization, bio-organic fertilizer and foliar application of a nano-organic extract (Nano Super Fifty). A factorial experiment (4 × 3) was conducted for the research using a split-plot design in a randomized complete block design (RCBD) with a total of three replicates. Three cucumber hybrids (Hybrid 182, Hybrid Falcon Star RZ and Hybrid 62 of betaalpha type) were considered under the first factor, and the second factor included four fertilization treatments (control, bio-organic fertilizer, Nano Super Fifty spray and a combination of bio-organic fertilizer with Nano Super Fifty). Results demonstrated that Hybrid 62 excelled in leaf width (1493 cm^2 $plant^{-1}$); fruit number (36.18 fruits $plant^{-1}$); total plastic house yield (4.427 tons $house^{-1}$); leaf chlorophyll content (42.72%); andaverage fruit weight (85.35 g), plant yield

(3.162 kg plant^{-1}), fruit protein content (4.392%), while Hybrid 182 exhibited the highest fruit nitrogen and phosphorus contents (1.279% and 0.328%, respectively). Among fertilization treatments, the integration of bio-organic fertilizer along with Nano Super Fifty significantly overall improved leaf width (1714 cm^2 plant^{-1}); fruit number (41.28 fruits plant^{-1}); average fruit weight (91.67 g); plant yield (3.028 kg plant^{-1}); leaf chlorophyll content (43.77 mg/100 g fresh weight); total plastic house yield (5.325 tons house^{-1}); and fruit phosphorus, potassium, and protein contents (0.372%, 3.047%, and 4.481%, respectively), whereas bio-organic fertilizer alone resulted in the highest fruit nitrogen content (1.293%). Interaction effects indicated that the combination of Hybrid 62 with bio-organic fertilizer plus Nano Super Fifty maximized growth, yield and quality traits, while Hybrid 182 with bio-organic fertilizer alone recorded the highest fruit nitrogen content [21].

Al-Waeli and Al-Janabi [10] conducted a study on a French black grape cultivar using a factorial randomized complete block design (RCBD) with a total of three replicates. The experiment assessed the impact of three factors: soil-applied biofertilizers (*Bacillus megaterium* and *Mycorrhiza glomus*) at 0- and 10-ml L^{-1}; liquid organic fertilizer at 0-, 15- and 30-ml L^{-1}; and foliar-applied nano-organic fertilizer at 0-, 0.1- and 3-ml L^{-1} on a total of 54 grapevines. Findings revealed that biofertilization at 10 ml L^{-1} significantly enhanced many chemical traits, including leaf phosphorus content (0.43–0.44%), total acidity (0.74–0.99%), TSS/TA ratio (27.6–32.5%), TSS/TA ratio (27.6–32.5%), total acidity (0.74–0.99%), berry juice total soluble solids (TSS; 14.46–15.19%), total acidity (0.74–0.99%) and reducing sugars in juice (7.24–8.02%). Moreover, soil-applied organic fertilizer at 30 ml L^{-1} positively impacted leaf phosphorus (0.392–0.475%) and enhanced the TSS/TA ratio in juice (30.3–37.6%) for both seasons, indicating that a combination of both bio- and organic fertilization can promptly enhance fruit quality of grapevine [10].

Kumar et al. [49–51] examined the impacts of organic manures, nano-nitrogen and nano-diammonium phosphate (DAP) on tomato and cauliflower cultivation. The results showed that mixing organic manure, biofertilizers and nanofertilizers had a synergistic impact, which led to yield increases of 12.6–14.9% in tomatoes and 12.2–15.4% in cauliflower. There were also clear economic benefits, with net returns going up by 25.8–31.7% in cauliflower and 21.4–25.0% in tomato compared to control treatments. Better growth factors, like plant height, number of main branches and average fruit weight, led to increased productivity. Soil tests after harvest showed higher levels of organic carbon, NPK and a lower C:N ratio, which means the soil is healthier. These results show that nano-nitrogen and nano-DAP are effective at increasing both the quality and quantity of crops, as well as the overall profitability of a farm [49–51].

6 Benefits of Smart-Delivering Nanofertilizers

6.1 *Agronomic Benefits (Yield, Quality, Stress Tolerance)*

Hamoda [39] experimented to study the impact of foliar spraying with nano-nitrogen fertilizer (with doses of 0, 250 and 500 mg/L) and salicylic acid (with doses of 0, 100 and 200 mg/L) on yield attributes of wheat. The experiments were conducted in a randomized complete block design (RCBD) using split-plot arrangement with a total of three replications. Findings demonstrated that nano-N had an expectational influence on enhancing yield attributes like the number of tillers/m^2; number of grain/spike; and grain, straw and biological yield of wheat, which was compared to the control treatment ($p < 0.05$), and was observed under high concentration (500 mg/L). The highest positive effect of salicylic acid (200 mg/L) application on the yield attributes of wheat was observed. Moreover, the interaction effect between nano-N at 500 mg/L and SA at 200 mg/L significantly increased the straw, grain and biological yield (7719, 6648 and 14,367 kg/ha, respectively) relative to control and/or other treatments ($p < 0.05$), whereas the maximal harvest index was observed for untreated plants [0 (nano-N) x 0 mg/L (SA)]. Grain yield had positive/direct interrelationship with no. of tillers/m2 ($r = 0.75$), spike length ($r = 0.85$), no. of grains/spike ($r = 0.69$) and no. of spikelet's ($r = 0.70$), under correlational analysis demonstrated that whereas the increase of plant height strongly contributed to the enhancement of biological yield by 86% [39].

Micronutrients and macronutrients play key roles in the defence of plants against disease. Using NPs, the nutritional status of plants can be increased to improve yields, tolerate stress and withstand attack from diseases. When utilized in agriculture, these nanoparticles display potential for boosting crop yield amid varied biotic and abiotic stress conditions [9]. Superior water retention capacities were observed for the nanofertilizer-enriched soil as compared with the control treatment, indicating that such soils can effectively help to gain more water for crops and delay the beginning of water stress. Numerous agricultural researchers also concentrated on the uses of nanotechnology, reporting significant characteristics such as improved abiotic and biotic stress resistance, greater plant uptake of nutrients and increased agricultural productivity [120]. It was shown that using PGPR to reduce crop growth and yield is sustainable in agriculture. For example, applying chitosan particles to immobilize *Methylobacterium oryzae* has improved tomato crop growth under salt stress [18]. Additionally, Zn oxide nanoparticles lessen the effects of stress on plants, such as those caused by salt, drought and cadmium [114].

6.2 *Economic Benefits (Cost-Effectiveness, Reduced Inputs)*

Technological improvements may enhance the production of agro-industrial, physiological and agronomical essential metallic NPs for making fertilizers with reduced nutritional losses and enhanced NUE with smart delivery systems [32]. Reduce nitrogen losses due to leaching and volatilization with nano-encapsulated nitrogen fertilizers, which ensures steady supply of nitrogen to crops. Moreover, nano-phosphorus fertilizers also enhance phosphorus solubilization and uptake, promoting proper root growth and higher crop yields [8].

Sun et al. [109] experimented to develop a high-performance nano-clinoptilolite-based N fertilizer capable of partially replacing urea via enhanced small-size properties of nano-clinoptilolite, particularly its increased meso porosity and elevated Si/Al ratio. The design of the experiment was a randomized complete block design, which included five treatments: 100% urea (CK) and nano-clinoptilolite-urea mixtures of 20%:80% (Z2N8), 30%:70% (Z3N7), 40%:60% (Z4N6) and 50%:50% (Z5N5). By analysing the data of both years, Z2N8, Z3N7, Z4N6 and Z5N5 decreased ammonia volatilization (AV) by 8.57%, 20.52%, 30.20% and 37.13%, respectively, and reduced N runoff losses by 23.29, 29.93, 39.66 and 43.76%. Moreover, they also enhanced N accumulation of the whole plant by 24.32, 16.84, 9.00 and 4.85%, while boosting rice yield by 15.28, 10.28, 6.99 and 5.05%. Although nano-clinoptilolite increased the N utilization efficiency, as the substitution ratio increased, its effectiveness declined. The surplus was reduced by 48.77%, 25.84% and 3.17% under Z2N8, Z3N7 and Z4N6, respectively, but increased by 9.61% under Z5N5 relative to CK. Moreover, economically, Z2N8 enhanced the income of the farm by 14.75%, lowered environmental damage costs by 8.77% and improved both net economic benefits (5.33%) and combined economic-ecological benefits (5.75%) [109].

Beeresha and Jayadev [15] studied the effects of nano-potassium (Nano-K) fertilizer on the growth and yield of maize crop grown in red sandy loam soil. The study was conducted in RCBD with a total of ten treatments and three replications. BRMH-1 was used as a cultivar. Findings revealed that soil application of 15 kg ha^{-1} K_2O in combination with foliar spraying of Nano-K at 2500 ppm gave significantly superior yield attributes, including greater cob length (17.9 cm), kernel weight cob^{-1} (188.4 g), test weight (34.8 g), cob girth (17.2 cm), rows cob^{-1} (16.8), kernels row^{-1} (41.3), kernels cob^{-1} (694.8), kernel yield (9051 kg ha^{-1}) and straw yield (11,667 kg ha^{-1}). The same treatment also resulted in high economic returns (gross: Rs. 138,330 ha^{-1}; net: Rs. 88,240 ha^{-1}), while the highest benefit-cost ratio (4.02) was observed under the recommended dose of fertilizer (RDF) [129].

Rajesh et al. [82] experimented to study the response of foliar-applied nano-N and nano-Zn on the overall growth and yield of sweet corn using RCBD comprising studied 13 different nutrient treatments including chemical nano-N, chemical nano-Zn, green nano-Zn, recommended fertilizer doses and an absolute control, revealed that applying 75% N + 100% P & K, combined with foliar sprays of chemically synthesized nano-N (4 mL L^{-1}) and nano-Zn (2 mL L^{-1}) at 25 and 50 DAS, produced the highest fresh cob yield (158.1 q ha^{-1}) and green fodder yield (214.5 q ha^{-1}), representing an 18% increase over the recommended NPK dose (150:75:37.5 kg ha^{-1}). The same treatment also significantly improved plant height (225.7 cm), SPAD (61.74), leaf number (12.60), leaf area (7130 cm^2 $plant^{-1}$), leaf area index (5.94), dry matter accumulation (295.2 g $plant^{-1}$), NDVI (0.80) and N uptake (246.2 kg ha^{-1}). Moreover, in economical perspective, it showed the highest gross returns (₹1,26,503 ha^{-1}), net returns (₹77,928 ha^{-1}) and a beneficial B:C ratio (2.60) [82].

Abd El-Azeim et al. [3] experimented to evaluate whether NPK nanofertilizers, applied at equivalent or reduced rates, can properly replace the recommended doses of conventional NPK fertilizers in the production of potato without compromising yield or quality. Compared to the conventional fertilizer control, NPK nanofertilizers (foliar application) at 50% or 25% significantly increased economic yield (23.59 t ha^{-1}); starch content (79.62%); nutrient use efficiency for N, P and K (67.74, 278.92 and 118.54 kg potato kg^{-1} nutrient, respectively) and harvest index (59.24%), moreover, while reducing tuber nitrate content (1.15 g kg^{-1}). Among all treatments, the greatest economic return, producing the highest yield, improved quality and maximum profit-to-cost ratio was shown by the treatment of foliar NPK nanofertilizers at 50% of the recommended rate. The experimental findings revealed that foliar-applied nanofertilizers at minimum doses offer an environmentally promising and economically viable alternative source of fertilizer compared to full-rate chemical fertilizers, while also enhancing potato productivity, as well as lowering chemical inputs [3].

6.3 Environmental Benefits (Soil Health, Biodiversity Conservation)

Jia et al. [43] experimented to examine the environmental implications of nano-hydroxyapatite (nano-HA), by growing maize crop for a duration of 4 weeks in soil amended with 150 mg P kg^{-1} soil. They also evaluated plant responses, soil properties and microbial dynamics. HA significantly increased maize fresh biomass, content of chlorophyll and bioavailable P, while also improving the quality of soil, as evidenced by a 12.7-fold increase in bioavailable P, a 28.7% rise in EC and a 15.6% increase in SOM. GC-MS-based soil metabolomics revealed elevated levels of low-molecular-weight metabolites, including sugars and sugar alcohols (arabinose, L-arabitol, palatinitol, sorbitol), amino acids (L-/D-tyrosine) and phenolic acids (catechin, scopoletin). Although microbial richness remained unchanged, 16S rRNA sequencing showed a shift in bacterial community structure, with increases in plant growth-promoting rhizobacteria (PGPR), particularly P-solubilizing groups such as *Burkholderiaceae*, *Massilia*, *Sinomonas* and *Streptomyces* [43].

Upadhyay [116] investigated the effectiveness of nano-N (nano-urea), nano-Zn and nano-Cu applied at different N levels [0, 50, 75 and 100% of the recommended rate of N (RRN)] in maize-wheat and pearl millet-mustard systems during 2019–20 and 2020–21. The study demonstrated that foliar application of nano-N + nano-Zn with 100% RRN produced the highest grain yields-maize (66.2–68.8%), wheat (61.9–62.6%), pearl millet (57.1–65.4%) and mustard (47.2–69.0%), surpassing both the absolute control and treatments combining all three nanofertilizers (nano-N + nano-Zn + nano-Cu), largely due to greater crop uptake of N and Zn. Moreover, 75% RRN + nano-N + nano-Zn responded with higher yields comparable to 100%

RRN, indicating that foliar nano-N and nano-Zn, when applied along with conventional urea, can help to reduce 25% N fertilizer requirements without compromising overall productivity. Similar trends were observed for profitability, soil mineral N, dehydrogenase activity (DHA) and soil microbial biomass C (SMBC) for both years [116].

Tao et al. [113] conducted the first systematic assessment of zinc-carbon dot nanofertilizers (Nano-ZCDs) on Bok choy (*Brassica chinensis* L.) growth and soil health under multi-batch, consecutive foliar applications across three cultivation cycles. Compared with the control (CK), Nano-ZCDs markedly increased fresh biomass by 75.5 ± 1.8%, 75.1 ± 0.2% and 74.6 ± 0.4% over the three successive harvests, while also enhancing nutritional attributes, including amino acids, vitamin C, soluble sugars, proteins and Zn accumulation. In the rhizosphere, Nano-ZCDs significantly improved soil water content, redox potential (Eh), soil organic C and available K and P. Furthermore, Nano-ZCD treatment increased bacterial community richness, stability and network complexity. According to the Cornell Soil Health Assessment, the soil health index in Nano-ZCD-treated soil was 8.1% higher than CK. Overall, the study demonstrates that repeated foliar application of Nano-ZCDs consistently promotes Bok choy productivity and nutritional quality without compromising and indeed enhancing soil health [113].

Das and Ghosh [22] experimented to create slow-release nanofertilizers by precisely impregnating four different biochars with micronutrient and secondary solutions through a three-step synthesis process to evaluate the suitability and response of biochar as a support matrix for controlled nutrient delivery. Transmission electron microscopy confirmed the nanoscale structure of the synthesized materials (8.33–24.33 nm), which possessed near-neutral pH values (5.94–6.61) and appropriate C, H, N and O contents. High water absorbency (32.46–45.76%), low density (0.731–0.762 g cm^{-3}), low salt index (0.04–0.07), good swelling capacity (4.15–4.61%) and high equilibrium water content (68–74%) were observed under nanofertilizers. After 45 days, they efficiently enhanced soil water retention capacity by 42.47–53.16%, which was far more effective than commercial compound fertilizer (22.43%) or untreated soil (19.52%), resulting in strong potential for water stress conditions. Slower release of Ca^{2+}, Mg^{2+}, Zn^{2+}, $Fe^{2+/3+}$ and Na_2O compared to commercial fertilizers were revealed in nutrient release studies; after 90 days, cumulative release reached 87.4–90.7% for Ca^{2+}, 96.2–99.1% for Na^+, 2.8–5.2% for Fe^{2+}, 81.8–85.3% for Mg^{2+} and 0.5–1.6% for Zn^{2+}, indicating the leaching hierarchy $Na_2O > Ca^{2+} > Mg^{2+} > Fe^{2+/3+} > Zn^{2+}$. Pine needle biochar exhibited the highest retention of Ca^{2+}, Mg^{2+}, $Fe^{2+/3+}$ and Na_2O among all the biochar treatments, whereas superior Zn^{2+} sorption was demonstrated by *Lantana camara* biochar. These results properly highlight the strong influence of feedstock type on nutrient release behaviour and also confirm the potential of biochar-based nanofertilizers to reduce nutrient leaching, improve soil water retention and enhance overall nutrient and water-use efficiency [22].

6.4 *Social Benefits (Food Security, Sustainable Livelihoods)*

The goal of employing nanomaterials (NMs) in the field of agriculture is to increase the efficiency and sustainability of agricultural processes by putting less input and generating less waste than current goods and procedures.

The precise function of nanofertilizers is to effectively enhance nutrient use efficiency as well as plant productivity by properly facilitating controlled and accurate nutrient delivery. They significantly reduce nutrient losses from different factors like leaching, volatilization or runoff, ensuring that essential elements such as nitrogen, phosphorus and potassium remain properly accessible to crops during their critical growth period. This precisely targeted application, therefore, enhances germination rates, root growth, chlorophyll levels in leaves and overall crop yield with higher productivity, while minimizing environmental pollution caused by conventional fertilizers. Nanoscale science and nanotechnology have the ability to alter the agriculture and food systems [71]. They enhance the efficiency of nutrient absorption, thereby directly supporting food security. Increased and steadier crop production on current farms addresses the dietary requirements of expanding populations without the need to expand agricultural areas. They enhance crop resilience to issues like drought, salinity or nutrient-poor soils, ensuring stable yields in vulnerable areas.

As a result, communities facing soil infertility or extreme weather conditions can yield larger crops, reducing the likelihood of hunger and malnutrition. It also provides significant economic advantages for farmers and rural livelihoods. Lower fertilizer needs decrease input expenses, while enhanced and steadier yields boost income reliability and market competitiveness. Enhanced crop quality can open doors to high-end markets, thus boosting family economies. The ecological advantages, such as diminished soil and water pollution, ensure sustainable agricultural practices, preserving essential natural resources crucial for future rural growth. It offers huge potential in agriculture upsurge, high reactivity, improved bioavailability, bioactivity and the surface impacts of NPs [42].

The application of NFs through soil irrigation ensures double advantages, i.e. soil improvement to optimize plant development productivity [56]. Furthermore, incorporating nanofertilizers into agricultural practices promotes sustainable farming by optimizing resource use and minimizing environmental harm. They conform to climate-smart agricultural practices by reducing greenhouse gas emissions associated with heavy fertilizer application. Ultimately, this fosters resilient agroecosystems, ensuring that both current and future generations can depend on fertile land for sustenance and income of the farming community.

7 Challenges and Risks

Although the prospect of smart-delivering nanofertilizers to transform the concept of sustainable agriculture is enormous, the safe and expansive adoption of this technique is largely dependent upon the successful surpassing of numerous scientific, regulatory, economic and social challenges. It is precisely due to the properties of nanoscale size, high reactivity and bio-interaction ability that make them effective that complex and largely unexplored risks exist that must be scrutinized rigorously. There is a need to have a cautious, risk-sensitive approach to avoiding unintended environmental or human health impacts that might derail the long-term sustainability objectives of the technology.

7.1 Ecotoxicity and Bioaccumulation Concerns

The most significant scientific problem is the overall evaluation of the ecotoxicity and environmental fate of engineered nanomaterials (ENMs) and their remnants. Due to their fine size and large surface-to-volume ratio, nanoparticles may have new biological and chemical behaviours different from those of their bulk counterparts. Concerns include the following.

7.1.1 Phytotoxicity

Research has demonstrated that, although numerous formulations of nanofertilizers have shown to be advantageous, concentrations or certain formulations of nanofertilizers (e.g. zinc oxide, titanium dioxide nanoparticles, etc.) can lead to concentration-dependent growth-inhibition and phytotoxicity in some crops or non-target plants. The type of formulation, size and surface coating of the ENM have a significant impact on its interaction with the plant tissues and soil microbes [126].

7.1.2 Soil and Water Contamination

When nanoparticles are incorporated into the soil, they could undergo transformation, aggregation or movement through the soil profile, which may lead to contamination of groundwater and surface water. The long-term implications of soil contamination on microbial communities in the soil, which are crucial in soil fertility, nutrient cycling and plant health, are yet to be well understood, although preliminary studies have shown that they can be disrupted [48].

7.1.3 Bioaccumulation and Food Chain Transfer

It accumulates in the food chain and undergoes biomagnification by transferring through the bioaccumulation process and food chain. The possibility of the bioaccumulation of nanoparticles in edible portions of plants and, consequently, into the food chain would result in the exposure of livestock and humans, perhaps the most significant public-health issue. Although the goal of smart delivery seeks to ensure specific uptake, the final destination of any leftover nanomaterial in the environment and the indirect consequences of the same on human health over decades require much, much more standardized analyses of the life cycle [17].

7.2 Regulatory and Policy Issues

Absence of a harmonized, nano-specific regulatory framework in the world is a major impediment to commercialization and safe uptake of smart nanofertilizers. The current laws that govern traditional fertilizers and chemicals are usually not applicable to the specific features and dangers of the products at the nanoscopic level. Major regulation issues are as follows.

7.2.1 Definition and Classification

Practically no legal standard exists as to what constitutes a nanomaterial in an agricultural product, which creates ambiguity in the necessary safety testing and labelling.

7.2.2 Risk Assessment Standards

In some cases (ex, EPA in the US or EFSA in the EU), regulatory bodies (especially) demand extensive risk testing before they approve, yet protocols to determine the toxicity, environmental transport and long-term fate of nano-agro-products continue to develop, and need international standardization [93].

7.2.3 Labelling and Public Trust

Nano-enabled products should be clearly and informatively labelled so that consumers will be able to make informed choices on product purchase. Regulatory ambiguity usually breeds social distrust and opposition to the acceptance of emerging technologies that are seen to have uncertainties about their long-term health effects. Responsible innovation can be impaired by the lack of clear safety information.

7.3 Cost and Scalability of Production

The challenge of the economic feasibility of smart nanofertilizers is a crucial bottleneck, especially in developing countries, where the necessity of a sustainable solution is the most urgent.

7.3.1 High Manufacturing Costs

Formulations that are complex and stimuli-responsive nanocarriers and nano-encapsulated preparations usually necessitate specialized equipment, accurate reaction parameters and costly raw materials. This leads to a unit cost being incurred for nanofertilizers as a higher than that of bulk conventional fertilizers [65].

7.3.2 Scaling up Challenges

Several of the synthesis mechanisms (primarily those that rely on bottom-up chemical synthesis or green synthesis) remain limited to laboratory scale. It is a complicated engineering task to effectively transfer these techniques to the industrial level of production and still achieve uniformity of particles, stability and intelligent behaviour of the particles [25].

7.3.3 Cost-Benefit Trade-Off

Although the great efficiency (NUE) and, possibly, a decrease in the overall dosage might result in economic benefits in the long term, the initial cost may be prohibitive. There are still very few, field-validated cost-benefit analyses that show net positive generation of revenue, especially on macronutrient nanofertilizers [33].

7.4 Farmer Awareness and Adoption Barriers

Despite the known scientific and economic advantages, the success or failure of any new agricultural technology depends on the end-user, who is the farmer.

7.4.1 Lack of Awareness and Training

A good deal of the farmers, especially those who are not involved in highly industrialized agricultural systems have little idea of the principles of nanotechnology, how nano-products should be handled and the time and method of application nanofertilizers should be applied foliarly (as is common) or applied to soil in a traditional way [84].

7.4.2 Need for Precision Infrastructure

The smart nanofertilizers are frequently made to perform best in a precision farming environment; this involves farmers incorporating data collected by sensors (IoT) and applying specialized application equipment. This infrastructure is a great capital outlay, and certain technical abilities that are not common everywhere are required [115].

7.4.3 Perceived Risk and Behavioural Change

The nature of farmers is risk-averse, and replacing an accustomed, albeit unproductive, traditional product with a new, high-tech one needs well-developed and localized field-trial data to show a steady payback and safety. Breaking the psychological and practical obstacles to the mass behaviour change is still a pivotal issue [76].

8 Recent Advances and Future Prospects

The nanofertilizer smart-delivering domain is evolving at an accelerated rate, shifting from proof-of-concept devices towards multi-purpose, highly specific applications. Recent advances are not only aimed at improving nutrient use efficiency (NUE) but also at integrating these nano-inputs into a fully data-driven and digital agricultural system (Fig. 2). The future potential of the technology is to provide high quality of safety, increased efficiency and smooth flexibility to meet the real-time needs of the farm.

8.1 Integration with Precision Farming and IoT

The most topical recent development is the convergence of smart nanofertilizers and precision farming (PF) with the help of the Internet of Things (IoT) and artificial intelligence (AI).

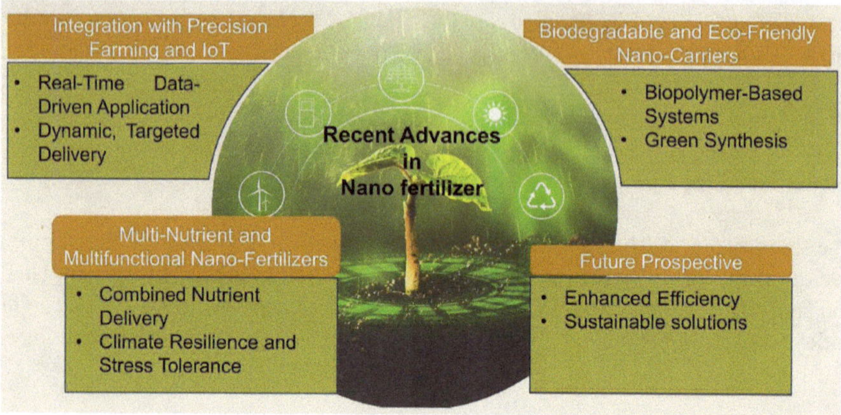

Fig. 2 An overview representation of recent advances and prospects of the nanofertilizer smart-delivering domain

8.1.1 Real-Time Data-Driven Application

Real-time data are produced by IoT sensors (measuring pH, moisture, temperature and nutrient levels in the soil) and drone/satellite cameras (monitoring the well-being of crops and their nutrient content). These big data are then processed by AI and machine-learning algorithms to create site-specific fertilization maps [81].

8.1.2 Dynamic, Targeted Delivery

These insights into the digital environment command the use of autonomous systems or drones to apply smart nanofertilizers on a variable rate basis. The nanofertilizer stimuli-response mechanism (e.g. pH or moisture sensitive release) will be best adjusted to the environmental factors measured by the sensors, and thus nutrients are only released when and where they are needed by the crop. Such a closed-loop system is more effective at maximizing NUE and minimizing environmental leakage that has been previously realized [77].

8.2 Biodegradable and Eco-friendly Nanocarriers

Responding to the severe ecotoxicity and bioaccumulation issues that are presented by some synthetic nanoparticles, it is applied with significant research effort on the production of biodegradable and biocompatible nanocarriers.

8.2.1 Biopolymer-Based Systems

Current studies focus on the use of biopolymers such as chitosan, alginate, pectin and starch derivatives as nano-encapsulators. Such natural materials offer a protective framework to the nutrient cargo, enabling controlled discharge and safely degrading into non-toxic substances in the soil to avoid the remnants of residual plastic or metallic nanocontaminants [125].

8.2.2 Green Synthesis

The green nanotechnology idea of using plant extracts, microbes or fungi to synthesize metallic and metal-oxide nanoparticles is becoming a popular one. This strategy replaces the use of harmful chemicals with natural sources that are harmless to the environment, and the ecological footprint of the nanofertilizer making process decreases dramatically [20, 74].

8.3 Multi-nutrient and Multifunctional Nanofertilizers

The future research directions are moving away (around single-nutrient delivery), e.g. nano-urea, to the development of multifunctional nano-products that tackle multiple agronomic problems at the same time.

8.3.1 Combined Nutrient Delivery

New formulations are designed as nanocomposites or hybrid nanofertilizers that can deliver more than one essential component (macro- and micronutrient, e.g. a nano-DAP formulation of nitrogen and phosphorus or one of NPK with zinc and copper). The approach will allow farmers to implement a holistic nutritional package in one effective application [64].

8.3.2 Climate Resilience and Stress Tolerance

The next important emerging field is the production of nanostimulants and multifunctional nanofertilizers that would provide nutrients as well as increase the resistance of the plant to biotic and abiotic stresses. These nanocarriers can co-encapsulate nutrients with plant hormones, stress-reducing compounds or even biocontrol agents, and so lead to better crop tolerance to drought, salinity and heat, an immediate requirement in the changing climate [108].

9 Conclusion

The integration of smart-delivering nanofertilizers into agricultural systems offers a potent, necessary pathway towards achieving truly sustainable and resilient agriculture. This technology fundamentally addresses the critical inefficiency of conventional fertilization, where low nutrient use efficiency (NUE) leads to massive resource waste and pervasive environmental pollution. By utilizing advanced stimuli-responsive nanocarriers (triggered by pH, moisture or root exudates), these fertilizers ensure that nutrients are released precisely when and where the crop needs them, resulting in superior uptake, reduced application rates and minimal nutrient runoff and volatilization. This precision not only boosts crop productivity and quality but also significantly contributes to soil health and drastically lowers greenhouse gas emissions, thus mitigating agriculture's climate impact. The successful widespread implementation of this technology, particularly when seamlessly integrated with precision farming operations and IoT data, will be instrumental in meeting global food demand sustainably and directly supports the UN Sustainable Development Goals (SDGs), particularly SDG 2 (Zero Hunger) and SDG 13 (Climate Action). Future efforts must focus on standardizing ecotoxicology testing, scaling up cost-effective, biodegradable carriers and harmonizing regulatory frameworks to ensure the safe, equitable and secure advancement of this transformative agricultural innovation.

References

1. Abay KA, Chamberlin J, Chivenge P, Spielman DJ (2025) Fertilizer, soil health, and economic shocks: a synthesis of recent evidence. Food Policy 133:102892. https://doi.org/10.1016/j.foodpol.2025.102892
2. Abbasi MR, Sepaskhah AR (2023) Nitrogen leaching and groundwater N contamination risk in saffron/wheat intercropping under different irrigation and soil fertilizers regimes. Sci Rep 13(1):6587. https://doi.org/10.1038/s41598-023-33817-5
3. Abd El-Azeim MM, Sherif MA, Hussien MS, Tantawy IAA, Bashandy SO (2020) Impacts of nano- and non-nanofertilizers on potato quality and productivity. Acta Ecol Sin 40(5):388–397. https://doi.org/10.1016/j.chnaes.2019.12.007
4. Abid N, Khan AM, Shujait S, Chaudhary K, Ikram M, Imran M, Haider J, Khan M, Khan Q, Maqbool M (2022) Synthesis of nanomaterials using various top-down and bottom-up approaches, influencing factors, advantages, and disadvantages: a review. Adv Colloid Interf Sci 300:102597. https://doi.org/10.1016/j.cis.2021.102597
5. Adhikary T, Basak P (2024) The paradigm shift towards utilizing nanofertilizers paves a promising future in industrial agriculture and soilless cultivation. Hybrid Adv 6:100255. https://doi.org/10.1016/j.hybadv.2024.100255
6. Agbna GHD, Zaidi SJ (2025) Hydrogel performance in boosting plant resilience to water stress—a review. Gels 11(4):276. https://doi.org/10.3390/gels11040276
7. Aguilar-Paredes A, Valdés G, Araneda N, Valdebenito E, Hansen F, Nuti M (2023) Microbial Community in the Composting Process and its Positive Impact on the soil biota in sustainable agriculture. Agronomy 13(2):542. https://doi.org/10.3390/agronomy13020542
8. Al-Juthery HWA, Al-Taee RAHG, Alhasan AS, Hassan DF, Al-Jassani NAA, Jarallah RS (2025) Nano-biofertilizers: a promising technology for sustainable soil fertility, soil health,

and environmental protection. Asian J Water Environ Pollut 22(3):15. https://doi.org/10.36922/AJWEP025160123

9. Al-Khayri JM, Rashmi R, Surya Ulhas R, Sudheer WN, Banadka A, Nagella P, Aldaej MI, Rezk AA-S, Shehata WF, Almaghasla MI (2023) The role of nanoparticles in response of plants to abiotic stress at physiological, biochemical, and molecular levels. Plants 12(2):292. https://doi.org/10.3390/plants12020292
10. Al-Waeli NHJ, Al-Janabi ASA (2023) Effect of bio fertilizer, Nano organic fertilizer and non-Nano organic fertilizer on the quantitative traits of French black grape cultivars. IOP Conf Ser: Earth Environ Sci 1262(4):042033. https://doi.org/10.1088/1755-1315/1262/4/042033
11. Almendro-Candel MB, Lucas IG, Navarro-Pedreño J, Zorpas AA (2018) Physical properties of soils affected by the use of agricultural waste. In: Agricultural waste and residues. InTech. https://doi.org/10.5772/intechopen.77993
12. Azam M, Bhatti HN, Khan A, Zafar L, Iqbal M (2022) Zinc oxide nano-fertilizer application (foliar and soil) effect on the growth, photosynthetic pigments and antioxidant system of maize cultivar. Biocatal Agric Biotechnol 42:102343. https://doi.org/10.1016/j.bcab.2022.102343
13. Babu S, Singh R, Yadav D, Rathore SS, Raj R, Avasthe R, Yadav SK, Das A, Yadav V, Yadav B, Shekhawat K, Upadhyay PK, Yadav DK, Singh VK (2022) Nanofertilizers for agricultural and environmental sustainability. Chemosphere 292:133451. https://doi.org/10.1016/j.chemosphere.2021.133451
14. Basavegowda N, Baek K-H (2021) Current and future perspectives on the use of nanofertilizers for sustainable agriculture: the case of phosphorus nanofertilizer. *3*. Biotech 11(7):357. https://doi.org/10.1007/s13205-021-02907-4
15. Beeresha KJ, Yadeva H (2020) Effect of nano-potassium fertilizer on yield and economics of maize. Mysore J Agric Sci 54:28–32
16. Bekah D, Boyjoo Y, Mistry Panpadoo R, White JC, Bhaw-Luximon A (2025) Nanostimulants and nanofertilizers for precision agriculture: transforming food production in the 21st century. Environ Sci Nano 12(3):1740–1766. https://doi.org/10.1039/D5EN00055F
17. Bolan S, Sharma S, Mukherjee S, Zhou P, Mandal J, Srivastava P, Hou D, Edussuriya R, Vithanage M, Truong VK, Chapman J, Xu Q, Zhang T, Bandara P, Wijesekara H, Rinklebe J, Wang H, Siddique KHM, Kirkham MB, Bolan N (2024) The distribution, fate, and environmental impacts of food additive nanomaterials in soil and aquatic ecosystems. Sci Total Environ 916:170013. https://doi.org/10.1016/j.scitotenv.2024.170013
18. Chanratana M, Joe MM, Roy Choudhury A, Anandham R, Krishnamoorthy R, Kim K, Jeon S, Choi J, Choi J, Sa T (2019) Physiological response of tomato plant to chitosan-immobilized aggregated Methylobacterium oryzae CBMB20 inoculation under salinity stress. *3*. Biotech 9(11):397. https://doi.org/10.1007/s13205-019-1923-1
19. Chinnakoti P, Kurdekar AD, Rao KM, Kamisetti V (2025) A review of the emerging role of engineered nanomaterials as nanoadsorbents in enhanced drinking water defluoridation. Discov Environ 3(1):151. https://doi.org/10.1007/s44274-025-00345-0
20. Chowdhury NK, Choudhury R, Gogoi B, Chang C-M, Pandey RP (2022) Microbial synthesis of gold nanoparticles and their application. Curr Drug Targets 23(7):752–760. https://doi.org/10.2174/1389450123666220128152408
21. Dahham AA, Manea AI, Attia Al-Juthery HW (2025) The effect of bio organic fertilizer and spraying with Nano-organic extract on the growth and yield of three hybrids of cucumbers for protected cultivation. IOP Conf Ser: Earth Environ Sci 1487(1):012039. https://doi.org/10.1088/1755-1315/1487/1/012039
22. Das SK, Ghosh GK (2023) Development and evaluation of biochar-based secondary and micronutrient enriched slow release nano-fertilizer for reduced nutrient losses. Biomass Convers Biorefinery 13(13):12193–12204. https://doi.org/10.1007/s13399-021-01880-5
23. Davarpanah S, Tehranifar A, Davarynejad G, Abadía J, Khorasani R (2016) Effects of foliar applications of zinc and boron nano-fertilizers on pomegranate (Punica granatum cv. Ardestani) fruit yield and quality. Sci Hortic 210:57–64. https://doi.org/10.1016/j.scienta.2016.07.003

24. Dhaliwal SS, Shukla S, Vivek AK (2022) Impact of micronutrients in mitigation of abiotic stresses in soils and plants—a progressive step toward crop security and nutritional quality:1–78. https://doi.org/10.1016/bs.agron.2022.02.001
25. Dikshit P, Kumar J, Das A, Sadhu S, Sharma S, Singh S, Gupta P, Kim B (2021) Green synthesis of metallic nanoparticles: applications and limitations. Catalysts 11(8):902. https://doi.org/10.3390/catal11080902
26. Ding Y, Zhao W, Zhu G, Wang Q, Zhang P, Rui Y (2023) Recent trends in foliar Nanofertilizers: a review. Nano 13(21):2906. https://doi.org/10.3390/nano13212906
27. Dutta P, Muthukrishnan G, Kutalingam Gopalasubramaiam S, Dharmaraj R, Karuppaiah A, Loganathan K, Periyasamy K, Arumugam Pillai M, Upamanya G, Boruah S, Deb L, Kumari A, Mahanta M, Heisnam P, Mishra A (2022) Plant growth-promoting rhizobacteria (PGPR) and its mechanisms against plant diseases for sustainable agriculture and better productivity. Biocell 46(8):1843–1859. https://doi.org/10.32604/biocell.2022.019291
28. El-Desouky HS, Islam KR, Bergefurd B, Gao G, Harker T, Abd-El-Dayem H, Ismail F, Mady M, Zewail RMY (2021) Nano iron fertilization significantly increases tomato yield by increasing plants' vegetable growth and photosynthetic efficiency. J Plant Nutr 1–15. https://doi.org/10.1080/01904167.2021.1871749
29. El-Saadony MT, ALmoshadak AS, Shafi ME, Albaqami NM, Saad AM, El-Tahan AM, Desoky E-SM, Elnahal ASM, Almakas A, Abd El-Mageed TA, Taha AE, Elrys AS, Helmy AM (2021) Vital roles of sustainable nano-fertilizers in improving plant quality and quantity-an updated review. Saudi J Biol Sci 28(12):7349–7359. https://doi.org/10.1016/j.sjbs.2021.08.032
30. Elrys AS, Elnahal AS, Abdo AI, Desoky E-SM, Selem E, Rady MM (2022) Traditional, modern, and molecular strategies for improving the efficiency of nitrogen use in crops for sustainable agriculture: a fresh look at an old issue. J Soil Sci Plant Nutr 22(3):3130–3156. https://doi.org/10.1007/s42729-022-00873-1
31. Farmand M, Jahanpeyma F, Gholaminejad A, Azimzadeh M, Malaei F, Shoaie N (2022) Carbon nanostructures: a comprehensive review of potential applications and toxic effects. *3.* Biotech 12(8):159. https://doi.org/10.1007/s13205-022-03175-6
32. Fellet G, Pilotto L, Marchiol L, Braidot E (2021) Tools for Nano-enabled agriculture: fertilizers based on calcium phosphate, silicon, and chitosan nanostructures. Agronomy 11(6):1239. https://doi.org/10.3390/agronomy11061239
33. Folina A, Kakabouki I, Baginetas K, Bilalis D (2025) Integration of bioresources for sustainable development in organic farming: a comprehensive review. Resources 14(7):102. https://doi.org/10.3390/resources14070102
34. Garg S, Rumjit NP, Roy S (2024) Smart agriculture and nanotechnology: technology, challenges, and new perspective. Adv Agrochem 3(2):115–125. https://doi.org/10.1016/j.aac.2023.11.001
35. Garg D, Sridhar K, Stephen Inbaraj B, Chawla P, Tripathi M, Sharma M (2023) Nano-biofertilizer formulations for agriculture: a systematic review on recent advances and prospective applications. Bioengineering 10(9):1010. https://doi.org/10.3390/bioengineering10091010
36. Govil S, Van Duc Long N, Escribà-Gelonch M, Hessel V (2024) Controlled-release fertiliser: recent developments and perspectives. Ind Crop Prod 219:119160. https://doi.org/10.1016/j.indcrop.2024.119160
37. Guo F, Aryana S, Han Y, Jiao Y (2018) A review of the synthesis and applications of polymer–Nanoclay composites. Appl Sci 8(9):1696. https://doi.org/10.3390/app8091696
38. Habib M, Singh S, Jan S, Jan K, Bashir K (2025) The future of the future foods: understandings from the past towards SDG-2. npj Sci Food 9(1):138. https://doi.org/10.1038/s41538-025-00484-x
39. Hamoda, asmaa. (2024) Effect of Nano-fertilizer and bio-growth regulator on yield attributes of wheat. J Plant Prod:101–109. https://doi.org/10.21608/jpp.2024.266830.1305
40. Harris LK, Theriot JA (2018) Surface area to volume ratio: a natural variable for bacterial morphogenesis. Trends Microbiol 26(10):815–832. https://doi.org/10.1016/j.tim.2018.04.008

41. Jacob S, Kather FS, Boddu SHS, Shah J, Nair AB (2024) Innovations in Nanoemulsion technology: enhancing drug delivery for Oral, parenteral, and ophthalmic applications. Pharmaceutics 16(10):1333. https://doi.org/10.3390/pharmaceutics16101333
42. Javier F, Luisa M, Gaton P, Rojo R (2012) Nanotechnology and food industry. In: Scientific, health and social aspects of the food industry. InTech. https://doi.org/10.5772/33458
43. Jia X, Shi N, Tang W, Su Z, Chen H, Tang Y, Sun B, Zhao L (2022) Nano-hydroxyapatite increased soil quality and boosted beneficial soil microbes. Plant Nano Biology 1:100002. https://doi.org/10.1016/j.plana.2022.100002
44. Joudeh N, Linke D (2022) Nanoparticle classification, physicochemical properties, characterization, and applications: a comprehensive review for biologists. J Nanobiotechnol 20(1):262. https://doi.org/10.1186/s12951-022-01477-8
45. Kashyap S, Kumar R, Ram H, Kumar A, Basak N, Sheoran P, Bhatacharjee S, Biswal B, Ali G, Kumar B, Bhakuni K, Hindoriya PS, Birbal, & Min, D. (2023) Quantitative and qualitative response of fodder maize to use of bulk and Nano-fertilizers in North Western Plains of India. Agronomy 13(7):1889. https://doi.org/10.3390/agronomy13071889
46. Kekeli MA, Wang Q, Rui Y (2025) The role of Nano-fertilizers in sustainable agriculture: boosting crop yields and enhancing quality. Plants 14(4):554. https://doi.org/10.3390/plants 14040554
47. Khan N, Sudhakar K, Mamat R (2024) Macroalgae farming for sustainable future: navigating opportunities and driving innovation. Heliyon 10(7):e28208. https://doi.org/10.1016/j.heliyon.2024.e28208
48. Khanna K, Kohli SK, Handa N, Kaur H, Ohri P, Bhardwaj R, Yousaf B, Rinklebe J, Ahmad P (2021) Enthralling the impact of engineered nanoparticles on soil microbiome: a concentric approach towards environmental risks and cogitation. Ecotoxicol Environ Saf 222:112459. https://doi.org/10.1016/j.ecoenv.2021.112459
49. Klimczyk M, Siczek A, Schimmelpfennig L (2021) Improving the efficiency of urea-based fertilization leading to reduction in ammonia emission. Sci Total Environ 771:145483. https://doi.org/10.1016/j.scitotenv.2021.145483
50. Kumar A, Ram H, Kumar S, Kumar R, Yadav A, Gairola A, Kumar V, Sharma T (2023a) A comprehensive review of Nano-urea vs. conventional urea. Int J Plant Soil Sci 35(23):32–40. https://doi.org/10.9734/ijpss/2023/v35i234212
51. Kumar N, Samota SR, Venkatesh K, Tripathi SC (2023b) Global trends in use of nano-fertilizers for crop production: advantages and constraints—a review. Soil Tillage Res 228:105645. https://doi.org/10.1016/j.still.2023.105645
52. Kumar A, Singh K, Sharma S, Mishra JP, Sharmila JS, Kumar Y, Lakshmanan A, Singh T, Singh S, Panwar A, Sivashankari L (2023c) Effect of organic manure, bio-fertilizer and Nano-fertilizer on yield and economics of different vegetable crops and soil nutrient status. https://doi.org/10.21203/rs.3.rs-2960032/v1
53. Lahiri D, Nag M, Sheikh HI, Sarkar T, Edinur HA, Pati S, Ray RR (2021) Microbiologically-synthesized nanoparticles and their role in silencing the biofilm Signaling Cascade. Front Microbiol 12. https://doi.org/10.3389/fmicb.2021.636588
54. Lewicka K, Szymanek I, Rogacz D, Wrzalik M, Łagiewka J, Nowik-Zając A, Zawierucha I, Coseri S, Puiu I, Falfushynska H, Rychter P (2024) Current trends of polymer materials' application in agriculture. Sustainability 16(19):8439. https://doi.org/10.3390/su16198439
55. Madzokere TC, Murombo LT, Chiririwa H (2021) Nano-based slow releasing fertilizers for enhanced agricultural productivity. Mater Today Proc 45:3709–3715. https://doi.org/10.1016/j.matpr.2020.12.674
56. Mahapatra DM, Satapathy KC, Panda B (2022) Biofertilizers and nanofertilizers for sustainable agriculture: Phycoprospects and challenges. Sci Total Environ 803:149990. https://doi.org/10.1016/j.scitotenv.2021.149990
57. Mahmoud NN, Selim MT (2025) Phytochemical analysis and antimicrobial activity of Silybum marianum L. via multi-solvent extraction. AMB Express 15(1):122. https://doi.org/10.1186/s13568-025-01925-2

58. Malik S, Chaudhary K, Malik A, Punia H, Sewhag M, Berkesia N, Nagora M, Kalia S, Malik K, Kumar D, Kumar P, Kamboj E, Ahlawat V, Kumar A, Boora K (2022) Superabsorbent polymers as a soil amendment for increasing agriculture production with reducing water losses under water stress condition. Polymers 15(1):161. https://doi.org/10.3390/polym15010161
59. Merghes P, Ilia G, Maranescu B, Varan N, Simulescu V (2024) The sol–gel process, a green method used to obtain hybrid materials containing phosphorus and zirconium. Gels 10(10):656. https://doi.org/10.3390/gels10100656
60. Millar N, Robertson GP, Grace PR, Gehl RJ, Hoben JP (2010) Nitrogen fertilizer management for nitrous oxide (N2O) mitigation in intensive corn (maize) production: an emissions reduction protocol for US Midwest agriculture. Mitig Adapt Strateg Glob Chang 15(2):185–204. https://doi.org/10.1007/s11027-010-9212-7
61. Mim JJ, Rahman SMM, Khan F, Paul D, Sikder S, Das HP, Khan S, Orny NT, Hossain Shuvo MR, Hossain N (2025) Towards smart agriculture through nano-fertilizer-a review. Mater Today Sustain 30:101100. https://doi.org/10.1016/j.mtsust.2025.101100
62. Mittal D, Kaur G, Singh P, Yadav K, Ali SA (2020) Nanoparticle-based sustainable agriculture and food science: recent advances and future outlook. Front Nanotechnol 2. https://doi.org/10.3389/fnano.2020.579954
63. Mohsen KH, Alrubaiee SHA, Alfarjawi TMK (2022) Response of wheat varieties, Triticum aestivum L., to spraying by iron nano-fertilizer. Casp J Environ Sci 20(4):775–783. https://doi.org/10.22124/cjes.2022.5764
64. Mridha D, Lamsal B, Antonangelo JA (2025) Nanotechnology in agriculture: innovations for sustainability and greenhouse gas mitigation-a review. Sci Total Environ 995:180065. https://doi.org/10.1016/j.scitotenv.2025.180065
65. Mustapha LS, Obayomi OV, Lau SY, Obayomi KS (2025) The role of nanotechnology in agricultural systems with emphasis on water management. Inorg Chem Commun 182:115507. https://doi.org/10.1016/j.inoche.2025.115507
66. Nandhini J, Karthikeyan E, Rajeshkumar S (2024) Eco-friendly bio-nanocomposites: pioneering sustainable biomedical advancements in engineering. Discov Nano 19(1):86. https://doi.org/10.1186/s11671-024-04007-7
67. Nandini B, Mawale KS, Giridhar P (2023) Nanomaterials in agriculture for plant health and food safety: a comprehensive review on the current state of agro-nanoscience. Biotech 13(3):73. https://doi.org/10.1007/s13205-023-03470-w
68. Narayan R, Nayak UY, Raichur AM, Garg S (2018) Mesoporous silica nanoparticles: a comprehensive review on synthesis and recent advances. Pharmaceutics 10(3):118. https://doi.org/10.3390/pharmaceutics10030118
69. Neale RE, Barnes PW, Robson TM, Neale PJ, Williamson CE, Zepp RG, Wilson SR, Madronich S, Andrady AL, Heikkilä AM, Bernhard GH, Bais AF, Aucamp PJ, Banaszak AT, Bornman JF, Bruckman LS, Byrne SN, Foereid B, Häder D-P et al (2021) Environmental effects of stratospheric ozone depletion, UV radiation, and interactions with climate change: UNEP environmental effects assessment panel, update 2020. Photochem Photobiol Sci 20(1):1–67. https://doi.org/10.1007/s43630-020-00001-x
70. Nongbet A, Mishra AK, Mohanta YK, Mahanta S, Ray MK, Khan M, Baek K-H, Chakrabartty I (2022) Nanofertilizers: a smart and sustainable attribute to modern agriculture. Plants 11(19):2587. https://doi.org/10.3390/plants11192587
71. Norman S, Hongda C (2013) IB IN DEPTH-special section on Nanobiotechnology, part 2. Ind Biotechnol 9:17-18
72. Nuruzzaman M, Bahar MM, Naidu R (2025) Diffuse soil pollution from agriculture: impacts and remediation. Sci Total Environ 962:178398. https://doi.org/10.1016/j.scitotenv.2025.178398
73. Osman AI, Fawzy S, Farghali M, El-Azazy M, Elgarahy AM, Fahim RA, Maksoud MIAA, Ajlan AA, Yousry M, Saleem Y, Rooney DW (2022) Biochar for agronomy, animal farming, anaerobic digestion, composting, water treatment, soil remediation, construction, energy storage, and carbon sequestration: a review. Environ Chem Lett 20(4):2385–2485. https://doi.org/10.1007/s10311-022-01424-x

74. Osman AI, Zhang Y, Farghali M, Rashwan AK, Eltaweil AS, Abd El-Monaem EM, Mohamed IMA, Badr MM, Ihara I, Rooney DW, Yap P-S (2024) Synthesis of green nanoparticles for energy, biomedical, environmental, agricultural, and food applications: a review. Environ Chem Lett 22(2):841–887. https://doi.org/10.1007/s10311-023-01682-3
75. Pagano M, Lunetta E, Belli F, Mocarli G, Cocozza C, Cacciotti I (2025) Advancements in agricultural nanotechnology: an updated review. Plants 14(18):2939. https://doi.org/10.3390/plants14182939
76. Patil V, Veettil PC (2024) Farmers' risk attitude, agricultural technology adoption and impacts in eastern India. Agric Food Secur 13(1):50. https://doi.org/10.1186/s40066-024-00497-x
77. Peladarinos N, Piromalis D, Cheimaras V, Tserepas E, Munteanu RA, Papageorgas P (2023) Enhancing smart agriculture by implementing digital twins: a comprehensive review. Sensors 23(16):7128. https://doi.org/10.3390/s23167128
78. Peng W, Li P, Zia M, Bhuiyan SA, Liu Y, Chen D, Tudi M, Gao Y, Yan X, Jia Y, Li Q (2024) A comparative review on carbon nanomaterials and their impact on plant growth with the lens of green chemistry principles. *Global.* Challenges 8(11). https://doi.org/10.1002/gch2.202400008
79. Pilotto L, Scalera F, Piccirillo C, Marchiol L, Zuluaga MYA, Pii Y, Cesco S, Civilini M, Fellet G (2025) Phosphorus release from Nano-hydroxyapatite derived from biowastes in the presence of phosphate-solubilizing bacteria: a soil column experiment. J Agric Food Chem 73(7):3918–3929. https://doi.org/10.1021/acs.jafc.4c09325
80. Ponnusamy P, Loganathan M, Krishnasamy S (2025) Mass transfer modelling and nitrogen release kinetics of nano-biochar dispersed PVA/PVP matrix encapsulated slow-release fertilizer for sustainable agriculture. Chem Eng J Adv 23:100817. https://doi.org/10.1016/j.ceja.2025.100817
81. Rajak P, Ganguly A, Adhikary S, Bhattacharya S (2023) Internet of things and smart sensors in agriculture: scopes and challenges. J Agric Food Res 14:100776. https://doi.org/10.1016/j.jafr.2023.100776
82. Rajesh H, Yadahalli GS, Chittapur B, Halepyati AS, Sharanugouda H (2021) Growth, yield and economics of sweet corn (Zea mays L. Saccarata) as influencedby foliar sprays of nano fertilisers. J Farm Sci 34(4):381–385
83. Ramalingappa PL, Shrivastava M, Dhar S, Bandyopadhyay K, Prasad S, Langyan S, Tomer R, Khandelwal A, Darjee S, Singh R (2023) Reducing options of ammonia volatilization and improving nitrogen use efficiency via organic and inorganic amendments in wheat (Triticum aestivum L.). PeerJ 11:e14965. https://doi.org/10.7717/peerj.14965
84. Rana L, Kumar M, Rajput J, Kumar N, Sow S, Kumar S, Kumar A, Singh SN, Jha CK, Singh AK, Ranjan S, Sahoo R, Samanta D, Nath D, Panday R, Raigar BL (2024) Nexus between nanotechnology and agricultural production systems: challenges and future prospects. Discov Appl Sci 6(11):555. https://doi.org/10.1007/s42452-024-06265-7
85. Reddy KS, Shivay YS, Kumar D, Pooniya V, Prasanna R, Mandi S, Nayak S, Baral K, Alekhya G, Borate RB (2025) Relative performance of granulated and nano urea on productivity and nitrogen use efficiency of wheat–rice sequence. Plant Nano Biol 11:100131. https://doi.org/10.1016/j.plana.2024.100131
86. Ribeiro C, Santos DMF (2025) Transitioning ammonia production: green hydrogen-based Haber–Bosch and emerging nitrogen reduction technologies. Clean Technol 7(2):49. https://doi.org/10.3390/cleantechnol7020049
87. Rockström J, Williams J, Daily G, Noble A, Matthews N, Gordon L, Wetterstrand H, DeClerck F, Shah M, Steduto P, de Fraiture C, Hatibu N, Unver O, Bird J, Sibanda L, Smith J (2017) Sustainable intensification of agriculture for human prosperity and global sustainability. Ambio 46(1):4–17. https://doi.org/10.1007/s13280-016-0793-6
88. Sadati Valojai ST, Niknejad Y, Fallah Amoli H, Barari Tari D (2021) Response of rice yield and quality to nano-fertilizers in comparison with conventional fertilizers. J Plant Nutr 44(13):1971–1981. https://doi.org/10.1080/01904167.2021.1884701
89. Sahoo BR, Dash AK, Mohapatra KK, Mohanty S, Sahu SG, Sahoo BR, Prusty M, Priyadarshini E (2024) Strategic management of nano-fertilizers for sustainable rice yield,

grain quality, and soil health. Front Environ Sci 12. https://doi.org/10.3389/fenvs.2024.1420505
90. Samir A, Ashour FH, Hakim AAA, Bassyouni M (2022) Recent advances in biodegradable polymers for sustainable applications. Npj Mater Degrad 6(1):68. https://doi.org/10.1038/s41529-022-00277-7
91. Sapkota TB, Bijay-Singh. (2025) India's fertilizer policies: implications for food security, environmental sustainability, and climate change. Reg Environ Chang 25(2):63. https://doi.org/10.1007/s10113-025-02395-9
92. Saritha GNG, Anju T, Kumar A (2022) Nanotechnology - big impact: how nanotechnology is changing the future of agriculture? J Agric Food Res 10:100457. https://doi.org/10.1016/j.jafr.2022.100457
93. Schoonjans R, Castenmiller J, Chaudhry Q, Cubadda F, Daskaleros T, Franz R, Gott D, Mast J, Mortensen A, Oomen AG, Rauscher H, Weigel S, Astuto MC, Cattaneo I, Barthelemy E, Rincon A, Tarazona J (2023) Regulatory safety assessment of nanoparticles for the food chain in Europe. Trends Food Sci Technol 134:98–111. https://doi.org/10.1016/j.tifs.2023.01.017
94. Seleiman MF, Almutairi KF, Alotaibi M, Shami A, Alhammad BA, Battaglia ML (2020) Nano-fertilization as an emerging fertilization technique: why can modern agriculture benefit from its use? Plants 10(1):2. https://doi.org/10.3390/plants10010002
95. Sembada AA, Lenggoro IW (2024) Transport of nanoparticles into plants and their detection methods. Nano 14(2):131. https://doi.org/10.3390/nano14020131
96. Sen A, Faseela P, Jacob J, Siju Thomas T (2025) Nano fertilizers for precision agriculture:153–178. https://doi.org/10.1007/978-3-031-90506-3_7
97. Shang Y, Hasan MK, Ahammed GJ, Li M, Yin H, Zhou J (2019) Applications of nanotechnology in plant growth and crop protection: a review. Molecules 24(14):2558. https://doi.org/10.3390/molecules24142558
98. Sher A, Li H, Ullah A, Hamid Y, Nasir B, Zhang J (2024) Importance of regenerative agriculture: climate, soil health, biodiversity and its socioecological impact. Discov Sustain 5(1):462. https://doi.org/10.1007/s43621-024-00662-z
99. Shi Y, Wang X, Feng C, Yang S (2024) Nano-clay montmorillonite removes tetracycline in water: factors and adsorption mechanism in aquatic environments. IScience 27(2):108952. https://doi.org/10.1016/j.isci.2024.108952
100. Singh S, Mali N, Jain A, Tailor G, P., N. J. (2025) A review: biogenic synthesis of ca, Mg, Si nano fertilizers and their foliar studies on legume crops production. Green Technol Sustainability 3(3):100206. https://doi.org/10.1016/j.grets.2025.100206
101. Singh A, Sharma A, Singh O, Rajput VD, Movsesyan H, Minkina T, Alexiou A, Papadakis M, Singh RK, Singh S, Sousa JR, El-Ramady HR, Zulfiqar F, Kumar R, Al-Ghamdi AA, Ghazaryan K (2024a) In-depth exploration of nanoparticles for enhanced nutrient use efficiency and abiotic stresses management: present insights and future horizons. Plant Stress 14:100576. https://doi.org/10.1016/j.stress.2024.100576
102. Singh S, Singh R, Singh K, Katoch K, Zaeen AA, Birhan DA, Singh A, Sandhu HS, Singh H, Sahrma LK (2024b) Smart fertilizer technologies: an environmental impact assessment for sustainable agriculture. Smart Agric Technol 8:100504. https://doi.org/10.1016/j.atech.2024.100504
103. Skiba UM, Rees RM (2014) Nitrous oxide, climate change and agriculture. CABI Rev 1–7. https://doi.org/10.1079/PAVSNNR20149010
104. Sousa de Almeida M, Susnik E, Drasler B, Taladriz-Blanco P, Petri-Fink A, Rothen-Rutishauser B (2021) Understanding nanoparticle endocytosis to improve targeting strategies in nanomedicine. Chem Soc Rev 50(9):5397–5434. https://doi.org/10.1039/D0CS01127D
105. Spasojević L, Ivanišević I, Dutour Sikirić M (2025) Stability and application of TiO 2 nanomaterials in aqueous suspensions: a review. RSC Adv 15(27):21341–21368. https://doi.org/10.1039/D5RA02499D
106. Srisha V, Prasad VM (2022) Effect of NPK and Nano fertilizers on plant growth, yield and quality of gladiolus (Gladiolus grandiflorus L.). Int J Plant Soil Sci:1529–1534. https://doi.org/10.9734/ijpss/2022/v34i2231528

107. Srivastava P, Singh S (2012) Conservation of soil, water and nutrients in surface runoff using riparian plant species. J Environ Biol 33(1):43–49
108. Stojanova M, Demiri S, Stojanova MT, Djukic DA, Kaya Y (2025) From cultivation to consumption: evaluating the effects of nano fertilizers on food quality and safety. Adv Agrochem 4(3):217–234. https://doi.org/10.1016/j.aac.2025.07.001
109. Sun Y, Zhou H, Sun Y, Wu Q, Chen H, Chi D (2024) The application of nano−clinoptilolite based nitrogen fertilizer mixed with urea promotes nitrogen balance and enhances economic and ecological benefits in paddy fields. J Clean Prod 453:142257. https://doi.org/10.1016/j.jclepro.2024.142257
110. Taha A, Ghazy M, Omar M (2024) Enhancing spinach performance: effectiveness of Nano-fertilizers in conjunction with conventional NPK fertilizers. Egypt J Soil Sci 64(3). https://doi.org/10.21608/ejss.2024.278586.1741
111. Tahat MM, Alananbeh K, Othman Y, Leskovar DI (2020) Soil health and sustainable agriculture. Sustainability 12(12):4859. https://doi.org/10.3390/su12124859
112. Tamrin SH, Phelps J, Nezhad AS, Sen A (2023) Critical considerations in determining the surface charge of small extracellular vesicles. J Extracell Vesicles 12(9). https://doi.org/10.1002/jev2.12353
113. Tao M, Zhang J, Ren Y, Zhang D, Cheng B, Wang C (2025) Multi-batch consecutive foliar spraying zinc–carbon dot Nano-fertilizer improving soil health for Bok choy cultivation production. Nano 15(22):1714. https://doi.org/10.3390/nano15221714
114. Taran M, Rad M, Alavi M (2017) Antibacterial activity of copper oxide (CuO) nanoparticles biosynthesized by bacillus sp. FU4: optimization of experiment design. Pharm Sci 23(3):198–206. https://doi.org/10.15171/PS.2017.30
115. Thilakarathne NN, Abu Bakar MS, Abas PE, Yassin H (2025) Internet of things enabled smart agriculture: current status, latest advancements, challenges and countermeasures. Heliyon 11(3):e42136. https://doi.org/10.1016/j.heliyon.2025.e42136
116. Upadhyay PK, Singh VK, Rajanna GA, Dwivedi BS, Dey A, Singh RK, Rathore SS, Shekhawat K, Babu S, Singh T, Kumar Y, Singh C, Rangot M, Kumar A, Sarkar S, Dash S, Rawat S (2023) Unveiling the combined effect of nano fertilizers and conventional fertilizers on crop productivity, profitability, and soil Well-being. Front Sustain Food Syst 7. https://doi.org/10.3389/fsufs.2023.1260178
117. Verma KK, Song X-P, Joshi A, Tian D-D, Rajput VD, Singh M, Arora J, Minkina T, Li Y-R (2022) Recent trends in Nano-fertilizers for sustainable agriculture under climate change for global food security. Nano 12(1):173. https://doi.org/10.3390/nano12010173
118. Vinukonda A, Bolledla N, Jadi RK, Chinthala R, Devadasu VR (2025) Synthesis of nanoparticles using advanced techniques. Next Nanotechnol 8:100169. https://doi.org/10.1016/j.nxnano.2025.100169
119. Vives-Peris V, de Ollas C, Gómez-Cadenas A, Pérez-Clemente RM (2020) Root exudates: from plant to rhizosphere and beyond. Plant Cell Rep 39(1):3–17. https://doi.org/10.1007/s00299-019-02447-5
120. Wani KA, Kothari R (2018) Agricultural nanotechnology: applications and challenges. Ann Plant Sci 7(3):2146. https://doi.org/10.21746/aps.2018.7.3.9
121. Xavier SJS, Klunk MA, Caetano NR (2025) Advances in the use of zeolites for industrial effluent treatment and carbon dioxide mitigation: a review. Appl Sci 15(24):13171. https://doi.org/10.3390/app152413171
122. Xu L, Niu H, Xu J, Wang X (2013) Nitrate-nitrogen leaching and Modeling in intensive agriculture farmland in China. Sci World J 2013(1). https://doi.org/10.1155/2013/353086
123. Xue J, Wu T, Dai Y, Xia Y (2019) Electrospinning and electrospun nanofibers: methods, materials, and applications. Chem Rev 119(8):5298–5415. https://doi.org/10.1021/acs.chemrev.8b00593
124. Yadav A, Yadav K, Abd-Elsalam K (2023) Nanofertilizers: types, delivery and advantages in agricultural sustainability. Agrochemicals 2(2):296–336. https://doi.org/10.3390/agrochemicals2020019

125. Yarahmadi A, Dousti B, Karami-Khorramabadi M, Afkhami H (2024) Materials based on biodegradable polymers chitosan/gelatin: a review of potential applications. Front Bioeng Biotechnol 12. https://doi.org/10.3389/fbioe.2024.1397668
126. Zafar S, Bilal M, Ali MF, Mahmood A, Kijsomporn J, Wong LS, Harshini M, Kumar V, Alotaibi SS (2024) Nano-biofertilizer an eco-friendly and sustainable approach for the improvement of crops under abiotic stresses. Environ Sustain Indic 24:100470. https://doi.org/10.1016/j.indic.2024.100470
127. Zahra Z, Habib Z, Hyun H, Shahzad HMA (2022) Overview on recent developments in the design, application, and impacts of Nanofertilizers in agriculture. Sustainability 14(15):9397. https://doi.org/10.3390/su14159397
128. Zamanian K, Taghizadeh-Mehrjardi R, Tao J, Fan L, Raza S, Guggenberger G, Kuzyakov Y (2024) Acidification of European croplands by nitrogen fertilization: consequences for carbonate losses, and soil health. Sci Total Environ 924:171631. https://doi.org/10.1016/j.scitotenv.2024.171631
129. Zea L (2020) Effect of Nano-potassium fertilizer on yield and economics o f maize. 54(1):28–32
130. Zhang W, Taheri-Ledari R, Ganjali F, Mirmohammadi SS, Qazi FS, Saeidirad M, KashtiAray A, Zarei-Shokat S, Tian Y, Maleki A (2023) Effects of morphology and size of nanoscale drug carriers on cellular uptake and internalization process: a review. RSC Adv 13(1):80–114. https://doi.org/10.1039/D2RA06888E
131. Zielińska A, Carreiró F, Oliveira AM, Neves A, Pires B, Venkatesh DN, Durazzo A, Lucarini M, Eder P, Silva AM, Santini A, Souto EB (2020) Polymeric nanoparticles: production, characterization, Toxicology and Ecotoxicology. Molecules 25(16):3731. https://doi.org/10.3390/molecules25163731
132. Zulfiqar F, Navarro M, Ashraf M, Akram NA, Munné-Bosch S (2019) Nanofertilizer use for sustainable agriculture: advantages and limitations. Plant Sci 289:110270. https://doi.org/10.1016/j.plantsci.2019.110270

Smart Nanocarriers for Delivering Pesticides Toward Sustainable Agriculture

Kinjal Mondal, Soham Barik, Trisha Majumder, Pradeep Kumar, Sukanya Dutta, Santanu Kundu, and Maksud Hasan Shah

Abstract Current advancements in nanotechnology have promisingly enabled both manufacturing and judicial deployment of smart nanocarriers for pesticide delivery, offering enhanced precision and sustainability in agriculture. These nanoscale delivery systems, including polymeric nanoparticles, mesoporous silica nanoparticles, solid-lipid nanoparticles, and nano-emulsions, possess high surface area-to-volume ratios, allowing for effective encapsulation, protection, and controlled release of active components. Functionalization using stimuli-responsive polymers enables triggered release under certain environmental conditions (e.g., pH, temperature, or enzymatic activity), assuring site-specific action while reducing off-target effects. Furthermore, the delayed and sustained release profile of such novel nanocarriers minimizes treatment frequency and pesticide runoff, hence reducing environmental pollution and insect resistance development. Surface-functionalized nanocarriers allow for targeted delivery by preferentially adhering to insect cuticles or plant surfaces. Integrating these smart carriers with precision agricultural technologies (remote sensing, GIS, and drone-based applications) improves targeted deployment and reduces input waste. Collectively, nanocarrier-based pesticide formulations offer a paradigm shift toward environmentally benign, resource-efficient crop protection measures, hence promoting the sustainable intensification of contemporary agriculture.

K. Mondal (✉) · P. Kumar
Rajasthan College of Agriculture, Maharana Pratap University of Agriculture and Technology, Udaipur, Rajasthan, India

S. Barik
Bidhan Chandra Krishi Viswavidyalaya, Mohanpur, West Bengal, India

T. Majumder
Department of Horticulture, Bangladesh Agricultural University, Maimansingh, Bangladesh

S. Dutta
Faculty of Agriculture, JIS University, Kolkata, West Bengal, India

S. Kundu
ICAR-Indian Agricultural Research Institute, New Delhi, India

M. H. Shah (✉)
CAE, Faculty of Agricultural Science, Aligar Muslim University, Aligarh, Uttar Pradesh, India

J.-T. Chen (ed.), *AI-Nano-Enabled Climate-Smart Agriculture and Sustainability*, Smart Nanomaterials Technology, https://doi.org/10.1007/978-981-92-1694-9_14

Keywords Nanocarrier · Nanoparticle · Encapsulation · Functionalization · Site-specific action

1 Introduction

The agriculture-food system is a complex socio-ecological and biochemical continuum that combines sustainable resource management, cultural practices, and modern agronomic research to guarantee food supply, ecosystem functioning, and global stability. Contemporary agriculture faces two major challenges: (i) maintaining or improving crop productivity as well as nutritional quality to fulfill the food demands of a growing population and (ii) ensuring the functional resilience, environmental sustainability, and biosecurity of food systems [12]. The attainment of these objectives is increasingly impeded by the multifaceted effects of climate change, including temperature extremes, alterations in precipitation patterns, and heightened CO_2 levels, alongside soil nutrient depletion, diminishing biodiversity, and rising metabolic and energetic expenses associated with agricultural inputs. The intensifying burden of insect pest infestations, exacerbated by climate warming, altered pest phenology, expanded geographic ranges, and increasing pesticide resistance, significantly amplifies crop losses and narrows the yield gap between potential and actual productivity [14]. By 2050, it is expected that the current 7.7 billion people on Earth will have grown to around 9.8 billion, with the majority of that growth occurring in emerging nations [26]. Presently, farmers are broadly practicing chemical farming, i.e., spraying of pesticides, herbicides, etc., to double their income, but this is not a sustainable approach. In agriculture, pesticides are a class of chemical compounds that are employed to control pests and other damaging pests like mites, nematodes, snails, and so on [10]. Globally, the annual usage of pesticides surpasses two million tons, with Europe representing approximately 45%, the USA accounting for 25%, and the remaining regions collectively contributing 25%; India constitutes merely 3.75% of worldwide consumption [11]. Nevertheless, even at these levels, pesticides produce significant environmental externalities. Residual agrochemicals infiltrate the ecosystem via aerosol dispersion, volatilization, surface discharge, and leaching, resulting in the contamination of air, groundwater, and surface water bodies. The persistent cycling of these chemical contaminants interferes with ecosystem homeostasis, leading to the degradation of soil's physicochemical properties, diminution of organic matter, reduction in microbial diversity, and a sustained decline in soil fertility [32]. Furthermore, sustained pesticide exposure has expedited the depletion of biodiversity, encompassing the mortality of non-target beneficial species such as aquatic fauna, pollinators, and keystone species, most notably honey bees, whose decline directly compromises crop pollination efficacy and the robustness of the agroecosystem [18].

The limitations of traditional pesticide application, characterized by limited target specificity, swift degradation, and significant environmental dissemination, underscore the necessity for more precise and controlled crop protection approaches.

In this context, nanotechnology provides a transformative platform to overcome these limitations. Nano-engineered materials are prepared in such a way that pesticide molecules of interest can easily be integrated into the nanocarriers, which further offer controlled release, increased stability, enhanced plant absorption, and targeted delivery to pest or pathogen locations, thereby augmenting bio-efficacy while decreasing chemical usage [68]. Complementing this, nanosensors can identify early signs of pest infestation and plant duress, facilitating targeted, site-specific interventions instead of uniform application [7]. Consequently, nanotechnology possesses the potential to transform pesticide application by transitioning agriculture from extensive, broad-spectrum chemical usage to precision-guided, need-based interventions that improve productivity and reduce ecological footprint [18]. Recently, the amity of nanotechnology with plant pathology has led to remarkable progress in pesticide application, by endowing conventional pesticides with prolonged efficacy, controlled release or degradation, and selective pest management.

Nanotechnology has markedly enhanced biogenic carrier systems through the development of nanoscale, biologically derived matrices capable of encapsulating, stabilizing, and accurately delivering agrochemicals, particularly pesticides, directly to insect pest targets [20]. Unlike traditional formulations that experience rapid photodegradation, volatilization, or nonspecific dispersion, nano-enabled biocarriers offer controlled-release kinetics, superior adhesion to plant surfaces, increased bioavailability, and the capacity to deliver active ingredients in response to environmental stimuli (e.g., pH, enzymatic activity, humidity). In targeted insect pest management, these biogenic nanocarrier systems can be crafted on a laboratory scale to detect specific physiological signals of pests, enhancing their localization to the insect intestine, epidermis, or feeding locations [62]. This elevates the local pesticide concentration at the target organism while decreasing off-target toxicity to beneficial insects and limiting wider ecological exposure. Their nanoscale dimension enables enhanced penetration through insect cuticles or into cellular compartments, thereby increasing pesticidal efficacy at considerably reduced concentrations. In this chapter, we have thoroughly discussed how the advent of the nanocarrier system revolutionized the field of plant protection through targeted and/or controlled delivery of pesticides.

2 Concept of Smart Nanocarrier System: Classification, Advantages, and Disadvantages

New methods for developing intelligent nanopesticides have been made possible by the development of nanocarriers and associated technology. Nanosized molecules having different physical as well as chemical characteristics from macroscopic materials show significant applications in fields including biology, medicine, and agricultural research. According to Ma and Wang [40], nanotechnology is thought of as a modernized form of the traditional application approach, offering more precise

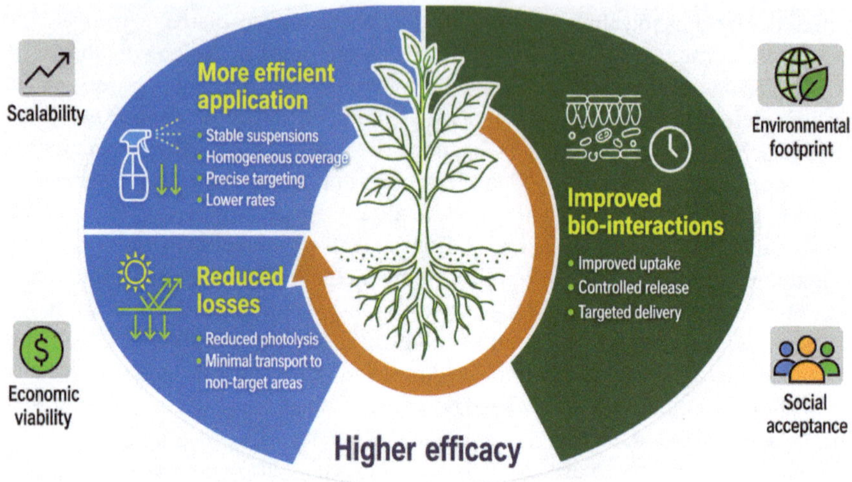

Fig. 1 Benefits of using nanoscale agrochemicals in contemporary agriculture

and regulated pesticide distribution. In agriculture, nanosized materials are mostly used for disease diagnostics, crop protection, and plant nutrition [15, 33]. Nanostructured systems, owing to their nanoscale dimensions and elevated surface-to-volume ratio, offer huge benefits in encapsulation of active substances into them, facilitating the slow and controlled delivery of potential agrochemicals. Current studies have reported that the utilization of active compound-containing nanostructures functions as a good controlled-release mechanism for these active ingredients. Current research on micro (10–100 μm), sub-micro (1–10 μm), and nano(<1 μm) sized particles in relation to regulated administration of agrochemicals revealed that the nanoscale particulates are potentially more effective (Fig. 1) than others [10].

Among currently developed nanoscale carrier systems, metal/metal-oxide nanocomposites (ZnO, TiO_2, Ag-based hybrids), dendrimeric constructs (PAMAM dendrimers), lipid-based structures (liposomes, solid-lipid nanoparticles, lipid carriers), inorganic nanocarriers (mesoporous silica nanoparticles, layered double hydroxides, nano-clays, nano-zeolites, iron oxide nanoparticles), and carbon-based nanomaterials (graphene oxide nanosheets, carbon nanotubes, carbon quantum dots) are very useful to deliver pesticidal chemicals (Fig. 2) in an ecologically friendly way [34]. Together, these nanocarrier systems maintain pesticidal efficacy within biologically effective thresholds while lowering input dosages by minimizing environmental residues, reducing off-target drift, improving photothermal and enzymatic stability, and guaranteeing regulated diffusion or stimuli-responsive release (pH, UV, enzyme-triggered). The slow-release of pesticide molecules is achieved on the basis of passive diffusion, capsule erosion, and osmotic-driven permeation [77].

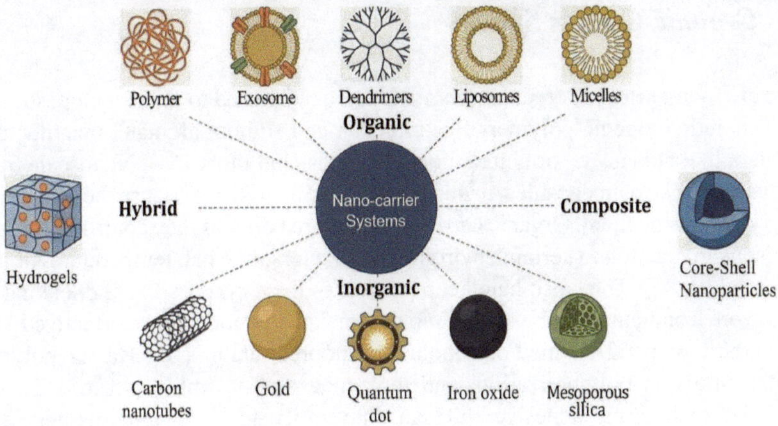

Fig. 2 Different kinds of nanocarrier systems useful in targeted pesticide delivery

2.1 Inorganic Carrier System

Incorporation or packaging of pesticide molecules into nanocarrier systems seems to be the most extensively explored approach in the development of nanopesticides [79, 80]. In this context, conventional inorganic carriers, such as mesoporous silica nanoparticles (MSNs), nano-clay, nanosized biochar, calcium carbonate, and graphene oxide, got ample success these days, likely due to superior physicochemical characteristics [6, 58]. Recently, MSNs have been reported to minimize the quick degradation and/or vaporization of active ingredients, sustaining prolonged impact on pest control. For instance, β-cyhalothrin-loaded MSNs were reported to prolong the larvicidal effect against *Ostrinia furnacalis* for more than one week due to sustained release of active agents [35].

Calcium carbonate nanoparticles ($CaCO_3$ NPs) are especially appropriate for acid-triggered release mechanisms. It is often manufactured by controlled precipitation through combining $CaCl_2$ and Na_2CO_3 at different surfactant levels in reverse-phase microemulsions, resulting in nanoscale crystal structures (50–200 nm) with ample loading capacity ($\approx$20%) [50]. The dry $CaCO_3$ NPs enwrap the typical insect parts (pulvilli, tarsi), hindering adherence to plant surfaces [52]. Similarly, graphene oxide (GO) was revealed to have potential carrying capacity for the transport of ß-cyfluthrin and imidacloprid, significantly improving their larvicidal effectiveness against lepidopteran insects [72]. The multilayer structure of GO was found to improve pesticide absorption by increasing adherence to insect cuticles and plant leaves. An in vitro investigation revealed a 1.5–1.8fold enhancement in acaricidal impacts to counter T. *urticae* with the application of GO-loaded pesticides, ascribed to improved coating and delayed environmental release [71].

2.2 *Organic Carrier System*

There are many kinds of organic polymers that can be used to make nanopesticides. These include biogenic polymers like chitosan and sodium alginate, manufactured polymers like polyesters, polyurea, polyurethanes, and more [64]. Nanoscale transport systems, like nanocapsules, nanomicelles, and nanogels [64], use these materials to carry AIs inside them. Organic carrier systems are better as they control the release of poisons in reaction to certain environmental factors, like pH, temperature, or light exposure [39, 83]. They also help keep pesticides from evaporating and leaking out into the environment, which cleans it up and makes treatment more effective [31].

Standard preparation methods include nanoprecipitation, interfacial polymerization, emulsion polymerization, and molecular self-assembly [45, 63]. Natural polymers, such as bio-adhesive chitosan, may infiltrate insect midguts and respiratory membranes, hence enhancing the concentrations of active ingredients. For example, laboratory-made camptothecin (CPT) micelles, formed by self-assembly with polyethylene glycol (PEG), had significant acaricidal efficacy against *Tetranychus cinnabarinus* [86]. In the same way, a star polymer (SPc) was used to naturally enclose dinotefuran by H-bonding as well as Van der Waals interactions. This made it much more effective at killing *Myzus persicae* [23].

2.3 *Inorganic–Organic Hybrid Carrier System*

Inorganic–organic hybrid carriers are fabricated through the integration of an inorganic core-shell with organic constituents through covalent interactions. A frequently employed strategy entails the modification of inorganic materials like MSNs with chitosan or dopamine to improve biological functionalities. In contrary, hybrid systems can be manufactured by assembling the inorganic compounds with organic ligands via coordination bonding [74]. For example, Xu et al. [81] synthesized dopamine-based MSNs and subsequently affixed a stimuli-responsive copolymer onto their surfaces by a free radical polymerization process. This approach used azodiisobutyronitrile (AIBN) to start the reaction and included monomers, i.e., tri(ethylene glycol) methyl ether methacrylate (MEO3MA) and N-vinyl caprolactam (NVCL). The synthetic MSN-based nanocarriers demonstrated a significant loading efficiency of 40.6% for the model insecticide imidacloprid. The stimuli-responsive copolymer (PVNM) maintains the core structure of the entire nanosystem and restricts premature insecticide discharge, thereby improving foliar retention. In another research, a dual-responsive metal-organic framework (MOF), i.e., zeolitic imidazolate framework-8 (ZIF-8), was reported to respond to both changes in pH and enzyme (amylase) activities by the in situ crystal growth method [85]. These hybrid systems demonstrate structural flexibility, increased loading capacity, and specific responses to particular stimuli, which make them promising for precise pesticide application.

Nano-delivery systems should ideally meet certain technological and economic criteria. The materials used as carriers must not elicit any negative reaction in the receiving organism. Furthermore, both the matrix material and its breakdown products must exhibit biocompatibility. Secondly, the mechanical qualities of the polymer must provide sustained protection for its contents, facilitating chemical stability over time. Third, the scalability of the manufacturing process must be technologically achievable and economically sustainable. The methods used for the production of nanocarriers must provide similar outcomes across batches regarding size, polydispersity, encapsulation efficiency, and stability. The materials designated as nanocarriers must be meticulously chosen, as they must satisfy technical specifications to comply with regulatory requirements before commercialization [66], and they must also exhibit favorable performance regarding cost-effectiveness and environmental compatibility. Table 1 summarizes the benefits and drawbacks of nanocarriers for the delivery of agrochemicals with prospective applications in agriculture. Pesticide nano-encapsulation fundamentally relies on the physicochemical characteristics of the encapsulation matrix, the active ingredient, and the methodology used in the process. Nonetheless, irrespective of the encapsulating matrix and the nature of the cargo or the fabrication technique used for the active ingredient-loaded nanocarriers, several papers affirm that some procedures for their elaboration provide the possibility for standardization due to their constant repeatability.

3 Mechanism of Smart Release and Targeting

Intelligent nanocarrier technologies are transforming pesticide application by facilitating targeted distribution and regulated release of active compounds. These nanoscale carriers, composed of components like polymers, lipids, or inorganic compounds, safeguard pesticides against environmental degradation and maintain their stability throughout delivery to the target location. They are generally described as protective enclosures for the material of interest against any detrimental biotic or abiotic challenges, hence enhancing their physicochemical stability. Their function in the intelligent release of essential molecules (specifically pesticides) is to (i) preserve the active compounds without premature release or degradation prior to reaching the target; (ii) enhance the specific activity of the compounds upon arrival at the target, such as facilitating penetration into the plant tissues; and (iii) regulate the active release mechanisms in adjacent environments [36, 65]. Their efficacy is fairly influenced by certain environmental or biological stimuli, such as changes in temperature, pH, light, or the presence of enzymes, prompting the nanocarrier to release its pesticide payload in a regulated way at the targeted site [19, 65]. For instance, pH-responsive nanocarriers may expedite the release of pesticides in the alkaline or acidic environments present in certain insect intestines or plant tissues, while temperature-responsive systems can be engineered to release upon thermal exposure, hence improving efficacy and minimizing off-target effects. Targeting is accomplished by engineering nanocarriers that may be absorbed passively via plant

Table 1 Merits and demerits of various nanocarrier systems in agriculture

Nanocarrier system	Merits	Demerits
MSNs	Stable structure, uniform pore size, tunable for targeted release phenomenon	Inorganic in nature, non-biodegradable
Solid lipidic nanoparticles (SLNs)	Highly soluble in water, cheaper in cost, biodegradable as well as biocompatible, easy to scale up	Low loading capacity, low encapsulation efficiency, chance of premature release of Ais during storage
Nano-emulsions	Highly soluble in water, highly stable to gravitational separation, biocompatible and biodegradable, suitable for carrying lipophilic materials	Requires huge amounts of surfactant to stabilize the nanometric size of the oil droplets
Dendrimers	Resistant to hydrolysis, suitable for carrying both lipophilic and lipophobic materials, enables targeted delivery of AIs	Difficult to maintain structural morphology as well as crystallinity, chance of cytotoxicity if used in high amount
Nanocrystals	Improves AI stability in liquid formulations, increases the bioavailability of water-soluble pesticide materials	Expensive, requires a high amount of stabilizers or solvents
Hydrogels	Highly bio- and eco-compatible, very cheap in cost, and easy to formulate	Variability exists from batch to batch as the nature of the polymer changes, and stability issues

Source [66]

stomata or root systems, or by conjugating ligands that engage with particular biological structures, so assuring accurate delivery of the pesticide to the required location [60]. This dual strategy mitigates environmental pollution, decreases pesticide use, and enhances pest control effectiveness.

Biodegradable polymers, including chitosan, gelatin, alginate, collagen, polyethylene glycol (PEG), etc., are generally used to form nanocapsules which can further be employed as nanocarriers [43]. Nanocapsules possess a core-shell structure that envelops the material of interest. Typically, release transpires by the diffusion of the active ingredient through the polymeric membrane into the nucleus, ultimately reaching the shell, where it is subjected to various external stimuli. The tensile strength and biological degradation of the coating material, coating thickness, active component density, and soil physiology and moisture content all play a significant role in the controlled-release mechanism of nanostructures [17]. The intricacy of the phenomena may encompass water transport through the coating, water condensation within the nanocapsule nucleus, osmotic pressure development, active compound dilution, granule swelling, and micropore modification. The transport of pesticide

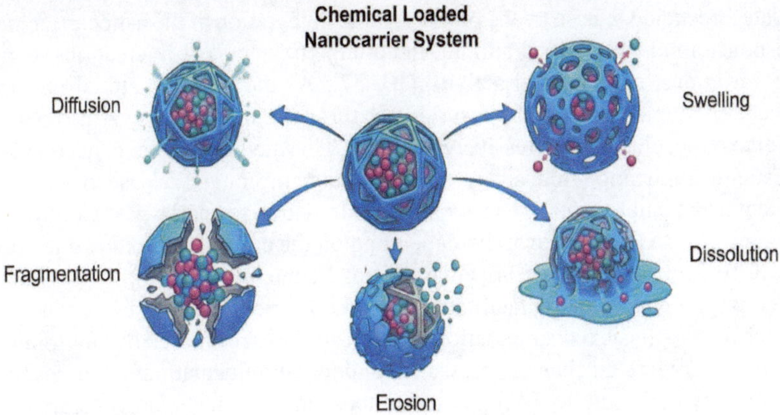

Fig. 3 Vital mechanisms for releasing active compounds from nanocarriers

chemicals integrated and/or adsorbed onto nanocarriers may transpire by (i) diffusion, (ii) dissolution, (iii) erosion, (iv) fragmentation, and (v) swelling (Fig. 3) [44].

The diffusion mechanism of pesticide release from smart nanocarriers entails the progressive translocation of pesticide molecules from the carrier matrix into the external environment, propelled by a concentration gradient. This passive mechanism relies on the structural characteristics of the nanocarrier—specifically polymer density, porosity, and molecular interactions—that govern the diffusion rate of pesticides. The release generally adheres to Fick's law of diffusion, facilitating a regulated, prolonged pesticide administration that minimizes environmental loss and improves target efficacy [42]. Moreover, diffusion-based release may be integrated with stimuli-responsive triggers (such as pH, temperature, or enzymes) to optimize the time and location of pesticide delivery, hence improving effectiveness and sustainability in agricultural activities [10, 37].

The dissolving mechanism of smart pesticide release from nanocarrier systems entails the slow degradation or solubilization of the nanocarrier matrix under certain environmental circumstances, such as variations in pH, temperature, or humidity. The carrier material, often made of biodegradable polymers or responsive organic compounds, dissolves or degrades, releasing the contained pesticide molecules into the surrounding environment [10, 37]. The release mechanism is regulated by the dissolution rate of the nanocarrier, which may be modified by altering the carrier's chemical composition and cross-linking density. Dissolution mode, in contrast to diffusion-controlled release, relies on the physical breakdown of the carrier rather than only on molecular movement [42]. This method facilitates accurate, prolonged, and stimulus-responsive pesticide distribution, enhancing stability and reducing premature loss or environmental contamination in agricultural contexts.

Erosion, in contrast to diffusion or dissolution, depends on the physical breakdown of the carrier rather than only on molecular mobility or solubilization, enabling predictable and sustained pesticide availability. The erosion mechanism facilitates regulated pesticide release by the progressive disintegration of the nanocarrier matrix in response to environmental stimuli, including enzymes, pH fluctuations, temperature variations, or microbial activity [10, 37]. As the carrier material undergoes physical erosion—either from its surface or throughout its entirety—the contained pesticide molecules are progressively liberated. This method enables exact regulation of pesticide application time and rate, guaranteeing prolonged release in accordance with environmental factors. The erosion mode decreases early pesticide loss and promotes targeted effectiveness by depending on the carrier's breakdown instead of mere diffusion, hence decreasing environmental contamination [42].

The fragmentation mode facilitates regulated pesticide release from nanocarrier systems by the physical disintegration of the carrier matrix caused by environmental stressors, including mechanical pressures, temperature fluctuations, chemical exposure, or enzymatic activity [42]. As the nanocarrier fragments, it disintegrates into smaller components, gradually releasing the enclosed pesticide molecules into the environment. This process continues until the carrier is entirely degraded, enabling a sustained and regulated release dependent upon the pace and degree of fragmentation [10, 37]. Fragmentation works in conjunction with other processes, such as erosion and dissolution, to optimize the time and quantity of pesticide application, enhancing targeted precision and minimizing premature loss or environmental pollution.

The swelling mode enables controlled-pesticide release from nanocarrier systems by the absorption of water or other solvents into the polymeric matrix, causing it to swell and expand. This swelling increases the internal pore size of the nanocarrier, creating pathways through which the pesticide molecules can diffuse out gradually [42]. The discharge rate depends primarily on how quickly the nanocarrier swells and the diffusion time of the pesticide through the swollen matrix [10, 37]. Environmental factors such as pH, temperature, and humidity influence the swelling behavior, allowing for stimuli-responsive and sustained pesticide delivery. This mechanism slows down pesticide release, enhancing stability and reducing premature loss in the field.

Pesticide deposition efficiency and adherence to the leaf surface are critical variables in reducing pesticide loss and increasing use efficiency [61]. NPs characterized by a high specific surface area and robust surface reactivity are extensively anticipated to be immobilized or aggregated on plant epidermis through mechanisms such as electrostatic attraction, mechanical adhesion, and hydrophobic interactions. The interaction between nanopesticides and plants primarily involves three steps: (1) NPs are adsorbed on plant surfaces; (2) NPs penetrate the cuticle and epidermis and migrate to the vascular tissues via symplast or exoplasm; and (3) NPs are transported to other parts of the plant via the vascular tissue [53, 69]. Although roots and leaves have comparable properties (an epidermis with stomata, mesophyll, and vascular tissue), the structural differences between roots and leaves would result in differing barriers to NP entrance. Meanwhile, nanoparticle form, application technique, and the type of plant tissues all influence NP accumulation and transport, with the phloem transport pathway facilitating NP translocation from leaf to root [70].

4 Real-Life Applications in Modern Agriculture

Nanotechnology has been extensively employed recently in the creation of novel formulations, including insecticides, pesticides, and insect repellents [54, 57]. Pesticide active ingredients (AIs) are often lipid-soluble, necessitating the use of many organic solvents, as well as emulsifiers and other additives [28]. These criteria may lead to excessive usage of organic solvents, endangering the environment and human health. Solid formulations need solid emulsifiers and a variety of inert fillers, and they are extensively employed owing to their ease of storage and transport. Furthermore, traditional pesticide products are becoming increasingly inefficient, with AIs using less than 0.1% of biological target absorption after dust drift and rainfall leaching [8, 29, 89]. In this context, potential nanocarrier systems can improve the bioavailability of AIs. For example, Avermectin (AV), a highly effective bio-pesticide, is susceptible to ultraviolet (UV) radiation and has a short half-life of 6 h [46]. Zhu et al. [91] similarly conjugated AV to the poly (ethylene glycol)-carboxymethyl cellulose (PEG-CMC) chain, yielding nanoparticles that significantly improved the stability of AV relative to the control, with a degradation rate of under 50% after 72 h of exposure to light. Similar to AV, some unstable pesticides, such as phoxim, fosthiazate, and quinalphos, which are prone to photolysis and degradation, would likely benefit from the use of potential nanocarriers [38, 90].

The nanocarrier system mitigates the wastage of active ingredients (AIs) and inadvertent damage to non-target organisms, while also diminishing the likelihood of resistance arising from suboptimal concentrations of AIs [75]. Xiang et al. [78] engineered a pH-responsive nano-mesh structure of polydopamine-modified attapulgite (PA) for the regulated release of chlorpyrifos (CPF). In alkaline circumstances, the outer layer of material deteriorated, resulting in the prolonged release of AIs. The safeguarding of the material enhanced the efficiency of the AIs and extended the period of pesticide efficacy. Pyraclostrobin has shown remarkable effectiveness against rice blast disease. This compound induces physiological changes in crops, enhancing their disease resistance and markedly augmenting yields [24].

NMs serve as carriers that enhance the dissolution of active ingredients while safeguarding them against volatilization and degradation [4]. Moreover, the particle size of pesticides affects their application [88]. Due to the small size effect and elevated specific surface area, a greater number of NMs containing active ingredients (AIs) of pesticides can adhere to the target per unit area, thereby significantly enhancing the contact area of AIs with the target and augmenting their bioavailability [87]. This results in a direct or indirect impact on the generation of abnormal reactive oxygen species (ROS) within insect bodies [76]. Table 2 illustrates pertinent studies on the preparation, deployment, and assessment of nanocarrier systems for pesticide chemicals.

Nonetheless, it is evident that the laboratory-simulated release environment is relatively simplistic, whereas the actual release environment is intricate and variable, as pH, salinity, light, microorganisms, and other factors may individually or

Table 2 Nanocarrier-based pesticide synthesis and application in crop pest management

Nanocarrier	Active ingredient (pesticide)	Synthesis process	Effect(s)	Reference
Mesoporous silica and trimethylammonium	2,4- dichlorophenoxy acetic acid	Sol-gel and nanosilica grafting	Slow/regulated release of pesticide agents to dicot plants	[6]
ZnO loaded silica gel microparticles and Cu NPs	–	Chemical reaction in aqueous dispersion	Excellent antimicrobial properties and high effectiveness against *Xanthomonas alfafae*, *pseudomonas syringae*, and *Clavibacter michiganensis*	[84]
2-nitrobenzylcarboxymethyl-chitosan succinate micelles	Diuron	Conjugation	Elevated rate of photo-responsive release (96.8%) for up to 8 h (at pH 7) under solar radiation	[82]
Hollow TiO_2 nanoparticles doped with Ag	–	Chemical reaction in alcoholic dispersion	Excellent antifungal activity against *Fusarium solani* and *Venturia inaequalis* under visible light	[2]
Chitosan and gum arabic nanoparticles	Carvacrol/linalool	Ionogenic gelation	Excellent insecticidal action against *Helicoverpa armígera* and *Tetranychus urticae*	[5]
Zein nanoparticles	Neem oil	Anti-solvent precipitation	Excellent antimicrobial activity and safe to apply soil biota	[48]
Microspheres of vinyl polyacetate (PVA)	Emamectin-benzoate (EMB)	Microemulsion polymerization	Superb photoprotective efficacy of active substances, structural stability and sustained EMB release beyond 200 h	[73]

interactively affect pesticide release. Therefore, it is crucial to examine the correlation between the laboratory-based release model and the real-world release model to enhance the precision and sophistication of pesticide delivery systems.

5 Safety and Environmental Impact

Nanotechnology offers novel approaches to enhance agricultural output and address several environmental challenges related to scarce water resources, soil degradation, energy shortages, and climate change [30]. Nanotechnology techniques have shown effective outcomes across several agricultural applications to improve crop output and yield [49]. Numerous studies have shown the significant impact of nanoscale agrochemicals on enhancing agricultural output, yet research remains in its nascent phase [25]. The research indicates that the advantageous impacts of nanoscale agrochemicals are mostly contingent upon their physicochemical qualities, duration of exposure, target species, and environmental variables [27]. Consequently, meticulous evaluation of characteristics like dosage, administration method, and experimental settings is essential for assessing these items.

Currently, researchers are concentrating on creating nanocarriers that possess safety attributes and effective mechanisms for the sustained and targeted release of nutrients and insecticides. The development of nanocarriers is expected to improve their effectiveness on the target organism and enable the controlled release of compounds or agrochemicals, hence minimizing losses due to premature degradation, leaching, and volatilization [3, 51]. Consequently, controlled-release systems using nanocarriers to enhance plant development and protection are anticipated to be a crucial instrument for addressing the environmental challenges posed by traditional agrochemicals in future agriculture [16]. The environmental safety and possible dangers of nanoscale agrochemicals to non-target creatures and human health must be evaluated to guarantee proper management [56]. To date, risk assessments of nanoscale agrochemicals have primarily focused on experimental evaluations performed in laboratory environments, while real-world scenarios within agricultural systems remain inadequately investigated. Consequently, novel strategies must be devised to enhance the delivery, absorption, targeting, and sustained impacts under field situations to assess efficacy and environmental hazards.

Future research should also look at how nanoscale agrochemicals interact with non-target species in the environment. The interplay of nanoscale agrochemicals applied to the foliage and roots of plants must be examined to identify possible hazards and adverse effects throughout the food chain. In this regard, proteomic, genomic, and metabolomic investigations might aid in understanding the processes underlying interactions with exposed species [9]. A thorough multidisciplinary investigation of nanoscale agrochemicals' influence on ecosystems and the food chain should be conducted to understand their consequences for environmental destiny, which is critical for making policy choices and determining market status. At present, there exist no established safety standards for regulatory bodies to authorize innovative nanoproducts; nonetheless, the European Food Safety Authority and OECD guidelines for assessing soil leaching and nanomaterial toxicity represent significant advancements in the validation of nanoscale agrochemicals [1, 21].

Consequently, to address the existing challenges stemming from insufficient understanding of the implications of nanoscale agrochemicals in agricultural systems and their true impact on ecosystem functionality, the utilization of nanoscale agrochemicals in agriculture necessitates significant effort across multiple scientific domains [25]. To critically evaluate advancements in the use of nanoscale agrochemicals, a number of views from the public, industry, regulators, and research must be combined. In the last several years, numerous initiatives have been executed into global management plans to assess the possible risks and dangers of nanoscale agrochemicals [22]. Furthermore, novel methods for evaluating nanoscale agrochemicals in the lab should be developed for soils and field research to show their effectiveness in actual farming settings. Although several businesses have created procedures for the manufacturing and use of nanoscale agrochemicals, which now provide a number of advantages, commercial goods still need additional development. Production at the industrial level has been constrained by the expense, complexity of the formulation, and high demand for skilled workers at the retail level. One of the main obstacles to scaling production is the low-margin business and high manufacturing costs [59]. Furthermore, proper assessment has been hampered by the absence of common procedures and standards for developing laws regarding nanoscale agrochemicals in food and agriculture. Consequently, the development of standardized techniques to assess the safety hazards for their long-term use in field settings is imperative. Lastly, because of their impact on food and agribusiness, as well as the lack of trustworthy data about their usage in actual agricultural settings, ethical concerns with the use of nanoscale agrochemicals must be taken into account.

6 Challenges and Future Perspectives

Nanopesticides are one of the most useful new-generation bioweapons in pest management. These nanometric systems are popularly known for prolonged and targeted delivery of active molecules inside the domain [55]. Consequently, nanodelivery techniques for pesticides have been thoroughly investigated to root out a fruitful course of action in order to counter recent challenges of environmental pollution, inadequate efficacy of traditional application systems, and reduced toxicity to users and food. The research indicates several benefits associated with the usage of these nanoparticles, suggesting a positive and more sustainable future for these nanostructures in comparison to conventional agriculture [10].

Nonetheless, some challenges that impede the use of nanopesticides in agriculture have constrained the development of industrial manufacturing and the establishment of these materials in the marketplace. Obstacles to the advancement of nanotechnology encompass inadequate attention in education, especially related to R&D infrastructure, high production costs of NMs, limited financial returns in agriculture, and issues within the agriculture sector to adopt NMs in small-scale farming [47]. Numerous technological constraints in the industrial manufacturing of nanopesticides hinder the progress in their use, such as the substantial energy requirements of the related processes [13]. Owing to their dimensions, nanoformulations often aggregate or disrupt the matrix, hence altering their surface chemistry. The aggregation of NPs results in the formation of non-nanometric substances, which contradicts their primary purpose of manufacturing [13]. Additional challenges to the progress of nanoscience include the uncertain dangers associated with the absorption and deposition of NPs in the food chain, thereby presenting acute health risks to humans, animals, and the environment [41, 67]. Therefore, additional research and more cohesive concepts are essential concerning health safety in the utilization of NMs, encompassing their mechanisms of functioning, bioaccumulation patterns, processing costs, economic feasibility of production, distribution logistics, stability of nanoformulation inventory, application methodologies in the field, and financial advantages to the producer.

7 Conclusion

Pesticides are essential for agricultural production, used to safeguard crops, and enhance their quality. Eventually, conventional pesticides are widely applied; however, they cause harmful impacts on non-target organisms and the natural surroundings that must not be disregarded. In this context, the development of regulated and long-term release nanocarriers for pesticides has become a crucial approach to reducing pesticide consumption. The retention characteristics of nanopesticides on plant surfaces and their efficient absorption by plants enhance the efficacy of pest control in agriculture. Nonetheless, numerous challenges related to their implementation persist, particularly regarding their placement within the ecosystem. Primarily, the question of whether nanocarriers can be broadly utilized across a category of AIs or universally for all AIs is essential for decreasing the cost of nanopesticides. Secondly, the responsiveness of the nanocarriers to external stimuli governs the precise release of the AIs. It is essential to integrate the switch for regulating pesticide release with stimuli produced by detrimental organisms or external environmental conditions to enhance response sensitivity and guarantee accurate agent release upon activation. The processes of absorption and translocation of nanocarriers in plants, as well as their ecological effects, need a comprehensive assessment prior to the commercialization of nanocarriers. However, it is highly expected that ongoing research advancements and enhancements in preparation methods would provide significant advantages for contemporary agriculture via nanocarrier-based tailored pesticide delivery.

References

1. Amenta V, Aschberger K, Arena M, Bouwmeester H, Moniz FB, Brandhoff P, Peters RJ (2015) Regulatory aspects of nanotechnology in the Agri/feed/food sector in EU and non-EU countries. Regul Toxicol Pharmacol 73(1):463–476
2. Boxi SS, Mukherjee K, Paria S (2016) Ag doped hollow TiO_2 nanoparticles as an effective green fungicide against *fusarium solani* and *Venturia inaequalis* phytopathogens. Nanotechnology 27(8):085103
3. Camara MC, Campos EVR, Monteiro RA, do Espirito Santo Pereira A, de Freitas Proença PL, Fraceto LF (2019) Development of stimuli-responsive nano-based pesticides: emerging opportunities for agriculture. J Nanobiotechnol 17(1):100
4. Campos EV, Proença PL, Oliveira JL, Bakshi M, Abhilash PC, Fraceto LF (2019) Use of botanical insecticides for sustainable agriculture: future perspectives. Ecol Indic 105:483–495
5. Campos EV, Proença PL, Oliveira JL, Pereira AE, de Morais Ribeiro LN, Fernandes FO, Fraceto LF (2018) Carvacrol and linalool co-loaded in β-cyclodextrin-grafted chitosan nanoparticles as sustainable biopesticide aiming pest control. Sci Rep 8(1):7623
6. Cao L, Zhou Z, Niu S, Cao C, Li X, Shan Y, Huang Q (2018) Positive-charge functionalized mesoporous silica nanoparticles as nanocarriers for controlled 2, 4-dichlorophenoxy acetic acid sodium salt release. J Agric Food Chem 66(26):6594–6603
7. Chaturvedi A, Tripathi D, Ranjan R (2025) Nano-enabled biosensors in early detection of plant diseases. Front Nanotechnol 7:1545792
8. Chen H, Zhi H, Liang J, Yu M, Cui B, Zhao X, Zeng Z (2021) Development of leaf-adhesive pesticide nanocapsules with pH-responsive release to enhance retention time on crop leaves and improve utilization efficiency. J Mater Chem B 9(3):783–792
9. Chi Y, Zhang G, Xiang Y, Cai D, Wu Z (2017) Fabrication of a temperature-controlled-release herbicide using a nanocomposite. ACS Sustain Chem Eng 5(6):4969–4975
10. de Oliveira C, Mulinari J, Reichert F, Júnior A (2020) Nano-delivery systems of pesticides active agents for agriculture applications-an overview. In: Ciência, Tecnologia e Inovação: Do Campo à Mesa, Proceeding of International Agribusiness Congress, Recife, Brazil, pp 25–27
11. De A, Bose R, Kumar A, Mozumdar S (2013) Worldwide pesticide use. In: Targeted delivery of pesticides using biodegradable polymeric nanoparticles. New Delhi, Springer India, pp 5–6
12. Devlet A (2021) Modern agriculture and challenges. Front Life Sci Relat Technol 2(1):21–29
13. Dimkpa CO, Bindraban PS (2017) Nanofertilizers: new products for the industry? J Agric Food Chem 66(26):6462–6473
14. Eigenbrode SD, Adhikari S (2023) Climate change and managing insect pests and beneficials in agricultural systems. Agron J 115(5):2194–2215
15. Ghormade V, Deshpande MV, Paknikar KM (2011) Perspectives for nano-biotechnology enabled protection and nutrition of plants. Biotechnol Adv 29(6):792–803
16. Gomez A, Narayan M, Zhao L, Jia X, Bernal RA, Lopez-Moreno ML, Peralta-Videa JR (2021) Effects of nano-enabled agricultural strategies on food quality: current knowledge and future research needs. J Hazard Mater 401:123385
17. Gutiérrez CA, Ledezma-Delgadillo A, Juárez-Luna G, Neri-Torres EE, Ibanez JG, Quevedo IR (2022) Production, mechanisms, and performance of controlled-release fertilizers encapsulated with biodegradable-based coatings. ACS Agric Sci Technol 2(6):1101–1125
18. Hou Q, Zhang H, Bao L, Song Z, Liu C, Jiang Z, Zheng Y (2021) NCs-delivered pesticides: a promising candidate in smart agriculture. Int J Mol Sci 22(23):13043
19. Huang B, Chen F, Shen Y, Qian K, Wang Y, Sun C, Cui H (2018) Advances in targeted pesticides with environmentally responsive controlled release by nanotechnology. Nano 8(2):102
20. Izuafa A, Chimbekujwo KI, Raji RO, Oyewole OA, Oyewale RO, Abioye OP (2025) Application of nanoparticles for targeted management of pests, pathogens and disease of plants. Plant Nano Biol 100177
21. Jain A, Das S (2021) Regulatory requirements for nanopesticides and nanofertilizers. In: Advances in Nano-fertilizers and Nano-pesticides in agriculture. Woodhead Publishing, pp 145–152

22. Jain A, Ranjan S, Dasgupta N, Ramalingam C (2018) Nanomaterials in food and agriculture: an overview on their safety concerns and regulatory issues. Crit Rev Food Sci Nutr 58(2):297–317
23. Jiang Q, Xie Y, Peng M, Wang Z, Li T, Yin M, Yan S (2022) A nanocarrier pesticide delivery system with promising benefits in the case of dinotefuran: strikingly enhanced bioactivity and reduced pesticide residue. Environ Sci Nano 9(3):988–999
24. Kabir MH, Tisha FS, Nayan HR, Islam MA, Kashem MA, Uddin MI, Meah MB (2021) Determining an effective and economic fungicide spray schedule for reducing blast of wheat. Int J Agric Res, Innov Technol (IJARIT) 11(1):10–16
25. Kah M (2015) Nanopesticides and nanofertilizers: emerging contaminants or opportunities for risk mitigation? Front Chem 3:64
26. Kah M, Tufenkji N, White JC (2019) Nano-enabled strategies to enhance crop nutrition and protection. Nature Nanotechnol 14(6):532–540
27. Kah M, Kookana RS, Gogos A, Bucheli TD (2018) A critical evaluation of nanopesticides and nanofertilizers against their conventional analogues. Nat Nanotechnol 13(8):677–684
28. Kaur R, Mavi GK, Raghav S, Khan I (2019) Pesticides classification and its impact on environment. Int J Curr Microbiol Appl Sci 8(3):1889–1897
29. Kole RK (2021) Improved pesticide formulation for sustainable crop protection. Ensuring Food Saf, Secur Sustain Crop Prot 5:50–55
30. Kumar A, Choudhary A, Kaur H, Mehta S, Husen A (2021) Smart nanomaterial and nanocomposite with advanced agrochemical activities. Nanoscale Res Lett 16(1):156
31. Kumar S, Bhanjana G, Sharma A, Sidhu MC, Dilbaghi N (2014) Synthesis, characterization and on field evaluation of pesticide loaded sodium alginate nanoparticles. Carbohydr Polym 101:1061–1067
32. Kundu S, Kumar SV, Pooniya V, Reddy BM, Basha NH, Das K, Barman A, Dutta S, Shah MH, Erapuram P (2025a) Soil conservation techniques and sustainable land management. In: Eco-resilience: climate change, land degradation and sustainable solutions. Cham, Springer Nature Switzerland, pp 381–404
33. Kundu S, Kumar SV, Reddy BM, Kasbe SS, Goud GMK, Pooniya V, Banoth SK, Shah MH, Ali S, Pramanik K (2025b) Effect of doses, particle shape, size, and composition of Nanofertilizers on yield of crops. In: Nanofertilizers in agriculture: synthesis, mechanisms, and effect on plants. Cham, Springer Nature Switzerland, pp 291–310
34. Li J, Tang S, Lu L, Zeng HC (2007) Preparation of nanocomposites of metals, metal oxides, and carbon nanotubes via self-assembly. J Am Chem Soc 129(30):9401–9409
35. Li M, Xue L, Gao T, Zhang Z, Zhao D, Li X, Kang Z (2025) Sustainable Pest management with hollow mesoporous silica nanoparticles loaded with β-Cypermethrin. Agronomy 15(3):737
36. Li N, Sun C, Jiang J, Wang A, Wang C, Shen Y, Wang Y (2021) Advances in controlled-release pesticide formulations with improved efficacy and targetability. J Agric Food Chem 69(43):12579–12597
37. Liechty WB, Kryscio DR, Slaughter BV, Peppas NA (2010) Polymers for drug delivery systems. Annu Rev Chem Biomol Eng 1(1):149–173
38. Liu B, Zhang J, Chen C, Wang D, Tian G, Zhang G, Wu Z (2021) Infrared-light-responsive controlled-release pesticide using hollow carbon microspheres@ polyethylene glycol/α-cyclodextrin gel. J Agric Food Chem 69(25):6981–6988
39. Ma Y, Li J, Yang Q, Li F, Wang L, Yan P, Wu X (2024) Redox-responsive nanopesticides based on natural polymers for environmentally safe delivery of pesticides with enhanced foliar dispersion and washout resistance. J Agric Food Chem 72(37):20343–20353
40. Ma X, Wang X (2022) Multigenerational exposure of plants to engineered nanoparticles (ENPs). In: Plant exposure to engineered nanoparticles. Academic Press, pp 195–207
41. Manchikanti P (2019) Bioavailability and environmental safety of nanobiopesticides. Nano-Biopestic Today Futur Perspect:207–222
42. McClements DJ (2014) Nanoparticle-and microparticle-based delivery systems: encapsulation, protection and release of active compounds. CRC press
43. Musyanovych A, Landfester K (2014) Polymer micro-and nanocapsules as biological carriers with multifunctional properties. Macromol Biosci 14(4):458–477

44. Nasif SO, Nuruzzaman M, Naidu R (2024) Porous silica nanocarriers: advances in structural orientation and modification to develop sustainable pesticide delivery systems. ACS Agric Sci Technol 4(2):144–172
45. Nuruzzaman MD, Rahman MM, Liu Y, Naidu R (2016) Nanoencapsulation, nano-guard for pesticides: a new window for safe application. J Agric Food Chem 64(7):1447–1483
46. Pang Y, Qin Z, Wang S, Yi C, Zhou M, Lou H, Qiu X (2020) Preparation and application performance of lignin-polyurea composite microcapsule with controlled release of avermectin. Colloid Polym Sci 298(8):1001–1012
47. Parisi C, Vigani M, Rodríguez-Cerezo E (2015) Agricultural nanotechnologies: what are the current possibilities? Nano Today 10(2):124–127
48. Pascoli M, Jacques MT, Agarrayua DA, Avila DS, Lima R, Fraceto LF (2019) Neem oil based nanopesticide as an environmentally-friendly formulation for applications in sustainable agriculture: an ecotoxicological perspective. Sci Total Environ 677:57–67
49. Qazi G, Dar FA (2020) Nano-agrochemicals: economic potential and future trends. In: Nanobiotechnology in agriculture: an approach towards sustainability. Springer International Publishing, Cham, pp 185–193
50. Qian K, Shi T, Tang T, Zhang S, Liu X, Cao Y (2011) Preparation and characterization of nano-sized calcium carbonate as controlled release pesticide carrier for validamycin against Rhizoctonia solani. Microchim Acta 173(1):51–57
51. Rajput VD, Singh A, Minkina T, Rawat S, Mandzhieva S, Sushkova S, Upadhyay SK (2021) Nano-enabled products: challenges and opportunities for sustainable agriculture. Plants 10(12):2727
52. Salerno G, Rebora M, Piersanti S, Saitta V, Kovalev A, Gorb E, Gorb S (2021) Reduction in insect attachment caused by different nanomaterials used as particle films (kaolin, zeolite, calcium carbonate). Sustainability 13(15):8250
53. Santiago EF, Pontes MS, Arruda GJ, Caires AR, Colbeck I, Maldonado-Rodriguez R, Grillo R (2020) Understanding the interaction of nanopesticides with plants. In: Nanopesticides: from research and development to mechanisms of action and sustainable use in agriculture. Springer International Publishing, Cham, pp 69–109
54. Shah MA, Wani SH, Khan AA (2016) Nanotechnology and insecticidal formulations. J Food Bioeng Nanoprocessing 1(3):285–310
55. Shah MH, Aktar SN, Barik S, Chowdhury A, Molla SR, Kundu S, Ahmad S, Mukherjee D, Mandal B, Hossain A (2025) Effect of various Nanofertilizers on physiological activities of plants. In: Nanofertilizers in agriculture: synthesis, mechanisms, and effect on plants. Cham, Springer Nature Switzerland, pp 311–354
56. Shahane SP, Kumar A (2022) Estimation of health risks due to copper-based nanoagrochemicals. Environ Sci Pollut Res 29(17):25046–25059
57. Shahid M, Naeem-Ullah U, Khan W, Saeed DS, Razzaq K (2021) Application of nanotechnology for insect pests management: a review. J Innov Sci 7:28–39
58. Sharma S, Sahu BK, Cao L, Bindra P, Kaur K, Chandel M, Shanmugam V (2021) Porous nanomaterials: Main vein of agricultural nanotechnology. Prog Mater Sci 121:100812
59. Singh H, Sharma A, Bhardwaj SK, Arya SK, Bhardwaj N, Khatri M (2021) Recent advances in the applications of nano-agrochemicals for sustainable agricultural development. Environ Sci: Processes Impacts 23(2):213–239
60. Smith AM, Gilbertson LM (2018) Rational ligand design to improve agrochemical delivery efficiency and advance agriculture sustainability. ACS Sustain Chem Eng 6(11):13599–13610
61. Su H, Liang Y, Gao A, Yang P (2023) Tackling pesticide overusing through foliar deposition and retention: Frontiers and challenges. Adv Agrochem 2(4):298–305
62. Su Y, Ashworth V, Kim C, Adeleye AS, Rolshausen P, Roper C, White J, Jassby D (2019) Delivery, uptake, fate, and transport of engineered nanoparticles in plants: a critical review and data analysis. Environ Sci Nano 6(8):2311–2331
63. Sun C, Wang Y, Zhao X, Cui B, Zhang L, Zeng Z, Cui H (2020b) Progress on categories and synergistic mechanism of nanopesticides. Chin J Pestic Sci 22(2):205–213

64. Sun C, Zeng Z, Cui H, Verheggen F (2020a) Polymer-based nanoinsecticides: current developments, environmental risks and future challenges-a review. Biotechnol Agron Soc Environ:24
65. Tao R, You C, Qu Q, Zhang X, Deng Y, Ma W, Huang C (2023) Recent advances in the design of controlled-and sustained-release micro/nanocarriers of pesticide. Environ Sci Nano 10(2):351–371
66. Vega-Vásquez P, Mosier NS, Irudayaraj J (2020) Nanoscale drug delivery systems: from medicine to agriculture. Front Bioeng Biotechnol 8:79
67. Vilchez-Aruani J, Cuello-Carrión FD, Valdez SR, Nadin SB (2020) Genomic effects of a nanostructured alumina insecticide in human peripheral blood lymphocytes in vitro. Heliyon 6(6)
68. Vishnu M, Kannan M, Soundararajan RP, Suganthi A, Subramanian A, Senthilkumar M, Govindaraju K (2024) Nano-bioformulations: emerging trends and potential applications in next generation crop protection. Environ Sci Nano 11(7):2831–2860
69. Wang D, Saleh NB, Byro A, Zepp R, Sahle-Demessie E, Luxton TP, Su C (2022) Nano-enabled pesticides for sustainable agriculture and global food security. Nature Nanotechnol 17(4):347–360
70. Wang X, Xie H, Wang P, Yin H (2023) Nanoparticles in plants: uptake, transport and physiological activity in leaf and root. Materials 16(8):3097
71. Wang X, Xie H, Wang Z, He K (2019b) Graphene oxide as a pesticide delivery vector for enhancing acaricidal activity against spider mites. Colloids Surf B: Biointerfaces 173:632–638
72. Wang X, Xie H, Wang Z, He K, Jing D (2019a) Graphene oxide as a multifunctional synergist of insecticides against lepidopteran insect. Environ Sci Nano 6(1):75–84
73. Wang Y, Wang A, Wang C, Cui B, Sun C, Zhao X, Cui H (2017) Synthesis and characterization of emamectin-benzoate slow-release microspheres with different surfactants. Sci Rep 7(1):12761
74. Wei Y, Chen J, Dong M, Yin M, Shen J, Gao L, Yan S (2025) Nano-enabled insecticides for efficient pest management: definition, classification, synergistic mechanism, and safety assessment. Nano 15(13):1050
75. Wu D, Qin M, Xu D, Wang L, Liu C, Ren J, Lu Y (2019) A bioinspired platform for effective delivery of protein therapeutics to the central nervous system. Adv Mater 31(18):1807557
76. Wu F, Harper BJ, Crandon LE, Harper SL (2020) Assessment of cu and CuO nanoparticle ecological responses using laboratory small-scale microcosms. Environ Sci Nano 7(1):105–115
77. Wu H, Li Z (2022) Nano-enabled agriculture: How do nanoparticles cross barriers in plants? Plant Commun 3(6)
78. Xiang Y, Zhang G, Chen C, Liu B, Cai D, Wu Z (2018) Fabrication of a pH-responsively controlled-release pesticide using an attapulgite-based hydrogel. ACS Sustain Chem Eng 6(1):1192–1201
79. Xiang Y, Zhang G, Chi Y, Cai D, Wu Z (2017) Fabrication of a controllable nanopesticide system with magnetic collectability. Chem Eng J 328:320–330
80. Xu C, Cao L, Zhao P, Zhou Z, Cao C, Li F, Huang Q (2018) Emulsion-based synchronous pesticide encapsulation and surface modification of mesoporous silica nanoparticles with carboxymethyl chitosan for controlled azoxystrobin release. Chem Eng J 348:244–254
81. Xu WS, Yang ZXS, Zhang GY, Liu H (2022) Light and temperature dual responsive pesticide release system based on mesoporous silica nanoparticles modified by dopamine. J Cent South Univ 29(2):397–409
82. Ye Z, Guo J, Wu D, Tan M, Xiong X, Yin Y, He G (2015) Photo-responsive shell cross-linked micelles based on carboxymethyl chitosan and their application in controlled release of pesticide. Carbohydr Polym 132:520–528
83. You C, Lin H, Ning L, Ma N, Wei W, Ji X, Wang F (2024) Advances in the design of functional cellulose based nanopesticide delivery systems. J Agric Food Chem 72(20):11295–11307
84. Young M, Ozcan A, Myers ME, Johnson EG, Graham JH, Santra S (2017) Multimodal generally recognized as safe ZnO/nanocopper composite: a novel antimicrobial material for the management of citrus phytopathogens. J Agric Food Chem 66(26):6604–6608

85. Zhang X, He Y, Yuan Z, Shen G, Zhang Z, Niu J, Qian K (2023) A pH-and enzymatic-responsive nanopesticide to control pea aphids and reduce toxicity for earthworms. Sci Total Environ 861:160610
86. Zhang ZJ, Shang XF, Yang L, Shi YB, Liu YQ, Li JC, Yang CJ (2020) Engineering of peglayted camptothecin into nanomicelles and supramolecular hydrogels for pesticide combination control. Front Chem 7:922
87. Zhao X, Cui H, Wang Y, Sun C, Cui B, Zeng Z (2017) Development strategies and prospects of nano-based smart pesticide formulation. J Agric Food Chem 66(26):6504–6512
88. Zheng L, Cao C, Cao L, Chen Z, Huang Q, Song B (2018) Bounce behavior and regulation of pesticide solution droplets on rice leaf surfaces. J Agric Food Chem 66(44):11560–11568
89. Zheng L, Cao C, Chen Z, Cao L, Huang Q, Song B (2021) Efficient pesticide formulation and regulation mechanism for improving the deposition of droplets on the leaves of rice (*Oryza sativa* L.). Pest Manag Sci 77(7):3198–3207
90. Zhou M, Xiong Z, Yang D, Pang Y, Wang D, Qiu X (2018) Preparation of slow release nanopesticide microspheres from benzoyl lignin. Holzforschung 72(7):599–607
91. Zhu H, Shen Y, Cui J, Li N, Wang C, Cui B, Cui H (2020) Avermectin loaded carboxymethyl cellulose nanoparticles with stimuli-responsive and controlled release properties. Ind Crop Prod 152:112497

Synergistic Application of Nanoparticles and Plant Growth-Promoting Rhizobacteria in Sustainable Agriculture

Bomba Dam, Sanchayita Saha, Susmita Akhuli, Prioty Bandhan Roop, and Sandip Debnath

Abstract Over time, to enhance agricultural productivity, the use of synthetic chemical pesticides has increased significantly. However, this approach has had detrimental effects on plants, soil, the environment, and human health, requiring immediate attention and prompt action. Among existing approaches, the use of natural microbial-based biofertilizers, i.e., plant growth-promoting rhizobacteria (PGPR), is a promising option to address the unrelenting challenge. PGPR enhances plant productivity through multiple direct and indirect pathways, including the most prominent ones: nitrogen fixation, potassium and phosphate solubilization, synthesis of plant growth-promoting hormones, siderophore production, and inhibition of pathogenic strains. Some strains are also demonstrating advanced traits, such as endophytic potential and nanoparticle production. Several PGPR are now reported to have the potential to produce nanoparticles, particularly silica, selenium, and silver nanoparticles, thereby enabling green synthesis. The trait also addresses the challenges associated with PGPR field application, including restricted diffusion, limited targeted delivery, and low permeability. Nanotechnology-based delivery systems are presently being introduced in agriculture due to their innovative formulation strategies and anticipated environmental and human health benefits. This book chapter discusses the application of nanoparticles used along with PGPR within the framework of sustainable agriculture.

Keywords Nanoparticles · Plant growth-promoting Rhizobacteria (PGPR) · Sustainable agriculture · Nanotechnology-based delivery systems · Biofertilizers · Green synthesis · Microbial-nanoparticle synergy

B. Dam (✉) · S. Saha · S. Akhuli
Microbiology Laboratory, Department of Botany (DST-FIST and UGC-DRS Funded), Institute of Science, Visva-Bharati (A Central University), Santiniketan, West Bengal, India
e-mail: bomba.dam@visva-bharati.ac.in

S. Saha
e-mail: sanchayitasaha.rs.bot@visvabharati.ac.in

P. B. Roop · S. Debnath (✉)
Department of Genetics and Plant Breeding, Palli-Siksha Bhavana (Institute of Agriculture), Visva-Bharati (A Central University), Sriniketan, West Bengal, India
e-mail: sandip.debnath@visva-bharati.ac.in

J.-T. Chen (ed.), *AI-Nano-Enabled Climate-Smart Agriculture and Sustainability*, Smart Nanomaterials Technology, https://doi.org/10.1007/978-981-92-1694-9_15

1 Introduction

The world's population is expected to surpass 9.7 billion by 2050 [1], thereby imposing substantial pressure on global food demand, which is projected to increase by 35–56% across a range of socio-economic and climate scenarios [2], which may not be met by conventional farming. To fulfill this demand, the present-day agricultural system faces challenges, including overwhelming environmental degradation, climate variability, declining soil fertility, and reduced soil microbial diversity resulting from the overuse of conventional pesticides, fungicides, herbicides, and fertilizers. In addition, crops are unavoidably affected by both biotic and abiotic stresses. Thus, there is an urgent need for sustainable, environmentally friendly technologies that enhance plant productivity and protection while minimizing environmental degradation. Among known systems, the most promising technology is the use of plant growth-promoting rhizobacteria (PGPR), which has diverse potential. PGPR are naturally occurring rhizospheric bacteria that enhance plant growth through multiple direct and indirect pathways, including nitrogen fixation, phosphate and potassium solubilization, synthesis of plant growth-promoting hormones, siderophore release, and induction of systemic resistance [3]. Traditionally, PGPRs have one or a few of these traits. But the recent trend is to identify PGPRs with multiple traits, such as the novel bacterium *Pantoea tagorei*, which possesses most of these traits [4]. These microbes improve nutrient acquisition, enhance plant growth by suppressing pathogens, and help plants adapt to abiotic stresses such as drought, salinity, and heavy metal toxicity. Over the past few decades, PGPR has been applied as biofertilizers and biocontrol agents in many cropping systems. However, the effectiveness of PGPR under environmental conditions is often unsustainable due to variable soil conditions, limited diffusion, non-targeted delivery, reduced permeability, competition with native microbes, and environmental stresses that reduce PGPR's performance [5]. Hence, improving stability and bioavailability and achieving the full potential of PGPR across diverse field conditions remain essential goals in sustainable agriculture.

A relatively new field, nanotechnology, offers new strategies to overcome many of these limitations. Nanoparticles (NPs) exhibit distinctive physicochemical properties, including a large surface-to-volume ratio, high reactivity, high solubility, and controlled release of their active ingredients [6]. In agricultural sciences, NPs of metals and metal oxides such as gold (Au), copper (Cu), silica (Si), zinc (Zn), zinc oxide (ZnO), iron oxide (Fe_3O_4), selenium (Se), silica (SiO_2), copper oxide (CuO), carbon-based nanomaterials, and organic form of NPs such as chitosan, and lignin have shown promising effect in enhancing seed germination, stimulating photosynthesis, providing targeted nutrient delivery, and protecting plants from biotic and abiotic stresses in the form of nanofertilizers, nanopesticides, and nanoherbicides [7]. Despite these advantages, concerns about nanoparticle toxicity, environmental persistence, and their effects on soil microbial communities persist, necessitating careful integration with biological approaches, such as optimizing concentration, shape, and size.

The co-application of PGPR and NPs represents an innovative approach that unites the strengths of both systems while potentially mitigating their limitations. In many cases, NPs can enhance PGPR performance by improving survival, root colonization, and stress tolerance. For example, nano-formulated nutrients can supplement PGPR by providing readily available micronutrients, such as Zn, Fe, Mn, and Mo, thereby enhancing metabolic activity and improving the efficiency of biofertilizers. NPs can also protect PGPR against environmental stresses through encapsulation technologies, creating nano-structured coatings that enhance microbial stability during storage, transport, and field application. Specific nanoparticles, particularly polymer-based nanocarriers, are now being used to encapsulate PGPR, enabling controlled release, improved root adhesion, and protection against UV radiation and desiccation [8].

Thus, while nanoparticles supplement PGPR's nutrition, activities, and survivability in soil, the later in turn, can mitigate the potential toxicity associated with some nanoparticles. Many PGPR possess metal tolerance, enabling them to detoxify nanoparticle formulations or regulate their dissolution rates in soil [9]. In fact, some PGPR strains can biosynthesize nanoparticles that are more stable and less toxic, thereby contributing to safer nanomaterial applications in the field [10]. This attribute is now even considered one of the sought-after advanced traits in PGPR, with reported potential to produce silica, selenium, and molybdenum nanoparticles, among others, thereby enabling green synthesis. This mutual relationship between PGPR and nanoparticles creates a positive feedback system in which the latter enhances microbial performance, while PGPR stabilizes NPs-induced toxicity and their prolonged presence in the soil ecosystem.

The synergistic application of nanoparticles and PGPR not only mitigates each other's drawbacks but also provides multiple benefits to the plants. First, co-application can enhance nutrient uptake by combining nanofertilizers with PGPR-mediated solubilization and mobilization processes. Second, plants exhibit improved photosynthetic efficiency, increased chlorophyll content, and enhanced antioxidant capacity under such synergistic field trials [11]. Third, the combined treatment enhances plant defense responses by activating antioxidant enzymes, modifying phytohormone profiles, and inducing systemic resistance against pathogens [11]. Fourth, nanoparticles may facilitate the delivery of PGPR to plant tissues, improving colonization. Numerous studies have reported enhanced root and shoot biomass, increased flowering rates, improved stress tolerance, and higher yields when both components are used together.

Thus, the combined application of nanoparticles and PGPR represents a highly promising, innovative, and multidisciplinary approach that could promote sustainable agricultural productivity. By combining the unique advantages of nanomaterials with the natural beneficial functions of PGPR, this strategy offers the potential for more efficient nutrient management, enhanced plant resilience, reduced chemical dependency, restoring soil health through synergistic interactions between nanomaterials and beneficial microbes, and improved crop productivity. This book chapter discusses the implications of conventional agrochemicals and how to mitigate them using new-generation technologies, namely nanoparticles, PGPR, and their synergetic applications. It also explores the production of nanoparticles through various modes, particularly via green synthesis, and their synergistic co-application.

2 Limitations of Conventional Farming

The widespread use of agrochemicals, including synthetic fertilizers, pesticides, herbicides, and fungicides, has revolutionized agriculture by increasing yields and supporting global food production. Additionally, the implementation of high-yielding varieties, and the use of highly mechanized and chemically intensive systems, improved irrigation, and synthetic inputs in modern agriculture has resulted in remarkable yield improvements. However, it has also initiated a cycle of environmental degradation and resource depletion, especially in relation to climate change. With a projected surge in global food demand of 60–70% by 2050 [12], intensifying pressure on agricultural systems already stressed by land degradation, water scarcity, and climate fluctuations, there is now a greater focus on modern and sustainable agricultural practices.

Excess fertilizers often leach into groundwater or run off into surface waters, causing eutrophication and disrupting aquatic ecosystems. Soil contamination, biodiversity loss, and disruption of beneficial soil microbiota also result from prolonged chemical use, leading to long-term declines in soil fertility and structure [13]. Moreover, nitrogen-based fertilizers are potential contributors to greenhouse gas emissions, raising serious environmental concerns. Despite the abundance of iron in soils, its bioavailability is low, especially in alkaline and calcareous soils. Additionally, synthetic iron chelators, such as EDDHA (Ethylenediamine-N, N'-bis(2-hydroxyphenylacetic acid)) and DTPA (Diethylenetriaminepentaacetic Acid), although effective for treating iron deficiency-induced chlorosis (IDIC) in plants, persist in the environment and pose risks for heavy metal mobilization [14]. Similarly, water-based fertilizers often lead to nutrient runoff, reducing nutrient use efficiency and contributing to pollution rather than plant uptake.

Residues from chemical pesticides and fertilizers pose numerous risks to human health. Chronic exposure has been linked to neurological disorders, reproductive complications, respiratory diseases, and developmental abnormalities, including congenital disabilities and methemoglobinemia in infants. Moreover, fertilizers often contain high levels of toxic, non-essential metals, such as aluminum, and heavy metals, such as cobalt, which accumulate in the food chain, causing neurological and bone-related disorders [15].

Farmers face rising input costs due to higher prices for chemical products and diminishing returns from degraded soils. The overuse of fungicides and pesticides also leads to resistance in pathogens and pests, requiring even more potent formulations that can reduce both costs and risks. This trend is especially challenging for smallholder farmers in developing regions where chemical regulation and access to training are limited. Given these limitations, the transition toward sustainable agricultural practices is urgent to reduce reliance on synthetic inputs, improve soil health, enhance nutrient availability, and strengthen and adapt plants for climate-resilient agriculture [16].

3 PGPR Attributes of Bacteria

PGPR-mediated effects on plants are achieved via both direct and indirect pathways. Direct mechanisms function internally in the plant [17] and unswervingly influence its growth, either by providing or improving the bioavailability of essential nutrients such as N, P, and K; serving as a source of phytohormones; or by producing growth regulators [18]. Indirect mechanisms function outside the plant, such as activating the plant's defensive metabolic processes through signals from PGPR.

3.1 Direct Plant Growth-Promoting Mechanisms of PGPR

3.1.1 Microbial Nitrogen Fixation

Nitrogen-fixing bacteria can be categorized into two distinct groups. It includes symbiotic bacteria associated with leguminous plants that selectively colonize roots and induce nodule formation, for example, *Rhizobium*. The other group includes free-living nitrogen fixers that do not show selectivity for plants [19], e.g., *Azotobacter, Azospirillum, Burkholderia, Herbaspirillum, Bacillus,* and *Paenibacillus* [20]. *Paenibacillus azotofixans* is the most proficient nitrogen fixer known so far among the genus *Paenibacillus* [21]. Thus, nitrogen fixation is considered a fundamental property of PGPR, as it enhances the direct availability of nitrogen to the plant.

3.1.2 Phosphate Solubilization

Despite abundant phosphorus reserves in soil, predominantly in insoluble forms, it is not readily accessible for plant uptake. The bioavailable forms for plants include soluble forms such as monophosphate and dibasic phosphate [22]. Microorganisms mineralize insoluble, complex structures, such as tricalcium phosphate, rock phosphate, and aluminum phosphate, into a soluble form, ultimately increasing phosphate availability to plants. Microorganisms in the rhizosphere use sugars from root exudates and convert them into organic acids [23]. These microbially derived acids act as chelators of Fe^{3+}, Al^{3+}, and Ca^{2+}, facilitating the conversion of insoluble phosphates into bioavailable forms [24]. Examples of some rhizospheric phosphate-solubilising bacteria are *Bacillus safensis*, *Falsibacillus pallidus* [25], *Sinorhizobium meliloti*, *Pseudomonas aeruginosa*, *Pseudomonas kribbensis* [26], *Pseudomonas fluorescens*, *Bacillus cereus*, *Bacillus licheniformis* [27], and *Bacillus subtilis* [28].

3.1.3 Phytohormone Synthesis

Indole-3-Acetic Acid (IAA): Plants perceive and react to phytohormones in the rhizosphere regardless of whether these compounds are externally supplied or produced by associated microbes. IAA, the predominant natural auxin, is a key phytohormone synthesized by numerous PGPR. Prolonged exposure of plants to IAA promotes extensive root development, enhancing nutrient absorption and supporting overall growth and productivity [29]. A large proportion (nearly 80%) of rhizospheric bacteria produce IAA, and their field application elevates endogenous IAA levels in plants [30]. Several IAA-producing bacterial species have been isolated, including *Bacillus subtilis* [31], *Arthrobacter pascens* [32], *Bacillus amyloliquefaciens* [33], *Burkholderia pyrrocinia* [34], and *Azospirillum brasilense* [35].

Cytokinins: Similar to IAA, the application of external cytokinins triggers various plant responses, including enhanced cell division, better root growth, increased root hair formation, initiation of shoots, expansion of leaves, promotion of branching and seed germination, increased chlorophyll production, development of embryonic vascular tissues, and a delay in the senescence process [36]. The ability of *Pseudomonas fluorescens* AK1 and *P. aeruginosa* AK2 to produce isopentenyladenosine (IPA), dihydroxyzeatin riboside (DHZR), and zeatin riboside (ZR) plays a significant role in promoting rice seedling growth [37]. Furthermore, nearly all naturally occurring cytokinins are produced by *Bacillus rugosus*, which promotes growth in *Arachis hypogaea* (L.) [38].

Gibberellins: An early investigation by Atzorn et al. (1988) used physicochemical techniques to characterize gibberellins in bacteria, demonstrating the presence of GA_1, GA_4, GA_9, and GA_{20} in gnotobiotic cultures of *Rhizobium meliloti* [3]. Subsequently, multiple gibberellins were isolated from *Leifsonia soli* SE134 [39]. In addition, *Promicromonospora* sp. SE188 was shown to produce both physiologically active and inactive gibberellins in varying amounts, as identified through GC/MS-SIM analysis, resulting in significant increases in shoot length and biomass of tomato plants [40].

3.1.4 Siderophore Production

In soils, iron predominantly exists as ferric hydroxide (Fe^{3+}), a form that is poorly accessible to plants. To overcome this limitation, certain bacteria produce and release low-molecular-weight compounds known as siderophores, which bind Fe^{3+} strongly and convert it into a soluble form that can be more easily taken up by plants. These siderophores possess a high affinity for iron, enabling the conversion of insoluble Fe^{3+} to the more soluble Fe^{2+} form, which can be readily transported into plant cells

[41]. Based on their chemical structure, siderophores are classified as catecholates, phenolates, hydroxamates, carboxylates, and mixed-type compounds [42]. The use of PGPR in managing plant diseases is effective because these microbes bind and utilize iron, reducing its availability to harmful pathogens, thereby promoting better crop growth and higher yields [43]. Siderophore-producing bacteria include *Bradyrhizobium, Rhizobium, Bacillus, Serratia, Enterobacter, Streptomyces, Pseudomonas,* and others [3].

3.2 Indirect Mechanisms of PGPR

3.2.1 Induced Systemic Resistance (ISR)

ISR is an elevated state of defensive readiness in plants that is stimulated by PGPR, including species of *Flavomonas, Pseudomonas, Serratia,* and *Bacillus* [44]. In this process, these beneficial microbes activate the plant's inherent defense mechanisms, enabling it to more effectively resist pathogen attack. Elicitors of both biological and non-biological origin, including polysaccharides from *Pseudomonas fluorescens* and *Burkholderia gladioli,* play a role in enhancing tomato plant resistance to tomato brown rugose fruit virus (ToBRFV) [45]. The *Pseudomonas*-derived high-affinity siderophore, pseudomanine, promotes salicylic acid accumulation in radish, which in turn improves the plant's defensive response [46].

3.2.2 Synthesis of 1-Aminocyclopropane-1-Carboxylic Acid (ACC) Deaminase

Ethylene is commonly regarded as a stress-related phytohormone because it negatively impacts plant growth and development, leading to symptoms such as flower wilting, chlorosis, leaf abscission, and leaf senescence. Several PGPR bacteria can synthesize the enzyme ACC deaminase, which degrades the ethylene precursor, ACC, thereby inhibiting its production. Under abiotic stress, plants accumulate higher levels of ACC. Bacterial ACC deaminase cleaves this compound into ammonia and α-ketobutyrate, thereby decreasing the ethylene concentration induced by stress [47]. Notably, ACC deaminase-producing *Pseudomonas fluorescens* YsS6 and *P. migulae* 8R6 have been demonstrated to efficiently postpone flower senescence in carnation [48].

3.2.3 Synthesis of Exopolysaccharide (EPS)

EPS produced by PGPR plays vital roles in various processes, including biofilm formation, bacterial protection against desiccation, maintenance of essential cellular processes, and antibacterial activity against predators. In addition, it effectively improves plant growth, physiology, and soil properties [49]. *Pseudomonas entomophila* FAP1 was selected as a better rhizospheric colonizer because it produced EPS, a trait related to biofilm, swarming, and swimming motility, and cell surface hydrophobicity [50]. EPS-producing rhizobacteria *Pseudomonas azotoformans* JRBHU5 and *Burkholderia seminalis* JRBHU6 exhibit efficient root colonization, boost early plant vigor, and thereby support enhanced wheat growth [51].

3.3 Next-Generation PGPR Attributes

Apart from these core attributes of PGPR, several emerging or next-generation plant growth-improving functional competencies of bacteria are being discovered, adding to the multifunctionality of the biofertilizer organisms. Endophytic potential and nanoparticle production are two such potentially rewarding competencies. The endophytic capability of PGPR can protect them from environmental stresses and microbial competitors, allowing for longer persistence within the plant. This, in turn, facilitates faster communication with plant cells and stronger induction of PGPR-mediated plant pathways, thereby overcoming biotic and abiotic stresses. Nanotechnology is another innovative approach employed in modern agriculture, where different plant-nutrient nanoparticles (such as silica or Mo nanoparticles) are provided either as a supplement along with PGPR, or the latter bacteria are searched for with nanoparticle-producing abilities to enhance agricultural sustainability. For example, PGPR (*Bacillus* sp. ZH16), along with the molybdenum NPs it produced, was co-applied to effectively improve wheat growth in arsenic (As)-contaminated soil [52].

4 Nanoparticle in Sustainable Agriculture

According to the International Organization for Standardization (ISO), nanoparticles (NPs) are defined as nano-objects in which all external dimensions fall within the nanoscale range, and the lengths of the longest and shortest axes are approximately equal [53]. When the dimensional difference exceeds three times, materials are often classified as nanofibers or nanoplates [53]. The size of the NPs can be between 1 and 100 nm, although there are certain NPs [53], such as polymer-coated inorganic NPs, certain polymeric NPs (PNPs), etc., which have a size of more than 100 nm [54]. NPs represent an advanced formulation in modern agriculture, encompassing nano-scale pesticides, fertilizers, fungicides, herbicides, and sensing devices. These innovations aim to transform agricultural practices by enabling controlled, sustained release of agrochemicals, incorporating intelligent sensing technologies, and enhancing plant resistance to diseases, ultimately improving crop productivity in an eco-friendly, sustainable way (Fig. 1).

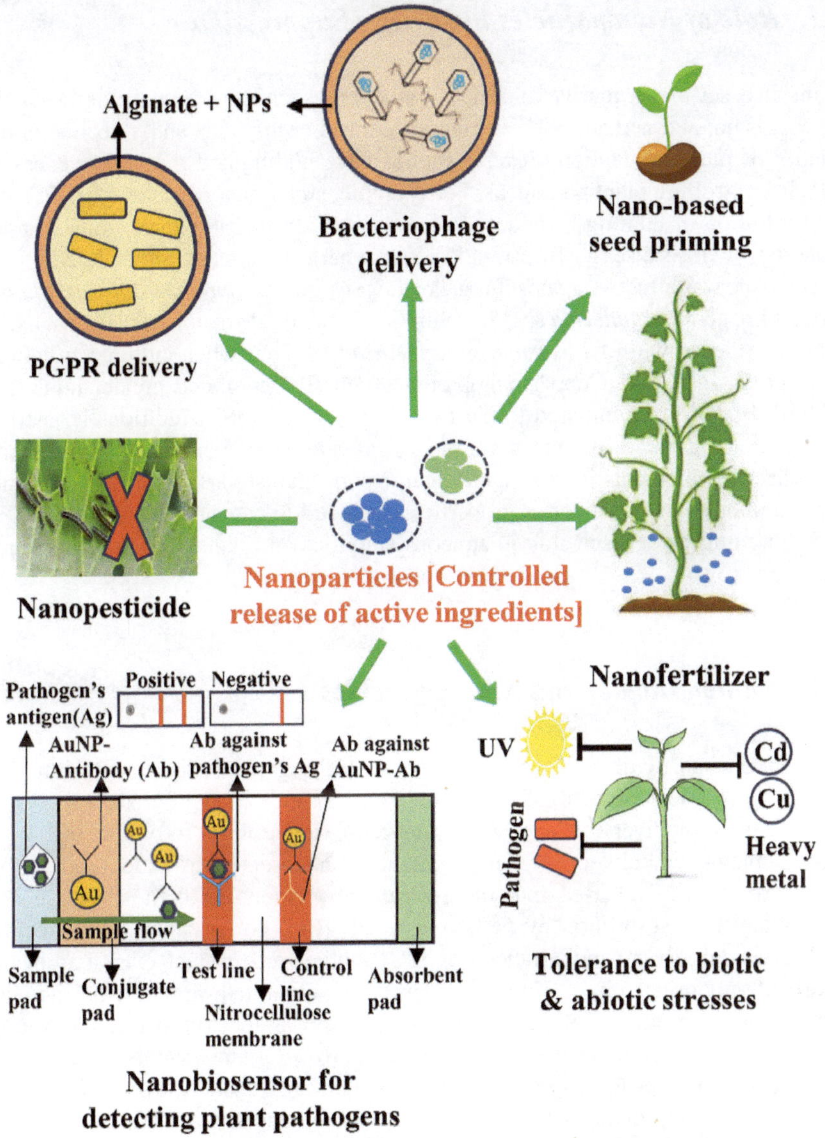

Fig. 1 Applications of nanoparticles in agriculture for improving plant productivity and their protection. Different types of NPs (Ag, Si, Au, Cu, Se, Zn) are used in agriculture in the form of nanofertilizers, nanopesticides, nanobiosensors, and for seed priming, plant tolerance to biotic and abiotic stresses, and delivery of plant growth-promoting rhizobacteria (PGPR) and bacteriophages. The NPs are generally encapsulated within a coating material (shown by dotted lines) for the controlled release of active ingredients at the site of action. For PGPR and phage delivery, alginate serves as a versatile protective matrix, encapsulating PGPR and phage within NPs and forming a gel-like shield that protects them from degradation

4.1 Role of Nanoparticles in Disease Suppression

Plant diseases are primarily caused by pathogens such as viruses, bacteria, fungi, mycoplasma, and nematodes, resulting in reduced crop yields and a decline in the quality of plant production. Several studies have highlighted the effectiveness of NPs in controlling plant pathogens. For example, silver nanoparticles (AgNPs) at a concentration of 200 mg/L reduced colony formation of pathogenic fungi responsible for rye grass diseases by 50% [55]. The synergistic application of AgNPs with fluconazole was effective against fungal pathogens such as *Candida albicans, Phoma glomerata*, and *Trichoderma* sp. [56]. Similarly, zinc oxide nanoparticles (ZnONPs) inhibited the growth of *Fusarium graminearum* by 26% when cultured in mungbean broth agar [57]. Copper nanoparticles (CuNPs) exhibited greater inhibition of fungal pathogen compared to the fungicide bavistin [58]. Additionally, oleoyl-chitosan nanoparticles produced via the oil-in-water (O/W) emulsification method have been evaluated as in vitro antifungal agents against various plant pathogenic fungi and have shown superior properties compared to conventional chitosan NPs, including improved dispersion in aqueous solutions and reduced sensitivity to pH variations [59].

4.2 Nanopesticides and Nanoherbicides

Conventional pesticides and herbicides have several negative environmental impacts, including the development of resistant pests, accumulation of toxic chemicals, reduced soil biodiversity, and their nitrogen-fixing ability [60]. On the other hand, innovative delivery systems containing nanoscale pesticides and herbicides are steadily released into the soil, thereby extending their shelf life and improving their absorption by pests and weeds [61]. For instance, pediculocidal and larvicidal silver nanoparticles (AgNPs) synthesized by leaf extract from the heartleaf moonseed plant, *Tinospora cordifolia*, exhibited strong lethal activity against the head louse *Pediculus humanus*, as well as the fourth instar larvae of *Anopheles subpictus* and *Culex quinquefasciatus* [62]. Organic nanoherbicides such as atrazine-poly (epsilon-caprolactone), metribuzin-poly(epsilon-caprolactone), and nano-encapsulated glyphosate, as well as inorganic nano-silica, ZnONPs, and titanium oxide NPs (TiO_2NPs), are very effective against various weed species, including *Chenopodium album, Echiniochloa crus-galli, Amaranthus viridis, Bidens pilosa*, and *Raphanus raphanistrum* [63].

4.3 Nanobiosensors

Nanobiosensors can detect a wide range of plant pathogens, as well as conventional chemical fertilizers, herbicides, pesticides, insecticides, and heavy metals, thereby supporting sustainable agriculture by enhancing crop productivity. A nanobiosensor comprises three key elements: a biological recognition component (e.g., aptamers, antibodies, peptides, or enzymes), a transducer, and a detection system [64]. Nanoparticles used in nanobiosensors include AuNPs, carbon nanotubes, titanium oxide NPs, and silicon oxide NPs, which are used to fabricate biosensors that produce an excellent, sensitive, and specific output [65]. Colorimetric, electrochemical, and lateral flow assays are popular types of nanobiosensors [65]. For example, a lateral flow assay-type nanobiosensor with an asymmetric PCR-mediated *Phytophthora infestans*-specific DNA probe and an AuNPs-based sensing assay can detect the potato late blight pathogen with its DNA load as low as 0.1 pg μL^{-1} [66] (Fig. 1).

5 Nanoparticle Production

5.1 Top-Down/Physical Approach

In the top-down/physical method, bulk materials (such as metals, ceramics, semiconductors, and polymers) are fragmented into smaller nanoparticles (Fig. 2) [67]. A range of methods, including mechanical milling, electrospinning, laser ablation, sputtering, electron explosion, pulsed wire discharge, arc discharge, probe sonication, and lithography, are utilized in the top-down approach (Table 1). Mechanical milling, including planetary and shaker mills, is a process that uses balls inside containers to grind materials through a combination of high-energy impact and attrition [68]. Using high-energy ball milling, ZnO nanoparticles are produced, which require temperatures above 800 °C to achieve high purity. These ZnONPs exhibit antibacterial activity [69]. Electrospinning is a commonly used technique for producing nanofibers from a range of materials, particularly polymers, into which NPs can be embedded [70]. In this process, electrically charged jets are drawn from polymer solutions to generate nanofibers [71]. The laser ablation method uses high-intensity laser irradiation to vaporize precursor materials, thereby forming NPs [72]. Sputtering involves the ejection of solid microparticles from a material's surface when it is bombarded with high-energy plasma or gas particles [67].

In the electron-explosion technique, a thin metal wire is exposed to a high-current pulse, which rapidly vaporizes and ionizes the metal. The resulting vapor expands, cools, and reacts with the surrounding gas or liquid medium, eventually condensing into NPs [73]. Probe sonication uses ultrasonic waves delivered directly into a liquid sample through an immersed probe, facilitating the breakdown of bulk materials into NPs [74]. Pulsed wire discharge is a widely used method for the synthesis of metal NPs, in which a high-intensity pulsed current vaporizes a metal wire, and the

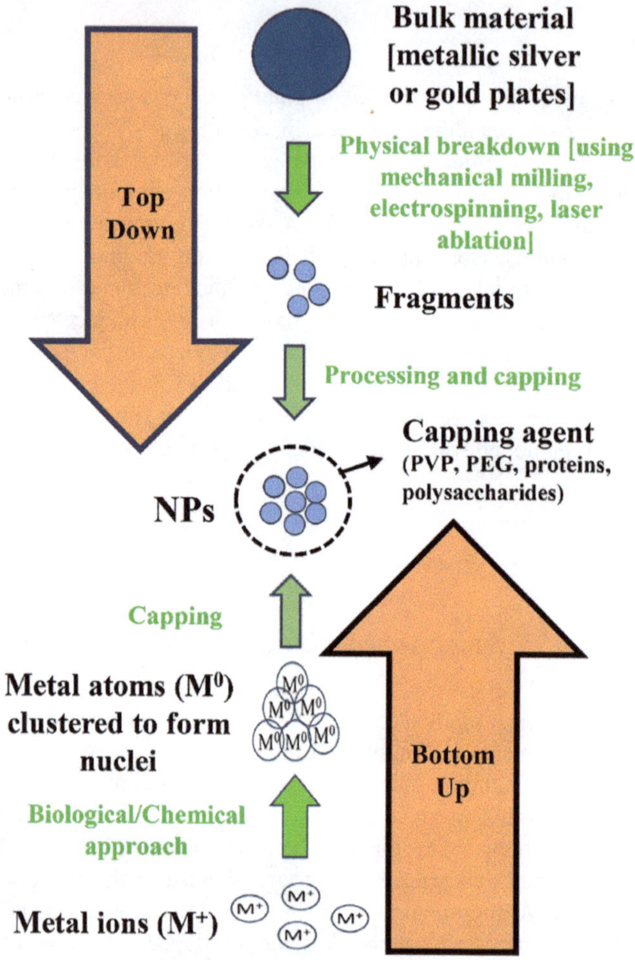

Fig. 2 Production of nanoparticles through top-down and bottom-up approaches. In the top-down method, bulk materials, such as metallic silver and gold plates, are fragmented into smaller pieces via mechanical milling, electrospinning, or laser ablation. These fragments are further processed to obtain the active ingredients, which are then encapsulated with capping agents, such as polyvinylpyrrolidone (PVP) and polyethene glycol (PEG). On the other hand, in the bottom-up approach, metal ions are reduced to metal atoms using biological/chemical methods, followed by their encapsulation within capping agents, such as PVP and PEG in chemical methods, and proteins and polysaccharides in biological methods

generated vapor subsequently condenses into NPs upon cooling in a surrounding gas atmosphere [75]. The arc-discharge method generates NPs by creating a high-temperature arc between two electrodes, vaporizing the material, and allowing it to condense into NPs [76]. Electron beam lithography generally employs a focused electron beam to fabricate NPs with high precision [75].

Table 1 Different techniques of nanoparticle production through physical and chemical approaches

A. Top-down/physical approach

Name of the technique	Example of NPs	Precursor molecules	Advantage/ Disadvantage	References
Mechanical milling	ZnO, Magnetite, Iron oxide	$ZnCl_2$ & Na_2CO_3 combination (ZnO-NPs), bulk hematite powder (magnetite NPs), $FeCl_3.H_2O$ & NaOH combination (iron oxide NPs)	Pros: Eco-friendly as it is devoid of toxic chemicals and solvents Cons: Contamination from milling media, loss of crystallinity due to high energy in the case of high-energy ball milling	[161]
Electrospining	Iron oxide NP-nanofiber composites	Iron salt, such as ferric chloride or ferrous nitrate typically mixed with a polymer solution of polyvinyl alcohol (PVA) or polyvinylpyrrolidone (PVP)	Pros: Precise fiber morphology production is useful for tissue engineering and drug delivery Cons: Need high voltage, low yield rate	[162]
Laser ablation	Ag, Au	Bulk metallic silver (Ag-NPs), bulk gold plate (99.99% purity) (Au-NPs)	Pros: Eco-friendly for metal NPs production Cons: High cost due to expensive lasers and low yield	[67, 163]
Sputtering	Ag	A bulk piece of silver, such as a silver plate	Pros: High purity, compact and similar to the target material Cons: Complex process, high cost and low yield rate	[67, 163]
Electron explosion of wire (EEW)	Cu, Al, Au, Ag, Zn, Fe, Ni metal, metal composite NPs	Pure metallic wire made up of either copper (CuNPs), aluminum (Al-NPs), gold (AuNPs), zinc (ZnNPs), iron (FeNPs), and nickel (NiNPs)	Pros: High purity, high yield rate, relatively simple method Cons: Fire and explosion risk	[164]

(continued)

Table 1 (continued)

Name of the technique	Example of NPs	Precursor molecules	Advantage/ Disadvantage	References
A. Top-down/physical approach				
Probe sonication	Ag-TiO2	Titanium (IV) isopropoxide (TTIP) and $AgNO_3$	Pros: Provide focused energy Cons: Metal contamination from the probe, overheating	[165]
Arc discharge	Carbon nanotubes, metal NPs, metal oxide NPs	Solid graphite (carbon) as the component of the electrode (carbon nanotubes), bulk metal electrodes (metal & metal oxide NPs)	Pros: low cost and eco-friendly method Cons: Size and uniformity not maintained	[76]
Electron beam lithography	Au	Dimethylgold (III) trifluoroacetylacetonate	Pros: High resolution Cons: Low yield and high cost, electron scattering	[166]
B. Bottom-up chemical approach				
Chemical vapor deposition	Ag, TiO_2, carbon nanotubes	$[Ag_5(O_2CC_2F_5)_5(H_2O)_3]$ (AgNPs), titanium (IV) isopropoxide (TTIP) (TiO_2NPs), various hydrocarbons (methane, acetylene) (carbon nanotubes)	Pros: High-purity and high-quantity materials Cons: High temperature, hazardous chemical gases, high cost	[167, 168]
Sol–gel method	TiO_2, ZnO, Ag, CuO	Titanium tetraisopropoxide (TTIP) (TIO_2NPs), zinc acetate dihydrate and zinc nitrate hexahydrate (ZnONPs), $AgNO_3$ (AgNPs), Copper (II) Nitrate trihydrate (CuONPs)	Pros: Uniform output quality maintained under low-temperature processing conditions Cons: Time-consuming due to slow gel formation, expensive	[79]

(continued)

Table 1 (continued)

A. Top-down/physical approach

Name of the technique	Example of NPs	Precursor molecules	Advantage/ Disadvantage	References
Co-precipitation	Fe_3O_4, Ag, ZnO	$FeCl_3.H_2O$ & NaOH combination (Fe_3O_4NPs), $AgNO_3$ (AgNPs), zinc nitrate hexahydrate, zinc acetate dihydrate, and zinc chloride (ZnONPs)	Pros: High quantities of water-soluble NPs Cons: Does not maintain the size of the NPs	[169, 170]
Inert gas condensation	Cu, Ag, Zn	Bulk elemental copper, either metal plate, wire, rods, or powder (CuNPs), bulk elemental silver (Ag metal) (AgNPs), bulk zinc metal powder (ZnNPs)	Pros: Controllable size of NPs with high purity Cons: High equipment cost, chances of NPs agglomeration	[171]

5.2 *Bottom-Up Approach*

In bottom-up methods, atoms and molecules are combined to form nanoscale particles (Fig. 2) [67]. These include chemical (Table 1) and biological approaches (Table 2).

5.2.1 Chemical Approaches

In chemical vapor deposition (CVD), thin films are deposited onto substrate surfaces using vapor-phase precursor compounds. This technique yields high-purity, mechanically robust, uniform, and homogeneous nanoparticles [77]. As a wet-chemical strategy, the sol–gel process is extensively used to fabricate nanomaterials [78]. In this method, metal precursor solutions undergo condensation, hydrolysis, and thermal decomposition, leading to the formation of a stable colloidal suspension, or sol [79].

Co-precipitation is another wet-chemical strategy, a solvent-displacement technique in which a polymer solution is mixed with a non-solvent, leading to rapid diffusion of polymer molecules, supersaturation, and subsequent precipitation as NPs [80]. In the inert gas condensation (or molecular condensation) technique, metal NPs are produced by evaporating metals, such as copper, into an inert gas environment, typically argon, helium, or neon. The vaporized metal atoms rapidly lose energy through cooling by the inert gas, reach supersaturation, and aggregate to form stable nuclei that grow by condensation, producing CuNPs typically in the 2–100 nm size range [75].

Table 2 Nanoparticle synthesis by green/biological approach

Name of the organism	Type of NP	Size of NP	Precursor/Substrate (Concentration applied)	Application	References
i. Biosynthesis of NPs by bacteria					
Bacillus subtilis ZBP4	ZnO	22–59 nm	$ZnSO_4 \cdot 7H_2O$	Broad antibacterial activity against food pathogens	[91]
Bacillus atrophaeus	CuO, Ag	4–28 nm (CuoNP), 15–38 nm (AgNP)	$CuSO_4$, $AgNO_3$	Strong antibacterial activity against *Escherichia coli* and *Staphylococcus aureus* was tested	[92]
Pseudomonas stutzeri	Cu	50–150 nm	Cu^{2+} heavy metal contaminated effluent wastewater	Not specified	[93]
Paenibacillus polymyxa Sx3	ZnO, MnO_2 and MgO	56.1–110.0 nm (ZnO), 10.1–18.8 nm (MnO_2), 19.8–63.9 nm (MgO)	ZnO, MgO and MnO2 (1 mM), respectively	Effectively suppressed the growth of *Xanthomonas oryzae* pv. *oryzae*	[94]
Proteus vulgaris ATCC-29905	Iron oxide	19.23–30.51 nm	$FeCl_3$ solution (3 mM)	Significant inhibition of *Staphylococcus aureus*, also displaying antioxidant potential and cytocompatibility	[172]
ii. Biosynthesis of NPs by fungi					
Aspergillus oryzae var. *viridis*	Ag	5–50 nm	$AgNO_3$ (1 mM)	Bactericidal Activity toward *Staphylococcus aureus* KCCM 12,256 was carried out	[173]

(continued)

Table 2 (continued)

Name of the organism	Type of NP	Size of NP	Precursor/Substrate (Concentration applied)	Application	References
Trichoderma viride and *Hypocrea lixii*	Au	20–30 nm	Not specified	Strong antimicrobial and biocatalytic activities	[174]
Aspergillus niger G3-1	CuO	14–47.4 nm	$CuSO_4 \cdot 5H_2O$ (2 mM)	Effectively controls wheat storage insects, causing 55–94.4% mortality in *Sitophilus granarius* and 70–90% in *Rhyzopertha dominica*	[175]
Trichoderma asperellum, *Fusarium incarnatum* and *Phialemoniopsis ocularis*	Iron oxide	2–16 nm	$FeCl_3$:$FeCl_2$ (2:1 mM)	Bioremediation of water containing chromium	[176]
iii. Biosynthesis of NPs by actinomycetes					
Streptomyces sp. LK-3	Ag	10 nm	$AgNO_3$	Antiparasitic activity against the ticks *Haemaphysalis bispinosa* and *Rhipicephalus microplus*	[98]
Streptomyces capillispiralis Ca-1	Cu	3.6–59 nm	Not specified	Showed significant antimicrobial as well as antifungal efficacy	[99]
Rhodococcus sp.	Au	5–15 nm	$HAuCl_4$ (10^{-3} M)	Not specified	[100]
iv. Biosynthesis of NPs by Algae					
Hormophysa triquetra	Ag	Not specified	$AgNO_3$	Significant antibacterial activity	[103]

(continued)

Table 2 (continued)

Name of the organism	Type of NP	Size of NP	Precursor/Substrate (Concentration applied)	Application	References
Padina sp.	Ag	25–60 nm	$AgNO_3$ (0.01 M)	Effectively suppressed the growth of pathogenic bacteria, *Staphylococcus aureus* and *Pseudomonas aeruginosa*	[104]
Euglena gracilis	Au	Not specified	$HAuCl_4$ (0.01 M)	Not specified	[105]
v. Biosynthesis of NPs by plant extract					
Eucalyptus globules	Zn	52–70 nm	Zinc nitrate hexahydrate (1 mM)	Protects apple orchards by targeting their primary fungal pathogens	[107]
Triticum aestivum	Ag	27–33 nm	$AgNO_3$ (1 mM)	Foliar application of AgNP regulates tolerance of stress in wheat plants	[108]

5.2.2 Green/Biological Method

Microorganisms, plant extracts, and other biological components, such as enzymes, tissues, and waste materials, are utilized in the biological synthesis of nanoparticles (Table 2). This method offers numerous benefits, such as low cost, simplicity, rapid production of highly stable nanoparticles, non-harmful byproducts, scalability, and environmental friendliness [81]. One of the most studied and widely used types of biological NPs in agricultural research is the AgNPs produced from plant extracts, which, at low concentrations, exhibit strong antimicrobial properties and could promote plant growth [82]. In addition, gold nanoparticles (AuNPs) are synthesized biologically and applied in diverse fields, including the formulation of bio-nanosensors [82].

Nanoparticle Synthesis Using Microorganisms

Microorganisms produce nanoparticles through the following steps: the capture or trapping of metal ions, their bioreduction by a reducing agent, such as an enzyme, the accumulation of reduced metal atoms to form a nucleus, and the stabilization or capping of the nucleus after reaching a critical size (Fig. 3) [83]. This can either be performed intracellularly or extracellularly, depending upon the microorganism. Intracellular nanoparticle production starts with the trapping of metal ions inside the cell and their reduction by enzymes or proteins to zero-valent metal atoms. In contrast, the extracellular method primarily captures metal ions and reduces them on the cell wall surface or by secreted extracellular components, which actually utilize the cell-free supernatant rather than the cells themselves [84]. Microbes produce various enzymes to detoxify heavy metals (precursors of NPs) in the environment, thereby reducing their toxicity [85]. These ensure their survival and growth in heavy-metal contaminated environments [85]. Enzymes such as NADH-dependent reductase, nitrate reductase, sulfite reductase, glutathione reductase [86], along with other enzymes, such as xylanases [87], and keratinases [88], are also described. These metal atoms further aggregate to form nuclei of standard size, which are then stabilized using capping agents. Capping agents are also synthesized by microbes, either intracellularly or extracellularly, including polysaccharides, proteins, lipids, and so on, ultimately forming NPs [84].

Bacterial NPs: Bacteria produce an array of nanoparticles that are highly adaptable to extreme conditions. The extracellular method of NP extraction is easier, faster, and more widely used than the intracellular method [89]. Several bacteria can synthesize NPs because they produce metabolites and enzymes responsible for NPs production. The bacterium *Rhodopseudomonas capsulata* releases the cofactor NADH and cofactor-dependent enzymes, which cause reduction of Au^{3+} to Au^{0} and the formation of gold NPs [90]. *Bacillus subtilis* ZBP4 was used to produce ZnONPs, which showed potent antibacterial activity against food pathogens [91]. *Bacillus atrophaeus* spores are used to synthesize AgNPs and CuONPs, which exhibit strong antibacterial activity against *Escherichia coli* and *Staphylococcus* [92]. CuNPs were

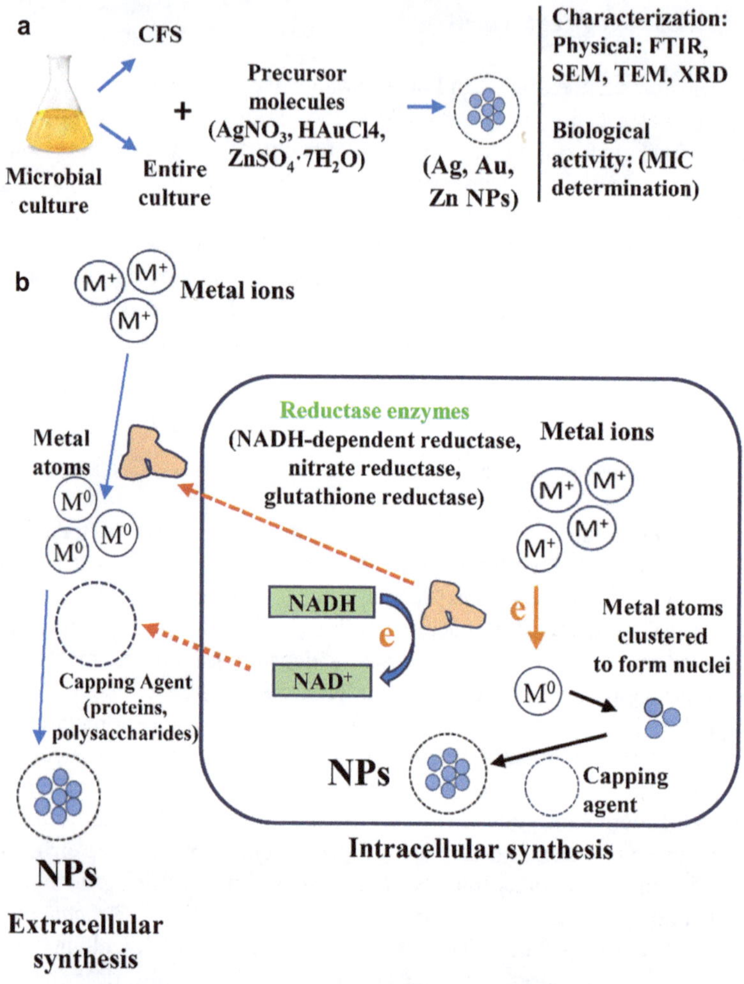

Fig. 3 **Mechanism of nanoparticles synthesis through microbial route.** **a** In the laboratory, NPs are synthesized using either the entire microbial culture or cell-free supernatant (CFS), which contains extracellular enzymes and some capping agents. In both methods, precursor molecules such as $AgNO_3$, $HAuCl_4$, and $ZnSO_4 \cdot 7H_2O$ should be provided externally for NPs synthesis. After synthesis, the size, shape, and further characterization should be performed using Fourier Transform Infrared Spectroscopy (FTIR), Scanning Electron Microscopy (SEM), Transmission Electron Microscopy (TEM), and X-ray diffraction (XRD). Then, their minimum inhibitory concentration (MIC) should be determined to avoid toxicity at higher concentrations, and they should finally be applied in agriculture. **b** In intracellular synthesis, the positively charged metal ions inside the cell are reduced to non-charged metal atoms by reductase enzymes like NADH-dependent reductase, nitrate reductase, and glutathione reductase. The reduced metal atoms are then accumulated to form nuclei and stabilized within capping agents, such as proteins and polysaccharides, once they reach a critical size, to generate the active form of NPs. In extracellular synthesis, the CFS is used as a source of the reductase enzyme and the capping agent, with the steps remaining the same

produced by the bacterium *Pseudomonas stutzeri* from a Cu^{2+}-contaminated effluent water sample, demonstrating bioremediation of heavy metals and the simultaneous production of CuNPs for further use [93]. Metal oxide NPs, ZnO, MgO, and MnO_2, were produced extracellularly using the cell-free supernatant of rhizospheric bacteria *Paenibacillus polymyxa* Sx3, and all three NPs had significant antibacterial activity against *Xanthomonas oryzae*, responsible for bacterial leaf blight of rice [94].

Fungal NPs: Fungi are ideal candidates for nanoparticle synthesis due to several factors: they produce a diverse range of intracellular enzymes, exhibit a remarkable ability to absorb metals within their cells, and can be easily cultivated on a large scale using solid-substrate fermentation, allowing the production of significant biomass for subsequent processing. Both viable and non-viable cells, along with the cell-free extract of *Aspergillus oryzae* var. *viridis*, were employed to synthesize nanocrystalline gold particles [95]. Extracellular synthesis of silver NPs from silver nitrate ($AgNO_3$) solution using fungal species, viz. *Rhizopus nigricans, Colletorichum gloeosporiodes, Fusarium semitectum,* and *Aspergillus nidulans* were reported [96]. A low-cost, green, and reproducible method for the biosynthesis of TiO_2NPs using *Saccharomyces cerevisiae* was also reported [97].

Actinomycetes NPs: Actinobacteria can also synthesize NPs via both extracellular and intracellular routes, as they possess NP-synthesizing metabolites similar to those found in bacteria and fungi. *Streptomyces* sp. LK-3 synthesizes AgNPs extracellularly through the action of NADH-dependent nitrate reductase, exhibiting strong antiparasitic effects against *Haemaphysalis bispinosa* and *Rhipicephalus microplus* [98]. Copper nanoparticles (CuNPs) were produced using the endophytic actinobacterium *Streptomyces capillispiralis* Ca-1, as indicated by a visible colour change of the biomass filtrate from light blue to greenish-brown. The synthesized CuNPs demonstrated significant antibacterial and antifungal activities [99]. Alkalotolerant bacteria, *Rhodococcus* sp., are able to withstand high concentrations of metal ions of 10^{-3} M $HAuCl_4$ during the production of AuNPs and produce AuNPs intracellularly [100].

Algal NPs: Algal cells generate biologically active compounds and secondary metabolites that serve as natural capping agents in NPs synthesis, transforming the algae into a distinctive "nanofactory" capable of producing various NPs [101]. Silver and gold are the most widely studied noble metals for NP synthesis using algae, which can occur either inside their cells (intracellularly) or in the surrounding medium (extracellularly) [102]. Marine macroalgae *Hormophysa triquetra* synthesized AgNPs possess significant antibacterial activity [103]. Another marine microalga, *Padina* sp., has been reported to biosynthesize uniform silver NPs (AgNPs) with strong antimicrobial activity against the pathogenic bacteria *Pseudomonas aeruginosa* and *Staphylococcus aureus* [104]. AuNPs were biosynthesized by the microalga *Euglena gracilis*. Growth accelerated by seven to eight times, which positively affected the NPs' kinetics, colloidal stability, and production yield [105].

Nanoparticle Synthesis Using Plant Extracts

Plant extracts contain a variety of phytochemicals, including phenols, flavonoids, alkaloids, and terpenoids, as well as other metabolic components such as proteins and carbohydrates. These components act as reducing agents, converting metal ions into metal atoms, and also serve as capping agents to inhibit NP's agglomeration [106]. Zinc oxide nanoparticles synthesized using *Eucalyptus globulus* extract show significant antifungal activity against key pathogens affecting apple orchards [107]. *Triticum aestivum* synthesizes AgNPs of 27–33 nm size, whose foliar application regulates the tolerance of stress in wheat plants [108].

6 Nanoparticle in Plant Growth Improvement

Nanoparticles are used in agriculture due to their distinctive physicochemical properties [53]. A variety of studies have focused on understanding the mechanisms by which nanoparticles (NPs) regulate plant growth and developmental processes. The metal nanoparticles enable targeted delivery of essential nutrients to specific sites, thereby enhancing plant growth and increasing overall productivity. The interaction of NPs with plants induces a range of physical and physiological alterations, which are largely influenced by their characteristics, such as size, surface area, chemical composition, reactivity, and the concentration at which they exhibit beneficial effects [109]. NPs exert a wide range of beneficial effects on plant growth and development (Fig. 1).

6.1 *Carbon Nanomaterials and Plant Growth*

Carbon-based nanomaterials have gained significant attention among nanoparticles due to their distinctive chemical, thermal, mechanical, and electrical characteristics. When used at low concentrations, it generally promotes plant growth by enhancing water and nutrient uptake, improving seed germination, and stimulating antioxidant enzyme activity. However, when used at high concentrations, it can cause adverse effects. Tomato seeds treated with carbon NPs showed a notable increase in germination rate and vegetative growth [109]. NPs facilitate penetration of the seed coat, leading to improved water absorption and consequently accelerating the germination process. Treatment with multi-walled carbon nanotubes (MWCNTs) at 50 and 200 μg/mL stimulated growth and enhanced flowering in *Lycopersicum esculentum* [110].

6.2 *Role of Metal-Based Nanoparticles in Modulating Plant Growth*

Metal nanoparticles are distinct from carbon-based nanomaterials because they are composed solely of metallic elements like Ag, Au, Si, Zn, etc., rather than carbon, and have been repeatedly demonstrated to be beneficial for plant growth. For example, treatment with AgNPs led to superior root growth in maize and cabbage compared to $AgNO_3$, precursor of AgNPs [111]. Studies have demonstrated that AuNPs play a crucial role in promoting seed germination and strengthening antioxidant defense activity in *Arabidopsis thaliana* [112]. In *Boswellia avalifoliata*, AuNPs also enhanced seed germination [113]. In *Bacopa monnieri*, AgNPs promoted protein and carbohydrate production but reduced total phenolics [114]. Exposure of spinach seeds to titanium NPs solutions promoted biomass accumulation (both fresh and dry) and improved nitrogen, chlorophyll, and protein concentrations in leaves [115]. Similarly, pearl millet treated with biosynthesized ZnNPs exhibited improved shoot growth, chlorophyll content, and grain yield [116]. In hydroponic spinach systems, iron NPs treatment enhanced plant development and biomass production, accompanied by an 11–21-fold increase in iron accumulation across various plant parts [117].

6.3 *Metal Oxide Nanoparticles and Plant Development*

Many studies have documented the influence of metal oxide nanoparticles on plant germination and development. Treatment of seeds with TiO_2NPs improved the germination rate of *Salvia* [118]. Additionally, TiO_2NPs positively influenced seed germination and boosted the elongation of radicle and plumule in canola seedlings [119]. Soybean plants treated with TiO_2NPs and SiO_2NPs together exhibited elevated nitrate reductase activity and improved antioxidant defense responses [120]. The presence of ZnONPs positively influenced seed germination and promoted root and shoot growth in peanut plants [121]. Treatment of *Cyamopsis tetragonoloba* with ZnONPs significantly promoted biomass yield, stimulated root and shoot growth, and elevated chlorophyll and protein levels, thereby improving overall plant growth [122].

6.4 *Nanoparticle-Based Seed Treatments*

Inorganic nanoparticles, such as ZnO, CuO, graphene oxide, Ag, and Si, as well as organic nanoparticles including chitosan, cellulose, and lignin, are used for seed treatment through seed coating or seed priming. Nano-iron-treated rice seeds showed significant seed germination and inhibition of the fungus *Fusarium verticillioides*, causing sheath rot and seed discolouration of rice [123]. Seed priming with ZnONPs helps to mitigate salinity stress in lupine (*Lupinus termis*) [124]. *Phyllanthus emblica* fruit extract stabilized Ag-NPs act as a growth promoter of wheat varieties by reducing reactive oxygen species toxicity, and seed treatment increased seed and seedling vigor [125]. Thiamine-incorporated chitosan NPs-treated seeds improved plant growth regulators in Chickpea (*Cicer arietinum* L.) [126].

6.5 Nanofertilizers

Nanofertilizers are nanomaterials that provide nutrients to plants, serving as an alternative to chemical fertilizers. Compared to chemical fertilizers, nanofertilizers deliver nutrients gradually and target them to the soil, thereby enhancing nutrient use efficiency, reducing environmental pollution, and maintaining soil integrity [127]. Application of 16 nm hydroxyapatite NPs ($Ca_5(PO_4)_3OH$) improved nutrient availability and supported soybean growth [128]. A substantial improvement in the growth of 15-day-old lettuce seedlings was observed in soil supplemented with metallic CuNPs [129]. The nanocomposite of *Gliricidia sepium* encapsulated with urea-modified hydroxyapatite nanoparticles exhibits controlled, prolonged nitrogen release over a range of pH values [130]. An alternative is the use of nanoporous zeolite as an N fertilizer to enhance N effectiveness in crop production [131]. Foliar application of MoS_2-based nanofertilizer significantly increased rice plant biomass and leaf chlorophyll content and regulated the expression of the aquaporin gene [132]. Molybdenum NP (MoNP) supplementation can improve nitrogen fixation, as the element is a cofactor of the key enzyme nitrogenase, making Mo available to nitrogen-fixers to efficiently synthesize nitrogenase and thus promote plant growth. MoNPs have been reported to boost plant performance by enhancing biomass production, increasing yield, and promoting nitrogen fixation, while simultaneously improving beneficial soil microbial communities. However, excessive doses can be toxic, so proper application is needed.

6.6 Role of Nanoparticles in Photosynthesis

Researchers are examining genetic modification alongside supportive strategies to boost photosynthetic performance in plants. The application of nanotechnology can effectively boost photosynthetic performance. For instance, incorporating single-walled carbon nanotubes into isolated chloroplasts led to a threefold increase in photosynthetic performance relative to the control, accompanied by a twofold increase in the maximum electron transport rate [133]. The application of silicon dioxide NPs enhances photosynthetic performance by modulating carbonic anhydrase activity and promoting pigment biosynthesis [134]. TiO_2NPs promote carbon assimilation during photosynthesis by activating Rubisco, thereby accelerating carboxylation reactions [135]. Magnesium nanoparticles (MgNPs) further improve photosynthesis by increasing chlorophyll content, enhancing light absorption, and critically improving magnesium bioavailability in plant tissues. Acting as an efficient magnesium source, MgNPs support carbon metabolism, stimulate gas exchange, and ultimately improve plant growth and yield [136]. Moreover, magnesium nanoparticles (MgNPs) have been shown to enhance plant growth and modify the structure of the rhizospheric microbial community in soybean (*Glycine max* L.) [136].

7 Nanoparticle Produced by PGPR Bacteria and Its Relation to Plant Growth

The main benefit of PGPR-derived nanoparticles is that these growth-promoting bacteria can also be employed in the field with the nanoparticle precursor, providing dual benefits if desired. Otherwise, if the soil is already enriched or is contaminated with the nanoparticle's precursor substances, application of the NP-producing PGPR will help in its proper removal and utilization, along with contributing to plant growth.

Several studies have been performed in which NPs were biosynthesized by PGPR and applied against plant pathogens, offering a sustainable alternative to chemical and physical methods of NPs production (Table 3). Under salinity stress, wheat treated with *Azospirillum lipoferum*, *Bacillus coagulans*, *B. circulans*, and *B. subtilis* together with ZnONPs showed a substantial increase in antioxidant enzymes, potassium uptake, and soil enzyme activities (urease and dehydrogenase up to 232%), while experiencing significant reductions in membrane damage, proline, and H_2O_2 accumulation [137]. In sugar beet irrigated with saline water, *Pseudomonas koreensis* and *B. coagulans* with SiNPs restored ionic balance (higher K^+/Na^+ ratio), stomatal conductance, photosynthetic pigments, and lowered oxidative markers as antioxidant enzymes rose to 2.5-fold [138]. Wheat subjected to drought and heat stress exhibited improved stress tolerance when combined with *Pseudomonas* sp. and ZnONPs, which elevated proline and antioxidant systems, while minimizing membrane injury and H_2O_2, ultimately preventing plant mortality [139]. For onion, *Bacillus pumilus* with AgNPs enhanced chlorophyll, carotenoids, sugars, and proline under salinity conditions, while reducing soil moisture retention, which reflects better osmotic regulation [140]. Cabbage treated with *Nocardiopsis* sp. and SeNPs showed greater glucosinolate accumulation and myrosinase activity [141], and okra exposed to *Pseudomonas libanensis* with SeNPs produced more phytochemicals, improved fruit quality, and greater shoot–root growth without harmful Se accumulation [142]. In galaxolide-contaminated soils, soybeans treated with SeNPs, and *Actinobacterium* sp. exhibited higher photosynthetic efficiency and CO_2 assimilation, while significantly reducing oxidative stress and enhancing detoxification molecules, such as phytochelatins and metallothioneins [143]. Cucumbers exposed to *Pseudomonas putida* or *P. stutzeri* with AgNPs showed increased protein, phenolics, sugars, and flavonoids levels [144]. Although AgNPs reduced root and shoot length, defense and antioxidant responses were significantly strengthened. Bitter gourd treated with *P.*

stutzeri plus AgNPs showed dramatic increases in carotenoids, flavonoids, proline, and antioxidant enzymes (increases up to 678%), indicating potent mitigation of oxidative stress [145]. The synergistic role of PGPR and NPs was prominent in heavy-metal-contaminated soils: FeNPs with chromium-resistant *Staphylococcus aureus* improved rice growth by transforming toxic Cr(VI) to Cr(III) and restricting its uptake [146]. CeO_2NPs with *S. aureus* enhanced sunflower growth, reduced membrane damage, and raised antioxidative activity under chromium stress [147]. *Pseudomonas fluorescens* combined with TiO_2NPs improved white clover growth while enhancing phytoremediation potential and chlorophyll levels in cadmium-polluted soil [148]. Collectively, these studies reveal a consistent pattern: PGPR–NP combinations enhance plant resilience, nutrient uptake, soil biochemical activity, detoxification capacity, and crop productivity, demonstrating a robust and sustainable strategy for future agriculture.

8 Nanoparticles as a Delivery System of PGPR

PGPR potential under environmental conditions is often inconsistent due to variable soil environments, limited diffusion, non-targeted delivery, permeability, competition with native microbes, and environmental stresses, all of which reduce PGPR performance. Hence, improving stability and bioavailability and achieving the full potential of PGPR across diverse field conditions remain essential goals in sustainable agriculture. To combat this problem, nanoparticles can be used to stabilize PGPR in the soil system.

Entrapping live plant growth-promoting rhizobacteria (PGPR) within polymer-based nanocarriers has gained considerable attention as an effective approach to enhance their performance in agricultural systems. In this method, viable microbial cells are incorporated into a polymer matrix, ensuring their stability while preserving their functional properties [149]. The gradual, sustained release of PGPR from these carriers enhances root colonization efficiency compared to traditional application methods, as the microorganisms are released only as the polymer slowly breaks down under natural soil conditions [149].

The functionality of these delivery systems may be improved by integrating plant-beneficial compounds into the encapsulating polymer matrix. For instance, *Pseudomonas* sp. DN18 immobilized in alginate beads enriched with salicylic acid (SA) and zinc oxide nanoparticles (ZnONPs) exhibited plant growth-promoting effects similar to those of free-living cells and successfully inhibited disease caused by *Sclerotium rolfsii*. Due to these benefits, the encapsulated *Pseudomonas* sp. DN18 formulation supplemented with ZnO nanoparticles (ZnONPs) and salicylic acid (SA) demonstrated significant promise as an effective biofertilizer or biopesticide for agricultural applications [150].

Similarly, nanoencapsulation of plant growth-promoting microorganisms (PGPM), *Bacillus* sp. CP4 and AHP3 using silica, copper oxide, and zinc oxide nanoparticles via physical adherence in phosphate-buffered saline (PBS) have been

Table 3 Nanoparticles produced by PGPR bacteria and their relation to plant growth

S. No.	PGPR bacteria	Type of nanoparticles	NP character	Concentration used	Plant species	Application	Effect	References
1	*Bacillus brevis, Bacillus megaterium, Pseudomonas fluorescens, Azotobacter*	Silica nanoparticles (SiNPs)	Amorphous silica, ~ 50 nm, > 99.9% purity	Different concentrations (exact values not specified; dose-dependent uptake)	Maize (*Zea mays*)	Soil application, along with comparison to sodium silicate and other bulk Si sources	Improved soil nutrient content and surface properties; increased silica and protein content in bacterial biomass; Promoted seed germination to 100%	[9]
2	Phosphate solubilizers, nitrogen fixers, silicate solubilizers (PGPR community)	Silica nanoparticles (SiNPs)	20–40 nm; high surface area; highly bioavailable; amorphous nano-silica	Not specified	Maize (*Zea mays*)	Soil/rhizosphere application	Significant enhancement of microbial populations, increased microbial biomass; Improved silica availability (14.75 mg mL^{-1}), better nutrient status and potential to promote maize growth	[177]
3	*Pseudomonas koreensis, Bacillus coagulans*	Silica nanoparticles (SiNPs)	Not specified	Not specified	Sugar beet (*Beta vulgaris*)	Seed inoculation (PGPR) + Foliar spraying (SiNPs)	Improved growth, yield, water status, K^+/Na^+ balance, chlorophyll content, and antioxidant enzyme activities; reduced oxidative stress under saline irrigation	[138]
4	*Paenibacillus polymyxa, Bacillus thuringiensis*	Titanium dioxide (TiO_2) nanoparticles	Biocompatible, nanoscale, produced via Sol–Gel	Not specified	Wheat	Seed/root inoculation	Enhanced root colonization and seedling biomass under drought, salt, and pathogen stress	[178]

(continued)

Table 3 (continued)

S. No.	PGPR bacteria	Type of nanoparticles	NP character	Concentration used	Plant species	Application	Effect	References
5	*Bacillus amyloliquefaciens*	Titanium dioxide (TiO_2) nanoparticles	Spherical, ~ 3–3.5 nm; biocompatible; negatively charged at neutral pH; produced via Sol–Gel method	Not explicitly stated (biologically safe concentrations implied; TNs are safe even at up to 30%)	Oilseed rape (*Brassica napus*) and *Arabidopsis thaliana*	Root inoculation with bacteria + nanoparticles	Improved plant growth and stress tolerance, enhanced bacterial adhesion and clustering on roots; protection against fungal pathogen *Alternaria brassicae*	[179]
6	*Pseudomonas fluorescens*	Titanium dioxide (TiO_2) nanoparticles	Not specified	0, 100, 250, 500, 1000 mg/kg soil	White clover (*Trifolium repens)*	Soil amendment + PGPR inoculation	Co-application enhanced plant growth, chlorophyll content, Cd uptake, and accumulation; Highest Cd accumulation (1235 µg/pot) at PGPR + 500 mg/kg TiO_2 NPs; Adverse effects at 1000 mg/kg TiO_2 NPs; Reduced TiO_2 requirement when combined with PGPR	[148]
7	*Pseudomonas libanesis*	Selenium nanoparticles (SeNPs)	Average size 93.31 nm;	75 ppm	Okra	Nanobiofertilizer (Se + PGPR)	30–40% increase in germination; 25–35% increase in shoot & root length; 25–35% increase in phytochemicals; increased fruit yield, Se uptake ~ 80%	[142]

(continued)

Table 3 (continued)

S. No.	PGPR bacteria	Type of nanoparticles	NP character	Concentration used	Plant species	Application	Effect	References
8	*Actinobacterium* sp.	Selenium nanoparticles (SeNPs)	Not specified (likely elemental or oxide form)	25 mg L^{-1}	Soybean	Soil/pot treatment (biofertilizer + SeNPs supplementation)	Reduced oxidative stress, enhanced reactive oxygen species detoxification (antioxidant enzymes, ascorbate–glutathione pathway), increased chlorophyll a, stomatal conductance, photosystem II efficiency, photorespiration, photosynthesis capacity, and biomass (+74%); Combined treatment (PGPB + Se) showed the strongest effects	[169]
9	*Nocardiopsis* sp.	Selenium nanoparticles (SeNPs)	Not specified	Not explicitly stated	*Brassica oleracea*	Seed priming individually or combined with SeNPs	Increased sprout growth (fresh and dry weights); enhanced glucosinolate accumulation, tyrosinase activity, and amino-acid-derived glucosinolates; reduced sulforaphane nitrile (linked to decreased epithiospecifier protein); improved antioxidant activity	[141]
10	*Pseudomonas monteilii*	Gold nanoparticles (AuNPs)	Spherical, hexagonal, triangular; size 12–32 nm; absorption peak 545 nm	50 µg/mL	Cowpea (*Vigna unguiculata*)	Seedling growth with PGPR + AuNPs	Enhanced plant growth and indole acetic acid (IAA) production	[180]

(continued)

Table 3 (continued)

S. No.	PGPR bacteria	Type of nanoparticles	NP character	Concentration used	Plant species	Application	Effect	References
11	*Staphylococcus aureus*	Iron nanoparticles (FeNPs)	Metal-derived, nano-scale	0–20 mg/L	Rice (*Oryza sativa*)	Foliar application	Reduced chromium uptake; enhanced chlorophyll content, growth, biomass, photosynthesis, and oxidative stress tolerance	[146]
12	*Pseudomonas* sp.	Zinc oxide nanoparticles (ZnONPs)	Green-synthesized using Papaya fruit extract;	10 ppm	Wheat (*Triticum aestivum* L.)	Pot experiment under drought, heat, and combined heat + drought stress	Improved plant growth and stress resistance via higher biomass, photosynthetic pigments, nutrient content, soluble sugars, protein, IAA, proline, antioxidant enzymes, and abscisic acid; reduced H_2O_2, and electrolyte leakage	[181]
13	*Azospirillum lipoferum, Bacillus coagulans, Bacillus circulans, Bacillus subtilis*	Zinc oxide nanoparticles (ZnONPs),	Not specified	500 mg L^{-1} (ZnO-NPs), 950 g ha^{-1} (PGPR)	Wheat (*Triticum aestivum* L.)	Foliar + soil application	Enhanced antioxidant enzymes, K^+ uptake, Zn content in grains/straw; reduced electrolyte leakage, proline, H_2O_2; improved soil urease and dehydrogenase activity under saline irrigation	[137]

(continued)

Table 3 (continued)

S. No.	PGPR bacteria	Type of nanoparticles	NP character	Concentration used	Plant species	Application	Effect	References
14	*Bacillus* spp.	Nanozeolite	Not specified	50 ppm	Maize (*Zea mays*)	Seeds treated with PGPR + nanozeolite	Increased plant height, leaf area, number of leaves, chlorophyll, total protein (up to 2 times); enhanced soil biochemical activities (dehydrogenase, alkaline phosphatase up to 3 times); higher phenolic and acid ester compounds; improved soil nutrient status (C, P, K, N)	[182]
15	*Bacillus* spp.	Nanochitosan	Not specified	Not specified	Maize (*Zea mays*)	Seeds treated with PGPR + nanochitosan	Enhanced seed germination (60 to 96.97%), increased plant height (1.5 times), leaf area (double); improved soil physicochemical parameters (pH, organic C, P, K, N); enhanced soil enzyme activities (dehydrogenase, alkaline phosphatase 2–3 times); higher alcohols, acid esters, aldehydes, and organic acids in plant indicating stress tolerance	[183]

(continued)

Table 3 (continued)

S. No.	PGPR bacteria	Type of nanoparticles	NP character	Concentration used	Plant species	Application	Effect	References
16	*Staphylococcus aureus*	Cerium dioxide (CeO_2) nanoparticles	Nanoscale particles (oxidative stress reducer, metal stress alleviator)	25 & 50 mg L^{-1}	Sunflower (*Helianthus annuus*)	Applied to plants grown in Cr-contaminated soil	Improved plant growth and biomass, reduced oxidative stress, enhanced enzymatic activity, increased chlorophyll content, reduced electrolyte leakage (121 to 104%), synergistic effect when combined with bacteria	[147]
17	*Pseudomonas* sp., *Pseudomonas fluorescence, Bacillus cereus*	Silver nanoparticles (AgNPs)	Not specified	Not specified	Maize (*Zea mays*)	Seeds inoculated with PGPR and irrigated with municipal wastewater + AgNPs	Enhanced root length/area, root-shoot ratio, peroxidase/catalase, modulated abscisic acid (34%), IAA (55%), gibberellic acid (82%), proline (70%), and augmented heavy metal (Pb, Cd, Ni) bioremediation	[184]
18	*Bacillus pumilus*	Silver nanoparticles (AgNPs)	Not specified	5 ppm	Onion	Rhizosphere soil	Maximally improved growth, chlorophyll, carotenoids, protein, sugar, proline, and bulb flavonoids; mitigated salt stress effects	[140]
19	*Bacillus cereus*	Silver nanoparticles (AgNPs)	Synthesized from *Tagetes erecta* (Marigold) leaf and flower extracts; ~ 60 nm size;	Not specified (standard for seed coating)	Maize (*Zea mays*)	Seed bacterization/coating	Increased germination (75% to 87.5%); maximized root and shoot growth; inhibited fungal pathogen *Macrophomina phaseolin*	[185]

(continued)

Table 3 (continued)

S. No.	PGPR bacteria	Type of nanoparticles	NP character	Concentration used	Plant species	Application	Effect	References
20	*Pseudomonas putida, Pseudomonas stutzeri*	Silver nanoparticles (AgNPs)	Not specified	5 ppm	Cucumber	Seed inoculation (PGPR), foliar spray (AgNPs)	Increased leaf protein, phenolics, flavonoids, chlorophyll, sugar, and antioxidant enzymes; reduced shoot/root growth but enhanced defense compounds	[179]
21	*Pseudomonas stutzeri*	Silver nanoparticles (AgNPs)	Not explicitly detailed (likely colloidal/ metallic)	Not fully specified; applied with PGPR	*Momordica charantia L*	Seeds soaked in a 73 h old culture of PGPR; AgNPs applied in combination	Reduced oxidative stress, increased carotenoids, flavonoids, proline, and antioxidant enzymes; enhanced germination and soil organic matter	[180]

reported. Greenhouse trials conducted on two wheat cultivars, HD3086 and black wheat NABI MG11, revealed significant improvements in plant biomass, height, nitrogen uptake, and photosynthetic pigment content, among other growth-related parameters [151].

9 Synergistic Activity of NPs and PGPR

The synergistic application of nanoparticles and PGPR is more effective than either input alone. Both can address their challenges individually; thus, their collaborative approach yields better results. Nanoparticles can enhance PGPR performance by improving survival, root colonization, and stress tolerance. Many PGPR possess metal-tolerant capabilities, enabling them to detoxify nanoparticle formulations or regulate their dissolution rates in soil.

The combined application of PGPR and nanoparticles has shown considerable potential in alleviating abiotic stress and enhancing plant performance. Synergistic treatment with PGPR *Bacillus mycoides* PM35 and TiO_2NPs substantially boosts morpho-physio-biochemical parameters in Cd-stressed barley (*Hordeum vulgare* L.) [139]. Exposure to cadmium adversely influences gas exchange, carbohydrate levels, photosynthetic pigments, and nutrient distribution in roots and shoots. Furthermore, the combined treatment with *Bacillus mycoides* PM35 and TiO_2NPs effectively alleviated these negative impacts by promoting plant growth and biomass production. This improvement was attributed to enhanced scavenging of reactive oxygen species, reduction of oxidative stress, and decreased cadmium accumulation in plant tissues [139].

Likewise, integrating PGPR strains *Bacillus pumilus* and *Pseudomonas moraviensis* with silver nanoparticles promoted improved growth in onion plants. The treated plants also showed greater resilience to salt stress than the untreated plants [140]. The combined application of zinc oxide nanoparticles and PGPR strains (*Azospirillum lipoferum* SP2, *Bacillus coagulans* NCAIM B.01123, *Bacillus circulans* NCAIM B.02324, and *Bacillus subtilis* MF497446) was found to enhance both salinity tolerance and yield of wheat in sodic–saline soils [137].

Wheat plants irrigated with saline water exhibited improved protection when treated with both PGPR and ZnONPs. This effect is mediated by elevated activities of antioxidant enzymes—peroxidase (102%), catalase (47%), and superoxide dismutase (106%)—along with a 27% increase in K^+ uptake relative to the control. Effective stress reduction was evident from the substantial decreases in malondialdehyde (56%), electrolyte leakage (62%), proline (39%), and H_2O_2 (60%). Wheat grains exhibited a 57% increase in nitrogen uptake, whereas zinc levels in the grains and straw increased by 117% and 72%, respectively. Application of PGPR combined with ZnONPs in saline-irrigated soils significantly increased soil enzymatic activity, with urease activity increasing by 80% and dehydrogenase activity by 232% [137].

In addition, the synergistic use of ZnO nanoparticles with PGPR such as *Priestia megaterium* (RGKP3) and *Pseudomonas songnenensis* (RG8) enhanced growth in groundnut (*Arachis hypogaea* L.) [152]. A bio-nanofertilizer formulation comprising ZnO and γ-Fe_2O_3 nanoparticles, PGPR (*Pseudomonas* spp.), and the microalga *Spirulina platensis* has also been reported to significantly improve overall plant productivity, highlighting the promise of integrated nano-bio approaches in sustainable agriculture [153].

10 Artificial Intelligence (AI) in Optimizing Nano-PGPR Applications

Artificial intelligence (AI) provides a powerful analytical framework for interpreting the multidimensional interactions among nanoparticles (NPs), plant growth-promoting rhizobacteria (PGPR), plants, and soil environments. By integrating large datasets generated from nanomaterial characterization, soil microbiome sequencing, and plant physiological monitoring, AI and machine-learning models can improve predictions of NP-PGPR compatibility, optimize concentration ranges, and identify early signatures of phytotoxicity or microbial disruption [154]. AI-enabled tools further support high-resolution microbiome analysis, improving our understanding of NP-induced ecological changes and community stability [155]. In parallel, deep learning and big data approaches provide enhanced capabilities for precision agriculture, enabling real-time monitoring and site-specific recommendations for NP–PGPR application [156, 157]. Moreover, machine-learning-based decision-support systems can optimize resource use and minimize environmental risks by integrating sensor-derived soil and crop data with predictive models [158]. Collectively, AI strengthens the design, optimization, and field translation of nano-PGPR technologies, supporting their safe and sustainable deployment in modern agriculture.

11 Conclusion and Future Perspectives

PGPR and nanoparticles have demonstrated remarkable abilities to boost plant growth. Both can enhance plant growth, either directly or indirectly, through nutrient mobilization, improved photosynthetic efficiency, increased phytohormone production, and enhanced stress tolerance. Their synergistic application is an emerging sustainable agriculture approach. When applied together, their interactions can amplify beneficial effects, creating a multifunctional system that supports plant performance more efficiently than either used alone. This synergy often arises from complementary mechanisms: NPs acting as micronutrient carriers can improve soil physicochemical properties, while PGPR colonizes the rhizosphere to facilitate production of growth hormones and protect plants from pathogens [11]. Collectively, these interactions can enhance crop yield, reduce dependency on chemical fertilizers, and contribute to ecological safety [11]. The combined application of

PGPR and NPs is still in its early stages. Further detailed investigation is required for NP–microbe compatibility, soil interactions, optimal dosages, and plant-specific responses. The introduction of nanomaterials alters soil microbial communities in unpredictable ways, influencing the effectiveness of PGPR and long-term soil health. Similarly, the physicochemical properties of NPs significantly influence their effects in biological systems and, therefore, their interactions with beneficial microbes [53].

Future research should prioritize mechanistic studies to elucidate how nanoparticles influence PGPR colonization, signaling pathways, and metabolic activities. Integrated multi-omics, including transcriptomics, metabolomics, and proteomics, together with recent AI-based developments in agriculture, can provide a more comprehensive understanding of plant–microbe–nanoparticle interactions at the molecular level [159]. Additionally, research must focus on developing biocompatible and biodegradable nanoparticles, such as green-synthesized or bio-inspired NPs, that minimize ecological risks while maximizing agricultural productivity.

Another important direction is nanoencapsulation of PGPR, in which nanoparticles serve as carriers, enhancing PGPR's survival, shelf life, and colonization capacity [160]. Controlled-release of PGPR and nanoparticles can revolutionize precision agriculture. Maximizing field-based applications is more important than relying solely on in vitro applications. Collaboration among agronomists, microbiologists, soil scientists, and nanotechnologists will accelerate the development of integrated nano-PGPR technologies for agroecosystems. Through continued interdisciplinary research and careful environmental assessment, this innovative approach has the potential to become a model for next-generation sustainable farming.

References

1. Nations U (2024) Global issues population. https://www.un.org/en/global-issues/population
2. Park J, Guan W, Yu G (2025) Smart hydrogels for sustainable agriculture. EcoMat 7(4):e70011
3. Khoso MA, Wagan S, Alam I, Hussain A, Ali Q, Saha S et al (2024) Impact of plant growth-promoting rhizobacteria (PGPR) on plant nutrition and root characteristics: current perspective. Plant Stress 11:100341
4. Biswas R, Misra A, Ghosh S, Chakraborty A, Mukherjee P, Dam B (2024) *Pantoea tagorei* sp. nov., a rhizospheric bacteria with plant growth-promoting activities. Indian J Microbiol 64(3):937–949
5. Copeland C, Schulze-Lefert P, Ma K-W (2025) Potential and challenges for application of microbiomes in agriculture. Plant Cell 37(8):koaf185
6. Bilia AR, Piazzini V, Risaliti L, Vanti G, Casamonti M, Wang M et al (2019) Nanocarriers: a successful tool to increase solubility, stability and optimise bioefficacy of natural constituents. Curr Med Chem 26(24):4631–4656
7. Mittal D, Kaur G, Singh P, Yadav K, Ali SA (2020) Nanoparticle-based sustainable agriculture and food science: recent advances and future outlook. Front Nanotechnol 2:579954
8. Kant R, Shukla DR, Mandal S (2025) Nanocarriers for delivering plant growth promoting microorganisms. Nanocarriers Plant Sci Agric 235–272
9. Khoshru B, Fallah Nosratabad A, Khosravi H, Asgharzadeh A, Faridian L (2025) Enhancing agricultural productivity using PGPR and nanoparticles: mechanisms, challenges, and future directions. J Soil Biol 2025(2)

10. Karunakaran G, Suriyaprabha R, Manivasakan P, Yuvakkumar R, Rajendran V, Prabu P et al (2013) Effect of nanosilica and silicon sources on plant growth promoting rhizobacteria, soil nutrients and maize seed germination. IET Nanobiotechnol 7(3):70–77
11. Verma KK, Joshi A, Song X-P, Singh S, Kumari A, Arora J et al (2024) Synergistic interactions of nanoparticles and plant growth promoting rhizobacteria enhancing soil-plant systems: a multigenerational perspective. Front Plant Sci 15:1376214
12. Falcon WP, Naylor RL, Shankar ND (2022) Rethinking global food demand for 2050. Popul Dev Rev 48(4):921–957
13. Ashitha A, Rakhimol K (2021) Fate of the conventional fertilizers in environment. In: Controlled release fertilizers for sustainable agriculture. Elsevier, pp 25–39
14. Ferreira CM, Soares HM, Soares EV (2019) Promising bacterial genera for agricultural practices: an insight on plant growth-promoting properties and microbial safety aspects. Sci Total Environ 682:779–799
15. Onakpa MM, Njan AA, Kalu OC (2018) A review of heavy metal contamination of food crops in Nigeria. Ann Glob Health 84(3):488
16. Giri S, Panwar H, Singh K, Kumar M (2025) Microbial contributions to crop adaptation: innovation for climate-resilient agriculture. In: Cutting edge technologies for developing future crop plants. Springer, pp 349–364
17. Siddikee MA, Chauhan PS, Anandham R, Han G-H, Sa T (2010) Isolation, characterization, and use for plant growth promotion under salt stress, of ACC deaminase-producing halotolerant bacteria derived from coastal soil. J Microbiol Biotechnol 20(11):1577–1584
18. Glick BR (2014) Bacteria with ACC deaminase can promote plant growth and help to feed the world. Microbiol Res 169(1):30–39
19. Oberson A, Frossard E, Bühlmann C, Mayer J, Mäder P, Lüscher A (2013) Nitrogen fixation and transfer in grass-clover leys under organic and conventional cropping systems. Plant Soil 371(1):237–255
20. Goswami D, Parmar S, Vaghela H, Dhandhukia P, Thakker JN (2015) Describing *Paenibacillus mucilaginosus* strain N3 as an efficient plant growth promoting rhizobacteria (PGPR). Cogent Food Agric 1(1):1000714
21. Govindasamy V, Senthilkumar M, Upendra-Kumar AK (2008) PGPR-biotechnology for management of abiotic and biotic stresses in crop plants. In: Potential microorganisms for sustainable agriculture, pp 26–48
22. Jha CK, Saraf M (2015) Plant growth promoting rhizobacteria (PGPR). J Agric Res Dev. 5(2):108–119
23. Goswami D, Patel K, Parmar S, Vaghela H, Muley N, Dhandhukia P et al (2015) Elucidating multifaceted urease producing marine *Pseudomonas aeruginosa* BG as a cogent PGPR and bio-control agent. Plant Growth Regul 75(1):253–263
24. Pan L, Cai B (2023) Phosphate-solubilizing bacteria: advances in their physiology, molecular mechanisms and microbial community effects. Microorganisms. 11(12):2904
25. Wang Z, Zhang H, Liu L, Li S, Xie J, Xue X et al (2022) Screening of phosphate-solubilizing bacteria and their abilities of phosphorus solubilization and wheat growth promotion. BMC Microbiol 22(1):296
26. Kiprotich K, Muoma J, Omayio DO, Ndombi TS, Wekesa C (2023) Molecular characterization and mineralizing potential of phosphorus solubilizing bacteria colonizing common bean (*Phaseolus vulgaris* L.) rhizosphere in Western Kenya. Int J Microbiol 2023(1):6668097
27. Ren J, Liu H, Wu X, Wang Q, Ren Y, Liu Y et al (2012) Screening, identification, and promoting effect of phosphate-solubilizing bacteria in rhizosphere of *Taxus chinensis* var. *mairei*. Wei Sheng wu xue bao = Acta Microbiologica Sinica 52(3):295–303
28. Gupta R, Kumari A, Sharma S, Alzahrani OM, Noureldeen A, Darwish H (2022) Identification, characterization and optimization of phosphate solubilizing rhizobacteria (PSRB) from rice rhizosphere. Saudi J Biol Sci. 29(1):35–42
29. Aeron A, Kumar S, Pandey P, Maheshwari D (2011) Emerging role of plant growth promoting rhizobacteria in agrobiology. In: Bacteria in agrobiology: crop ecosystems. Springer, pp 1–36

30. Lata DL, Abdie O, Rezene Y (2024) IAA-producing bacteria from the rhizosphere of chickpea (*Cicer arietinum* L.): isolation, characterization, and their effects on plant growth performance. Heliyon 10(21)
31. Swain MR, Naskar SK, Ray RC (2007) Indole-3-acetic acid production and effect on sprouting of yam (*Dioscorea rotundata* L.) minisetts by *Bacillus subtilis* isolated from culturable cowdung microflora. Pol J Microbiol 56(2):103
32. Li M, Guo R, Yu F, Chen X, Zhao H, Li H et al (2018) Indole-3-acetic acid biosynthesis pathways in the plant-beneficial bacterium *Arthrobacter pascens* ZZ21. Int J Mol Sci 19(2):443
33. Shao J, Li S, Zhang N, Cui X, Zhou X, Zhang G et al (2015) Analysis and cloning of the synthetic pathway of the phytohormone indole-3-acetic acid in the plant-beneficial *Bacillus amyloliquefaciens* SQR9. Microb Cell Fact 14(1):130
34. Liu W-H, Chen F-F, Wang C-E, Fu H-H, Fang X-Q, Ye J-R et al (2019) Indole-3-acetic acid in *Burkholderia pyrrocinia* JK-SH007: enzymatic identification of the indole-3-acetamide synthesis pathway. Front Microbiol 10:2559
35. Castro-Guerrero J, Romero A, Aguilar JJ, Xiqui ML, Sandoval JO, Baca BE (2012) The hisC1 gene, encoding aromatic amino acid aminotransferase-1 in *Azospirillum brasilense* Sp7, expressed in wheat. Plant Soil 356(1):139–150
36. Amara U, Khalid R, Hayat R (2015) Soil bacteria and phytohormones for sustainable crop production. In: Bacterial metabolites in sustainable agroecosystem. Springer, pp 87–103
37. Karnwal A, Kaushik P (2011) Cytokinin production by fluorescent *Pseudomonas* in the presence of rice root exudates. Arch Phytopathol Plant Prot 44(17):1728–1735
38. Rabari A, Ruparelia J, Jha CK, Perveen K, Debnath A, Behera M et al (2025) Evaluating the phytohormone proficiencies of multifarious *Bacillus rugosus* for growth promotion in *Arachis hypogaea* (L.). J Basic Microbiol 65(6):e70011
39. Kang S-M, Khan AL, You Y-H, Kim J-G, Kamran M, Lee I-J (2014) Gibberellin production by newly isolated strain *Leifsonia soli* SE134 and its potential to promote plant growth. J Microbiol Biotechnol 24(1):106–112
40. Kang S-M, Khan AL, Hamayun M, Hussain J, Joo G-J, You Y-H et al (2012) Gibberellin-producing *Promicromonospora* sp. SE188 improves *Solanum lycopersicum* plant growth and influences endogenous plant hormones. J Microbiol 50(6):902–909
41. Passari AK, Ruiz-Villafan B, Cruz-Bautista R, Diaz-Dominguez V, Rodriguez-Sanoja R, Sanchez S (2023) Opportunities and challenges of microbial siderophores in the medical field. Appl Microbiol Biotechnol 107(22):6751–6759
42. Timofeeva AM, Galyamova MR, Sedykh SE (2022) Bacterial siderophores: classification, biosynthesis, perspectives of use in agriculture. Plants 11(22):3065
43. Shen X, Hu H, Peng H, Wang W, Zhang X (2013) Comparative genomic analysis of four representative plant growth-promoting rhizobacteria in *Pseudomonas*. BMC Genomics 14(1):271
44. Annapurna K, Kumar A, Kumar LV, Govindasamy V, Bose P, Ramadoss D (2012) PGPR-induced systemic resistance (ISR) in plant disease management. In: Bacteria in agrobiology: disease management. Springer, pp 405–425
45. Zohoursoleimani R, Aeini M, Ghodoum Parizipour MH (2025) Effects of abiotic and biotic elicitors on tomato resistance to tomato brown rugose fruit virus. Sci Rep 15(1):27216
46. Van Loon L, Bakker P (2006) Induced systemic resistance as a mechanism of disease suppression by rhizobacteria. PGPR: Biocontrol Biofertilization 6:39–66
47. del Carmen O-M, Glick BR, Santoyo G (2020) ACC deaminase in plant growth-promoting bacteria (PGPB): an efficient mechanism to counter salt stress in crops. Microbiol Res 235:126439
48. Ali S, Charles T, Glick B (2012) Delay of flower senescence by bacterial endophytes expressing 1-aminocyclopropane-1-carboxylate deaminase. J Appl Microbiol 113(5):1139–1144
49. Dar A, Zahir ZA, Iqbal M, Mehmood A, Javed A, Hussain A et al (2021) Efficacy of rhizobacterial exopolysaccharides in improving plant growth, physiology, and soil properties. Environ Monit Assess 193(8):515

50. Ansari FA, Ahmad I (2018) Biofilm development, plant growth promoting traits and rhizosphere colonization by *Pseudomonas entomophila* FAP1: a promising PGPR. Adv Microbiol 8(3):235–251
51. Prasad J, Dey R, Raghuwanshi R (2022) Exopolysaccharide-producing rhizospheric bacteria enhance yield via promoting wheat (*Triticum aestivum* L.) growth at early stages. Microbiology 91(6):757–769
52. Ahmed T, Noman M, Rizwan M, Ali S, Ijaz U, Nazir MM et al (2022) Green molybdenum nanoparticles-mediated bio-stimulation of *Bacillus* sp. strain ZH16 improved the wheat growth by managing in planta nutrients supply, ionic homeostasis and arsenic accumulation. J Hazard Mater 423:127024
53. Joudeh N, Linke D (2022) Nanoparticle classification, physicochemical properties, characterization, and applications: a comprehensive review for biologists. J Nanobiotechnology 20(1):262
54. Zielińska A, Carreiró F, Oliveira AM, Neves A, Pires B, Venkatesh DN et al (2020) Polymeric nanoparticles: production, characterization, toxicology and ecotoxicology. Molecules 25(16):3731
55. Jo Y-K, Kim BH, Jung G (2009) Antifungal activity of silver ions and nanoparticles on phytopathogenic fungi. Plant Dis 93(10):1037–1043
56. Gajbhiye M, Kesharwani J, Ingle A, Gade A, Rai M (2009) Fungus-mediated synthesis of silver nanoparticles and their activity against pathogenic fungi in combination with fluconazole. Nanomedicine: Nanotechnol Biol Med 5(4):382–386
57. Dimkpa CO, McLean JE, Britt DW, Anderson AJ (2013) Antifungal activity of ZnO nanoparticles and their interactive effect with a biocontrol bacterium on growth antagonism of the plant pathogen *Fusarium graminearum*. Biometals 26:913–924
58. Kanhed P, Birla S, Gaikwad S, Gade A, Seabra AB, Rubilar O et al (2014) *In vitro* antifungal efficacy of copper nanoparticles against selected crop pathogenic fungi. Mater Lett 115:13–17
59. Xing K, Shen X, Zhu X, Ju X, Miao X, Tian J et al (2016) Synthesis and *in vitro* antifungal efficacy of oleoyl-chitosan nanoparticles against plant pathogenic fungi. Int J Biol Macromol 82:830–836
60. Ghormade V, Deshpande MV, Paknikar KM (2011) Perspectives for nano-biotechnology enabled protection and nutrition of plants. Biotechnol Adv 29(6):792–803
61. Gupta A, Tripathy DB, Kumar G, Agarwal P, Ghosal A (2022) Nanopesticides, nanoherbicides, and nanofertilizers. CRC Press, Taylor and Francis
62. Jayaseelan C, Rahuman AA, Rajakumar G, Vishnu Kirthi A, Santhoshkumar T, Marimuthu S et al (2011) Synthesis of pediculocidal and larvicidal silver nanoparticles by leaf extract from heartleaf moonseed plant, *Tinospora cordifolia* Miers. Parasitology Res 109:185–194
63. Ghosh S, Sarkar B, Thongmee S (2022) Nanoherbicides for field applications. In: Agricultural nanobiotechnology. Elsevier, pp 439–463
64. Naresh V, Lee N (2021) A review on biosensors and recent development of nanostructured materials-enabled biosensors. Sensors 21(4):1109
65. Mondal R, Dam P, Chakraborty J, Paret ML, Katı A, Altuntas S et al (2022) Potential of nanobiosensor in sustainable agriculture: the state-of-art. Heliyon8(12).
66. Zhan F, Wang T, Iradukunda L, Zhan J (2018) A gold nanoparticle-based lateral flow biosensor for sensitive visual detection of the potato late blight pathogen, *Phytophthora infestans*. Anal Chim Acta 1036:153–161
67. Baig N, Kammakakam I, Falath W (2021) Nanomaterials: a review of synthesis methods, properties, recent progress, and challenges. Mater Adv 2(6):1821–1871
68. Gorrasi G, Sorrentino A (2015) Mechanical milling as a technology to produce structural and functional bio-nanocomposites. Green Chem 17(5):2610–2625
69. Wirunchit S, Gansa P, Koetniyom W (2021) Synthesis of ZnO nanoparticles by Ball-milling process for biological applications. Mater Today: Proc 47:3554–3559
70. Ostermann R, Cravillon J, Weidmann C, Wiebcke M, Smarsly BM (2011) Metal–organic framework nanofibers via electrospinning. Chem Commun 47(1):442–444

71. Chronakis IS (2010) Micro-/nano-fibers by electrospinning technology: processing, properties and applications. Micromanufacturing Eng Technol 2010:264–286
72. Amendola V, Meneghetti M (2009) Laser ablation synthesis in solution and size manipulation of noble metal nanoparticles. Phys Chem Chem Phys 11(20):3805–3821
73. Joh D-W, Jung T-K, Lee H-S, Kim D-H (2013) Synthesis of nanoparticles using electrical explosion of Ni wire in Pt solution. J Nanosci Nanotechnol 13(9):6092–6094
74. Zheng Z, Zhang X, Carbo D, Clark C, Nathan C-A, Lvov Y (2010) Sonication-assisted synthesis of polyelectrolyte-coated curcumin nanoparticles. Langmuir 26(11):7679–7681
75. Patil N, Bhaskar R, Vyavhare V, Dhadge R, Patil Y (2021) Overview on methods of synthesis of nanoparticles. Int J Curr Pharm Res 11–16
76. Hontanon E, Palomares JM, Stein M, Guo X, Engeln R, Nirschl H et al (2013) The transition from spark to arc discharge and its implications with respect to nanoparticle production. J Nanopart Res 15(9):1957
77. Ago H (2015) CVD growth of high-quality single-layer graphene. Front Graphene Carbon Nanotub: Devices Appl 3–20
78. Das S, Srivasatava VC (2016) Synthesis and characterization of ZnO–MgO nanocomposite by co-precipitation method. Smart Sci 4(4):190–195
79. Parashar M, Shukla VK, Singh R (2020) Metal oxides nanoparticles via sol–gel method: a review on synthesis, characterization and applications. J Mater Sci Mater Electron 31(5):3729–3749
80. Das S, Srivastava VC (2016) Hierarchical nanostructured ZnO-CuO nanocomposite and its photocatalytic activity. J Nano Res 35:21–26
81. Malhotra SPK, Alghuthaymi MA (2022) Biomolecule-assisted biogenic synthesis of metallic nanoparticles. In: Agri-waste and microbes for production of sustainable nanomaterials, pp 139–163
82. Castillo-Henríquez L, Alfaro-Aguilar K, Ugalde-Álvarez J, Vega-Fernández L, Montes de Oca-Vásquez G, Vega-Baudrit JR (2020) Green synthesis of gold and silver nanoparticles from plant extracts and their possible applications as antimicrobial agents in the agricultural area. Nanomaterials 10(9):1763
83. Ghosh S, Ahmad R, Zeyaullah M, Khare SK (2021) Microbial nano-factories: synthesis and biomedical applications. Front Chem 9:626834
84. Liu S, Pan L, Chen J, Wang Z, Li Z, Gao C et al (2024) Mechanisms and applications of microbial synthesis of metal nanoparticles in agri-sectors. Environ Sci Nano 11(7):2803–2830
85. Iravani S, Varma RS (2020) Bacteria in heavy metal remediation and nanoparticle biosynthesis. ACS Sustain Chem Eng 8(14):5395–5409
86. Koul B, Poonia AK, Yadav D, Jin J-O (2021) Microbe-mediated biosynthesis of nanoparticles: applications and future prospects. Biomolecules 11(6):886
87. Elegbede JA, Lateef A, Azeez MA, Asafa TB, Yekeen TA, Oladipo IC et al (2018) Fungal xylanases-mediated synthesis of silver nanoparticles for catalytic and biomedical applications. IET Nanobiotechnol 12(6):857–863
88. Abada E, Al-Faifi Z, Osman M (2017) Enzymes and nanoparticles produced by microorganisms and their applications in biotechnology. In: Fungal nanotechnology: applications in agriculture, industry, and medicine. Springer, pp 119–150
89. Tsekhmistrenko S, Bityutskyy V, Tsekhmistrenko O, Horalskyi L, Tymoshok N, Spivak M (2020) Bacterial synthesis of nanoparticles: a green approach. Biosyst Divers 28(1):9–17
90. He S, Guo Z, Zhang Y, Zhang S, Wang J, Gu N (2007) Biosynthesis of gold nanoparticles using the bacteria *Rhodopseudomonas capsulata*. Mater Lett 61(18):3984–3987
91. Hamk M, Akçay FA, Avcı A (2023) Green synthesis of zinc oxide nanoparticles using *Bacillus subtilis* ZBP4 and their antibacterial potential against foodborne pathogens. Prep Biochem Biotechnol 53(3):255–264
92. Alali A, Hosseini-Abari A, Bahrami A, Yazdan MM (2023) Biosynthesis of copper oxide and silver nanoparticles by *Bacillus* spores and evaluation of the feasibility of their use in antimicrobial paints. Materials 16(13):4670

93. Varshney R, Bhadauria S, Gaur MS, Pasricha R (2011) Copper nanoparticles synthesis from electroplating industry effluent. Nano Biomed Eng. 3(2):115–119
94. Ogunyemi SO, Zhang M, Abdallah Y, Ahmed T, Qiu W, Ali MA et al (2020) The bio-synthesis of three metal oxide nanoparticles (ZnO, MnO_2, and MgO) and their antibacterial activity against the bacterial leaf blight pathogen. Front Microbiol 11:588326
95. Binupriya A, Sathishkumar M, Vijayaraghavan K, Yun S-I (2010) Bioreduction of trivalent aurum to nano-crystalline gold particles by active and inactive cells and cell-free extract of *Aspergillus oryzae* var. *viridis*. J Hazard Mater 177(1–3):539–545
96. Ravindra B, Rajasab A (2014) A comparative study on biosynthesis of silver nanoparticles using four different fungal species
97. Jha AK, Prasad K, Kulkarni A (2009) Synthesis of TiO_2 nanoparticles using microorganisms. Colloids Surf, B 71(2):226–229
98. Karthik L, Kumar G, Kirthi AV, Rahuman A, Bhaskara Rao K (2014) *Streptomyces* sp. LK3 mediated synthesis of silver nanoparticles and its biomedical application. Bioprocess Biosyst Eng 37(2):261–267
99. Hassan SED, Salem SS, Fouda A, Awad MA, El-Gamal MS, Abdo AM (2018) New approach for antimicrobial activity and bio-control of various pathogens by biosynthesized copper nanoparticles using endophytic actinomycetes. J Radiat Res Appl Sci 11(3):262–270
100. Ahmad A, Senapati S, Khan MI, Kumar R, Ramani R, Srinivas V et al (2003) Intracellular synthesis of gold nanoparticles by a novel alkalotolerant actinomycete, *Rhodococcus* species. Nanotechnology 14(7):824
101. Fawcett D, Verduin JJ, Shah M, Sharma SB, Poinern GEJ (2017) A review of current research into the biogenic synthesis of metal and metal oxide nanoparticles via marine algae and seagrasses. J Nanosci 2017(1):8013850
102. Dahoumane SA, Mechouet M, Wijesekera K, Filipe CD, Sicard C, Bazylinski DA et al (2017) Algae-mediated biosynthesis of inorganic nanomaterials as a promising route in nanobiotechnology–a review. Green Chem 19(3):552–587
103. Bibi S, Abu-Dieyeh MH, Al-Ghouti MA (2025) Biosynthesis of silver nanoparticles from macroalgae *Hormophysa triquetra* and investigation of its antibacterial activity and mechanism against pathogenic bacteria. Sci Rep 15(1):2476
104. Bhuyar P, Rahim MHA, Sundararaju S, Ramaraj R, Maniam GP, Govindan N (2020) Synthesis of silver nanoparticles using marine macroalgae Padina sp. and its antibacterial activity towards pathogenic bacteria. Beni-Suef Univ J Basic Appl Sci 9(1):1–15
105. Dahoumane SA, Yéprémian C, Djédiat C, Couté A, Fiévet F, Coradin T et al (2016) Improvement of kinetics, yield, and colloidal stability of biogenic gold nanoparticles using living cells of *Euglena gracilis* microalga. J Nanopart Res 18(3):79
106. Mohamad NAN, Arham NA, Jai J, Hadi A (2014) Plant extract as reducing agent in synthesis of metallic nanoparticles: a review. Adv Mater Res 832:350–355
107. Ahmad H, Venugopal K, Rajagopal K, De Britto S, Nandini B, Pushpalatha HG et al (2020) Green synthesis and characterization of zinc oxide nanoparticles using *Eucalyptus globules* and their fungicidal ability against pathogenic fungi of apple orchards. Biomolecules 10(3):425
108. Wahid I, Kumari S, Ahmad R, Hussain SJ, Alamri S, Siddiqui MH et al (2020) Silver nanoparticle regulates salt tolerance in wheat through changes in ABA concentration, ion homeostasis, and defense systems. Biomolecules 10(11):1506
109. Khodakovskaya M, Dervishi E, Mahmood M, Xu Y, Li Z, Watanabe F et al (2009) Carbon nanotubes are able to penetrate plant seed coat and dramatically affect seed germination and plant growth. ACS Nano 3(10):3221–3227
110. Khodakovskaya MV, Kim BS, Kim JN, Alimohammadi M, Dervishi E, Mustafa T et al (2013) Carbon nanotubes as plant growth regulators: effects on tomato growth, reproductive system, and soil microbial community. Small 9(1):115–123
111. Pokhrel LR, Dubey B (2013) Evaluation of developmental responses of two crop plants exposed to silver and zinc oxide nanoparticles. Sci Total Environ 452:321–332

112. Kumar V, Guleria P, Kumar V, Yadav SK (2013) Gold nanoparticle exposure induces growth and yield enhancement in *Arabidopsis thaliana*. Sci Total Environ 461:462–468
113. Savithramma N, Ankanna S, Bhumi G (2012) Effect of nanoparticles on seed germination and seedling growth of *Boswellia ovalifoliolata* an endemic and endangered medicinal tree taxon. Nano Vision 2(1):2
114. Krishnaraj C, Jagan E, Ramachandran R, Abirami S, Mohan N, Kalaichelvan P (2012) Effect of biologically synthesized silver nanoparticles on *Bacopa monnieri* (Linn.) Wettst. plant growth metabolism. Process Biochemistry 47(4):651–658
115. Yang F, Liu C, Gao F, Su M, Wu X, Zheng L et al (2007) The improvement of spinach growth by nano-anatase TiO_2 treatment is related to nitrogen photoreduction. Biol Trace Elem Res 119:77–88
116. Tarafdar J, Raliya R, Mahawar H, Rathore I (2014) Development of zinc nanofertilizer to enhance crop production in pearl millet (*Pennisetum americanum*). Agric Res 3(3):257–262
117. Almeelbi T, Bezbaruah A (2014) Nanoparticle-sorbed phosphate: iron and phosphate bioavailability studies with *Spinacia oleracea* and *Selenastrum capricornutum*. ACS Sustain Chem Eng 2(7):1625–1632
118. Feizi H, Kamali M, Jafari L, Moghaddam PR (2013) Phytotoxicity and stimulatory impacts of nanosized and bulk titanium dioxide on fennel (*Foeniculum vulgare* Mill). Chemosphere 91(4):506–511
119. Mahmoodzadeh H, Nabavi M, Kashefi H (2013) Effect of nanoscale titanium dioxide particles on the germination and growth of canola (*Brassica napus*)
120. Lu ChangMei LC, Zhang ChaoYing ZC, Wen JunQiang WJ, Wu GuoRong WG, Tao MingXuan TM (2002) Research of the effect of nanometer materials on germination and growth enhancement of *Glycine max* and its mechanism
121. Singh A, Singh N, Hussain I, Singh H, Singh S (2015) Plant-nanoparticle interaction: an approach to improve agricultural practices and plant productivity. Int J Pharm Sci Invent. 4(8):25–40
122. Raliya R, Tarafdar JC (2013) ZnO nanoparticle biosynthesis and its effect on phosphorous-mobilizing enzyme secretion and gum contents in Clusterbean (*Cyamopsis tetragonoloba* L.). Agric Res 2:48–57
123. Ahuja R, Sidhu A, Bala A (2019) Synthesis and evaluation of iron (ii) sulfide aqua nanoparticles (FeS-NPs) against *Fusarium verticillioides* causing sheath rot and seed discoloration of rice. Eur J Plant Pathol 155(1):163–171
124. Abdel Latef AAH, Abu Alhmad MF, Abdelfattah KE (2017) The possible roles of priming with ZnO nanoparticles in mitigation of salinity stress in lupine (*Lupinus termis*) plants. J Plant Growth Regul 36(1):60–70
125. Kannaujia R, Srivastava CM, Prasad V, Singh BN, Pandey V (2019) *Phyllanthus emblica* fruit extract stabilized biogenic silver nanoparticles as a growth promoter of wheat varieties by reducing ROS toxicity. Plant Physiol Biochem 142:460–471
126. Muthukrishnan S, Murugan I, Selvaraj M (2019) Chitosan nanoparticles loaded with thiamine stimulate growth and enhances protection against wilt disease in Chickpea. Carbohyd Polym 212:169–177
127. Naderi M, Danesh-Shahraki A (2013) Nanofertilizers and their roles in sustainable agriculture
128. Liu R, Lal R (2014) Synthetic apatite nanoparticles as a phosphorus fertilizer for soybean (*Glycine max*). Sci Rep 4(1):5686
129. Shah V, Belozerova I (2009) Influence of metal nanoparticles on the soil microbial community and germination of lettuce seeds. Water Air Soil Pollut 197:143–148
130. Kottegoda N, Munaweera I, Madusanka N, Karunaratne V (2011) A green slow-release fertilizer composition based on urea-modified hydroxyapatite nanoparticles encapsulated wood. Curr Sci 73–78
131. Manikandan A, Subramanian K (2014) Fabrication and characterisation of nanoporous zeolite based N fertilizer. Afr J Agric Res 9(2):276–284
132. Li Y, Jin Q, Yang D, Cui J (2018) Molybdenum sulfide induce growth enhancement effect of rice (*Oryza sativa* L.) through regulating the synthesis of chlorophyll and the expression of aquaporin gene. J Agric Food Chem 66(16):4013–4021

133. Giraldo JP, Landry MP, Faltermeier SM, McNicholas TP, Iverson NM, Boghossian AA et al (2014) Plant nanobionics approach to augment photosynthesis and biochemical sensing. Nat Mater 13(4):400–408
134. Siddiqui MH, Al-Whaibi MH, Faisal M, Al Sahli AA (2014) Nano-silicon dioxide mitigates the adverse effects of salt stress on *Cucurbita pepo* L. Environ Toxicol Chem 33(11):2429–2437
135. Gao F, Hong F, Liu C, Zheng L, Su M, Wu X et al (2006) Mechanism of nano-anatase TiO_2 on promoting photosynthetic carbon reaction of spinach: inducing complex of rubisco-rubisco activase. Biol Trace Elem Res 111:239–253
136. Hu Y, Zhong X, Wang X, Yi J, Luo J, Tang G (2025) Magnesium nanoparticles enhance growth and reshape the rhizosphere microbial community in soybean (*Glycine max* L.). Plant Physiol Biochem 110460
137. Alharbi K, Hafez EM, Omara AE-D, Rashwan E, Alshaal T (2023) Zinc oxide nanoparticles and PGPR strengthen salinity tolerance and productivity of wheat irrigated with saline water in sodic-saline soil. Plant Soil 493(1):475–495
138. Alharbi K, Hafez E, Omara AE-D, Awadalla A, Nehela Y (2022) Plant growth promoting rhizobacteria and silica nanoparticles stimulate sugar beet resilience to irrigation with saline water in salt-affected soils. Plants 11(22):3117
139. Ma J, Li Y, Chen F, Sun Y, Zhu Y, Wang L (2023) *Bacillus mycoides* PM35 in combination with titanium dioxide (TiO_2)—nanoparticles enhanced morpho-physio-biochemical attributes in Barley (*Hordeum vulgare* L.) under cadmium stress. Chemosphere 323:138224
140. Jahangir S, Javed K, Bano A (2020) Nanoparticles and plant growth promoting rhizobacteria (PGPR) modulate the physiology of onion plant under salt stress. Pak J Bot 52(4):1473–1480
141. AbdElgawad H, Magdy Korany S, Reyad AM, Zahid I, Akhter N, Alsherif E et al (2023) Synergistic impacts of plant-growth-promoting bacteria and selenium nanoparticles on improving the nutritional value and biological activities of three cultivars of *Brassica* sprouts. ACS Omega 8(29):26414–26424
142. Sonali JMI, Gayathri KV, Rangasamy G, Kumar PS, Rajagopal R (2024) Efficacy of biogenic selenium nanoparticles from *Pseudomonas libanesis* towards growth enhancement of Okra. Waste and Biomass Valorization 15(3):1793–1806
143. Halawani RF, Aloufi FA (2023) Galaxolide-contaminated soil and tolerance strategies in soybean plants using biofertilization and selenium nanoparticle supplementation. Front Plant Sci 14:1221780
144. Nawaz S, Bano A (2020) Effects of PGPR (*Pseudomonas* sp.) and Ag-nanoparticles on enzymatic activity and physiology of cucumber. Recent PatS Food Nutr Agric 11(2):124–136
145. Tariq S, Bano A (2023) Role of PGPR and silver nanoparticles on the physiology of *Momordica charantia* L. irrigated with polluted water comprising high Fe and Mn. Int J Phytoremediation 25(12):1643–1655
146. Alharby HF, Ali S (2022) Combined role of Fe nanoparticles (Fe NPs) and *Staphylococcus aureus* L. in the alleviation of chromium stress in rice plants. Life 12(3):338
147. Ma J, Alshaya H, Okla MK, Alwasel YA, Chen F, Adrees M et al (2022) Application of cerium dioxide nanoparticles and chromium-resistant bacteria reduced chromium toxicity in sunflower plants. Front Plant Sci 13:876119
148. Zand AD, Mikaeili Tabrizi A, Vaezi HA (2020) Application of titanium dioxide nanoparticles to promote phytoremediation of Cd-polluted soil: contribution of PGPR inoculation. Bioremediat J 24(2–3):171–189
149. Young CC, Rekha P, Lai WA, Arun A (2006) Encapsulation of plant growth-promoting bacteria in alginate beads enriched with humic acid. Biotechnol Bioeng 95(1):76–83
150. Panichikkal J, Prathap G, Nair RA, Krishnankutty RE (2021) Evaluation of plant probiotic performance of *Pseudomonas* sp. encapsulated in alginate supplemented with salicylic acid and zinc oxide nanoparticles. Int J Biol Macromol 166:138–143
151. Karunakaran A, Fathima Y, Singh P, Beniwal R, Singh J, Ramakrishna W (2024) Next-generation biofertilizers: nanoparticle-coated plant growth-promoting bacteria biofertilizers for enhancing nutrient uptake and wheat growth. Agriculture 14(4):517

152. Vasant GH, Raghav R (2024) Co-application of metal oxide nanoparticle (s) and plant growth promoting Rhizobacteria on the growth of groundnut plant (*Arachis hypogaea* L.). Department of Microbiology Faculty of Science Atmiya University
153. Guardiola-Márquez CE, López-Mena ER, Segura-Jiménez ME, Gutierrez-Marmolejo I, Flores-Matzumiya MA, Mora-Godínez S et al (2023) Development and evaluation of zinc and iron nanoparticles functionalized with plant growth-promoting rhizobacteria (PGPR) and microalgae for their application as bio-nanofertilizers. Plants 12(20):3657
154. Epa VC, Burden FR, Tassa C, Weissleder R, Shaw S, Winkler DA (2012) Modeling biological activities of nanoparticles. Nano Lett 12(11):5808–5812
155. Knight R, Vrbanac A, Taylor BC, Aksenov A, Callewaert C, Debelius J et al (2018) Best practices for analysing microbiomes. Nat Rev Microbiol 16(7):410–422
156. Wolfert S, Ge L, Verdouw C, Bogaardt M-J (2017) Big data in smart farming–a review. Agric Syst 153:69–80
157. Kamilaris A, Prenafeta-Boldú FX (2018) Deep learning in agriculture: a survey. Comput Electron Agric 147:70–90
158. Liakos KG, Busato P, Moshou D, Pearson S, Bochtis D (2018) Machine learning in agriculture: a review. Sensors 18(8):2674
159. Liu W, Wang A, Li X, Wang J, Liu X, Zhao Y et al (2024) Multi-omics analysis for mechanistic understanding of microbial-mediated synthesis of silver nanoparticles. Chem Eng J 497:154410
160. Kumar P, Vashistha H, Kumar P, Chandra H, Dubey RC (2025) Plant growth promoting Rhizobacteria (PGPR)-based nanoparticles in agriculture. In: Nanofertilizers in agriculture: synthesis, mechanisms, and effect on plants. Springer, pp 117–140
161. Yadav TP, Yadav RM, Singh DP (2012) Mechanical milling: a top down approach for the synthesis of nanomaterials and nanocomposites. Nanosci Nanotechnol 2(3):22–48
162. Mortimer CJ, Wright CJ (2017) The fabrication of iron oxide nanoparticle-nanofiber composites by electrospinning and their applications in tissue engineering. Biotechnol J 12(7):1600693
163. Kratochvíl J, Kuzminova A, Kylián O, Biederman H (2015) Comparison of magnetron sputtering and gas aggregation nanoparticle source used for fabrication of silver nanoparticle films. Surf Coat Technol 275:296–302
164. Pervikov A (2021) Metal, metal composite, and composited nanoparticles obtained by electrical explosion of wires. Nanobiotechnology Rep 16(4):401–420
165. Phromma S, Injun K, Srisuda A, Wutikhun T, Supcharoengoon U, Kangwansupamonkon W et al (2023) The influence of ultra-probe sonication and polyvinylpyrrolidone on dispersion stability and photocatalytic activity of Ag-TiO_2 nanoparticles. J Mol Liq 390:123032
166. Corbierre MK, Beerens J, Lennox RB (2005) Gold nanoparticles generated by electron beam lithography of gold (I)—thiolate thin films. Chem Mater 17(23):5774–5779
167. Zhao X, Wei C, Gai Z, Yu S, Ren X (2020) Chemical vapor deposition and its application in surface modification of nanoparticles. Chem Pap 74(3):767–778
168. Robotnik K, Zieliński T, Walczak-Skierska J, Sibińska E, Rudzik P, Piszczek P et al (2025) Synthesis of Silver nanoparticles by chemical vapor deposition method and its application in laser desorption/ionization techniques. Nanomaterials 15(13):973
169. Arteaga-Díaz SJ, Meramo-Hurtado SI, León-Pulido J, Zuorro A, González-Delgado AD (2019) Environmental assessment of large scale production of magnetite (Fe_3O_4) nanoparticles via coprecipitation. Appl Sci 9(8):1682
170. Gatou M-A, Lagopati N, Vagena I-A, Gazouli M, Pavlatou EA (2022) ZnO nanoparticles from different precursors and their photocatalytic potential for biomedical use. Nanomaterials 13(1):122
171. Gracia-Pinilla M, Martínez E, Vidaurri GS, Pérez-Tijerina E (2009) Deposition of size-selected Cu nanoparticles by inert gas condensation. Nanoscale Res Lett 5(1):180
172. Majeed S, Danish M, Mohamad Ibrahim MN, Sekeri SH, Ansari MT, Nanda A et al (2021) Bacteria mediated synthesis of iron oxide nanoparticles and their antibacterial, antioxidant, cytocompatibility properties. J Cluster Sci 32(4):1083–1094

173. Binupriya AR, Sathishkumar M, Yun S-I (2010) Myco-crystallization of silver ions to nanosized particles by live and dead cell filtrates of *Aspergillus oryzae* var. *viridis* and its bactericidal activity toward *Staphylococcus aureus* KCCM 12256. Ind Eng Chem Res 49(2):852–858
174. Mishra A, Kumari M, Pandey S, Chaudhry V, Gupta K, Nautiyal C (2014) Biocatalytic and antimicrobial activities of gold nanoparticles synthesized by *Trichoderma* sp. Biores Technol 166:235–242
175. Badawy AA, Abdelfattah NA, Salem SS, Awad MF, Fouda A (2021) Efficacy assessment of biosynthesized copper oxide nanoparticles (CuO-NPs) on stored grain insects and their impacts on morphological and physiological traits of wheat (*Triticum aestivum* L.) plant. Biology 10(3):233
176. Mahanty S, Bakshi M, Ghosh S, Chatterjee S, Bhattacharyya S, Das P et al (2019) Green synthesis of iron oxide nanoparticles mediated by filamentous fungi isolated from Sundarban mangrove ecosystem, India. BioNanoScience 9(3):637–651
177. Rangaraj S, Gopalu K, Rathinam Y, Periasamy P, Venkatachalam R, Narayanasamy K (2014) Effect of silica nanoparticles on microbial biomass and silica availability in maize rhizosphere. Biotechnol Appl Biochem 61(6):668–675
178. Timmusk S, Seisenbaeva G, Behers L (2018) Titania (TiO_2) nanoparticles enhance the performance of growth-promoting rhizobacteria. Sci Rep 8(1):617
179. Palmqvist N, Bejai S, Meijer J, Seisenbaeva G, Kessler V (2015) Nano titania aided clustering and adhesion of beneficial bacteria to plant roots to enhance crop growth and stress management. Sci Rep 5(1):10146
180. Panichikkal J, Thomas R, John JC, Radhakrishnan E (2019) Biogenic gold nanoparticle supplementation to plant beneficial *Pseudomonas monteilii* was found to enhance its plant probiotic effect. Curr Microbiol 76(4):503–509
181. Azmat A, Tanveer Y, Yasmin H, Hassan MN, Shahzad A, Reddy M et al (2022) Coactive role of zinc oxide nanoparticles and plant growth promoting rhizobacteria for mitigation of synchronized effects of heat and drought stress in wheat plants. Chemosphere 297:133982
182. Khati P, Parul, Bhatt P, Nisha, Kumar R, Sharma A (2018) Effect of nanozeolite and plant growth promoting rhizobacteria on maize. 3 Biotech 8(3):141
183. Khati P, Chaudhary P, Gangola S, Bhatt P, Sharma A (2017) Nanochitosan supports growth of Zea mays and also maintains soil health following growth. 3 Biotech 7(1):81
184. Khan N, Bano A (2016) Role of plant growth promoting rhizobacteria and Ag-nano particle in the bioremediation of heavy metals and maize growth under municipal wastewater irrigation. Int J Phytorem 18(3):211–221
185. Kumar P, Pahal V, Gupta A, Vadhan R, Chandra H, Dubey RC (2020) Effect of silver nanoparticles and *Bacillus cereus* LPR2 on the growth of *Zea mays*. Sci Rep 10(1):20409

Smart Nano-Delivering Bio-Stimulants in Sustainable Agriculture

Pravallika Sree Rayanoothala, Desetty JMSNK Sri Veda, Sipika Devi, Debasmita Adak, and Palla Sahithi

Abstract Bio-stimulants represent a promising innovation in modern agriculture, offering sustainable alternatives to conventional agrochemicals for enhancing crop growth, yield, and resilience. Derived from diverse natural sources such as microbes, humic substances, amino acids, and seaweed extracts, these bio-based formulations support plant nutrition, improve stress tolerance, and promote soil health. Rccent progress in nanotechnology has further expanded this field through the development of nano-bio-stimulants, which enhance stability, bioavailability, and targeted delivery of active compounds, thereby increasing efficiency and reducing input costs. Despite their potential, the widespread adoption of bio-stimulants faces several challenges, including inconsistency in product quality, variable field performance, and lack of global regulatory harmonization. Natural variability in raw materials and differences in extraction or formulation processes often lead to inconsistent outcomes, highlighting the need for product standardization and clear application guidelines. Additionally, fragmented regulatory frameworks across regions hinder commercialization and international trade. Addressing these limitations requires coordinated efforts among researchers, policymakers, and industry stakeholders to establish science-based standards, testing protocols, and quality control measures. As research advances and regulations evolve, bio-stimulants, particularly in nano-formulations, are expected to play a transformative role in achieving eco-efficient, resilient, and climate-smart agriculture.

Keywords Nano-delivery · Precision agriculture · Sustainable agriculture · Plant growth promotion · Soil microbiome

---

P. S. Rayanoothala (✉) · D. J. S. Veda · S. Devi
Department of Plant Pathology, M S Swaminathan School of Agriculture, Centurion University of Technology and Management, Paralakhemundi, Gajapati, Odisha, India
e-mail: rpravallikasree@gmail.com

D. Adak
Centurion School of Smart Agriculture, Centurion University of Technology and Management, Vizianagaram, Andhra Pradesh, India

P. Sahithi
M S Swaminathan School of Agriculture, Centurion University of Technology and Management, Paralakhemundi, Odisha, India

J.-T. Chen (ed.), *AI-Nano-Enabled Climate-Smart Agriculture and Sustainability*, Smart Nanomaterials Technology, https://doi.org/10.1007/978-981-92-1694-9_16

1 Introduction

Bio-stimulants represent an emerging field in agricultural science that offers environmentally sound alternatives to conventional agrochemicals for enhancing plant vigor and productivity. These agents improve plant performance under various environmental conditions and include diverse categories such as beneficial microorganisms, humic and fulvic acids, amino acids, and seaweed derivatives, each contributing uniquely to plant development. The growing interest in this domain reflects a shift toward sustainable agricultural practices aimed at addressing global production challenges [1, 2]. In addition to improving crop yield and quality, bio-stimulants play a vital role in minimizing chemical inputs, thereby supporting long-term food security and environmental protection. Their adoption can significantly promote eco-friendly and resilient farming systems worldwide [3]. Recent progress in nanotechnology has enabled the development of nano-bio-stimulants, advanced formulations that utilize nanomaterials or nanoencapsulation to enhance the stability, uptake efficiency, and targeted delivery of active ingredients to plants. These systems allow for controlled release and site-specific action, thereby increasing the effectiveness of conventional bio-stimulants while minimizing wastage and required application doses [4, 5]. Ongoing research on bio-stimulants has revealed their multifaceted roles, ranging from improving nutrient assimilation to regulating plant stress tolerance mechanisms. Such insights underscore their potential to revolutionize sustainable agriculture, as these natural compounds can significantly contribute to improving crop productivity and resilience. By utilizing these multifunctional agents, scientists, and cultivators are adopting innovative, eco-friendly strategies that support efficient and environmentally responsible farming systems [6].

Nevertheless, the rapidly expanding biostimulant sector faces certain obstacles. The diversity in composition and functional mechanisms among different formulations poses difficulties in defining standardized doses and application protocols. This variability hampers seamless adoption in existing agricultural systems, necessitating further scientific validation and optimization for consistent and effective field use. Getting consistent results and wide use of bio-stimulants in farming is still tricky. This shows there's a need for more research to create clear rules for using these products effectively in agriculture [7]. The rules about bio-stimulants differ a lot around the world, making it hard for them to be used globally. These differing regulations are a big hurdle to spreading bio-stimulants everywhere. On the bright side, some progress is being made. For example, the European Union has new rules that show a growing understanding of how valuable bio-stimulants can be. These changes are helping to create clearer standards for how bio-stimulants are classified, checked for quality, and used safely and sustainably in farming, which could lead to wider acceptance in farming [8]. This review looks at how bio-stimulants have developed, their different types, and how they are used. By examining research, rules, and upcoming ideas, the article aims to give a thorough look at bio-stimulants and their crucial role in making agriculture stronger and more sustainable [9]. As rules change and new biostimulant traits are found, there's a hopeful outlook for improving global farming.

This shows the need for ongoing research and discussion in this fast-changing area and its importance for the future of farming.

2 Limitations of Conventional Biostimulant Application

Bio-stimulants are a blend of old farming methods, new biotech, and eco-friendly practices, providing special advantages for farming. This unique mix shows how they could change farming for the better, but some problems hold back their use. These include tough regulations, mixed results from different weather conditions, and the need for more research to understand how they work. The lack of clear definitions and testing methods makes it harder to use them widely in farming. Overcoming these issues is important to fully benefit from bio-stimulants, helping crops grow more while being friendly to the environment.

2.1 Consistency and Reliability Issues

Natural products are quite different from each other, which makes researching and using bio-stimulants tricky. Bio-stimulants are made from a mix of natural ingredients like plants, seaweed, and microbes. This variety is helpful for farming, but it also makes it hard to create standard products, leading to inconsistent research results and on-the-ground use. Differences in where raw materials come from, the seasons, how they are extracted and processed, and the types of microbes used can all change what active ingredients are in the final product. Each of these differences can affect how well the bio-stimulants work, making it tough to ensure they perform the same way every time. Fixing these issues is important for making trustworthy bio-stimulants that farmers can reliably use to improve crop growth and sustainability [4, 10].

Making biostimulant products standard is crucial for their success in farming. This means creating strict rules for how to get and process the raw materials and improving ways to measure the active ingredients. Recent progress in chemistry and genetics has helped identify and describe the important compounds in bio-stimulants. These new methods help produce bio-stimulants that are uniform in quality and effectiveness. By refining these products, researchers, and producers can better understand and predict how bio-stimulants will affect plant growth and health, making them more useful in farming [7].

Making bio-stimulants easier to use and fitting them to different farming situations can help farmers grow more crops sustainably and use fewer chemicals [6, 11, 12]. To tackle issues with changing quality and consistency in bio-stimulants, we need teamwork among scientists, businesses, and regulators. By creating standard ways to test bio-stimulants, improving quality checks [13], and running thorough field tests, the farming industry can tackle these issues and make the most of bio-stimulants [1, 14].

2.2 Regulatory Gaps and Standardization Needs

The rules for bio-stimulants are different in many parts of the world. This inconsistency can make it hard for these products to be sold internationally. It's important to create clear and science-based rules to support new ideas while keeping products safe and effective [9]. In the European Union, bio-stimulants are regulated under specific laws (EU 2019/1009) that outline how they can be sold and used. However, in places like the United States, the rules are more relaxed and scattered, which often creates confusion when trying to register and approve products [6, 14]. A major problem is that there isn't a clear definition of what counts as a biostimulant. They include various products, like microbes, plant extracts, and natural mixes, each working differently and used in different ways. This variety makes it tough to create standard rules for all bio-stimulants [5, 15]. Efforts are being made to bring these regulations closer together in different regions. Unclear rules can slow down progress in the biostimulant industry. Companies may hesitate to spend money on creating new products if they don't know how to get them approved for sale. Also, mixed regulations can make it hard for new businesses to enter the market and limit the types of bio-stimulants available to farmers [16]. In the end, these efforts will help the agricultural sector make the best use of bio-stimulants, promoting sustainable farming and better crop yields [17, 18].

2.3 Mechanistic and Efficacy Limitations

Even though bio-stimulants are known to help plants grow better, figuring out exactly how they work is tough. This is because bio-stimulants can be made of many different substances, and they behave differently with various plants and in different environments. This makes it hard to understand how they help, so more research is needed on how plants react to these treatments at a tiny level. To use bio-stimulants effectively, we need thorough studies that look at how these products influence plant growth in different situations. Their effectiveness can change a lot depending on what type of crop it is, how mature the plant is, the current weather conditions, and what the biostimulant contains. Knowing these factors will help farmers use bio-stimulants in the best way possible. This need for a better understanding shows why more research and teamwork among scientists is important to create clear guidelines for testing and using bio-stimulants. This will help clarify how they work and make them more useful, leading to better and more sustainable farming practices [19, 20].

Bio-stimulants help plants grow and stay healthy in different ways. They can make it easier for plants to take in nutrients, affect how plants manage their metabolic processes and hormones, help them handle stress better, and boost their natural defenses against diseases. Each of these functions is important for improving how well crops perform. However, understanding how these processes work can be tough for researchers because there are many different ways bio-stimulants interact with plants. These interactions can change based on the type of biostimulant used, the plant species, the environment, and the plant's growth stage, making it hard to draw broad conclusions. To tackle this, researchers are using advanced tools like genomics, proteomics, and metabolomics to study how bio-stimulants impact plants at a detailed level. This deeper understanding helps scientists pinpoint how bio-stimulants work, allowing for better and more specific use of these products in farming. This research can lead to developing stronger biostimulant products and smarter ways to apply them for maximum benefit in different farming situations [21–23].

2.4 *Cost and Market Challenges*

The moneymaking potential of bio-stimulants is really important for how widely they are used in farming. The costs for researching, developing, and bringing bio-stimulants to market can be high, which makes it tough for companies and farmers to use them. Farmers usually need to see that bio-stimulants save them money compared to regular products like fertilizers and pesticides before they decide to use them. Researchers are putting more effort into looking at the money side of bio-stimulants. This means checking market trends, figuring out if the benefits outweigh the costs, and how using them affects farmers' profits. These money checks are key to seeing if bio-stimulants can be used widely in farming. By showing how bio-stimulants can help grow better crops, lower costs, and make farms more resilient to climate change, studies aim to highlight the financial advantages for farmers. Also, with more people wanting eco-friendly farming, bio-stimulants could become more popular, leading to more farmers trying them out. Knowing these financial factors is vital for everyone in agriculture, from producers to consumers, to make wise decisions about using bio-stimulants in their farms [24, 25].

2.5 *Expensive Innovation Process*

Developing bio-stimulants takes a lot of time and money, as it involves a lot of research and testing. Finding and improving effective ingredients requires not just money but also a strong dedication to science. It's complicated to understand how these bio-stimulants work and the best ways to use them, making it tough for one organization to cover all the costs alone. That's why teamwork between universities, government bodies, and private companies is really important. These partnerships let them share resources, knowledge, and skills to make biostimulant research better and more efficient. They can collaborate in many ways, like joint research projects or shared funding, taking advantage of each other's strengths. For example, universities provide high-level scientific research and can run independent tests, while private companies offer market knowledge and production skills. Government agencies can help with rules and provide funding, like grants or tax cuts, to support this work. By joining forces, these groups can advance biostimulant technologies more effectively and deal with the high costs and technical issues of launching new products. Such teamwork not only speeds up development but also helps integrate bio-stimulants into regular farming practices, benefiting agriculture overall [26].

2.6 Compliance and Legal Limitations

Dealing with rules for bio-stimulants can be expensive and complicated because different countries have their own guidelines. Each place has specific requirements for these products, which can differ a lot, making it hard for companies to sell their products around the world. This means businesses need special plans for each location, which increases costs and complexity. By cutting down on the hassle and cost of following the rules, clearer regulations would help more biostimulant products reach the market, supporting innovation and wider use of these eco-friendly farming solutions [27]. This would benefit manufacturers, farmers, and the whole agricultural industry by providing more options to boost productivity and promote sustainability [6, 28].

2.7 Challenges in Market Penetration

Getting bio-stimulants available for sale involves costs for making them, promoting them, teaching people about them, and getting them to stores. It's important to convince farmers to use bio-stimulants, especially in places where they are used to the usual farming products. Clear communication about their benefits is needed. Showing clear results like better crop yields, quality, and strength is key to getting farmers to try them [29].

2.8 Limited Cost–Benefit for Farmers

Farmers decide to use bio-stimulants mainly based on whether they save money compared to traditional farming methods. It's important to show that bio-stimulants can increase crop production, lower input costs, or do both. By looking at costs and benefits and how these products affect farm profits, farmers can get important information to help them decide if they should use bio-stimulants in their farming [30].

2.9 Uncertain Market Prospects

The market for bio-stimulants is growing fast because more people want eco-friendly farming methods, and bio-stimulants help crops grow better and handle stress. This growth could lead to lower costs for bio-stimulants in future, making them more affordable for all types of farmers. The business side of bio-stimulants is complicated, involving their development, legal rules, launching in the market, and how farmers start using them. While these challenges exist, the benefits of bio-stimulants for better crop growth and sustainable farming make them important in today's farming practices. As the market for bio-stimulants grows, ongoing research, better rules, and working closely with farmers are needed to maximize their benefits [7, 18].

2.10 Challenges in Global Market Expansion

The global market for bio-stimulants is growing quickly, with estimates saying it could grow by 11–13% each year to reach over $5 billion by 2030 [1, 7]. This growth is driven by more awareness of environmental issues, helpful regulations like the EU's Fertilizing Products Regulation [6], and new technologies in biotech and nanotech [31, 32]. But there are still challenges, like high costs to produce and register these products, different regulations across regions, and varying effectiveness of bio-stimulants on different crops and environments [5, 6]. Europe leads the market because of strict environmental rules and strong investment in research, while the Asia–Pacific area is likely to grow the fastest due to increasing farming activity and government support for sustainability [7, 33]. North America is also becoming a major market thanks to new techniques in precision farming (Nkebiwe PM, Lekfeldt JDS, Mäder P, Cozzolino V, Di SE, Dobczinski S, et al.). To really take advantage of bio-stimulants, we need to tackle economic challenges like lowering production costs and improving supply chains, as well as encouraging their use with subsidies or financial help [5, 34].

2.11 Environmental Challenges

Weather and environment can really change how well bio-stimulants work, making it tricky to use them. Things like soil health, changes in climate, and issues like drought or salt in the soil can hurt how bio-stimulants perform [33]. In tough weather areas, these products might not work as well, which makes them less worth the costs. Also, how well bio-stimulants do often depends on weather factors like temperature and rain, plus soil traits like pH and the amount of organic matter and microbes. For example, a biostimulant meant to help plants take in water might not work well in too wet soils or places with a lot of rain. While bio-stimulants can help plants cope with challenges like drought and high salt levels, they may not be very effective when conditions get really bad [35, 36]. Studies show that even strong bio-stimulants may not help much in extreme drought, pointing out the need to create bio-stimulants that are designed for specific tough environments [37].

2.12 Crop-Specific Limitations

How well bio-stimulants work depends on the environment and the type of crop being treated. Different plants react differently because each has its own way of functioning. Studies show that while bio-stimulants can help one crop grow, they might not work well or even harm another. This difference shows the need for focused research to improve biostimulant products for specific crops, making them more useful and appealing to farmers [38–40]. To get the most out of bio-stimulants, it's crucial to tailor them to the needs and growing conditions of each crop type. This requires a good understanding of how bio-stimulants interact with the unique workings of different plants, which means a lot of research is needed to make sure these products really support plant health and growth [41, 42].

2.13 Compatibility Challenges

Adding bio-stimulants to current farming methods comes with some challenges. How they work with things like crop rotation, watering, and pest control can really affect how well they work and how healthy the crops are. For bio-stimulants to be useful, they need to fit well with other farm inputs and methods. If they don't, they might not work well and could harm plant health and crop output. So, it's important to carefully look at and adjust how bio-stimulants are used to make sure they help rather than hurt farming productivity [20, 43]. When done right, bio-stimulants can improve eco-friendly farming practices, help keep the ecological balance, and enhance crop health. To make the most of these benefits, detailed planning and a good understanding of how bio-stimulants interact with other farming practices is necessary. This approach helps crops stay strong and productive while promoting green farming methods, leading to a more sustainable agricultural system [44, 45].

2.14 Cost-Scalability Constraints

Many farmers are looking to bio-stimulants, but their use depends a lot on how much they cost to develop and sell. This can be a big problem, especially for new products, making it harder for farmers to get them. Researchers point out that these costs need to be lower, and the bio-stimulants should really help improve crops to convince farmers to use them. Also, making bio-stimulants for large farms has its own set of challenges [46, 47]. To keep quality and effectiveness when making more, we need better manufacturing and strict quality checks [48]. Experts are exploring new ways to tackle these issues while also considering the environment and sustainability, which are important for balancing nature [49–51].

3 Overview of Nanotechnology in Agriculture

Nanotechnology is helping many fields grow quickly, including science and technology. It's making big changes in areas like electronics and computers. Although it's not widely used in agriculture yet, its advancements suggest a bright future for farming. It can aid sustainable farming methods like precision agriculture, managing soil nutrients, studying plant biology, and creating new tools for agriculture. There are already nanosensors being used in food production. For example, nanotechnology is being tested for breaking down pesticides, detecting biomolecules, and turning vegetable oils into biofuels, all with good results. Additionally, nanoparticles help in detecting pesticides and reducing environmental harm, and they can be used for self-cleaning surfaces in cars and boats. Research is also ongoing to see if nanopaints could help reduce how much soil sticks to tools. Among all these applications, precision agriculture has the most potential, as it can use smart GPS systems to deliver fertilizers, pesticides, or herbicides more effectively, with nanosolvents improving their absorption. Similar uses of nanotechnology are expected to grow in other agricultural areas, too.

4 Types of Nanodelivery System Used for Bio-Stimulants

In farming, bio-stimulants are a mixed group of products that help plants grow better by boosting their natural processes. These processes help plants take in nutrients, use them efficiently, withstand tough conditions, and enhance crop quality. Bio-stimulants are categorized (Fig. 1) by where they come from and what ingredients they contain, showing the many types available. This variety shows how important these products are for sustainable farming.

4.1 Nanoencapsulation

Nanoencapsulation is a cutting-edge method in farming that helps deliver plant boosters more effectively. It involves covering a helpful substance—such as a plant hormone, a natural extract, or a tiny form of a microbe—inside a very small shell.

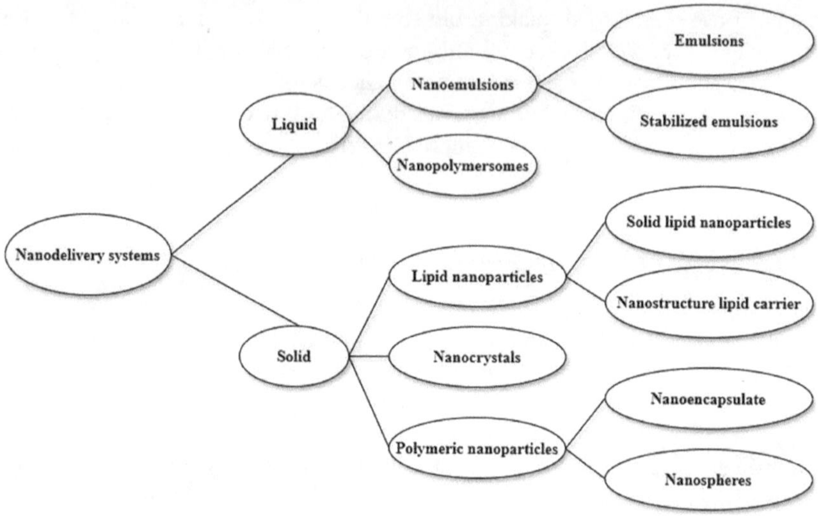

Fig. 1 Types of nano-delivery systems

This makes the plant booster more stable, easier to dissolve, and work better, allowing for smarter and more precise use [52, 53].

4.1.1 Why Use Nanoencapsulation for Bio-Stimulants

Using bio-stimulants the usual way can be hard because the active ingredients can be affected by the environment and plants may struggle to take them in. Nanoencapsulation helps fix these problems using various methods:

4.1.2 Protection and Degradation

Lots of bio-stimulants can break down due to sunlight, air, or tiny organisms in the soil. The nanoshell works like a shield, keeping the main ingredient safe and making it last longer and work better after being used [54].

4.1.3 Improving Solubility and Bioavailability

Many natural plant helpers don't dissolve well in water. Using tiny capsules can help them spread out better, so plants can take them in through their roots and leaves more easily. This means you need less of the product to get the same results [55].

4.1.4 Controlled and Sustained Release

The nanocapsule can be made to slowly let out what's inside over time. This steady release gives the plant a constant amount of the helpful substance, which works better than one big dose and means you don't have to keep using it often [56].

4.1.5 Targeted Delivery

We can change the outside of the tiny capsule to focus on certain parts of a plant, like the roots or leaves, making the treatment more accurate and reducing waste [57].

4.2 Nanoemulsion

Nanoemulsions are special liquids made by mixing oil and water, held together by substances called surfactants. This results in a clear or slightly cloudy liquid with tiny droplets, usually between 20 and 200 nm [54, 55]. When it comes to plant boosters, nanoemulsions are an advanced way to deliver these products effectively, providing better stability, dissolving capabilities, and effectiveness than regular formulas.

4.3 Advantages of Nanoemulsion in Bio-Stimulants

4.3.1 Enhanced Bioavailability

Many biostimulant ingredients don't mix well with water because they like oil more. This makes it hard for plants to take them in from water. However, breaking these ingredients into tiny droplets helps them mix better with water. This makes it easier for plants to absorb and use these ingredients effectively [53, 56].

4.3.2 Improved Stability

Biostimulant materials can break down when they are exposed to things like sunlight and high heat. Nanoemulsions help keep these materials safe inside tiny oil droplets, protecting them from breaking down and making them last longer after being used [57].

4.3.3 Increased Efficiency at Low Doses

Nanoemulsions work better and last longer, so you need less of the active biostimulant to get the same or even better results than regular products. This is good for your wallet and the environment, as it means less active substance is put into nature [7].

4.4 Nanogels

Nanogels are soft and bendy materials that can soak up a lot of water. Imagine them as tiny, superabsorbent sponges [54]. Their special design lets them hold and let go of a lot of things, which makes them great for helping plants grow.

4.5 Biopolymeric Nanocarriers

Biopolymeric nanocarriers are tiny delivery systems made from natural materials called biodegradable polymers, like starch and cellulose. These small carriers help move and protect important compounds, making them work better while also being eco-friendly [52, 55].

5 Mechanisms of Smart Delivery and Plant Uptake

Bio-stimulants smart delivery uses techniques like nanoencapsulation, which controls the release of active ingredients, while plant uptake is done through root and leaf level mechanisms involving passive or negative energy movement, active transport, and beneficial microorganisms (Fig. 2).

5.1 Nanoparticle Plant Interactions

When nanoparticles are inserted into root or leaves, they move through different pathways, apoplastic or symplastic, then reach the xylem and phloem for distribution throughout the plant or absorbed through stomatal pores to reach the mesophyll and vascular tissues.

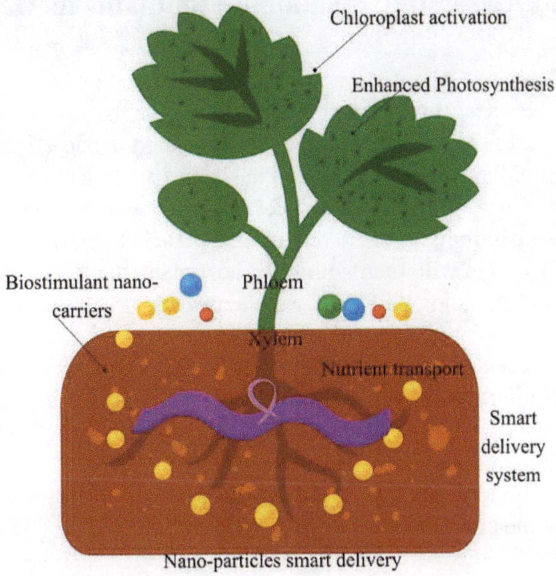

Fig. 2 Nano-enabled smart delivery of bio-stimulants

5.2 *Plant Uptake*

Nanoparticles applied as foliar sprays enter the plant leaves through several points, like stomata, cuticle, hydathodes, and wounds. During the daytime, the particles smaller than the stomatal opening size can pass through. The leaf surface is covered by a waxy cuticle that acts as a protective barrier. However, extremely small nanoparticles can penetrate through hydrophilic or lipophilic pores in the cuticle. Hydathodes, located at leaf tips and margins, secrete water and can provide entry sites for nanoparticles. Wounds created by insects, pathogens, or mechanical damage also create openings that allow nanoparticles to enter plant tissues.

5.3 *Nutrient Transport*

Efficiency of nanoparticles is more than conventional fertilizers, ensuring that nutrients are given directly to where they are required within the plant [58]. Effective nutrient utilization is achieved by using this targeted delivery system. Plant uptake and translocation of NPs into plant cells is prompted by the chemical composition of NPs.

6 Case Study: Nanobio Stimulants in Major Crops

Nano bio stimulants are the integrated product of nanoparticles and bio-stimulants. The encapsulation technique is used for the same. Nano-bio-stimulants lower the process to release nutrients but mitigate the bad effects of biotic and abiotic stress. Simultaneously, it increases the capacity of plant nutrient uptake [59].

Case study 1: When salicylic acid-coated bioFeNPs (SI) NCs (Nanocomposites) were introduced in the watermelon plant, which was under the stressful condition, i.e., Fusarium wilt, then within the host plant, the fungal growth was inhibited. Salicylic acid signaling pathways were activated, which improved the antioxidative property and systemic acquired response [60].

Case study 2: When nano ZnFe oxide plus Azotobacter, Azospirillum, and Pseudomonas was incorporated in a wheat plant that faced drought, it increased the concentration of proline, soluble sugars, chlorophyll, and carotenoid, and it increased the grain yield of wheat [61].

Case study 3: The growth rates of Mint (Mentha longifolia L.) plants cultivated in the control soil without the addition of biochar (BC), Spirulina (SP), or both (BC + SP) were significantly reduced by water salinity stress. When biochar was not used, the plant showed the detrimental effects of water salinity (5 dS m^{-1}) on the growth and physiological performance of Mentha longifolia cultivated in sandy soil [62]. When biochar and spirulina were added either separately or in combination, they reduced the detrimental effects of salty wastewater irrigation on the mint plants [62]. Therefore, the combined application of biochar and the biostimulant Spirulina platensis significantly reduced salinity stress, resulting in an increase in mint biomass compared to untreated saline controls (Table 1).

7 Advantages Over Traditional Delivery Methods

The advantages of nanodelivering bio-stimulants over traditional methods are primarily due to improved effectiveness, stability, and controlled delivery. Nanotechnology offers a more efficient and sustainable approach by addressing many of the limitations of conventional biostimulant formulations. Plant growth, resilience, and productivity are increased by this innovative approach [66]. Plant growth and development are stimulated by nano-bio-stimulants, which comprise nano-encapsulated hormones and growth-promoting substances [67]. The nanoscale size and high surface area of these materials enable precise and efficient delivery of active compounds at lower dosages, thereby reducing their environmental footprint. Nano-bio-stimulants have also demonstrated the ability to alleviate abiotic stresses, such as drought and salinity, thereby contributing to sustainable agricultural systems, as observed by Guerriero et al. in studies on tomato plants. These nanoformulations improve the bioavailability and functional efficiency of active ingredients, facilitating

Table 1 Influence of nano-bio-stimulants on plants under stress

Nano-bio-stimulants	Plants	Stress	Impact on plants	References
Salicylic acid coated bioFeNPs (SI) NCs (Nanocomposites)	Watermelon	Fusarium wilt	Host plants are inhibited by fungal invasive growth	[60]
Nano ZnFe Oxide + Azotobacter, *Azospirillum*, and *Pseudomonas*	Wheat	Drought	Grain yield of wheat increased, and proline, soluble sugar, etc., increased	[61]
Zerovalent iron nanoparticles (nZVI) + compost, biochar	Mustard	Heavy metal toxicity	Heavy metals and metalloids are immobilized from the soil to the plant	[63]
ZnONPs + farmyard manure and compost	Wheat	Cd toxicity	Plant biomass, chlorophyll concentrations increased. Decreased electrolyte leak	[64]
Agnanoparticles + PGPR (*Bacillus pumilus* and *Pseudomonas*)	Onion	Salinity	Salt-induced retention is decreased, and chlorophyll content is increased	[65]

targeted nutrient uptake and enhanced stress resilience, which together promote superior crop growth with minimal ecological disturbance. Furthermore, nanomaterials have been reported to stimulate plant metabolism, optimize nutrient absorption, and boost photosynthetic activity, resulting in healthier and more vigorous plants [68]. The integration of nano-bio-stimulants into crop management practices has been linked to improved yield and quality outcomes, highlighting their promising role across diverse farming systems [69].

8 Toxicological and Environmental Concerns

Nanomaterials' origin varies from natural processes like volcanic explosions, forest fires [70], or other anthropogenic activities, including fossil fuel burning, mining, automobile traffic, and nanomaterial production, and wastewater release from industries [71].

8.1 Nanomaterials of Different Substances and Their Toxicity

Numerous nanomaterials are used in daily life, like zinc oxide nanoparticles (ZnONPs), silica nanoparticles (SiO_2NPs), titanium dioxide nanoparticles (TiO2NPs), silver nanoparticles (AgNPs), gold NPs (AuNPs), and polymeric nanoparticles (PNPs) [72]. Toxicological effects harmfully occur when they change after interacting with the physicochemical and systemic properties [73].

8.2 Environmental Impact of Nanoparticles

Nanomaterials are increasingly entering the environment through industrial discharges, waste streams, and the contamination of soil, water, and air. As nanotechnology advances, these particles display distinct behaviors across various environmental compartments, influenced by their composition, size, and surface characteristics.

8.2.1 Nanoparticles in the Air

In the atmospheric environment, nanoparticles can lead to uncontrolled human exposure and exert toxic effects on both terrestrial and aquatic organisms [74]. Their minute size allows them to remain suspended for long periods, facilitating widespread dispersion and bioaccumulation.

8.2.2 Nanoparticles in the Soil

Soil contamination by nanoparticles commonly occurs through agricultural inputs such as fertilizers and pesticides, or via waste disposal and wastewater treatment byproducts, including sludge and biosolids [75]. Once introduced, these particles can alter soil microbial activity, influence nutrient cycling, and potentially impact plant and soil health.

8.3 Challenges in Environmental Toxicology of Nanoparticles

The toxicological assessment of nanomaterials remains a major challenge due to limited scientific understanding and inadequate regulatory frameworks [76]. One key difficulty lies in hazard and exposure evaluation, as the absence of standardized testing protocols makes it difficult to determine the true risks posed by different nanoparticles [77]. Moreover, the scarcity of comprehensive data on the chemical composition, physical properties, and environmental behavior of nanomaterials restricts effective monitoring and testing [78].

9 Biostimulant Regulations in India

Bio-stimulants are substances or microorganisms that enhance plant growth, nutrient absorption, stress tolerance, and yield without functioning as fertilizers, pesticides, or plant growth regulators. These include seaweed extracts, humic and fulvic acids, amino acids, protein hydrolysates, microbial inoculants, and other organic formulations. Recognizing their agricultural importance, the Government of India established a regulatory mechanism for bio-stimulants under the Fertilizer Control Order (FCO), 1985 (Amendment 2021). This framework mandates product registration, field efficacy testing, and postmarket monitoring to ensure quality and safety.

9.1 Evolution of Biostimulant Regulation in India

Initially, the biostimulant market in India operated without clear regulatory oversight. To address this, the government progressively introduced a structured policy framework under the FCO, 1985.

2019: Informal consultations began to define and standardize biostimulant regulation.

2021: The amendment officially recognized bio-stimulants as a distinct product category under the FCO.

2023: Draft guidelines were circulated to refine registration and quality norms.

2024: The "live microorganisms excluding biofertilizers and biopesticides" category was added, along with 11 new entries in Schedule VI.

2025: The government extended provisional registration timelines and authorized State and Union Territory governments to oversee sales under the FCO framework.

9.2 Regulatory Framework and Key Components

9.2.1 Central Biostimulant Committee (CBC)

The Central Biostimulant Committee (CBC), operating under the Department of Agriculture and Farmers' Welfare, is the apex body for reviewing product applications, issuing guidelines, and granting exemptions. It ensures that bio-stimulants meet safety, quality, and efficacy standards, thereby reinforcing confidence among farmers and manufacturers.

9.2.2 Approved Biostimulant Categories Under Indian Regulations

Bio-stimulants recognized by the FCO are classified into the following groups:

1. Seaweed and Plant Extracts: Derived from marine and terrestrial sources with proven stimulatory properties.
2. Protein Hydrolysates and Amino Acids: Contain peptides and amino compounds beneficial for growth and stress recovery.
3. Vitamins: Support enzymatic and metabolic processes.
4. Humic and Fulvic Acids: Enhance soil fertility and nutrient uptake.
5. Antioxidants: Help mitigate oxidative damage in plants.
6. Antitranspirants: Reduce water loss and improve drought resistance.
7. Biochemicals: Naturally occurring compounds that promote plant vigor.
8. Cell-free Microbial Products: Include enzymes, metabolites, and signaling molecules derived from microorganisms.
9. Chitosan and Other Biopolymers: Biopolymeric agents that strengthen plant immunity and growth.
10. Inorganic Compounds: Mineral-based substances with growth-promoting effects.
11. Beneficial Microorganisms (NonBiofertilizer/NonBiopesticide): Promote plant growth through nonnutritional mechanisms.
12. The Fertilizer (Inorganic, Organic, or Mixed) (Control) (Third) Amendment Order, 2024, issued by the Ministry of Agriculture and Farmers' Welfare on May 8, 2024, further streamlined the regulation of bio-stimulants in India.

9.2.3 Key Provisions

Tracer Elements: All product specifications must include tracer elements alongside active ingredients for compositional transparency.

New Category Addition: Introduction of a category for "live microorganisms excluding biofertilizers and biopesticides" in Schedule VI.

Revised Pesticide Limits: The permissible pesticide residue level was increased from 0.01 ppm to 1 ppm, ensuring both safety and manufacturing feasibility.

Toxicity Testing: Manufacturers must submit toxicity data; however, low-risk categories (such as seaweed extracts and humic acids) may receive exemptions upon providing a safety affidavit.

9.3 Registration Procedure

A biostimulant may be manufactured or imported unless it is listed in Schedule VI and approved by the Central Government. The process involves the following steps:

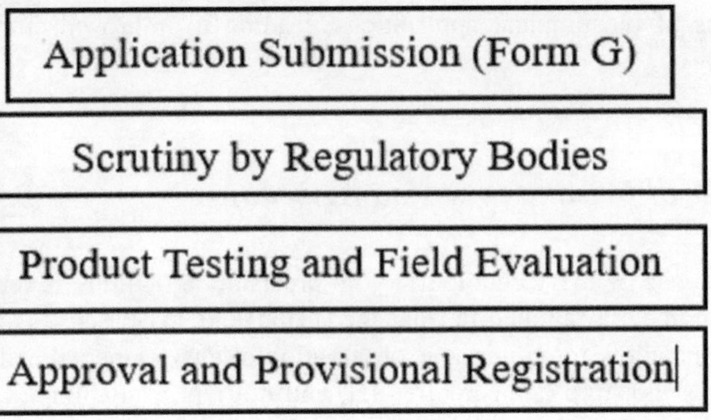

9.4 Compliance and Quality Assurance

Biostimulant manufacturers must adhere to strict quality control and compliance protocols to maintain efficacy and safety. Regular audits are conducted at production facilities to ensure conformity with FCO guidelines. Product labels must clearly display information such as trade name, composition, category, manufacturing and expiry dates, dosage, application method, and safety instructions. Random postmarket testing is performed to monitor ongoing compliance.

9.5 Challenges in Biostimulant Registration

The Indian biostimulant industry faces several challenges, including lengthy approval durations, high compliance expenses, and misalignment with international standards. The registration process may take 12–24 months, significantly delaying product commercialization. Moreover, the costs associated with field trials, toxicity assessments, and certification impose a heavy financial burden, particularly on small and medium enterprises (SMEs) [79].

10 Integration with Digital and Precision Farming Technologies

The integration of bio-stimulants with digital and precision farming tools, including Internet of Things (IoT) systems and Artificial Intelligence (AI), is reshaping modern agriculture into a data-driven, efficient, and sustainable practice. These technologies enable real-time decision making, optimize resource use, and enhance the overall effectiveness of biostimulant applications, leading to improved crop health and productivity.

10.1 Role of IoT and Precision Agriculture

The convergence of IoT technologies with precision agriculture is revolutionizing how bio-stimulants are applied and managed in the field. IoT-based sensors and monitoring systems allow for continuous observation of environmental and soil parameters, such as moisture, temperature, pH, and nutrient availability. The real-time data generated helps farmers make informed decisions and customize biostimulant application according to crop-specific needs, thereby improving efficiency, nutrient uptake, and yield potential [22, 80–83].

The adoption of drones and automated sprayers has further advanced precision farming. Equipped with high-resolution imaging and spraying systems, drones can uniformly apply bio-stimulants across diverse and hard-to-reach terrains, ensuring consistent coverage while reducing labor requirements and minimizing human error. When coupled with IoT-generated data, automated sprayers can dynamically adjust application rates based on real-time field variability, thus maximizing precision, resource efficiency, and crop response [84–86].

This synergistic integration between IoT devices, drones, and automated systems fosters sustainable agricultural practices by reducing input wastage and promoting responsible resource use. It aligns with global agricultural goals of boosting productivity while minimizing environmental impact. As digital technologies continue to evolve, their seamless incorporation into crop management systems can help address rising food demands and the challenges of environmental conservation, driving the transition toward smarter, adaptive, and resilient farming ecosystems [31, 87–91].

11 Future Perspectives and Challenges

The future of biostimulant research and application lies in their potential to contribute to sustainable and climate-resilient agriculture. They play a key role in improving nutrient use efficiency, enhancing crop tolerance to abiotic stresses, and mitigating the effects of climate change. However, several challenges persist. These include variability in product performance, nonuniform formulation standards, fragmented regulatory frameworks, and limited mechanistic understanding of how bio-stimulants interact with plants, soil, and other inputs. Additionally, economic constraints and the need for comprehensive, multidisciplinary research remain major hurdles to their widespread adoption. Future efforts should focus on establishing standardized evaluation protocols, expanding scientific validation, and promoting the integration of bio-stimulants with digital and precision farming systems to achieve maximum efficiency and sustainability in agricultural production.

11.1 Regulatory Harmonization and Market Access

Market access and safe usage can be improved by efforts to standardize biostimulant regulations on a global scale. The innovation and commercialization of bio stimulants are supported by the establishment of transparent, science-based regulatory frameworks [92, 93].

11.2 Research Innovations and Technical Advancements

Next-generation bio stimulants need to be developed through advancements in molecular biology and bioinformatics. The intricate interactions between bio-stimulants and plant physiology are deciphered by focusing on these fields. Researchers are leveraging cutting-edge technologies to elucidate the precise mechanisms through which bio-stimulants exert their effects. To unravel the specific pathways bio-stimulants influence, emerging research in genomics and proteomics is instrumental. This knowledge increases our ability to tailor exact and effective biostimulant formulations that work effectively under various agricultural conditions [94, 95].

11.3 Sustainable Agricultural Practices and Climate Resilience

Bio stimulants play an important role in promoting sustainable agriculture practices by improving plant efficiency and reducing the dependency on chemical inputs. This is integrated with comprehensive agroecological systems, which play a vital role in building resilience against climate variability. Bio-stimulants' future lies in their strategic integration into holistic farming systems, including crop rotation, organic farming, and integrated pest management. With this integration, this enhances crop health and productivity and aligns with global sustainable development and food security objectives [86, 96, 97].

11.4 Advanced Biotechnological Approaches in Biostimulant Development

Agricultural productivity and sustainability can be significantly enhanced by the application of advanced biotechnological approaches. A prime example of such innovation is synthetic biology, which employs genetically engineered microorganisms designed to produce specific bio-stimulants. Engineered bio-stimulants are used because they can be customized for particular crops, which improves their absorption rates and efficacy under various environmental conditions [21, 98, 99].

11.5 Challenges in Large-Scale Applications and Economic Viability

Bio-stimulants require several logistical and economic challenges in their large-scale application, although bio-stimulants offer a more sustainable alternative to traditional agrochemicals. One of the key barriers is cost-effectiveness compared to conventional approaches. Though they are environmentally friendly, their high production cost is a shortcoming since it is derived from natural or organic sources. There is a concern rising among the farmers because of such high costs and leading to a dilemma for affording to integrate bio-stimulants into their farming practices without incurring substantial costs [7]. Farmers have to adjust their management practices as the bio-stimulants are most effective when applied at specific stages of plant growth or under certain environmental conditions. For example, to increase root development, seaweed extracts are often applied during early plant growth stages, but if applied later in the crop cycle or under drought conditions, this reduces their efficacy. This variability in timing and conditions of application includes another layer of complexity for farmers, particularly those managing large-scale operations [100].

12 Conclusion

Smart nanodelivery systems are significantly advancing in the field of sustainable agriculture. The processes regulating this intelligent delivery, featuring improved penetration via foliar and root routes, highlight how these nanoscale methods can surpass conventional obstacles to plant absorption. Research on primary crops has offered strong proof of enhanced nutrient use efficiency, better resilience to stress, and increased yields, highlighting the real advantages of this innovation for agricultural methods. Nonetheless, like any new technology, the journey ahead has its obstacles. The essential conversations about toxicological and environmental issues underscore the importance of ethical research and development, especially concerning lasting effects on soil quality, nontarget species, and possible bioaccumulation in the food chain. The regulatory framework is still developing, necessitating strong, science-driven systems to evaluate risk, guarantee safety, and foster public confidence in these products. Moreover, tackling the economic feasibility and access for small-holder farmers is essential to ensure this technology does not worsen the current inequalities in the agricultural sector. In conclusion, the intelligent nano-delivery of bio-stimulants is set to become fundamental to future sustainable farming. It pledges to harmonize the necessity of increasing food output for a rising global population with the pressing requirement to safeguard and rejuvenate our natural resources. By skillfully managing the obstacles and seizing the advantages, we can utilize this transformative technology to foster a more robust, efficient, and eco-friendly food system for future generations.

References

1. Calvo P, Nelson L, Kloepper JW (2014) Agricultural uses of plant Bio-stimulants. Plant Soil 383:3–41
2. Sharma HSS, Fleming C, Selby C, Rao JR, Martin T (2014) Plant Bio-stimulants: a review on the processing of macroalgae and use of extracts for crop management to reduce abiotic and biotic stresses. J Appl Phycol 26:465–490
3. Halpern M, BarTal A, Ofek M, Minz D, Muller T, Yermiyahu U (2015) The use of bio-stimulants for enhancing nutrient uptake. Elsevier Inc., Amsterdam
4. Mannino G (2023) A new era of sustainability: plant Bio-stimulants. Int J Mol Sci. https://doi.org/10.3390/ijms242216329
5. Ruzzi M, Colla G, Rouphael Y (2024) Editorial: bio-stimulants in agriculture II: towards a sustainable future. Front Plant Sci 15:1–7
6. Berendsen RL, Pieterse CMJ, Bakker PAHM (2012) The rhizosphere microbiome and plant health. Trend Plant Sci. 17:478–486
7. du Jardin P (2015) Plant bio-stimulants: definition, concept, main categories and regulation. Sci Hortic 196:3–14. https://doi.org/10.1016/j.scienta.2015.09.021
8. Ricci M, Tilbury L, Daridon B, Sukalac K (2019) General principles to justify plant biostimulant claims. Front Plant Sci 10:1–8
9. Rouphael Y, Colla G (2020) Toward a sustainable agriculture through plant bio-stimulants: from experimental data to practical applications. Agronomy. https://doi.org/10.3390/agronomy10101461
10. Sharp RG (2013) A review of the applications of chitin and its derivatives in agriculture to modify plantmicrobial interactions and improve crop yields. Agronomy 3:757–793
11. Epstein E (2009) Silicon: its manifold roles in plants. Ann Appl Biol 155:155–160
12. Kumar N, Kumar V, Kishor K, Singh AK (2024) Optimizing nutrient application and nitrogen transformation to maximize the growth and productivity of bud chip transplanted sugarcane. J Plant Nutr 47:2583–2596
13. Suwanchaikasem P, Idnurm A, SelbyPham J, Walker R, Boughton BA (2024) The impacts of chitosan on plant root systems and its potential to be used for controlling fungal diseases in agriculture. J Plant Growth Regul 43:3424–3445. https://doi.org/10.1007/s00344024113561
14. Alengebawy A, Abdelkhalek ST, Qureshi SR, Wang MQ (2021) Heavy metals and pesticides toxicity in agricultural soil and plants: ecological risks and human health implications. Toxics. 9:1–34
15. Khoulati A, Saalaoui E (2021) Impact of Moroccan *Crocus sativus* L. tepals, corms, and stigmas extract on growth and photosynthetic pigments in tomato seedling. Adv Hortic Sci. 35:207–213
16. Sarov A, Boevsky I (2021) Economic efficiency of Bio-stimulants in agriculture. Trakia J Sci 19:176–180
17. Fauteux F, RémusBorel W, Menzies JG, Bélanger RR (2005) Silicon and plant disease resistance against pathogenic fungi. FEMS Microbiol Lett 249:1–6
18. Rinaudo M (2006) Chitin and chitosan: properties and applications. Prog Polym Sci 31:603–632

19. Sammauria R, Kumawat S, Kumawat P, Singh J, Jatwa TK (2020) Microbial inoculants: potential tool for sustainability of agricultural production systems. Arch Microbiol 202:677–693. https://doi.org/10.1007/s0020301901795w
20. Alori ET, Dare MO, Babalola OO (2017) Microbial inoculants for soil quality and plant health. In: Lichtfouse E (ed) Sustainable agriculture reviews. Springer International Publishing, Cham, pp 281–307
21. Martins EM, Pillajo JQ, Jones ML (2024) Humic and fulvic acids promote growth and flowering in petunias at low and optimal fertility. HortScience 59:235–244
22. Alsudays IM, Alshammary FH, Alabdallah NM, Alatawi A, Alotaibi MM, Alwutayd KM et al (2024) Applications of humic and fulvic acid under saline soil conditions to improve growth and yield in barley. BMC Plant Biol BioMed Central. 24:1–16. https://doi.org/10.1186/s12870024048636
23. Tahoun AMMA, ElEnin MMA, Mancy AG, Sheta MH, Shaaban A (2022) Integrative soil application of humic acid and foliar plant growth stimulants improves soil properties and wheat yield and quality in nutrientpoor sandy soil of a semiarid region. J Soil Sci Plant Nutr. https://doi.org/10.1007/s42729022008517
24. Thamvithayakorn P, Phosri C, Robinsonboyer L, Limnonthakul P, Doonan JH, Suwannasai N (2024) The synergistic impact of a novel plant growthpromoting rhizobacterial consortium and ascophyllum nodosum seaweed extract on rhizosphere microbiome dynamics and growth enhancement in *Oryza sativa* L. RD79. Agronomy. https://doi.org/10.3390/agronomy14112698
25. Nanda S, Kumar G, Hussain S (2022) Utilization of seaweedbased Bio-stimulants in improving plant and soil health: current updates and future prospective. Int J Environ Sci Technol 19:12839–12852. https://doi.org/10.1007/s13762021035689
26. Nathaniel H, Franzén D, Lingegård S, Franzén F, Söderqvist T, Gröndahl F (2023) Using stakeholder perceptions to deepen the understanding of beachcast governance and management practices on Gotland, Sweden. Ocean Coast Manag. https://doi.org/10.1016/j.ocecoaman.2023.106583
27. Alliance BPI (2023) United States biostimulant industry recommendations to assess the efficacy, composition, and safety of plant biostimulant products. J Regul Sci. 10:1–29
28. Jakubas MS, Nowak BH (2022) Protective, biostimulating, and eliciting effects of chitosan and its derivatives on crop plants. Molecules 27:1–17
29. OrozcoMosqueda MD, Flores A, RojasSánchez B, UrtisFlores CA, MoralesCedeño LR, ValenciaMarin MF et al (2021) Plant growthpromoting bacteria as bioinoculants: attributes and challenges for sustainable crop improvement. Agronomy 11:1–15
30. Murugan AC, Perumal V, Kannan S (2022) Production, cost analysis, and marketing of agricultural effective microorganisms. In: Amaresan N, Dharumadurai D, Cundell DR (eds) Industrial microbiology based entrepreneurship: making money from microbes, vol 42. Springer Nature Singapore, Singapore
31. Nkebiwe PM, Lekfeldt JDS, Mäder P, Cozzolino V, Di SE, Dobczinski S et al (2024) Effectiveness of bioeffectors on maize, wheat and tomato performance and phosphorus acquisition from greenhouse to fi eld scales in Europe and Israel : a metaanalysis. Front Plant Sci. https://doi.org/10.3389/fpls.2024.1333249
32. Brazienė Z, Paltanavičius V, Avižienytė D (2019) The influence of fulvic acid on seed germination and plant productivity. B Abstr Fifth Int Conf CIS IHSS Humic Innov Technol Humic Subst living Syst. 22:59–59
33. Hayashi K (2023) Partitioning product footprint changes into yield and environmental improvement effects: toward life cycle assessment of agricultural system transitions. Int J Life Cycle Assess 28:336–348. https://doi.org/10.1007/s11367022021122
34. Nephali L, Piater LA, Dubery IA, Patterson V, Huyser J, Burgess K et al (2020) Bio-stimulants for plant growth and mitigation of abiotic stresses: a metabolomics perspective. Metabolites 10:1–26
35. Deng Q, Yu T, Zeng Z, Ashraf U, Shi Q, Huang S et al (2021) Silicon application modulates the growth, rhizosphere soil characteristics, and bacterial community structure in sugarcane. Front Plant Sci 12:1–15

36. MajkowskaGadomska J, Dobrowolski A, Jadwisieńczak KK, Kaliniewicz Z, Francke A (2021) Effect of bio-stimulants on the growth, yield and nutritional value of *Capsicum annuum* grown in an unheated plastic tunnel. Sci Rep 11:1–14. https://doi.org/10.1038/s41598021 01834x
37. Battacharyya D, Babgohari MZ, Rathor P, Prithiviraj B (2015) Seaweed extracts as bio-stimulants in horticulture. Sci Hortic 196:39–48. https://doi.org/10.1016/j.scienta.2015. 09.012
38. Chaudhari V, Patel N, Barot D, Nadoda N (2023) Role of bio-stimulants in solanaceous vegetable crops: a review. Int J Adv Biochem Res. 7:178–183
39. Rodrigues LA, Alves CZ, Rego CHQ, da Silva TRB, da Silva JB (2017) Humic acid on germination and vigor of corn seeds. Rev Caatinga. 30:149–154
40. HernándezHerrera RM, GonzálezGonzález MF, VelascoRamírez AP, VelascoRamírez SF, SantacruzRuvalcaba F, ZamoraNatera JF (2023) Seaweed extract components are correlated with the seeds germination and growth of tomato seedlings. Seeds 2:436–448
41. O'Callaghan M, Ballard RA, Wright D (2022) Soil microbial inoculants for sustainable agriculture: limitations and opportunities. Soil Use Manag 38:1340–1369
42. Sun W, Shahrajabian MH, Kuang Y, Wang N (2024) Amino acids bio-stimulants and protein hydrolysates in agricultural sciences. Plants. https://doi.org/10.3390/plants13020210
43. Abdelkader M, Voronina L, Baratova L, Shelepova O, Zargar M, Puchkov M et al (2023) Bio-stimulantsbased amino acids augment physiobiochemical responses and promote salinity tolerance of lettuce plants (*Lactuca sativa* L.). Horticulturae. https://doi.org/10.3390/hortic ulturae9070807
44. Jindo K, Olivares FL, da Paixão Malcher DJ, SánchezMonedero MA, Kempenaar C, Canellas LP (2020) From lab to field: role of humic substances under openfield and greenhouse conditions as biostimulant and biocontrol agent. Front Plant Sci 11:1–10
45. Jain S, Nidhi N, Kale S, Rathod M, Dhurve L, Mehara H et al (2023) A comprehensive review on role of bio regulators in the growth and development of fruit and vegetable crops. Int J Environ Clim Chang. 13:2879–2892
46. Rathore SS, Chaudhary DR, Boricha GN, Ghosh A, Bhatt BP, Zodape ST et al (2009) Effect of seaweed extract on the growth, yield and nutrient uptake of soybean (glycine max) under rainfed conditions. South African J Bot 75:351–355. https://doi.org/10.1016/j.sajb. 2008.10.009
47. Kularathne MAMN, Srikrishnah S, Sutharsan S (2021) Effect of seaweed extracts on ornamental plants: article review. Curr Agric Res J. 9:149–160
48. Behera B, Kolli Venkata Supraja BP (2021) Integrated microalgal biorefinery for the production and application of Bio-stimulants in circular bioeconomy. Bioresour Technol. https://doi. org/10.1016/j.biortech.2021.125588
49. RojasRodríguez ML, RamírezGil JG, GonzálezConcha LF, BalagueraLópez HE (2023) Bio-stimulants improve yield and quality in preharvest without impinging on the postharvest quality of hass avocado and mango fruit: evaluation under organic and traditional systems. Agronomy. https://doi.org/10.3390/agronomy13071917
50. Drobek M, Frąc M, Cybulska J (2019) Plant Bio-stimulants: importance of the quality and yield of horticultural crops and the improvement of plant tolerance to abiotic stressa review. Agronomy. https://doi.org/10.3390/agronomy9060335
51. Bulgari R, Cocetta G, Trivellini A (2015) Bio-stimulants and crop responses: a review. Biol Agric Hortic. https://doi.org/10.1080/01448765.2014.964649
52. Kah M et al (2018) Nanotechnology for plant protection and production: a critical review. J Agric Food Chem 66(50):13182–13192
53. Adisa AR et al (2020) Nanoencapsulation of plant Bio-stimulants for sustainable agriculture. J Agric Food Chem 68(19):5275–5286
54. Rai M et al (2017) Nanoparticles in plant protection and agriculture. J Appl Microbiol 123(3):577586
55. Sabir A et al (2014) Encapsulation of plant growth promoting bacteria: a potential for sustainable agriculture. Crit Rev Biotechnol 34(3):263277

56. Giraldo JP et al (2019) Nanotechnologyenabled tools for sustainable agriculture. Nat Nanotechnol 14(7):643652
57. Kumari S et al (2020) Nanoenabled strategies for sustainable crop production. J Nanosci Nanotechnol 20(3):16931707
58. Guo H, White JC, Wang Z, Xing B (2018) Nanoenabled fertilizers to control the release and use efficiency of nutrients. Curr Opin Environ Sci Health 6:7783. https://doi.org/10.1016/j.coesh.2018.07.009
59. Eliaspour S, Seyed Sharifi R, Shirkhani A, Farzaneh S (2020) Effects of biofertilizers and iron nano-oxide on maize yield and physiological properties under optimal irrigation and drought stress conditions. Food Sci Nutr 8(11):5985–5998. https://doi.org/10.1002/fsn3.1884
60. Noman M, Ahmed T, Shahid M, Nazir MM, Azizullah Li D, Song F (2023) Salicylic acid-doped iron nanoBio-stimulants potentiate defense responses and suppress Fusarium wilt in watermelon. J Adv Res. https://doi.org/10.1016/j.jare.2023.06.011
61. Sharifi RS, Khalilzadeh R, Pirzad A, Anwar S (2020) Effects of biofertilizers and nano zinc-iron oxide on yield and physicochemical properties of wheat under water deficit conditions. Commun Soil Sci Plant Anal 51:2511–2524
62. Khoulati A, Ouahhoud S, Taibi M et al (2025) Harnessing biostimulants for sustainable agriculture: innovations, challenges, and future prospects. Discov Agric 3:56. https://doi.org/10.1007/s44279-025-00177-9
63. Baragano D, Forján R, FernandezPerez B, Ayala J, Afif E, Gallego JLR (2020) Application of biochar, compost and ZVI nanoparticles for the remediation of As, Cu, Pb and Zn polluted soil. Environ Sci Pollut Reserch 27:33681–33691
64. Bashir A, Rizwan M, Ali S, Adrees M, Qayyum MF (2020) Effect of composted organic amendments and zinc oxide nanoparticles on growth and cadmium accumulation by wheat; a life cycle study. Environ Sci Pollut Res 27:23926–23936
65. Jahangir S, Javed K, Bano A (2020) Nanoparticles and plant growth promoting rhizobacteria (PGPR) modulate the physiology of onion plant under salt stress. Pak J Bot 52:1473–1480
66. Kohatsu MY, Lange CN, Pelegrino MT et al (2021) Foliar spraying of biogenic CuO nanoparticles protects the defence system and photosynthetic pigments of lettuce (*Lactuca sativa*). J Clean Prod 324:129264
67. Guerriero G, Sutera FM, Hoffmann J et al (2023) Nanoporous quercetinloaded siliconstabilized hybrid lipid nanoparticles alleviate salt stress in tomato plants. ACS Appl Nano Mater. 6:3647–3660
68. Magnabosco P, Masi A, Shukla R et al (2023) Advancing the impact of plant Bio-stimulants to sustainable agriculture through nanotechnologies. Chem Biol Technol Agric 10:117. https://doi.org/10.1186/s40538023004918
69. Bahuguna A, Sharma S, Rai A et al (2022) Chapter 17 Advance technology for Bio-stimulants in agriculture. In: Singh HB, Vaishnav A (eds). Elsevier, pp 393–412
70. Bakshi S, He ZL, Harris WG (2015) Natural nanoparticles: implications for the environment and human health. Crit Rev Environ Sci Technol 45:861904. https://doi.org/10.1080/10643389.2014.921975
71. Kabir E, Kumar V, Kim KH, Yip ACK, Sohn JR (2018) Environmental impacts of nanomaterials. J Environ Manag 225:261271. https://doi.org/10.1016/j.jenvman.2018.07.087
72. Sajid M, Ilyas M, Basheer C, Tariq M, Daud M, Baig N, Shehzad F (2015) Impact of nanoparticles on human and environment: review of toxicity factors, exposures, control strategies, and prospects. Environ Sci Pollut Res 22:41224143. https://doi.org/10.1007/s1135601439941
73. Gatoo MA, Naseem S, Arfat MY, Mahmood Dar A, Qasim K, Zubair S (2014) Physicochemical properties of nanomaterials: implication in associated toxic manifestations [WWW Document]. BioMed Res Int 48:842. https://doi.org/10.1155/2014/498420
74. Ghadimi M, Zangenehtabar S, Homaeigohar S (2020) An overview of the water remediation potential of nanomaterials and their ecotoxicological impacts. Water 12:1150. https://doi.org/10.3390/w12041150
75. Courtois P, Rorat A, Lemiere S, Guyoneaud R, Attard E, Longepierre M, Rigal F, Levard C, Chaurand P, Grosser A, Grobelak A, Kacprzak M, Lors C, Richaume A, Vandenbulcke F

(2020) Mediumterm effects of Ag supplied directly or via sewage sludge to an agricultural soil on Eisenia fetida earthworm and soil microbial communities. Chemosphere 171:128761. https://doi.org/10.1016/j.chemosphere.2020.128761
76. Lai RWS, Yeung KWY, Yung MMN, Djurišı AB, Giesy JP, Leung KMY (2018) Regulation of engineered nanomaterials: current challenges, insights and future directions. Environ Sci Pollut Res 25:30603077. https://doi.org/10.1007/s1135601794890
77. Weinberg H, Galyean A, Leopold M (2011) Evaluating engineered nanoparticles in natural waters: characterization, analysis, and risks of nanomaterials in environmental and food samples. TrAC Trends Anal Chem 30:7283. https://doi.org/10.1016/j.trac.2010.09.006
78. Lead JR, Batley GE, Alvarez PJJ, Croteau MN, Handy RD, McLaughlin MJ, Judy JD, Schirmer K (2018) Nanomaterials in the environment: behavior, fate, bioavailability, and effects—An updated review. Environ Toxicol Chem 37:20292063. https://doi.org/10.1002/etc.4147
79. https://www.bioagworlddigest.com/2025/03/Bio-stimulantsindianregulatorylandscape/
80. Baghdadi A, Della Lucia MC, Borella M, Bertoldo G, Ravi S, ZegadaLizarazu W et al (2022) A dualomics approach for profiling plant responses to biostimulant applications under controlled and field conditions. Front Plant Sci 13:1–19
81. Mendes R, de Barros I, D'Andréa PA, D'AndréaKühl MSC, Rodrigues GS (2023) A multiattribute approach to evaluating the impact of Bio-stimulants on crop performance. Front Plant Sci 14:1–12
82. Naikoo NB, Chesti MH, Bhat MA, Mir AH, Bashir O, Bhat TA et al (2022) Bio-stimulants towards soil health improvement: a review. Agric Rev I:1–8
83. Hawes C, Iannetta PPM, Squire GR (2021) Agroecological practices for wholesystem sustainability. CAB Rev Perspect Agric Vet Sci Nutr Nat Resour 16:2
84. Asif A, Ali M, Qadir M, Karthikeyan R, Singh Z, Khangura R et al (2023) Enhancing crop resilience by harnessing the synergistic effects of Bio-stimulants against abiotic stress. Front Plant Sci 14:1–16
85. Webster LJ, Villagomez D, Brown R, Clarke W, Schenk PM (2024) A synthetic biology approach for the treatment of pollutants with microalgae. Bioeng Biotechnol. https://doi.org/10.3389/fbioe.2024.1379301
86. Rouphael Y, Spíchal L, Panzarová K, Casa R, Colla G (2018) Highthroughput plant phenotyping for developing novel Bio-stimulants: from lab to field or from field to lab? Front Plant Sci 9:1–6
87. Goold HD, Wright P, Hailstones D (2018) Emerging opportunities for synthetic biology in agriculture. Genes (Basel). https://doi.org/10.3390/genes9070341
88. Erdoğan İ, CevherKeskin B, Bilir Ö, Hong Y, Tör M (2023) Recent developments in CRISPR/Cas9 genomeediting technology related to plant disease resistance and abiotic stress tolerance. Biology (Basel). https://doi.org/10.3390/biology12071037
89. Wang Y, Zafar N, Ali Q, Manghwar H, Wang G, Yu L et al (2022) CRISPR/Cas genome editing technologies for plant limitations, and future perspectives. Cells 11:3928
90. Idbella M, Giusti D, Gulli G, Bonanomi G (2023) Structure, functionality, compatibility with pesticides and beneficial microbes, and potential applications of a new delivery system based on inkjet technology. Sensors. https://doi.org/10.3390/s23063053
91. Ayaz M, AmmadUddin M, Sharif Z, Mansour A, Aggoune EHM (2019) Internetofthings (IoT)based smart agriculture: toward making the fields talk. IEEE Access 7:129551–129583
92. Yakhin OI, Lubyanov AA, Yakhin IA, Brown PH (2017) Bio-stimulants in plant science: a global perspective. Front Plant Sci 7:2049
93. Mauceri A, Puccio G, Faddetta T, Abbate L, Polito G, Caldiero C et al (2024) Integrated omics approach reveals the molecular pathways activated in tomato by Kocuria rhizophila, a soil plant growthpromoting bacterium. Plant Physiol Biochem 210:108609. https://doi.org/10.1016/j.plaphy.2024.108609
94. FAO (2021) The FAO action plan on antimicrobial resistance 2021–2025
95. EBIC (2020) Position papers and regulatory documents

96. Pechsiri JS, Thomas JBE, El BN, Fernandez FGA, Chaouki J, Chidami S et al (2023) Comparative life cycle assessment of conventional and novel microalgae production systems and environmental impact mitigation in urbanindustrial symbiosis. Sci Total Environ 854:158445
97. El Chami D, Santagata R, Moretti S, Moreschi L, Del Borghi A, Gallo M (2023) A life cycle assessment to evaluate the environmental benefits of applying the circular economy model to the fertiliser sector. Sustainability 15:15468
98. Sible CN, Seebauer JR, Below FE (2021) Plant Bio-stimulants: a categorical review, their implications for row crop production, and relation to soil health indicators. Agronomy. https://doi.org/10.3390/agronomy11071297
99. Hamid B, Zaman M, Farooq S, Fatima S, Sayyed RZ, Baba ZA et al (2021) Bacterial plant Bio-stimulants: a sustainable way towards improving growth, productivity, and health of crops. Sustain 13:1–24
100. Sujeeth N, Petrov V, Guinan KJ, Rasul F, O'Sullivan JT, Gechev TS (2022) Current insights into the molecular mode of action of seaweedbased Bio-stimulants and the sustainability of seaweeds as raw material resources. Int J Mol Sci. https://doi.org/10.3390/ijms23147654

Smart Nanosensors in Precision Agriculture

K. K. Vishnupriya, Aatika Nizam, Vasantha Veerappa Lakshmaiah, and Praveen Nagella

Abstract The integration of nanotechnology along with precision agriculture has emerged as an advanced concept for sustainable crop production and resource management. With respect to advanced innovations in the field of nanotechnology, nanosensors play an important role in real-time, sensitive, and specific monitoring of soil as well as plant health parameters, allowing for precise agricultural practices. Nanosensors equipped for the monitoring of soil health continuously facilitate the assessment of soil moisture, salinity, pH, nutrient monitoring, presence of heavy metals, detection of pesticides, and herbicides. Nanomaterials, including nanoparticles, carbon nanotubes, and various other nanostructures, allow prior identification of soil and plant health. To continue, the application of nanosensors makes it possible to monitor the use of chemicals in the agriculture industry, thus assuring a safer environment and better agrochemical practices, which are mostly sustainable. Alongside, the incorporation of soil nanosensors with smart irrigation systems leads to maximum use of water efficiently, which in turn helps in saving water and increasing crop yield. Besides, the use of nanosensors in monitoring plant health is quite significant as they allow the early detection of pathological changes occurring in the plant biome. Nanosensor technology enables the detection of plant hormones like abscisic acid, salicylic acid, auxins, etc., thus providing information about plant growth and stress conditions. In addition, the detection of plant diseases is made feasible by the use of nanosensors. The pairing of nanosensor technology with precision agriculture promotes effective and efficient crop management, hence minimizing yield losses and adopting sustainable farming practices. In conclusion, the use of nanosensors marks a new, promising area for soil–plant-environment interactions and the attainment of proper agricultural ecosystems.

Keywords Nanotechnology · Precision agriculture · Nanosensors · Soil health · Plant health

---

K. K. Vishnupriya · V. V. Lakshmaiah (✉) · P. Nagella
Department of Life Sciences, School of Sciences, Christ University, Bengaluru, India
e-mail: vasantha.vl@christuniversity.in

A. Nizam
Department of Chemistry, School of Sciences, Christ University, Bengaluru, India

J.-T. Chen (ed.), *AI-Nano-Enabled Climate-Smart Agriculture and Sustainability*, Smart Nanomaterials Technology, https://doi.org/10.1007/978-981-92-1694-9_17

1 Introduction

The introduction of nanotechnology has led to significant changes in one of the most crucial domains, which is agriculture. The agriculture industry has reached a very high peak, especially in the areas of irrigation, soil, plant health, and greenhouse management, where sensor-based solutions have been invented. Nanosensors play a crucial role in maintaining humidity, temperature, and moisture levels in Greenhouses, among others [48]. Besides that, the use of nanosensors is the next-generation innovation because of their incredibly high sensitivity and real-time analysis for detecting and monitoring the critical changes in agricultural fields [19]. Moreover, the use of nanosensors can also help to identify the differences in soil profiles and environmental conditions that are suitable for crop improvement and management [58]. Nanosensors in agriculture have attracted the attention of many, as they have been the reason behind the transformation of conventional farming into modern farming and have also helped in the optimization of resource management for better crop production [57]. These tiny nanosensors are non-invasive tools that assist in the fast evaluation of both non-living and living stress factors acting on plants [25]. AI integrated with smart nanosensor systems offers molecular-level data-driven decisions that allow for measuring pH, nutrients, light, temperature, moisture, etc., in a vast way. Besides that, the changes in soil profile, environmental conditions, and plant growth are also being tracked. The use of nanosensors in conjunction with smart agriculture opens up new possibilities for the efficient use of resources and also allows farmers to foresee weather patterns and become more resilient to them, thus leading to an increase in crop yield and better management [26].

2 Precision Agriculture

Precision agriculture is the comprehensive adoption of management and administrative principles for large-scale farming. One of the primary goals of precision farming is to raise the crop yield to the maximum level by using pesticides, fertilizers, and herbicides in only the amounts that will not have a negative impact on the environment [56]. Besides these, a number of different environmental factors are kept track of as controlled variables. The use of technologies like the geographical information system (GIS), remote sensing, and the geosynchronous positioning system (GPS) not only measures but also gives more accuracy to the conditions that are essential to the process of precision agriculture. These systems collect satellite/UAV-driven data continuously for surveillance and to accurately predict the changes in soil at a fixed GPS location at any given point in time. Precision agriculture takes the path of sustainability in agriculture and plans to meet the crop inputs based on the plant's needs. Farmers will be able to take site-specific and time-specific decision to ensure sustainable agriculture with reduced use of agrochemicals [41]. It checks the mineral needs of the plants, their water absorption, and total crop growth in a more centralized way. Variable-rate technology (VRT) is a key concept in precision agriculture, which integrates various technological dimensions for different field devices. VRT brings down the costs related to labor, crop losses, and ensures marketing strategies [53]. Given that the control of diseases and insects is among the major problems that limit plant growth, these technologies will eventually assist in the differentiation of these issues and make agricultural methods more efficient and less intensive [21]. Besides that, precision agriculture also employs systems empowered by nanotechnology that entail the deployment of sensors that interact with the target analyte and generate primary data connected to GPS. GIS and VRT for effective crop management and soil monitoring. The incorporation of nano-based smart delivery systems and nanosensors will direct the farmers regarding the right resource management within the farms through different sensing mechanisms [73].

3 Nanosensors in Smart Agriculture

Nanotechnology is one of the key contributors that has elevated the global industry to new heights. Alongside the widely known applications in the food sector and healthcare, agriculture emphasizes nanotechnology as a key player in environmental stewardship. Among the different types of top-down and bottom-up nanotechnological processes, one of the most important and probably the least evoked is the advancement in the detection of plant pathogens and plant biotic stress factors [24]. The journey from producer to consumer has been made easier and safer by utilizing nanotechnology and its various applications for the development of nanosensors that can provide real-time detection of contaminants in the worldwide food supply [11]. Nanosensors are unique materials at the nano level that are specifically molded to which the changes (chemical, biological, or physical) at very low frequency are captured and amplified. Gold, silver, and many other metallic, inorganic, and organic nanomaterials are functionalized for their potential application in nanosensors [22]. Nanosensors have been the best in the Agri-food industries in terms of detection of pesticide, pathogen, toxin, and allergen residues for eco-friendly agriculture. Supremely developed nanotechnology has given rise to next-generation advanced sensors that work efficiently in tandem with portable devices and with AI and ML to keep a check on a broad spectrum of environmental factors, including soil moisture, humidity, atmospheric gases, and salinity [9]. Working in a combination of nanomaterials with sensing technology, these nanosensors are able to determine the physicochemical changes in soil, identify plant diseases, and monitor plant-associated hormones even in the surrounding flora [14]. These nanosensors have the ability to indicate the presence of target molecules by visual or digital means, such as color changes, which is a big step forward in smart agriculture [62]. Smart agriculture represents the combination of advanced mechanisms and technologies, especially the use of nanotechnology in the sensing system and data analytics, which are very sensitive, productive, and efficient [17]. Nanosensors were incorporated into conventional farming in order to improve farming practices; thus, they have been seen to be the revolution in the agriculture sector. Within this framework, the role of nanosensors as revolutionary tools for transforming the agriculture domain in monitoring soil and crop health, water usage, and various environmental factors has been recognized [73]. The distinctive attributes of these nanosensors, like fast responsiveness, sensing of target molecules, low detection limits, and sustainability, make them highly qualified for precision agriculture and smart irrigation systems [30]. This review has brought forth the minute improvements in the role of nanosensors in the production of food farming technology that is more resilient and of superior quality. In summary, nanosensors indicated a huge rise in monitoring plant and soil health, as well as smart agriculture practices.

4 Nanosensors in Monitoring Soil Health

Soil health is defined by its potential to support plant growth, microbial community,maintain moisture content, soil biodiversity, nutrient cycling, and fertility. Sustainable farming and a healthy plant ecosystem are supported by soil health [12]. Additionally, it represents a safe and healthy environment for the long-term sustainable agriculture-based decomposition of organic and inorganic matter. Physical, biological, and chemical components interact efficiently, thus giving birth to soil health. Good soil with optimum porosity, ideal pH, and water retention capacity, balanced quantity of macro and micro nutrients means better crop health [68]. Real-time sensitive detection of major soil health parameters is provided by nanosensors. They play an important role in monitoring soil moisture, measuring salinity, variations in pH, identifying heavy metals, major plant nutrients and their deficiencies, detecting herbicide residues, in the development of optimized irrigation systems, and hence, giving insights into improving soil quality and plant productivity.

4.1 Assessment of Soil Moisture

Soil moisture measurement is a significant factor for monitoring both plant growth and irrigation efficiency. In fact, soil moisture monitoring by conventional means has been described as slow and not very accurate. However, continuous soil moisture monitoring by electrical and optical nanosensors is not only more accurate but also sensitive and stable, which positively affects the environment and the precision in agriculture [18]. Another way to measure soil moisture and its related properties is through soil apparent electrical conductivity (ECa). However, according to some studies, even the ECa values of soil can be inaccurate due to different potential error factors such as positional offset, sensor drift, and environmental variability. One way of getting rid of these inaccuracies is by the application of more sensitive and reliable nanosensors equipped with state-of-the-art technology, ultimately leading to real-time soil moisture monitoring. Nanosensors can effectively counteract environmental variability that usually hampers conventional ECa-based moisture monitoring in precision agriculture, highlighting the urgent need for trustworthy and consistent soil moisture data across diverse conditions [63]. Among the various methods available, capacitive sensors have emerged as the most precise for soil moisture measurement compared to resistive sensors that are normally cheaper but less accurate. Resistive sensors are said to be improved in terms of temperature and salinity sensitivity, while capacitive sensors continue to deliver stable soil-moisture assessments [1]. An innovative single-probe heat pulse sensor, which was represented in the form of a bipolar transistor, showed excellent temperature sensitivity for soil moisture assessment, and thus, the development of the top-notch nanosensor-based moisture detection technology has been accelerated [20]. Optical fiber-based sensors through changes in refractive index and spectral variations provide higher sensitivity, and this characteristic serves as a strongpoint of the optical fiber technology for precise soil water measurement at the nanoscale [36].

4.2 *Assessment of Salinity*

Nanosensors have been shown to be a suitable candidate for the monitoring of salinity conditions of water environments. The technology is based on measuring changes in refractive index that are closely associated with fresh and salt water mixing. This research presented a reliable measurement of the shifts in hydrology of the adjacent ecosystem [50]. Nanosensors are applied in the proper management of salt in the soil, and regulate the non-plant growth-promoting toxic ions of salts in the ecosystem. This area of application encourages plant forbearance, and high planting and reducing salt stress in the crops' development gives rise to the best use of nanosensors [67]. Moreover, the use of nanosensors is allowing a better understanding of salt-stress responses and supporting the plant towards this stress tolerance. The levels of glycine betaine are natural indicators whose in vivo concentration accumulates as an osmoprotectant when plants experience high salinity conditions, indicating FRET-based nanosensors are optical biosensors that can effectively capture levels of glycine betaine and generate proportional signal measure Förster Resonance Energy Transfer (FRET)-based nanosensors [2]. However, the salinity stress can be resolved for plants by irrigation and salt flushing, which requires close monitoring of the salinity. Nanosensors measure the electrical conductivity of pore water, wherein the conductivity is proportional to ions. Nanosensors with carbon nanotubes or graphene with high surface areas and conductivity as excellent choice to measure soil conductivity [35]. Nanosensors working with IoT have greatly improved underground water detection by providing accurate readings of water salinity, pH, water level, and TDS in real-time. ZnO nanotubes have been used as nanostructure sensing materials due to their high sensitivity and ability to detect water sources, thus preventing unnecessary drilling for borewells in water-scarce areas [51].

4.3 Heavy Metal Detection

The process of industrialization has contributed to the accumulation of heavy metals in the environment, particularly in cities, and this situation has made health risks inevitable. To this end, the development of portable nanosensors that possess excellent sensitivity and can control the levels of toxic heavy metal ions in the environment for public health benefits is a must [70]. The new generation of nanosensors, made up of graphene, are excellent for the precise detection and quantification of heavy metal ions. However, there are optical sensing devices for heavy metal detection customized with graphene, graphene oxide, and graphene quantum dots for their better performance [74]. The use of nanomaterials science involving graphene, MOFs, metal oxides, and polymer-based nanostructures results in prior heavy metal detection through electrochemical principle ensures to reduce the risks of heavy metal pollution [45]. An optical nanosensor made with carbon quantum dots (CQDs) with its fluorescence property has excellent sensitivity and spatiotemporal regulation capability. The application of machine learning is ubiquitous in signal extraction and processing. Furthermore, it is a great help in the exact detection of ions like Hg^{2+}, Cu^{2+}, Fe^{3+}, Pb^{2+}, and Cr^{6+} [65, 66]. Thiol-capped CdZnTe quantum dots with a thiol group, which ensures stability and water solubility, were intended for the accurate binding and detection of a broad spectrum of heavy metals at low concentrations. MPA-capped QDs responded to the quenching effect in correlation to the presence of Fe ions, present in fertilizers and water samples [49]. An ultra-thin MXene/bacterial cellulose (M/BC) composite paper was the source of a triboelectric nanosensor (TENS) for heavy metal detection. The device utilizes the properties of MXene and cellulose, where MXene contributes conductivity and cellulose contributes flexibility, thus yielding a nanosensor that is set for the detection of the heavy metal ions of Zn^{2+}, Cu^{2+}, and Cr^{3+} with limits that offer a cost-effective platform for the application of nanosensing [70].

4.4 Detection of Soil pH

Nanosensors made of carbon-based nanomaterials and polyaniline have been utilized for monitoring the quality of water samples, not only by determining the pH level but also by indicating organic matter from water with exceptional sensitivity. Real-time water quality assessment has experienced a shift from conventional methods to these types of nanosensors because of their low detection limits and fast response times [15]. Nanosensors with carbon nanopipette electrodes in a potentiometric system measure the pH level and show Nernstian response, hence fast and stable detection. This cellular-level application of the nanosensor increases the value of pH-related physiological study [8]. Fiber-optic nanosensors have been made with nanoporous silica-titania coatings. Different pH ranges showed the same characteristics of speed diffusion, enhanced porosity, and fast responses of the sensor. The durability and sensitivity of the produced sensor make it practical for long-term pH monitoring [28]. A nanocomposite-based optical pH sensor strip made with polyaniline on PET films presented color changes and detected various pH levels from pH 2 to 12. Its stable and UV–Vis-based responses made it suitable for natural water monitoring and illustrated the wider potential of the nanosensor [47].

4.5 Assessment of Soil Nutrients

Nanosensors are while to the point detectors of soil nutrients and organic matter, a real-time dirt quality and fertility indicator. Carbon and metal-based nanosensors are qualified for microorganisms and nutrients, exclusive to the soil profile detection. Nanosensors are engaged in smart agriculture, together with crop health benefits and sustainability [43]. Electrochemical nanosensors merge the detection of nitrate levels in a variety of soils with carbon nanotubes to achieve great sensitivity, reduced detection limits, and thereby improve nutrient monitoring and crop management [33]. Ammonium presence in soils was monitored using a chemiresistive nanosensor consisting of polyaniline/Ti_3C_2Tx hybrid films. This nanosensor has made it possible to achieve very fast and accurate gas detection, thus enabling real-time crop management [37]. Metal-oxide nanostructures (ZnO, TiO_2, SnO_2) are the perfect candidates for gas-based (especially ammonia) nutrient detection in smart agriculture because of their wide surface area, reactivity, and sensitivity. Different methods of synthesis result in the building of high-performance nanosensors for smart nutrient monitoring [7]. Electrodes modified with activated carbon nanotubes combined with graphene oxide and nickel hydroxide have been created for the selective detection of glucose. The glucose nanosensor has been reported to have very low interference, a low detection limit, and a fast response time [42].

4.6 Detection of Pesticides and Herbicides

The detection of pesticides by means of nanosensors has appeared as the application of nanotechnology in a new way and has beaten the conventional methods. Gas nanosensors made of nanoparticles have the great ability of the latter, hence, very sensitive detection of pesticides by means of the former. The main types of sensors involved in the monitoring of pesticides and their metabolites are optical and electrochemical nanosensors, which offer a real-time, smarter, and safer approach to better agricultural management and practices [52]. A working nanosensor made from polyaniline graphite and polyaniline silver nanostructure has demonstrated its ability to identify insecticides such as malathion and malaoxon by their respective vapor selectively. The pesticides have been detected owing to the low detection limits and the precise functionality of the pesticide monitoring system in agriculture [60]. A nonenzymatic electrical nanosensor, which is biomimetic in nature, applied the use of reduced graphene oxide and imprinting polymer films to identify atrazine by the exclusive binding of target molecules, and it has demonstrated remarkable durability and trustworthy sensitivity in both food and water samples [34]. A fluorescent pyrene-modified carbon nanodot-based nanosensor allowed for the detection of nitroaniline herbicides in soil through a "turn off" mechanism. It was mainly based on the photoinduced electron transfer process. The achieved nanosensor yielded great linearity and sensitivity, thus marking a superb research work in the area of herbicide detection with its further application in hydrogel-based kits [46]. Ratiometric nanosensor in conjunction with N-doped carbon dots and gold nanoclusters generated a "signal on–off–on" through copper-driven technique and fluorescence, thereby distinguishing the presence of glyphosate with ultra-sensitivity even in a mixture of compounds [39].

4.7 Nanosensors in the Irrigation System

Nanosensors placed in water irrigation systems are contributing to saving water. Nanosensors give accurate readings of soil moisture, humidity, and temperature, therefore, indirectly aiding irrigation management. Nanosensors are responsible for the very delicate observation of different factors using different signals, which makes it possible for them to turn into measurable outputs. Nanosensors are not just merely monitoring soil moisture but also hydraulic modeling and thus, determining the water level and making decisions about cutting off the inflow [54]. Nanosensors are given signals indicating the discharge of water that is to be optimized, thus enhancing surface irrigation systems [10]. Table 1 representing few examples for nanosensors in irrigation system.

5 Nanosensors in Monitoring Plant Health

Monitoring of plant health and monitoring of soil health are equally important. For increased crop yield, early disease diagnosis, and sustainable agriculture, plant health monitoring is no less important [64]. The classical methods for evaluating plant physiology, hormone level, plant diseases, and stress responses are very often considered to be time consuming owing to the lack of their real-time sensing capabilities, and hence their application is limited or even altogether abandoned [38]. The use of various biorecognition methods, such as antibodies, DNA probes, and microbial markers, makes this process possible. It is very much helping in the timely management of diseases before they show their symptom.

[40]. Nanosensors, among them nanobiosensors, are valuable tools for assessing abiotic stress caused by factors like temperature extremes, drought, salinity, etc., and for identifying biotic stress symptoms due to the presence of pests and microbiological infections by detecting the accumulation of metabolites and biomarkers [5].

Table 1 Overview of nanoparticle-based sensors: Types, target analytes, and detection mechanisms

Nanoparticle type	Sensor type	Target analyte	Detection mechanism	Reference
Pristine $LaFeO_3$ (LFO1, LFO2) and Cr-, Cu-, Mn-doped $LaFeO_3$ nanoparticle	Humidity sensor	Water vapour	Adsorption–desorption of water molecules	[32]
Si nanosphere core with corrugated Ag shell, modified with poly(diallyldimethylammonium chloride) polymer	Plasmonic nanosensor	Salicylic acid, extracellular ATP, cruciferous phytoalexin, glutathione (stress signaling molecules)	SERS signal enhancement through plasmonic hotspots and analyte interaction with polymer modified with Ag Shell; near IR Raman excitation	[61]
Au nano-urchins coated with molecularly imprinted sol–gel (MISG) solution	Plasmonic VOC sensor	cis-Jasmone, limonene, α-pinene, γ-terpinene (VOC biomarkers of plants)	RI-based sensing, molecular imprinting for terpene binding; PCA for analyte discrimination	[16]
Ferrocenecarboxylic-acid–functionalized MOF (Fc-NH_2-UiO-66) integrated with graphene oxide (trGNO)	Ratiometric electrochemical sensor	Cd^{2+}, Pb^{2+}, and Cu^{2+} (heavy metal ions)	Heavy metals are adsorbed by porous MOF; conductivity and electron transport are improved by trGNO and Fc	[71]
ZnO nanoparticles with ionic liquid ([EMIM][Otf]) and chitosan forming a nanocomposite membrane on an Au electrode	Electrochemical DNA biosensor	Identifies DNA sequences of *Trichoderma harzianum* (soil-borne fungi)	ssDNA probe immobilized on ZnO–CHIT/AuE; methylene blue used as hybridization indicator; using DPV sensitive detection of target DNA sequence	[59]
ZnO nanorods (NRs) grown hydrothermally on ZnO seed–layer–coated interdigitated electrodes (IDEs)	pH sensor	Hydrogen ion concentration (pH 4–10)	Capacitance variation caused by surface charge variation of ZnO NRs with pH	[31]

The combination of nanosensors with a wireless system is bringing about precision agriculture, which has great potential to uplift the modern agriculture that practices sustainable agriculture systems.

5.1 Detection of Plant Hormones

Nanosensors enable the use of advanced nanomaterials to gain very sensitive, very selective, very fast, and very real-time monitoring of the main plant hormones. Through the use of smart nanosensors made up of quantum dots, nanoparticles, and carbon nanotubes, the traditional methods of sensing have been replaced altogether with these new techniques, which not only increase the efficiency of the signal but also improve the accuracy of the measurements, thus helping to deal with the problem of poor stability and scalability. These sensors have a major role in remotely monitoring plant growth, stress response, and physiological changes, which is very useful in precision agriculture [23]. Nanosensors are able to monitor the plants in real time and notify when any stress signals are emitted. A double sensor made of salicylic acid (SA) and hydrogen peroxide (H_2O_2) has been developed, which is capable of tracking light, heat, and wounding stresses as well as pathogen attacks in real time and generating unique fluorescence-based response patterns. The use of nanotechnology in plant growth and stress monitoring has led to the development of new methods that are suitable for climate-resilient crop management [6]. Studies have reported the ability of polymer-based carbon nanotube sensors to detect very low levels of gibberellin hormones (GA_3 and GA_4) by means of selective binding. The system employing near-infrared fluorescence and Raman-referencing techniques has been successful in determining the GA level in plant roots in real time. The aforementioned nanosensors have primarily revealed the fluctuations of the hormones coinciding with the plant growth and stress periods [13]. The pyrite FeS_2 nanosensor on flexible paper has singled out the plant stress hormone methyl jasmonate. The sensors operate on the principle of electrochemical oxidation that the conductive and large surface area nanosheets provide implanting with wireless data transmission via Bluetooth, facilitating a low-cost remote monitoring of hormones for agricultural applications [55].

5.2 Detection of Pathogens

The fluorescent nanosensor, which was made with GelGreen–polyethylenimine silica nanoparticles, was designed for the real-time detection of extracellular DNA mainly released by bacterial invasion. The extracellular DNA that was formed acted as a bridge between the nanoparticles and the resulting bright fluorescence, thus indicating sensitivity and confirming the presence of pathogens for non-invasive early bacterial detection in plants [75]. Carbon nanotube, along with gold nanoparticle-based electrochemical nanosensors, has been invented to detect the quinoline metabolite linked to bacterial activity and hence, microbial markers. The conductivity of the nanosensors, the synergistic interactions, and fast and ultra-sensitive detection are the reasons for their exposure, and the analyte oxidation levels have been increased by identifying them for the assessment of related bacterial metabolites [3]. A functionalized multiwalled carbon nanotube (MWCNT)-based electrochemical nanosensor with gold nanoparticles and chitosan is able to detect the metabolic changes in plants caused by the *Ganoderma boninense* infection in oil palm. The nanosensor can tell sick plant tissues from the healthy ones through electrocatalytic oxidation, ultra-sensitivity, and low detection limits [27]. 3D SERS nanosensors built with gold nanoparticles on turnip yellow mosaic virus (TYMV) capsids, wherein a nano-bio-hybrid is produced. The functionalized AuNP-TYMV nanostructure emits Raman signals due to its efficient plasmonic effects that facilitate the identification of viral components [44]. Nano-immunosensors can reveal the existence of biological contaminants, particularly viruses. The major concept behind nano-immunosensors is the antigen–antibody interactions that, in the end, amplify their presence, thus leading to the rapid detection of viral particles. They can provide real-time, on-site detection of contaminants, which is a highly accurate sensing method to support public health and food safety [4].

5.3 *Monitoring of Abiotic and Biotic Stress*

Plants suffer from both abiotic and biotic stresses to a great extent, which decreases the yields of crops. The use of advanced nanosensors powered by nanotechnology, however, allows for the continuous monitoring of plant stress. The application of carbon nanotubes, SERS, and nanoparticle systems makes it possible to manage plant stress effectively through their strategic use in precision agriculture [5]. Abscisic acid (ABA) is a key player in the regulation of plant responses to different environmental stresses, particularly in the case of high salinity and drought. Traditional techniques mainly focus on measuring the average ABA levels in plant tissues. In contrast, the genetically engineered nanosensor proteins provide real-time and accurate detection of the levels of ABA in a single plant cell and its corresponding cellular sites [69]. Fluorescent-based ABACUS nanosensors are made to bind ABA levels reversibly. Changes in fluorescence ratios revealed the direct ABA concentration, and these sensors provided information on the ABA transport pathway and the root degradation mechanism. The sensor showed that the root cells always keep a lower internal ABA level, leading to the prevention of excess hormone inside the root cells [29]. A SERS (Surface Enhanced Raman Scattering) nanosensor made of silver decahedral nanoparticles is highly sensitive to the detection of hydrogen peroxide (H_2O_2), which is a biomarker for plant stress. This biochemical sensing helps detect the H_2O_2 build-up under different heat, cold, and wound stress conditions, thereby allowing the early detection and warning of abiotic and biotic stress long before the actual symptoms

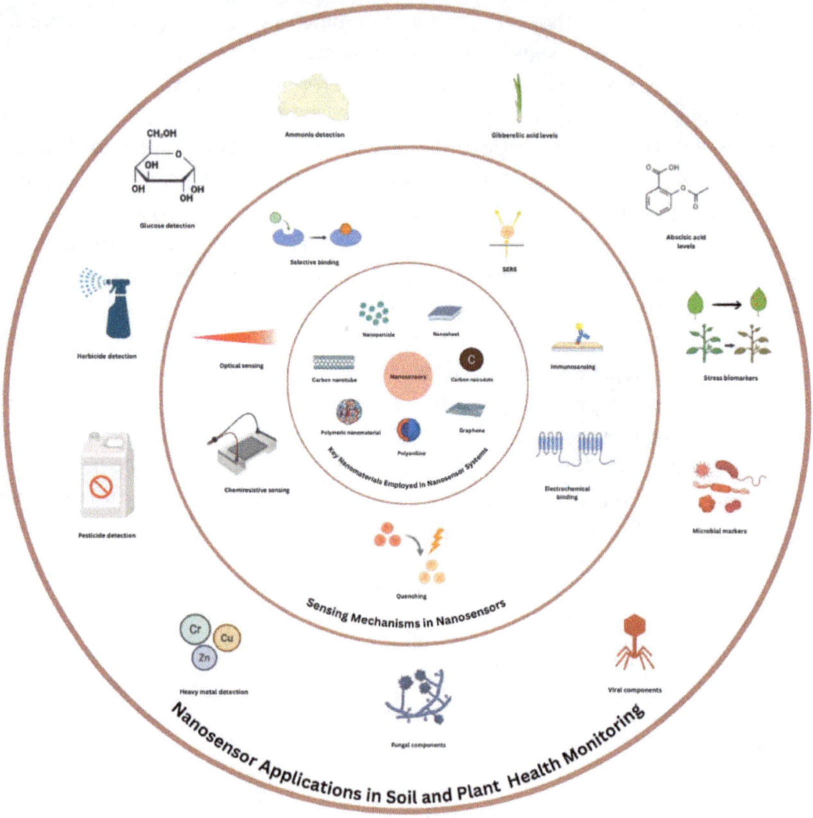

Fig. 1 Schematic representation of nanosensor applications in the soil and plant system

show up in the crops [65, 66]. Copper nanoparticle immobilized on gold electrodes is potential in the detection of salicylic acid, which is a major plant stress hormone. The main principle of electrocatalytic oxidation is at play, which in turn allows real-time monitoring of salicylic acid levels. The nanosensor was even more effective than HPLC in revealing the stress-induced dynamics of infected plants [72]. Figure 1 representing wide spectrum of nanosensor applications in precise agricultre.

6 Conclusion and Future Perspectives

The advent of nano sensors has revolutionized agriculture to meet the challenges like food scarcity, plant diseases, and climate change. The advances in the nano sensors ensure real-time analysis of plant phenotypes with high sensitivity and versatility at low cost compared to conventional methods. There are diverse transducers with the potential to selectively record and monitor significant signals designed on a nanoscale dimension. Nanosensors assist in precise analyte sensing and measurement. Research focusing on the synthesis, selection, fabrication, and construction of nano sensors is crucial for impactful, sustainable, and precise agriculture. This integration of nano sensors to plant sciences can gear up the production by optimizing the resources required and making it resilient to ever-changing environmental conditions. Some of the major hindrances to nanosensors are their toxicity, as they utilize highly reactive and persistent nanomaterials in the environment, which are major concerns affecting human health, soil microbes, soil health, and plants. Furthermore, there is enormous data generated, which requires robust AI and Machine learning tools. By integrating AI and ML, we can unlock the highest potential of the nano sensors by neutralizing the interfering background noises for accurate and real-time estimation of parameters. Another challenge towards implication of nanosensors is their scalability, and educating the public and farmers while conducting hands-on training and awareness campaigns will equip the farmers for their effective use in precise agriculture.

References

1. Adla S, Rai NK, Karumanchi SH, Tripathi S, Disse M, Pande S (2020) Laboratory calibration and performance evaluation of low-cost capacitive and very low-cost resistive soil moisture sensors. Sensors 20(2):363. https://doi.org/10.3390/S20020363
2. Ahmad M, Ameen S, Siddiqi TO, Khan P, Ahmad A (2016) Live cell monitoring of glycine betaine by FRET-based genetically encoded nanosensor. Biosens Bioelectron 86:169–175. https://doi.org/10.1016/j.bios.2016.06.049
3. Akanbi FS, Yusof NA, Abdullah J, Sulaiman Y, Hushiarian R (2017) Detection of quinoline in G. boninense-infected plants using functionalized multi-walled carbon nanotubes: a field study. Sensors 17(7):1538. https://doi.org/10.3390/s17071538
4. Akbari-Alavijeh S, Shaddel R, Lee CC, Pourjafar H, Ansari F, Alizadeh Sani M, Ajili N, Assadpour E, Zhang F, Jafari SM (2025) Nano-immunosensors for the rapid and sensitive detection of foodborne toxins: recent advances. Ind Crops Prod 228:120879. https://doi.org/10.1016/j.indcrop.2025.120879
5. Amjad S, Serajuddin M (2022) Role of nanosensors and bionanosensors in crop abiotic stress. In: Augmenting crop productivity in stress environment, pp 1–12. Springer, Singapore. https://doi.org/10.1007/978-981-16-6361-1_1
6. Ang MCY, Saju JM, Porter TK, Mohaideen S, Sarangapani S, Khong DT, Wang S, Cui J, Loh SI, Singh GP, Chua NH, Strano MS, Sarojam R (2024) Decoding early stress signaling waves in living plants using nanosensor multiplexing. Nat Commun 15(1):47082. https://doi.org/10.1038/s41467-024-47082-1
7. Arafat MM, Dinan B, Akbar SA, Haseeb AS (2012) Gas sensors based on one-dimensional nanostructured metal-oxides: a review. Sensors 12(6):7207–7258. https://doi.org/10.3390/s120607207
8. Aref M, Ranjbari E, García-Guzmán JJ, Hu K, Lork A, Crespo GA, Ewing AG, Cuartero M (2021) Potentiometric pH nanosensor for intracellular measurements: real-time and continuous assessment of local gradients. Anal Chem 93(47):15744–15751. https://doi.org/10.1021/acs.analchem.1c0387
9. Arora K (2018) Advances in nano based biosensors for food and agriculture. In: Advances in nano-based biosensors for food and agriculture, pp 1–52. Springer, Cham. https://doi.org/10.1007/978-3-319-70166-0_1

10. Aslam MA, Cheema MJ, Saleem S, Basit A, Hussain S, Waqas MS (2023) Application of sensor-based precision irrigation methods for improving water use efficiency of maize crop. Environ Sci Proc 23(1):38. https://doi.org/10.3390/environsciproc2022023038
11. Awlqadr FH, Altemimi AB, Qadir SA, Hama Salih TA, Alkanan ZT, AlKaisy QH, Mohammed OA, Hesarinejad MA (2025) Emerging trends in nano-sensors: a new frontier in food safety and quality assurance. Heliyon 11(1):e41181. https://doi.org/10.1016/j.heliyon.2024.e41181
12. Bhaduri D, Sihi D, Bhowmik A, Verma BC, Munda S, Dari B (2022) A review on effective soil health bio-indicators for ecosystem restoration and sustainability. Front Microbiol 13:938481. https://doi.org/10.3389/fmicb.2022.938481
13. Boonyaves K, Ang MCY, Park M, Cui J, Khong DT, Singh GP, Koman VB, Gong X, Porter TK, Choi SW, Chung K, Chua NH, Urano D, Strano MS (2023) Near-infrared fluorescent carbon nanotube sensors for the plant hormone family gibberellins. Nano Lett 23(3):916–924. https://doi.org/10.1021/acs.nanolett.2c04128
14. Carrillo-Lopez LM, Villanueva-Verduzco C, Villanueva-Sánchez E, Fajardo-Franco ML, Aguilar-Tlatelpa M, Ventura-Aguilar RI, Soto-Hernández RM (2024) Nanomaterials for plant disease diagnosis and treatment: a review. Plants 13(18):2634. https://doi.org/10.3390/plants13182634
15. Chajanovsky I, Cohen S, Shtenberg G, Suckeveriene RY (2020) Development and characterization of integrated nano-sensors for organic residues and pH field detection. Sensors 21(17):5842. https://doi.org/10.3390/s21175842
16. Chen B, Liu C, Shang L, Guo H, Qin J, Ge L, Jing CJ, Feng C, Hayashi K (2019) Electric-field enhancement of molecularly imprinted sol–gel-coated Au nano-urchin sensors for vapor detection of plant biomarkers. J Mater Chem C 8(1):262–269. https://doi.org/10.1039/C9TC05522
17. Christopher Selvam D, Devarajan Y (2025) Investigation of emerging technologies in agriculture: an in-depth look at smart farming, nano-agriculture, AI, and big data. J Biosyst Eng 50(2):170–192. https://doi.org/10.1007/s42853-025-00258-z
18. Cipriano PE, Siueia Júnior M, da Silva DF, Mars G, da Silva RF, Rabêlo FHS (2025) Nanosensors and their role in soil quality, moisture, pH, contaminants, and other molecules detection. In: Husen A (eds) Nanobiosensors for agricultural and other related sectors. Smart nanomaterials technology. Springer, Singapore. https://doi.org/10.1007/978-981-96-6499-3_5
19. Darwish MA, Abd-Elaziem W, Elsheikh A, Zayed AA (2024) Advancements in nanomaterials for nanosensors: a comprehensive review. Nanoscale Adv 6(16):4015–4046. https://doi.org/10.1039/d4na00214h
20. Dias PC, Roque W, Ferreira EC, Siqueira Dias JA (2013) A high sensitivity single-probe heat pulse soil moisture sensor based on a single NPN junction transistor. Comput Electron Agric 96:139–147. https://doi.org/10.1016/j.compag.2013.05.003
21. El-Saadony MT, Saad AM, Soliman SM, Salem HM, Ahmed AI, Mahmood M, El-Tahan AM, Ebrahim AAM, Abd El-Mageed TA, Negm SH, Selim S, Babalghith AO, Elrys AS, El-Tarabily KA, AbuQamar SF (2022) Plant growth-promoting microorganisms as biocontrol agents of plant diseases: mechanisms, challenges and future perspectives. Front Plant Sci 13:923880. https://doi.org/10.3389/fpls.2022.923880
22. Feng L, Song S, Li H, He R, Chen S, Wang J, Zhao G, Zhao X (2023) Nano-biosensors based on noble metal and semiconductor materials: Emerging trends and future prospects. Metals 13(4):792. https://doi.org/10.3390/met13040792
23. Giri S, Virk V (2025) Nanobiosensors: advanced detection and monitoring of plant hormones. In: Chen JT (eds) Nanobiosensors for crop monitoring and precision agriculture. Sustainability sciences in Asia and Africa, pp 189–206. Springer, Singapore. https://doi.org/10.1007/978-981-96-8335-2_9
24. Gul MZ, Beedu SR (2023) Integrated approaches to agri-nanotechnology: applications, challenges, and future perspectives. In: Microbiomes for the management of agricultural sustainability, pp 1–28. Springer, Cham. https://doi.org/10.1007/978-3-031-32967-8_1
25. Gulia V, Dhull SS, Kamboj H, Kumar S, Deepali Dhanda N, Rani J (2025) Pushing limits: integrating smart nanosensors and the Internet of Things in agriculture. Rendiconti Lincei. Scienze Fisiche e Naturali 36(2), 479–497. https://doi.org/10.1007/s12210-025-01337-1

26. Ibrahim AA, Abd-Ellatif S, Abdel Razik ESS, Hamedo HA, Salem KFM (2024) Nanosensors for enhancing plant growth and productivity. In: Al-Khayri JM, Alnaddaf LM, Jain SM, Penna S (eds) Innovative methods in horticultural crop improvement. Advances in Plant Breeding Strategies, pp 25–50. Springer, Cham. https://doi.org/10.1007/978-3-031-61095-0_2
27. Isha A, Akanbi FS, Yusof NA, Osman R, Mui-Yun W, Abdullah SNA (2019) An NMR metabolomics approach and detection of Ganoderma boninense-infected oil palm leaves using MWCNT-based electrochemical sensor. J Nanomater 2019:4729706. https://doi.org/10.1155/2019/4729706
28. Islam S, Bidin N, Riaz S, Krishnan G, Naseem S (2016) Sol–gel based fiber optic pH nanosensor: structural and sensing properties. Sens Actuators, A 238:8–18. https://doi.org/10.1016/j.sna.2015.12.003
29. Jones AM, Danielson JÅ, Manojkumar SN, Lanquar V, Grossmann G, Frommer WB (2014) Abscisic acid dynamics in roots detected with genetically encoded FRET sensors. eLife 3:e01741. https://doi.org/10.7554/eLife.01741
30. Kaushal M, Wani SP (2017) Nanosensors: frontiers in precision agriculture. In: Nanotechnology: an agricultural paradigm, pp 279–291. Springer, Singapore. https://doi.org/10.1007/978-981-10-4573-8_13
31. Kumar AA, Kumar NS, Ayele Aniley A, Edwin Fernandez R, Bhansali S, Ayele A, Edwin R (2019) Hydrothermal growth of zinc oxide (ZnO) nanorods on screen-printed interdigitated electrodes for pH measurement application. J Electrochem Soc 166(9):B324–B330. https://doi.org/10.1149/2.0431909jes
32. Kumari M, Dhariwal N, Yadav P, Kumar V, Thakur O (2025) Urea-modified Cu-doped $LaFeO_3$ nanoparticles for humidity sensing with contactless moisture detection for medical and agricultural applications. Sens Actuators, B Chem 423:136749. https://doi.org/10.1016/j.snb.2024.136749
33. Kundu M, Krishnan P, Chobhe KA, Manjaiah KM, Pant RP, Chawla G (2022) Fabrication of electrochemical nanosensor for detection of nitrate content in soil extract. J Soil Sci Plant Nutr 22(3):2777–2792. https://doi.org/10.1007/s42729-022-00845-5
34. Lakavath K, Kafley C, Sukumaran RA, Umadevi D, Kotagiri YG (2025) Advanced molecularly imprinted polymer-based electrochemical nanosensor for the sensitive detection of atrazine in agricultural fields. J Environ Chem Eng 13(5):117673. https://doi.org/10.1016/j.jece.2025.117673
35. Lee JR, Bechstein DJB, Ooi CC, Patel A, Gaster RS, Ng E, Gonzalez LC, Wang SX (2016) Magneto-nanosensor platform for probing low-affinity protein–protein interactions and identification of a low-affinity PD-L1/PD-L2 interaction. Nat Commun 7:12220. https://doi.org/10.1038/ncomms12220
36. Leone M (2022) Advances in fiber optic sensors for soil moisture monitoring: a review. Results in Optics 7:100213. https://doi.org/10.1016/j.rio.2022.100213
37. Li X, Xu J, Jiang Y, He Z, Liu B, Xie H, Li H, Li Z, Wang Y, Tai H (2020) Toward agricultural ammonia volatilization monitoring: a flexible polyaniline/$Ti_3C_2T_X$ hybrid sensitive film-based gas sensor. Sens Actuators, B Chem 316:128144. https://doi.org/10.1016/j.snb.2020.128144
38. Lowe A, Harrison N, French AP (2017) Hyperspectral image analysis techniques for the detection and classification of the early onset of plant disease and stress. Plant Methods 13:80. https://doi.org/10.1186/s13007-017-0233-z
39. Ma R, Fu L, Long N, Guo H, Hou Y, Li Y, Li P, Wang J, Zhou L, Kong W (2023) Gold nanoclusters and silica-coated carbon dots-assisted ratiometric fluorescent nanosensors for ultrasensitive detection of glyphosate. ACS Sustain Chem Eng 11(13):5093–5104. https://doi.org/10.1021/acssuschemeng.2c07077
40. Martinelli F, Scalenghe R, Davino S, Panno S, Scuderi G, Ruisi P, Villa P, Stroppiana D, Boschetti M, Goulart LR, Davis CE, Dandekar AM (2014) Advanced methods of plant disease detection: a review. Agron Sustain Dev 35(1):1–25. https://doi.org/10.1007/s13593-014-0246-1
41. Mgendi G (2024) Unlocking the potential of precision agriculture for sustainable farming. Discover Agriculture 2(1):87. https://doi.org/10.1007/s44279-024-00078-3

42. Mohammadi S, Taheri A, Rezayati-Zad Z (2018) Ultrasensitive and selective non-enzymatic glucose detection based on a Pt electrode modified by carbon nanotubes@graphene oxide/nickel hydroxide–Nafion hybrid composite in alkaline media. PC Biochemistry Research 1(1):1–8
43. Narang C, Kant R, Mandal S (2024) Nanosensors for soil and nutrients analytics. In: Nanotechnology in the life sciences, pp 177–200. Springer, Cham. https://doi.org/10.1007/978-3-031-76000-6_8
44. Nguyen HA, Jupin I, Decorse P, Lau-Truong S, Ammar S, Ha-Duong NT (2019) Assembly of gold nanoparticles using turnip yellow mosaic virus as an in-solution SERS sensor. RSC Adv 9(55):32296–32307. https://doi.org/10.1039/c9ra08015e
45. Numan A, Gill AA, Rafique S, Guduri M, Zhan Y, Maddiboyina B, Li L, Singh S, Nguyen Dang N (2021) Rationally engineered nanosensors: A novel strategy for the detection of heavy metal ions in the environment. J Hazard Mater 409:124493. https://doi.org/10.1016/j.jhazmat.2020.124493
46. Öztürk D, Sanko V, Khataee A, Demirbaş E, Durmuş M, Yeşilot S, Kılıç Z, Demirbaş Ü, Şenocak A, Tümay SO (2025) A pyrene-modified carbon nanodot as a fluorescence nanosensor for nitroaniline herbicide detection and smartphone-assisted RGB analysis using hydrogel detection kits. Analyst 150(19):4320–4331. https://doi.org/10.1039/d5an00624d
47. De Paula Herrmann PS (2011) Nanosensors applied to water quality: developing a low-cost pH sensor for natural water, and application of other techniques. Int J Nucl Desalin 4(3):218–229. https://doi.org/10.1504/IJND.2011.040124
48. Pramanik P, Krishnan P, Maity A, Mridha N, Mukherjee A, Rai V (2020) Application of nanotechnology in agriculture. In: Dasgupta N, Ranjan S, Lichtfouse E (eds) Environmental nanotechnology volume 4. Environmental chemistry for a sustainable world, vol 32, pp 317–348, Springer, Cham. https://doi.org/10.1007/978-3-030-26668-4_9
49. Qian J, Lu X, Wang C, Cui H, An K, Long L, Hao N, Wang K (2020) Controlling over the terminal functionalities of thiol-capped CdZnTe QDs to develop fluorescence nanosensor for selective discrimination and determination of Fe(II) ions. Sens Actuators, B Chem 322:128636. https://doi.org/10.1016/j.snb.2020.128636
50. Rakib A, Muttaqi MR, Siddique ATB, Siddique MAB, Sagor RH (2023) Oxidation-resistant nanosensor: a label free, on-chip sensing approach of protein concentration, salinity, tissue classification. Results Phys 53:106956. https://doi.org/10.1016/j.rinp.2023.106956
51. Rathi, P. A. (2020) Nanotechnology and Nanosensors, Part-1 Coursera Mini-Project IoT based Underground Water Detection with Purity Measurement. https://doi.org/10.13140/RG.2.2.21483.49447
52. Saini RK, Bagri LP, Bajpai AK (2017) Smart nanosensors for pesticide detection. New Pestic Soil Sens 519–559. https://doi.org/10.1016/B978-0-12-804299-1.00015-1
53. Saleem SR, Zaman QU, Schumann AW, Abbas Naqvi SMZ (2023) Variable rate technologies: development, adaptation, and opportunities in agriculture. Precis Agric: Evol, Insights Emerg Trends 103–122. https://doi.org/10.1016/B978-0-443-18953-1.00010-6
54. Selvaraj SK, Sanjeev Kumar R, Balamurugan K, Joany RM, Dorothy R, Nguyen TA, Rajendran S (2022) Wireless nanosensor network for irrigation control. Nanosensors Smart Agric 463–478https://doi.org/10.1016/B978-0-12-824554-5.00005-7
55. Sha R, Kadu A, Matsumoto K, Uno S, Badhulika S (2020) Ultra-low cost, smart sensor based on pyrite FeS_2 on cellulose paper for the determination of vital plant hormone methyl jasmonate. Eng Res Express 2(2). https://doi.org/10.1088/2631-8695/ab8b
56. Shafi U, Mumtaz R, Hassan SA, Zaidi SA, Iqbal N (2018) Precision agriculture techniques and practices: from considerations to applications. Sensors 19(17):3796. https://doi.org/10.3390/s19173796
57. Shah F, Nagaraj K, Papriwal S, Selwal G, Mistry R, Maity P, Kamalesu S, Radhakrishnan L, Venkatachalapathy R (2024) Innovative nanotechnological solutions for sustainable agriculture amidst rising population demands. Food and Humanity 3:100425. https://doi.org/10.1016/j.foohum.2024.100425

58. Sharma V, Bag AG, Changade NM, Parihar NS (2025) Precision agriculture and nanosensor. In: Husen A (eds) Nanobiosensors for agricultural and other related sectors. Smart nanomaterials technology, pp 33–48. Springer, Singapore. https://doi.org/10.1007/978-981-96-6499-3_2
59. Siddiquee S, Rovina K, Yusof NA, Rodrigues KF, Suryani S (2014) Nanoparticle-enhanced electrochemical biosensor with DNA immobilization and hybridization of *Trichoderma harzianum* gene. Sens Bio-Sens Res 2:16–22. https://doi.org/10.1016/j.sbsr.2014.06.002
60. Silva DM, Machado EP, Fernandes IA, Steffens J, Steffens C (2025) Characterization and performance of gas nanosensors for the detection of malathion and malaoxon and application in a kale (*Brassica oleracea*) environment. Microchem J 219:115916. https://doi.org/10.1016/j.microc.2025.115916
61. Son WK, Choi YS, Han YW, Shin DW, Min K, Shin J, Lee MJ, Son H, Jeong DH, Kwak SY (2023) In vivo surface-enhanced Raman scattering nanosensor for the real-time monitoring of multiple stress signalling molecules in plants. Nat Nanotechnol 18(2):205–216. https://doi.org/10.1038/s41565-022-01274-2
62. Srivastava AK, Dev A, Karmakar S (2017) Nanosensors and nanobiosensors in food and agriculture. Environ Chem Lett 16(1):161–182. https://doi.org/10.1007/s10311-017-0674-7
63. Sudduth K, Drummond S, Kitchen N (2001) Accuracy issues in electromagnetic induction sensing of soil electrical conductivity for precision agriculture. Comput Electron Agric 31(3):239–264. https://doi.org/10.1016/S0168-1699(00)00185-X
64. Tahat MM, Alananbeh KM, Othman YA, Leskovar DI (2020) Soil health and sustainable agriculture. Sustainability 12(12):4859. https://doi.org/10.3390/su12124859
65. Tian Y, Cong L, Sun H, Wei W, Wu F, Li F, Xu G, Niu W (2024) An in vivo surface-enhanced Raman scattering nanosensor for detecting hydrogen peroxide in plants under abiotic stresses. ACS Applied Nano Materials 7(17):20308–20316. https://doi.org/10.1021/acsanm.4c03271
66. Tian C, Lee Y, Song Y, Elmasry MR, Yoon M, Kim DH, Cho SY (2024) Machine-learning-enhanced fluorescent nanosensor based on carbon quantum dots for heavy metal detection. ACS Applied Nano Materials 7(5):5576–5586. https://doi.org/10.1021/acsanm.4c00359
67. Tiwari S, Lata C, Prasad V (2021) Application of nanobiotechnology in overcoming salinity stress. In: Nanobiotechnology: mitigation of abiotic stress in plants, pp 375–398. Springer, Cham. https://doi.org/10.1007/978-3-030-73606-4_16
68. Usharani KV, Roopashree KM, Naik D (2019) Role of soil physical, chemical and biological properties for soil health improvement and sustainable agriculture. J Pharmacogn Phytochem 8(5):1256–1267
69. Waadt R, Hitomi K, Nishimura N, Hitomi C, Adams SR, Getzoff ED, Schroeder JI (2014) FRET-based reporters for the direct visualization of abscisic acid concentration changes and distribution in *Arabidopsis*. eLife 3:e01739. https://doi.org/10.7554/eLife.01739
70. Wang E, Cao Y, Bai Y, Gai Y, Shan Y, Li Q, Jiang T, Feng H, Li Z (2022) A triboelectric nanosensor based on ultra-thin MXene composite paper for heavy metal ion detection. J Micromech Microeng 32(4):044003. https://doi.org/10.1088/1361-6439/ac542b
71. Wang X, Qi Y, Shen Y, Yuan Y, Zhang L, Zhang C, Sun Y (2020) A ratiometric electrochemical sensor for simultaneous detection of multiple heavy metal ions based on ferrocene-functionalized metal–organic framework. Sens Actuators, B Chem 321:127756. https://doi.org/10.1016/j.snb.2020.127756
72. Wang Z, Wei F, Liu SY, Xu Q, Huang JY, Dong XY, Yu JH, Yang Q, Zhao YD, Chen H (2010) Electrocatalytic oxidation of phytohormone salicylic acid at copper nanoparticles-modified gold electrode and its detection in oilseed rape infected with fungal pathogen *Sclerotinia sclerotiorum*. Talanta 80(3):1277–1281. https://doi.org/10.1016/j.talanta.2009.09.023
73. Yadav A, Yadav K, Ahmad R, Kumar A (2023) Emerging frontiers in nanotechnology for precision agriculture: advancements, hurdles and prospects. Agrochemicals 2(2):220–256. https://doi.org/10.3390/agrochemicals2020016
74. Zhang L, Peng D, Liang R, Qiu J (2018) Graphene-based optical nanosensors for detection of heavy metal ions. TrAC, Trends Anal Chem 102:280–289. https://doi.org/10.1016/j.trac.2018.02.010

75. Zhu C, Kang S, Wu S, Sun Q, Zhang Z, Hu J, Liu J, Li C, Wang F (2024) A self-reporting nanosensor for detecting bacterial exploitation in plants through extracellular DNA. Nano Lett 24(48):15464–15472. https://doi.org/10.1021/acs.nanolett.4c05168

Smart Nano-delivering Genetic Engineering for Agricultural Sustainability

Elshan Musazade, Nargiz Bayramova, Samra Mirzayeva, Javaid Akhter Bhat, and Xianzhong Feng

Abstract Nanotechnology is revolutionizing plant genetic engineering by overcoming persistent challenges in gene delivery that have impeded advances in crop enhancement, functional genomics, and precise genome editing. Traditional transformation techniques, such as *Agrobacterium*-mediated transfer, polyethylene glycol (PEG)-based transfection, electroporation, biolistic bombardment, and viral vectors, are limited by species specificity, low efficiency, tissue damage, and the labor-intensive nature of regeneration. These issues are particularly problematic for CRISPR-Cas applications, which require effective intracellular transport of nucleic acids and ribonucleoproteins (RNPs). Recent findings suggest that nanomaterials can penetrate plant cell walls, shield biomolecules from degradation, and facilitate the delivery of DNA, RNA, proteins, and CRISPR/Cas-gRNA complexes in a transgene-free manner. Key platforms, such as mesoporous silica nanoparticles (MSNs), Au nanoparticles (AuNPs), carbon nanotubes (CNTs), and layered double hydroxides (LDHs), have achieved transient expression, stable genomic integration, germline modification, and targeted editing of plastids and mitochondria. Their adjustable physicochemical properties enable controlled cargo loading, low cytotoxicity, and species-independent uptake, offering advantages over conventional methods. This chapter synthesizes recent advances in nano-enabled plant genetic engineering, encompassing gene delivery for transformation, gene silencing, and genome editing, with emphasis on structure–function relationships, rational

---

E. Musazade · X. Feng (✉)
Key Laboratory of Soybean Molecular Design Breeding, Northeast Institute of Geography and Agroecology, Chinese Academy of Sciences, Changchun, China
e-mail: fengxianzhong@iga.ac.cn

N. Bayramova
LANDAU School, Baku, Azerbaijan

S. Mirzayeva
Science, Medicine and Education Department, Azerbaijan National Anti-Doping Agency, Baku, Azerbaijan

J. A. Bhat (✉)
Zhejiang Lab, Research Center for Life Science Computing, Hangzhou, China

J.-T. Chen (ed.), *AI-Nano-Enabled Climate-Smart Agriculture and Sustainability*, Smart Nanomaterials Technology, https://doi.org/10.1007/978-981-92-1694-9_18

nanocarrier (NC) design, and integration with CRISPR-based technologies. Collectively, these developments underscore the transformative potential of nanotechnology to enable efficient, precise, and broadly applicable genetic manipulation for next-generation crop improvement.

Keywords Nanocarriers · Biomolecule delivery · Nanotechnology-enabled transformation · CRISPR-Cas genome editing · Plant genetic engineering

1 Introduction

Plants are fundamental to human well-being and environmental stability, underpinning natural ecosystems and serving as primary sources of food, medicine, fragrances, flavors, sustainable biomaterials, and recombinant bioproducts [1, 2]. However, rapid population growth, projected to exceed 10 billion by 2050, coupled with climate change, land degradation, pollution, and the emergence of new pests and diseases, increasingly threatens agricultural productivity and global food security [3, 4]. These converging pressures highlight the urgent need for innovative strategies to enhance crop resilience, improve plant performance, and enable sustainable bioproduction.

Modern plant biotechnology has enabled significant advances in molecular farming and genetic engineering, supporting the development of high-yield and stress-resistant crops [5, 6]. Although the Green Revolution substantially increased agricultural productivity through conventional breeding [7], traditional approaches remain constrained by limited precision, low throughput, and an inability to introduce traits beyond natural gene pools [8]. Pioneering transformation technologies, including *Agrobacterium tumefaciens*-mediated gene transfer [9] and biolistic particle delivery [10], significantly accelerated crop improvement and recombinant protein production in plants [11, 12]. However, these methods are often species- and genotype-dependent, labor-intensive, and inefficient for non-nuclear genome engineering, with transformation and regeneration remaining major bottlenecks [6, 13].

To overcome these limitations, a range of alternative or derivative delivery approaches, such as polyethylene glycol (PEG)-mediated transfection, viral vectors, electroporation, and pollen-tube-mediated transfer, have been explored across diverse crop species [14, 15]. Nevertheless, each method presents inherent drawbacks: PEG transfection and electroporation are often associated with cytotoxicity and low regeneration efficiency, viral vectors are constrained by cargo capacity and host specificity, and pollen-tube–mediated approaches suffer from poor reproducibility and efficiency [16, 17]. These technical challenges are further compounded by intrinsic structural barriers of plant cells, including the cell wall, composed primarily of cellulose, hemicellulose, and pectin, which restricts the passive entry of biomolecules larger than ~ 20 nm, as well as the plasma membrane, nuclear envelope, and organellar membranes [18, 19]. Collectively, these constraints have contributed to the comparatively slower development of plant transformation technologies relative to animal systems.

The introduction of CRISPR-Cas genome-editing technologies has revolutionized plant genetic engineering by enabling precise, targeted DNA modifications through the delivery of plasmid DNA, RNA, or ribonucleoprotein (RNP) complexes [20, 21]. The delivery of CRISPR-Cas components as RNA or RNP complexes enables permanent genome editing without stable transgene integration, and in many regions, exempts such plants from GMO (Genetically Modified Organism) regulatory frameworks [22], thereby accelerating trait improvement and modern breeding programs [23, 24]. In parallel, RNA- and protein-based delivery platforms enable RNA interference (RNAi)-mediated gene silencing and support a broad range of precision engineering approaches, including miRNA-based regulation, zinc-finger nucleases, TALENs, and CRISPR systems [25, 26]. Despite these advances, the efficiency of DNA-, RNA-, and protein-based editing and silencing technologies remains constrained by extracellular and intracellular transport barriers, which limit the effective delivery of cargo to target tissues and subcellular compartments.

Nanotechnology offers a versatile framework for overcoming delivery barriers in plant genetic engineering by enabling controlled transport of genetic cargos across plant cellular boundaries. By tailoring physicochemical properties such as size, shape, surface charge, and surface chemistry, nanomaterials can be engineered to penetrate rigid plant cell walls, protect nucleic acids from enzymatic degradation, and deliver diverse biomolecules into intact plant cells [27, 28]. Nanoscale platforms have been applied to molecular sensing [29], stress priming [30, 31], genetic engineering [32, 33], and post-transcriptional regulation [34, 35]. Notably, these systems enable species-independent delivery, facilitate access to organellar genomes, and support efficient transport of DNA, RNA, and RNPs without external physical intervention [18, 36]. In parallel, agricultural nanotechnologies, including nanosensors [37], nanopesticides [38], and nanofertilizers [39], demonstrate the broad compatibility of nanomaterials with plant systems. Their high biocompatibility, biodegradability, and low immunogenicity [40, 41] further position nanomaterials as effective next-generation carriers for CRISPR-, RNAi-, and protein-based applications.

This review provides an in-depth analysis of the latest advancements in plant genetic engineering, combining traditional transformation techniques with innovative nanotechnologies. It begins by discussing conventional delivery methods and their drawbacks. Then it proceeds to the design and engineering principles of nanocarriers (NCs), detailing strategies for loading and delivering DNA, RNA, proteins, and CRISPR RNP complexes. This review focuses on the mechanisms that facilitate gene delivery, genome editing, and regeneration through nanomaterials in various plant systems, as well as biosafety issues and unresolved mechanistic questions in this rapidly advancing field. The final sections emphasize the significant technical challenges that currently limit the widespread use of nano-enabled transformation and highlight key opportunities to enhance delivery precision, efficiency, and species independence, establishing smart nano-delivery systems as a cornerstone for future sustainable agriculture.

2 Conventional Approaches for Plant Genetic Transformation

Genome engineering is a key approach for enhancing crops sustainably and is set to become increasingly vital in future breeding initiatives [42]. Traditionally, transformation systems such as *Agrobacterium*-mediated infection, particle bombardment, PEG-induced protoplast transfection, and viral vector-based delivery have facilitated significant progress in functional genomics and crop improvement [43, 44]. Plant gene delivery methods can be categorized into two main types: direct and indirect techniques [45]. Direct methods introduce DNA into cells through physical or chemical approaches, whereas indirect methods rely on biological systems such as *Agrobacterium* for DNA transfer. The selection of an appropriate delivery platform depends on the biological context, desired expression profile, and the experimental goals. Although these conventional methods are widely used, their overall effectiveness is limited by the complex, multilayered structure of the plant cell wall, species-specific regeneration challenges, and the inherent limitations of each transformation system [46]. Consequently, achieving efficient and predictable genetic modifications across various plant species remains a considerable challenge. A comparative overview of plant gene-delivery systems, encompassing both conventional and nanomaterial-mediated approaches, is presented in Table 1, summarizing their key characteristics, strengths, and limitations.

2.1 Direct Physical Transformation Methods

Direct gene delivery techniques utilize physical or chemical approaches to insert DNA into plant cells, including particle bombardment, microinjection, electroporation, and PEG-mediated protoplast transfection. Among these, particle bombardment, also known as biolistic transformation, is one of the most widely used methods in plant biotechnology. This technique involves using compressed gas to propel gold (Au) or wolfram (W) microprojectiles, which are coated with DNA, RNA, proteins, or peptides, through the plant cell wall and plasma membrane into the cytoplasm and nucleus, facilitating gene transfer [47–49]. Biolistics is compatible with a variety of tissues and cargo types and is particularly useful for transforming mitochondria and chloroplasts [50]. Owing to the high-speed microprojectiles disrupting cellular turgor balance and potentially causing leakage, target tissues are often placed on osmotic media during bombardment. Subsequently, DNA is released from the particles to facilitate genomic integration, although the exact mechanism of release is not fully understood. The broad applicability of this method across different plant species and tissue types makes it worthwhile wherever effective regeneration systems are available. However, high-speed acceleration can lead to tissue damage, random and multiple transgene insertions, and structural genomic changes, such as deletions, inversions, and translocations, which can limit its suitability for commercial breeding

Table 1 Comparison between various gene-delivery systems used for plant genetic engineering

Classification	Systems	Characteristics	Strengths	Limitations
Traditional genetic transformation	*Agrobacterium*-mediated transformation	*A. tumefaciens* infects wounded plant tissues and transfers T-DNA into the plant genome	• High transformation efficiency • Low operational cost • Capable of generating stable transgenic lines	• Narrow host range, with reduced efficiency in recalcitrant species • Random genomic integration of T-DNA
	Biolistic (particle bombardment) delivery	DNA-coated heavy-metal microparticles are accelerated into plant cells at high velocity	• Species-independent delivery • Simple technical workflow • Capable of introducing large genetic constructs	• Lower efficiency compared to *Agrobacterium* • Potential for severe cellular or tissue damage • High equipment and consumable costs • Random integration events
	Electroporation	High-voltage electric pulses transiently permeabilize the cell membrane, allowing DNA to enter the cytoplasm	• Rapid and cost-effective • High transformation efficiency in suitable materials	• Limited applicability to a narrow range of plant species • Difficult to apply in intact, walled cells • Risk of electrical or physiological damage

(continued)

Table 1 (continued)

Classification	Systems	Characteristics	Strengths	Limitations
	PEG-mediated delivery	PEG destabilizes the plasma membrane of protoplasts, thereby facilitating the entry of DNA into the cytoplasm	• Highly efficient for protoplast transformation	• Restricted to protoplast systems • Random genomic integration • Requires regeneration, often a major bottleneck
Nanomaterial-mediated gene delivery systems	CNTs, MSNs, liposomes, and other NCs	Nanomaterials transport exogenous nucleic acids into the cytoplasm or organelles through endocytic or non-endocytic pathways	• Species-independent delivery • Simple operational workflow • High biocompatibility • High cargo-loading capacity • Potential for high transformation efficiency	• Limited availability of optimized NCs • Delivery outcome strongly influenced by physical and chemical properties of carriers

[51, 52]. The need for specialized equipment and expensive materials further limits its widespread use, although its ability to distribute cargo nonspecifically remains beneficial for targeting the nuclear genome [18].

Microinjection is another physical technique in which DNA is introduced using fine needles that range in diameter from 0.5 to 10 μm, depending on the size of the cell. Although it can be applied to whole cells, it is most commonly conducted in protoplasts. These protoplasts can be stabilized using holding capillaries or substrates such as poly-L-lysine (PLL) and agarose to maintain the correct orientation during intranuclear injection [53, 54].

Electroporation facilitates the introduction of foreign DNA by applying short, intense electric pulses that cause temporary pores to form in the plasma membrane [55]. Initially demonstrated in plant protoplasts in 1985, this technique enabled the transformation of both monocots and dicots, such as carrot (*Daucus carota*), tobacco (*Nicotiana tabacum*), and maize (*Zea mays*), by optimizing the pulse amplitude, duration, and buffer conditions [56]. Electric pulses induce localized membrane curvature and the formation of transient hydrophobic pores in plant protoplast membranes, which can subsequently transition into hydrophilic pores [57], permitting the entry of macromolecules and enabling efficient recovery of electropermeabilized cells, as demonstrated by the delivery of Cre recombinase in *Arabidopsis thaliana* [58]. Although electroporation is rapid, cost-effective, and efficient, its application is constrained by species specificity, limited performance in cells with thick or lignified cell walls, and the risk of nucleic acid damage that may result in aberrant translation and cytotoxic effects [59]. Moreover, exposure to high-voltage pulses can increase cell mortality, further restricting its practical applicability.

Protoplast transfection using PEG is a commonly employed direct delivery method. When PEG is used in conjunction with Mg^{2+} or Ca^{2+}, it facilitates the formation of DNA-PEG complexes that temporarily increase membrane permeability, enabling the endocytic uptake of naked DNA into the cytoplasm [60]. This method of transfection is straightforward, cost-effective, and highly efficient for introducing plasmids, RNA, and CRISPR-Cas RNPs [20, 61]. It is frequently used for transient gene expression, protein localization studies, and assessment of genome editing. Nonetheless, its use is limited to protoplasts, and regenerating entire plants from transfected protoplasts remains a challenging, genotype-dependent, and inefficient process [35, 62].

Although physical delivery techniques are useful, they have significant drawbacks. Particle bombardment can lead to severe tissue damage, diminished regenerative ability, and the integration of multiple transgene copies, which increases the likelihood of gene silencing and unpredictable segregation patterns [63]. Similarly, electroporation carries the risk of cytotoxicity and decreased cell viability. Additionally, all physical methods require specialized equipment and technical skills, limiting their accessibility and widespread application in plant transformation research [64].

2.2 Indirect Vector-Mediated Transformation Method

Indirect gene delivery utilizes biological systems, mainly *Agrobacterium* or plant viruses, to insert foreign DNA into the plant cells. *Agrobacterium*-mediated transformation is the most widely used indirect technique in plant biotechnology [65]. During its natural infection process, *Agrobacterium* transfers genes responsible for phytohormone biosynthesis and opine production into the host genome, leading to uncontrolled cell growth and providing the bacterium with specific nutrients [66]. These genes, along with the regulatory elements necessary for DNA transfer, are located on tumor-inducing (Ti) or root-inducing (Ri) plasmids. In engineered binary vectors, the T-DNA region is substituted with a desired gene, and the T-DNA is processed and incorporated into the host genome as a single DNA strand [67]. Since dicots naturally activate *Agrobacterium* virulence (Vir) genes through phenolic signals induced by wounds, transformation is generally more effective in dicots, whereas monocot transformation often requires external acetosyringone (AS) or strains with constant *Vir* gene expression [68, 69].

Although *Agrobacterium*-mediated transformation is known for its procedural simplicity, high stability, and tendency for single-copy insertions, it is highly genotype-dependent. Even closely related cultivars can exhibit varying levels of susceptibility due to differences in their genetic makeup, hormone responsiveness, regenerative ability, and metabolic conditions [70]. Recent advancements in the transient or stable overexpression of developmental regulators, such as *BBM*, *WUS*, *GRF5*, and *GRF4-GIF1*, have enhanced transformation in genotypes that were previously difficult to transform by promoting cell fate reprogramming [71–73]. Specific endogenous metabolites, such as ethylene (ET), salicylic acid (SA), and γ-aminobutyric acid (GABA), can inhibit bacterial growth and *Vir* gene activation, thereby decreasing transformation efficiency [74, 75], while incompatibilities between the host and *Agrobacterium* may lead to symptoms resembling disease [76]. Additionally, *Agrobacterium* is unable to deliver RNAs, RNPs, or oligonucleotides, which restricts its application in transgene-free genome editing [77, 78].

There are two primary methods for *Agrobacterium* transformation. The in vitro method involves using sterile plant parts, such as leaves, petioles, hypocotyls, stem internodes, roots, calli, and cell suspensions. However, it necessitates that regenerated plants develop from individually transformed cells [79–81]. In contrast, *in planta* transformation avoids extensive tissue culture by focusing on reproductive tissues, meristems, or the pollen tube pathway. This approach, first introduced in 1983, involves introducing DNA into the ovary post-pollination, where it is absorbed through pollen tube channels [82]. Although widely applicable, a narrow host range limits *Agrobacterium* [83, 84], the potential for random genomic insertions that could disrupt native genes [85, 86], and regulatory challenges due to its classification as a GMO technology in many areas, including regulation by the USDA in the United States [87] and restricted use in Europe [88].

Plant viruses present an alternative vector system that can encapsulate, replicate, and systemically transport foreign genes during infection [89]. Viral vectors are extensively utilized in functional genomics, primarily through virus-induced gene silencing (VIGS) [90]. *Agrobacterium*-mediated viral vectors have been employed to alter crop traits via foliar application [91]. Viral systems have been used to deliver sgRNAs into Cas9-expressing plants, such as tobacco, cotton (*Gossypium hirsutum*), *Arabidopsis*, tomato (*Solanum lycopersicum*), wheat (*Triticum aestivum*), and maize, to produce precisely edited and virus-free offspring in subsequent generations [92]. However, viral vectors face significant biosafety and technical challenges. Many of these are seed-borne, easily transmissible, or capable of causing developmental issues. Antiviral RNA-silencing responses destabilize vectors in most species, making tobacco, where antiviral silencing is weak, the preferred host [93]. Host specificity limits their applicability, as demonstrated by the tobacco rattle virus (TRV), which cannot infect rice (*Oryza sativa*) or other non-host species [94]. Effective systemic dissemination also requires meticulous handling and substantial technical expertise. Although viral vectors allow rapid and high-copy expression, they do not integrate into the nuclear genome and thus cannot produce heritable transgenic plants [95]. Ongoing challenges, including species dependency, variable transformation efficiency, and technical or biological constraints, continue to restrict both *Agrobacterium*- and virus-based methods, highlighting the need for more adaptable delivery platforms [96].

3 Nanocarrier-Mediated Delivery Systems in Plants

Nanoparticles (NPs) with sizes ranging from 1 to 100 nm exhibit unique thermal, optical, catalytic, magnetic, and mechanical properties owing to their high surface area-to-volume ratio and quantum effects [127]. These physicochemical traits, along with adjustable size, shape, surface charge, and concentration, render NPs highly efficient carriers for biomolecules, such as DNA, RNA, proteins, and CRISPR reagents, in plant genetic engineering. The behavior of NPs within plant tissues is influenced by both their inherent properties and external factors, such as pH and ionic composition, which collectively impact NP stability, mobility, and biological interactions [128]. NPs can enter through stomata, cuticular gaps, and root epidermal structures, allowing them to move via both apoplastic and symplastic pathways, which enables many to penetrate the cell wall and plasma membrane to access the cytoplasm [129].

NP-based approaches offer unique benefits over traditional transformation techniques. One of these is minimal cytotoxicity, straightforward procedures that eliminate the need for cell wall removal, applicability across various species, and the capability to transport a range of cargos, including nucleic acids, imaging agents, and regulatory molecules. Numerous studies have indicated that effective gene delivery is closely linked to the efficient internalization of NPs, with the most successful carriers having at least one dimension smaller than 20 nm, including quantum dots (QDs) [130], Au nanoclusters [115], carbon-based nanomaterials [104, 131], and nanowires [132]. Their nanoscale size also allows targeting within subcellular organelles, such as chloroplasts [104], mitochondria, and nuclei [32]. Collectively, these NCs improve molecular stability, lower species-specific transformation barriers, and facilitate the delivery of pDNA, siRNA, dsRNA, proteins, and RNP complexes.

A comparative overview of representative nanomaterial-mediated delivery systems, including carrier materials, genetic cargos, target tissues, plant species, and delivery strategies, is summarized in Table 2. Ongoing advances in nanomaterial design and functionalization are expected further to improve delivery efficiency, specificity, and biocompatibility, positioning nanocarrier-based platforms as promising tools for precise and environmentally sustainable plant genome engineering.

3.1 Classes of Nanocarriers

Numerous types of NPs have been investigated for delivering genetic material to plants. These include bio-inspired liposomes [133], carbon-based nanomaterials such as carbon nanotubes (CNTs) [32] and carbon dots (CDs) [34], conjugated polymeric NPs [134], and metal-based systems such as Au [35] and iron oxide [135]. Other categories of NCs, such as DNA nanostructures, peptide NPs, silicon-based materials, clay nanosheets, and various polymeric platforms, possess distinct physicochemical properties that affect cargo loading, cellular uptake, and whether the delivered material facilitates transient or stable transformation [136, 137].

3.1.1 Carbon-Based Nanomaterials

Carbon-based nanomaterials, such as CNTs, graphene derivatives, and CDs, have emerged as highly effective platforms for delivering biomolecules to plants. This is due to their remarkable tensile strength, thermal stability, electrical conductivity, biocompatibility, and nanoscale size, which allow them to penetrate plant cell walls [138]. The ability of carbon nanomaterials (CNs) to enter plant cells was first demonstrated in 2009 [139], but their use for transporting genetic material was only realized a decade later. Currently, CNTs and CDs are the primary carbon-based NCs used for plant transformation.

CNTs, which are essentially graphene sheets rolled into tubes, can be found as single-walled (SWCNTs) or multi-walled (MWCNTs) structures. Both types are capable of transporting nucleic acids, proteins, and other biomolecules through the plant cell walls [140, 141]. CNTs grafted with polyethyleneimine (PEI) have been shown to penetrate benth (*N. benthamiana*), arugula (*Eruca sativa*), wheat, and cotton, successfully delivering a 4.2-kb plasmid for temporary GFP expression [32]. However, delivering larger plasmids ($\geq$ 10 kb) is challenging and requires optimized surface chemistries and precise plasmid-to-carrier ratios [120]. The toxicity and biocompatibility of CNTs are significantly influenced by the polymer used for grafting [142], highlighting the need for a thoughtful structure–function design. Increasing the aspect ratio of CNTs can further improve their plasmid loading

Table 2 Overview of nanomaterial-based platforms for the delivery of nucleic acids and genome-editing components in plants across diverse species and target tissues

Materials/ nanoparticles	Cargo	Plant species	Target site	Delivery method	References
Casein nanoparticles	pDNA	*N.benthamiana*	Leaves	Syringe infiltration	[97]
CDs	pDNA	*N. benthamiana*	Leaves	Syringe infiltration	[98]
Silica nanopowder	dsDNA	*N. benthamiana*	Leaves	Syringe infiltration	[99]
Silica nanopowder	dsDNA	*N. benthamiana*	Roots	Passive diffusion/culture uptake	[99]
MSNs	dsDNA	*N. benthamiana*	Leaves and roots	Syringe infiltration	[100]
SWCNTs	pDNA	*N. tabacum*	Root cells	Co-culture	[101]
Graphene quantum dots (GQDs)	dsDNA	*T. aestivum*	Spikes	Spray method	[102]
Carbon quantum dots (CQDs)	dsDNA	*N. benthamiana*	Leaves	Syringe infiltration	[99]
CQDs	dsRNA	N. benthamiana	Leaves	Spray method	[99]
Lipid NPs	CRISPR-Cas9 plasmid	Sweet orange (*Citrus sinensis*)	Protoplasts	PEG- and lipofectamine-mediated delivery	[103]
Chitosan-based nanoparticles (ChNPs)	dsDNA	*N. benthamiana*	Leaves	Syringe infiltration	[99]
ChNPs	dsRNA	N. benthamiana	Leaves	Spray method	[99]

(continued)

Table 2 (continued)

Materials/ nanoparticles	Cargo	Plant species	Target site	Delivery method	References
SWCNTs	pDNA	*E. sativa*, watercress (*Nasturtium officinale*), *N. tabacum*, Spinach (*Spinacia oleracea*)	Leaves	Injection	[104]
DNA nanostructures	siRNA	*N. benthamiana, N. tabacum, E. sativa, N. officinale*	Leaf tissues	Infiltration with subsequent active or passive internalization	[83]
LDH nanoparticles	siRNA	*N. benthamiana*	Leaves	Syringe infiltration	[105]
LDH nanosheets	microRNA	*S. lycopersicum*	Leaves	Spray method	[106]
LDH nanosheets	dsRNA	*N. tabacum*	Leaves	Spray method	[107]
LDH nanoparticles	dsRNA	*S. lycopersicum*	Flower buds	Cellular uptake	[108]
LDH nanoparticles	dsRNA	*S. lycopersicum*	Flower buds	Syringe injection	[108]
Liposomes	Tobacco mosaic virus (TMV) RNA	*N. tabacum*	Protoplasts	Co-culture	[109]
MSNs	pDNA	*S. lycopersicum*	Leaves	Spray method	[110]
MSNs	pDNA	*S. lycopersicum*	Shoot	Syringe injection	[110]
LDH NPs	dsRNA	*S. lycopersicum*	Pollen	Co-incubation/co-culture	[111]
MSNs	siRNA	*N. benthamiana*	Leaf tissues	Syringe infiltration	[112]
Virus-like NCs	pDNA	*A. thaliana*	Protoplasts	Passive diffusion/culture uptake	[113]
AuNPs	DNA	Hemp (*Cannabis sativa*)	Leaf tissues	Infiltration	[114]
AuNPs	siRNA	*N. benthamiana*	Leaf tissues	Infiltration	[115]

(continued)

Table 2 (continued)

Materials/ nanoparticles	Cargo	Plant species	Target site	Delivery method	References
AuNPs	amiRNA (artificial miRNA)	*A. thaliana*	Leaf tissues	Infiltration	[116]
CDs	pDNA	*O. sativa*, *T. aestivum*, Mung bean (*Phaseolus radiatus*)	Leaf tissues	Infiltration	[117]
MSNs	pDNA	*N. tabacum*	suspension cell culture	Ultrasound-assisted delivery	[118]
CDs	dsRNA	*C. sativus*	Leaves	Spray method	[119]
CDs	siRNA	*N. benthamiana*	Leaves	Spray method	[34]
CDs	siRNA	*S. lycopersicum*	Leaves	Spray method	[34]
CNTs	pDNA	*N. benthamiana*	Leaves	Syringe infiltration	[120]
AuNPs	siRNA	*A. thaliana*	Leaves	Syringe infiltration	[121]
SWCNTs	pDNA	*N. benthamiana*, *E. sativa*, *T. aestivum*, *G. hirsutum*	Leaf tissues	Passive diffusion//cellular uptake	[32]
SWCNTs	pDNA	*A. thaliana*	Seedlings	Vacuum infiltration	[122]
Ag and Au microcarriers	pDNA	*N. tabacum*	Leaves	Biolistic bombardment	[123]
SWNTs	siRNA	*N. benthamiana*	Leaves	Syringe infiltration	[124]
Metal–organic framework NPs (MOF NPs)	pDNA	*N. benthamiana*	Leaves	Syringe infiltration	[125]
MOF NPs	pDNA	*A. thaliana*	Roots	Passive diffusion/culture uptake	[125]
Quantum dots (QDs)	pDNA	*A. thaliana*	Leaves	Spray method	[126]

capacity and delivery efficiency [143]. SWCNTs have proven particularly effective in transporting siRNAs for post-transcriptional gene silencing, facilitating efficient siRNA uptake and activity in tobacco leaves [124]. Additionally, CNTs can deliver small RNAs into intact plant cells through electrostatic adsorption, thereby protecting against nuclease degradation [32]. DNA-coated CNTs, when conjugated with engineered peptides, can further enhance their ability to traverse plant cell walls efficiently [101].

CDs serve as additional NC platforms. CDs functionalized with PEI have facilitated siRNA delivery across different species [34] and enabled temporary DNA expression in the leaves and roots of rice and wheat [117]. Their effectiveness is strongly dependent on size. CDs larger than 8.7 nm exhibit limited siRNA transport capacity, whereas MWCNTs with diameters of approximately 12.1 nm can still deliver plasmid DNA [32, 34], illustrating the combined influence of morphology and cargo type on delivery efficiency. Carbon nanoparticles (CNPs) loaded with plasmids have also been used to express GFP (Green Fluorescent Protein), YFP (Yellow Fluorescent Protein), and β-glucuronidase (GUS) in rice embryos and leaves [144].

Subcellular targeting represents a significant advancement in the field of CNs. Using the LEEP model, chitosan-coated CNTs were designed to penetrate chloroplasts, achieving a zeta potential of approximately 35 mV and a diameter of approximately 10 nm. These characteristics are expected to improve organelle uptake and have been shown to facilitate chloroplast genome transformation in various plant species, as evidenced by GFP fluorescence [104]. However, direct comparisons with biolistic delivery methods have not been conducted, leaving the relative efficiencies undetermined. Mitochondrial targeting has also been achieved: CNT-polymer hybrids modified with targeting peptides resulted in a 30-fold increase in delivery efficiency in *Arabidopsis* compared to peptides alone, likely due to improved cargo stability and protection [122]. The general applicability of these mitochondrial delivery techniques across different plant species requires further assessment.

The combination of adjustable surface chemistry, substantial loading capacity, and proven ability to deliver specifically to organelles makes carbon-based nanomaterials promising candidates for the future of plant genetic engineering.

3.1.2 Metallic and Metal Oxide Nanoparticles

Metallic NCs, especially AuNPs, have been extensively used for genetic delivery in both animal and plant systems. In plant biology, Au particles at the microscale have historically been used as biolistic carriers, facilitating gene transfer with minimal tissue damage [145]. The ongoing use of Au is attributed to its strong affinity for thiolated nucleic acids, which helps stabilize genetic material and allows for precise control over their loading [146]. Initial research using nanoscale Au particles in biolistic delivery demonstrated approximately 40% higher transformation efficiency and less tissue damage compared to traditional micro-Au particles. Additionally, nanogold-coated carbon bullets have been used to deliver binary vectors into rice and tobacco, illustrating the compatibility of nanoscale metallic carriers with biolistic transformation methods [147].

Although biolistic methods have been extensively explored, the use of AuNPs in non-biolistic plant applications remains relatively underdeveloped, with a primary focus on delivering siRNA. In *Arabidopsis* protoplasts, polydisperse spherical AuNPs have achieved approximately 80% gene knockdown [148], and the Au nanoclusters have been effective in delivering siRNA to *N. benthamiana* leaves [115]. Interestingly, subsequent research has challenged the assumption that cellular internalization is required for gene silencing, demonstrating that spherical AuNPs (5–15 nm) retained within the cell wall can still mediate effective siRNA-induced silencing [35]. Conversely, Au nanorods (~ 13 nm in diameter) can penetrate plant cells, likely because of their higher aspect ratio. Similar to their behavior in animal systems, rod-shaped NCs tend to align parallel to membrane surfaces, thereby enhancing their penetration and uptake compared to spherical particles [149, 150], which highlights the significance of NP shape in plant transport dynamics. Other metallic systems, such as Au-capped MSNs [151] and PEG-coated magnetic AuNPs used for transforming canola protoplasts [152], demonstrate the adaptability of AuNP surface chemistry.

Silver nanoparticles (AgNPs) exhibit potent antimicrobial properties that help minimize mechanical stress during gene transfer, thereby enhancing the integration efficiency [153, 154]. In specific scenarios, AgNPs surpass Au microparticles in biolistic transformation, achieving significantly higher gene delivery efficiency [123]. Their adjustable size, surface characteristics, and ability to bind biomolecules bolster their effectiveness as part of the larger group of metallic NCs.

Magnetic iron oxide nanoparticles (MNPs) have been investigated for their role in plant transformation, particularly through a method known as magnetofection. In this process, DNA-coated MNPs are directed into the pollen using an external magnetic field [135]. This approach circumvents the need for tissue culture by allowing genetically altered pollen to pass on its changes directly to the offspring. Although the initial study successfully delivered a GUS reporter, subsequent attempts failed to replicate these results, likely because of the limitations of the GUS assay [155]. A later study on maize using a GFP reporter confirmed successful transformation [156] and suggested that earlier inconsistencies were due to species-specific differences in pollen morphology, particularly the aperture structure, which significantly affects NP entry. The inherent surface charge of pollen [157] may also interact electrostatically with charged NPs, hindering their movement through the germination pores. Fe_3O_4 (magnetite) is the most commonly used MNP because of its biocompatibility, ability to penetrate plant cell walls, and role in enhancing root development without causing phytotoxicity [158]. Despite this, magnetofection remains a contentious topic, and comprehensive cross-species studies are necessary to understand NP-pollen interactions better and establish consistent transformation efficiencies.

Metallic and metal oxide NPs have adjustable surface chemistries, robust biomolecule attachment, and distinct optical or magnetic characteristics that facilitate various plant gene delivery applications. However, further investigation is needed to clarify their uptake mechanisms, species-specific effectiveness, and reproducibility to achieve consistent and predictable transformation results.

3.1.3 Silicon and Silica Nanoparticles

Initially developed for delivering biomedical substances in animal models [159, 160], silicon-based NCs were later adapted for plant use. MSNs were among the first silicon-based materials used in plant systems. They are known for their honeycomb-like porous structures, extensive surface areas, and low toxicity [161]. Early experiments employed MSNs measuring 100–200 nm to introduce plasmid DNA into *N. tabacum* through biolistic bombardment [151]. Successful delivery necessitated the use of AuNP capping, as MSNs without caps were ineffective in transferring plasmids. Building on this foundation, Au-functionalized MSNs were subsequently used to co-deliver nucleic acids and proteins [162]. The direct delivery of Cre recombinase protein facilitated DNA-free genome modification with efficiencies of up to 20% [163].

In addition to genetic modification, MSNs can breach the plant cell wall and access internal structures, such as the endodermal and vascular layers, which have been linked to enhanced growth in plants like wheat and lupin (*Lupinus* spp.) [163]. Their adaptability is highlighted by their substantial DNA-loading capacity and robust protection of biomolecules. Another category of silicon-based NCs is ceramic nanoparticles (CeNPs), which are created from inorganic substances such as phosphates, carbides, carbonates, and oxides through high-temperature methods. Compounds such as calcium triphosphate [$Ca_3(PO_4)_2$] and nanohydroxyapatite [$Ca_{10}(PO_4)_6(OH)_2$] are highly biocompatible and offer significant protection to enzymes, DNA, and RNA against degradation caused by pH and temperature changes, making them ideal for delivering biomolecules to plants [164, 165].

Despite these advancements, the use of silicon-based platforms for plant transformation remains limited. Most successful applications have been limited to young maize tissues and onion (*Allium cepa*) epidermis. Since 2015, progress in silicon-mediated delivery has been slow. The ongoing dependence on biolistic methods and the rise of more efficient NCs that do not rely on biolistics, especially those made from carbon and metals, which offer better delivery efficiency and compatibility with a broader range of species, have led to a decline in the use of silicon and silica nanoparticles (SiNPs) in modern plant genetic engineering [4].

3.1.4 Bio-inspired and Biomimetic Nanocarriers

Bio-inspired NCs, such as liposomes, synthetic vesicles, and organic NPs made from proteins, polysaccharides, lipids, or biomimetic analogs, are highly biocompatible, biodegradable, and exhibit low toxicity, making them ideal for genetic delivery in plants. Lipid-based nanoparticles (LNPs), which consist of phospholipids, cationic lipids, cholesterol, and PEGylated lipids, utilize their membrane-like structure to enable effective cellular uptake and controlled release of cargoes, such as plasmid DNA, mRNA, and other nucleic acids, within cells [166]. Polymer-based bio-inspired NCs also enable ligand conjugation, such as peptides or antibodies, to minimize off-target delivery and achieve targeted delivery to specific tissues or organelles [167]. Their primary non-covalent assembly enables biodegradation and removal from plant tissues, thereby reducing the risk of long-term accumulation.

In plant biotechnology, liposome-mediated transfection has been predominantly used for protoplasts or young tissues. Liposomes have facilitated the delivery of plasmid DNA encoding Cas9 [103] and the direct transfer of the Cas9 protein [106]. Employing Lipofectamine 3000, DNA-free Cas9 RNPs achieved an editing efficiency of 6%, surpassing the 3% efficiency of macro-Au biolistics [106]. A supplementary vesicle system adorned with cell-penetrating peptides (CPPs) also introduced Cas9 RNPs into *Arabidopsis* callus, albeit with efficiencies of less than 1% [168]. Beyond genetic material, liposomes have also been employed for nutrient transport in whole plants. Lecithin-based liposomes containing 16–18 carbon chains can penetrate the cuticle and translocate systemically to roots and distal leaves, enabling efficient iron delivery [169]. These results highlight the potential of liposome-based NCs for applications across the entire plant, although further refinement of lipid composition and enhanced performance in mature tissues are still needed.

Bio-inspired and biomimetic NCs are at the forefront of advances in nano-enabled plant genetic engineering. Their natural compatibility with biological systems, along with their customizable and flexible nature, makes them attractive alternatives to inorganic vectors, especially for applications that require safe, sustainable, and environmentally friendly delivery methods [170].

3.2 Mechanisms of Nanoparticle Uptake and Transport

The absorption and movement of NPs within plants are influenced by several factors, including the type of plant, NP concentration, NP size, surface charge of the NPs, and exposure duration [129]. NPs mainly enter through the leaves and roots, where they encounter unique anatomical and physiological barriers that affect their transport [137, 171]. The cuticle, a hydrophobic, waxy barrier, limits the leaf uptake of NPs larger than approximately 10 nm, except through stomatal openings. As stomatal density and size differ among species and change in response to environmental conditions, these characteristics significantly impact NP absorption [172]. For instance, a decrease in stomatal diameter in wheat resulted in a 33.2 and 8.3% reduction in ZnO NP accumulation in the chloroplasts and cytoplasm, respectively [173]. Similarly, root uptake is restricted by the epidermis and Casparian strip, allowing NPs to pass mainly through damaged areas, sites of lateral root emergence, and specialized passage cells.

To overcome the cuticle barrier, some studies have employed methods such as petiole feeding or direct injection of NPs into leaves or stems [171]. However, these techniques are labor-intensive and unsuitable for large-scale applications [137].

There is a growing preference for injection-free methods, with recent studies indicating that certain NPs can independently penetrate the cuticle [174]. Once NPs enter leaf tissues, they must navigate through the mesophyll to reach the vascular system, allowing for systemic distribution throughout the plant [175]. The primary obstacle to internalization within mesophyll cells is the cell wall, with the plasma membrane posing a lesser challenge [176]. The suggested mechanisms for overcoming these barriers include passive diffusion, endocytosis, movement through plasmodesmata, and localized disruption of the cell wall. Adjusting the osmotic composition of NP suspensions can alter the water potential of the cell wall, thereby loosening the microfibrillar network and facilitating NP entry [137].

Some NCs, such as DNA-coated Au microparticles, still require physical force, achieved through biolistic propulsion, to penetrate tissues and access intracellular targets [177, 178]. To achieve intracellular transport without assistance, a deeper understanding of NP-membrane interactions is necessary, as the roles of passive penetration and endocytosis remain debated. Interestingly, certain studies have observed gene silencing effects without any detectable NP internalization, indicating potential extracellular or cell wall-related mechanisms [35, 179].

The morphology and charge of NPs also affect their uptake by cells. Rod-shaped AuNPs are absorbed and internalized more effectively in *Arabidopsis* leaves than spherical particles of the same size [35]. Positively charged NPs tend to adhere more to plant tissues because of the electrostatic attraction to the negatively charged cell wall matrix [180]. Within mesophyll tissues, NPs travel through both apoplastic (cell walls, intercellular spaces, and xylem) and symplastic (cytoplasm connected by plasmodesmata) routes. The apoplastic pathway generally faces less hydraulic resistance, and the introduction of NPs might further decrease apoplastic conductivity, indicating that the apoplast could be a primary route for translocation. Nonetheless, the predominant pathway remains uncertain, and the mechanisms by which NPs interact with cell walls require further investigation [175].

An interdisciplinary, multiscale strategy is crucial for identifying NP transport pathways and directing NP movement to targeted cellular or subcellular locations [171]. Mathematical models suggest that the aspect ratio and stiffness of NPs significantly affect their capacity to penetrate the cell wall [174, 177, 181]. Ultimately, the transport of NPs is determined by the interaction between the physicochemical characteristics of the NPs and the anatomical and physiological features specific to each species [182]. Understanding these mechanisms is vital for enhancing NP-based delivery systems in plant genetic engineering and broader applications in agricultural nanotechnology [177].

3.3 Delivery Methods in Nanoparticle-Mediated Transformation

NP-mediated transformation involves a wide range of physical, chemical, and physiology-based methods for delivering nucleic acids and proteins into plant tissues to achieve temporary expression, gene silencing, and genome editing in plants. These techniques avoid the challenges of low efficiency, laborious regeneration, and genotype dependency associated with stable transformation [183, 184]. By incorporating nanotechnology, delivery methods have broadened to include syringe and vacuum infiltration, force-assisted biolistic transformation, magnetofection, ultrasound- and spray-based non-invasive systems, and passive uptake pathways.

Syringe infiltration is one of the most prevalent methods for delivering NPs because of its procedural simplicity, accuracy, and adaptability to various NCs. This technique involves using a needleless syringe to gently push the infiltration solution through the underside of the leaf [185–187], allowing for either a uniform distribution across the leaf or targeted delivery of multiple constructs [188–190]. NPs that have been pre-incubated for 30–60 min with plasmid DNA, siRNA, dsRNA, or CRISPR components [32] can be introduced without causing damage, facilitating quick transient expression, gene silencing, and genome editing [191, 192]. Notable high-performance instances include β-cyclodextrin-functionalized QDs, which achieved approximately 75% targeting of chloroplasts in *Arabidopsis* [126], chitosan-SWCNTs, which enabled chloroplast delivery in leafy greens [104], layered double hydroxide (LDH) NPs, which achieved over 70% mRNA knockdown in *N. tabacum* [105], and AuNP-PEI-siRNA complexes, which reached approximately 80% NPR1 silencing in *Arabidopsis* [148]. Despite these achievements, the range of species covered remains limited, highlighting the need for NC engineering tailored to specific crops and their anatomical features.

Vacuum infiltration offers an alternative method driven by negative pressure to improve access to intercellular space [193, 194]. There are two prevalent techniques for this. For smaller tissues, such as leaves, seedlings, and pollen, these are immersed in NP-cargo solutions inside a syringe barrel, where manipulating the plunger facilitates the entry of the solution [195–197]. In contrast, larger organs undergo vacuum infiltration in sealed chambers [198–200], where negative pressure expels air from the apoplastic spaces, and the return of atmospheric pressure aids in deep NP penetration. Successful applications include PEI-CDs delivering GUS plasmids into rice calli [117], CNT-plasmid complexes expressing reporters in isolated rice embryos [144], and nano-biomimetic plasmid carriers facilitating transient activation across various species and seed tissues with optimized vacuum cycles [201]. However, unlike syringe infiltration, vacuum infiltration has not yet been used for siRNA or dsRNA delivery, and hybrid pressure-based methods have not been investigated.

Biolistic transformation, also known as particle bombardment, propels DNA-coated microprojectiles into plant cells in a vacuum [106, 202]. Using Au or W particles coated with spermidine and $CaCl_2$-assisted adsorption [203], this method can transform leaves, calli, embryos, and seeds regardless of their genotype [204,

205]. Biolistics can deliver plasmid DNA, siRNA, dsRNA, proteins, and Cas9-RNP complexes [206, 207]. Innovations utilizing nanomaterials include MSNs loaded with plasmids and capped with Au for the dual delivery of DNA and small molecules [151], as well as Au-functionalized MSNs for concurrent DNA–protein transfer [163]. AgNPs (100 nm) have surpassed 0.6 μm Au microcarriers by enhancing efficiency, reducing rupture pressures, and cutting costs by approximately 37.5 times in tobacco plants [123]. Ongoing challenges involve optimizing DNA adsorption and controlled release, as spermidine is expensive and chemically intricate [208, 209].

Magnetofection utilizes MNPs composed of Fe_3O_4, ranging from 5 to 14 nm, to introduce nucleic acids into plant cells when subjected to a magnetic field [117, 210]. The process involves incubating nucleic acid-coated MNPs with tissues and then applying a magnetic field of 0.1–1 T for 10–60 min to facilitate internalization [156]. Pollen magnetofection is notably revolutionary: DNA attached to MNPs was successfully integrated into pollen, resulting in heritable transgene incorporation [135], and a system that does not require culture and is independent of genotype has been established for maize [156]. The reported success rates include 22% RFP-positive grains and 55% GUS-positive embryos in maize, along with 85.73% pollen germination and 88.32% transformation efficiency in Regal lily (*Lilium regale*) [116]. Beyond pollen, magnetofection has facilitated transient GUS expression in *N. tabacum* protoplasts [211] and GFP delivery in okra (*Abelmoschus esculentus*) using green-synthesized MNPs [212]. To expand their application, optimization of NP size, functionalization, magnetic strength, exposure duration, and tissue-specific conditions is necessary.

Ultrasound-based delivery is a non-invasive technique that utilizes acoustic cavitation to create temporary membrane pores, enabling nucleic acids to enter [213]. When ultrasound is combined with NCs, it enhances the ability to penetrate the cell wall and plasma membrane [214]. Successful transfection has been observed at frequencies of 20–50 kHz [215], although naked nucleic acids must be encapsulated in NPs to prevent degradation by ultrasound [27]. Synergistic systems using ultrasound and NCs, such as PEI-MSNs and hyperbranched PAMAM-G2 dendrimers, have been shown to aid DNA uptake in suspension cells while safeguarding the nucleic acids [118, 216]. Silica-coated barium titanate nanoparticles (BTNPs, 60 nm) have been found to boost GUS activity by 56.9% in dicots and 72.6% in monocots under optimal conditions [100]. Current applications primarily focus on plasmid DNA, highlighting the need to expand delivery methods to include siRNA, dsRNA, and CRISPR reagents.

Spray-based delivery offers an entirely non-destructive and field-suitable method in which DNA or RNA, combined with surfactants or adjuvants, infiltrates through stomata or cuticular pores [217–219]. Atomized NP-cargo complexes pass through the cell wall and deliver their contents to the cell [107]. This technique is particularly effective for RNAi, as NP-RNA complexes with neutralized or adjustable charges can bypass the electrostatic barriers that limit the uptake of naked RNA [35, 115]. Examples include CD-dsRNA complexes that inhibit *Phytophthora* pathogenicity genes [220], LDH-dsRNA formulations that shield *N. benthamiana* from Tomato Yellow Leaf Curl Virus (TYLCV) [221], polymer-functionalized graphene oxide achieving up to 97.2% siRNA silencing [222], and MSNs facilitating up to 98% gene knockdown with minimal phytotoxicity [112].

Passive NP delivery utilizes natural uptake mechanisms, including stomata, hydathodes, and root hairs, to internalize nanomaterials without requiring an external force [223]. Successful entry relies on aligning the NP size (10–40 nm) with the dimensions of the natural pores and ensuring compatibility in terms of physico-chemical properties. Metal oxide NPs, such as Zn (Zinc), Au, Ce (Cerium), and Ti (Titanium), and carbon-based nanomaterials, are particularly well-suited for passive uptake because of their small size and biocompatibility [224]. Although the cell wall limits particles larger than approximately 5–20 nm [18, 156], autonomous permeation has been observed: 5 nm CDs complexed with plasmid DNA penetrated *O. sativa* root tip cells within 12 h, resulting in strong mCherry expression. In protoplasts, where the cell wall is absent, CNT-plasmid complexes achieved a 76% transformation rate, with CNTs crossing plastid membranes within seconds [143]. Passive pathways also facilitate RNA-based silencing. For example, dip-inoculated dsRNA induced PDS knockdown in rice, while CPP6-complexed dsRNA enhanced silencing efficiency, with similar effects observed in Arabidopsis and cucumber (*Cucumis sativus*) [119, 224]. The combination of syringe and vacuum infiltration, biolistic transformation, magnetofection, ultrasound- and spray-based delivery, and passive nano-uptake methods constitutes a comprehensive toolkit for NP-facilitated gene transfer in plants [4]. Ongoing advancements in NC chemistry, delivery parameters tailored to specific species, and compatibility with CRISPR-based tools are essential for achieving reliable, high-efficiency, and widely applicable genetic engineering across a variety of crop species.

4 Nano-Enabled Genetic Engineering Applications

Nanotechnology is revolutionizing plant genetic engineering by enabling the precise and species-independent introduction of DNA, RNA, proteins, and RNPs into plant cells with cell walls, thereby overcoming the persistent challenges of traditional transformation methods. As depicted in Fig. 1, NCs can penetrate the plant cell wall, protect biomolecular cargos from degradation, and direct them to specific intracellular locations, facilitating transient expression, stable integration, RNAi, and genome editing. A diverse range of nanomaterials, including MSNs, metal, and magnetic nanostructures, CNTs, and peptide-based complexes, offers adjustable physicochemical properties that enable controlled cargo loading, minimal cytotoxicity, and programmable release kinetics [4]. These platforms have expanded the potential of plant biotechnology by enabling targeted delivery to nuclear, chloroplast, mitochondrial, and germline tissues in both model and challenging plant species. Together, these advancements establish nanotechnology as a fundamental tool for the next generation of plant transformation, functional genomics, and precision genome engineering.

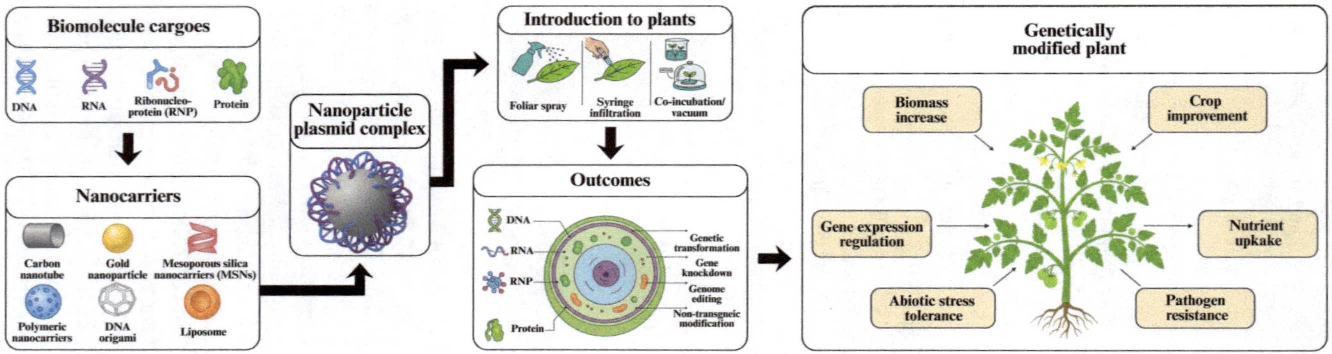

Fig. 1 **Schematic representation of nanocarrier-mediated biomolecule delivery in a plant**. The diagram illustrates the overall workflow of NP-assisted delivery of DNA, RNA, RNPs, and proteins into plant tissues. Various NCs, including CNTs, AuNPs, MSNs, polymeric NCs, DNA-origami structures, and liposomes, form stable NC-cargo complexes that are introduced into plants via foliar spray, syringe infiltration, or co-incubation/vacuum infiltration methods. Delivered biomolecules mediate genetic transformation, gene knockdown, genome editing, and non-transgenic modification within plant cells, ultimately producing genetically enhanced plants with improved biomass, crop performance, nutrient uptake, pathogen resistance, stress tolerance, and regulated gene expression

4.1 Nanotechnology-Assisted Plant Genetic Transformation

Over the last decade, the field of NC-facilitated gene delivery has witnessed significant progress, offering promising alternatives to traditional transformation methods by enhancing cargo protection, facilitating cell wall penetration, and enabling gene transfer across diverse species. Various nanomaterials, such as MSNs, metal-based, magnetic, and CNT NPs, have been developed for transporting DNA or RNA for either temporary expression or permanent genomic integration. Recent research has demonstrated the successful nano-assisted delivery of DNA to chloroplasts and mitochondria [225, 226], as well as the direct modification of the germline genome in pollen [135, 156], highlighting the potential of NCs to address the long-standing challenges in plant biotechnology. However, despite these advancements, many NC systems initially designed for mammalian cells require significant modifications for use in plant biology. For instance, cationic liposomes, which are commonly used in mammalian transfection, perform poorly in walled plant cells compared to protoplasts [227, 228], highlighting the difficulty posed by the rigid cell wall.

MSNs are pioneering nanoplatforms used for plant transformation. Their notable attributes, including high stability, extensive surface area, porous structure, and biocompatibility, which are endorsed by the GRAS (Generally Recognized as Safe) program, facilitate the effective encapsulation and controlled release of biomolecules [161]. Torney et al. illustrated transient GFP expression in tobacco protoplasts by employing plasmid-loaded MSNs with 3-nm pores that could simultaneously encapsulate DNA, proteins, and small chemical inducers, while safeguarding the contents from nucleases. For whole tissues, MSN-DNA complexes need to be capped with AuNPs to enhance the density for biolistic delivery, with subsequent uncapping prompting the synchronized release of DNA and inducers [151]. Expanding on this method, Au-functionalized MSNs were used to deliver plasmid DNA and proteins, such as FITC-BSA, TRITC-BSA, and GFP, into onion epidermal cells, where only the plasmid-loaded MSNs produced fluorescence, confirming the specificity of the delivery [162].

MSN-based delivery techniques were later modified to apply to entire plants, eliminating the need for mechanical forces. Chang et al. demonstrated, using a co-culture method, that DNA-coated MSNs can infiltrate the *Arabidopsis* cell wall, penetrate the cortical and endodermal layers, accumulate in vascular bundles and various organelles, and promote gene expression throughout the tissues [229]. Hajiahmadi et al. accomplished stable genomic integration in tomatoes by injecting functionalized plasmid-MSN complexes, likely through homologous recombination between pPZP122-derived sequences and chromosome 11 [110]. Subsequent studies reported a 10.71% transformation frequency and T_2 inheritance of a cryIAB transgene that conferred resistance to tomato pinworm (*Tuta absoluta*) [230]. The use of ultrasound to aid the uptake of DNA-loaded MSNs enhances cell wall penetration but also poses a risk of cellular damage [118]. Although MSNs provide high loading capacity and controlled-release capabilities, effective transformation in many tissues still necessitates biolistic force, which limits their scalability.

Metal NPs present alternative methods for delivering DNA owing to their adjustable size, straightforward synthesis, and optical characteristics, such as surface plasmon resonance (SPR) [231]. Au and Ag NPs are particularly appealing because of their compatibility with biological systems and affinity for nucleic acids [36]. Biolistic systems utilizing nanogold can carry up to 2.4 μg of plasmid DNA per carbon projectile, facilitating the delivery of the GUS gene into tobacco and rice cells [147]. PEI-functionalized AuNPs enabled swift gene expression in *Arabidopsis* and Madagascar periwinkle (*Catharanthus roseus*), with nuclear-localized fluorescence observable within hours [35]. In contrast, AgNPs demonstrated higher efficiency in delivering GUS to *Nicotiana* leaves compared to Au [123].

MNPs offer a unique method for gene delivery via magnetically directed manipulation. These NPs, when coated with PEI, can electrostatically attach to plasmid DNA and be transported across the cell membrane using an external magnetic field. Hao et al. introduced PEG-functionalized magnetic gold nanoparticles (mGNPs) that successfully delivered FITC-labeled DNA into canola (*Brassica napus*) protoplasts and whole cells [152]. A significant advancement was achieved with pollen magnetofection, which utilizes the natural 5–10 μm openings in pollen grains to introduce DNA-loaded MNPs, facilitating both temporary expression and stable integration without the need for tissue culture [135]. This technique has led to the production of transgenic seeds and fully transformed plants in cotton and maize [129, 156], and has been applied to okra [212] and cucumber [232]. However, despite its potential, the reproducibility of this method remains variable, especially in monocots, highlighting the need for a deeper understanding of DNA release, pollen physiology, and integration mechanisms [155].

CNPs, such as CDs and CNTs, have gained recognition as effective gene delivery systems because of their adjustable surface properties and ability to penetrate plant cell walls. CDs functionalized with PEI create positively charged complexes that tightly bind DNA and release their cargo under alkaline intracellular conditions, facilitating the transfection of tobacco leaves [98]. Both single-walled (SWCNTs) and multi-walled (MWCNTs) CNTs can cross the cell walls [139]. Surface modifications influence organelle targeting. CNTs coated with chitosan exhibit efficient chloroplast entry [233] and have been used for chloroplast-targeted DNA delivery across multiple plant species [104]. In contrast, CNTs coated with lipids or PVA have limited chloroplast penetration. CNTs modified with cellulase further improve entry by forming localized nanoholes [234].

Effective DNA loading is essential for transformation using CNTs. Electrostatic loading offers a greater DNA loading capacity than dialysis while ensuring effective delivery in both monocots and dicots [32]. CNTs bound with PEI have successfully delivered ssDNA-FITC into German chamomile (*Matricaria recutita*) cells, with ultrasound treatment enhancing cell wall penetration, and CNTs protecting DNA from enzymatic degradation [235]. Imidazolium-functionalized SWCNTs have achieved an approximately 35% transformation efficiency in oil palm (*Elaeis guineensis*) pollen, outperforming biolistic techniques [236]. CNT platforms facilitate organelle-targeted DNA delivery in line with the lipid exchange envelope penetration (LEEP) model [104, 177]. Nonetheless, concerns about CNT-induced cytotoxicity, such as membrane disruption and reactive oxygen species (ROS) production [225], persist, and biocompatibility is heavily reliant on the surface polymer [142]. However, uncertainties remain regarding the mechanisms of CNT uptake pathways, intracellular movement, and DNA release processes.

Peptide-based nanocomplexes offer a flexible approach for genetic transformation in plants. These peptides are carefully designed to merge DNA-binding cationic motifs with CPPs or sequences that target specific organelles [237, 238]. They create electrostatic nanostructures that shield DNA from nucleases [239, 240] and facilitate the delivery of genetic material to nuclei [227, 241], plastids [242], and mitochondria [243]. Despite their hydrodynamic size often surpassing the traditional cell wall exclusion limit, they are absorbed, likely through wall gaps or more minor complex subpopulations, and this process is energy-dependent and occurs via endocytosis. These complexes can facilitate homologous recombination-based insertion into chloroplasts and mitochondria [244, 245], although their efficiency remains significantly lower than that required for homoplasmy [246]. The efficiency of delivery has been enhanced by engineered peptides that loosen cell walls [247], trigger macropinocytosis-like uptake [248], disrupt endosomes [249], or degrade in response to stimuli [250]. Hybrid systems that combine peptides with NCs or chemical permeabilizers, such as zwitterionic ionic liquids (ZILs) or CNT-peptide hybrids, have further improved delivery to chloroplasts and mitochondria [225, 227].

SiNPs altered with PLL (SiNP-PLL) offer an alternative method for delivering DNA biolistically. According to Fu et al., SiNP-PLL protects plasmid DNA from degradation by DNase and facilitates its stable incorporation into the tobacco genome, resulting in transformation efficiencies of approximately 47%. Penetration was more successful in the callus than in the leaf tissue, which is attributed to variations in the cell wall structure [251].

Collectively, nanotechnology-based transformation strategies encompass a diverse suite of NCs, including silica-, metal-, magnetic-, carbon-, and peptide-based carriers, each offering distinct mechanisms for cellular entry, organelle targeting, and transgene expression. Despite substantial progress, challenges remain regarding scalability, cytotoxicity, reproducibility, controlled cargo release, and species-specific optimization. Overcoming these limitations will be essential for establishing reliable, efficient, and regeneration-independent nano-enabled transformation platforms suited for next-generation crop engineering.

4.2 Nanocarrier-Mediated Gene Silencing

NC systems have become a practical approach for introducing small RNAs into plant cells, facilitating efficient and temporary gene silencing without the need for DNA integration or permanent genetic changes. RNA-based strategies utilize RNAi, in which external small interfering RNAs (siRNAs) guide sequence-specific cleavage through the RNA-induced silencing complex (RISC) [252]. RNAi has been extensively utilized to investigate gene functions, enhance crop characteristics [253], and provide protection against pests and diseases [254, 255]. Traditional methods, such as hairpin RNA (hpRNA) and VIGS [256, 257], primarily depend on DNA vectors delivered via *Agrobacterium*, which have limitations including a lack of uniformity, species specificity, and biosafety issues. NC-facilitated RNA delivery offers a direct, transgene-free approach for delivering siRNA or double-stranded RNA (dsRNA) into cells, eliminating the need for genomic integration. While some NCs initially designed for DNA delivery, such as CNTs and CPPs, can also carry RNA, recent developments have increasingly focused on specialized RNAi nanoplatforms [96].

Initial experiments with NC-facilitated RNAi utilized cationic dendrimers and synthetic peptide complexes. Jiang et al. demonstrated that siRNA, when electrostatically bound to cationic dendrimers, could induce gene silencing and affect the development of meristems and roots in *Arabidopsis* seedlings [258]. Numata et al. also found that a cell-penetrating polycationic peptide successfully delivered dsRNA into *A. thaliana* leaves, effectively downregulating both foreign (YFP) and native (Chalcone synthase (CHS)) genes, with significantly greater efficiency than unmodified dsRNA [259]. These CPP-dsRNA complexes were later adapted into a spray-based system for scalable foliar application [219]. However, their relatively large size (> 100 nm) might restrict their ability to pass through the size-exclusion limit of the cell wall.

LDHs are among the most sophisticated RNAi NC systems used in plants for gene silencing. These lamellar nanoclays, which carry a positive charge, possess a high capacity for anion exchange and are inherently biocompatible, allowing them to bind dsRNA securely [96]. According to Mitter et al., LDH-dsRNA formulations can shield RNA from degradation by nucleases for more than a month after application to leaves. The gradual dissolution of LDHs under mildly acidic conditions on leaf surfaces facilitates the controlled release of dsRNA and its uptake by cells. It provides long-lasting antiviral protection for at least 20 days. The delivered dsRNA can move systemically to untreated parts of the plant, traveling through the xylem and spongy mesophyll and reaching the apical leaves. LDH-dsRNA systems have been shown to effectively confer virus resistance in *Arabidopsis*, cowpea (*Vigna unguiculat*a), and tobacco [107], and offer complete protection against the transmission of Bean Common Mosaic Potyvirus (BCMV) by aphids [260]. Beyond viral protection, MgFe-LDH nanosheets carrying dsRNA targeting whiteflies have been shown to induce significant gene silencing and high mortality rates [261]. Although simple spraying results in limited LDH penetration, infiltration studies indicate that particles as large as 50 nm can penetrate leaves and pollen [105, 111], suggesting that optimizing the particle size and charge could enhance intracellular uptake. However, the design of LDHs is still limited by an incomplete understanding of their uptake pathways, trafficking behaviors, and systemic mobility.

Au nanomaterials have emerged as effective platforms for RNA delivery. Au nanoclusters functionalized with PEI have been shown to successfully transport siRNA and trigger gene silencing [115]. Interestingly, the shape of AuNPs significantly influences their ability to enter plant cells. Au nanorods with a high aspect ratio (13 × 68 nm) are more adept at penetrating cells than spherical AuNPs, which often get stuck in the cell wall [35]. Surprisingly, 10 nm spherical AuNPs, although they remained outside the cells, demonstrated the highest silencing efficiency, suggesting that siRNA released from these extracellular carriers might enter cells through unknown uptake pathways that do not rely on NP internalization. Recently, the AuNP-mediated delivery of artificial microRNAs targeting ATG6 enabled rapid gene silencing in *Arabidopsis* and *Nicotiana*, thereby decreasing their resistance to *Pseudomonas syringae* [116]. While Au nanomaterials can be extensively modified in terms of size, shape, and surface chemistry, the exact connections between their physical characteristics and biological silencing capabilities are not yet fully understood.

Carbon-based nanomaterials, including CNTs, CDs, and graphene oxide nanoparticles (GONs), have emerged as highly adaptable RNA carriers. CNTs can penetrate plant cell walls and deliver RNA via electrostatic interactions or π-π stacking [262]. In a study by Demirer et al. [124], single-walled CNTs (SWNTs) were used to introduce siRNA into intact *N. benthamiana* leaves, resulting in approximately 95% GFP transcript knockdown within 6 h, marking one of the quickest and most effective RNAi responses observed in walled plant cells [124]. CDs, which are less than 20 nm in size, offer high biocompatibility and intrinsic fluorescence, making them safer options. CDs derived from PEI can deliver siRNA via foliar spraying, effectively silencing magnesium chelatase genes involved in chlorophyll production [34]. Additionally, functionalized graphene oxide NPs enhance the design of carbon-based NCs. PEI/PEG-modified GONs create near-spheroidal siRNA complexes with significant potential for plant delivery [222].

Collectively, these strategies utilizing nanotechnology position RNAi as a flexible, species-neutral, and transgene-free method for enhancing crop performance. Ongoing refinement of NC chemistry, particle dimensions, surface charge, and delivery methods is crucial for enhancing targeted gene silencing and fully leveraging the potential of RNA-guided regulation in contemporary plant biotechnology.

4.3 Nanotechnology-Facilitated Genome Editing

Genome editing facilitates the precise alteration of specific genomic sites using programmable nucleases that create targeted double-strand breaks (DSBs) [263]. Initially, plant transformation techniques depended on random DNA insertion. However, the development of engineered nuclease systems, such as meganucleases [264], zinc finger nucleases (ZFNs) [265], transcription activator-like effector nucleases (TALENs) [266], and CRISPR-Cas platforms [267, 268], has enabled sequence-specific cutting followed by repair through non-homologous end joining (NHEJ) or homologous recombination (HR). This advancement allows both gene disruption and precise sequence replacement [25, 269]. Among these technologies, CRISPR-Cas9 has gained the most widespread use owing to its flexibility and ease of retargeting. However, its practical application in plants is dependent on the coordinated delivery of Cas nucleases and guide RNA within cells, which remains a significant challenge in many species [270].

Nanotechnology has introduced innovative approaches to address these delivery obstacles. NCs can enhance cargo stability, boost intracellular uptake, and decrease the risk of unintended transgene integration compared to methods using plasmids or *Agrobacterium*. For instance, CNTs have facilitated the temporary delivery of CRISPR plasmids into plant tissues, promoting nuclease expression while reducing stable DNA integration [271]. More generally, the use of NPs for CRISPR delivery offers benefits such as high loading capacity, adjustable surface chemistry, low cytotoxicity, and compatibility with a wide range of species [272, 273]. Liposomal NCs exemplify these advantages: cationic and multilamellar liposomes can encapsulate CRISPR cargos in DNA, RNA, or RNP formats and deliver them efficiently due to their biocompatibility and membrane-fusion capabilities [274–276].

Polymeric NCs broaden the possibilities of plant genome editing. Chitosan nanoparticles (CS NPs) have facilitated the delivery of CRISPR constructs targeting the *coilin* gene in potato (*S. tuberosum*), resulting in edited plants with improved resistance to potato virus Y and better tolerance to salinity [277]. Similarly, CS NPs were employed to introduce two *Arabidopsis thionin* genes into *S. tuberosum*, producing lines resistant to Alternaria stem canker of tomato (*Alternaria alternata*) and Rhizoctonia aerial blight (*Rhizoctonia solani*) [278]. These studies highlight the capability of biopolymer NCs to aid intracellular uptake and stable genomic integration.

Recent advancements in synthetic nanomaterials have introduced novel methods for delivering CRISPR components. A nanofiber matrix composed of self-assembling peptides and polycaprolactone (SAP/PCL) facilitated substantial plasmid loading and gradual release of CRISPR-dCas9 through hydrophobic interactions [279]. PLGA NPs, approximately 160 nm in size, encapsulated CRISPR-Cas9 plasmids with an efficiency of approximately 80% and released them within a day, allowing for swift gene-editing actions [280]. Alginate NPs created through electrospraying, measuring 228 nm with a charge of − 4.42 mV, encapsulated dual CRISPR plasmids with 99% efficiency and achieved high expression levels in HepG2 cells, indicating promising applications in plant systems [281]. Further advancements include Arg-functionalized NPs that deliver Cas9En (E-tag) RNPs, boosting editing efficiency by up to 30% in cultured cells [282].

Simultaneously, systems utilizing NPs specifically designed for genome editing without DNA or transgenes are gaining traction. Recent research indicates that nanotubes and MSNs offer a high surface area and significant capacity for carrying programmable nucleases, guide RNAs, and Cas-gRNA RNPs [18]. RNPs can be attached to NP surfaces using enzymatically cleavable linkers, allowing controlled release and minimizing off-target effects. Initial examples include MSN-Au hybrids employed to deliver Cre recombinase into maize cells, achieving approximately 20% recombination [162], and the delivery of preassembled Cas9-gRNA complexes into protoplasts of *Arabidopsis*, tobacco, and rice, resulting in heritable edits after regeneration. DNA nanostructures with various shapes, such as nanostring-SWCNT hybrids, have demonstrated effective uptake and facilitated siRNA delivery into tobacco leaves without genome integration [279]. Functionalized single-walled CNTs have also been utilized to introduce plasmid DNA into intact cells, enabling transient Cas9 expression and stable editing without DNA integration [32]. In contrast, CDs carrying plasmid DNA have allowed for spray-based genome editing [283].

Delivering RNPs is still the most promising approach for achieving transgene-free editing. However, current techniques largely depend on particle bombardment or PEG-mediated protoplast transfection [284]. NC systems, such as CNTs, AuNPs, MSNs, nanoclews, nanocapsules, hydrogels, and metallic NPs, present a viable alternative as they facilitate direct intracellular transport and safeguard RNPs from degradation [285]. Initially, CPPs served as RNP carriers. However, their high positive charge can destabilize Cas9-gRNA complexes, leading to efforts to develop optimized variants. Vesicular NCs, such as polyion complex vesicles (PICsomes), create semipermeable protective environments for RNPs. CPP-modified PICsomes successfully delivered Cas9 RNPs into *Arabidopsis* calli, demonstrating DNA-free editing in a proof-of-concept study [33]. PEG-free, sarcosine-based PICsomes further enhance vesicle design and improve RNP encapsulation [286].

Despite these advances, broadly applicable design principles for NP-mediated genome editing in plants remain limited. NCs encounter a suite of plant-specific barriers, including the hydrophobic cuticle, restricted entry through stomata and hydathodes, aggregation driven by root exudates, cell-wall porosity constraints, and the Casparian strip. Addressing these challenges will require coordinated progress across plant physiology, colloid science, nanomaterial engineering, and cellular biophysics. As these fields continue to converge, nanotechnology-enabled genome editing is poised to evolve into a core platform for next-generation crop improvement, supporting precise, efficient, and species-independent genetic modification without reliance on stable DNA integration.

4.4 Nanotechnology-Supported Plant Regeneration

Germline transformation is the most effective nano-enabled approach to circumvent regeneration and reduce chimerism. However, nanomaterials can also directly enhance the regeneration process. The fate of plant cells during regeneration is mainly controlled by the balance between auxin and cytokinin: high auxin-to-cytokinin ratios encourage totipotency, whereas lower ratios favor differentiation and shoot development [287]. Regeneration protocols depend on the precise spatial and temporal regulation of these hormones in calli or embryogenic tissues, and NCs offer a potent method for achieving this control. Systems based on nanomaterials have been employed to deliver and monitor essential plant growth regulators [288], with chitosan NCs and bidirectional pH-responsive supramolecular nanovalves facilitating the efficient delivery of gibberellic acid [289]. Similarly, Au-capped MSNs carrying the auxin 2,4-dichlorophenoxyacetic acid into common flax (*Linum usitatissimum*) callus resulted in a three-fold increase in embryogenic cell formation compared to using exogenous auxin alone [290]. In addition to hormone delivery, NPs have been utilized in tissue culture to decrease microbial contamination, promote callus induction, enhance organogenesis and rooting, and boost secondary metabolite accumulation [291].

Nanotechnology has the potential to enable the delivery of key developmental regulators, such as WUSCHEL, PLETHORA, and BABY BOOM, which promote embryogenesis and significantly enhance regeneration in recalcitrant species, including sorghum (*Sorghum bicolor*) and indica rice [292–294], while growth-regulating factors and their associated cofactors have likewise improved regeneration efficiency across both monocots and dicots [72, 73]. Insights from mammalian systems, where nanoparticles have been used to deliver [295] or mimic [296] transcription factors, provide valuable conceptual guidance for developing analogous strategies in plants. However, in practice, plant developmental regulators are still predominantly introduced through transgene-based expression, which requires subsequent removal by breeding or genome-editing strategies. Direct, DNA-free delivery of transcription factors remains technically challenging due to their large size, charge, and structural instability. To date, only a single study has demonstrated the successful DNA-free delivery of a plant developmental regulator, WUSCHEL, using CPPs [297]. Nevertheless, many crucial plant developmental transcription factors have molecular weights under 70 kDa, indicating that they might be more suitable for NP-mediated intracellular transport than larger RNP complexes [4].

5 Nanoplatforms for Protein Delivery

Direct protein delivery is a potent, transgene-free method for influencing plant cellular activities and developing traits without the need for DNA-based transformation. This approach circumvents obstacles related to gene transfer, regeneration challenges, random integration of transgenes, and regulatory limitations, offering a swift and secure alternative for functional research and trait enhancement in plants. While protein delivery is well established in mammalian systems, its use in plants is relatively underdeveloped, with most current techniques depending on particle bombardment or PEG-mediated protoplast transfection [298]. These techniques are hindered by low efficiency, limited throughput, and restricted applicability across species because of the low regeneration potential of many crops.

Within NC systems, CPPs are one of the most sophisticated methods for introducing bioactive proteins into intact plant tissues. Initial research showed that covalent attachment of arginine-rich CPPs could effectively transport fluorescent proteins into living plant cells [299]. Subsequent studies found that even simple co-incubation, without the need for covalent attachment, allowed for the swift internalization of proteins such as GFP [300–302]. Building on these insights, the Numata group developed synthetic peptides that combine a CPP with a polycationic domain, facilitating the efficient delivery of a wide array of negatively charged proteins, such as yellow fluorescent protein (27 kDa), bovine serum albumin (66 kDa), and alcohol dehydrogenase (150 kDa), without requiring specialized equipment or protoplast isolation [136].

Despite these advancements, several barriers hinder the widespread application of these technologies. The delivery of CPPs is significantly affected by protein charge and size. Large peptide-protein complexes often become trapped in the cell wall, and the highly positive charge of many CPPs can disrupt electrostatic interactions with proteins, thereby reducing their compatibility with certain enzymes. Additionally, the mechanisms of intracellular trafficking are not yet fully understood, making rational design even more challenging [45].

A significant advancement was achieved with the creation of the macropinocytosis-inducing peptide, dTat-Sar-EED4. This synthetic peptide, which features multiple domains, includes a D-arginine-rich CPPs, a sarcosine linker, and an endosomolytic motif. This enables highly effective co-incubation-based delivery of large active proteins, such as the tetrameric ~290 kDa GUS, across various plant systems. Its delivery efficiency is approximately 20 times greater than that of conventional CPPs. Mechanistic investigations revealed that dTat-Sar-EED4 initiates a distinct uptake pathway similar to macropinocytosis, marked by actin polymerization, PI3K activity, membrane ruffling, and the formation of large vesicles (>200 nm), a phenomenon not previously observed in plants. Despite its effectiveness, high concentrations of the peptide are necessary to counteract strong electrostatic interactions with the negatively charged cell wall, leaving only a small portion available for membrane interactions [227].

Protein delivery nanoplatforms offer substantial promise for rapid, transgene-free modulation of plant traits. However, continued progress is required to improve protein compatibility, maintain functional stability during delivery, and enable scalable, species-independent deployment across diverse crop systems.

6 Nanocarrier-Based Organelle Targeting Strategies

Subcellular organelles serve as crucial sites for cargo delivery via NCs, playing a vital role in managing essential metabolic and genetic activities. The nucleus is a key focus for delivering DNA and RNP, while chloroplasts and mitochondria are equally significant due to the roles of their genomes in photosynthesis, respiration, and traits important for agriculture [202, 303]. These three compartments are surrounded by selective membranes that limit the passage of macromolecules, posing significant challenges to NC-mediated transport.

The nucleus is enclosed in a double-layered membrane that contains nuclear pore complexes (NPCs). These NPCs manage the exchange between the nucleus and cytoplasm through FG-nucleoporins (FG-Nups), which have phenylalanine-glycine repeats that create a dynamic barrier believed to block macromolecules larger than approximately 60 kDa [304]. Nonetheless, the size-exclusion limits of NPCs are more permissive than previously thought [305, 306]. Transport efficiency is strongly influenced by surface chemistry, as residues such as Arg, His, and Cys enhance interactions with FG-Nups and facilitate nuclear passage [305]. FG-Nup-dependent transport has also been shown for phenylalanine-derivatized heparin and maltooligosaccharide-modified liposomes in mammalian cells [307, 308], although similar findings have not been reported in plants. While nuclear localization signals (NLSs) allow for active import through importin complexes [309] and have been employed to direct DNA to the nucleus [241], the processes that control nuclear complex-mediated nuclear entry and whether DNA enters as free cargo or as part of an intact NC-DNA complex remain unclear.

Chloroplasts are surrounded by outer and inner envelope membranes that house the TOC/TIC import system, which facilitates the post-translational import of approximately 95% of chloroplast proteins through N-terminal chloroplast transit peptides (CTPs) [310]. CTPs have been employed to transport various NCs, such as QDs [126], CDs, CNTs [225], and peptide-based complexes [242, 245], into chloroplasts. Even peptide/DNA complexes larger than 100 nm can penetrate the stroma and enable transgene expression, despite exceeding the estimated 3–5 nm diameter of TOC/TIC pores. Live-cell imaging has revealed an alternative process in which outer membrane protrusions capture and engulf incoming complexes [244], allowing their entry into chloroplasts and other plastid types, such as chromoplasts and amyloplasts. The LEEP model also suggests passive diffusion [177], although CNT diffusion may cause chlorophyll loss and ROS accumulation due to membrane disruption [225]. Peptide-based nanocomplexes have shown potential for chloroplast delivery. However, their synthesis and assembly remain technically challenging. NPs, such as CDs and SWCNTs linked to chloroplast-targeting peptides, provide a biocompatible alternative and have enabled effective delivery of agrochemicals and DNA with minimal toxicity [225]. The physicochemical properties of NCs significantly affect chloroplast targeting, and functionalized SWCNTs combined with chitosan or lipid-PEG have been demonstrated to cross the cell wall and chloroplast envelopes under the LEEP model, facilitating transient YFP expression [104, 311]. Beyond terrestrial plants, COOH-SWCNTs coated with DNA-PEI complexes and covalently linked to fluorophores have facilitated the transport of DNA into algal chloroplasts [312].

Mitochondria pose unique challenges due to their dual membranes, compact size, and limited permeability, with pore diameters generally ranging from 5 to 20 nm, and the inner membrane restricting hydrophilic diffusion [313, 314]. Initial attempts at mitochondrial transformation involved electroporation, which facilitated the expression of the *coxII* gene in wheat embryos [315], and particle bombardment, which enabled transformation in *Chlamydomonas* and yeast (*Saccharomyces cerevisiae*) [316, 317], although these methods had low efficiencies [243]. Peptide-based carriers offer a promising alternative. Yoshizumi et al. successfully introduced foreign genes into the mitochondria of *Arabidopsis* cotyledons and meristematic tissues using the mitochondrial targeting peptide Cytcox-KH fused to DNA. This peptide-DNA complex (~ 356 nm, – 25 mV) was delivered through syringe-assisted vacuum infiltration, and PCR and reporter assays confirmed successful mitochondrial transformation [245]. Expanding on this approach, Law et al. linked Cytcox-KH peptides to polymer-coated SWCNTs, creating complexes that transported GFP-encoding plasmids driven by a mitochondrial-specific promoter [122]. Western blotting and fluorescence assays confirmed strong mitochondrial expression, underscoring the potential of CNT-peptide hybrids for targeted organelle engineering. Since these complexes are significantly larger than TOM/TIM pore diameters [318], their entry likely occurs via nontraditional mechanisms, highlighting the nascent and exploratory stages of mitochondrial NC delivery.

One of the main challenges in organelle engineering is the plant cell wall, which limits permeability based on the size, shape, and surface chemistry of the NCs. To address this issue, researchers have developed cellulase-coated CNTs that can locally degrade cellulose, forming nanoholes that facilitate the entry of NCs and enhance gene delivery [234]. Similarly, a zwitterionic liquid (ZIL) that disrupts hydrogen bonds to dissociate cellulose was used to improve the uptake of peptide-DNA complexes, significantly boosting chloroplast transformation in *Arabidopsis* [227]. These studies highlight that specific enzymatic or chemical alterations to the cell wall can significantly enhance the delivery of NCs to organelles.

Recent advances in NC-mediated subcellular targeting underscore the potential and intricacy of organelle engineering. Understanding the mechanisms by which NCs cross nuclear, chloroplast, and mitochondrial membranes, as well as addressing the ongoing challenge posed by the plant cell wall, is crucial for achieving accurate, effective, and species-independent manipulation of organelle genomes in plants.

7 Challenges and Limitations

The use of NCs for biomolecule delivery has broadened the scope of plant genetic engineering by facilitating temporary gene manipulation, stable transformation, genome editing, and non-genetic modification. However, despite these advancements, several core challenges limit the effectiveness, dependability, and scalability of NC-based transformations.

A significant challenge is the difficulty in achieving cargo delivery that is not species-dependent. Most NC platforms have primarily been evaluated in model species, such as *A. thaliana* and *N. benthamiana*, creating uncertainty about their effectiveness in a variety of crops. This issue is further complicated by species-specific regeneration needs, which limit the applicability of traditional transformation methods. The mechanistic understanding of NC transport through various plant barriers at multiple scales remains incomplete. NCs must traverse the surfaces of leaves and roots, the vascular system, cell walls, plasma membrane, and organelle membranes. However, the mechanisms that control their movement, uptake, and intracellular trafficking are not well understood. This limited understanding of these interactions hampers the rational design of NCs and obstructs their widespread application. Additionally, environmental and procedural stress during delivery can harm plant tissues or destabilize biomolecules, thereby decreasing the transformation efficiency [4].

Achieving precise and controlled cargo release at specific target sites remains a significant challenge. While many NC platforms facilitate DNA transport, most deliveries only result in temporary expression, underscoring the challenge of achieving stable genomic integration [19]. Cargo release is influenced by the binding affinity between the NP and biomolecules, the dynamics of intracellular trafficking, and the size of DNA fragments, all of which vary across different delivery systems. The maximum size of DNA fragments that NPs can transport remains unknown [16], and the processes governing nuclear import and chromosomal integration are not well understood. Factors such as NP concentration, surface functionalization, and physicochemical properties, including the charge, hydrophobicity, and density of functional groups, are crucial for delivery efficiency. Inadequate modification can weaken DNA binding, impede uptake, and cause aggregation [285].

Germline and organelle transformation present further obstacles. While direct germline transformation through pollen magnetofection shows potential [135, 155], achieving consistent, heritable changes necessitates a more comprehensive understanding of NC-pollen interactions. Organelle transformation poses even greater challenges; stable plastid transformation is limited to a few dicot species, and stable mitochondrial transformation in plants remains unaccomplished [202]. Although organelle-targeting peptides can guide DNA to specific compartments, the current efficiency levels are low, and there is a need for enhanced selection and regeneration methods [227].

Delivering large DNA constructs, such as multigene assemblies and genome-scale pathways, remains technically challenging due to difficulties in stabilizing and releasing high-molecular-weight cargos. RNA-based methods also have limitations. Although NC-mediated delivery of dsRNA and siRNA holds significant promise for crop protection [107, 261], important questions remain regarding RNA stability, membrane transport, long-distance movement, and the effective use against plant-pathogen interactions. Similarly, NC-based delivery of miRNAs, CRISPR-Cas reagents, and RNAi constructs is still in the early stages of development, with only a few systems showing consistent and scalable results.

Protein delivery is an emerging field in plant research. Current methods predominantly use cationic carriers, such as CPPs, which possess substantial positive charges that can destabilize the interactions between Cas and guide RNA, potentially affecting the integrity of the Cas-RNP. To address this challenge, it is essential to develop NCs that can transport proteins and protein-RNA complexes without compromising their structural or functional stability [96].

Ensuring the stability of NPs in plant-suitable environments remains a significant challenge. The colloidal properties of NPs are highly influenced by factors such as ionic strength, osmotic conditions, and the composition of common plant culture media, many of which can lead to the destabilization of NP suspensions and encourage aggregation of NPs. Additionally, the lack of thorough biosafety evaluations, which should include considerations of environmental fate, ecological risks, and effects on human and animal health, underscores the necessity for enhanced formulations and detailed safety assessments before these technologies can be widely implemented in agriculture [96].

In summary, although NP-mediated gene transfer offers a flexible approach for plant transformation, its effectiveness is still significantly lower than that of microbial and animal systems. To achieve reliable, high-efficiency, and widely applicable NC-enabled plant biotechnology, ongoing improvements in NC design, surface functionalization, delivery methods, and buffer formulations are crucial.

8 Future Directions

Nanomaterials offer significant potential as future delivery systems for plant genetic engineering. However, numerous mechanistic, technical, and biosafety issues must be resolved before their capabilities can be fully harnessed. A key focus is to ascertain whether DNA delivered by NCs can consistently integrate into plant genomes. Unlike *Agrobacterium*, which employs Vir proteins to aid T-DNA integration, nanovectors do not possess similar machinery, and foreign DNA is frequently degraded during replication. However, the use of MNPs for delivery into cotton and maize pollen has resulted in stable genomic integration, with GUS and GFP signals observed in offspring. While these findings suggest the possibility of heritable modification, the specific genomic locations and border sequences of nano-facilitated insertions remain unidentified. Uncovering the mechanisms behind integration will be a primary goal of future studies.

Despite the successful delivery of DNA by NCs, the current levels of expression are often inadequate for applications that require a high concentration of intracellular cargo, such as genome editing. For instance, plasmid-loaded SWCNTs resulted in only faint, localized GFP expression compared to delivery by *Agrobacterium*. CRISPR-Cas editing encounters further challenges because Cas proteins and nucleic acids are susceptible to degradation or inactivation during transport within cells. While strong constitutive or viral promoters can enhance editing efficiency, comprehensive optimization of NC chemistry, vector design, and cargo dosage is essential to achieve effective nano-enabled genome editing.

Another significant challenge is the reproducibility of the results. Variations between laboratories, resulting from differences in the synthesis of nanomaterials, the types of explants used, and physiological conditions, make comparisons challenging and hinder their wider acceptance. Therefore, standardized nanomaterial-based delivery systems are urgently needed. Cytotoxicity is another crucial factor to consider. Although many nanomaterials applied to plants exhibit low toxicity and can even enhance stress tolerance, specific surface ligands, such as high-molecular-weight PEI, are toxic in animal systems. For plant applications, safer options such as chitosan, arginine, and low-molecular-weight PEI are preferred.

A multidisciplinary approach is crucial for advancing plant engineering through the application of nanotechnology. Machine learning (ML) techniques, such as supervised, semi-supervised, and unsupervised algorithms, can be employed to analyze extensive genetic and phenotypic datasets, thereby enhancing nanomaterial design, optimizing delivery conditions, and selecting appropriate targets. More sophisticated models, including convolutional neural networks (CNNs), Deep belief networks (DBNs), multivariate Poisson deep neural network (MPDN), multilayer perceptrons, Probabilistic Neural Networks (PNNs), and Radial Basis Function (RBF) networks, can integrate diverse datasets and facilitate predictive modeling for engineering guided by nanotechnology. Concurrently, callus-based systems may serve as effective intermediaries for NP delivery, potentially overcoming species-specific transformation barriers that vary by species.

Long-term safety and environmental effects remain vital factors to consider. It is crucial to thoroughly analyze off-target genome edits that occur during transient expression to understand their biological implications. As agricultural nanomaterials see increased use, environmental monitoring systems similar to those for GM crops, such as Europe's decade-long post-market surveillance, will be necessary. Given the prevalence of both natural and engineered NPs in ecosystems and their potentially harmful impacts, ongoing and transparent risk evaluations are essential. Moreover, it is important to address issues related to the accumulation of NPs in soil, water, and edible plant tissues, as well as the possible disruption of beneficial microbiota.

Progressing in the field of NC chemistry remains a crucial priority. The creation of biocompatible nanomaterials with adjustable reactivity, enhanced stability, controlled cargo release, and minimal unintended effects is crucial for reliable and species-independent delivery. Future initiatives should also aim to facilitate direct germline modification, including methods based on pollen, to eliminate the need for tissue culture, as well as to develop NCs capable of targeting organelles for genome editing within plastids and mitochondria. Increasing the delivery capacity to handle large DNA constructs will accelerate the application of synthetic biology, such as the assembly of multigene pathways and genome-scale engineering. Together, these advancements will help overcome persistent challenges in plant biotechnology, enabling more precise, efficient, and sustainable crop improvement through genetic engineering powered by nanotechnology.

9 Conclusion

Efficient delivery of DNA, RNA, proteins, and CRISPR-Cas RNP complexes remains a central challenge in plant genetic engineering. Although conventional methods, such as Agrobacterium-mediated transformation, PEG-based transfection, and biolistics, have enabled significant advances, they are constrained by species dependence, random genome integration, tissue damage, and regeneration bottlenecks, limitations that are particularly problematic for DNA-free CRISPR applications, which are increasingly favored by regulatory frameworks.

Nanotechnology offers a transformative solution to these barriers. Diverse nanomaterials, including MSNs, CNTs, LDHs, AuNPs, and engineered peptide complexes, have demonstrated the ability to traverse plant cell walls, protect genetic cargos from degradation, and enable cross-species delivery to cellular and subcellular targets, including organelles. Experimental evidence across multiple crop systems highlights their potential for transient expression, RNA-guided silencing, organelle targeting, and transgene-free genome editing.

Despite this progress, critical challenges persist, including an incomplete understanding of nanomaterial transport mechanisms, the reliable delivery of fragile CRISPR RNPs, precise organelle targeting, and a comprehensive evaluation of biocompatibility and environmental impact. Addressing these issues will require close integration of plant physiology, nanomaterials science, and biophysics. Continued development of biocompatible, targetable, and stimuli-responsive nanocarriers positions nanotechnology as a foundational platform for next-generation, species-independent, and sustainable plant genome engineering.

Acknowledgements We thank the editors and reviewers for their critical comments on the manuscript.

Author Contributions X.F. and J.A.B. conceptualized the manuscript; E.M. prepared the initial draft; E.M., N.B., S.M., J.A.B., and X.F. reviewed and revised the manuscript. All authors have read and agreed to the published version of the manuscript.

Funding This research was supported by the Biological Breeding-National Science and Technology Major Project of China (2023ZD0403203) and the National Natural Science Foundation of China (U21A20215 and 32488102).

Institutional Review Board Statement Not Applicable.

Informed Consent Statement Not Applicable.

Data Availability Statement Not Applicable.

Conflicts of Interest The authors declare no conflict of interest.

References

1. Waring B, Neumann M, Prentice IC, Adams M, Smith P, Siegert M (2020) Forests and decarbonization—roles of natural and planted forests. Front For Glob Chang 3. https://doi.org/10.3389/ffgc.2020.00058
2. Dahl TW, Arens SKM (2020) The impacts of land plant evolution on earth's climate and oxygenation state—an interdisciplinary review. Chem Geol 547:119665. https://doi.org/10.1016/j.chemgeo.2020.119665
3. Dasgupta S, Robinson EJZ (2022) Attributing changes in food insecurity to a changing climate. Sci Rep 12:4709. https://doi.org/10.1038/s41598-022-08696-x
4. Squire HJ, Tomatz S, Voke E, González-Grandío E, Landry M (2023) The emerging role of nanotechnology in plant genetic engineering. Nat Rev Bioeng 1:314–328. https://doi.org/10.1038/s44222-023-00037-5

5. Mora C, McKenzie T, Gaw IM, Dean JM, von Hammerstein H, Knudson TA, Setter RO, Smith CZ, Webster KM, Patz JA et al (2022) Over half of known human pathogenic diseases can be aggravated by climate change. Nat Clim Chang 12:869–875. https://doi.org/10.1038/s41558-022-01426-1
6. Schillberg S, Finnern R (2021) Plant molecular farming for the production of valuable proteins—critical evaluation of achievements and future challenges. J Plant Physiol 258–259:153359. https://doi.org/10.1016/j.jplph.2020.153359
7. Pingali PL (2012) Green revolution: impacts, limits, and the path ahead. Proc Natl Acad Sci 109:12302–12308. https://doi.org/10.1073/pnas.0912953109
8. Breseghello F, Coelho ASG (2013) Traditional and modern plant breeding methods with examples in rice (*Oryza Sativa* L.). J Agric Food Chem 61:8277–8286. https://doi.org/10.1021/jf305531j
9. Bevan M (1984) Binary agrobacterium vectors for plant transformation. Nucleic Acids Res 12:8711–8721. https://doi.org/10.1093/nar/12.22.8711
10. Klein TM, Wolf ED, Wu R, Sanford JC (1987) High-velocity microprojectiles for delivering nucleic acids into living cells. Nature 327:70–73. https://doi.org/10.1038/327070a0
11. Buyel JF (2019) Plant molecular farming—integration and exploitation of side streams to achieve sustainable biomanufacturing. Front Plant Sci 9. https://doi.org/10.3389/fpls.2018.01893
12. Chung YH, Church D, Koellhoffer EC, Osota E, Shukla S, Rybicki EP, Pokorski JK, Steinmetz NF (2021) Integrating plant molecular farming and materials research for next-generation vaccines. Nat Rev Mater 7:372–388. https://doi.org/10.1038/s41578-021-00399-5
13. McNulty MJ, Xiong Y (Mary), Yates K, Karuppanan K, Hilzinger JM, Berliner AJ, Delzio J, Arkin AP, Lane NE, Nandi S et al (2021) Molecular pharming to support human life on the moon, mars, and beyond. Crit Rev Biotechnol 41:849–864. https://doi.org/10.1080/07388551.2021.1888070
14. Venugopal PD, Dively GP (2017) Climate change, transgenic corn adoption and field-evolved resistance in corn earworm. R Soc Open Sci 4:170210. https://doi.org/10.1098/rsos.170210
15. Altpeter F, Springer NM, Bartley LE, Blechl A, Brutnell TP, Citovsky V, Conrad L, Gelvin SB, Jackson D, Kausch AP et al (2016) Advancing crop transformation in the era of genome editing. Plant Cell. tpc.00196.2016. https://doi.org/10.1105/tpc.16.00196
16. Lv Z, Jiang R, Chen J, Chen W (2020) Nanoparticle-mediated gene transformation strategies for plant genetic engineering. Plant J 104:880–891. https://doi.org/10.1111/tpj.14973
17. Khan I, Saeed K, Khan I (2019) Nanoparticles: properties, applications and toxicities. Arab J Chem 12:908–931. https://doi.org/10.1016/j.arabjc.2017.05.011
18. Cunningham FJ, Goh NS, Demirer GS, Matos JL, Landry MP (2018) Nanoparticle-mediated delivery towards advancing plant genetic engineering. Trends Biotechnol 36:882–897. https://doi.org/10.1016/j.tibtech.2018.03.009
19. Wu K, Xu C, Li T, Ma H, Gong J, Li X, Sun X, Hu X (2023) Application of nanotechnology in plant genetic engineering. Int J Mol Sci 24:14836. https://doi.org/10.3390/ijms241914836
20. Woo JW, Kim J, Kwon SII, Corvalán C, Cho SW, Kim H, Kim S-G, Kim S-T, Choe S, Kim J-S (2015) DNA-free genome editing in plants with preassembled CRISPR-Cas9 ribonucleoproteins. Nat Biotechnol 33:1162–1164. https://doi.org/10.1038/nbt.3389
21. Gao C (2021) Genome engineering for crop improvement and future agriculture. Cell 184:1621–1635. https://doi.org/10.1016/j.cell.2021.01.005
22. Duensing N, Sprink T, Parrott WA, Fedorova M, Lema MA, Wolt JD, Bartsch D (2018) Novel features and considerations for ERA and regulation of crops produced by genome editing. Front Bioeng Biotechnol 6. https://doi.org/10.3389/fbioe.2018.00079
23. Sharma A, Toepfer CN, Ward T, Wasson L, Agarwal R, Conner DA, Hu JH, Seidman CE (2018) CRISPR/Cas9-mediated fluorescent tagging of endogenous proteins in human pluripotent stem cells. Curr Protoc Hum Genet 96. https://doi.org/10.1002/cphg.52
24. Lu Y, Tian Y, Shen R, Yao Q, Wang M, Chen M, Dong J, Zhang T, Li F, Lei M et al (2020) Targeted, efficient sequence insertion and replacement in rice. Nat Biotechnol 38:1402–1407. https://doi.org/10.1038/s41587-020-0581-5

25. Nandy D, Maity A, Mitra AK (2020) Target-specific gene delivery in plant systems and their expression: insights into recent developments. J Biosci 45:30. https://doi.org/10.1007/s12038-020-0008-y
26. Rajam M (2020) V RNA silencing technology: a boon for crop improvement. J Biosci 45:118. https://doi.org/10.1007/s12038-020-00082-x
27. Sanzari I, Leone A, Ambrosone A (2019) Nanotechnology in plant science: to make a long story short. Front Bioeng Biotechnol 7. https://doi.org/10.3389/fbioe.2019.00120
28. Buschmann MD, Carrasco MJ, Alishetty S, Paige M, Alameh MG, Weissman D (2021) Nanomaterial delivery systems for MRNA vaccines. Vaccines 9:65. https://doi.org/10.3390/vaccines9010065
29. Li H, Merkl P, Sommertune J, Thersleff T, Sotiriou GA (2022) SERS hotspot engineering by aerosol self-assembly of plasmonic Ag nanoaggregates with tunable interparticle distance. Adv Sci 9. https://doi.org/10.1002/advs.202201133
30. El-Shetehy M, Moradi A, Maceroni M, Reinhardt D, Petri-Fink A, Rothen-Rutishauser B, Mauch F, Schwab F (2021) Silica nanoparticles enhance disease resistance in arabidopsis plants. Nat Nanotechnol 16:344–353. https://doi.org/10.1038/s41565-020-00812-0
31. Zarattini M, Corso M, Kadowaki MA, Monclaro A, Magri S, Milanese I, Jolivet S, de Godoy MO, Hermans C, Fagard M et al (2021) LPMO-oxidized cellulose oligosaccharides evoke immunity in arabidopsis conferring resistance towards necrotrophic fungus B. Cinerea Commun Biol 4:727. https://doi.org/10.1038/s42003-021-02226-7
32. Demirer GS, Zhang H, Matos JL, Goh NS, Cunningham FJ, Sung Y, Chang R, Aditham AJ, Chio L, Cho M-J et al (2019) High aspect ratio nanomaterials enable delivery of functional genetic material without DNA integration in mature plants. Nat Nanotechnol 14:456–464. https://doi.org/10.1038/s41565-019-0382-5
33. Odahara M, Watanabe K, Kawasaki R, Tsuchiya K, Tateishi A, Motoda Y, Kigawa T, Kodama Y, Numata K (2021) Nanoscale polyion complex vesicles for delivery of cargo proteins and Cas9 ribonucleoprotein complexes to plant cells. ACS Appl Nano Mater 4:5630–5635. https://doi.org/10.1021/acsanm.1c00695
34. Schwartz SH, Hendrix B, Hoffer P, Sanders RA, Zheng W (2020) Carbon dots for efficient small interfering RNA delivery and gene silencing in plants. Plant Physiol 184:647–657. https://doi.org/10.1104/pp.20.00733
35. Zhang H, Goh NS, Wang JW, Pinals RL, González-Grandío E, Demirer GS, Butrus S, Fakra SC, Del Rio Flores A, Zhai R et al (2022) Nanoparticle cellular internalization is not required for RNA delivery to mature plant leaves. Nat Nanotechnol 17:197–205. https://doi.org/10.1038/s41565-021-01018-8
36. Jat SK, Bhattacharya J, Sharma MK (2020) Nanomaterial based gene delivery: a promising method for plant genome engineering. J Mater Chem B 8:4165–4175. https://doi.org/10.1039/D0TB00217H
37. Voke E, Pinals RL, Goh NS, Landry MP (2021) In planta nanosensors: understanding biocorona formation for functional design. ACS Sens 6:2802–2814. https://doi.org/10.1021/acssensors.1c01159
38. Nisha Raj S, Anooj ES, Rajendran K, Vallinayagam S (2021) A comprehensive review on regulatory invention of nano pesticides in agricultural nano formulation and food system. J Mol Struct 1239:130517. https://doi.org/10.1016/j.molstruc.2021.130517
39. Zulfiqar F, Navarro M, Ashraf M, Akram NA, Munné-Bosch S (2019) Nanofertilizer use for sustainable agriculture: advantages and limitations. Plant Sci 289:110270. https://doi.org/10.1016/j.plantsci.2019.110270
40. Yin H, Kanasty RL, Eltoukhy AA, Vegas AJ, Dorkin JR, Anderson DG (2014) Non-viral vectors for gene-based therapy. Nat Rev Genet 15:541–555. https://doi.org/10.1038/nrg3763
41. Hu Q, Kou E, Liao X, Qiu R, Tang Q, Zhang H, Zheng Y, Yang J, Zheng B (2025) Nanoparticle delivery of antisense MiR162 inhibits invasive habitat adaption of alternanthera philoxeroides. Adv Sci 12. https://doi.org/10.1002/advs.202416747
42. Steinwand MA, Ronald PC (2020) Crop biotechnology and the future of food. Nat Food 1:273–283. https://doi.org/10.1038/s43016-020-0072-3

43. Jähne A, Becker D, Lörz H (1995) Genetic engineering of cereal crop plants: a review. Euphytica 85:35–44. https://doi.org/10.1007/BF00023928
44. Behzadi S, Serpooshan V, Tao W, Hamaly MA, Alkawareek MY, Dreaden EC, Brown D, Alkilany AM, Farokhzad OC, Mahmoudi M (2017) Cellular uptake of nanoparticles: journey inside the cell. Chem Soc Rev 46:4218–4244. https://doi.org/10.1039/C6CS00636A
45. Liu BR, Chen C-W, Huang Y-W, Lee H-J (2023) Cell-penetrating peptides for use in development of transgenic plants. Molecules 28:3367. https://doi.org/10.3390/molecules28083367
46. Shivashakarappa K, Marriboina S, Dumenyo K, Taheri A, Yadegari Z (2025) Nanoparticle-mediated gene delivery techniques in plant systems. Front Nanotechnol 7. https://doi.org/10.3389/fnano.2025.1516180
47. Liang Z, Chen K, Li T, Zhang Y, Wang Y, Zhao Q, Liu J, Zhang H, Liu C, Ran Y et al (2017) Efficient DNA-free genome editing of bread wheat using CRISPR/Cas9 ribonucleoprotein complexes. Nat Commun 8:14261. https://doi.org/10.1038/ncomms14261
48. Elosua-Bayes M, Nieto P, Mereu E, Gut I, Heyn H (2021) SPOTlight: seeded NMF regression to deconvolute spatial transcriptomics spots with single-cell transcriptomes. Nucleic Acids Res 49:e50–e50. https://doi.org/10.1093/nar/gkab043
49. Zhang Y, Liang Z, Zong Y, Wang Y, Liu J, Chen K, Qiu J-L, Gao C (2016) Efficient and transgene-free genome editing in wheat through transient expression of CRISPR/Cas9 DNA or RNA. Nat Commun 7:12617. https://doi.org/10.1038/ncomms12617
50. Baltes NJ, Gil-Humanes J, Voytas DF (2017) Genome engineering and agriculture: opportunities and challenges. pp 1–26
51. Banakar R, Eggenberger AL, Lee K, Wright DA, Murugan K, Zarecor S, Lawrence-Dill CJ, Sashital DG, Wang K (2019) High-frequency random DNA insertions upon co-delivery of CRISPR-Cas9 ribonucleoprotein and selectable marker plasmid in rice. Sci Rep 9:19902. https://doi.org/10.1038/s41598-019-55681-y
52. Liu J, Nannas NJ, Fu F, Shi J, Aspinwall B, Parrott WA, Dawe RK (2019) Genome-scale sequence disruption following biolistic transformation in rice and maize. Plant Cell 31:368–383. https://doi.org/10.1105/tpc.18.00613
53. Nomura K, Komamine A (1986) Embryogenesis from microinjected single cells in a carrot cell suspension culture. Plant Sci 44:53–58. https://doi.org/10.1016/0168-9452(86)90168-8
54. Toyoda H, Matsuda Y, Utsumi R, Ouchi S (1988) Intranuclear microinjection for transformation of tomato callus cells. Plant Cell Rep 7:293–296. https://doi.org/10.1007/BF00269921
55. Fromm ME, Taylor LP, Walbot V (1986) Stable transformation of maize after gene transfer by electroporation. Nature 319:791–793. https://doi.org/10.1038/319791a0
56. Fromm M, Taylor LP, Walbot V (1985) Expression of genes transferred into monocot and dicot plant cells by electroporation. Proc Natl Acad Sci 82:5824–5828. https://doi.org/10.1073/pnas.82.17.5824
57. Weaver JC (1993) Electroporation: a general phenomenon for manipulating cells and tissues. J Cell Biochem 51:426–435. https://doi.org/10.1002/jcb.2400510407
58. Furuhata Y, Sakai A, Murakami T, Morikawa M, Nakamura C, Yoshizumi T, Fujikura U, Nishida K, Kato Y (2019) A method using electroporation for the protein delivery of cre recombinase into cultured arabidopsis cells with an intact cell wall. Sci Rep 9:2163. https://doi.org/10.1038/s41598-018-38119-9
59. Rakoczy-Trojanowska M (2002) Alternative methods of plant transformation—a short review. Cell Mol Biol Lett 7:849–858
60. Wu Q, Miao W, Zhang Y, Gao H, Hui D (2020) Mechanical properties of nanomaterials: a review. Nanotechnol Rev 9:259–273. https://doi.org/10.1515/ntrev-2020-0021
61. Lin C, Hsu C, Yang L, Lee L, Fu J, Cheng Q, Wu F, Hsiao HC-W, Zhang Y, Zhang R et al (2018) Application of protoplast technology to CRISPR/Cas9 mutagenesis: from single-cell mutation detection to mutant plant regeneration. Plant Biotechnol J 16:1295–1310. https://doi.org/10.1111/pbi.12870

62. Maćkowska K, Jarosz A, Grzebelus E (2014) Plant regeneration from leaf-derived protoplasts within the daucus genus: effect of different conditions in alginate embedding and phytosulfokine application. Plant Cell Tissue Organ Cult 117:241–252. https://doi.org/10.1007/s11240-014-0436-1
63. Altpeter F, Baisakh N, Beachy R, Bock R, Capell T, Christou P, Daniell H, Datta K, Datta S, Dix PJ et al (2005) Particle bombardment and the genetic enhancement of crops: myths and realities. Mol Breed 15:305–327. https://doi.org/10.1007/s11032-004-8001-y
64. Ozyigit II (2020) Gene transfer to plants by electroporation: methods and applications. Mol Biol Rep 47:3195–3210. https://doi.org/10.1007/s11033-020-05343-4
65. Abdallah NA, Prakash CS, McHughen AG (2015) Genome editing for crop improvement: challenges and opportunities. GM Crops Food 6:183–205. https://doi.org/10.1080/21645698.2015.1129937
66. Tiwari M, Mishra AK, Chakrabarty D (2022) Agrobacterium-mediated gene transfer: recent advancements and layered immunity in plants. Planta 256:37. https://doi.org/10.1007/s00425-022-03951-x
67. Gordon JE, Christie PJ (2014) The agrobacterium Ti plasmids. Microbiol Spectr 2. https://doi.org/10.1128/microbiolspec.PLAS-0010-2013
68. Gelvin SB (2017) Integration of agrobacterium T-DNA into the plant genome. Annu Rev Genet 51:195–217. https://doi.org/10.1146/annurev-genet-120215-035320
69. Wu C-F, Smith DA, Lai E-M, Chang JH (2018) The agrobacterium type VI secretion system: a contractile nanomachine for interbacterial competition, pp 215–231
70. Das D, Reddy M, Upadhyaya K, Sopory S (2022) An efficient leaf-disc culture method for the regeneration via somatic embryogenesis and transformation of grape (Vitis Vinifera L.). Plant Cell Rep 20:999–1005. https://doi.org/10.1007/s00299-002-0441-4
71. Aesaert S, Impens L, Coussens G, Van Lerberge E, Vanderhaeghen R, Desmet L, Vanhevel Y, Bossuyt S, Wambua AN, Van Lijsebettens M et al (2022) Optimized transformation and gene editing of the B104 public maize inbred by improved tissue culture and use of morphogenic regulators. Front Plant Sci 13. https://doi.org/10.3389/fpls.2022.883847
72. Debernardi JM, Tricoli DM, Ercoli MF, Hayta S, Ronald P, Palatnik JF, Dubcovsky J (2020) A GRF–GIF chimeric protein improves the regeneration efficiency of transgenic plants. Nat Biotechnol 38:1274–1279. https://doi.org/10.1038/s41587-020-0703-0
73. Kong J, Martin-Ortigosa S, Finer J, Orchard N, Gunadi A, Batts LA, Thakare D, Rush B, Schmitz O, Stuiver M et al (2020) Overexpression of the transcription factor growth-regulating Factor5 improves transformation of dicot and monocot species. Front Plant Sci 11:572319. https://doi.org/10.3389/fpls.2020.572319
74. Nonaka S, Ezura, Plantâ H (2014) Agrobacterium interaction mediated by ethylene and super-agrobacterium conferring efficient gene transfer. Front Plant Sci 5. https://doi.org/10.3389/fpls.2014.00681
75. Nonaka S, Someya T, Zhou S, Takayama M, Nakamura K, Ezura H (2017) An agrobacterium tumefaciens strain with gamma-aminobutyric acid transaminase activity shows an enhanced genetic transformation ability in plants. Sci Rep 7:42649. https://doi.org/10.1038/srep42649
76. Baker CJ, Mock NM, Whitaker BD, Roberts DP, Rice CP, Deahl KL, Aver'yanov AA (2005) Involvement of acetosyringone in plant–pathogen recognition. Biochem Biophys Res Commun 328:130–136. https://doi.org/10.1016/j.bbrc.2004.12.153
77. Gu X, Liu L, Zhang H (2021) Transgene-free genome editing in plants. Front Genome Ed 3:805317. https://doi.org/10.3389/fgeed.2021.805317
78. Park J, Choi S, Park S, Yoon J, Park AY, Choe S (2019) DNA-free genome editing via ribonucleoprotein (RNP) delivery of CRISPR/Cas in Lettuce, pp 337–354
79. Kiryushkin AS, Ilina EL, Guseva ED, Pawlowski K, Demchenko KN (2021) Hairy CRISPR: genome editing in plants using hairy root transformation. Plants 11:51. https://doi.org/10.3390/plants11010051
80. Moniruzzaman M, Zhong Y, Huang Z, Yan H, Yuanda L, Jiang B, Zhong G (2021) Citrus cell suspension culture establishment, maintenance, efficient transformation and regeneration to complete transgenic plant. Plants 10:664. https://doi.org/10.3390/plants10040664

81. Wen S-S, Ge X-L, Wang R, Yang H-F, Bai Y-E, Guo Y-H, Zhang J, Lu M-Z, Zhao S-T, Wang L-Q (2022) An efficient agrobacterium-mediated transformation method for hybrid poplar 84K (Populus Alba × P. Glandulosa) Using Calli as Explants. Int J Mol Sci 23:2216. https://doi.org/10.3390/ijms23042216
82. Zhou G, Weng J, Zeng Y, Huang J, Qian S, Liu G (1983) Introduction of exogenous DNA into cotton embryos, pp 433–481
83. Zhang H, Demirer GS, Zhang H, Ye T, Goh NS, Aditham AJ, Cunningham FJ, Fan C, Landry MP (2019) DNA nanostructures coordinate gene silencing in mature plants. Proc Natl Acad Sci 116:7543–7548. https://doi.org/10.1073/pnas.1818290116
84. Zhang H, Zhang H, Demirer GS, González-Grandío E, Fan C, Landry MP (2020) Engineering DNA nanostructures for SiRNA delivery in plants. Nat Protoc 15:3064–3087. https://doi.org/10.1038/s41596-020-0370-0
85. Landry MP, Mitter N (2019) How nanocarriers delivering cargos in plants can change the GMO landscape. Nat Nanotechnol 14:512–514. https://doi.org/10.1038/s41565-019-0463-5
86. Yan Y, Zhu X, Yu Y, Li C, Zhang Z, Wang F (2022) Nanotechnology strategies for plant genetic engineering. Adv Mater 34. https://doi.org/10.1002/adma.202106945
87. Kuzma J, Kokotovich A (2011) Renegotiating GM crop regulation. EMBO Rep 12:883–888. https://doi.org/10.1038/embor.2011.160
88. Ichim MC (2021) The more favorable attitude of the citizens toward GMOs supports a new regulatory framework in the European Union. GM Crops Food 12:18–24. https://doi.org/10.1080/21645698.2020.1795525
89. Ji X, Zhang H, Zhang Y, Wang Y, Gao C (2015) Establishing a CRISPR–Cas-like immune system conferring DNA virus resistance in plants. Nat Plants 1:15144. https://doi.org/10.1038/nplants.2015.144
90. Dinesh-Kumar SP, Anandalakshmi R, Marathe R, Schiff M, Liu Y (2003) Virus-induced gene silencing. In: Plant functional genomics. Humana Press, New Jersey, pp 287–294
91. Torti S, Schlesier R, Thümmler A, Bartels D, Römer P, Koch B, Werner S, Panwar V, Kanyuka K, von Wirén N et al (2021) Transient reprogramming of crop plants for agronomic performance. Nat Plants 7:159–171. https://doi.org/10.1038/s41477-021-00851-y
92. Ma X, Zhang X, Liu H, Li Z (2020) Highly efficient DNA-free plant genome editing using virally delivered CRISPR–Cas9. Nat Plants 6:773–779. https://doi.org/10.1038/s41477-020-0704-5
93. Shi G, Hao M, Tian B, Cao G, Wei F, Xie Z (2021) A methodological advance of tobacco rattle virus-induced gene silencing for functional genomics in plants. Front Plant Sci 12. https://doi.org/10.3389/fpls.2021.671091
94. Ratcliff F, Martin-Hernandez AM, Baulcombe DC (2001) Technical advance: tobacco rattle virus as a vector for analysis of gene function by silencing. Plant J 25:237–245. https://doi.org/10.1046/j.0960-7412.2000.00942.x
95. Mahmood MA, Naqvi RZ, Rahman SU, Amin I, Mansoor S (2023) Plant virus-derived vectors for plant genome engineering. Viruses 15:531. https://doi.org/10.3390/v15020531
96. Mujtaba M, Wang D, Carvalho LB, Oliveira JL, do Espirito Santo Pereira A, Sharif R, Jogaiah S, Paidi MK, Wang L, Ali Q et al (2021) Nanocarrier-mediated delivery of MiRNA, RNAi, and CRISPR-Cas for plant protection: current trends and future directions. ACS Agric Sci Technol 1:417–435. https://doi.org/10.1021/acsagscitech.1c00146
97. Ben-Haim AE, Feldbaum RA, Belausov E, Zelinger E, Maria R, Nativ-Roth E, Mani KA, Barda O, Sionov E, Mechrez G (2024) DNA delivery to intact plant cells by casein nanoparticles with confirmed gene expression. Adv Funct Mater 34. https://doi.org/10.1002/adfm.202314756
98. Liu Y, Yang F, Jing X, Liu X, Wang G, Jian-Ping A, Huo H, Zhang Z, Wang X-F, Zhang S et al (2023) A biomimetic nanoparticle for PDNA delivery and expression in plant cells in a PH-dependent manner. ACS Agric Sci Technol 3:631–641. https://doi.org/10.1021/acsagscitech.3c00068
99. Xu X, Yu T, Zhang D, Song H, Huang K, Wang Y, Shen L, Li Y, Wang F, Zhang S et al (2023) Evaluation of the anti-viral efficacy of three different DsRNA nanoparticles against potato

virus Y using various delivery methods. Ecotoxicol Environ Saf 255:114775. https://doi.org/10.1016/j.ecoenv.2023.114775

100. Sangwan A, Gupta D, Singh OW, Roy A, Mukherjee SK, Mandal B, Singh N (2023) Size variations of mesoporous silica nanoparticle control uptake efficiency and delivery of AC2-derived DsRNA for protection against tomato leaf curl new Delhi virus. Plant Cell Rep 42:1571–1587. https://doi.org/10.1007/s00299-023-03048-z
101. Golestanipour A, Nikkhah M, Aalami A, Hosseinkhani S (2018) Gene delivery to tobacco root cells with single-walled carbon nanotubes and cell-penetrating Fusogenic peptides. Mol Biotechnol 60:863–878. https://doi.org/10.1007/s12033-018-0120-5
102. Gyawali B, Rahimi R, Alizadeh H, Mohammadi M (2024) Graphene Quantum Dots (GQD)-mediated DsRNA delivery for the control of fusarium head blight disease in wheat. ACS Appl Bio Mater 7:1526–1535. https://doi.org/10.1021/acsabm.3c00972
103. Mahmoud LM, Kaur P, Stanton D, Grosser JW, Dutt M (2022) A cationic lipid mediated CRISPR/Cas9 technique for the production of stable genome edited citrus plants. Plant Methods 18:33. https://doi.org/10.1186/s13007-022-00870-6
104. Kwak S-Y, Lew TTS, Sweeney CJ, Koman VB, Wong MH, Bohmert-Tatarev K, Snell KD, Seo JS, Chua N-H, Strano MS (2019) Chloroplast-selective gene delivery and expression in planta using chitosan-complexed single-walled carbon nanotube carriers. Nat Nanotechnol 14:447–455. https://doi.org/10.1038/s41565-019-0375-4
105. Yong J, Wu M, Zhang R, Bi S, Mann CWG, Mitter N, Carroll BJ, Xu ZP (2022) Clay nanoparticles efficiently deliver small interfering RNA to intact plant leaf cells. Plant Physiol 190:2187–2202. https://doi.org/10.1093/plphys/kiac430
106. Liu Q, Li Y, Xu K, Li D, Hu H, Zhou F, Song P, Yu Y, Wei Q, Liu Q et al (2020) Clay nanosheet-mediated delivery of recombinant plasmids expressing artificial MiRNAs via leaf spray to prevent infection by Plant DNA viruses. Hortic Res 7:179. https://doi.org/10.1038/s41438-020-00400-2
107. Mitter N, Worrall EA, Robinson KE, Li P, Jain RG, Taochy C, Fletcher SJ, Carroll BJ, Lu GQ, Xu ZP (2017) Clay nanosheets for topical delivery of RNAi for sustained protection against plant viruses. Nat Plants 3:16207. https://doi.org/10.1038/nplants.2016.207
108. Molesini B, Pennisi F, Cressoni C, Vitulo N, Dusi V, Speghini A, Pandolfini T (2022) Nanovector-mediated exogenous delivery of DsRNA induces silencing of target genes in very young tomato flower buds. Nanoscale Adv 4:4542–4553. https://doi.org/10.1039/D2NA00478J
109. Nagata T, Okada K, Takebe I, Matsui C (1981) Delivery of tobacco mosaic virus RNA into plant protoplasts mediated by reverse-phase evaporation vesicles (liposomes). Mol Gen Genet MGG 184:161–165. https://doi.org/10.1007/BF00272899
110. Hajiahmadi Z, Shirzadian-Khorramabad R, Kazemzad M, Sohani MM (2019) Enhancement of tomato resistance to Tuta Absoluta using a new efficient mesoporous silica nanoparticle-mediated plant transient gene expression approach. Sci Hortic (Amsterdam). 243:367–375. https://doi.org/10.1016/j.scienta.2018.08.040
111. Yong J, Zhang R, Bi S, Li P, Sun L, Mitter N, Carroll BJ, Xu ZP (2021) Sheet-like clay nanoparticles deliver RNA into developing pollen to efficiently silence a target gene. Plant Physiol 187:886–899. https://doi.org/10.1093/plphys/kiab303
112. Cai Y, Liu Z, Wang H, Meng H, Cao Y (2024) Mesoporous silica nanoparticles mediate SiRNA delivery for long-term multi-gene silencing in intact plants. Adv Sci 11. https://doi.org/10.1002/advs.202301358
113. Islam MR, Youngblood M, Kim H-I, González-Gamboa I, Monroy-Borrego AG, Caparco AA, Lowry GV, Steinmetz NF, Giraldo JP (2024) DNA delivery by virus-like nanocarriers in plant cells. Nano Lett 24:7833–7842. https://doi.org/10.1021/acs.nanolett.3c04735
114. Ahmed S, Gao X, Jahan MA, Adams M, Wu N, Kovinich N (2021) Nanoparticle-based genetic transformation of Cannabis sativa. J Biotechnol 326:48–51. https://doi.org/10.1016/j.jbiotec.2020.12.014
115. Zhang H, Cao Y, Xu D, Goh NS, Demirer GS, Cestellos-Blanco S, Chen Y, Landry MP, Yang P (2021) Gold-nanocluster-mediated delivery of SiRNA to intact plant cells for efficient gene knockdown. Nano Lett 21:5859–5866. https://doi.org/10.1021/acs.nanolett.1c01792

116. Zhang B, Huang S, Meng Y, Chen W (2023) Gold nanoparticles (AuNPs) can rapidly deliver artificial microRNA (AmiRNA)-ATG6 to silence ATG6 expression in Arabidopsis. Plant Cell Rep 42:1191–1201. https://doi.org/10.1007/s00299-023-03026-5
117. Wang B, Huang J, Zhang M, Wang Y, Wang H, Ma Y, Zhao X, Wang X, Liu C, Huang H et al (2020) Carbon dots enable efficient delivery of functional DNA in plants. ACS Appl Bio Mater 3:8857–8864. https://doi.org/10.1021/acsabm.0c01170
118. Zolghadrnasab M, Mousavi A, Farmany A, Arpanaei A (2021) Ultrasound-mediated gene delivery into suspended plant cells using polyethyleneimine-coated mesoporous silica nanoparticles. Ultrason Sonochem 73:105507. https://doi.org/10.1016/j.ultsonch.2021.105507
119. Delgado-Martín J, Delgado-Olidén A, Velasco L (2022) Carbon dots boost DsRNA delivery in plants and increase local and systemic SiRNA production. Int J Mol Sci 23:5338. https://doi.org/10.3390/ijms23105338
120. Ali Z, Serag MF, Demirer GS, Torre B, di Fabrizio E, Landry MP, Habuchi S, Mahfouz M (2022) DNA-carbon nanotube binding mode determines the efficiency of carbon nanotube-mediated DNA delivery to intact plants. ACS Appl Nano Mater 5:4663–4676. https://doi.org/10.1021/acsanm.1c03482
121. Qi J, Li Y, Yao X, Li G, Xu W, Chen L, Xie Z, Gu J, Wu H, Li Z (2024) Rational design of ROS scavenging and fluorescent gold nanoparticles to deliver SiRNA to improve plant resistance to pseudomonas Syringae. J Nanobiotechnology 22:446. https://doi.org/10.1186/s12951-024-02733-9
122. Law SSY, Liou G, Nagai Y, Giménez-Dejoz J, Tateishi A, Tsuchiya K, Kodama Y, Fujigaya T, Numata K (2022) Polymer-coated carbon nanotube hybrids with functional peptides for gene delivery into plant mitochondria. Nat Commun 13:2417. https://doi.org/10.1038/s41467-022-30185-y
123. Rajkumari N, Alex S, Soni KB, Anith KN, Viji MM, Kiran AG (2021) Silver nanoparticles for biolistic transformation in nicotiana Tabacum L. 3 Biotech 11:497. https://doi.org/10.1007/s13205-021-03043-9
124. Demirer GS, Zhang H, Goh NS, Pinals RL, Chang R, Landry MP (2020) Carbon nanocarriers deliver SiRNA to intact plant cells for efficient gene knockdown. Sci Adv 6. https://doi.org/10.1126/sciadv.aaz0495
125. Yu P, Zheng X, Alimi LO, Al-Babili S, Khashab NM (2024) Metal-organic framework-mediated delivery of nucleic acid across intact plant cells. ACS Appl Mater Interfaces 16:18245–18251. https://doi.org/10.1021/acsami.3c19571
126. Santana I, Wu H, Hu P, Giraldo JP (2020) Targeted delivery of nanomaterials with chemical cargoes in plants enabled by a biorecognition motif. Nat Commun 11:2045. https://doi.org/10.1038/s41467-020-15731-w
127. Guisbiers G, Mejía-Rosales S, Leonard Deepak F (2012) Nanomaterial properties: size and shape dependencies. J Nanomater. https://doi.org/10.1155/2012/180976
128. Báez-Cruz R, Baptista LA, Ntim S, Manidurai P, Espinoza S, Ramanan C, Cortes-Huerto R, Sulpizi M (2021) Role of PH in the synthesis and growth of gold nanoparticles using L-asparagine: a combined experimental and simulation study. J Phys Condens Matter 33:254005. https://doi.org/10.1088/1361-648X/abf6e3
129. Wang X, Xie H, Wang P, Yin H (2023) Nanoparticles in plants: uptake, transport and physiological activity in leaf and root. Materials (Basel). 16:3097. https://doi.org/10.3390/ma16083097
130. Pagano L, Maestri E, White JC, Marmiroli N, Marmiroli M (2018) Quantum dots exposure in plants: minimizing the adverse response. Curr Opin Environ Sci Heal 6:71–76. https://doi.org/10.1016/j.coesh.2018.09.001
131. Li Y, Xu X, Wu Y, Zhuang J, Zhang X, Zhang H, Lei B, Hu C, Liu Y (2020) A review on the effects of carbon dots in plant systems. Mater Chem Front 4:437–448. https://doi.org/10.1039/C9QM00614A
132. Majumdar S, Long RW, Kirkwood JS, Minakova AS, Keller AA (2021) Unraveling metabolic and proteomic features in soybean plants in response to copper hydroxide nanowires compared

to a commercial fertilizer. Environ Sci Technol 55:13477–13489. https://doi.org/10.1021/acs.est.1c00839
133. Liu W, Rudis MR, Cheplick MH, Millwood RJ, Yang J-P, Ondzighi-Assoume CA, Montgomery GA, Burris KP, Mazarei M, Chesnut JD et al (2020) Lipofection-mediated genome editing using DNA-free delivery of the Cas9/GRNA ribonucleoprotein into plant cells. Plant Cell Rep 39:245–257. https://doi.org/10.1007/s00299-019-02488-w
134. Silva AT, Nguyen A, Ye C, Verchot J, Moon JH (2010) Conjugated polymer nanoparticles for effective SiRNA delivery to tobacco BY-2 protoplasts. BMC Plant Biol 10:291. https://doi.org/10.1186/1471-2229-10-291
135. Zhao X, Meng Z, Wang Y, Chen W, Sun C, Cui B, Cui J, Yu M, Zeng Z, Guo S et al (2017) Pollen magnetofection for genetic modification with magnetic nanoparticles as gene carriers. Nat Plants 3:956–964. https://doi.org/10.1038/s41477-017-0063-z
136. Guo B, Itami J, Oikawa K, Motoda Y, Kigawa T, Numata K (2019) Native protein delivery into rice callus using ionic complexes of protein and cell-penetrating peptides. PLoS ONE 14:e0214033. https://doi.org/10.1371/journal.pone.0214033
137. Hubbard JD, Lui A, Landry MP (2020) Multiscale and multidisciplinary approach to understanding nanoparticle transport in plants. Curr Opin Chem Eng 30:135–143. https://doi.org/10.1016/j.coche.2020.100659
138. Maiti D, Tong X, Mou X, Yang K (2019) Carbon-based nanomaterials for biomedical applications: a recent study. Front Pharmacol 9. https://doi.org/10.3389/fphar.2018.01401
139. Liu X, Gao C, Xing D (2009) A non-invasive and rapid seed vigor biosensor based on quantitative measurement of superoxide generated by aleurone cell in intact seeds. Biosens Bioelectron 24:1537–1542. https://doi.org/10.1016/j.bios.2008.06.040
140. Khodakovskaya MV, de Silva K, Nedosekin DA, Dervishi E, Biris AS, Shashkov EV, Galanzha EI, Zharov VP (2011) Complex genetic, photothermal, and photoacoustic analysis of nanoparticle-plant interactions. Proc Natl Acad Sci 108:1028–1033. https://doi.org/10.1073/pnas.1008856108
141. Majeed N, Panigrahi KCS, Sukla LB, John R, Panigrahy M (2020) Application of carbon nanomaterials in plant biotechnology. Mater Today Proc 30:340–345. https://doi.org/10.1016/j.matpr.2020.01.618
142. González-Grandío E, Demirer GS, Jackson CT, Yang D, Ebert S, Molawi K, Keller H, Landry MP (2021) Carbon nanotube biocompatibility in plants is determined by their surface chemistry. J Nanobiotechnology 19:431. https://doi.org/10.1186/s12951-021-01178-8
143. Demirer GS, Zhang H, Goh NS, González-Grandío E, Landry MP (2019) Carbon nanotube-mediated DNA delivery without transgene integration in intact plants. Nat Protoc 14:2954–2971. https://doi.org/10.1038/s41596-019-0208-9
144. Dunbar T, Tsakirpaloglou N, Septiningsih EM, Thomson MJ (2022) Carbon nanotube-mediated plasmid DNA delivery in rice leaves and seeds. Int J Mol Sci 23:4081. https://doi.org/10.3390/ijms23084081
145. Lévy R, Shaheen U, Cesbron Y, Sée V (2010) Gold nanoparticles delivery in mammalian live cells: a critical review. Nano Rev 1:4889. https://doi.org/10.3402/nano.v1i0.4889
146. Ding Y, Jiang Z, Saha K, Kim CS, Kim ST, Landis RF, Rotello VM (2014) Gold nanoparticles for nucleic acid delivery. Mol Ther 22:1075–1083. https://doi.org/10.1038/mt.2014.30
147. Vijayakumar PS, Abhilash OU, Khan BM, Prasad BLV (2010) Nanogold-loaded sharp-edged carbon bullets as plant-gene carriers. Adv Funct Mater 20:2416–2423. https://doi.org/10.1002/adfm.200901883
148. Lei W-X, An Z-S, Zhang B-H, Wu Q, Gong W-J, Li J-M, Chen W-L (2020) Construction of gold-SiRNA NPR1 nanoparticles for effective and quick silencing of NPR1 in Arabidopsis thaliana. RSC Adv 10:19300–19308. https://doi.org/10.1039/D0RA02156C
149. Huang C, Zhang Y, Yuan H, Gao H, Zhang S (2013) Role of nanoparticle geometry in endocytosis: laying down to stand up. Nano Lett 13:4546–4550. https://doi.org/10.1021/nl402628n
150. Yu M, Wang J, Yang Y, Zhu C, Su Q, Guo S, Sun J, Gan Y, Shi X, Gao H (2016) Rotation-facilitated rapid transport of nanorods in mucosal tissues. Nano Lett 16:7176–7182. https://doi.org/10.1021/acs.nanolett.6b03515

151. Torney F, Trewyn BG, Lin VS-Y, Wang K (2007) Mesoporous silica nanoparticles deliver DNA and chemicals into plants. Nat Nanotechnol 2:295–300. https://doi.org/10.1038/nnano.2007.108
152. Hao Y, Yang X, Shi Y, Song S, Xing J, Marowitch J, Chen J, Chen J (2013) Magnetic gold nanoparticles as a vehicle for fluorescein isothiocyanate and DNA delivery into plant cells. Botany 91:457–466. https://doi.org/10.1139/cjb-2012-0281
153. Loo YY, Rukayadi Y, Nor-Khaizura M-A-R, Kuan CH, Chieng BW, Nishibuchi M, Radu S (2018) In vitro antimicrobial activity of green synthesized silver nanoparticles against selected gram-negative foodborne pathogens. Front Microbiol 29. https://doi.org/10.3389/fmicb.2018.01555
154. Gur T (2022) Green synthesis, characterizations of silver nanoparticles using sumac (Rhus Coriaria L.) plant extract and their antimicrobial and DNA damage protective effects. Front Chem 10. https://doi.org/10.3389/fchem.2022.968280
155. Vejlupkova Z, Warman C, Sharma R, Scheller HV, Mortimer JC, Fowler JE (2020) No evidence for transient transformation via pollen magnetofection in several monocot species. Nat Plants 6:1323–1324. https://doi.org/10.1038/s41477-020-00798-6
156. Wang Z, Zhang Z, Zheng D, Zhang T, Li X, Zhang C, Yu R, Wei J, Wu Z (2022) Efficient and genotype independent maize transformation using pollen transfected by DNA-coated magnetic nanoparticles. J Integr Plant Biol 64:1145–1156. https://doi.org/10.1111/jipb.13263
157. Vercoulen PHW, Roos RA, Marijnissen JCM, Scarlett B (1992) Measuring electric charge on pollen. J Aerosol Sci 23:377–380. https://doi.org/10.1016/0021-8502(92)90428-X
158. Sun X-D, Ma J-Y, Feng L-J, Duan J-L, Xie X-M, Zhang X-H, Kong X, Ding Z, Yuan X-Z (2023) Magnetite nanoparticle coating chemistry regulates root uptake pathways and iron chlorosis in plants. Proc Natl Acad Sci 120. https://doi.org/10.1073/pnas.2304306120
159. Kneuer C, Sameti M, Bakowsky U, Schiestel T, Schirra H, Schmidt H, Lehr C-M (2000) A nonviral DNA delivery system based on surface modified silica-nanoparticles can efficiently transfect cells in vitro. Bioconjug Chem 11:926–932. https://doi.org/10.1021/bc0000637
160. Radu DR, Lai C-Y, Jeftinija K, Rowe EW, Jeftinija S, Lin VS-Y (2004) A polyamidoamine dendrimer-capped mesoporous silica nanosphere-based gene transfection reagent. J Am Chem Soc 126:13216–13217. https://doi.org/10.1021/ja046275m
161. Bharti C, Gulati N, Nagaich U, Pal A (2015) Mesoporous silica nanoparticles in target drug delivery system: a review. Int J Pharm Investig 5:124. https://doi.org/10.4103/2230-973X.160844
162. Martin-Ortigosa S, Valenstein JS, Lin VS-Y, Trewyn BG, Wang K (2012) Gold functionalized mesoporous silica nanoparticle mediated protein and DNA codelivery to plant cells via the biolistic method. Adv Funct Mater 22:3576–3582. https://doi.org/10.1002/adfm.201200359
163. Martin-Ortigosa S, Peterson DJ, Valenstein JS, Lin VS-Y, Trewyn BG, Lyznik LA, Wang K (2014) Mesoporous silica nanoparticle-mediated intracellular Cre protein delivery for maize genome editing via LoxP site excision. Plant Physiol 164:537–547. https://doi.org/10.1104/pp.113.233650
164. Bhardwaj P, Singh B, Behera SP (2021) Green approaches for nanoparticle synthesis: emerging trends. In: Nanomaterials. Elsevier, pp 167–193
165. Ghaffari A, Abazari M, Moghimi HR (2023) Wound healing and nanotechnology: opportunities and challenges. In: Bioengineered nanomaterials for wound healing and infection control. Elsevier, pp 115–174
166. Mehta M, Bui TA, Yang X, Aksoy Y, Goldys EM, Deng W (2023) Lipid-based nanoparticles for drug/gene delivery: an overview of the production techniques and difficulties encountered in their industrial development. ACS Mater Au 3:600–619. https://doi.org/10.1021/acsmaterialsau.3c00032
167. Amreddy N, Babu A, Muralidharan R, Munshi A, Ramesh R (2017) Polymeric nanoparticle-mediated gene delivery for lung cancer treatment. Top Curr Chem 375:35. https://doi.org/10.1007/s41061-017-0128-5
168. Cai Q, He B, Wang S, Fletcher S, Niu D, Mitter N, Birch PRJ, Jin H (2021) Message in a bubble: shuttling small RNAs and proteins between cells and interacting organisms using

extracellular vesicles. Annu Rev Plant Biol 72:497–524. https://doi.org/10.1146/annurev-arplant-081720-010616
169. Karny A, Zinger A, Kajal A, Shainsky-Roitman J, Schroeder A (2018) Therapeutic nanoparticles penetrate leaves and deliver nutrients to agricultural crops. Sci Rep 8:7589. https://doi.org/10.1038/s41598-018-25197-y
170. Lowry GV, Giraldo JP, Steinmetz NF, Avellan A, Demirer GS, Ristroph KD, Wang GJ, Hendren CO, Alabi CA, Caparco A et al (2024) Towards realizing nano-enabled precision delivery in plants. Nat Nanotechnol 19:1255–1269. https://doi.org/10.1038/s41565-024-01667-5
171. Su Y, Ashworth V, Kim C, Adeleye AS, Rolshausen P, Roper C, White J, Jassby D (2019) Delivery, uptake, fate, and transport of engineered nanoparticles in plants: a critical review and data analysis. Environ Sci Nano 6:2311–2331. https://doi.org/10.1039/C9EN00461K
172. Ali S, Mehmood A, Khan N (2021) Uptake, translocation, and consequences of nanomaterials on plant growth and stress adaptation. J Nanomater 2021:1–17. https://doi.org/10.1155/2021/6677616
173. Zhu J, Li J, Shen Y, Liu S, Zeng N, Zhan X, White JC, Gardea-Torresdey J, Xing B (2020) Mechanism of zinc oxide nanoparticle entry into wheat seedling leaves. Environ Sci Nano 7:3901–3913. https://doi.org/10.1039/D0EN00658K
174. Banerjee A, Sarkar A, Acharya K, Chakraborty N (2021) Nanotechnology: an emerging hope in crop improvement. Lett Appl NanoBioScience 10:2784–2803. https://doi.org/10.33263/LIANBS104.27842803
175. Avellan A, Yun J, Zhang Y, Spielman-Sun E, Unrine JM, Thieme J, Li J, Lombi E, Bland G, Lowry GV (2019) Nanoparticle size and coating chemistry control foliar uptake pathways, translocation, and leaf-to-rhizosphere transport in wheat. ACS Nano 13:5291–5305. https://doi.org/10.1021/acsnano.8b09781
176. Demirer GS, Silva TN, Jackson CT, Thomas JB, Ehrhardt WD, Rhee SY, Mortimer JC, Landry MP (2021) Nanotechnology to advance CRISPR–Cas genetic engineering of plants. Nat. Nanotechnol 16:243–250. https://doi.org/10.1038/s41565-021-00854-y
177. Lew TTS, Wong MH, Kwak S, Sinclair R, Koman VB, Strano MS (2018) Rational design principles for the transport and subcellular distribution of nanomaterials into plant protoplasts. Small 14. https://doi.org/10.1002/smll.201802086
178. Zhu Z-J, Wang H, Yan B, Zheng H, Jiang Y, Miranda OR, Rotello VM, Xing B, Vachet RW (2012) Effect of surface charge on the uptake and distribution of gold nanoparticles in four plant species. Environ Sci Technol 46:12391–12398. https://doi.org/10.1021/es301977w
179. Chiu YTE, Choi CHJ (2021) Enabling transgenic plant cell–derived biomedicines with nanotechnology. Adv NanoBiomed Res 1. https://doi.org/10.1002/anbr.202000028
180. Zhu J, Wang J, Zhan X, Li A, White JC, Gardea-Torresdey JL, Xing B (2021) Role of charge and size in the translocation and distribution of zinc oxide particles in wheat cells. ACS Sustain Chem Eng 9:11556–11564. https://doi.org/10.1021/acssuschemeng.1c04080
181. Banerjee K, Pramanik P, Maity A, Joshi DC, Wani SH, Krishnan P (2019) Methods of using nanomaterials to plant systems and their delivery to plants (mode of entry, uptake, translocation, accumulation, biotransformation and barriers). In: Advances in phytonanotechnology. Elsevier, pp 123–152
182. Zhang P, Guo Z, Zhang Z, Fu H, White JC, Lynch I (2020) Nanomaterial transformation in the soil–plant system: implications for food safety and application in agriculture. Small 16. https://doi.org/10.1002/smll.202000705
183. Guidarelli M, Baraldi E (2015) Transient transformation meets gene function discovery: the strawberry fruit case. Front Plant Sci 6. https://doi.org/10.3389/fpls.2015.00444
184. Canto T (2016) Transient expression systems in plants: potentialities and constraints, pp 287–301
185. Schöb H, Kunz C, Meins F (1997) Silencing of transgenes introduced into leaves by agroinfiltration: a simple, rapid method for investigating sequence requirements for gene silencing. Mol Gen Genet MGG 256:581–585. https://doi.org/10.1007/s004380050604

186. Wroblewski T, Tomczak A, Michelmore R (2005) Optimization of agrobacterium-mediated transient assays of gene expression in lettuce, tomato and arabidopsis. Plant Biotechnol J 3:259–273. https://doi.org/10.1111/j.1467-7652.2005.00123.x
187. Orzaez D, Mirabel S, Wieland WH, Granell A (2006) Agroinjection of tomato fruits. A tool for rapid functional analysis of transgenes directly in fruit. Plant Physiol 140:3–11. https://doi.org/10.1104/pp.105.068221
188. Santi L, Batchelor L, Huang Z, Hjelm B, Kilbourne J, Arntzen CJ, Chen Q, Mason HS (2008) An efficient plant viral expression system generating orally immunogenic Norwalk virus-like particles. Vaccine 26:1846–1854. https://doi.org/10.1016/j.vaccine.2008.01.053
189. Xu K, Huang X, Wu M, Wang Y, Chang Y, Liu K, Zhang J, Zhang Y, Zhang F, Yi L et al (2014) A rapid, highly efficient and economical method of agrobacterium-mediated in planta transient transformation in living onion epidermis. PLoS ONE 9:e83556. https://doi.org/10.1371/journal.pone.0083556
190. Vaghchhipawala Z, Rojas CM, Senthil-Kumar M, Mysore KS (2011) Agroinoculation and agroinfiltration: simple tools for complex gene function analyses, pp 65–76
191. Přibylová A, Fischer L, Pyott DE, Bassett A, Molnar A (2022) DNA methylation can alter CRISPR/Cas9 editing frequency and DNA repair outcome in a target-specific manner. New Phytol 235:2285–2299. https://doi.org/10.1111/nph.18212
192. Kaur M, Manchanda P, Kalia A, Ahmed FK, Nepovimova E, Kuca K, Abd-Elsalam KA (2021) Agroinfiltration mediated scalable transient gene expression in genome edited crop plants. Int J Mol Sci 22:10882. https://doi.org/10.3390/ijms221910882
193. Yan H, Fu D, Zhu B, Liu H, Shen X, Luo Y (2012) Sprout vacuum-infiltration: a simple and efficient agroinoculation method for virus-induced gene silencing in diverse solanaceous species. Plant Cell Rep 31:1713–1722. https://doi.org/10.1007/s00299-012-1285-1
194. Bian Y, Zhang X, Chen C, Wang G, Zhang P, Liu G, Ma F, Bao Z (2024) Establish an efficient inoculation system of tomato yellow leaf curl virus to assist tomato resistance breeding. Sci Hortic (Amsterdam) 324:112567. https://doi.org/10.1016/j.scienta.2023.112567
195. Zheng L, Yang J, Chen Y, Ding L, Wei J, Wang H (2021) An improved and efficient method of agrobacterium syringe infiltration for transient transformation and its application in the elucidation of gene function in poplar. BMC Plant Biol 21:54. https://doi.org/10.1186/s12870-021-02833-w
196. Chen J, Wang Y, Wang F, Yang J, Gao M, Li C, Liu Y, Liu Y, Yamaji N, Ma JF et al (2015) The rice CK2 kinase regulates trafficking of phosphate transporters in response to phosphate levels. Plant Cell 27:711–723. https://doi.org/10.1105/tpc.114.135335
197. Matsuo K, Fukuzawa N, Matsumura T (2016) A simple agroinfiltration method for transient gene expression in plant leaf discs. J Biosci Bioeng 122:351–356. https://doi.org/10.1016/j.jbiosc.2016.02.001
198. Simmons CW, VanderGheynst JS, Upadhyaya SK (2009) A model of agrobacterium tumefaciens vacuum infiltration into harvested leaf tissue and subsequent in planta transgene transient expression. Biotechnol Bioeng 102:965–970. https://doi.org/10.1002/bit.22118
199. Ranjbaran M, Datta AK (2019) Pressure-driven infiltration of water and bacteria into plant leaves during vacuum cooling: a mechanistic model. J Food Eng 246:209–223. https://doi.org/10.1016/j.jfoodeng.2018.10.032
200. Simmons CW, Nitin N, VanderGheynst JS (2012) Rapid, in situ detection of agrobacterium tumefaciens attachment to leaf tissue. Biotechnol Prog 28:1321–1328. https://doi.org/10.1002/btpr.1608
201. Izuegbunam CL, Wijewantha N, Wone B, Ariyarathne MA, Sereda G, Wone BWM (2021) A nano-biomimetic transformation system enables in planta expression of a reporter gene in mature plants and seeds. Nanoscale Adv 3:3240–3250. https://doi.org/10.1039/D1NA00107H
202. Maliga P (2022) Engineering the plastid and mitochondrial genomes of flowering plants. Nat Plants 8:996–1006. https://doi.org/10.1038/s41477-022-01227-6
203. Macklin MD, Drape RJ, Swain WF (2000) Preparations for particle-mediated gene transfer using the Accell® gene gun. In: DNA vaccines. Humana Press, New Jersey, pp 297–304

204. Zhang K, Liu J, Zhang Y, Yang Z, Gao C (2015) Biolistic genetic transformation of a wide range of Chinese Elite Wheat (Triticum Aestivum L.) varieties. J Genet Genom 42:39–42. https://doi.org/10.1016/j.jgg.2014.11.005
205. Lacroix B, Citovsky V (2020) Biolistic approach for transient gene expression studies in plants, pp 125–139
206. Ueki S, Lacroix B, Krichevsky A, Lazarowitz SG, Citovsky V (2009) Functional transient genetic transformation of arabidopsis leaves by biolistic bombardment. Nat Protoc 4:71–77. https://doi.org/10.1038/nprot.2008.217
207. Wu H, Acanda Y, Canton M, Zale J (2019) Efficient biolistic transformation of immature citrus rootstocks using phosphomannose-isomerase selection. Plants 8:390. https://doi.org/10.3390/plants8100390
208. Fadeev VS, Blinkova OV, Gaponenko AK (2006) Optimization of biological and physical parameters for biolistic genetic transformation of common wheat (Triticum Aestivum L.) using a particle inflow gun. Russ J Genet 42:402–411. https://doi.org/10.1134/S1022795406040077
209. Wang K, Frame B (2009) Biolistic gun-mediated maize genetic transformation, pp 29–45
210. Gong J, Tian Z, Qu X, Meng Q, Guan Y, Liu P, Chen C, Deng X, Guo W, Cheng Y et al (2021) Illuminating the cells: transient transformation of citrus to study gene functions and organelle activities related to fruit quality. Hortic Res 8:175. https://doi.org/10.1038/s41438-021-00611-1
211. George ACS (2018) Magnetofection of tobacco protoplasts: a novel mechanism for plant transformation
212. Farooq N, Ather L, Shafiq M, Nawaz-ul-Rehman MS, Haseeb M, Anjum T, Abbas Q, Hussain M, Ali N, Asad Abbas SAA et al (2022) Magnetofection approach for the transformation of okra using green iron nanoparticles. Sci Rep 12:16568. https://doi.org/10.1038/s41598-022-20569-x
213. Choudhary ML, Chin CK (1995) Ultrasound mediated delivery of compounds into petunia protoplasts and cells. J Plant Biochem Biotechnol 4:37–39. https://doi.org/10.1007/BF03262948
214. Lo C-W, Desjouy C, Chen S-R, Lee J-L, Inserra C, Béra J-C, Chen W-S (2014) Stabilizing in vitro ultrasound-mediated gene transfection by regulating cavitation. Ultrason Sonochem 21:833–839. https://doi.org/10.1016/j.ultsonch.2013.10.017
215. Liu Y, Yang H, Sakanishi A (2006) Ultrasound: mechanical gene transfer into plant cells by sonoporation. Biotechnol Adv 24:1–16. https://doi.org/10.1016/j.biotechadv.2005.04.002
216. Amani A, Zare N, Asadi A, Asghari-Zakaria R (2018) Ultrasound-enhanced gene delivery to alfalfa cells by HPAMAM dendrimer nanoparticles. TURKISH J Biol 42:63–75. https://doi.org/10.3906/biy-1706-6
217. Wang M, Jin H (2017) Spray-induced gene silencing: a powerful innovative strategy for crop protection. Trends Microbiol 25:4–6. https://doi.org/10.1016/j.tim.2016.11.011
218. Rank AP, Koch A (2021) Lab-to-field transition of RNA spray applications—how far are we? Front Plant Sci 12. https://doi.org/10.3389/fpls.2021.755203
219. Thagun C, Horii Y, Mori M, Fujita S, Ohtani M, Tsuchiya K, Kodama Y, Odahara M, Numata K (2022) Non-transgenic gene modulation via spray delivery of nucleic acid/peptide complexes into plant nuclei and chloroplasts. ACS Nano 16:3506–3521. https://doi.org/10.1021/acsnano.1c07723
220. Wang Z, Li Y, Zhang B, Gao X, Shi M, Zhang S, Zhong S, Zheng Y, Liu X (2023) Functionalized carbon dot-delivered RNA nano fungicides as superior tools to control phytophthora pathogens through plant RdRP1 mediated spray-induced gene silencing. Adv Funct Mater 33. https://doi.org/10.1002/adfm.202213143
221. Hernandez ES (2021) Layered double hydroxide (LDH)-mediated topical delivery of DsRNA for protection against tomato yellow leaf curl virus (TYLCV) in nicotiana benthamiana
222. Li S, Li J, Du M, Deng G, Song Z, Han H (2022) Efficient gene silencing in intact plant cells using SiRNA delivered by functional graphene oxide nanoparticles. Angew Chemie Int Ed 61. https://doi.org/10.1002/anie.202210014

223. Sembada AA, Lenggoro IW (2024) Transport of nanoparticles into plants and their detection methods. Nanomaterials 14:131. https://doi.org/10.3390/nano14020131
224. Pal G, Ingole KD, Yavvari PS, Verma P, Kumari A, Chauhan C, Chaudhary D, Srivastava A, Bajaj A, Vemanna RS (2024) Exogenous application of nanocarrier-mediated double-stranded RNA manipulates physiological traits and defence response against bacterial diseases. Mol Plant Pathol 25. https://doi.org/10.1111/mpp.13417
225. Santana I, Jeon S-J, Kim H-I, Islam MR, Castillo C, Garcia GFH, Newkirk GM, Giraldo JP (2022) Targeted carbon nanostructures for chemical and gene delivery to plant chloroplasts. ACS Nano 16:12156–12173. https://doi.org/10.1021/acsnano.2c02714
226. Chuah J-A, Horii Y, Numata K (2016) Peptide-derived method to transport genes and proteins across cellular and organellar barriers in plants. J Vis Exp. https://doi.org/10.3791/54972
227. Miyamoto T, Tsuchiya K, Toyooka K, Goto Y, Tateishi A, Numata K (2022) Relaxation of the plant cell wall barrier via Zwitterionic liquid pretreatment for micelle-complex-mediated DNA delivery to specific plant organelles. Angew Chemie 134. https://doi.org/10.1002/ange.202204234
228. Caboche M (1990) Liposome-mediated transfer of nucleic acids in plant protoplasts. Physiol Plant 79:173–176. https://doi.org/10.1111/j.1399-3054.1990.tb05882.x
229. Chang F-P, Kuang L-Y, Huang C-A, Jane W-N, Hung Y, Hsing YC, Mou C-Y (2013) A simple plant gene delivery system using mesoporous silica nanoparticles as carriers. J Mater Chem B 1:5279. https://doi.org/10.1039/c3tb20529k
230. Hajiahmadi Z, Shirzadian-Khorramabad R, Kazemzad M, Sohani MM, Khajehali J (2020) A novel, simple, and stable mesoporous silica nanoparticle-based gene transformation approach in solanum lycopersicum. Biotech 10:370. https://doi.org/10.1007/s13205-020-02359-2
231. Chandrakala V, Aruna V, Angajala G (2022) Review on metal nanoparticles as nanocarriers: current challenges and perspectives in drug delivery systems. Emergent Mater 5:1593–1615. https://doi.org/10.1007/s42247-021-00335-x
232. Park C-W, Choi J-Y, Son Y-J, Kim D-H, Li H, Liang W, Lee C, Jung K-H, Kim Y-J (2024) Magnetofected pollen gene delivery system could generate genetically modified cucumis sativus. Hortic Res 11. https://doi.org/10.1093/hr/uhae179
233. Giraldo JP, Landry MP, Faltermeier SM, McNicholas TP, Iverson NM, Boghossian AA, Reuel NF, Hilmer AJ, Sen F, Brew JA et al (2014) Plant nanobionics approach to augment photosynthesis and biochemical sensing. Nat Mater 13:400–408. https://doi.org/10.1038/nmat3890
234. Burlaka OM, Pirko YV, Yemets AI, Blume YB (2015) Plant genetic transformation using carbon nanotubes for DNA delivery. Cytol Genet 49:349–357. https://doi.org/10.3103/S009545271506002X
235. Ghaghelestany AB, Jahanbakhshi A, Taghinezhad E (2020) Gene transfer to german chamomile (L Chamomilla M) using cationic carbon nanotubes. Sci Hortic (Amsterdam) 263:109106. https://doi.org/10.1016/j.scienta.2019.109106
236. Lew TTS, Park M, Wang Y, Gordiichuk P, Yeap W-C, Mohd Rais SK, Kulaveerasingam H, Strano MS (2020) Nanocarriers for transgene expression in pollen as a plant biotechnology tool. ACS Mater Lett 2:1057–1066. https://doi.org/10.1021/acsmaterialslett.0c00247
237. Numata K, Horii Y, Oikawa K, Miyagi Y, Demura T, Ohtani M (2018) Library screening of cell-penetrating peptide for BY-2 cells, leaves of arabidopsis, tobacco, tomato, poplar, and rice callus. Sci Rep 8:10966. https://doi.org/10.1038/s41598-018-29298-6
238. Watanabe K, Odahara M, Miyamoto T, Numata K (2021) Fusion peptide-based biomacromolecule delivery system for plant cells. ACS Biomater Sci Eng 7:2246–2254. https://doi.org/10.1021/acsbiomaterials.1c00227
239. Mann A, Thakur G, Shukla V, Singh AK, Khanduri R, Naik R, Jiang Y, Kalra N, Dwarakanath BS, Langel U et al (2011) Differences in DNA condensation and release by lysine and arginine homopeptides govern their DNA delivery efficiencies. Mol Pharm 8:1729–1741. https://doi.org/10.1021/mp2000814
240. Adami RC, Collard WT, Gupta SA, KwoK KY, Bonadio J, Rice KG (1998) Stability of peptide-condensed plasmid DNA formulations. J Pharm Sci 87:678–683. https://doi.org/10.1021/js9800477

241. Ni R, Feng R, Chau Y (2023) Peptide-DNA co-assembled nanoparticles as a nonviral vector for gene delivery with high transgene expression. ACS Appl Nano Mater 6:3191–3201. https://doi.org/10.1021/acsanm.2c04279
242. Thagun C, Chuah J, Numata K (2019) Targeted gene delivery into various plastids mediated by clustered cell-penetrating and chloroplast-targeting peptides. Adv Sci 6. https://doi.org/10.1002/advs.201902064
243. Chuah J-A, Yoshizumi T, Kodama Y, Numata K (2015) Gene introduction into the mitochondria of arabidopsis thaliana via peptide-based carriers. Sci Rep 5:7751. https://doi.org/10.1038/srep07751
244. Oikawa K, Tateishi A, Odahara M, Kodama Y, Numata K (2021) Imaging of the entry pathway of a cell-penetrating peptide–DNA complex from the extracellular space to chloroplast nucleoids across multiple membranes in arabidopsis leaves. Front Plant Sci 12. https://doi.org/10.3389/fpls.2021.759871
245. Yoshizumi T, Oikawa K, Chuah J-A, Kodama Y, Numata K (2018) Selective gene delivery for integrating exogenous DNA into plastid and mitochondrial genomes using peptide–DNA complexes. Biomacromol 19:1582–1591. https://doi.org/10.1021/acs.biomac.8b00323
246. Odahara M, Horii Y, Itami J, Watanabe K, Numata K (2022) Functional peptide-mediated plastid transformation in tobacco, rice, and kenaf. Front Plant Sci 13. https://doi.org/10.3389/fpls.2022.989310
247. Tsuchiya K, Yilmaz N, Miyamoto T, Masunaga H, Numata K (2020) Zwitterionic polypeptides: chemoenzymatic synthesis and loosening function for cellulose crystals. Biomacromol 21:1785–1794. https://doi.org/10.1021/acs.biomac.9b01700
248. Miyamoto T, Tsuchiya K, Numata K (2020) Dual peptide-based gene delivery system for the efficient transfection of plant callus cells. Biomacromol 21:2735–2744. https://doi.org/10.1021/acs.biomac.0c00481
249. Miyamoto T, Tsuchiya K, Numata K (2021) Endosome-escaping micelle complexes dually equipped with cell-penetrating and endosome-disrupting peptides for efficient DNA delivery into intact plants. Nanoscale 13:5679–5692. https://doi.org/10.1039/D0NR08183C
250. Chuah J-A, Numata K (2018) Stimulus-responsive peptide for effective delivery and release of DNA in plants. Biomacromol 19:1154–1163. https://doi.org/10.1021/acs.biomac.8b00016
251. Fu Y, Li L, Wang H, Jiang Y, Liu H, Cui X, Wang P, Lü C (2015) Silica nanoparticles-mediated stable genetic transformation in nicotiana tabacum. Chem Res Chinese Univ 31:976–981. https://doi.org/10.1007/s40242-015-5088-0
252. Das PR, Sherif SM (2020) Application of exogenous DsRNAs-induced RNAi in agriculture: challenges and triumphs. Front Plant Sci 11. https://doi.org/10.3389/fpls.2020.00946
253. Frizzi A, Huang S (2010) Tapping RNA silencing pathways for plant biotechnology. Plant Biotechnol J 8:655–677. https://doi.org/10.1111/j.1467-7652.2010.00505.x
254. Yan S, Ren B, Shen J (2021) Nanoparticle-mediated double-stranded RNA delivery system: a promising approach for sustainable pest management. Insect Sci 28:21–34. https://doi.org/10.1111/1744-7917.12822
255. Zhao Y, Yang X, Zhou G, Zhang T (2020) Engineering plant virus resistance: from RNA silencing to genome editing strategies. Plant Biotechnol J 18:328–336. https://doi.org/10.1111/pbi.13278
256. Small I (2007) RNAi for revealing and engineering plant gene functions. Curr Opin Biotechnol 18:148–153. https://doi.org/10.1016/j.copbio.2007.01.012
257. Watson JM, Fusaro AF, Wang M, Waterhouse PM (2005) RNA silencing platforms in plants. FEBS Lett 579:5982–5987. https://doi.org/10.1016/j.febslet.2005.08.014
258. Jiang L, Ding L, He B, Shen J, Xu Z, Yin M, Zhang X (2014) Systemic gene silencing in plants triggered by fluorescent nanoparticle-delivered double-stranded RNA. Nanoscale 6:9965–9969. https://doi.org/10.1039/C4NR03481C
259. Numata K, Ohtani M, Yoshizumi T, Demura T, Kodama Y (2014) Local gene silencing in plants via synthetic Ds RNA and carrier peptide. Plant Biotechnol J 12:1027–1034. https://doi.org/10.1111/pbi.12208

260. Worrall EA, Bravo-Cazar A, Nilon AT, Fletcher SJ, Robinson KE, Carr JP, Mitter N (2019) Exogenous application of RNAi-inducing double-stranded RNA inhibits aphid-mediated transmission of a plant virus. Front Plant Sci 10. https://doi.org/10.3389/fpls.2019.00265
261. Jain RG, Fletcher SJ, Manzie N, Robinson KE, Li P, Lu E, Brosnan CA, Xu ZP, Mitter N (2022) Foliar application of clay-delivered RNA interference for whitefly control. Nat Plants 8:535–548. https://doi.org/10.1038/s41477-022-01152-8
262. Chakraborty S, Bera D, Roy L, Ghosh CK (2023) Biomimetic and bioinspired nanostructures. In: Bioinspired and green synthesis of nanostructures. Wiley, pp 353–404
263. Grohmann L, Keilwagen J, Duensing N, Dagand E, Hartung F, Wilhelm R, Bendiek J, Sprink T (2019) Detection and identification of genome editing in plants: challenges and opportunities. Front Plant Sci 10. https://doi.org/10.3389/fpls.2019.00236
264. Tröder SE, Zevnik B (2022) History of genome editing: from meganucleases to CRISPR. Lab Anim 56:60–68. https://doi.org/10.1177/0023677221994613
265. Baltes NJ, Voytas DF (2015) Enabling plant synthetic biology through genome engineering. Trends Biotechnol 33:120–131. https://doi.org/10.1016/j.tibtech.2014.11.008
266. Becker S, Boch J (2021) TALE and TALEN genome editing technologies. Gene Genome Ed. 2:100007. https://doi.org/10.1016/j.ggedit.2021.100007
267. Mao Y, Botella JR, Liu Y, Zhu J-K (2019) Gene editing in plants: progress and challenges. Natl Sci Rev 6:421–437. https://doi.org/10.1093/nsr/nwz005
268. Van Eck J (2020) Applying gene editing to tailor precise genetic modifications in plants. J Biol Chem 295:13267–13276. https://doi.org/10.1074/jbc.REV120.010850
269. Sukegawa S, Saika H, Toki S (2021) Plant genome editing: ever more precise and wide reaching. Plant J 106:1208–1218. https://doi.org/10.1111/tpj.15233
270. Gerashchenkov GA, Rozhnova NA, Kuluev BR, Kiryanova OY, Gumerova GR, Knyazev AV, Vershinina ZR, Mikhailova EV, Chemeris DA, Matniyazov RT et al (2020) Design of guide RNA for CRISPR/Cas plant genome editing. Mol Biol 54:24–42. https://doi.org/10.1134/S0026893320010069
271. Wang JW, Grandio EG, Newkirk GM, Demirer GS, Butrus S, Giraldo JP, Landry MP (2019) Nanoparticle-mediated genetic engineering of plants. Mol Plant 12:1037–1040. https://doi.org/10.1016/j.molp.2019.06.010
272. Xu X, Liu C, Wang Y, Koivisto O, Zhou J, Shu Y, Zhang H (2021) Nanotechnology-based delivery of CRISPR/Cas9 for cancer treatment. Adv Drug Deliv Rev 176:113891. https://doi.org/10.1016/j.addr.2021.113891
273. Song X, Liu C, Wang N, Huang H, He S, Gong C, Wei Y (2021) Delivery of CRISPR/Cas systems for cancer gene therapy and immunotherapy. Adv Drug Deliv Rev 168:158–180. https://doi.org/10.1016/j.addr.2020.04.010
274. Liu S, Cheng Q, Wei T, Yu X, Johnson LT, Farbiak L, Siegwart DJ (2021) Membrane-destabilizing ionizable phospholipids for organ-selective MRNA delivery and CRISPR–Cas gene editing. Nat Mater 20:701–710. https://doi.org/10.1038/s41563-020-00886-0
275. Lin Y, Wu J, Gu W, Huang Y, Tong Z, Huang L, Tan J (2018) Exosome–liposome hybrid nanoparticles deliver CRISPR/Cas9 system in MSCs. Adv Sci 5. https://doi.org/10.1002/advs.201700611
276. Farbiak L, Cheng Q, Wei T, Álvarez-Benedicto E, Johnson LT, Lee S, Siegwart DJ (2021) All-in-one dendrimer-based lipid nanoparticles enable precise HDR-mediated gene editing in vivo. Adv Mater 33. https://doi.org/10.1002/adma.202006619
277. Makhotenko AV, Khromov AV, Snigir EA, Makarova SS, Makarov VV, Suprunova TP, Kalinina NO, Taliansky ME (2019) Functional analysis of coilin in virus resistance and stress tolerance of potato solanum tuberosum using CRISPR-Cas9 editing. Dokl Biochem Biophys 484:88–91. https://doi.org/10.1134/S1607672919010241
278. Abdel-Razik AB, Hammad IA, Tawfik E (2017) Transformation of thionin genes using chitosan nanoparticle into potato plant to Be resistant to fungal infection. IOSR J Biotechnol Biochem 03:01–13. https://doi.org/10.9790/264X-03030113
279. Zhang K, Chooi WH, Liu S, Chin JS, Murray A, Nizetic D, Cheng D, Chew SY (2020) Localized delivery of CRISPR/DCas9 via layer-by-layer self-assembling peptide coating on

nanofibers for neural tissue engineering. Biomaterials 256:120225. https://doi.org/10.1016/j.biomaterials.2020.120225

280. Jo A, Ringel-Scaia VM, McDaniel DK, Thomas CA, Zhang R, Riffle JS, Allen IC, Davis RM (2020) Fabrication and characterization of PLGA nanoparticles encapsulating large CRISPR–Cas9 plasmid. J Nanobiotechnol 18:16. https://doi.org/10.1186/s12951-019-0564-1
281. Alallam B, Altahhan S, Taher M, Mohd Nasir MH, Doolaanea AA (2020) Electrosprayed alginate nanoparticles as CRISPR plasmid DNA delivery carrier: preparation, optimization, and characterization. Pharmaceuticals 13:158. https://doi.org/10.3390/ph13080158
282. Mout R, Ray M, Yesilbag Tonga G, Lee Y-W, Tay T, Sasaki K, Rotello VM (2017) Direct cytosolic delivery of CRISPR/Cas9-ribonucleoprotein for efficient gene editing. ACS Nano 11:2452–2458. https://doi.org/10.1021/acsnano.6b07600
283. Doyle C, Higginbottom K, Swift TA, Winfield M, Bellas C, Benito-Alifonso D, Fletcher T, Galan MC, Edwards K, Whitney HM (2019) A simple method for spray-on gene editing in planta
284. Ramakrishnan M, Kaul R, Sharma A, Ahmad Z, Vijayakanth V, Keerthana K, Gao Z, Zhou M, Wei Q (2025) CRISPR RNP-mediated transgene-free genome editing in plants: advances, challenges and future directions for tree species. Plant Cell Environ
285. Vats S, Kumawat S, Brar J, Kaur S, Yadav K, Magar SG, Jadhav PV, Salvi P, Sonah H, Sharma S et al (2022) Opportunity and challenges for nanotechnology application for genome editing in plants. Plant Nano Biol. 1:100001. https://doi.org/10.1016/j.plana.2022.100001
286. Fujita S, Tsuchiya K, Numata K (2021) All-peptide-based polyion complex vesicles: facile preparation and encapsulation of the protein in active form. ACS Polym Au 1:30–38. https://doi.org/10.1021/acspolymersau.1c00008
287. Zhao XY, Su YH, Cheng ZJ, Zhang XS (2008) Cell fate switch during in vitro plant organogenesis. J Integr Plant Biol 50:816–824. https://doi.org/10.1111/j.1744-7909.2008.00701.x
288. An C, Sun C, Li N, Huang B, Jiang J, Shen Y, Wang C, Zhao X, Cui B, Wang C et al (2022) Nanomaterials and nanotechnology for the delivery of agrochemicals: strategies towards sustainable agriculture. J Nanobiotechnol 20:11. https://doi.org/10.1186/s12951-021-01214-7
289. Li X, Han J, Wang X, Zhang Y, Jia C, Qin J, Wang C, Wu J-R, Fang W, Yang Y-W (2019) A triple-stimuli responsive hormone delivery system equipped with pillararene magnetic nanovalves. Mater Chem Front 3:103–110. https://doi.org/10.1039/C8QM00509E
290. Kokina I, Jahundoviča I, Mickeviča I, Jermaļonoka M, Strautiņš J, Popovs S, Ogurcovs A, Sledevskis E, Polyakov B, Gerbreders V (2017) Target transportation of auxin on mesoporous Au/SiO_2 nanoparticles as a method for somaclonal variation increasing in flax (L. Usitatissimum). J Nanomater 1–9. https://doi.org/10.1155/2017/7143269
291. Kim DH, Gopal J, Sivanesan I (2017) Nanomaterials in plant tissue culture: the disclosed and undisclosed. RSC Adv 7:36492–36505. https://doi.org/10.1039/C7RA07025J
292. Lowe K, La Rota M, Hoerster G, Hastings C, Wang N, Chamberlin M, Wu E, Jones T, Gordon-Kamm W (2018) Rapid genotype "independent" Zea Mays L. (Maize) transformation via direct somatic embryogenesis. Vitr Cell Dev Biol Plant 54:240–252. https://doi.org/10.1007/s11627-018-9905-2
293. Bouchabké-Coussa O, Obellianne M, Linderme D, Montes E, Maia-Grondard A, Vilaine F, Pannetier C (2013) Wuschel overexpression promotes somatic embryogenesis and induces organogenesis in cotton (Gossypium Hirsutum L.) tissues cultured in vitro. Plant Cell Rep 32:675–686. https://doi.org/10.1007/s00299-013-1402-9
294. Maher MF, Nasti RA, Vollbrecht M, Starker CG, Clark MD, Voytas DF (2020) Plant gene editing through de novo induction of meristems. Nat Biotechnol 38:84–89. https://doi.org/10.1038/s41587-019-0337-2
295. Liu Y, Wang H, Kamei K, Yan M, Chen K, Yuan Q, Shi L, Lu Y, Tseng H (2011) Delivery of intact transcription factor by using self-assembled supramolecular nanoparticles. Angew Chemie Int Ed 50:3058–3062. https://doi.org/10.1002/anie.201005740

296. Patel S, Jung D, Yin PT, Carlton P, Yamamoto M, Bando T, Sugiyama H, Lee K-B (2014) NanoScript: a nanoparticle-based artificial transcription factor for effective gene regulation. ACS Nano 8:8959–8967. https://doi.org/10.1021/nn501589f
297. Wang JW, Goh N, Squire H, Ni M, Lien E, González-Grandío E, Landry MP (2022) Fluorescence complementation enables quantitative imaging of cell penetrating peptide-mediated protein delivery in plants including WUSCHEL transcription factor
298. Svitashev S, Schwartz C, Lenderts B, Young JK, Mark Cigan A (2016) Genome editing in maize directed by CRISPR–Cas9 ribonucleoprotein complexes. Nat Commun 7:13274. https://doi.org/10.1038/ncomms13274
299. Chang M, Chou J-C, Lee H-J (2005) Cellular internalization of fluorescent proteins via arginine-rich intracellular delivery peptide in plant cells. Plant Cell Physiol 46:482–488. https://doi.org/10.1093/pcp/pci046
300. Chang M, Chou J, Chen C, Liu BR, Lee H (2007) Noncovalent protein transduction in plant cells by macropinocytosis. New Phytol 174:46–56. https://doi.org/10.1111/j.1469-8137.2007.01977.x
301. Jain A, Yadav BK, Chugh A (2015) Marine antimicrobial peptide tachyplesin as an efficient nanocarrier for macromolecule delivery in plant and mammalian cells. FEBS J 282:732–745. https://doi.org/10.1111/febs.13178
302. Lu S-W, Hu J-W, Liu BR, Lee C-Y, Li J-F, Chou J-C, Lee H-J (2010) Arginine-rich intracellular delivery peptides synchronously deliver covalently and noncovalently linked proteins into plant cells. J Agric Food Chem 58:2288–2294. https://doi.org/10.1021/jf903039j
303. Sharma AK, Sharma MK (2009) Plants as bioreactors: recent developments and emerging opportunities. Biotechnol Adv 27:811–832. https://doi.org/10.1016/j.biotechadv.2009.06.004
304. Hampoelz B, Andres-Pons A, Kastritis P, Beck M (2019) Structure and assembly of the nuclear pore complex. Annu Rev Biophys 48:515–536. https://doi.org/10.1146/annurev-biophys-052118-115308
305. Frey S, Rees R, Schünemann J, Ng SC, Fünfgeld K, Huyton T, Görlich D (2018) Surface properties determining passage rates of proteins through nuclear pores. Cell 174:202-217.e9. https://doi.org/10.1016/j.cell.2018.05.045
306. Timney BL, Raveh B, Mironska R, Trivedi JM, Kim SJ, Russel D, Wente SR, Sali A, Rout MP (2016) Simple rules for passive diffusion through the nuclear pore complex. J Cell Biol 215:57–76. https://doi.org/10.1083/jcb.201601004
307. Inui M, Hamada Y, Sejima N, Ueda N, Katayama T, Ono K, Nagahama K (2022) Exportin-inspired artificial cell nuclear-exporting nanosystems. Nanoscale Adv 4:2637–2641. https://doi.org/10.1039/D2NA00050D
308. Akita H, Masuda T, Nishio T, Niikura K, Ijiro K, Harashima H (2011) Improving in vivo hepatic transfection activity by controlling intracellular trafficking: the function of GALA and maltotriose. Mol Pharm 8:1436–1442. https://doi.org/10.1021/mp200189s
309. Kosugi S, Hasebe M, Matsumura N, Takashima H, Miyamoto-Sato E, Tomita M, Yanagawa H (2009) Six classes of nuclear localization signals specific to different binding grooves of importin α. J Biol Chem 284:478–485. https://doi.org/10.1074/jbc.M807017200
310. Bruce BD (2000) Chloroplast transit peptides: structure, function and evolution. Trends Cell Biol 10:440–447. https://doi.org/10.1016/S0962-8924(00)01833-X
311. Wong MH, Misra RP, Giraldo JP, Kwak S-Y, Son Y, Landry MP, Swan JW, Blankschtein D, Strano MS (2016) Lipid exchange envelope penetration (LEEP) of nanoparticles for plant engineering: a universal localization mechanism. Nano Lett 16:1161–1172. https://doi.org/10.1021/acs.nanolett.5b04467
312. Newkirk GM, Jeon S-J, Kim H-I, Sivaraj S, De Allende P, Castillo C, Jinkerson RE, Giraldo JP (2023) DNA delivery by high aspect ratio nanomaterials to algal chloroplasts. Environ Sci Nano 10:2890–2903. https://doi.org/10.1039/D3EN00268C
313. Daniell H, Chase C (eds) (2004) Molecular biology and biotechnology of plant organelles. Springer Netherlands, Dordrecht. ISBN 978-1-4020-2713-0
314. Schwab F, Zhai G, Kern M, Turner A, Schnoor JL, Wiesner MR (2016) Barriers, pathways and processes for uptake, translocation and accumulation of nanomaterials in plants-critical review. Nanotoxicology 10:257–278. https://doi.org/10.3109/17435390.2015.1048326

315. Farre J-C (2001) Gene expression in isolated plant mitochondria: high fidelity of transcription, splicing and editing of a transgene product in electroporated organelles. Nucleic Acids Res 29:2484–2491. https://doi.org/10.1093/nar/29.12.2484
316. Fox TD, Sanford JC, McMullin TW (1988) Plasmids can stably transform yeast mitochondria lacking endogenous MtDNA. Proc Natl Acad Sci 85:7288–7292. https://doi.org/10.1073/pnas.85.19.7288
317. Randolph-Anderson BL, Boynton JE, Gillham NW, Harris EH, Johnson AM, Dorthu M-P, Matagne RF (1993) Further characterization of the respiratory deficient Dum-1 mutation of chlamydomonas reinhardtii and its use as a recipient for mitochondrial transformation. Mol Gen Genet MGG 236–236:235–244. https://doi.org/10.1007/BF00277118
318. Murcha MW, Kmiec B, Kubiszewski-Jakubiak S, Teixeira PF, Glaser E, Whelan J (2014) Protein import into plant mitochondria: signals, machinery, processing, and regulation. J Exp Bot 65:6301–6335. https://doi.org/10.1093/jxb/eru399

AI-Nano Platforms for Enhancing CRISPR/Cas-Mediated Genome Editing in Precision Agriculture

Vikas and Rajiv Ranjan

Abstract The convergence of artificial intelligence (AI), nanotechnology, and CRISPR/Cas-based genome editing presents a transformative approach to address the complex challenges facing modern agriculture, especially in the context of climate change and sustainability. This chapter explores the development and application of AI-nano platforms to enhance the precision, efficiency, and scalability of CRISPR/Cas-mediated genome editing in crops. Nanotechnology plays a pivotal role in delivering CRISPR components into plant cells by overcoming barriers such as the rigid cell wall and enabling targeted, non-toxic, and efficient gene editing. Meanwhile, AI offers powerful tools for optimizing single-guide RNA (sgRNA) design, predicting editing outcomes, and managing large-scale genomic data to support decision-making in crop improvement. The synergy between AI and nanotechnology accelerates precision agriculture by enabling real-time monitoring, trait prediction, and adaptive responses to environmental stress. Through case studies and recent advances, the chapter highlights how AI-nano-CRISPR systems have been successfully used to develop climate-resilient crops with enhanced yield, disease resistance, and stress tolerance. Additionally, the chapter discusses the ethical, regulatory, and biosafety considerations of applying such advanced technologies in agriculture, emphasizing the need for transparent governance and public engagement. The prospects of integrating AI-nano-CRISPR platforms with other digital agriculture tools, such as IoT and blockchain, are also explored to envision a more sustainable, climate-smart agricultural paradigm. This chapter provides researchers, policymakers, and industry stakeholders with a comprehensive understanding of the potential and implications of these converging technologies in revolutionizing plant genome editing for global food security.

Keywords AI-nanotechnology integration · CRISPR/Cas genome editing · Precision agriculture · Climate-smart crops · Sustainable biotechnology

Vikas · R. Ranjan (✉)
Plant Biotechnology Lab, Department of Botany, Faculty of Science, Dayalbagh Educational Institute, Agra, India
e-mail: rajivranjan@dei.ac.in

J.-T. Chen (ed.), *AI-Nano-Enabled Climate-Smart Agriculture and Sustainability*, Smart Nanomaterials Technology, https://doi.org/10.1007/978-981-92-1694-9_19

1 Introduction

The escalating challenges of global food security, exacerbated by climate change, population growth, and diminishing arable land, necessitate transformative approaches in agriculture. Traditional breeding methods, while foundational, often fall short in addressing the rapid pace of environmental and biological stresses affecting crop productivity. In this context, the convergence of advanced biotechnologies, namely, CRISPR/Cas genome editing, nanotechnology, and artificial intelligence (AI), offers unprecedented opportunities to revolutionize precision agriculture and promote sustainability.

CRISPR/Cas Systems: Precision Tools for Crop Improvement

CRISPR/Cas systems have emerged as powerful tools for precise genome editing, enabling targeted modifications to enhance desirable traits in crops. Recent advancements have demonstrated the efficacy of CRISPR/Cas9 in improving agronomic characteristics such as yield, stress tolerance, and nutritional content. For instance, CRISPR/Cas9-mediated editing of the eukaryotic translation initiation factor eIF4E gene has conferred resistance to various viruses in crops like barley, melon, and wheat [10]. Similarly, the knockout of α-gliadin genes in wheat has resulted in low gluten varieties with reduced immunoreactivity, addressing health concerns related to gluten consumption [34].

Furthermore, CRISPR/Cas systems have been instrumental in enhancing the nutritional profile of staple crops. The development of vitamin D-rich tomatoes through CRISPR-mediated gene editing exemplifies the potential of this technology to address micronutrient deficiencies prevalent in human populations [3]. Such innovations underscore the versatility of CRISPR/Cas systems in tailoring crops to meet specific nutritional and agronomic needs.

Nanotechnology: Enhancing Delivery and Efficacy of Genome Editing

Despite the precision of CRISPR/Cas systems, efficient delivery of genome editing components into plant cells remains a significant challenge. Nanotechnology offers promising solutions by facilitating the delivery of CRISPR/Cas reagents through nanoscale carriers. Nanoparticles such as carbon nanotubes, mesoporous silica nanoparticles, and chitosan-based systems have been explored for their ability to traverse plant cell walls and deliver genetic material effectively [33].

The integration of nanotechnology not only improves the delivery efficiency of CRISPR/Cas components but also minimizes potential off-target effects by enabling controlled release and targeted delivery. Moreover, nanocarriers can be engineered to respond to specific environmental stimuli, providing a responsive system for genome editing in plants under varying conditions.

Artificial Intelligence: Optimizing Genome Editing and Agricultural Practices

The vast and complex datasets generated in genomics and agriculture necessitate advanced analytical tools for meaningful interpretation. AI, particularly machine learning algorithms, has been increasingly employed to enhance the precision and efficiency of genome editing. AI-driven models assist in designing optimal single guide RNAs (sgRNAs) for CRISPR/Cas systems, predicting off-target effects, and modeling gene editing outcomes (Lee M., 2023).

Beyond genome editing, AI applications extend to various aspects of precision agriculture, including crop monitoring, pest and disease prediction, and resource optimization. By analyzing data from diverse sources such as satellite imagery, sensor networks, and genomic databases, AI enables informed decision-making, leading to improved crop management and sustainability [24].

Synergistic Integration for Climate Smart Agriculture

The convergence of CRISPR/Cas genome editing, nanotechnology, and AI holds transformative potential for developing climate-smart agricultural practices. This integrated approach facilitates the rapid development of crop varieties with enhanced resilience to abiotic stresses such as drought, salinity, and extreme temperatures, as shown in Fig. 1. For example, CRISPR/Cas-mediated editing of ABA induced transcription repressors (AITRs) has been employed to elucidate mechanisms behind salt tolerance in crops, contributing to the development of stress-resilient varieties [2].

Moreover, the integration of AI and nanotechnology enhances the precision and efficiency of genome editing, enabling the development of crops tailored to specific environmental conditions and agricultural practices. This holistic approach aligns with the goals of sustainable agriculture by promoting resource efficiency, reducing environmental impact, and ensuring food security in the face of climate change.

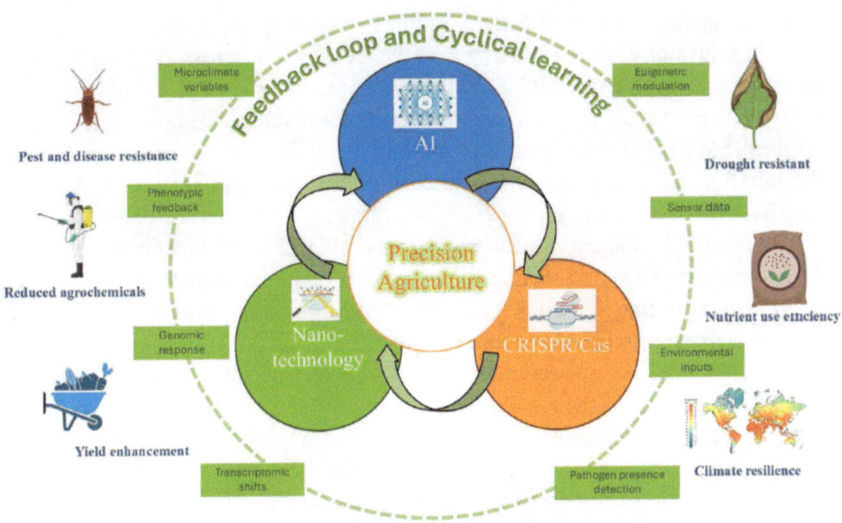

Fig. 1 Synergistic integration of AI, nanotechnology, and CRISPR/Cas in precision agriculture

2 Nanotechnology's Role in Enhancing CRISPR/Cas Delivery

Nanotechnology plays a pivotal role in overcoming one of the major limitations of CRISPR/Cas systems in plants: efficient and targeted delivery. Traditional methods struggle with plant cell walls and often involve transgenic approaches. Nanocarriers such as carbon nanotubes, silica nanoparticles, and liposomes offer a non-integrative, versatile alternative, capable of transporting CRISPR components directly into plant cells with high precision and minimal toxicity. Their customizable properties allow for tissue-specific and stimuli-responsive release, aligning well with the goals of precision agriculture. This section explores how nanotechnology is transforming CRISPR delivery, enabling safer, more effective genome editing for climate-smart and sustainable crop improvement.

2.1 *Overcoming Delivery Barriers*

One of the primary obstacles in applying CRISPR/Cas-mediated genome editing to plants is the difficulty of delivering biomolecules across the rigid plant cell wall, which is composed mainly of cellulose, hemicellulose, and lignin. This complex structure hinders the uptake of large biomolecules such as plasmid DNA, RNA, or ribonucleoprotein (RNP) complexes [5]. Traditional delivery methods like Agrobacterium-mediated transformation or biolistics often show limited efficiency, species-specificity, and can cause DNA damage or tissue necrosis [5, 37].

Nanomaterials offer a promising solution by enabling the transport of CRISPR components through passive diffusion or active penetration across the plant cell wall. Their small size (typically < 100 nm) allows them to bypass the physical constraints posed by the cell wall and plasma membrane [5]. For example, single-walled carbon nanotubes (SWCNTs) have been used to deliver plasmid DNA and CRISPR/Cas9 RNPs into *Nicotiana benthamiana* and *Arabidopsis thaliana* with high efficiency and without DNA integration [5, 12].

Mesoporous silica nanoparticles (MSNs), another class of nanocarriers, provide customizable pore sizes and surface functionalization, making them ideal for encapsulating and releasing CRISPR components within plant cells [25]. Similarly, chitosan-based nanoparticles offer biodegradable and biocompatible properties while facilitating electrostatic interaction with negatively charged nucleic acids, improving uptake and protection of the payload [17].

Notably, these nanocarriers allow for transient expression of genome editing tools, avoiding transgene integration and addressing biosafety concerns [5]. By optimizing nanoparticle composition and functionalization, it is possible to enhance the targeted delivery and genome editing efficiency across a range of plant species, thus overcoming one of the most significant barriers in precision agriculture [12, 37].

2.2 Types of Nanocarriers for CRISPR Delivery

A variety of nanocarriers have been explored for the efficient delivery of CRISPR/Cas components into plant cells, each offering unique physicochemical properties tailored to overcome species-specific and cellular barriers. These nanocarriers serve as vehicles for transporting DNA, RNA, or protein-based genome editing tools through the plant cell wall and membrane while protecting them from degradation [5].

Carbon-based nanomaterials, particularly single-walled carbon nanotubes (SWCNTs), have shown high efficacy in delivering CRISPR/Cas plasmids and ribonucleoproteins (RNPs) into plant tissues without causing genomic integration [5]. Their high aspect ratio and tunable surface chemistry allow for functionalization with polymers or peptides to facilitate cellular uptake and nuclear targeting [12]. SWCNTs also exhibit minimal cytotoxicity and can enter cells through endocytosis or passive diffusion.

Mesoporous silica nanoparticles (MSNs) provide a versatile platform with adjustable pore size and surface charge, enabling the encapsulation of large biomolecules such as Cas9 and guide RNA [25]. Functionalized MSNs can respond to pH or enzymatic triggers to release cargo in a controlled manner within the plant cell environment [25, 37].

Lipid-based nanoparticles (LNPs), widely used in animal systems, are being adapted for plant applications. They form stable complexes with nucleic acids and proteins, enabling efficient endosomal escape and intracellular trafficking [37].

Cationic polymer nanoparticles, such as chitosan or polyethyleneimine (PEI), interact electrostatically with negatively charged DNA or RNA molecules, enhancing stability and uptake through plant membranes [17]. Chitosan, in particular, offers biodegradability and low toxicity, making it suitable for agricultural uses.

The choice of nanocarrier depends on the type of plant tissue, the editing payload (plasmid DNA vs. RNP), and the desired outcome (transient vs. stable expression). Each nanomaterial provides different advantages in terms of targeting specificity, biocompatibility, and editing efficiency [5, 12, 37] (Fig. 2).

2.3 Advantages over Traditional Methods

Nanocarrier-mediated delivery systems offer several advantages over traditional methods such as Agrobacterium-mediated transformation and biolistics, which are often limited by host specificity, low transformation efficiency, and potential genomic instability [5]. Agrobacterium, though effective in dicots, performs poorly in monocots and has regulatory concerns due to its integration of foreign DNA. Biolistics, while more broadly applicable, causes mechanical damage to plant tissues and often results in random and multiple insertions of transgenes [37].

Nanocarriers enable species-independent, non-integrative, and transient delivery of CRISPR/Cas components, particularly in the form of ribonucleoproteins (RNPs),

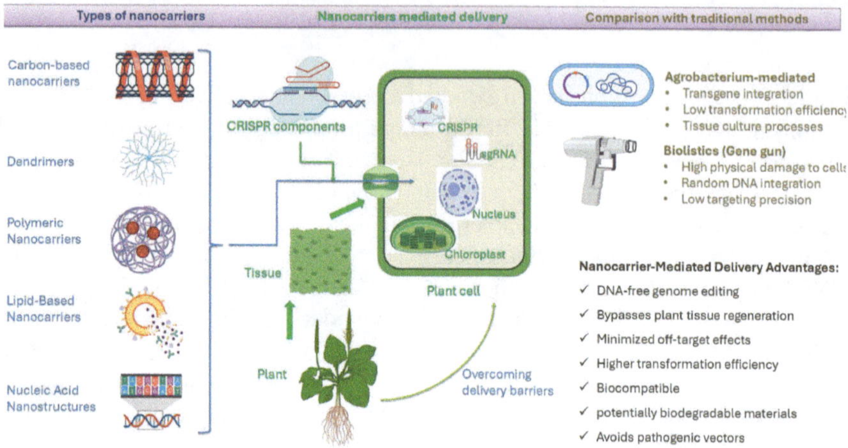

Fig. 2 Nanocarrier types and their delivery mechanisms in plants

which significantly reduce off-target effects and avoid insertional mutagenesis [5, 12]. They allow for precision targeting through surface functionalization and protect nucleic acids or proteins from enzymatic degradation during delivery [25].

Furthermore, nanocarriers offer minimal cytotoxicity, biodegradability, and scalable application, making them more suitable for field-level implementation in precision agriculture. Unlike gene guns or Agrobacterium, nanomaterials can be adapted for topical application, allowing genome editing without tissue culture or regeneration steps [5, 17]. These features make nano-enabled CRISPR delivery a promising alternative for next-generation sustainable crop improvement.

3 AI Integration in CRISPR/Cas Genome Editing

Artificial intelligence (AI) is transforming CRISPR/Cas genome editing by enabling smarter, data-driven decisions across the editing workflow. Traditional sgRNA design and prediction relied on rule-based methods, often limited by off-target effects and inefficiencies. AI models such as deep learning and ensemble algorithms now optimize sgRNA design, simulate editing outcomes, and manage complex genomic data with higher accuracy. In agriculture, this integration allows for precise trait targeting, multiplex editing, and real-time data analysis. This section explores how AI enhances CRISPR/Cas applications, improving both efficiency and predictability in plant genome editing for climate-resilient, high-performance crops.

3.1 *Optimizing sgRNA Design Using AI*

One of the critical challenges in CRISPR/Cas-based genome editing is the design of single-guide RNAs (sgRNAs) with high specificity and efficacy. sgRNAs guide the Cas enzyme to the target DNA locus, but improper design can lead to off-target effects, reducing precision and raising biosafety concerns. Traditional sgRNA design relied on manually curated rules or simple scoring matrices, but these approaches lacked predictive robustness across diverse genomes.

Artificial intelligence (AI), particularly deep learning and ensemble models, has revolutionized this step by enabling data-driven optimization of sgRNA sequences. Machine learning (ML) algorithms such as support vector machines (SVMs), random forests (RFs), and neural networks have been successfully employed to learn complex features that influence sgRNA activity, including GC content, thermodynamic stability, chromatin accessibility, and sequence context [1, 6]. These models outperform traditional tools by integrating multidimensional data and updating iteratively based on large training datasets derived from high-throughput CRISPR screens.

For instance, deep convolutional neural networks (CNNs) have been implemented in tools like DeepCRISPR and DeepSpCas9, which learn hierarchical features from raw nucleotide sequences without manual feature engineering [35]. These tools have demonstrated higher predictive accuracy than earlier rule-based tools like CCTop or CHOPCHOP, especially for novel sgRNA sequences not represented in training datasets.

Recent studies also introduce graph neural networks (GNNs) for off-target prediction and sgRNA activity scoring [20]. These models treat DNA as graph structures rather than flat sequences, enabling a more nuanced interpretation of spatial relationships between nucleotides and epigenomic features [31]. By incorporating topological data, GNNs further improve specificity predictions for complex plant genomes.

Moreover, attention-based models and transformer architectures, inspired by natural language processing, have been adapted to "read" DNA sequences, offering a new frontier for sgRNA design optimization [26]. AI-enhanced sgRNA design holds the promise of minimizing false-positive edits, increasing on-target activity, and accelerating the pipeline from hypothesis to functional genome modification [19] (Fig. 3).

3.2 *Predictive Modeling of Gene Editing Outcomes*

Beyond sgRNA design, AI also plays a pivotal role in predicting the functional consequences of CRISPR edits. Predictive modeling refers to the computational simulation of possible outcomes post-editing, ranging from mutation profiles to phenotypic expressions based on genome context, sequence characteristics, and biological pathways.

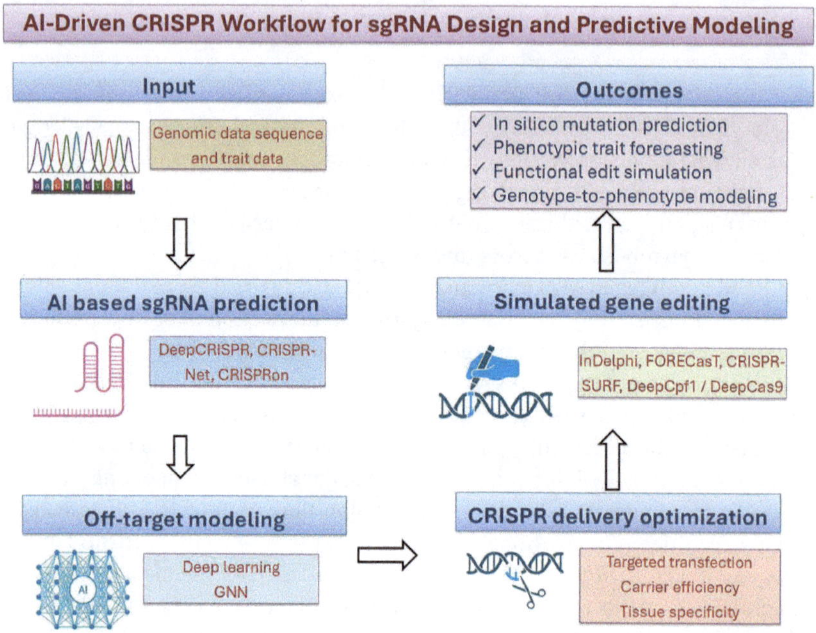

Fig. 3 AI-driven CRISPR workflow illustrating sgRNA design, off-target modeling, gene-editing simulation, and delivery optimization for precision agriculture applications

Machine learning models are now routinely used to simulate the likelihood and impact of indels generated by non-homologous end joining (NHEJ) or homology-directed repair (HDR). For example, algorithms like FORECasT and InDelphi predict the types and frequencies of insertions or deletions (indels) likely to occur after Cas9 cleavage [13].

These models utilize sequence features from the protospacer and flanking regions, and are trained on large datasets of experimentally validated edits. By simulating these outcomes in silico, researchers can select target sites that maximize desired mutations while avoiding deleterious effects.

In agricultural contexts, where the goal is often to knock-in or knock-out specific traits like disease resistance or stress tolerance, such preemptive modeling tools are invaluable. They reduce experimental burden, improve predictability, and enhance trait stability across generations (Table 1).

AI has also been applied in multiplex editing scenarios, where several genes are edited simultaneously. Models like CRISPRater and CRISPRon assess potential crosstalk or competitive binding effects between sgRNAs, allowing for better coordination of multiplexed approaches [7].

Additionally, ensemble learning, which combines predictions from multiple algorithms, has shown increased robustness in variable genomic backgrounds such as polyploid plants or transgenic systems [21].

Table 1 AI models for sgRNA design, efficiency prediction, and off-target analysis

AI tool	Algorithm type	Key features	Accuracy (AUC/F1)	Primary application	Reference
DeepCRISPR	CNN + RNN	Sequence embedding, chromatin info, multi-task learning	AUC ~0.93	sgRNA efficacy + off-target scoring	[4]
DeepSpCas9	Convolutional neural network (CNN)	Raw sequence-based prediction with end-to-end deep learning	AUC ~0.92	sgRNA activity scoring	[11]
CRISPR-net	LSTM (RNN)	Captures sequential nucleotide dependencies	AUC ~0.88	Off-target detection	[35]
InDelphi	Probabilistic ML model	Predicts insertions/deletions from double-strand breaks	N/A	Repair outcome modeling	[22]
CRISPRon	Random forest ensemble	Integrates epigenetic, thermodynamic, and sequence features	AUC ~0.94	sgRNA efficiency ranking	[35]
FORECasT	Decision trees	Outcome probability estimation for NHEJ editing	R^2 ~ 0.80	Mutation spectrum prediction	Liu et al. [14]
CRISPR-DIP	Attention-based model	Interpretable off-target prediction using attention layers	F1-score ~ 0.91	Explainable AI (XAI) for CRISPR	[11]
GUIDES	Gradient boosting	Integrates gene essentiality and sequence rules	AUC ~0.90	sgRNA library optimization	[30]
CRISPR-AI	Hybrid CNN-LSTM	Sequence-to-function modeling with contextual awareness	AUC ~ 0.91	Multi-species sgRNA design	Toufikuzzaman et al. [26]

3.3 AI for Genomic Data Analysis and Management

CRISPR genome editing in precision agriculture relies heavily on genomic data ranging from raw sequence data to epigenomic annotations and transcriptomic profiles. AI has proven indispensable in analyzing, integrating, and managing these data types.

First, AI facilitates variant calling and annotation, essential for identifying candidate loci for editing. Using unsupervised learning, AI algorithms can cluster phenotypic traits with genomic regions, identifying targets for precision trait engineering [29].

Second, in functional genomics, AI models correlate gene expression patterns with environmental stress responses, aiding in the selection of genes whose editing could improve drought or pest resistance [36]. Tools integrating AI with transcriptomic data, such as CRISPR-VAE, simulate how edits affect transcriptional networks.

Third, AI enhances metadata management across genome editing experiments. By employing automated classification and tagging systems, researchers can trace the provenance, parameters, and outcomes of experiments crucial for reproducibility and regulatory documentation.

In plant genome editing, where large-scale phenotyping generates terabytes of data, AI-assisted image analysis and sensor fusion are becoming standard. AI models process imaging data from drones or satellites to associate phenotypic outcomes with specific genomic edits [8].

Finally, explainable AI (XAI) is gaining traction. These models provide interpretable outputs (e.g., SHAP values or saliency maps), helping regulators and researchers understand why certain sgRNA sequences were predicted to be effective or safe [23].

4 Applications in Precision Agriculture

The integration of artificial intelligence (AI), nanotechnology, and CRISPR/Cas genome editing is revolutionizing precision agriculture. This triad, often referred to as AI-nano-CRISPR platforms, offers highly targeted and efficient methods to enhance crop traits, respond to real-time environmental signals, and implement adaptive interventions. These systems not only address traditional agronomic goals like yield and pest resistance but also contribute to climate resilience and sustainable practices.

4.1 Trait Improvement Through AI-Nano-CRISPR Platforms

CRISPR/Cas gene editing enables precise modifications of plant genomes, but its efficacy largely depends on delivery mechanisms and predictive design. Nanotechnology facilitates this delivery by encapsulating CRISPR components (e.g., sgRNAs and Cas9 proteins) into nanoscale carriers that traverse plant cell walls with minimal damage. AI further enhances this process by optimizing sgRNA design, predicting off-target effects, and modeling phenotype-genotype correlations for desired traits.

One prominent application is drought tolerance. For instance, researchers have edited the *DREB* and *AREB* gene families' transcription factors that regulate stress response pathways in crops like rice and maize using CRISPR-Cas systems delivered via carbon nanotubes (CNTs) and layered double hydroxide (LDH) nanoparticles. AI models helped predict promoter regions with high expression stability under water-limited conditions, thereby enhancing success rates [1].

In the domain of pest resistance, AI-guided CRISPR has been used to target susceptibility (S) genes that facilitate pathogen invasion. For example, in tomato and wheat, gene targets such as *MLO* and *eIF4E* were successfully knocked out using AI-curated sgRNA libraries, reducing powdery mildew and viral susceptibility, respectively. Nano-formulations such as mesoporous silica nanoparticles enabled these edits without relying on Agrobacterium-based methods, minimizing regulatory barriers associated with transgenic vectors [7].

Yield enhancement is another frontier. By multiplexing CRISPR edits, simultaneously modifying genes associated with grain size (*GS3*), tillering (*IPA1*), and nutrient uptake, scientists have engineered high-yield rice cultivars. AI was instrumental in simulating the combined phenotypic effects of these edits, allowing for targeted trait stacking with fewer trial-and-error cycles. Nanocarriers like chitosan-coated liposomes improved delivery efficacy and biosafety.

4.2 Real-Time Monitoring with AI-Powered Nano Sensors

While CRISPR edits optimize genetic potential, AI-nanosensors support the phenotypic realization of that potential through dynamic environmental monitoring. These sensors are embedded in or around plants and are capable of detecting abiotic and biotic stressors, including drought, salinity, nutrient deficiency, and pest activity at early stages.

Nanosensors, often made from quantum dots, gold nanoparticles, or graphene oxide, can detect specific biomarkers such as reactive oxygen species (ROS), plant hormones (like abscisic acid), or volatile organic compounds (VOCs). AI algorithms, particularly convolutional neural networks and decision-tree ensembles, process this sensor data to trigger alerts or automated responses. For example, a nanosensor might detect elevated ethylene levels as a stress signal and activate irrigation or prompt a CRISPR-induced response if engineered bioswitches are present.

Recent advancements have seen the development of "smart leaves" plant tissues embedded with biocompatible nanosensors that communicate wirelessly with decision-support systems. AI algorithms analyze multi-modal inputs (spectral, thermal, chemical) and deliver actionable insights for farmers via mobile platforms. These systems allow real-time adjustments in irrigation, pesticide application, and nutrient delivery, reducing environmental footprints and input costs.

4.3 Case Studies of Integrated Deployment

Several field and lab-scale case studies exemplify the effectiveness of AI-nano-CRISPR platforms:

- **Maize Drought Tolerance in Sub-Saharan Africa**: A pilot project used lipid nanoparticle-encapsulated CRISPR-Cas9 to edit *ARGOS8*, a gene associated with ethylene response, which improved drought tolerance by 15%. AI models guided sgRNA design, while embedded soil-moisture nanosensors allowed AI systems to schedule irrigation efficiently [21].
- **Rice Yield Enhancement in China**: A multi-institutional collaboration used deep learning to identify gene clusters responsible for panicle architecture. Simultaneous editing via AI-guided CRISPR, delivered by LDH nanosheets, led to a 25% yield increase. Monitoring drones equipped with nanosensors tracked physiological parameters during the growing season.
- **Smart Greenhouse in the Netherlands**: A fully automated greenhouse used AI-integrated nanobiosensors to detect aphid infestations. CRISPR-edited lettuce lines with pest-resistant traits were monitored in real-time, enabling autonomous responses such as targeted release of biocontrol agents.

5 Ethical, Environmental, and Regulatory Considerations

The integration of artificial intelligence (AI), nanotechnology, and CRISPR/Cas genome editing into agriculture presents transformative possibilities for sustainable food production. However, these innovations also raise profound ethical, environmental, and regulatory concerns that must be addressed to ensure responsible implementation. Societal acceptance and long-term ecological compatibility are critical, particularly as the release of gene-edited crops and the deployment of nanomaterials intersect with global biodiversity, biosafety, and public trust.

5.1 *Biosafety and Bioethics*

The use of CRISPR/Cas to modify crop genomes prompts critical ethical and ecological questions, particularly regarding the release of edited organisms into the environment. While genome editing is more precise than traditional genetic modification (GM), the long-term effects of these changes on ecosystems are not fully understood. Unintended off-target mutations, though minimized by AI optimization, may still occur, raising concerns about horizontal gene flow to wild relatives or non-target species [1].

Nanocarriers used for CRISPR delivery, such as carbon nanotubes, quantum dots, or layered double hydroxides, further complicate biosafety. Many of these materials are not biodegradable and may accumulate in soil or aquatic systems, affecting microbial communities, plant physiology, and food safety [21]. For example, nanosilver particles have shown toxicity to beneficial soil organisms such as *Rhizobium* spp., potentially disrupting nitrogen fixation cycles. The ethical use of such technologies demands rigorous assessment not only of their efficacy but also of their ecological persistence, degradability, and toxicity profiles.

From a bioethical standpoint, there's also concern about equitable access and techno-economic disparity. Technologies that rely on high-cost AI infrastructures and proprietary nano-formulations may be inaccessible to smallholder farmers, thereby exacerbating global inequities in agricultural innovation. Ethical frameworks must address not only environmental risk but also social justice, farmer autonomy, and the right to informed participation.

5.2 *Regulatory Landscape*

The regulatory frameworks governing genome editing and nanotechnology in agriculture vary widely across regions, reflecting different risk paradigms, scientific norms, and public values. While genome-edited organisms may not always be classified as genetically modified organisms (GMOs), their regulatory treatment often overlaps.

In the United States, the USDA considers gene-edited crops exempt from regulation if they do not contain foreign DNA and mimic naturally occurring mutations. This approach emphasizes the product, not the process. Similarly, the FDA assesses food safety based on composition rather than the editing method used [32].

Conversely, the European Union takes a more precautionary stance. The 2018 ruling by the European Court of Justice declared that genome-edited organisms fall under the GMO Directive, meaning they require full risk assessments, traceability, and labeling regardless of whether foreign DNA is introduced [7]. This has led to significant regulatory hurdles for CRISPR-derived crops in Europe.

For nanomaterials, the regulatory landscape is even more fragmented. While the EU has introduced REACH (Registration, Evaluation, Authorisation and Restriction of Chemicals) and the NanoDefine project to assess nanomaterials, these frameworks primarily address chemical safety rather than biological applications in agriculture. In contrast, India and China, both major adopters of nanotech in agriculture, have limited formal regulatory structures tailored to nano-enabled genome editing. Their oversight often falls under broader biosafety or food standards authorities, which may lack the technical capacity to assess nanomaterial-specific risks (Table 2).

Table 2 Regulatory frameworks for genome editing and nanotechnology in agriculture (global comparison)

Region/ Country	CRISPR regulatory status	Nanotech agricultural regulation	Regulatory notes	Reference
United States	Product-based evaluation; CRISPR crops without foreign DNA are often exempt from GMO regulation	EPA regulates nanomaterials under TSCA and FIFRA for environmental safety	No regulatory trigger if the end-product is not transgenic	Wolt et al. [32]
European Union	CRISPR-regulated as GMO under 2001/18/EC directive following 2018 ECJ ruling	Subject to REACH and EFSA risk assessments	Strict precautionary principle; requires full GMO compliance	[28]
China	Case-by-case approval based on product and risk profile; high investment in gene editing R&D	Active promotion, but no comprehensive nano-agri policy yet	Guidelines released for genome-edited crops (2022); nanotech is regulated under general safety laws	[36]
India	CRISPR is regulated under existing GMO rules via GEAC (genetic engineering appraisal committee)	No specific nano-agriculture law; governed by BIS and FSSAI guidelines	Exemption of the genome-edited plants falling under the categories of SDN1 and SDN2	[16]
Brazil	Favors "mutagenesis equivalence"—CRISPR crops may be non-GMO if no foreign DNA is inserted	Limited formal regulation; NANoREG pilot participation	EMBRAPA promotes gene-edited crops for national food security	[9]
Japan	Exempts CRISPR if no transgenes are present; process and product assessed together	Nano-regulated under the chemical substances control law	Encourages innovation while ensuring food and environmental safety	[15]
South Africa	CRISPR is regulated following the GMO rule; full biosafety approval is required	No distinct nano-agri regulatory framework yet	Aligned with the cartagena protocol; regulatory updates underway	[18]

In addition, international treaties like the Cartagena Protocol on Biosafety govern transboundary movement of living modified organisms, yet they do not fully address new breeding techniques (NBTs) or engineered nanocarriers, creating legal ambiguities. As such, harmonizing definitions and risk-assessment protocols for AI-nano-CRISPR platforms remains a global policy challenge.

5.3 Public Perception and Acceptance

Technological acceptance in agriculture is shaped not only by science and regulation but also by societal narratives, trust, and cultural values. Historically, public resistance to GMOs stemmed from concerns over unnaturalness, corporate control, and environmental unpredictability. Similar skepticism may arise around CRISPR and nanotechnology, particularly if transparency and inclusive dialogue are lacking.

Public perception studies show that terminology matters; terms like "gene editing" are viewed more favorably than "genetic engineering" or "GMO" [8]. Moreover, when the benefits of genome editing, such as climate-resilient or pesticide-free crops, are framed in the context of public good, acceptance increases, especially among environmentally conscious consumers.

AI-driven systems, especially those with opaque algorithms or "black box" decision-making, pose unique risks for public distrust. Thus, embedding explainable AI (XAI) models in genome editing pipelines can promote transparency, allowing stakeholders to understand and audit decisions related to gene targets or delivery strategies [23].

Active stakeholder engagement, including farmers, consumers, scientists, and indigenous communities, is critical. Mechanisms such as public consultations, participatory risk assessments, and transparent labeling can foster legitimacy. Educational initiatives, particularly in low- and middle-income countries, should emphasize risk–benefit literacy, demystifying both nanotech and genome editing.

6 Future Perspectives and Research Directions

As agriculture increasingly depends on biotechnology and data-driven systems, the convergence of nanotechnology, artificial intelligence (AI), and CRISPR/Cas genome editing offers a future-proof strategy for sustainable and resilient food systems. However, these platforms are still evolving, and their full potential is yet to be realized. This section outlines key research directions and technological frontiers likely to shape the next generation of AI-nano-CRISPR applications in precision agriculture.

6.1 Advancements in Nanocarriers

While current nanocarriers like carbon nanotubes, mesoporous silica nanoparticles, and lipid-based vesicles have demonstrated efficacy in CRISPR delivery, their limitations in specificity, stability, and biodegradability necessitate further innovation. The future lies in developing smart, targeted nanocarriers that respond to internal or external stimuli, such as pH, temperature, or enzyme activity, to release CRISPR payloads precisely at the desired site [7].

Emerging materials such as DNA nanostructures, metal-organic frameworks (MOFs), and peptide-based nanocarriers are being explored for their biocompatibility and customizable properties. For example, MOFs can encapsulate Cas9-sgRNA complexes and release them in response to cellular redox conditions, thereby improving editing efficiency while minimizing systemic exposure [13]. Additionally, multi-functional nanocarriers capable of delivering multiple sgRNAs simultaneously could enable multiplex editing, a vital requirement for complex trait enhancement in polygenic crops.

The integration of nanocarriers with biosensing and actuation capabilities is another exciting frontier. These "theranostic" carriers could not only deliver gene-editing machinery but also monitor gene expression changes post-delivery, closing the loop between intervention and outcome.

6.2 AI Enhancements

Artificial intelligence will continue to play a transformative role in genome editing, particularly as data volume and complexity escalate. One immediate direction is the improvement of AI models that predict off-target effects, guide sgRNA design, and simulate editing outcomes. Deep learning architectures such as transformers, already successful in natural language processing, are being repurposed for genomic sequence interpretation with impressive accuracy [23].

Another anticipated advancement is the development of explainable AI (XAI) systems for genome editing. These models will not only make accurate predictions but also provide interpretable insights into why certain gene targets or delivery routes are optimal. This is crucial for regulatory acceptance and farmer trust.

Moreover, generative AI models (e.g., GANs or diffusion models) may soon be used to propose novel gene-editing strategies by learning from vast genotype-phenotype datasets. Combined with reinforcement learning, these systems could enable autonomous experimental planning in genome editing labs or field stations [21].

6.3 Integration with Other Technologies

Looking ahead, the AI-nano-CRISPR triad is expected to converge with other emerging technologies, particularly the Internet of Things (IoT), synthetic biology, and edge computing. In smart agriculture, this integration would create closed-loop, self-optimizing systems capable of sensing, editing, and adapting in real time.

For example, IoT-enabled nanosensors embedded in crops could detect early stress signals or pathogen presence and relay the data to an AI system that determines the need for CRISPR-mediated intervention. This could trigger the activation of inducible gene-editing constructs delivered via environmentally responsive nanocarriers, thus achieving real-time genome adaptation [8].

Moreover, integration with cloud-based genomic databases and edge AI chips will enable on-site decision-making without the need for high-speed internet, a game-changer for remote or resource-poor agricultural regions.

The potential also exists to combine AI-nano-CRISPR with precision fermentation, biofoundries, and autonomous robotics, paving the way for a decentralized but interconnected global food system where gene-editing interventions are tailored, traceable, and sustainable.

7 Conclusion

The convergence of artificial intelligence (AI), nanotechnology, and CRISPR/Cas genome editing presents a paradigm shift in precision agriculture. These technologies, when synergistically integrated, offer transformative capabilities from designing highly specific gene edits and delivering them efficiently into plant cells to monitoring environmental cues in real time. Together, they enable unprecedented precision in trait improvement, environmental responsiveness, and sustainable crop management.

AI has proven invaluable in optimizing sgRNA design, modeling editing outcomes, and managing vast genomic datasets. Nanotechnology enhances the precision and safety of CRISPR/Cas delivery, overcoming traditional biological barriers. Meanwhile, CRISPR enables rapid, accurate genetic modifications tailored to evolving climate and agricultural challenges. Applications in drought tolerance, pest resistance, and real-time stress monitoring showcase how this triad is already reshaping modern agriculture.

However, the adoption of AI-nano-CRISPR platforms is not without complexity. Ethical considerations surrounding biosafety, environmental impact, and equitable access must be rigorously addressed. Regulatory frameworks remain fragmented, and public perception plays a crucial role in determining the social license to operate these technologies.

To fully realize the potential of these innovations, a call to action is necessary. Interdisciplinary collaboration among plant scientists, data engineers, nanotechnologists, ethicists, and policymakers must be intensified. Investment in open-access platforms, transparent risk assessment models, and public engagement campaigns will help bridge the gap between innovation and implementation.

Future research should focus on developing biodegradable, target-specific nanocarriers, explainable AI models, and integrative platforms combining IoT, robotics, and bioinformatics. The journey toward climate-smart, high-efficiency agriculture depends on our collective ability to align technological progress with ecological integrity and social trust.

The AI-nano-CRISPR nexus is not just a technical advancement, it is a foundation for building resilient, inclusive, and sustainable agricultural systems for the decades ahead.

References

1. Abbasi AF, Asim MN, Dengel A (2025) Transitioning from wet lab to artificial intelligence: a systematic review of AI predictors in CRISPR. J Transl Med 23(1):153. https://doi.org/10.1186/s12967-024-06013-w
2. Chen S, Zhang N, Zhou G, Hussain S, Ahmed S, Tian H, Wang S (2021) Knockout of the entire family of AITR genes in Arabidopsis leads to enhanced drought and salinity tolerance without fitness costs. BMC Plant Biol 21:1–15
3. Choi S, You MK, Jeon YA, Lee J, Kim J, Park YJ, Choe S (2023) Metabolic engineering to enhance provitamin D3 accumulation in edible tomatoes. GEN Biotechnol. 2(3):219–227
4. Chuai G, Ma H, Yan J, Chen M, Hong N, Xue D et al (2018) DeepCRISPR: optimized CRISPR guide RNA design by deep learning. Genome Biol 19(1):80. https://doi.org/10.1186/s13059-018-1459-4
5. Demirer GS, Zhang H, Matos JL, Goh NS, Cunningham FJ, Sung Y, Landry MP (2019) High aspect ratio nanomaterials enable delivery of functional genetic material without DNA integration in mature plants. Nat Nanotechnol 14(5):456–464. https://doi.org/10.1038/s41565-019-0382-5
6. Dhanjal JK, Dammalapati S, Pal S, Sundar D (2020) Evaluation of off-targets predicted by sgRNA design tools. Genomics 112(5):3609–3614. https://doi.org/10.1016/j.ygeno.2020.04.024
7. Dimauro G, Barletta VS, Catacchio CR, Colizzi L, Maglietta R, Ventura M (2022) A systematic mapping study on machine learning techniques for the prediction of CRISPR/Cas9 sgRNA target cleavage. Comput Struct Biotechnol J 20:5813–5823. https://doi.org/10.1016/j.csbj.2022.10.013
8. Dixit S, Kumar A, Srinivasan K, Vincent PDR, Ramu Krishnan N (2024) Advancing genome editing with artificial intelligence: opportunities, challenges, and future directions. Front Bioeng Biotechnol 11:1335901. https://doi.org/10.3389/fbioe.2023.1335901
9. Ishii T, Araki M (2017) A future scenario of the global regulatory landscape regarding genome-edited crops. GM Crop. Food 8(1):44–56. https://doi.org/10.1080/21645698.2016.1261787
10. Kan J, Cai Y, Cheng C, Chen S, Jiang C, He Z, Yang P (2023) CRISPR/Cas9-guided knockout of eIF4E improves wheat yellow mosaic virus resistance without yield penalty. Plant Biotechnol J 21(5):893
11. Kim N, Choi S, Kim S, Song M, Seo JH, Min S, Kim HH (2024) Deep learning models to predict the editing efficiencies and outcomes of diverse base editors. Nat Biotechnol 42(3):484–497

12. Kwak SY, Lew TTS, Sweeney CJ, Koman VB, Wong MH, Bohmert-Tatarev K et al (2019) Chloroplast-selective gene delivery and expression in planta using chitosan-complexed single-walled carbon nanotube carriers. Nat Nanotechnol 14(5):447–455. https://doi.org/10.1038/s41565-019-0375-4
13. Lee M (2023) Deep learning in CRISPR-Cas systems: a review of recent studies. Front Bioeng Biotechnol 11:1226182. https://doi.org/10.3389/fbioe.2023.1226182
14. Liu X, Wang S, Ai D (2022) Predicting CRISPR/Cas9 repair outcomes by attention-based deep learning framework. Cells 11(11):1847
15. Menz J, Modrzejewski D, Hartung F, Wilhelm R, Sprink T (2020) Genome edited crops touch the market: a view on the global development and regulatory environment. Front Plant Sci 11:586027
16. Ministry of Environment, Forest and Climate Change (2022, March 30) Exemption of the genome edited plants falling under the categories of SDN1 and SDN2 from the provisions of the rules, 1989 (Office Memorandum C-12013/3/2020-CS-III). Government of India. https://forest.delhi.gov.in/sites/default/files/circulars-orders/exemption_of_the_genome_edited_plants_falling_under_the_categories_of_sdn1_and_sdn2_from_the_provisions_of_the_rules_1989.pdf
17. Mitter N, Worrall EA, Robinson KE, Li P, Jain RG, Taochy C, Xu ZP (2017) Clay nanosheets for topical delivery of RNAi for sustained protection against plant viruses. Nat Plants 3(2):1–10. https://doi.org/10.1038/nplants.2016.207
18. Muzhinji N, Ntuli V (2021) Genetically modified organisms and food security in southern Africa: conundrum and discourse. GM Crop. Food 12(1):25–35
19. Patel J, Patel D, Raval A (2025) Artificial intelligence for predictive modeling in CRISPR/Cas9 gene editing: a survey of methods and design strategies. J Gene Med 27(12):e70061. https://doi.org/10.1002/jgm.70061
20. Pindi C, Palermo G (2026) Computation and deep-learning-driven advances in CRISPR genome editing. Nat Struct Mol Biol 1–12. https://doi.org/10.1038/s41594-025-01739-7
21. Rafiq S, Macha MA, Assad A (2025) Machine learning and deep learning for genomic data: a data-centric approach to CRISPR/Cas9 gene editing. Int J Data Sci Anal:1–30. https://doi.org/10.1007/s41060-025-00773-4
22. Shen MW, Arbab M, Hsu JY, Worstell D, Culbertson SJ, Krabbe O, Corn JE (2018) Predictable and precise template-free CRISPR editing of pathogenic variants. Nature 563(7733):646–651. https://doi.org/10.1038/s41586-018-0686-x
23. Sherkatghanad Z, Abdar M, Charlier J, Makarenkov V (2023) Using traditional machine learning and deep learning methods for on-and off-target prediction in CRISPR/Cas9: a review. Brief Bioinform 24(3):bbad131. https://doi.org/10.1093/bib/bbad131
24. Singh A, Mehrotra R, Rajput VD, Dmitriev P, Singh AK, Kumar P, Singh AK (2022) Geoinformatics, artificial intelligence, sensor technology, big data: emerging modern tools for sustainable agriculture. Sustain. Agric. Syst. Technol.:295–313
25. Torney F, Trewyn BG, Lin VSY, Wang K (2007) Mesoporous silica nanoparticles deliver DNA and chemicals into plants. Nat Nanotechnol 2(5):295–300. https://doi.org/10.1038/nnano.2007.108
26. Toufikuzzaman M, Hassan Samee MA, Sohel Rahman M (2024) CRISPR-DIPOFF: an interpretable deep learning approach for CRISPR Cas-9 off-target prediction. Brief Bioinform 25(2):bbad530
27. Toufikuzzaman M, Samee MAH, Rahman MS (2024) CRISPR-DIPOFF: an interpretable deep learning approach for CRISPR Cas-9 off-target prediction. Brief Bioinform 25(2):bbad530. https://doi.org/10.1093/bib/bbad530
28. Turnbull C, Lillemo M, Hvoslef-Eide TA (2021) Global regulation of genetically modified crops amid the gene edited crop boom–a review. Front Plant Sci 12:630396
29. Vadapalli S, Abdelhalim H, Zeeshan S, Ahmed Z (2022) Artificial intelligence and machine learning approaches using gene expression and variant data for personalized medicine. Brief Bioinform 23(5):bbac191

30. Vinceti A, Iannuzzi RM, Boyle I, Trastulla L, Campbell CD, Vazquez F, Iorio F (2024) A benchmark of computational methods for correcting biases of established and unknown origin in CRISPR-Cas9 screening data. Genome Biol 25(1):192
31. Vinodkumar PK, Ozcinar C, Anbarjafari G (2021) Prediction of sgRNA off-target activity in CRISPR/Cas9 gene editing using graph convolution network. Entropy 23(5):608. https://doi.org/10.3390/e23050608
32. Wolt JD, Wolf C (2018) Policy and governance perspectives for regulation of genome edited crops in the United States. Front Plant Sci 9:1606
33. Yin H, Kauffman KJ, Anderson DG (2017) Delivery technologies for genome editing. Nat Rev Drug Discov 16(6):387–399
34. Yu Z, Yunusbaev U, Fritz A, Tilley M, Akhunova A, Trick H, Akhunov E (2024) CRISPR-based editing of the ω-and γ-gliadin gene clusters reduces wheat immunoreactivity without affecting grain protein quality. Plant Biotechnol J 22(4):892–903
35. Zhang G, Luo Y, Dai X, Dai Z (2023) Benchmarking deep learning methods for predicting CRISPR/Cas9 sgRNA on-and off-target activities. Brief Bioinform 24(6):bbad333. https://doi.org/10.1093/bib/bbad333
36. Zhang X, Ibrahim Z, Khaskheli MB, Raza H, Zhou F, Shamsi IH (2024) Integrative approaches to abiotic stress management in crops: combining bioinformatics educational tools and artificial intelligence applications. Sustainability 16(17):7651
37. Zhang Y, Iaffaldano B, Qi Y (2021) CRISPR ribonucleoprotein-mediated genetic engineering in plants. Plant Commun 2(2). https://doi.org/10.1016/j.xplc.2021.100168

AI-Nano Strategies for Water Remediation in Agriculture

Aditya Pratap Singh, Sumit Kumar Mishra, Arjun Kumar Agarwal, Longjam Boris Singh, Juhi Rani, and Fatima Araish

Abstract Water is one of the most important resources to sustain life on our planet and an essential input for increasing agricultural productivity. Over-exploitation and contamination of the water bodies lead to an increased depth of the water table, which threatens crop health, soil fertility, and human health. Conventional approaches like sedimentation, evaporation, and filtration for water remediation are time-consuming and are inefficient in removing pollutants completely. Contamination of heavy metals like arsenic, nitrate, agrochemicals, wastewater, and various pesticide residues is entering the food chain, posing a direct threat to human and wildlife. Nanotechnology is the game changer in the field of water remediation because it utilizes advanced nano materials like carbon nanotubes, TiO_2, nano-filters, and nano-iron, which help in detection, monitoring, and removal of specific contaminants (e.g., heavy metals, pesticides) more efficiently than conventional methods. Moreover, integration of artificial intelligence (AI) with nanotechnology can greatly increase water use efficiency by utilizing an analysis of data from different sources, such as climate, vegetation, soil types, crop stages, and soil moisture index. AI can do real-time analysis, which enables it to check various water quality parameters instantly (e.g., pH, nitrate, salinity, heavy metals) and pass the information to all the stakeholders. By the use of AI models and nanotechnology, a normal farm can be converted to a smart farm in which everything is automatic, from irrigation scheduling to adjusting filtration systems. This integrated approach helps achieve resilient and sustainable agriculture in the face of growing environmental problems.

A. P. Singh (✉)
Department of Genetics and Plant Breeding, Bidhan Chandra Krishi Viswavidyalaya, Mohanpur, Nadia, West Bengal, India

S. K. Mishra · A. K. Agarwal · F. Araish
Department of Genetics and Plant Breeding, Birsa Agricultural University, Kanke, Ranchi, Jharkhand, India

L. B. Singh
Department of Agriculture, Krishi Vigyan Kendra Thoubal, Manipur, Umiam, Meghalaya, India

J. Rani
Department of Horticulture, Birsa Agricultural University, Kanke, Ranchi, Jharkhand, India

J.-T. Chen (ed.), *AI-Nano-Enabled Climate-Smart Agriculture and Sustainability*, Smart Nanomaterials Technology, https://doi.org/10.1007/978-981-92-1694-9_20

Keywords Artificial intelligence · Nanotechnology · Smart farming · Sustainable agriculture · Water management · Water remediation

1 Introduction

Life depends on water, which is essential for nurturing processes and sustaining the productivity of Earth's ecosystems. 97% of the water on Earth is seawater, making up 71% of the planet's surface [6]. Only 0.06% of the 3% of freshwater that is readily accessible is made up of groundwater, rivers, and the frozen polar ice cap or glaciers [6, 96]. Water is required primarily for the cultivation of crops and animal husbandry as well. The productivity and quality of produce to a larger degree are influenced by the quality of water that is employed in the irrigation of crops or provided to livestock. The source of this water not only requires it to be clean but also free from heavy metals measured at TDS (total dissolved solutes) to sustain the internal ecosystem of plants and livestock. Water is required in many physiological and chemical processes occurring inside the plants, and also the quality of water helps in maintaining soil equilibrium, which enables plants to absorb essential nutrients and minerals from the soil. A slight imbalance in composition can hinder the absorption of one mineral compared to others. Plants thrive on the availability of water for photosynthesis, nutrient transport through phloem, and temperature regulation. Water provides support to plant cell structure via osmotic pressure and turgor pressure to enable metabolic functions. Water has complex traits that determine both its allocation and requirement as an input for agriculture. Use of water for irrigation is itself contingent on land resources. Currently, agriculture uses over 70% of total freshwater withdrawals, while irrigated agricultural production uses nearly 85% of removed resources [34]. It has long been known that irrigation serves "productive" and "protective" purposes.

For all stages in agriculture, a crucial role is that of water, starting with animal husbandry, where it is used for maintaining hygiene and sanitation of livestock and cooling livestock to prevent heat stress; it helps in dissipating body heat of the livestock and curbing mortality. In crops also water is required for many agronomical interventions required for healthy upkeep of crops on fields, like protection from frost by forming a protective ice layer, which in turn prevents thermal shock to plants; also constant water is maintained in frost periods for field crops to help the crops regulate temperature and crop cooling through evaporative irrigation to prevent heat damage, soil leaching to remove excess salts and as a carrier for fertigation of crops. Water has an extensive role to play in post-harvest operation, and is also used extensively for washing and cleaning produce immediately, rapid cooling to extend the shelf life of produce, transportation of certain crops like cranberries grown in flooded fields, and various food processing activities, including cooking, blanching, and parboiling. This comprehensive use of water helps in maintaining the optimum quality of crops and livestock and enables food safety, highlighting its indispensable role in sustainable agricultural productivity.

Freshwater supplies are shrinking because of population growth, urbanization, industry, and climate change. This makes it hard for the world to meet the rising need for clean water [87]. Both population growth and climate change are making it harder to find water for agricultural production. Water scarcity occurs when the amount of water extracted is insufficient to meet needs and competition [96]. The UN estimated that in 2009, about 1.2 billion people across over 80 countries faced water shortages; by 2025, this number is predicted to increase to 2.7 billion [6]. Water shortages can cause starvation, death, and conflict both within and outside a nation. Numerous confrontations have already been reported between or within national borders [40, 85].

In order to meet agricultural needs without jeopardizing the sustainability of water supplies, these issues call for creative and effective water management techniques [97]. Despite being widely used, traditional irrigation techniques like flood irrigation frequently cause significant water losses from runoff and evaporation [100]. Modern irrigation methods can significantly improve crop production and water use efficiency. According to studies, drip irrigation increases crop productivity by using 50% less water than traditional techniques [116]. By simulating natural rainfall patterns using pressurized pipes and sprinkler heads, sprinkler irrigation systems increase the uniformity of water distribution and reduce losses from evaporation and deep percolation [81]. According to Maurya et al. [76], technologies like remote sensing, GIS, and soil moisture sensors are crucial for modern agriculture, as they enable real-time monitoring and precise management of water resources and improve irrigation efficiency. Rainwater collection and water recycling are also becoming popular sustainable methods to augment freshwater resources in farming. In order to lessen the demand on freshwater resources, water recycling entails cleaning agricultural wastewater and using it again for irrigation [77]. Rainfall harvesting systems provide an alternate water supply that lessens reliance on surface and groundwater by collecting and storing rainfall for agricultural usage during dry spells [89].

Water pollution is a serious environmental issue caused by the introduction of hazardous substances into water sources that harm both ecosystems and humans. In recent decades, agricultural pollution, mainly from agrochemicals, organic waste, and fertilizers, has become the leading cause of water quality decline worldwide, surpassing industrial and urban sources in many regions [45]. The heavy use of chemical fertilizers in farming leads to water pollution by causing nutrient runoff into rivers and lakes. This runoff results in harmful algal blooms and degrades water quality [36]. Addressing water pollution requires integrated management practices focusing on pollution source control, wastewater treatment, and sustainable agricultural techniques to protect both human health and aquatic ecosystems. For the ecosystem of our planet to be preserved and restored, environmental remediation is essential. Urbanization, industrialization, and intensification of agriculture over the past few centuries have exacerbated soil, water, and air pollution, endangering ecosystems, human health, and biodiversity [12, 13].

1.1 Limitations of Traditional Remediation Methods

Traditional water treatment methods have a number of disadvantages that may limit their effectiveness and safety, despite being widely accessible and fairly priced. Among the main restrictions are:

- Inadequate Removal of Contaminants: Conventional methods such as boiling, sedimentation, and filtering using local materials can remove some contaminants. Boiling eliminates infections but not chemicals or heavy metals; sedimentation eliminates only large particles; and filtering using local materials lowers turbidity but leaves out microorganisms and tiny particles [33, 15].
- Consistency: Recontamination may result from improper practices; effectiveness varies based on user knowledge and maintenance.
- Possible Health Risks: Poor filtering can add impurities, and boiling can concentrate hazardous elements.
- Limited Chemical and Microbial Control: Techniques like boiling and sedimentation are unable to eliminate pesticides, industrial pollutants, and dissolved microorganisms in areas with chemical pollutants [112].
- Environmental Impact: Conventional methods may have an impact on the environment, particularly in areas where they are widely used. Due to air pollution and deforestation, the common use of firewood to boil water can worsen environmental damage.

2 Sources of Contaminants

Water contamination is a major worldwide problem affecting environmental, human, and animal health. Multiple sources contribute to pollution, such as poor agricultural practices, discharge from industries, and municipal or domestic waste. Each of these categories contributes different types of pollutants into water bodies, impacting the chemical, physical, and biological quality of water [32]. Industrial activities play a big role in polluting water bodies. The main industry includes petroleum refineries, textile, pharmaceutical, power plants, and iron and steel companies discharging waste and polluted water directly into water bodies without any proper treatment, which contains large amounts of toxic chemicals, heavy metals (mercury, lead, arsenic, cadmium), oils, and dyes [104]. These pollutants are dangerous and pose a threat to animal and human life. Industries that use water as a coolant also cause thermal pollution by altering the temperature and amount of dissolved oxygen in the aquatic system.

Globally, agriculture has become a major cause of pollution due to an increase in intensive farming practices and growing demand for food and irrigation. Nitrogen and phosphorus are the primary pollutants sourced from organic manure and inorganic chemical fertilizers. They may reach ponds, lakes, and water reservoirs through leaching or through runoff after rainfall and cause eutrophication and toxic algal blooms. Pesticides, which include herbicides, insecticides, and fungicides, may accumulate in soil, air, and water and cause problems to human and aquatic life [92–9]. Disease-causing pathogens, including fungi, bacteria, and viruses, are found in animal manures and also pose a risk to human and animal health.

Water bodies are also contaminated with untreated sewage from industries, domestic waste, and gray water from households are discharged directly into freshwater bodies. Poorly managed sewage contains many harmful chemicals, detergents, organic materials, and pathogens, leading to depletion of dissolved oxygen, development of algal bloom, and spread of infectious diseases like typhoid, hepatitis, Cholera, and diarrhea [65]. Harmful chemicals from household products, leakage in septic tanks, and untreated wastes impact the quality of drinking water. Runoff after rainfall carries waste and other harmful chemicals into the natural water bodies, contaminating the quality of water.

3 Types of Pollutants

Among major types of pollutants, heavy metals, nitrates, and pesticides are the main concerns for environmental degradation, human and animal health. These are mainly produced from industries, intensive farming, and poor urban activities, which result in loss of quality of soil, air, and water. Heavy metals have high atomic weight and density and remain in the environment for a long time due to their non-biodegradable nature. They cause toxicity even at minute concentrations. Heavy metals generally accumulate in the ecosystem through bio-magnification and bioaccumulation. They can interact with cellular components, enzymes, and pose a threat to serious health issues, including nervous system damage, kidney dysfunction, heart problems, and decreased immunity [101]. Different group of pesticides used in agriculture mainly belongs to organophosphates, carbamates, organochlorines, and pyrethroids contain several toxic ingredients that persist in the environment and contaminate soil, air, and water. These chemicals are highly toxic to humans and wildlife, affecting their respiratory, nervous, and immune systems. They also disrupt ecosystem balance by eliminating beneficial microbes, flora, and fauna. Nitrates are another form of toxicants that originate mainly from the overuse of nitrogenous fertilizers. Every year, many children under six months of age lose their lives due to blue baby syndrome (methemoglobinemia) due to toxic concentrations of nitrate in drinking water.

3.1 Effect on Crop and Soil

Contaminants accumulate over time in the environment, significantly affecting the quality of air and water, and soil, agricultural output, and overall functioning of the ecosystem. Depending on the crop species, method of cultivation, and quality of soil, heavy metals concentrate and reach up to toxic levels that can influence the growth and development of plants and ultimately reduce overall output [113]. These toxic heavy metals interact with the plant biological system, disrupt various physiological processes, transport of food and water, and opening and closing of stomata [101, 113]. The toxicity of these molecules induces oxidative stress in plants and also generates reactive oxygen species in large amounts, which further affects various enzymatic reactions and metabolic pathways. These heavy metals are generally stored in the edible parts of plant and crop species, which adversely affect the health and well-being of humans and animals when consumed. Excessive use of copper fungicide in citrus and other crop plants develops symptoms of phytotoxicity in crops and degraded soil quality [113].

Pesticides used in different forms for crop protection contain multiple toxic ingredients, heavy metals, and compounds that increase the level of pollution in soil, air, and water, causing direct and indirect damage to the environment, plants, and animal life [8]. It is reported from a study that continuous use of insecticide thiamethoxam can reduce the number of nitrifying bacteria by 42% and enzyme phosphatase activity by 6.5%. Application of herbicides for control of weeds also negatively affects main crops by reducing growth, increasing flowering duration, altering biological pathways, and pigment discoloration, which ultimately results in lowered production. Pesticides enter in soil and damage soil microflora and fauna, which play a major role in important soil biological processes such as decomposition of wastes, mineralization, and denitrification of important mineral nutrients.

Soil acidification due to the leaching of nitrogenous fertilizer in soil disrupts many soil activities, such as nutrient mobilization, microbial activity, which collectively affect soil health. Acidification also reduces acid phosphatase and nitrate reductase enzyme activity through activation of metal ions, which interferes with mineral and water absorption by roots [94]. Studies reported that a decline in soil pH from 6.5 to 4.5 results in a decrease in absorption of phosphorus and nitrogen in plants by 19.2 and 21.6% respectively [8]. Soil acidification also affects the nitrogen fixation process, thus reducing soil nitrogen availability for crop plants in subsequent crop cycles.

3.2 *Risk to Human and Animal Health*

Exposure to heavy metals can occur via breathing, the food chain, or consumption of contaminated food or drinking water, causing serious health disorders that can damage various body organs and cause serious diseases. These contaminants also interact with enzymes, proteins, RNA, and DNA of the cell, damaging their molecular structure and subsequently affecting metabolic processes linked with these biomolecules. Prolonged exposure to these toxicants can result in the development of cancer affecting multiple body organs [113]. Pesticide residues, which persist in the environment for long durations, decrease the quality of air, cause various respiratory disorders, and also contaminate food and water that making them unfit for consumption, causing acute and chronic health issues. Long-term exposure to these pesticide residues can lead to the development of many genetic disorders and sometimes heart problems. Infants and children are particularly more vulnerable because their immune system and body organs are underdeveloped and cannot fight these toxicants effectively [52]. Nitrate and other related contaminants mix with groundwater and make drinking water unsuitable for consumption. Nitrate interacts with other molecules and forms nitrosamines, which results in the development of various forms of cancers such as cancer of the digestive tracts, bladder, ovary, respiratory tract, and other health issues [10].

4 Nanotechnology in Water Remediation

Nanotechnology is a novel field that can address the current water treatment issues by utilizing materials with at least one dimension less than 100 nanometers. With the emergence of nanoscience and nanotechnology over the past few decades, it has played a key role in shaping current environmental remediation science by providing innovative materials that can detect, remove, or degrade contaminants that are frequently missed by traditional approaches ([12, 13]; Lu et al. 2018). Nanoparticles effectively interact with contaminants at the molecular level due to precise control of adsorption, detection, and degradation processes, removing both inorganic and organic contaminants from wastewater and natural systems [107].

4.1 Common Nanomaterials in Water Treatment

4.1.1 Carbon Nanotubes (CNTs)

CNTs are cylindrical allotropes of carbon with very high mechanical strength, thermal stability, and large surface areas. Their adsorption efficiency is due to strong electrostatic forces that attract and retain pollutants and CNTs with carboxyl or hydroxyl groups exhibit increased affinity toward heavy metals like lead, arsenic, and mercury. Moreover, they have been used for the removal of dyes such as methylene blue, malachite green, and rhodamine B from industrial wastewater, achieving over 95% removal efficiency [22].

4.1.2 Titanium Dioxide (TiO_2) Nanoparticles

TiO_2 nanoparticles are recognized for their photocatalytic properties, which enable them to generate reactive oxygen species (ROS), including hydroxyl radicals and superoxide ions, when subjected to UV or solar light. This process facilitates the breakdown of organic compounds into carbon dioxide and water [37]. The primary application of TiO_2 is in the treatment of agricultural wastewater, specifically for the degradation of pesticides like atrazine and imidacloprid [29]. TiO_2 works remarkably well for breaking down harmful substances such as pesticides, phenols, pharmaceutical residues, and even disease-causing microbes. Through modifications of TiO_2 with elements like silver, nitrogen, or graphene, scientists have managed to make it active under visible light, making it viable for large-scale water purification under natural sunlight [46].

4.1.3 Nano-Zero Valent Iron (nZVI)

nZVI has strong reducing capacity and magnetic properties, and its magnetic recoverability reduces environmental release risk as well. nZVI detoxifies chlorinated hydrocarbons, nitroaromatics, and heavy metals [73]. Due to its nanoscale size, nZVI can be incorporated into contaminated aquifers, offering in situ remediation potential. Coating nZVI with stabilizers like starch, polyacrylic acid, or graphene oxide improves its dispersion and reduces oxidation. nZVI reduces Cr(VI) to Cr(III) and As(V) to less toxic forms, making it invaluable for groundwater treatment [48].

4.1.4 Nanofilters

Nanofilters are one of the most advanced innovations in modern water purification systems. These are membranes embedded with nanosized pores or nanomaterials like graphene oxide and zeolites that selectively remove contaminants. They can efficiently eliminate bacteria, viruses, and dissolved salts while maintaining high water flux [125]. They don't just strain contaminants like an ordinary mechanical sieve, but they actively interact with them through chemical and electrostatic forces. This lets them capture even the tiniest traces of pesticides, drugs, and heavy metals that conventional treatments often overlook, all without using extra chemicals or high energy (Table 1).

4.2 Functions of Nanomaterials in Water Remediation

Nanomaterials are tiny yet powerful materials that have outperformed all traditional water treatment technologies by doing far more than simply filtering. They have three fundamental functions: detection, adsorption, and degradation, which work in harmony to cleanse water at the molecular level, forming a system of precision and adaptability.

Table 1 Overview of nanomaterials used in water remediation

Nanomaterial	Composition/form	Primary function	Key application in agriculture
Carbon nanotubes (CNTs)	Cylindrical carbon nanostructures	Adsorption of heavy metals and dyes	Removal of cd, Pb from irrigation water
Titanium dioxide (TiO_2)	Metal oxide nanoparticle	Photocatalytic degradation	Decomposition of pesticides and organics
Nano-zero valent iron (nZVI)	Fe^0 nanoparticle suspension	Reduction of chlorinated and nitrate compounds	Pesticide and nitrate removal in irrigation channels
Graphene oxide (GO)	Layered carbon oxide sheets	Adsorption and filtration	Removal of arsenic and phosphate from agricultural runoff
Iron oxide nanoparticles	Fe_3O_4 or γ-Fe_2O_3 nanoparticles	Magnetic adsorption of heavy metals	Elimination of arsenic, cadmium from groundwater
Zeolite nanocomposites	Aluminosilicate nanostructures	Ion exchange, adsorption	Reduction of salinity and nitrate levels
Gold nanoparticles (AuNPs)	Functionalized gold nanosensors	Detection of ions and nutrients	Real-time monitoring of water salinity and Pb^{2+}

4.2.1 Detection

In a water purification process, before pollutants can be removed, they should be identified first, even if they exist in extremely small amounts and are invisible to the naked eye. Nanomaterials enhance this pollutant detection through nano sensors and biosensors capable of detecting extremely low levels of contaminants at the ppb or ppt level. Gold nanoparticles (AuNPs), quantum dots, and graphene-based composites are commonly used for colorimetric and electrochemical sensing [57]. For example, AuNP-based sensors change color when they come in contact with lead or mercury ions and further enabling quick and easy field detection [43]. These nano sensors make real-time water quality monitoring a reality for communities like researchers and farmers, allowing them to respond swiftly before pollutants spread and further mark a transformative shift from reactive to preventive environmental management, where contamination can be identified before it becomes a major environmental problem.

4.2.2 Adsorption

If detection is about awareness, adsorption is about action. Once the contaminants are identified, nanomaterials work as molecular magnets by immobilizing them efficiently. Carbon nanotubes (CNTs) and graphene oxide are a great example. Their intricate, layered structures allow pollutants to nestle between sheets or bind to reactive sites, leading to efficient removal. The process efficiency depends on pH, temperature, and nanomaterial concentration. Studies have shown that CNTs can remove over 95% of dyes like methylene blue and rhodamine B from industrial effluents in mere hours [74].

4.2.3 Degradation

After detection and adsorption, the next step is degradation. It is the stage where pollutants are not just trapped but broken down into harmless substances. Nano catalysts like titanium dioxide (TiO_2), zinc oxide (ZnO), and silver nanoparticles convert harmful organic pollutants into safe byproducts. When these photocatalysts are excited by light, they create electron–hole pairs or reactive oxygen species (ROS), which in turn produce radicals that facilitate the oxidation of pollutants [99]. For example, TiO_2–SiO_2 nanocomposites have demonstrated over 90% degradation of chlorpyrifos pesticide in 3 hours [46]. Hybrid nanomaterials, such as Ag–TiO_2 and Fe_3O_4–graphene, further enhance photocatalytic performance under visible light [48].

4.3 Advantages Over Conventional Methods

Conventional methods for treating water, like coagulation, chlorination, and sedimentation, usually encounter problems such as poor removal of contaminants and high energy consumption, and production of dangerous disinfection by-products. Nanotechnology can solve these problems by developing more efficient, selective, and sustainable alternatives [87].

a. **Enhanced Efficiency**: Due to high surface-to-volume ratios and active functional surfaces, nanoparticles can capture pollutants present at very low concentrations (parts-per-billion), something that many standard methods struggle with [74].
b. **High Selectivity**: Surface modification enables nanomaterials to target specific pollutants, minimizing interference from other ions.
c. **Eco-friendliness**: Photocatalytic processes use sunlight as energy, avoiding harmful chemicals.

d. **Regeneration and Reuse**: Many nanomaterials can be reactivated using simple UV light or gentle washing, enabling multiple reuse cycles with little to no drop in efficiency.
e. **Integration Capability**: Nanocomposites can be embedded into existing filtration systems, enhancing performance without completely changing the infrastructure [12, 13].

4.4 Applications in Agriculture

Since agriculture is a predominant user of freshwater, it has benefited immensely from nanotechnology. Nanotechnology enables sustainable management of water resources by purifying irrigation water, recycling agricultural runoff, and detecting contaminants that can reduce soil fertility, damage crops, and threaten food safety.

4.4.1 Removal of Pesticides and Fertilizers

TiO_2 and ZnO nanoparticles are widely used to degrade pesticides like atrazine, carbaryl, and chlorpyrifos [46]. Photocatalysis ensures complete mineralization without generating harmful residues. This solar-powered mechanism is particularly beneficial for rural regions where high-cost treatment systems are not feasible, also making it a cost-efficient approach to protecting crops and groundwater, as the removal only requires sunlight as a catalyst.

4.4.2 Heavy Metal Remediation

CNTs, graphene oxide, and iron oxide nanoparticles effectively capture harmful metals such as cadmium, arsenic, and chromium from irrigation water [105]. Graphene oxide (GO)-based membranes can eliminate more than 95% of arsenic and cadmium ions from contaminated irrigation water [12, 13]. Magnetic nanoparticles add an eco-friendly advantage that they can be separated easily with a magnet and reused, minimizing waste while restoring soil fertility and safeguarding food quality.

4.4.3 Recycling Agricultural Runoff

Nanofiltration systems embedded with graphene oxide can remove nitrates, phosphates, and organic residues, allowing water reuse for irrigation [77]. Such recycling reduces dependency on freshwater sources and supports circular water use in agriculture.

4.4.4 Nano Sensors in Irrigation

Nano-biosensors made from gold and graphene give farmers real-time insight into water quality. They help track salinity and nutrient levels instantly, making fertigation more precise and efficient. Such smart tools support sustainable farming and better resource management [57].

5 Sustainable Transformation: Toward Cleaner and Smarter Agriculture

The integration of nanotechnology in agricultural water management represents more than technological progress for developing countries, where access to safe water remains limited and shifts toward sustainability and resilience.

5.1 Role of Artificial Intelligence in Water Management

Water is a crucial resource for maintaining life on Earth. Most of our planet is covered by 2/3 water, out of which available fresh water is only 2.5% which is also getting contaminated by human activities and modern challenges [7]. Conventional water testing methods are not a real solution which as highly laborious, less accurate, and time-consuming method. To deal with this, innovative computing and machine-learning methods have emerged. In this, data collection is very easy and accurate as it uses advanced optical sensors, near-infrared spectroscopy, and remote sensing technology, and after the collection of these data, different machine-learning models are used, such as SVM, Random forest, which gives accurate results [111].

5.2 AI Tools for Water Quality Monitoring

5.2.1 Support Vector Machine (SVM)

One of the finest ways to use machine learning to figure out and predict how excellent water is is via the Support Vector Machine (SVM). It produces the best hyperplane that divides water samples into groups or guesses the values of continuous parameters. In classification tasks, SVM sorts water samples into quality groups like excellent, middling, or bad depending on things like pH, dissolved oxygen (DO), biochemical oxygen demand (BOD), turbidity, and electrical conductivity. It uses Support Vector Regression (SVR) to predict continuous outputs like DO, BOD, or Water Quality Index (WQI). The algorithm's kernel functions are linear, polynomial, and radial basis function (RBF), which can find both linear and nonlinear connections between parameters. Numerous studies indicate that SVM and SVR models achieve great predictive accuracy, frequently exhibiting correlation values over 0.90 for principal indicators [44, 108]. The SVM technique is effective for small to medium-sized datasets and nonlinear water quality systems [122]. But the model's performance relies on the kernel you use and how you configure the parameters. With bigger datasets, the calculation might take a lengthy time [59].

5.2.2 Decision Tree

The Decision Tree (DT) approach is another commonly used machine-learning tool for water quality monitoring and prediction. It functions by recursively dividing the dataset into subsets based on the values of input parameters, forming a tree-like structure where core nodes indicate decision rules and leaf nodes identify projected outcomes. DT models may perform both classification and regression problems, making them useful for classifying water quality into multiple groups or forecasting continuous variables such as DO, BOD, or WQI. The method's interpretability is one of its biggest features, since it gives unambiguous decision rules relating characteristics such as pH, turbidity, conductivity, and nutrient concentration to water quality status. Studies have demonstrated that Decision Tree–based models produce robust performance in water quality prediction, with accuracy equivalent to more sophisticated ensemble approaches when the dataset is well-structured and noise-free [98, 70]. Moreover, DT models require minimal data preparation and may easily handle missing or categorical data. However, solo Decision Trees may suffer from overfitting and instability, which can be addressed using ensemble techniques such as Random Forest or Gradient Boosting [56].

5.3 Use of Sensors and Remote Data in Water Management

The combined use of sensors and remote data has transformed modern water management by offering continuous, immediate analysis of water quality, quantity, and distribution. Advanced sensor networks, including Internet of Things (IoT)–based systems, are implemented in rivers, reservoirs, irrigation channels, and groundwater wells to record parameters such as pH, dissolved oxygen, turbidity, conductivity, and temperature with high temporal resolution. These sensors communicate data wirelessly to centralized platforms for processing and analysis. Simultaneously, remote sensing technologies, leveraging satellite and drone-based observations, give large-scale geographic information on surface water extent, evapotranspiration, soil moisture, and vegetation health. Data products from satellites such as Landsat, Sentinel, and MODIS are being utilized for analyzing water availability, pollutant dispersion, and drought conditions. When integrated, in situ sensor networks and distant data enhance data-driven decision making, early warning of pollution or shortage, and optimization of irrigation and reservoir operations. Machine-learning models can further strengthen these systems by identifying anomalies, anticipating trends, and enabling sustainable water resource management [73, 54].

5.3.1 Applications in Irrigation Scheduling

The implementation of Artificial Intelligence (AI) in irrigation scheduling has considerably increased the efficiency of agricultural water management by optimizing the timing and amount of water given to crops. AI methods like as Artificial Neural Networks (ANNs), Support Vector Machines (SVMs), Fuzzy Logic Systems, and Decision Trees are being employed to anticipate agricultural water requirements based on environmental and soil characteristics. Data from soil moisture sensors, meteorological stations, remote sensing platforms, and crop growth models are linked to build predictive algorithms that prescribe ideal watering schedules. Machine-learning algorithms assess complicated interactions between meteorological parameters (temperature, humidity, rainfall, solar radiation), soil conditions, and plant responses to calculate irrigation demands dynamically. IoT-enabled smart irrigation systems further automate this process, allowing real-time control of valves and pumps based on AI predictions. Such AI-driven technologies minimize water loss, boost agricultural output, and enable sustainable water resource usage [54, 23]. Recent studies have also proven that hybrid models integrating machine learning with optimization approaches (e.g., evolutionary algorithms or reinforcement learning) may further increase irrigation precision and adapt to changing environmental circumstances [24].

5.3.2 Real-time Decision-making and Automation

The combination of Artificial Intelligence (AI) with real-time data collecting and automation technologies is changing water resource management by allowing continuous, data-driven, and adaptive decision-making. AI algorithms, when integrated with IoT-enabled sensor networks, edge computing, and cloud-based platforms, provide real-time monitoring of hydrological, climatic, and water quality parameters [88, 86]. Data acquired from field sensors, satellite imaging, and weather models are processed using machine-learning (ML) and deep learning algorithms to discover abnormalities, improve operations, and anticipate water demand and supply in dynamic contexts [54]. Advanced AI techniques such as reinforcement learning, fuzzy logic controllers, and expert systems can autonomously alter irrigation schedules, regulate canal gates, or activate treatment systems based on feedback loops [17, 14].

6 Prospects and Challenges of AI–Nano Strategies for Water Remediation

The integration of Artificial Intelligence and nanotechnology presents a transformative approach to sustainable water remediation, particularly in the agricultural sector, where efficient water use and quality maintenance are critical. The combined strength of these technologies lies in their complementarity: AI contributes predictive intelligence and automation, while nanotechnology provides the precision tools necessary for pollutant detection and removal. Together, they form an advanced, adaptive, and data-driven system capable of addressing water contamination, scarcity, and inefficiency in real time. As these technologies mature, their practical applications are expected to expand rapidly, offering a framework for smart water governance and circular water use in agriculture (Fig. 1).

One of the most promising prospects of AI–Nano strategies is the development of autonomous water monitoring and treatment systems. These systems combine nanosensors, IoT devices, and AI analytics to detect and remove pollutants without

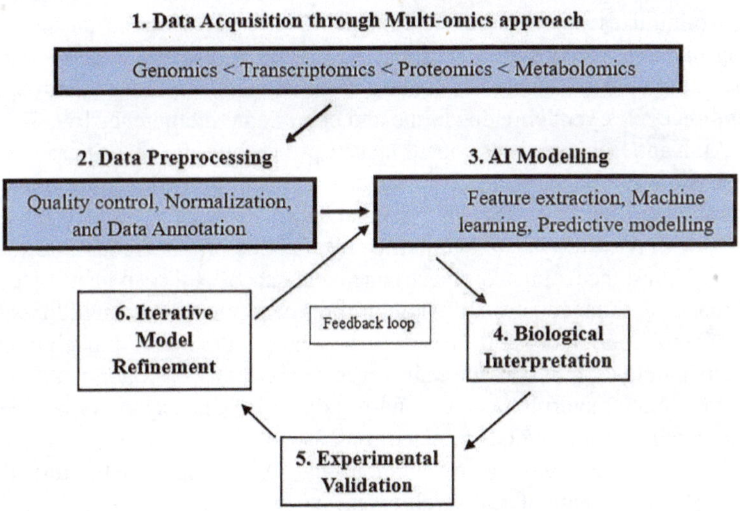

Fig. 1 Dynamic model of AI–nano integration for resilient water remediation, illustrating adaptive feedback between AI analytics and nanomaterial processes for real-time optimization of water quality in agricultural systems. [Schematic representation of an AI–nanotechnology closed-loop for agricultural water remediation. **a** Field system showing runoff capture by a GO-membrane collector, evaporation, and sensor feedback to an AI decision module that recommends nanomaterial adaptation for downstream treatment. **b** Stylized modeled water-quality trajectories across a crop season for four scenarios: baseline (no treatment), nano only, AI only, and combined AI + nano management. (Note: the figure shows conceptual model outputs; the 20% reduction claimed in the text refers to model validation reported in the Results section and is not plotted here.) **c** AI-predicted efficacy network among representative nanomaterials (nZVI, CNT, TiO_2) with arrow styles indicating direction and type of interaction (black = mechanistic link; green = AI-recommended optimization; orange = photochemical driver)]

human intervention. In such setups, nanosensors embedded in irrigation channels, reservoirs, or pipelines detect trace contaminants—such as heavy metals, pesticides, or nitrates—and transmit continuous data streams to an AI-based monitoring platform. The AI system interprets the signals, predicts contamination risks, and activates appropriate remediation measures, such as adjusting the flow to nanofiltration membranes or switching to alternative water sources. Over time, reinforcement learning algorithms refine their responses, enhancing efficiency and reducing energy consumption. The result is a closed-loop system that not only purifies water but also prevents contamination from spreading, ensuring consistent water quality for agricultural use [57].

Another major advantage of AI–Nano integration is its potential to reduce operational costs and environmental impact compared to traditional water treatment methods. Conventional systems often rely on chemical coagulants, chlorination, and sedimentation, which can generate harmful by-products and require significant energy inputs. Nanomaterials, in contrast, can function under mild environmental conditions, often using sunlight or ambient temperature for catalytic reactions. When paired with AI, these systems achieve optimized operation by automatically adjusting variables such as light intensity, catalyst loading, and hydraulic retention time. This precision minimizes material waste and extends the lifespan of nanocatalysts, thereby lowering maintenance costs. In addition, AI models trained on large datasets can forecast equipment wear, predict fouling rates in nanomembranes, and schedule regeneration cycles, reducing downtime and operational inefficiency [69, 72].

The AI–Nano framework also opens up new possibilities for real-time and predictive water quality management. Traditional water testing methods are often time-consuming, requiring laboratory analysis that delays decision-making. Nanosensors integrated with AI overcome this limitation by enabling instantaneous detection and response. For instance, optical or electrochemical nanosensors can identify multiple contaminants simultaneously, and AI algorithms can process this multidimensional data to classify pollutants and predict their sources. Predictive modeling further allows the anticipation of contamination events by correlating water quality parameters with climatic, hydrological, or land-use data [3]. This capability is especially valuable in agricultural regions affected by seasonal runoff and fertilizer leaching, where timely interventions can prevent large-scale water quality deterioration.

The use of AI in nanomaterial development and optimization also holds significant promise. The synthesis of nanomaterials with specific surface properties, catalytic activity, or selectivity typically requires numerous experiments and characterization steps. AI accelerates this process by analyzing large databases of previous syntheses and experimental results to identify patterns that link structure with performance. Machine learning models can thus predict optimal synthesis conditions or suggest modifications to enhance activity and stability. For example, AI can recommend dopants for TiO_2 or ZnO nanoparticles to improve visible-light photocatalytic efficiency or propose composite structures that combine adsorption and degradation properties. This data-driven approach not only speeds up innovation but also reduces waste in material development [3].

Despite these promising prospects, the practical implementation of AI–Nano strategies faces several technical, environmental, and ethical challenges. One of the most significant concerns is the potential toxicity and ecological impact of nanomaterials. While they are highly effective in pollutant removal, their small size and high reactivity can pose risks to non-target organisms and soil microbiota if released unintentionally into the environment. Nanoparticles such as silver or zinc oxide may interfere with microbial communities essential for nutrient cycling, while carbon-based nanomaterials might persist in soil or water for extended periods [113]. Long-term ecological studies and standardized risk assessment protocols are therefore essential to ensure that nanotechnology-based remediation remains environmentally safe.

Another challenge lies in the economic feasibility and scalability of AI–Nano systems. The synthesis of advanced nanomaterials, installation of nanosensor networks, and integration with AI platforms require substantial initial investment. In many developing regions where water contamination is most severe, limited financial resources, poor digital infrastructure, and low technical capacity restrict adoption. Furthermore, the maintenance of nanosensors and calibration of AI models demand specialized skills that may not be readily available in rural agricultural communities [97]. Developing cost-effective, modular, and user-friendly systems will be key to ensuring that these technologies benefit smallholder farmers and resource-limited regions.

From a technological standpoint, the integration of AI with nanotechnology presents challenges related to data quality, interoperability, and model transparency. AI algorithms depend heavily on large, high-quality datasets for accurate predictions. However, environmental data often contain gaps, noise, or inconsistencies due to varying sensor sensitivity and external influences such as temperature or humidity. This can reduce the reliability of model outputs. Moreover, many AI models, particularly deep learning networks, function as "black boxes," providing accurate results without clear explanations of the underlying reasoning. For water management applications where safety and regulatory compliance are critical, model interpretability and explainability become essential [116]. Addressing these issues requires developing transparent AI frameworks that balance performance with accountability.

Data security and privacy also pose emerging concerns. With IoT-enabled nanosensor networks transmitting continuous data on water quality and environmental conditions, ensuring the protection of sensitive information becomes a priority. Unauthorized access or manipulation of data could lead to false readings or system malfunctions, undermining trust in AI-based water management. Establishing robust cybersecurity protocols, encrypted communication systems, and regulatory frameworks for data governance will be necessary for large-scale implementation.

From an environmental ethics perspective, responsible innovation, and governance are vital. Policies must guide the production, deployment, and disposal of nanomaterials to prevent ecological harm. Similarly, AI algorithms used in environmental management must adhere to principles of fairness, transparency, and accountability. Collaborative frameworks involving scientists, policymakers, and local stakeholders can ensure that AI–Nano systems are applied in ways that align with sustainability goals and public welfare. Additionally, life-cycle assessments of nanomaterials and the energy costs of AI computation should be incorporated into overall sustainability evaluations to ensure that technological progress does not inadvertently increase environmental footprints [40].

Despite these challenges, the future outlook for AI–Nano strategies in water remediation is highly promising. Continued research and cross-disciplinary collaboration will likely produce more efficient, environmentally safe, and economically viable systems. The miniaturization of sensors, the advancement of low-cost nanomaterial synthesis methods, and the increasing availability of open-source AI frameworks are all expected to accelerate adoption. Integration with renewable energy systems, such as solar-powered photocatalytic reactors and self-sustaining nanosensor networks, could further enhance sustainability. Moreover, as governments and international organizations recognize the importance of digital and nanotechnological solutions for environmental protection, policy incentives and funding initiatives are likely to support wider deployment in the near future.

Ultimately, the successful implementation of AI–nano water remediation strategies will depend on balancing innovation with responsibility. Ensuring safety, affordability, and inclusiveness will make these technologies not just scientific achievements but practical tools for sustainable development. When applied thoughtfully, the convergence of Artificial Intelligence and nanotechnology holds the potential to revolutionize water management in agriculture—delivering cleaner water, healthier ecosystems, and greater resilience in the face.

7 Conclusion

Water is the foundation of agricultural productivity and ecosystem stability, yet its availability and quality are under increasing threat due to pollution, climate change, and unsustainable use. Conventional water treatment methods, though effective to some extent, often fail to address the complex mix of contaminants that characterizes modern agricultural and industrial wastewater. The emergence of nanotechnology and Artificial Intelligence has opened new possibilities for overcoming these limitations through precision, adaptability, and real-time control. Together, these technologies offer a transformative pathway for water remediation, integrating material innovation with digital intelligence to ensure both efficiency and sustainability.

Nanotechnology provides the means to detect and eliminate pollutants at the molecular level through materials with unique properties such as large surface area, high reactivity, and tunable selectivity. Nanomaterials like titanium dioxide, zinc oxide, carbon nanotubes, and zero-valent iron have proven highly effective in removing pesticides, dyes, heavy metals, and other toxic compounds. Their photocatalytic and adsorptive functions allow the breakdown or immobilization of harmful substances into non-toxic forms, often powered by renewable energy sources such as sunlight. When embedded in membranes, filters, or coatings, nanomaterials enhance purification systems far beyond the capabilities of conventional treatment techniques.

Artificial Intelligence complements these advances by introducing predictive power, automation, and learning capability to water management. Machine learning models analyze vast datasets from sensors, satellites, and field devices to identify contamination patterns, optimize treatment processes, and prevent waste. AI-driven control systems continuously adjust parameters like flow rate, catalyst dosage, and light intensity to maintain optimal purification performance. In irrigation management, AI algorithms integrate data from soil moisture sensors and weather forecasts to minimize water loss and ensure that crops receive adequate yet efficient watering. These intelligent systems transform water management from reactive to proactive, allowing decisions based on evidence and foresight rather than routine or guesswork.

The integration of AI and nanotechnology has created a new paradigm in water remediation—one where detection, diagnosis, and purification occur simultaneously within a unified framework. AI-guided nanosensors monitor water quality in real time, identifying contaminants before they reach harmful levels. Data analytics predict potential contamination events and recommend corrective measures, while nanomaterials carry out on-site remediation through adsorption or catalytic degradation. This dynamic, self-regulating system significantly enhances responsiveness and efficiency, offering a sustainable solution for agricultural and industrial water challenges. The concept of smart nanoremediation, powered by AI, thus bridges the gap between technological innovation and environmental management.

The prospects of these combined technologies extend beyond laboratory success. Field-deployable nanosensors, AI-enabled irrigation networks, and solar-driven photocatalytic reactors are already demonstrating practical viability in pilot projects across different parts of the world. These innovations not only conserve water but also improve soil health, safeguard crop quality, and protect ecosystems. In developing regions, where access to clean water remains a pressing challenge, portable AI–Nano systems can provide affordable and decentralized solutions, helping communities achieve self-reliant water security. As research advances, integrating renewable energy sources and recyclable nanomaterials will further improve sustainability and cost-effectiveness, making these systems accessible to a wider range of users.

Nevertheless, widespread implementation requires addressing several critical issues, including the potential toxicity of nanomaterials, their safe disposal, and the need for standardized environmental regulations. Ethical and governance frameworks must guide AI applications to ensure transparency, accountability, and data protection. Strengthening interdisciplinary collaboration between material scientists, environmental engineers, computer scientists, and policymakers will be essential for translating laboratory innovations into field-scale practices. Investment in capacity building and infrastructure will also play a vital role, particularly in developing countries where digital and nanotechnological adoption remains limited.

References

1. Adeleye AS, Conway JR, Garner K, Huang Y, Su Y, Keller AA (2016) Engineered nanomaterials for water treatment and remediation: costs, benefits, and applicability. Chem Eng J 286:640–662. https://doi.org/10.1016/j.cej.2015.10.105
2. Adeleye AS, Keller AA (2014) Long-term colloidal stability and metal leaching of single wall carbon nanotubes: implications for environmental release. Environ Sci Technol 48(21):12530–12539. https://doi.org/10.1021/es502470p
3. Aghbashlo M, Peng W, Tabatabaei M, Kalogirou SA, Soltanian S, Hosseinzadeh-Bandbafha H, Lam SS (2021) Machine learning technology in biodiesel research: a review. Prog Energy Combust Sci 85:100904. https://doi.org/10.1016/j.pecs.2021.100862
4. Ahmad R, Kumar R (2020) Nanomaterial-based adsorbents for removal of heavy metals and dyes from water: a review. J Environ Chem Eng 8(5):104503. https://doi.org/10.1016/j.jece.2020.104503
5. Ahmaruzzaman M, Gupta VK (2020) Synthesis and photocatalytic activity of porous ZnO stabilized by TiO_2 and Fe_3O_4 nanoparticles: investigation of pesticide degradation reaction in water treatment. J Environ Chem Eng 8(5):104012. https://pubmed.ncbi.nlm.nih.gov/33131041/
6. Ahuja S (2009) Handbook of water quality and purity. Elsevier
7. Ajakwe SO, Ihekoronye VU, Ajakwe IU, Jun T, Kim D-S, Lee JM (2022) Connected intelligence for smart water quality monitoring system in IIoT. In: 2022 13th international conference on information and communication technology convergence (ICTC). IEEE, pp 2386–2391. https://doi.org/10.1109/ICTC55196.2022.9952785
8. Alengebawy A, Abdelkhalek ST, Qureshi SR, Wang MQ (2021) Heavy metals and pesticides toxicity in agricultural soil and plants: ecological risks and human health implications. Toxics 9(3):42. https://doi.org/10.3390/toxics9030042
9. Alengebawy A, Abdelkhalek ST, Qureshi SR, Wang M-Q (2021) Heavy metals and pesticides toxicity in agricultural soil and plants: ecological risks and human health implications. Toxics 9(3):42. https://doi.org/10.3390/toxics9030042
10. Angon PB, Islam MS, Das A, Anjum N, Poudel A, Suchi SA (2024) Sources, effects and present perspectives of heavy metals contamination: soil, plants and human food chain. Heliyon 10(7):e28357. https://doi.org/10.1016/j.heliyon.2024.e28357
11. Azam A, Ahmed AS, Oves M, Khan MS, Habib SS, Memic A (2012) Antimicrobial activity of metal oxide nanoparticles against gram-positive and gram-negative bacteria: a comparative study. Int J Nanomedicine 7:6003–6009. https://doi.org/10.2147/IJN.S35347
12. Bediako JK, El Ouardi Y, Massima Mouele ES, Mensah B, Repo E (2023a) Polyelectrolyte and polyelectrolyte complex-incorporated adsorbents in water and wastewater remediation: a review of recent advances. Chemosphere 325:138418. https://doi.org/10.1016/j.chemosphere.2023.138418
13. Bediako JK, Wei W, Kim S, Yun YS, Choi J (2023b) Metal-organic frameworks and derivatives for sustainable removal of heavy metal ions from aqueous environments. Environ Res 220:115167. https://doi.org/10.1016/j.envres.2022.115167
14. Bukhari M, Athar SO, Ullah M, Aman MN (2024) A fuzzy-logic-based smart irrigation controller for precision agriculture. IEEE Internet Things J 11(22):37257–37268. https://doi.org/10.1109/JIOT.2024.3440200
15. Centers for Disease Control and Prevention (2024) Making water safe in an emergency. https://www.cdc.gov/healthywater/emergency/
16. Chakraborty P, Mukherjee A (2021) Toxicity and safety concerns of nanomaterials in the aquatic environment: a critical review. Environ Toxicol Pharmacol 84:103651. https://doi.org/10.1016/j.etap.2021.103651
17. Chandel AK, Ribeiro CJ, da Silva SS, Singh OV, Pereira CS (2023) State-of-the-art AI-enabled mobile device for real-time water stress detection of field crops. Comput Electron Agric 212:108123. https://doi.org/10.1016/j.compag.2023.108123

18. Chatterjee S, Woo SH (2009) The removal of reactive black 5 by chitosan hydrogel beads: batch and column studies. Bioresour Technol 100(23):6019–6026. https://doi.org/10.1016/j.biortech.2009.06.045
19. Chen H, Chen G (2015) Advanced oxidation processes for wastewater treatment: recent developments and trends. Environ Sci: Water Res & Technol 1(1):13–38. https://doi.org/10.1039/C4EW00066J
20. Chen C, Ding S, Yang H (2022) Recent advances in nano-enabled membranes for water purification: mechanisms and challenges. Prog Mater Sci 129:100967. https://doi.org/10.1016/j.pmatsci.2022.100967
21. Chen Q, Zhao X, Li Z (2021) Gold nanoparticle-based sensors for heavy metal detection in water: a review. Environ Sci Technol 55(14):9784–9803. https://doi.org/10.1021/acs.est.0c08673
22. De Volder MFL, Tawfick SH, Baughman RH, Hart AJ (2013) Carbon nanotubes: present and future commercial applications. Science 339(6119):535–539. https://doi.org/10.1126/science.1222453
23. Del-Coco M, Leo M, Carcagnì P (2024) Machine learning for smart irrigation in agriculture: how far along are we? Information 15(6):306. https://doi.org/10.3390/info15060306
24. Dhanke JA, Srivastava D, Menaga D, Raj R, Kumar KV, Jangir P, Mani P (2025) Climate-Based AI-Powered precision irrigation: sustainably smart agriculture frameworks for maximum crop yields. Remote Sens Earth Syst Sci 8(1):161–172. https://doi.org/10.1007/s41976-024-00174-4
25. Doelman JC, Beier FD, Stehfest E, Bodirsky BL, Beusen AH, Humpenöder F, Kram T (2022) Quantifying synergies and trade-offs in the global water-land-food-climate nexus using a multi-model scenario approach. Environ Res Lett 17(4):045004
26. Duan J, Yu Y, Li Y, Yu Y, Sun Z, Li Y, Sun H (2022) Nanotechnology for water treatment: from basic principles to applications. Chem Eng J 430:132883. https://doi.org/10.1016/j.cej.2021.132883
27. Dutta T, Kim KH, Deep A, Szulejko JE, Vellingiri K, Kumar S, Kwon EE (2018) Recovery of nanomaterials for reuse in environmental remediation and clean energy systems: a review. Environ Int 112:314–333. https://doi.org/10.1016/j.envint.2017.12.006
28. El-Saeid MH, Alotaibi MO, Alshabanat M, Alharbi K, Altowyan AS, Al-Anazy M (2021) Photo-catalytic remediation of pesticides in wastewater using UV/TiO_2. Water 13(21):3080. https://doi.org/10.3390/w13213080
29. El-Saeid MH, Alotaibi MA, Alshammari NA (2021) Photocatalytic degradation of pesticides in wastewater using UV/TiO_2: a review. Water 13(21):3080. https://doi.org/10.3390/w13213080
30. El-Sheikh SM, Zhang G (2018) Recent developments of nanomaterials for heavy metal removal from water and wastewater: a review. Environ Sci Pollut Res 25(32):31963–31978. https://doi.org/10.1007/s11356-018-3248-1
31. Esmail AA, Ibrahim MA, Abdallah SM, Radwan AE, Elsayed MA, Elnakeib NA, ... Moawad IF (2023, October) Smart irrigation system using IoT and machine learning methods. In 2023 5th Novel Intelligent and Leading Emerging Sciences Conference (NILES) (pp 362–367). IEEE. https://doi.org/10.1109/NILES59815.2023.10296736
32. Evans AE, Mateo-Sagasta J, Qadir M, Boelee E, Ippolito A (2019) Agricultural water pollution: key knowledge gaps and research needs. Curr Opin Environ Sustain 36:20–27. https://doi.org/10.1016/j.cosust.2018.10.003
33. Eze F (2016) Traditional water treatment in Nigeria. Heritage Publishers
34. Food and Agriculture Organization (FAO) (2021) The state of the world's land and water resources for food and agriculture—systems at breaking point: synthesis report 2021. FAO. https://doi.org/10.4060/cb7654en
35. Ganguly P, Byrne C, Breen A, Pillai SC (2018) Antimicrobial activity of photocatalysts: fundamentals, mechanisms, kinetics, and recent advances. Appl Catal B Environ 225:51–75. https://doi.org/10.1016/j.apcatb.2017.11.018

36. Garnache C, Smith CM, Swinton SM, Roth M, Segerson K (2016) Sustainable management of Great Lakes watersheds through market-based policies. Ecol Econ 124:63–74. https://doi.org/10.1016/j.ecolecon.2016.01.019
37. Gatou M-A, Syrrakou A, Lagopati N, Pavlatou EA (2024) Photocatalytic TiO_2-based nanostructures as a promising material for diverse environmental applications: a review. Reactions 5(1):135–194. https://doi.org/10.3390/reactions5010007
38. Gautam RK, Mudhoo A, Lofrano G, Chattopadhyaya MC (2014) Biomass-derived biosorbents for metal ions sequestration: adsorbent modification and activation methods and adsorbent regeneration. J Environ Chem Eng 2(1):239–259. https://doi.org/10.1016/j.jece.2013.12.019
39. Ghaedi M, Shokrollahi A (2006) Adsorption of brilliant green from aqueous solution onto carbon nanotubes. J Hazard Mater 137(1):226–231. https://doi.org/10.1016/j.jhazmat.2006.01.054
40. Gleick PH (2018) The world's water: the report on freshwater resources. Pac Inst Stud Dev, Environ, Secur
41. Grandjean P, Landrigan PJ (2014) Neurobehavioural effects of developmental toxicity. Lancet Neurol 13(3):330–338. https://doi.org/10.1016/S1474-4422(13)70278-3
42. Guo Z, Liang S, Tian Q, Wu C, Wang Z (2016) Enhanced visible-light photocatalytic degradation of methylene blue by ag nanoparticles loaded on TiO_2 nanotubes. J Mater Sci Mater Electron 27:5891–5898. https://doi.org/10.1007/s10854-016-4465-5
43. Guo Y, Wang Z, Qu W, Shao H, Jiang X (2011) Colorimetric detection of mercury, lead and copper ions simultaneously using protein-functionalized gold nanoparticles. Biosens Bioelectron 26(10):4064–4069. https://doi.org/10.1016/j.bios.2011.03.033
44. Hamid NA, Bhuiyan MAH, Zhang J (2019) Water quality prediction model based on support vector machine for ungauged river catchment under dual scenarios. Water 11(6):1231. https://doi.org/10.3390/w11061231
45. Hansen NC, Allen BL, Baumhardt RL, Lyon DJ (2012) Research achievements and adoption of practices for water conservation in dryland cropping. J Soil Water Conserv 67(5):130A–134A. https://doi.org/10.2489/jswc.67.5.130A
46. Hashimoto K, Irie H, Fujishima A (2005) TiO_2 photocatalysis: a historical overview and future prospects. Jpn J Appl Phys 44(12):8269–8285. https://doi.org/10.1143/JJAP.44.8269
47. He J, Wang L, Huang L, Liu Y (2020) Nanotechnology in wastewater treatment: a review on nanomaterials for environmental applications. Environ Chem Lett 18:1091–1110. https://doi.org/10.1007/s10311-019-00952-7
48. He F, Zhao D, Paul C (2005) Preparation and characterization of a new class of starch-stabilized bimetallic nanoparticles for degradation of chlorinated hydrocarbons in water. Environ Sci Technol 39(9):3314–3320. https://doi.org/10.1021/es048743y
49. Homaeigohar S, Elbahri M (2014) Graphene membranes for water desalination. NPG Asia Mater 6(9):e100. https://doi.org/10.1038/am.2014.76
50. Hu X, Zhang Y, Ding T, Liu J, Zhao H (2019) Multifunctional nanomaterials for advanced water treatment: adsorption, catalysis, and disinfection. Environ Sci Nano 6(11):3319–3347. https://doi.org/10.1039/C9EN00859E
51. Huang Y, Zhao R, Zhang R (2021) Nanocomposite hydrogels for water treatment: fabrication, performance, and prospects. Chem Eng J 420:130471. https://doi.org/10.1016/j.cej.2021.130471
52. Kim K-H, Kabir E, Jahan SA (2017) Exposure to pesticides and the associated human health effects. Sci Total Environ 575:525–535. https://doi.org/10.1016/j.scitotenv.2016.09.009
53. Klaine SJ, Alvarez PJ, Batley GE, Fernandes TF, Handy RD, Lyon DY, Lead JR (2008) Nanomaterials in the environment: behavior, fate, bioavailability, and effects. Environ Toxicol Chem 27(9):1825–1851. https://doi.org/10.1897/08-090.1
54. Krishnan SR, Nallakaruppan MK, Chengoden R, Koppu S, Iyapparaja M, Sadhasivam J, Sethuraman S (2022) Smart water resource management using Artificial Intelligence—a review. Sustainability 14(20):13384. https://doi.org/10.3390/su142013384
55. Kumar V, Shahi SK, Singh J (2020) Biogenic synthesis of nanoparticles and their role in environmental remediation: a review. Environ Chem Lett 18(5):1491–1510. https://doi.org/10.1007/s10311-020-00988-2

56. Kumar D, Singh A, Gupta V (2021) Evaluation of machine learning algorithms for river water quality classification. Ecol Inform 63:101305. https://doi.org/10.1016/j.ecoinf.2021.101305
57. Kusuma SAF, Harmonis JA, Pratiwi R, Hasanah AN (2023) Gold nanoparticle-based colorimetric sensors: properties and application in detection of heavy metals and biological molecules. Sensors 23(19):8172. https://doi.org/10.3390/s23198172
58. Lee J, Choi W, Yoon J (2018) Enhanced degradation of micropollutants by a hybrid nanomaterial-based photo-Fenton-like process. Appl Catal B Environ 237:111–122. https://doi.org/10.1016/j.apcatb.2018.05.010
59. Lee J-Y (2025) Low-cost sensor-based algal bloom labeling: a comparative study of SVM and logic methods. Environ Monit Assess 197(4):405. https://doi.org/10.1007/s10661-025-13815-y
60. Li J, Zhang Y (2020) Recent advances in photocatalytic water treatment: mechanisms and applications. Catal Today 352:196–207. https://doi.org/10.1016/j.cattod.2019.04.055
61. Li H, Zhang M, Ma Y, Wang C, Zhang H (2018) Nanocomposites for enhanced wastewater treatment: synergistic effects and mechanisms. Adv Mater Interfaces 5(20):1800899. https://doi.org/10.1002/admi.201800899
62. Li Y, Zhang X, Wang G (2020) Development of TiO_2-based photocatalysts for water purification: from morphology control to applications. Catal Today 355:276–288. https://doi.org/10.1016/j.cattod.2019.04.065
63. Li X, Zhang L, Wang J, Liu J (2021) Nanotechnology-based materials for environmental applications: a review of recent advances and future prospects. J Clean Prod 312:127730. https://doi.org/10.1016/j.jclepro.2021.127730
64. Li P, Karunanidhi D, Subramani T, Srinivasamoorthy K (2021) Sources and consequences of groundwater contamination. Archiv Environ Contam Toxicol 80(1):1–10. https://doi.org/10.1007/s00244-020-00805-z
65. Lin L, Yang H, Xu X (2022) Effects of water pollution on human health and disease heterogeneity: a review. Front Environ Sci 10:880246. https://doi.org/10.3389/fenvs.2022.880246
66. Ling CJ, Mohammad AW, Benamor A, Hilal N, Shon HK (2019) Surface modifications of PVDF membranes for antifouling and antibacterial applications: a review. J Membr Sci 575:158–188. https://doi.org/10.1016/j.memsci.2019.01.012
67. Liu S, Wei H, Zhang Z, Zhang Y (2022) Metal oxide nanomaterials for water treatment: design, mechanisms, and challenges. Environ Sci Nano 9(7):2290–2314. https://doi.org/10.1039/D2EN00361B
68. Liu X, Xu J, Liu F, Pan C, Li Y, Wu Y (2016) The impact of pesticides on soil microbial community composition and function in China: a review. Sci Total Environ 565:1025–1038. https://doi.org/10.1016/j.scitotenv.2016.05.003
69. Lowe M, Qin R, Mao X (2022) A review on machine learning, artificial intelligence, and smart technology in water treatment and monitoring. Water 14(9):1384. https://doi.org/10.3390/w14091384
70. Lu H, Ma X (2020) Hybrid decision tree-based machine learning models for short-term water quality prediction. Chemosphere 249:126169. https://doi.org/10.1016/j.chemosphere.2020.126169
71. Maleki M, Ghaedi M, Dashtian K, Hajati S, Ghaedi AM (2017) The performance of nanorods material as efficient adsorbent for removal of dyes and heavy metal ions: modelling and optimization of process conditions. J Mol Liq 238:630–640. https://doi.org/10.1016/j.molliq.2017.05.077
72. Mandal S, Yadav A, Panwar R, Kumar SS, Karthick A, Priya A, ... Ganesh SS (2025) Smart water management for SDG 6: a review of AI and iot-enabled solutions. Water Conserv Sci Eng 10(2):83. https://doi.org/10.1007/s41101-025-00407-7
73. Mangayayam MC, Perez JPH, Alonso-de-Linaje V, Dideriksen K, Benning LG, Tobler DJ (2022) Sulfidation extent of nanoscale zerovalent iron controls selectivity and reactivity with mixed chlorinated hydrocarbons in natural groundwater. J Hazard Mater 431:128534. https://doi.org/10.1016/j.jhazmat.2022.128534

74. Mathur J, Goswami P, Gupta A, Srivastava S, Minkina T, Shan, S, Rajput VD (2022) Nanomaterials for water remediation: an efficient strategy for prevention of metal(loid) hazard. Water 14(24):3998. https://doi.org/10.3390/w14243998
75. Maurer-Jones MA, Gunsolus IL, Murphy CJ, Haynes CL (2013) Toxicity of engineered nanoparticles in the environment. Anal Chem 85(6):3036–3049. https://doi.org/10.1021/ac303636s
76. Maurya DK, Maurya SK, Kumar M, Chaubey C, Gupta D, Patel KK, ... Yadav R (2024) A review on precision agriculture: an evolution and prospect for the future. Int J Plant Soil Sci 36(5):363–374. https://doi.org/10.9734/ijpss/2024/v36i54508
77. Mishra RK (2023) Fresh water availability and its global challenge. Br J Multidiscip Adv Stud 4(3):1–78. https://doi.org/10.37745/bjmas.2022.0208
78. Mubarak NM, Sahu JN, Abdullah EC, Jayakumar NS (2014) Removal of heavy metals from wastewater using carbon nanotubes. Sep & Purif Rev 43:311–338. https://doi.org/10.1080/15422119.2012.748671
79. Naseem T, Durrani T (2021) The role of some important metal oxide nanoparticles for wastewater and antibacterial applications: a review. Environ Chem Lett 19(2):1297–1313. https://doi.org/10.1007/s10311-020-01138-7
80. Nasrollahzadeh M, Sajadi SM, Iravani S, Varma RS (2019) Green synthesis of metal nanoparticles using plant extracts and their applications in environmental remediation. J Nanobiotechnol 17:1–30. https://doi.org/10.1186/s12951-019-0495-9
81. Nikolaou G, Neocleous D, Christou A, Kitta E, Katsoulas N (2020) Implementing sustainable irrigation in water-scarce regions under the impact of climate change. Agronomy 10(8):1120. https://doi.org/10.3390/agronomy10081120
82. Noman M, Shahid M, Ahmed T, Tahir M, Naqqash T, Islam W (2020) Nanotechnology for environmental remediation: recent advances and future prospects. Front Environ Sci 8:88. https://doi.org/10.3389/fenvs.2020.00088
83. Paerl HW, Otten TG (2013) Harmful cyanobacterial blooms: causes, consequences, and controls. Microb Ecol 65(4):995–1010. https://doi.org/10.1007/s00248-012-0159-y
84. Patil SS, Shedbalkar UU, Truskewycz A, Chopade BA, Ball AS (2016) Nanoparticles for environmental clean-up: a review of potential risks and emerging solutions. Environ Technol Innovation 5:10–21. https://doi.org/10.1016/j.eti.2015.11.001
85. Phelps EB (2007) Stream sanitation. John Wiley & Sons, New York
86. Preite L, Vignali G (2024) Artificial intelligence to optimize water consumption in agriculture: a predictive algorithm-based irrigation management system. Comput Electron Agricult 223:109126. https://doi.org/10.1016/j.compag.2024.109126
87. Qu X, Alvarez PJJ, Li Q (2013) Applications of nanotechnology in water and wastewater treatment. Water Res 47(12):3931–3946. https://doi.org/10.1016/j.watres.2012.09.058
88. Rahman A, Chung GC, Ng YH (2026) Applications, challenges, and future trends of Artificial Intelligence of Things (AIoT)-enabled water quality and resource management. Water 18(8):919. https://doi.org/10.3390/w18080919
89. Raimondi A, Quinn R, Abhijith GR, Becciu G, Ostfeld A (2023) Rainwater harvesting and treatment: state of the art and perspectives. Water 15(8):1518. https://doi.org/10.3390/w15081518
90. Rajput S, Pittman CU, Mohan D (2016) Magnetic magnetite (Fe_3O_4) nanoparticle synthesis and applications for lead (Pb^{2+}) and chromium (Cr^{6+}) removal from water. J Colloid Interface Sci 468:334–346. https://doi.org/10.1016/j.jcis.2015.12.008
91. Rana AK, Kalaichelvan PT (2021) A review on recent advances in nanomaterials for wastewater treatment. Environ Technol Innovation 23:101620. https://doi.org/10.1016/j.eti.2021.101620
92. Rasheed T, Bilal M, Nabeel F, Adeel M, Iqbal HMN (2019) Environmentally-related contaminants of high concern: potential sources and analytical modalities for detection, quantification, and treatment. Environ Int 122:52–66. https://doi.org/10.1016/j.envint.2018.11.038
93. Rashid MI, Mujawar LH, Almeelbi T (2021) Green and sustainable remediation technologies for emerging pollutants using nanomaterials. Environ Technol Innovation 22:101471. https://doi.org/10.1016/j.eti.2021.101471

94. Rashid A, Schutte BJ, Ulery A, Deyholos MK, Sanogo S, Lehnhoff EA, Beck L (2023) Heavy metal contamination in agricultural soil: environmental pollutants affecting crop health. Agronomy 13(6):1521. https://doi.org/10.3390/agronomy13061521
95. Reddy DHK, Yun YS (2016) Spinel ferrite magnetic adsorbents: current status and prospects for future applications. Chem Eng J 289:508–527. https://doi.org/10.1016/j.cej.2015.10.082
96. Rijsberman FR (2006) Water scarcity: fact or fiction? Agric Water Manage 80(1–3):5–22. https://doi.org/10.1016/j.agwat.2005.07.001
97. Ringler C, Agbonlahor M, Barron J, Baye K, Meenakshi JV, Mekonnen DK, Uhlenbrook S (2022) The role of water in transforming food systems. Glob Food Sec 33:100639. https://doi.org/10.1016/j.gfs.2022.100639
98. Sakaa B, Elbeltagi A, Boudibi S, Chaffaï H, Islam ARMT, Kulimushi LC, ... Wong YJ (2022) Water quality index modeling using random forest and improved SMO algorithm for support vector machine in Saf-Saf river basin. Environ Sci Pollut Res 29(32):48491–48508. https://doi.org/10.1007/s11356-022-18644-x
99. Saljooqi A, Shamspur T, Mostafavi A (2021) Synthesis and photocatalytic activity of porous ZnO stabilized by TiO2 and Fe3O4 nanoparticles: investigation of pesticide degradation reaction in water treatment. Environ Sci Pollut Res 28(8):9146–9156. https://doi.org/10.1007/s11356-018-3731-9
100. Santos E, Carvalho M, Martins S (2023) Sustainable water management: understanding the socioeconomic and cultural dimensions. Sustainability 15(17):13074. https://doi.org/10.3390/su151713074
101. Senila M, Cadar O (2024) Modification of natural zeolites and their applications for heavy metal removal from polluted environments: challenges, recent advances, and perspectives. Heliyon 10(3):e25303. https://doi.org/10.1016/j.heliyon.2024.e25303
102. Sharma R, Jha P, Kothari SL (2020) Nanotechnology for sustainable agriculture and food safety: recent developments and future prospects. Environ Nanotechnol, Monit & Manag 14:100333. https://doi.org/10.1016/j.enmm.2020.100333
103. Singh P, Kim YJ (2021) Nanoparticles in wastewater treatment: a review on their synthesis, applications, and toxicity. Environ Nanotechnol, Monit & Manag 16:100562. https://doi.org/10.1016/j.enmm.2021.100562
104. Singh V, Ahmed G, Vedika S, Kumar P, Chaturvedi SK, Rai SN, Vamanu E, Kumar A (2024) Toxic heavy metal ions contamination in water and their sustainable reduction by eco-friendly methods: isotherms, thermodynamics and kinetics study. Sci Rep 14:7595. https://doi.org/10.1038/s41598-024-58061-3
105. Tawabini BS, Al-Khaldi SF, Khaled MM, Atieh MA (2011) Removal of arsenic from water by iron oxide nanoparticles impregnated on carbon nanotubes. J Environ Sci Health Part A 46(3):215–223. https://doi.org/10.1080/10934529.2011.535389
106. Thakur R, Reddy CRK (2021) Sustainable remediation of contaminated environments using green nanotechnology: current trends and future prospects. J Environ Manag 290:112597. https://doi.org/10.1016/j.jenvman.2021.112597
107. Theron J, Walker JA, Cloete TE (2008) Nanotechnology and water treatment: applications and emerging opportunities. Crit Rev Microbiol 34(1):43–69. https://doi.org/10.1080/10408410701710442
108. Vellingiri J, Kalaivanan K, Shanmugaiah K, Shobana Bai FJJ (2024) AO-SVM: a machine learning model for predicting water quality in the Cauvery River. Environ Res Commun 6(7):075025. https://doi.org/10.1088/2515-7620/ad6061
109. Wang Z, Ma L, Zhao J (2019) Nanomaterials for environmental remediation: a review on recent progress. Chem Eng J 373:914–931. https://doi.org/10.1016/j.cej.2019.05.079
110. Wang S, Sun H, Ang HM, Tade MO (2013) Adsorptive remediation of environmental pollutants using novel graphene-based nanomaterials. Chem Eng J 226:336–347. https://doi.org/10.1016/j.cej.2013.04.070
111. Whaiduzzaman M, Barros A, Chanda M, Barman S, Sultana T, Rahman MS, Roy S, Fidge CJ (2022) A review of emerging technologies for IoT-based smart cities. Sensors 22(23):9271. https://doi.org/10.3390/s22239271

112. World Health Organization (2022) Guidelines for drinking-water quality (4th ed., incorporating the first and second addenda). World Health Organization
113. Xu W, Jin Y, Zeng G (2024) Introduction of heavy metals contamination in the water and soil: a review on source, toxicity and remediation methods. Green Chem Lett Rev 17(1):2404235. https://doi.org/10.1080/17518253.2024.2404235
114. Yadav D, Chandra R (2020) Green synthesis and characterization of metal nanoparticles: their applications in environmental remediation. Environ Nanotechnol, Monit & Manag 14:100346. https://doi.org/10.1016/j.enmm.2020.100346
115. Yang Y, Li X, Li X, Liu J, Zhang L (2020) Advances in metal–organic frameworks for water treatment: adsorption and photocatalytic degradation. Coord Chem Rev 415:213319. https://doi.org/10.1016/j.ccr.2020.213319
116. Yang P, Wu L, Cheng M, Fan J, Li S, Wang H, Qian L (2023) Review on drip irrigation: impact on crop yield, quality, and water productivity in China. Water 15(9):1733. https://doi.org/10.3390/w15091733
117. Yaqoob AA, Parveen T, Umar K, Mohamad IS (2020) Recent advances in metal decorated nanomaterials and their various biological applications: a review. Front Chem 8:341. https://doi.org/10.3389/fchem.2020.00341
118. Yaqoob AA, Umar K, Parveen T (2021) Role of nanomaterials in the treatment of industrial wastewater: an overview. Environ Chem Lett 19(5):3787–3808. https://doi.org/10.1007/s10311-021-01217-0
119. Zhang L, Fang M (2010) Nanomaterials in pollution trace detection and environmental improvement. Nano Today 5(2):128–142. https://doi.org/10.1016/j.nantod.2010.03.006
120. Zhang D, Ma Y, Zhang Y, Chen L, Liu J (2022) Emerging trends of nanomaterials in water and wastewater treatment: bibliometric analysis and research frontiers. Environ Sci Pollut Res 29(19):27863–27879. https://doi.org/10.1007/s11356-021-17677-3
121. Zhang W, Wang C, Zhang H (2021) Multifunctional nanomaterials for wastewater treatment and energy recovery: a review. Nano Today 36:101017. https://doi.org/10.1016/j.nantod.2020.101017
122. Zhang Z, Yang C, Qiao Q, Li X, Wang F, Li C (2023) Application of improved particle swarm pptimization SVM in water quality evaluation of Ming Cui Lake. Sustainability 15(12):9835. https://doi.org/10.3390/su15129835
123. Zhao R, Huang Y, Zhao C (2020) Nanotechnology-based water treatment strategies: challenges and future perspectives. Environ Int 138:105627. https://doi.org/10.1016/j.envint.2020.105627
124. Zhou Y, Jiang JQ, Gao Y (2017) Nanomaterials in water treatment: development and application. Front Environ Sci Eng 11(1):5. https://doi.org/10.1007/s11783-016-0858-7
125. Zunita M (2021) Graphene oxide-based nanofiltration for Hg removal from wastewater: a mini review. Membranes 11(4):269. https://doi.org/10.3390/membranes11040269

MIX
Papier | Fördert gute Waldnutzung
FSC® C083411

Kolibri 360 GmbH
Ferdinand-Jühlke-Straße 7
99095 Erfurt, Deutschland
produktsicherheit@kolibri360.de